Table of Meas

RIVERSIDE COMMUNITY COLLEGE 1916
A, WITH
992
P9-DWU-522
JUE

Length		Capacity		Weight		Area	
in.	inch	oz	ounce			in.^{2}	square inches
ft	feet	c	cup			ft^2	square feet
yd	yard	qt	quart				
mi	mile	gal	gallon				

Metric System

Length		Capacity		Weight/Mass		Area	
mm	millimeter (0.001 m)	ml	milliliter (0.001 L)	mg	milligram (0.001 g)	cm^2	square centimeters
cm	centimeter (0.01 m)	cl	centiliter (0.01 L)	cg	centigram (0.01 g)	m^2	square meters
dm	decimeter (0.1 m)	dl	deciliter (0.1 L)	dg	decigram (0.1 g)		
m	meter	L	liter	g	gram		
dam	decameter (10 m)	dal	decaliter (10 L)	dag	decagram (10 g)		
hm	hectometer (100 m)	hl	hectoliter (100 L)	hg	hectogram (100 g)		
km	kilometer (1000 m)	kl	kiloliter (1000 L)	kg	kilogram (1000 g)		

Time

h	hours	min	minutes	s	seconds

Table of Symbols

$+$	add
$-$	subtract
$\cdot$, $\times$, $(a)(b)$	multiply
$\frac{a}{b}$, $\div$	divide
( )	parentheses, a grouping symbol
[ ]	brackets, a grouping symbol
π	pi, a number approximately equal to $\frac{22}{7}$ or 3.14
$-a$	the opposite, or additive inverse, of a
$\frac{1}{a}$	the reciprocal, or multiplicative inverse, of a
$=$	is equal to
$\approx$	is approximately equal to
$\neq$	is not equal to
$<$	is less than
$\leq$	is less than or equal to
$>$	is greater than
$\geq$	is greater than or equal to
(a, b)	an ordered pair whose first component is a and whose second component is b
$^\circ$	degree (for angles and temperature)
$\sqrt{a}$	the principal square root of a
$\varnothing$, $\{\ \}$	the empty set
$\lvert a \rvert$	the absolute value of a
$\cup$	union of two sets
$\cap$	intersection of two sets
$\in$	is an element of (for sets)
$\notin$	is not an element of (for sets)
	reference to THE COMPUTER TUTOR™
	reference to the Math ACE Disk
	indicates calculator topics
	reference to the Video Tapes

INTERMEDIATE ALGEBRA

with Applications

INTERMEDIATE ALGEBRA

with Applications

THIRD EDITION

Richard N. Aufmann
Palomar College, California

Vernon C. Barker
Palomar College, California

Joanne S. Lockwood
Plymouth State College, New Hampshire

HOUGHTON MIFFLIN COMPANY Boston Toronto

Dallas Geneva, Illinois Palo Alto Princeton, New Jersey

Riverside Community College
Library
MAY '93
4800 Magnolia Avenue
Riverside, California 92506

Sponsoring Editor: Maureen O'Connor
Development Editor: Erika Desuk
Project Editor: Erika Desuk
Assistant Design Manager: Karen Rappaport
Production Coordinator: Frances Sharperson
Manufacturing Coordinator: Sharon Pearson
Marketing Manager: Michael Ginley

Most Point of Interest Art designed and illustrated by Daniel P. Derdula, with airbrushing by Linda Phinney on Chapters 2 and 9, and calligraphy by Susan Fong on Chapter 4. Linda Phinney is credited for illustrating Chapter 6.

All Interior Math Figures rendered by Network Graphics (135 Fell Court, Hauppauge, New York 11788).

Cover concept and design by Daniel P. Derdula and Libby Plaisted.

Interior design by George McLean.

Copyright © 1992 by Houghton Mifflin Company. All rights reserved.

No part of the format or content of this work may be reproduced or transmitted in any form or by any means, electronic or mechanical, including photocopying and recording, or by any information storage or retrieval system without the prior written permission of Houghton Mifflin Company unless such copying is expressly permitted by federal copyright law. Address inquiries to College Permissions, Houghton Mifflin Company, One Beacon Street, Boston, MA 02108.

Printed in the U.S.A.

Library of Congress Catalog Card Number: 91-72001

ISBN Numbers:
Text: 0-395-58889-8
Instructor's Annotated Edition: 0-395-58890-1
Solutions Manual: 0-395-58892-8
Student's Solutions Manual: 0-395-58893-6
Instructor's Resource Manual with Chapter and Cumulative Tests: 0-395-58891-X
Test Bank: 0-395-58894-4
Transparencies: 0-395-60130-4

CDEFGHIJ-D-95432

CONTENTS

PREFACE

The third edition of *Intermediate Algebra with Applications* provides mathematically sound and comprehensive coverage of the topics considered essential in an intermediate algebra course. Our strategy in preparing this revision has been to build on the successful features of the second edition, features designed to enhance the student's mastery of math skills. *Intermediate Algebra with Applications* provides a complete, integrated learning system organized by objectives and linked to the ancillary package. All of the components of the package were written by the authors.

Features

The Interactive Approach

Instructors have long recognized the need for a text that requires the student to use a skill as it is being taught. *Intermediate Algebra with Applications* uses an interactive technique that meets this need. Each section is divided into objectives, and every objective contains one or more sets of matched-pair examples. The first example in each set is worked out; the second example is not. By solving this second problem, the student interacts with the text. The complete worked-out solutions to these problems are provided in an appendix at the end of the book, so the student can obtain immediate feedback on and reinforcement of the skill being learned.

Emphasis on Problem-Solving Strategies

Intermediate Algebra with Applications features a carefully developed approach to problem solving that emphasizes developing strategies to solve problems. For each type of word problem contained in the text, the student is prompted to use a "strategy step" before performing the actual manipulation of numbers and variables. By developing problem-solving strategies, the student will know better how to analyze and solve those word problems encountered in an intermediate algebra course.

Applications

The traditional approach to teaching or reviewing algebra covers only the straightforward manipulation of numbers and variables and thereby fails to teach students the practical value of algebra. By contrast, *Intermediate Algebra with Applications* emphasizes applications. Wherever appropriate, the last objective in each section presents applications that require the student to use the skills covered in that section to solve practical problems. Most of Chapter 2, *"First-Degree Equations and Inequalities,"* and portions of several other chapters are devoted entirely to applications. This carefully integrated applied approach generates awareness on the student's part of the value of algebra as a real-life tool.

Complete, Integrated Learning System Organized by Objectives

Each chapter begins with a list of the learning objectives included within that chapter. Each of the objectives is then restated in the chapter to remind the student of the current topic of discussion. The same objectives that organize the text organize each ancillary. The Solutions Manual, Student's Solutions Manual, Computerized Test Generator, Computer Tutor™, Videos, Transparencies, Test Bank, and the Printed Testing Program have all been prepared so that both the student and instructor can easily connect all of the different aids.

Exercises

There are more than 6000 exercises in the text, grouped in the following categories:

- **End-of-section exercise sets,** which are keyed to the corresponding learning objectives, provide ample practice and review of each skill.
- **Supplemental exercise sets,** designed to increase the student's ability to solve problems requiring a combination of skills, have been added at the end of each section.
- **Chapter review exercises,** which appear at the end of each chapter, help the student integrate all of the skills presented in the chapter.
- **Chapter tests,** which appear at the end of each chapter, are typical one-hour exams that the student can use to prepare for an in-class test.
- **Cumulative review exercises,** which appear at the end of each chapter (beginning with Chapter 2), help the student retain math skills learned in earlier chapters.
- **The final exam,** which follows the last chapter, can be used as a review item or practice final.

Calculator and Computer Enrichment Topics

Each chapter also contains optional calculator or computer enrichment topics. Calculator topics provide the student with valuable key-stroking instructions and practice in using a hand-held calculator. Computer topics correspond directly to the programs found on the Math ACE (Additional Computer Exercises) Disk. These topics range from solving first-degree equations to graphing an ellipse or a hyperbola.

New To This Edition

Topical Coverage

In Chapter 2, the material on inequalities and absolute value equations has been placed at the end of the chapter, with the result that all applications of first-degree equations follow the section on solving first-degree equations. The introductory applications material has been changed to allow for a better development of problem-solving skills, and the presentation of integer problems postponed until Section 2.2 to allow for a better understanding of this material.

In Chapter 3, the reorganization of the material on integer exponents has negative exponents defined prior to division of monomials; therefore, a single rule for division of monomials is stated. Scientific notation is now presented in Chapter 3 in order to reinforce the material on integer exponents. The material on division of polynomials has also been included in this chapter, thus completing the material on operations on polynomials. Factoring by grouping is presented earlier in Chapter 3, enabling trinomials of the form $ax^2 + bx + c$ to be factored by grouping as well as by trial and error. Factoring trinomials that are quadratic in form is presented here, rather than postponed until Chapter 7.

In Chapter 4, proportions are included in the objective on solving equations containing fractions.

In Chapter 6, the material on graphing the solution set of an inequality in two variables is placed at the end of the chapter, enabling students to use all the graphing skills presented in the chapter to graph these inequalities.

Chapter 7 includes solving inequalities by factoring (previously in Chapter 3) and solving rational inequalities (previously in Chapter 4).

In Chapter 8, the material on composite functions has been expanded, and applications of functions (maximum and minimum problems) have been included.

Chapter 9 introduces the midpoint formula, which is then used in the objectives involving equations of circles.

In Chapter 10, the expansion of determinants by cofactors has been added.

The application problems throughout the text have been rewritten. Many have been updated to reflect contemporary situations.

Graphing Calculator

Material on using a graphing calculator has been included in the Calculators and Computers feature of chapters that present material on graphing.

New Applications Feature

Each chapter now includes an expanded application feature entitled Something Extra. Topics include such concepts as Venn diagrams, trajectories, cryptography, linear programming, and the Fibonacci Sequence.

Challenge Exercises

Challenge exercises are now denoted in the Instructor's Annotated Edition. An asterisk (*) is printed next to those Supplemental Exercises which require more analytical thought.

Chapter Review and Chapter Test

The Chapter Review at the end of each chapter has been expanded to include nearly twice the number of review exercises, and a Chapter Test has been added. The objective references for each of these features are provided in the Answer Section at the back of the book so that the student can determine which objectives require restudy.

New Testing Program

Both the Computerized Testing Program and the Printed Testing Program have been completely rewritten to provide instructors with the option of creating countless new tests. The computerized test generator contains high quality graphics and editing capabilities for all nongraphic questions.

New Transparencies

Approximately 150 transparencies containing the worked-out solutions to the "student problems" in the text have been added.

Supplements for the Student

Two computerized study aids, the Computer Tutor™ and the Math ACE (Additional Computer Exercises) Disk have been carefully designed for the student.

The COMPUTER TUTOR™

The Computer Tutor™ is an interactive instructional microcomputer program for student use. Each learning objective in the text is supported by a lesson on the Computer Tutor™. As a reminder of this, a small computer icon appears to the right of each objective title in the text. Lessons on the tutor provide additional instruction and practice and can be used in several ways: (1) to cover material the student missed because of absence from class; (2) to repeat instruction on a skill or concept that the student has not yet mastered; or (3) to review material in preparation for examinations. This tutorial program is available for the IBM PC and compatible computers, the Apple II family of computers, and the Macintosh. The IBM and Macintosh versions of the Computer Tutor™ have been expanded to include nine "you-try-it" examples for each lesson.

Math ACE (Additional Computer Exercises) Disk

The Math ACE Disk contains a number of computational and drill-and-practice programs that correspond to selected Calculator and Computer Enrichment Topics in the text. These programs are available for the Apple II family of computers and the IBM PC and compatible computers.

Student's Solutions Manual

The Student's Solutions Manual contains the complete worked-out solutions for all the odd-numbered exercises in the text. Also included are the complete solutions to the Chapter Reviews, Chapter Tests, and Cumulative Reviews.

Videotapes

Over 50 half-hour videotape lessons accompany *Intermediate Algebra with Applications*. These lessons follow the format and style of the text and are closely tied to specific sections of the text.

Supplements for the Instructor

Intermediate Algebra with Applications has an unusually complete set of teaching aids for the instructor.

Instructor's Annotated Edition

The Instructor's Annotated Edition is an exact replica of the student text except that the answers to all of the exercises are printed in color next to the problems.

Solutions Manual

The Solutions Manual contains worked-out solutions for all end-of-section exercise sets, chapter reviews, chapter tests, cumulative reviews, and the final exam.

Instructor's Resource Manual with Chapter and Cumulative Tests

The Instructor's Resource Manual/Testing Program contains the printed testing program, which is the first of three sources of testing material available to users of *Intermediate Algebra with Applications.* Eight printed tests (in two formats—free response and multiple choice) are provided for each chapter, as are cumulative and final exams. In addition, the Instructor's Manual includes the documentation for all the software ancillaries (ACE, the Computer Tutor™, and the Instructor's Computerized Test Generator) as well as suggested course sequences.

Instructor's Computerized Test Generator

The Instructor's Computerized Test Generator is the second source of testing material for use with *Intermediate Algebra with Applications.* The database contains over 1900 new test items. These questions are unique to the test generator and do not repeat items provided in the Instructor's Resource Manual/Testing Program. Organized according to the keyed objectives in the text, the Test Generator is designed to produce an unlimited number of tests for each chapter of the text, including cumulative tests and final exams. It is available for the Apple II family of computers and the Macintosh. It is also available for the IBM PC or compatible computers with editing capabilities for all nongraphic questions.

Test Bank

The Printed Test Bank, the third component of the testing materials, is a printout of all items in the Instructor's Computerized Test Generator. Instructors using the Test Generator can use the test bank to select specific items from the database. Instructors who do not have access to a computer can use the test bank to select items to be included on a test being prepared by hand.

Transparencies

Approximately 150 transparencies accompany *Intermediate Algebra with Applications.* These transparencies contain the complete solution to every "student problem" in the text.

Acknowledgments

The authors would like to thank the people who have reviewed this manuscript and provided many valuable suggestions:

Barbara Brook
Camden County College, NJ

Patricia Confort
Roger Williams College, RI

Sharon Edgmon
Bakersfield College, CA

Ervin Eltze
Fort Hays State University, KS

Gerald D. Fischer
Northeast Iowa Community College, IA

Carol L. Grover
Carlow College, PA

Frank Gunnip
Oakland Community College, MI

Tim Hall
Central Texas College, TX

John A. Heublein
Kansas College of Technology, KS

Katherine J. Huppler
St. Cloud State University, MN

Buddy A. Johns
The Wichita State University, KS

Ellen Milosheff
Triton College, IL

Allan Newhart
West Virginia University at Parkersburg, WV

Doris Nice
University of Wisconsin-Parkside, WI

Donald Perry
Lee College, TX

Judith A. Pokrop
Cardinal Stritch College, WI

Diane Shores
Phillips County Community College, AR

Dean Stowers
Nicolet Area Technical College, WI

James M. Sullivan
Massachusetts Bay Community College, MA

Lana Taylor
Siena Heights College, MI

Robert A. Tolar
College of the Canyons, CA

Beverly Weatherwax
Southwest Missouri State University, MI

Warren Wise
Blue Ridge Community College, WA

Wayne Wolfe
Orange Coast College, CA

TO THE STUDENT

Many students feel that they will never understand math while others appear to do very well with little effort. Oftentimes what makes the difference is that successful students take an active role in the learning process.

Learning mathematics requires your *active* participation. Although doing homework is one way you can actively participate, it is not the only way. First, you must attend class regularly and become an active participant. Second, you must become actively involved with the textbook.

Intermediate Algebra with Applications was written and designed with you in mind as a participant. Here are some suggestions on how to use the features of this textbook.

There are 12 chapters in this text. Each chapter is divided into sections and each section is subdivided into learning objectives. Each learning objective is labeled with a number from 1–5.

First, read each objective statement carefully so you will understand the learning goal that is being presented. Next, read the objective material carefully, being sure to note each bold word. These words indicate important concepts that you should familiarize yourself with. Study each in-text example carefully, noting the techniques and strategies used to solve the example.

You will then come to the key learning feature of this text, the paired Examples and Problems. These Examples and Problems have been designed to assist you in a very specific way. Notice that the Examples are completely worked-out and explanations are given for certain steps within the solutions. The solutions to the Problems are not given; *you* are expected to work these Problems, thereby testing your understanding of the material you have just studied.

Study the Examples carefully by working through each step presented. Then use the worked-out example as a model for solving the Problems. When you have completed your solution, check your work by turning to the page in the Appendix where the complete solution is given. The page number on which the solution appears is printed on the solution line below the Problem statement. By checking your solution, you will know immediately whether or not you fully understand the skill just studied.

When you have completed studying an objective, do the exercises in the exercise set that correspond with that objective. The exercises are labeled with the same number as the objective. Algebra is a subject that needs to be learned in small sections and practiced continually in order to be mastered. Doing the exercises in each exercise set will help you master the problem-solving techniques necessary for success.

Once you have completed the exercises for an objective, you should check your answers to the odd-numbered exercises with those found in the back of the book.

After completing a chapter, read the Chapter Summary. This summary highlights the important topics covered in the chapter. Following the Chapter Summary are Chapter Review Exercises, a Chapter Test, and a Cumulative Review (beginning with Chapter 2). Doing the review exercises is an important way of testing your understanding of the chapter. The answer to each review exercise is in an appendix at the back of the book. Each answer is followed by a reference that tells which objective that exercise was taken from. For example, (4.2.2) means Section 4.2, Objective 2. After checking your answers, restudy any objective that corresponds to an exercise you answered incorrectly. It may be very helpful to retry some of the exercises for that objective to reinforce your problem-solving techniques.

The Chapter Test should be used to prepare for an exam. We suggest that you try the Chapter Test a few days before your actual exam. Take the test in a quiet place and try to complete the test in the same amount of time you will be allowed for your exam. When taking the Chapter Test, practice the strategies of successful test takers: 1) scan the entire test to get a feel for the questions; 2) read the directions carefully; 3) work the problems that are easiest for you first; and perhaps most importantly, 4) try to stay calm.

When you have completed the Chapter Test, check your answers. If you missed a question, review the material in that objective and rework some of the exercises from that objective. This will strengthen your ability to perform the skills in that objective.

The Cumulative Review allows you to refresh the skills you have learned in previous chapters. This is very important in mathematics. By consistently reviewing previous materials, you will retain the previous skills as you build new ones.

Remember, to be successful, attend class regularly; read the textbook carefully; actively participate in class; work with your textbook using the Examples and Problems for immediate feedback and reinforcement of each skill; do all the homework assignments; review constantly; and work carefully.

INTERMEDIATE ALGEBRA

with Applications

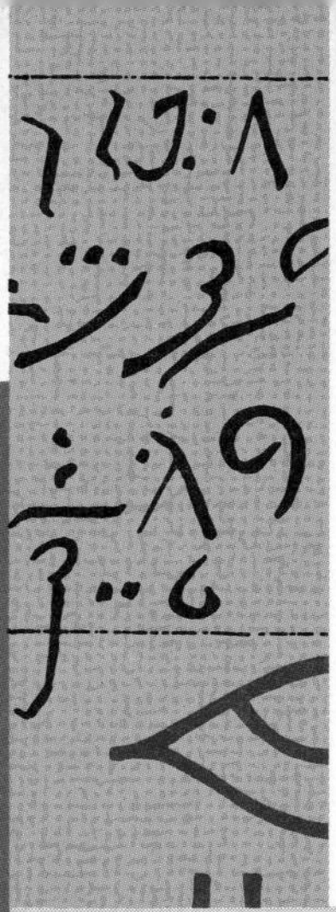

1

Review of Real Numbers

Objectives

- Absolute value and additive inverse
- Operations on rational numbers
- Exponential expressions
- The Order of Operations Agreement
- Evaluate variable expressions
- The Properties of the Real Numbers
- Simplify variable expressions
- Translate a verbal expression into a variable expression and simplify the resulting expression
- The union and intersection of sets
- Graph the solution set of an inequality in one variable

POINT OF INTEREST

Early Egyptian Number System

The early Egyptian type of picture writing shown at the right is known as hieroglyphics.

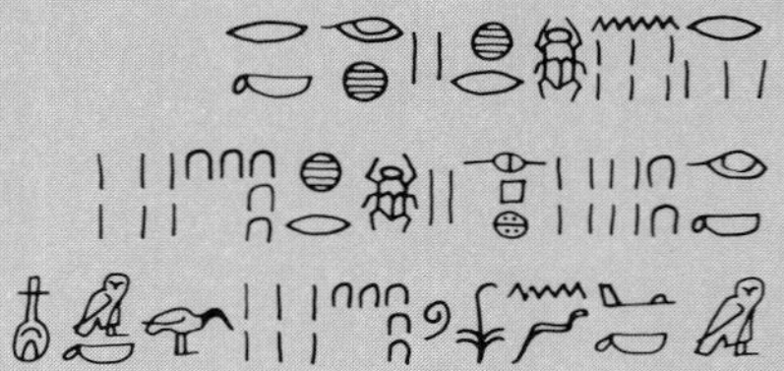

The Egyptian hieroglyphic method of representing numbers differs from our modern version in an important way. For the hieroglyphic number, the symbol indicated the value. For example, the symbol ∩ meant 10. The symbol was repeated to get larger values.

The markings at the right represent the number 743. Each vertical stroke represents 1, each ∩ represents 10, and each 9 represents 100.

There are 3 ones, 4 tens, and 7 hundreds representing the number 743.

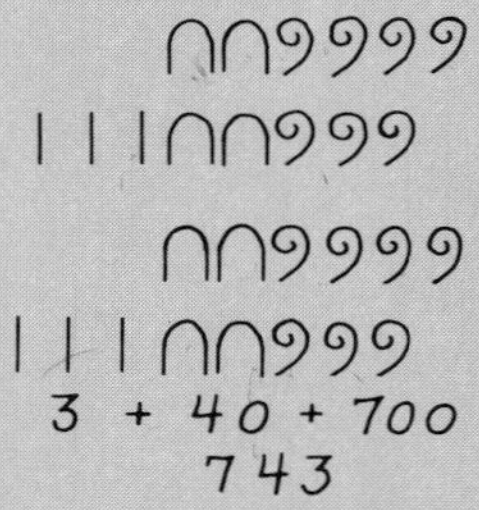

In our system, the position of a number is important. The number 5 in 356 means 5 tens but the number 5 in 3517 means 5 hundreds. Our system is called a positional number system.

Consider the hieroglyphic number at the right. Notice that the ∩ is missing. For the early Egyptians, when a certain group of ten was not needed, it was just omitted.

There are 4 ones and 5 hundreds. Thus, the number 504 is represented by this group of markings.

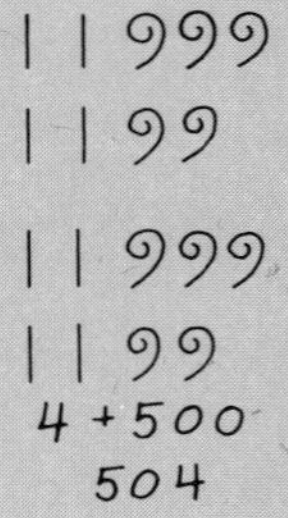

In a positional system of notation like ours, a zero is used to show that a certain group of ten is not needed. This may seem like a fairly simple idea, but it was not until near the end of the 7th century that a zero was introduced to the number system.

SECTION 1.1

Operations on the Real Numbers

1 Absolute value and additive inverse

The **integers** are . . ., −4, −3, −2, −1, 0, 1, 2, 3, 4, . . .

The three dots before and after the list of integers mean the list continues without end, and that there is no smallest integer and there is no largest integer.

The integers can be shown on the number line. The integers to the left of zero on the number line are **negative integers.** The integers to the right of zero are **positive integers.** Zero is neither a positive nor a negative integer.

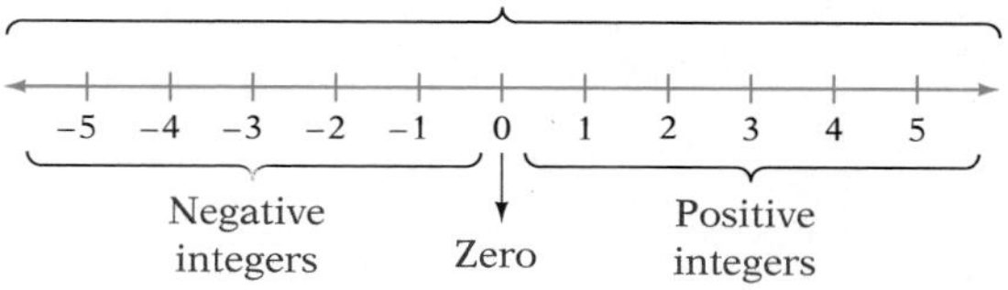

The positive integers are also called the **natural numbers.** 1, 2, 3, 4, . . .

The positive integers and zero are called the **whole numbers.** 0, 1, 2, 3, . . .

Just as the word *it* is used in language to stand for an object, a letter of the alphabet can be used in mathematics to stand for a number. Such a letter is called a **variable.**

A **rational number** is the quotient of two integers. Therefore, a rational number is a number that can be written in the form $\frac{a}{b}$, where a and b are integers, and $b \neq 0$ (b is not equal to zero). A rational number written in this way is commonly called a fraction.

Every integer is a rational number since an integer can be written as the quotient of the integer and 1; for example, $6 = \frac{6}{1}$. A number written in decimal notation is also a rational number; for example, $0.7 = \frac{7}{10}$.

Every rational number can be written as a terminating or repeating decimal. For example, the rational number $\frac{3}{8}$ can be written as 0.375 and the rational number $\frac{3}{11}$ can be written as $0.\overline{27}$. The bar over the 27 means that the block of digits 27 repeats without end, that is, 0.272727. . . .

Some numbers, for example $\sqrt{7}$ and π, have decimal representations that never terminate nor repeat. These numbers are called **irrational numbers.**

$\sqrt{7} = 2.6457513\ldots$ $\qquad$ $\pi = 3.141592654\ldots$

The rational numbers and the irrational numbers taken together are called the **real numbers.**

- Real Numbers
  - Rational Numbers
    - Integers
      - Whole Numbers
        - Zero
        - Positive Integers (Natural Numbers)
      - Negative Integers
    - Non-Integer Fractions Terminating Decimals Repeating Decimals
  - Irrational Numbers

A number line can be used to show the relative order of two numbers a and b. If a is to the left of b on the number line, then a is less than b ($a < b$). If a is to the right of b on the number line, then a is greater than b ($a > b$).

Negative 2 is greater than negative 4.

$-2 > -4$

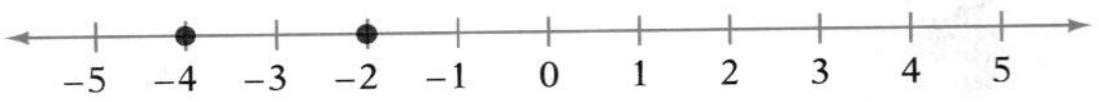

Negative 1 is less than 2.

$-1 < 2$

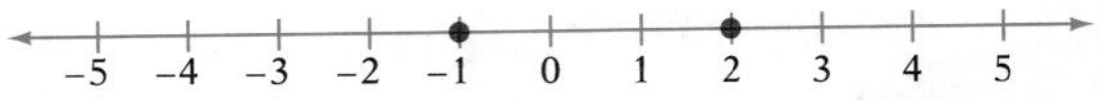

Two numbers that are the same distance from zero on the number line but on opposite sides of zero are **opposite numbers,** or **opposites.** The opposite of a number is called its **additive inverse.**

-3 is the additive inverse of 3.

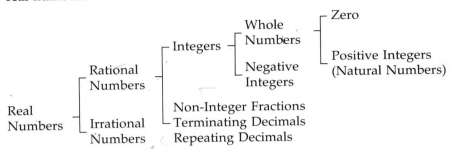

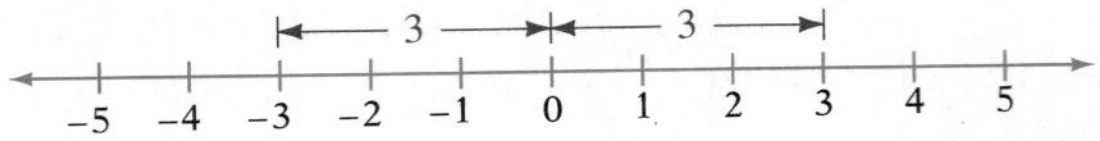

The additive inverse of -4 is 4.

Example 1 Find the additive inverse.

A. -16 $\qquad$ B. $\frac{3}{4}$

Solution A. 16 $\qquad$ B. $-\frac{3}{4}$

Problem 1 Find the additive inverse.

A. 18 B. -5.2

Solution See page A7.

The **absolute value** of a number is a measure of its distance from zero on the number line. Therefore, the absolute value of a number is a positive number or zero. The symbol for absolute value is $|\ |$.

The absolute value of a positive number is the number itself. $|7| = 7$

The absolute value of a negative number is its additive inverse. $|-7| = 7$

The absolute value of zero is zero. $|0| = 0$

Example 2 Evaluate.

A. $|-8|$ B. $-|-2.9|$

Solution A. 8 B. -2.9

▶ The absolute value sign does not affect the negative sign in front of the absolute value sign.

Problem 2 Evaluate.

A. $|-11|$ B. $-\left|-\frac{4}{5}\right|$

Solution See page A7.

2 Operations on rational numbers

To add two numbers with the same sign, add the absolute values of the numbers. Then attach the sign of the addends.

$$-15 + (-27) = -42$$

To add two numbers with different signs, find the difference between the absolute values of the numbers. Then attach the sign of the number with the greater absolute value.

$$-18.42 + 6.3 = -12.12$$

Subtraction is defined as the addition of the additive inverse.

Definition of Subtraction

> If a and b are real numbers, then $\boldsymbol{a - b = a + (-b)}$.

$$-17 - (-23) = -17 + 23 = 6$$

Addition of Fractions

> The sum of two fractions with the same denominators is the sum of the numerators over the common denominator.
>
> $$\frac{a}{c} + \frac{b}{c} = \frac{a+b}{c}$$

To add or subtract fractions with different denominators, first rewrite the fractions as equivalent fractions using the least common multiple (LCM) of the denominators as the common denominator. Then perform the indicated operations.

$$\frac{5}{8} + \frac{7}{12} = \frac{15}{24} + \frac{14}{24} = \frac{15+14}{24} = \frac{29}{24}$$

$$\frac{1}{4} - \frac{5}{6} = \frac{1}{4} + \left(-\frac{5}{6}\right) = \frac{3}{12} + \left(-\frac{10}{12}\right) = \frac{3+(-10)}{12} = \frac{-7}{12} = -\frac{7}{12}$$

To multiply two real numbers with the same sign, multiply the absolute values of the factors. The product is positive.

$$-31.7 \times (-0.05) = 1.585$$

To multiply two real numbers with different signs, multiply the absolute values of the factors. The product is negative.

$$-24.8 \times 0.02 = -0.496$$

Multiplication of Fractions

> The product of two fractions is the product of the numerators over the product of the denominators.
>
> $$\frac{a}{b} \cdot \frac{c}{d} = \frac{ac}{bd}$$

$$-\frac{5}{12} \cdot \frac{8}{15} = -\frac{5 \cdot 8}{12 \cdot 15} = -\frac{\overset{1}{\cancel{5}} \cdot \overset{1}{\cancel{2}} \cdot \overset{1}{\cancel{2}} \cdot 2}{\underset{1}{\cancel{2}} \cdot \underset{1}{\cancel{2}} \cdot 3 \cdot 3 \cdot \underset{1}{\cancel{5}}} = -\frac{2}{9}$$

The quotient of two numbers with the same sign is positive.

$$-1.5 \div (-0.05) = 30$$

The quotient of two numbers with different signs is negative.

$$3.8 \div (-0.19) = -20$$

The **reciprocal** of a fraction is the fraction with the numerator and denominator interchanged. The process of interchanging the numerator and denominator of a fraction is called **inverting.**

Division of Fractions

To divide two fractions, multiply by the reciprocal of the divisor.

$$\frac{a}{b} \div \frac{c}{d} = \frac{a}{b} \cdot \frac{d}{c}$$

$$-\frac{3}{8} \div \frac{9}{16} = -\frac{3}{8} \cdot \frac{16}{9} = -\frac{\overset{1}{\cancel{3}} \cdot \overset{1}{\cancel{2}} \cdot \overset{1}{\cancel{2}} \cdot \overset{1}{\cancel{2}} \cdot 2}{\underset{1}{\cancel{2}} \cdot \underset{1}{\cancel{2}} \cdot \underset{1}{\cancel{2}} \cdot \underset{1}{\cancel{3}} \cdot 3} = -\frac{2}{3}$$

Example 3 Simplify.

A. $\frac{3}{8} + \frac{5}{12} - \frac{9}{16}$ B. $6.329 - 12.49$ C. $14 - |3 - 18|$

Solution A. $\frac{3}{8} + \frac{5}{12} - \frac{9}{16} = \frac{18}{48} + \frac{20}{48} + \left(-\frac{27}{48}\right) = \frac{18 + 20 + (-27)}{48} = \frac{11}{48}$

B. $6.329 - 12.49 = 6.329 + (-12.49) = -6.161$

C. $14 - |3 - 18|$

$14 - |-15|$ ▶ Perform the indicated operation inside the absolute value symbols.

$14 - 15$ ▶ Rewrite $|-15|$ as 15.

-1

Problem 3 Simplify.

A. $\frac{5}{8} \div \left(-\frac{15}{40}\right)$ B. $-8.729 + 12.094$

Solution See page A7.

Example 4 Simplify: $0.0527 \div (-0.27)$ Round to the nearest hundredth.

Solution

```
          0.195 ≈ 0.20
0.27.)0.05.270
       -2 7
        2 57
       -2 43
          140
         -135
            5
```

▶ The symbol ≈ is used to indicate that the quotient is an approximate value after being rounded off.

$0.0527 \div (-0.27) \approx -0.20$

Problem 4 Simplify: $-4.027(0.49)$ Round to the nearest hundredth.

Solution See page A7.

3 Exponential expressions

Repeated multiplication of the same factor can be written using an exponent.

$2 \cdot 2 \cdot 2 \cdot 2 \cdot 2 \cdot 2 = 2^6$ ← Exponent; 2 ← Base

$b \cdot b \cdot b \cdot b \cdot b = b^5$ ← Exponent; b ← Base

The **exponent** indicates how many times the factor, called the **base,** occurs in the multiplication. The multiplication $2 \cdot 2 \cdot 2 \cdot 2 \cdot 2 \cdot 2$ is in **factored form.** The exponential expression 2^6 is in **exponential form.**

2^1 is read "the first power of two" or just "two." → Usually the exponent 1 is not written.

2^2 is read "the second power of two" or "two squared."

2^3 is read "the third power of two" or "two cubed."

2^4 is read "the fourth power of two."

2^5 is read "the fifth power of two."

b^5 is read "the fifth power of b."

If b is a number and n is a positive integer, the nth power of b is defined as the product of n factors of b.

$$b^n = \underbrace{b \cdot b \cdot b \ldots b}_{n \text{ factors}}$$

To evaluate an exponential expression, write each factor as many times as indicated by the exponent. Then multiply.

$$5^4 = 5 \cdot 5 \cdot 5 \cdot 5 = 625$$

$$3^2 \cdot 5^3 = (3 \cdot 3)(5 \cdot 5 \cdot 5) = 9 \cdot 125 = 1125$$

Example 5 Evaluate $(-3)^4$ and -3^4.

Solution $(-3)^4 = (-3)(-3)(-3)(-3) = 81$

$-3^4 = -(3 \cdot 3 \cdot 3 \cdot 3) = -81$

▶ The negative of a number is taken to a power only when the negative sign is *inside* the parentheses.

Problem 5 Evaluate $(-2)^4$ and -2^4.

Solution See page A7.

Example 6 Evaluate $\left(-\frac{2}{3}\right)^2 \cdot 3^3$.

Solution $\left(-\frac{2}{3}\right)^2 \cdot 3^3 = \left(-\frac{2}{3}\right)\left(-\frac{2}{3}\right) \cdot (3)(3)(3) = 12$

Problem 6 Evaluate $-\left(\frac{2}{5}\right)^3 \cdot 5^2$.

Solution See page A7.

4 The Order of Operations Agreement

In order to prevent more than one answer to the same numerical expression, an Order of Operations Agreement is followed.

The Order of Operations Agreement

Step 1	Perform operations inside grouping symbols. Grouping symbols include parentheses ( ), brackets [ ], and the fraction bar.
Step 2	Simplify exponential expressions.
Step 3	Do multiplication and division as they occur from left to right.
Step 4	Do addition and subtraction as they occur from left to right.

Simplify: $8 - \frac{12-2}{4+1} \div 2^2$

$$8 - \frac{12-2}{4+1} \div 2^2$$

Perform operations above and below the fraction bar. $\quad 8 - \frac{10}{5} \div 2^2$

Simplify exponential expressions. $\quad 8 - \frac{10}{5} \div 4$

Do multiplication and division as they occur from left to right. Note that a fraction bar can be read "÷." $\quad 8 - 2 \div 4$

$$8 - \frac{1}{2}$$

Do addition and subtraction as they occur from left to right. $\quad \frac{15}{2}$

One or more of the above steps may not be needed to simplify an expression. In that case, proceed to the next step in the Order of Operations Agreement.

When an expression has grouping symbols inside grouping symbols, perform the operations inside the inner grouping symbols first.

Simplify: $4.3 - [(25 - 9) \div 2]^2$ — $4.3 - [(25 - 9) \div 2]^2$

Perform operations inside grouping symbols. — $4.3 - [16 \div 2]^2$; $4.3 - [8]^2$

Simplify exponential expressions. — $4.3 - 64$

Perform addition and subtraction as they occur from left to right. — -59.7

Example 7 Simplify: $(-1.2)^3 - 8.4 \div 2.1$

Solution

$(-1.2)^3 - 8.4 \div 2.1$

$-1.728 - 8.4 \div 2.1$ ▶ Simplify exponential expressions.

$-1.728 - 4$ ▶ Do multiplication and division.

-5.728 ▶ Do addition and subtraction.

Problem 7 Simplify: $(3.81 - 1.41)^2 \div 0.036 - 1.89$

Solution See page A7.

Example 8 Simplify: $\left(\frac{1}{2}\right)^3 - \left[\left(\frac{2}{3} + \frac{1}{4}\right) \div \frac{5}{6}\right]$

Solution

$\left(\frac{1}{2}\right)^3 - \left[\left(\frac{2}{3} + \frac{1}{4}\right) \div \frac{5}{6}\right]$

$\left(\frac{1}{2}\right)^3 - \left[\frac{11}{12} \div \frac{5}{6}\right]$ ▶ Perform operations inside the inner grouping symbols.

$\left(\frac{1}{2}\right)^3 - \frac{11}{10}$ ▶ Perform operations inside the grouping symbols.

$\frac{1}{8} - \frac{11}{10}$ ▶ Simplify exponential expressions.

$-\frac{39}{40}$ ▶ Do addition and subtraction.

Problem 8 Simplify: $\frac{1}{3} + \frac{5}{8} \div \frac{15}{16} - \frac{7}{12}$

Solution See page A7.

A **complex fraction** is a fraction whose numerator or denominator contains one or more fractions. Examples of complex fractions are shown below.

$$\frac{\frac{2}{3}}{\frac{1}{3}} \qquad \frac{\frac{2}{5} + 1}{\frac{7}{8}} \longleftarrow \text{Main Fraction Bar}$$

To simplify a complex fraction, perform operations above and below the main fraction bar as the first step in the Order of Operations Agreement.

Simplify: $\dfrac{\frac{3}{4}-\frac{1}{2}}{\frac{2}{3}+\frac{1}{4}}$

Perform operations above and below the main fraction bar.

$$\frac{\frac{3}{4}-\frac{1}{2}}{\frac{2}{3}+\frac{1}{4}} = \frac{\frac{3}{4}-\frac{2}{4}}{\frac{8}{12}+\frac{3}{12}}$$

Multiply the numerator of the complex fraction by the reciprocal of the denominator of the complex fraction.

$$= \frac{\frac{1}{4}}{\frac{11}{12}} = \frac{1}{4}\cdot\frac{12}{11}$$

$$= \frac{1\cdot 12}{4\cdot 11}$$

$$= \frac{3}{11}$$

Example 9 Simplify: $9\cdot\dfrac{\frac{5}{6}-2}{\frac{3}{8}}\div\dfrac{7}{6}$

Solution $9\cdot\dfrac{\frac{5}{6}-2}{\frac{3}{8}}\div\dfrac{7}{6}$

$9\cdot\dfrac{-\frac{7}{6}}{\frac{3}{8}}\div\dfrac{7}{6}$ ▶ Do operations above the main fraction bar.

$9\cdot\left(-\dfrac{7}{6}\cdot\dfrac{8}{3}\right)\div\dfrac{7}{6}$ ▶ Multiply the numerator of the complex fraction by the reciprocal of the denominator of the complex fraction.

$9\cdot\left(-\dfrac{28}{9}\right)\div\dfrac{7}{6}$

$-28\div\dfrac{7}{6}$ ▶ Do multiplication and division as they occur from left to right.

$-28\cdot\dfrac{6}{7}$

-24

Problem 9 Simplify: $\dfrac{11}{12}-\dfrac{\frac{5}{4}}{2-\frac{7}{2}}\cdot\dfrac{3}{4}$

Solution See page A7.

EXERCISES 1.1

1 Find the additive inverse.

1. 83
2. 51
3. -75
4. -126
5. 9.3
6. 2.7
7. -6.4
8. -43.9
9. $-\frac{11}{12}$
10. $-\frac{8}{9}$

Evaluate.

11. $-|126|$
12. $-|89|$
13. $|-436|$
14. $|-502|$
15. $-|-16|$
16. $-|-22|$
17. $|-4.93|$
18. $-|72.1|$
19. $-\left|-\frac{7}{8}\right|$
20. $\left|-\frac{15}{16}\right|$

2 Simplify.

21. $\frac{7}{12} + \frac{5}{16}$
22. $\frac{3}{8} - \frac{5}{12}$
23. $-\frac{5}{9} - \frac{14}{15}$
24. $\frac{1}{2} + \frac{1}{7} - \frac{5}{8}$
25. $-\frac{1}{3} + \frac{5}{9} - \frac{7}{12}$
26. $\frac{1}{3} + \frac{19}{24} - \frac{7}{8}$
27. $\frac{2}{3} - \frac{5}{12} + \frac{5}{24}$
28. $-\frac{7}{10} + \frac{4}{5} + \frac{5}{6}$
29. $\frac{5}{8} - \frac{7}{12} + \frac{1}{2}$
30. $-\frac{1}{3} \cdot \frac{5}{8}$
31. $\left(\frac{6}{35}\right)\left(-\frac{5}{16}\right)$
32. $\frac{2}{3}\left(-\frac{9}{20}\right) \cdot \frac{5}{12}$
33. $-\frac{8}{15} \div \frac{4}{5}$
34. $-\frac{2}{3} \div \left(-\frac{6}{7}\right)$
35. $-\frac{11}{24} \div \frac{7}{12}$
36. $\frac{7}{9} \div \left(-\frac{14}{27}\right)$
37. $\left(-\frac{5}{12}\right)\left(\frac{4}{35}\right)\left(\frac{7}{8}\right)$
38. $\frac{6}{35}\left(-\frac{7}{40}\right)\left(-\frac{8}{21}\right)$
39. $-14.27 + 1.296$
40. $-0.4355 + 172.5$
41. $1.832 - 7.84$
42. $(3.52)(4.7)$
43. $(0.03)(10.5)(6.1)$
44. $(1.2)(3.1)(-6.4)$
45. $5.418 \div (-0.9)$
46. $-0.2645 \div (-0.023)$
47. $-0.4355 \div 0.065$
48. $|12 - 14|$
49. $|-25| + |-10|$
50. $4 - |6 - 12|$
51. $6 - |8 + 5|$
52. $|24 \div (-4)|$
53. $3 - |2 \cdot (-5)|$

3 Simplify.

54. -2^3
55. -4^3
56. $(-5)^3$
57. $(-8)^2$
58. $2^2 \cdot 3^4$
59. $4^2 \cdot 3^3$
60. $-2^2 \cdot 3^2$
61. $-3^2 \cdot 5^3$
62. $(-2)^3(-3)^2$
63. $(-4)^3(-2)^3$
64. $-4(-3)^2(4^2)$
65. $2^2(-10)(-2)^2$
66. $\left(-\frac{3}{4}\right)^2(2)^4$
67. $\left(\frac{2}{3}\right)^3(-9)^2$
68. $-\left(\frac{4}{5}\right)^2(5)^3$
69. $-\left(\frac{3}{5}\right)^2(10)^2$

4 Simplify.

70. $5 - 3(8 \div 4)^2$
71. $4^2 - (5 - 2)^2 \cdot 3$
72. $16 - \frac{2^2 - 5}{3^2 + 2}$
73. $\frac{4(5 - 2)}{4^2 - 2^2} \div 4$

74. $\dfrac{3+\frac{2}{3}}{\frac{11}{16}}$

75. $\dfrac{\frac{11}{14}}{4-\frac{6}{7}}$

76. $5[(2-4)\cdot 3-2]$

77. $2[(16 \div 8)-(-2)]+4$

78. $16-4\left(\frac{8-2}{3-6}\right) \div \frac{1}{2}$

79. $25 \div 5\left(\frac{16+8}{-2^2+8}\right)-5$

80. $6[3-(-4+2) \div 2]$

81. $12-4[2-(-3+5)-8]$

82. $\frac{1}{2}-\left(\frac{2}{3} \div \frac{5}{9}\right)+\frac{5}{6}$

83. $\left(-\frac{3}{5}\right)^2-\frac{3}{5}\cdot\frac{5}{9}+\frac{7}{10}$

84. $\frac{1}{2}-\dfrac{\frac{17}{25}}{4-\frac{3}{5}} \div \frac{1}{5}$

85. $\frac{3}{4}+\dfrac{3-\frac{7}{9}}{\frac{5}{6}}\cdot\frac{2}{3}$

86. $0.4(1.2-2.3)^2+5.8$

87. $5.4-(0.3)^2 \div 0.09$

SUPPLEMENTAL EXERCISES 1.1

Classify each of the following numbers as a natural number, an integer, a positive integer, a negative integer, a rational number, an irrational number, and a real number.

88. -1

89. 0

90. 65

91. $-\frac{19}{21}$

92. -8.43

93. $\sqrt{5}$

94. $-6.\overline{3}$

95. $0.232332333\ldots$

96. 3.14

97. π

Complete.

98. If a is a positive number, then $-a$ is a ______ number.

99. If a is a negative number, then $-a$ is a ______ number.

100. The product of an even number of negative factors is a ______ number.

101. The product of an odd number of negative factors is a ______ number.

102. A number that is its own additive inverse is ______.

103. A number that is its own reciprocal is ______.

Solve.

104. What is the sum of the three largest prime numbers less than 100?

105. How many positive integers less than 10 have an odd number of positive integral divisors?

106. What is the tens' digit in 11^{22}?

107. What is the ones' digit in 11^{22}?

108. What is the ones' digit in 7^{18}?

109. Does $(2^3)^4 = 2^{(3^4)}$?

SECTION 1.2 Variable Expressions

1 Evaluate variable expressions

An expression that contains one or more variables is called a **variable expression.**

A variable expression is shown at the right. The expression has 4 addends, which are called **terms** of the expression. The variable expression has 3 variable terms and 1 constant term.

4 terms

$$4x^2y \quad - \quad 2z \quad - \quad x \quad + \quad 4$$

Variable terms   Constant term

Each variable term is composed of a **numerical coefficient** and a **variable part.** When the numerical coefficient is 1 or −1, the 1 is usually not written.

numerical coefficient

$$4x^2y \quad - \quad 2z \quad - \quad 1x \quad + \quad 4$$

variable part

Replacing the variable in a variable expression by a numerical value and then simplifying the resulting expression is called **evaluating the variable expression.**

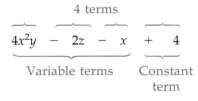

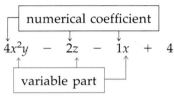

Example 1 Evaluate $a^2 - (ab - c)$ when $a = -2$, $b = 3$, and $c = -4$.

Solution

$a^2 - (ab - c)$

$(-2)^2 - [(-2)(3) - (-4)]$ ▶ Replace each variable in the expression with its value.

$(-2)^2 - [-6 - (-4)]$ ▶ Use the Order of Operations Agreement to simplify the resulting numerical expression.

$(-2)^2 - [-2]$

$4 - [-2]$

6

Problem 1 Evaluate $(b - c)^2 \div ab$ when $a = -3$, $b = 2$, and $c = -4$.

Solution See page A8.

2 The Properties of the Real Numbers

The Properties of the Real Numbers describe the way operations on numbers can be performed. Here is a list of some of the real number properties and an example of each property.

The Commutative Property of Addition

If a and b are real numbers, then $a + b = b + a$.

$$3 + 2 = 2 + 3$$
$$5 = 5$$

The Commutative Property of Multiplication

If a and b are real numbers, then $a \cdot b = b \cdot a$.

$$(3)(-2) = (-2)(3)$$
$$-6 = -6$$

The Associative Property of Addition

If a, b, and c are real numbers, then $(a + b) + c = a + (b + c)$.

$$(3 + 4) + 5 = 3 + (4 + 5)$$
$$7 + 5 = 3 + 9$$
$$12 = 12$$

The Associative Property of Multiplication

If a, b, and c are real numbers, then $(a \cdot b) \cdot c = a \cdot (b \cdot c)$.

$$(3 \cdot 4) \cdot 5 = 3 \cdot (4 \cdot 5)$$
$$12 \cdot 5 = 3 \cdot 20$$
$$60 = 60$$

The Addition Property of Zero

If a is a real number, then $a + 0 = 0 + a = a$.

$$3 + 0 = 0 + 3 = 3$$

The Multiplication Property of Zero

If a is a real number, then $a \cdot 0 = 0 \cdot a = 0$.

$$3 \cdot 0 = 0 \cdot 3 = 0$$

The Multiplication Property of One

If a is a real number, then $a \cdot 1 = 1 \cdot a = a$.

$$5 \cdot 1 = 1 \cdot 5 = 5$$

The Inverse Property of Addition

If a is a real number, then $a + (-a) = (-a) + a = 0$.

$$4 + (-4) = (-4) + 4 = 0$$

$-a$ is called the **additive inverse** of a. Because $-(-a) = a$, the additive inverse of $-a$ is a. The sum of a number and its additive inverse is 0.

The Inverse Property of Multiplication

> If a is a nonzero real number, then $a \cdot \frac{1}{a} = \frac{1}{a} \cdot a = 1$.

$$(4)\left(\frac{1}{4}\right) = \left(\frac{1}{4}\right)(4) = 1$$

$\frac{1}{a}$ is called the **reciprocal** or **multiplicative inverse** of a. The product of a number and its multiplicative inverse is 1.

The Distributive Property

> If a, b, and c are real numbers, then
> $a(b + c) = ab + ac$ and $(b + c)a = ba + ca$.

$$3(4 + 5) = 3 \cdot 4 + 3 \cdot 5$$
$$3 \cdot 9 = 12 + 15$$
$$27 = 27$$

$$(4 + 5)2 = 4 \cdot 2 + 5 \cdot 2$$
$$9 \cdot 2 = 8 + 10$$
$$18 = 18$$

Example 2 Complete the statement by using the Inverse Property of Addition.
$3x + ? = 0$

Solution $3x + (-3x) = 0$

Problem 2 Complete the statement by using the Commutative Property of Multiplication.
$(x)\left(\frac{1}{4}\right) = (?)(x)$

Solution See page A8.

Example 3 Identify the property that justifies the statement.
$3(x + 4) = 3x + 12$

Solution The Distributive Property

Problem 3 Identify the property that justifies the statement.
$(a + 3b) + c = a + (3b + c)$

Solution See page A8.

It is important to recognize the following facts concerning division by zero.

Zero divided by any nonzero number is zero. For $a \neq 0$, $\frac{0}{a} = 0$.

Division by zero is not defined. $\frac{a}{0}$ is not defined.

Any nonzero number divided by itself is 1. For $a \neq 0$, $\frac{a}{a} = 1$.

$\frac{0}{0}$ is not a real number.

3 Simplify variable expressions

Like terms of a variable expression are the terms with the same variable part.

Constant terms are like terms.

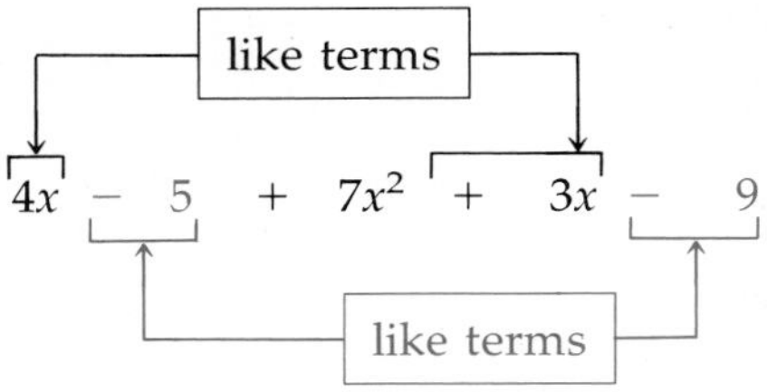

To **combine** like terms, use the Distributive Property $ba + ca = (b + c)a$ to add the coefficients.

$3x + 2x$
$(3 + 2)x$
$5x$

Example 4 Simplify: $2(x + y) + 3(y - 3x)$

Solution $2(x + y) + 3(y - 3x)$

$2x + 2y + 3y - 9x$ ▶ Use the Distributive Property to remove parentheses.

$(2x - 9x) + (2y + 3y)$ ▶ Use the Commutative and Associative Properties of Addition to rearrange and group like terms.

$-7x + 5y$ ▶ Combine like terms.

Problem 4 Simplify: $(2x + xy - y) - (5x - 7xy + y)$

Solution See page A8.

Example 5 Simplify: $4y - 2[x - 3(x + y) - 5y]$

Solution $4y - 2[x - 3(x + y) - 5y]$

$4y - 2[x - 3x - 3y - 5y]$ ▶ Use the Distributive Property to remove parentheses.

$4y - 2[-2x - 8y]$ ▶ Combine like terms.

$4y + 4x + 16y$ ▶ Use the Distributive Property to remove brackets.

$4x + 20y$ ▶ Combine like terms.

Problem 5 Simplify: $2x - 3[y - 3(x - 2y + 4)]$

Solution See page A8.

4 Translate a verbal expression into a variable expression and simplify the resulting expression

One of the major skills required in applied mathematics is the translation of a verbal expression into a variable expression. This requires recognizing the verbal phrases that translate into mathematical operations. Some of the verbal phrases used to indicate the different mathematical operations are given below.

Addition	added to	5 added to x	$x + 5$
	more than	2 more than t	$t + 2$
	the sum of	the sum of s and r	$s + r$
	increased by	z increased by 7	$z + 7$
	the total of	the total of 6 and b	$6 + b$
Subtraction	minus	y minus 8	$y - 8$
	less than	5 less than p	$p - 5$
	decreased by	n decreased by 1	$n - 1$
	the difference between	the difference between t and 4	$t - 4$
Multiplication	times	9 times y	$9y$
	of	one third of p	$\frac{1}{3}p$
	the product of	the product of x and y	xy
	multiplied by	t multiplied by 43	$43t$
	twice	twice n	$2n$
Division	divided by	a divided by 6	$\frac{a}{6}$
	the quotient of	the quotient of s and t	$\frac{s}{t}$
	the ratio of	the ratio of r to 5	$\frac{r}{5}$
Power	the square of	the square of b	b^2
	the cube of	the cube of x	x^3

In most applications that involve translating phrases into variable expressions, the variable to be used is not given. To translate these phrases, a variable must be assigned to an unknown quantity before the variable expression can be written. After translating a verbal expression into a variable expression, simplify the variable expression by using the Addition, Multiplication, and Distributive Properties.

Example 6 Translate and simplify "the total of five times a number and twice the difference between the number and three."

Solution

the unknown number: n ► Assign a variable to one of the unknown quantities.

five times the number: $5n$ ► Use the assigned variable to write an expression for any other unknown quantity.

the difference between the number and three: $n - 3$

twice the difference between the number and three: $2(n - 3)$

$5n + 2(n - 3)$ ► Use the assigned variable to write the variable expression.

$5n + 2n - 6$ ► Simplify the variable expression.

$7n - 6$

Problem 6 Translate and simplify "a number decreased by the difference between eight and twice the number."

Solution See page A8.

Example 7 Translate and simplify "fifteen minus one half the sum of a number and ten."

Solution

the unknown number: n ► Assign a variable to one of the unknown quantities.

the sum of the number and ten: $n + 10$ ► Use the assigned variable to write an expression for any other unknown quantity.

one half the sum of the number and ten: $\frac{1}{2}(n + 10)$

$15 - \frac{1}{2}(n + 10)$ ► Use the assigned variable to write the variable expression.

$15 - \frac{1}{2}n - 5$ ► Simplify the variable expression.

$-\frac{1}{2}n + 10$

Problem 7 Translate and simplify "the sum of three eighths of a number and five twelfths of the number."

Solution See page A8.

EXERCISES 1.2

1 Evaluate the variable expression when $a = 2$, $b = 3$, $c = -1$, and $d = -4$.

1. $ab + dc$

2. $2ab - 3dc$

3. $4cd \div a^2$

4. $b^2 - (d - c)^2$

5. $(b - 2a)^2 + c$

6. $(b - d)^2 \div (b - d)$

7. $(bc + a)^2 \div (d - b)$

8. $\frac{1}{3}b^3 - \frac{1}{4}d^3$

9. $\frac{1}{4}a^4 - \frac{1}{6}bc$

10. $2b^2 \div \frac{ad}{2}$

11. $\frac{3ac}{-4} - c^2$

12. $\frac{2d - 2a}{2bc}$

13. $\frac{3b - 5c}{3a - c}$

14. $\frac{2d - a}{b - 2c}$

15. $\frac{a - d}{b + c}$

16. $|a^2 + d|$

17. $-a|a + 2d|$

18. $d|b - 2d|$

19. $\frac{2a - 4d}{3b - c}$

20. $\frac{3d - b}{b - 2c}$

21. $-3d \div \left|\frac{ab - 4c}{2b + c}\right|$

22. $-2bc + \left|\frac{bc + d}{ab - c}\right|$

23. $2(d - b) \div (3a - c)$

24. $(d - 4a)^2 \div c^3$

Evaluate the variable expression when $a = 1.5$, $b = -2.4$, and $c = -0.5$.

25. $b^2 - ac$

26. $(a - b)^2 \div c^2$

27. $\frac{a^2}{c} - (b - a)^2$

2 Use the given Property of the Real Numbers to complete the statement.

28. The Commutative Property of Multiplication
$3 \cdot 4 = 4 \cdot ?$

29. The Commutative Property of Addition
$7 + 15 = ? + 7$

30. The Associative Property of Addition
$(3 + 4) + 5 = ? + (4 + 5)$

31. The Associative Property of Multiplication
$(3 \cdot 4) \cdot 5 = 3 \cdot (? \cdot 5)$

32. The Inverse Property of Addition
$2 + ? = 0$

33. The Multiplication Property of Zero
$5 \cdot ? = 0$

34. The Distributive Property
$3(x + 2) = 3x + ?$

35. The Distributive Property
$5(y + 4) = ? \cdot y + 20$

36. The Commutative Property of Multiplication
$b(2a) = ?b$

37. The Inverse Property of Addition
$(x + y) + ? = 0$

38. The Inverse Property of Multiplication
$\frac{1}{mn}(mn) = ?$

39. The Multiplication Property of One
$? \cdot 1 = x$

40. The Associative Property of Multiplication
$2(3x) = ? \cdot x$

41. The Commutative Property of Addition
$ab + bc = bc + ?$

Identify the property that justifies the statement.

42. $5(a6) = 5(6a)$

43. $-8 + 8 = 0$

44. $(-12)\left(-\frac{1}{12}\right) = 1$

45. $(3 \cdot 4) \cdot 2 = 2 \cdot (3 \cdot 4)$

46. $y + 0 = y$

47. $2x + (5y + 8) = (2x + 5y) + 8$

48. $3(b + a) = 3(a + b)$

49. $(x + y)z = xz + yz$

50. $6(x + y) = 6x + 6y$

51. $(-12y)(0) = 0$

52. $(ab)c = a(bc)$

53. $(x + y) + z = (y + x) + z$

3 Simplify.

54. $5x + 7x$

55. $3x + 10x$

56. $-8ab - 5ab$

57. $-2x + 5x - 7x$

58. $3x - 5x + 9x$

59. $-2a + 7b + 9a$

60. $5b - 8a - 12b$

61. $12\left(\frac{1}{12}x\right)$

62. $\frac{1}{3}(3y)$

63. $-3(x - 2)$

64. $-5(x - 9)$

65. $(x + 2)5$

66. $-(x + y)$

67. $-(-x - y)$

68. $3(-2a + 3a - 5)$

69. $3(x - 2y) - 5$

70. $4x - 3(2y - 5)$

71. $-2a - 3(3a - 7)$

72. $3x - 2(5x - 7)$

73. $2x - 3(x - 2y)$

74. $3[a - 5(5 - 3a)]$

75. $5[-2 - 6(a - 5)]$

76. $3[x - 2(x + 2y)]$

77. $5[y - 3(y - 2x)]$

78. $-2(x - 3y) + 2(3y - 5x)$

79. $4(-a - 2b) - 2(3a - 5b)$

80. $5(3a - 2b) - 3(-6a + 5b)$

81. $-7(2a - b) + 2(-3b + a)$

82. $3x - 2[y - 2(x + 3[2x - y])]$

83. $2x - 4[x - 4(y - 2[5y + 3])]$

84. $4 - 2(7x - 2y) - 3(-2x + 3y)$

85. $3x + 8(x - 4) - 3(2x - y)$

86. $\frac{1}{3}[8x - 2(x - 12) + 3]$

87. $\frac{1}{4}[14x - 3(x - 8) - 7x]$

4 Translate into a variable expression. Then simplify.

88. a number minus the sum of the number and two

89. a number decreased by the difference between five and the number

90. the sum of one third of a number and four fifths of the number

91. the difference between three eighths of a number and one sixth of the number

92. five times the product of eight and a number

93. a number increased by two thirds of the number

94. the difference between the product of seventeen and a number and twice the number

95. one half the total of six times a number and twenty-two

96. three times a number plus the difference between the number and five

97. sixteen minus the difference between five times a number and four

98. twice a number plus the product of three more than the number and five

99. two thirds of a number increased by five eighths of the number

100. one third of the total of six times a number and twelve

101. three times the quotient of twice a number and six

102. the sum of five times a number and twelve added to the product of fifteen and the number

103. four less than twice the sum of a number and eleven

104. twice a number plus the product of two more than the number and eight

105. twenty minus the product of four more than a number and twelve

106. a number added to the product of five plus the number and four

107. a number plus the product of the number minus twelve and three

SUPPLEMENTAL EXERCISES 1.2

Name the property that justifies each lettered step used in simplifying the expression.

108. $3(x + y) + 2x$
- **a.** $(3x + 3y) + 2x$
- **b.** $(3y + 3x) + 2x$
- **c.** $3y + (3x + 2x)$
- **d.** $3y + (3 + 2)x$
  $3y + 5x$

109. $3a + 4(b + a)$
- **a.** $3a + (4b + 4a)$
- **b.** $3a + (4a + 4b)$
- **c.** $(3a + 4a) + 4b$
- **d.** $(3 + 4)a + 4b$
  $7a + 4b$

110. $y + (3 + y)$
- **a.** $y + (y + 3)$
- **b.** $(y + y) + 3$
- **c.** $(1y + 1y) + 3$
- **d.** $(1 + 1)y + 3$
  $2y + 3$

111. $5(3a + 1)$
- **a.** $5(3a) + 5(1)$
- **b.** $(5 \cdot 3)a + 5(1)$
  $15a + 5(1)$
- **c.** $15a + 5$

Write a variable expression.

112. The length of a rectangle is 5 m more than the width. Write a variable expression for the length of the rectangle in terms of the width.

113. A mixture contains three times as many peanuts as cashews. Write a variable expression for the amount of peanuts in terms of the amount of cashews.

114. One cyclist rode 4 mph faster than a second cyclist. Write a variable expression for the speed of the first cyclist in terms of the speed of the second.

115. In a triangle, the measure of one angle is one third the measure of the largest angle. Write a variable expression for the measure of the smaller angle in terms of the largest angle.

116. In a coin bank, the number of nickels is 3 more than twice the number of dimes. Write a variable expression for the number of nickels in terms of the number of dimes.

117. The age of a gold coin is 15 years less than twice the age of a silver coin. Write a variable expression for the age of the gold coin in terms of the age of the silver coin.

SECTION 1.3 Sets

1 The union and intersection of sets

A **set** is a collection of objects. The objects in a set are the **elements** of the set.

A set can be written in various ways. The **roster method** of writing a set encloses the list of the elements of the set in braces.

The set of the three planets nearest the sun is written {Mercury, Venus, Earth}.

The set of even natural numbers less than 10 is written $\{2, 4, 6, 8\}$. This is an example of a **finite set.** All the elements of the set can be listed.

The set of natural numbers greater than 5 is written $\{6, 7, 8, 9, 10, 11, \ldots\}$. The three dots mean that the pattern of numbers continues without end. This is an example of an **infinite set.** It is impossible to list all of the elements of the set.

Example 1 Use the roster method to write the set of positive prime numbers less than 15.

Solution $A = \{2, 3, 5, 7, 11, 13\}$ ▶ A set is usually designated by a capital letter.

Problem 1 Use the roster method to write the set of even natural numbers less than 12.

Solution See page A8.

The symbol $\in$ means *is an element of.*

$9 \in B$ is read "9 is an element of set B."

Given $A = \{-1, 5, 9\}$, then $-1 \in A$, $5 \in A$, and $9 \in A$. $12 \notin A$ is read "12 is not an element of A."

The **empty set,** or **null set,** is the set that contains no elements. The symbol $\emptyset$ or $\{\ \}$ is used to represent the empty set. (Note: It is incorrect to write $\{\emptyset\}$ as the empty set.)

The set of trees over 1000 ft tall is the empty set.

A second method of representing a set is **set builder notation.** The set of all integers greater than −3 would be written:

$$\{x|x > -3, x \text{ is an integer}\}$$

and is read "the set of all x such that x is greater than −3 and x is an integer."

Set builder notation can be used to describe almost any set, but it is especially useful when writing infinite sets.

Using set builder notation, the set of real numbers less than 5 is written

$$\{x|x < 5, x \in \text{real numbers}\}$$

and is read "the set of all x such that x is less than 5 and x is an element of the real numbers." This is an infinite set. It is impossible to list all the elements in this set.

Example 2 Write $\{x|x < 10, x \text{ is a natural number}\}$ by using the roster method.

Solution $A = \{1, 2, 3, 4, 5, 6, 7, 8, 9\}$

Problem 2 Write $\{x|x < 10, x \text{ is a positive odd integer}\}$ by using the roster method.

Solution See page A8.

Example 3 Is $-7 \in \{x|x > 5, x \in \text{real numbers}\}$?

Solution −7 is not greater than 5.
No, −7 is not an element of the set.

Problem 3 Is $0.35 \in \{x|x < 2, x \in \text{real numbers}\}$?

Solution See page A8.

Just as operations such as addition and multiplication are performed on real numbers, operations are performed on sets.

The **union** of two sets, written $A \cup B$, is the set of all elements which belong to either A **or** B. In set builder notation, this is written

$$A \cup B = \{x|x \in A \text{ or } x \in B\}.$$

Given $A = \{2, 3, 5, 7\}$ and $B = \{0, 1, 2, 3, 4\}$, the union of A and B contains all the elements which belong to either A or B. The elements which belong to both sets are listed only once.

$A \cup B = \{0, 1, 2, 3, 4, 5, 7\}$

The **intersection** of two sets, written $A \cap B$, is the set of all elements which are common to both A **and** B. In set builder notation, this is written

$$A \cap B = \{x|x \in A \text{ and } x \in B\}.$$

Given $A = \{2, 3, 5, 7\}$ and $B = \{0, 1, 2, 3, 4\}$, the intersection of A and B contains all the elements which are common to both A and B.

$A \cap B = \{2, 3\}$

Example 4 Find $C \cup D$ given $C = \{1, 5, 9, 13, 17\}$ and $D = \{3, 5, 7, 9, 11\}$.

Solution $C \cup D = \{1, 3, 5, 7, 9, 11, 13, 17\}$

Problem 4 Find $A \cup C$ given $A = \{-2, -1, 0, 1, 2\}$ and $C = \{-5, -1, 0, 1, 5\}$.

Solution See page A8.

Example 5 Find $A \cap B$ given $A = \{x|x \text{ is a natural number}\}$ and $B = \{x|x \text{ is a negative integer}\}$.

Solution There are no natural numbers that are also negative numbers.
$A \cap B = \varnothing$

Problem 5 Find $E \cap F$ given $E = \{x|x \text{ is an odd integer}\}$ and $F = \{x|x \text{ is an even integer}\}$.

Solution See page A8.

2 Graph the solution set of an inequality in one variable

An **inequality** expresses the relative order of two mathematical expressions. The symbols $>$, $<$, $\leq$, and $\geq$ are used to write inequalities. The symbol $\leq$ means is less than or equal to. The symbol $\geq$ means is greater than or equal to.

$7 > -2$
$2x < 4$
$2x - 3y \geq 6$
$x^2 - 2x - 3 \leq 0$

The **solution set of an inequality** is the set of real numbers that makes the inequality true. The solution set can be graphed on the number line.

The graph of the solution set of $x > -2$ is shown below. The solution set is the real numbers greater than -2. The circle on the graph indicates that -2 is not included in the solution set.

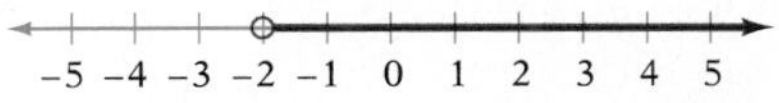

The solution set is written $\{x|x > -2, x \in \text{real numbers}\}$.

The graph of the solution set of $x \geq -2$ is shown below. The dot at -2 indicates that -2 is included in the solution set.

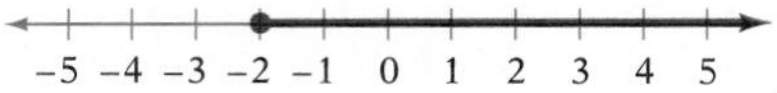

The solution set is written $\{x|x \geq -2, x \in \text{real numbers}\}$.

For the remainder of this section, all variables will represent real numbers. Using this convention, the solution set above would be written $\{x|x \geq -2\}$.

Example 6 Graph the solution set of $x \leq 3$.

Solution The solution set is $\{x|x \leq 3\}$.

Problem 6 Graph the solution set of $x > -3$.

Solution See page A8.

The union of two sets is the set of all elements belonging to either one or the other of the two sets.

The set $\{x|x \leq -1\} \cup \{x|x > 3\}$ is the set of all numbers that are either less than or equal to -1 or greater than 3.

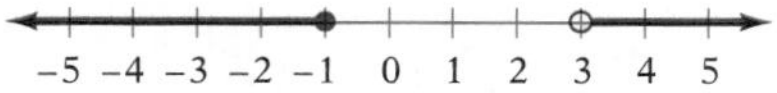

The set is written $\{x|x \leq -1 \text{ or } x > 3\}$.

The set $\{x|x > 2\} \cup \{x|x > 4\}$ is the set of all numbers that are either greater than 2 or greater than 4.

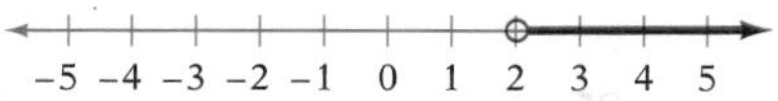

The set is written $\{x|x > 2\}$.

The intersection of two sets is the set that contains the elements common to both sets.

The set $\{x|x > -2\} \cap \{x|x < 5\}$ is the set of numbers that are greater than -2 and less than 5.

The set can be written $\{x|x > -2 \text{ and } x < 5\}$. However, it is more commonly written $\{x|-2 < x < 5\}$.

The set $\{x|x < 4\} \cap \{x|x < 5\}$ is the set of numbers that are less than 4 and less than 5.

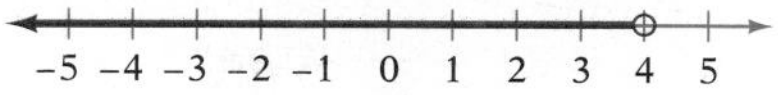

The set is written $\{x|x < 4\}$.

Example 7 Graph the set $\{x|x < 0\} \cap \{x|x > -3\}$.

Solution The set is $\{x|-3 < x < 0\}$.

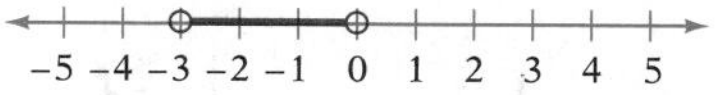

Problem 7 Graph the set $\{x|x \geq 1\} \cup \{x|x \leq -3\}$.

Solution See page A8.

EXERCISES 1.3

1 Use the roster method to write the set.

1. the integers between −3 and 5

2. the integers between −4 and 0

3. the even natural numbers less than 13

4. the odd natural numbers less than 13

5. the prime numbers between 2 and 5

6. the prime numbers between 30 and 40

7. the positive integers less than 20 that are divisible by 3

8. the perfect square integers less than 100

Use set builder notation to write the set.

9. the integers greater than 4

10. the integers less than −2

11. the real numbers greater than or equal to 1

12. the real numbers less than or equal to −3

13. the integers between −2 and 5

14. the integers between −5 and 2

15. the real numbers between 0 and 1

16. the real numbers between −2 and 4

Find $A \cup B$.

17. $A = \{1, 4, 9\}$, $B = \{2, 4, 6\}$

18. $A = \{-1, 0, 1\}$, $B = \{0, 1, 2\}$

19. $A = \{2, 3, 5, 8\}$, $B = \{9, 10\}$

20. $A = \{1, 3, 5, 7\}$, $B = \{2, 4, 6, 8\}$

21. $A = \{-4, -2, 0, 2, 4\}$, $B = \{0, 4, 8\}$

22. $A = \{-3, -2, -1\}$, $B = \{-2, -1, 0, 1\}$

23. $A = \{1, 2, 3, 4, 5\}$, $B = \{3, 4, 5\}$

24. $A = \{2, 4\}$, $B = \{0, 1, 2, 3, 4, 5\}$

Find $A \cap B$.

25. $A = \{6, 12, 18\}$, $B = \{3, 6, 9\}$

26. $A = \{-4, 0, 4\}$, $B = \{-2, 0, 2\}$

27. $A = \{1, 5, 10, 20\}$, $B = \{5, 10, 15, 20\}$

28. $A = \{-9, -5, 0, 7\}$, $B = \{-7, -5, 0, 5, 7\}$

29. $A = \{1, 2, 4, 8\}$, $B = \{3, 5, 6, 7\}$

30. $A = \{-3, -2, -1, 0\}$, $B = \{1, 2, 3, 4\}$

31. $A = \{2, 4, 6, 8, 10\}$, $B = \{4, 6\}$

32. $A = \{1, 3, 5, 7, 9\}$, $B = \{1, 9\}$

2 Graph the set.

33. $\{x|x \geq 3\}$

34. $\{x|x \leq -2\}$

35. $\{x|x > 0\}$

36. $\{x|x < 4\}$

37. $\{x|x > 1\} \cup \{x|x < -1\}$

38. $\{x|x \leq 2\} \cup \{x|x > 4\}$

39. $\{x|x \leq 2\} \cap \{x|x \geq 0\}$

40. $\{x|x > -1\} \cap \{x|x \leq 4\}$

41. $\{x|x > 1\} \cap \{x|x \geq -2\}$

42. $\{x|x < 4\} \cap \{x|x \leq 0\}$

43. $\{x|x > 2\} \cup \{x|x > 1\}$

44. $\{x|x < -2\} \cup \{x|x < -4\}$

SUPPLEMENTAL EXERCISES 1.3

Graph the set.

45. $\left\{x|x > \frac{3}{2}\right\} \cup \left\{x|x < -\frac{1}{2}\right\}$

46. $\{x|x \leq -2.5\} \cup \{x|x > 1.5\}$

47. $\left\{x|x > -\frac{5}{2}\right\} \cap \left\{x|x \leq \frac{7}{3}\right\}$

48. $\{x|x > -0.5\} \cap \{x|x \geq 3.5\}$

Graph the solution set.

49. $|x| < 2$

50. $|x| < 5$

51. $|x| > 3$

52. $|x| > 4$

Use set builder notation to write $A \cup B$.

53. $A = \{1, 3, 5, 7, \ldots\}$
$B = \{2, 4, 6, 8, \ldots\}$

54. $A = \{\ldots, -6, -4, -2\}$
$B = \{\ldots, -5, -3, -1\}$

Use set builder notation to write $A \cap B$.

55. $A = \{15, 17, 19, 21, \ldots\}$
$B = \{11, 13, 15, 17, \ldots\}$

56. $A = \{-12, -10, -8, -6, \ldots\}$
$B = \{-4, -2, 0, 2, \ldots\}$

Solve.

57. Given that a, b, c, and d are real numbers, which of the following will ensure that $a + c < b + d$?

a. $a < b$ and $c < d$ **b.** $a > b$ and $c > d$
c. $a < b$ and $c > d$ **d.** $a > b$ and $c < d$

58. Given that a and b are real numbers, which of the following will ensure that $a^2 < b^2$?

a. $a < b$ **b.** $a > b$
c. $0 < a < b$ **d.** $a < b < 0$

59. Given that a, b, c, and d are positive real numbers, which of the following will ensure that $\frac{a - b}{c - d} \leq 0$?

a. $a \geq b$ and $c > d$ **b.** $a \leq b$ and $c > d$
c. $a \geq b$ and $c < d$ **d.** $a \leq b$ and $c < d$

60. Given that a, b, and c are nonzero real numbers and $a < b < c$, which of the following will ensure that $\frac{1}{a} > \frac{1}{b} > \frac{1}{c}$?

a. $a > 0$ **b.** $c < 0$
c. $a < 0, c > 0$ **d.** $b < c$

Calculators and Computers

Euclidean Algorithm

An algorithm is a method or procedure that repeats the same sequence of steps over and over. For example, to divide two whole numbers, the division algorithm is used.

$$\begin{array}{r} 2 \\ 3\overline{)747} \\ \underline{-6} \\ 14 \end{array}$$

1. Estimate the quotient. $\left(3\overline{)7}^{\,2}\right)$
2. Multiply the quotient times the divisor. $(2 \times 3 = 6)$
3. Subtract. $(7 - 6 = 1)$
4. "Bring down" the next digit. (4)

$$\begin{array}{r} 249 \\ 3\overline{)747} \\ \underline{-6} \\ 14 \\ \underline{-12} \\ 27 \\ \underline{-27} \\ 0 \end{array}$$

The division problem is completed by repeating these four steps until there are no numbers to bring down. The algorithm is the four steps: estimate, multiply, subtract, and bring down.

The Euclidean Algorithm is a procedure to find the greatest common divisor (GCD) of two numbers.

Find the GCD of 15 and 25 using the Euclidean Algorithm.

$$\begin{array}{r} 1 \\ 15\overline{)25} \\ \underline{-15} \\ 10 \end{array}$$

1. Divide the smaller number into the larger number.

$$\begin{array}{r} 1 \\ 10\overline{)15} \\ \underline{-10} \\ 5 \end{array}$$

2. Divide the remainder (10) into the divisor (15).

$$\begin{array}{r} 2 \\ 5\overline{)10} \\ \underline{-10} \\ 0 \end{array}$$

3. Repeat Step 2 until the remainder is zero.
4. The divisor in the last division is the GCD.

The GCD is 5.

The Euclidean Algorithm is the four listed steps. A computer program can be written to carry out this procedure for any two positive integers. One such program is on the Math ACE Disk.

Find the GCD of 5 and 8 using the Euclidean Algorithm. You may use the program on the Math ACE Disk to check your answer.

Something Extra

Venn Diagrams

In working with sets, diagrams can be very useful. These diagrams are called Venn diagrams.

In the Venn diagram at the right, the rectangle represents the set U and the circles, A and B, represent subsets of set U. The common area of the two circles represents the intersection of A and B.

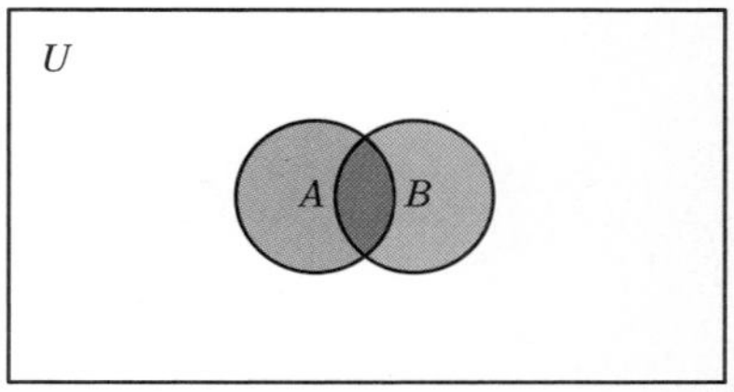

This Venn Diagram represents the set $U = \{1, 2, 3, 4, 5, 6, 7, 8\}$. The sets $A = \{2, 3, 4, 5\}$ and $B = \{4, 5, 6, 7\}$ are subsets of U. Note that 1 and 8 are in set U but not in A or B. The numbers 4 and 5 are in both set A and set B.

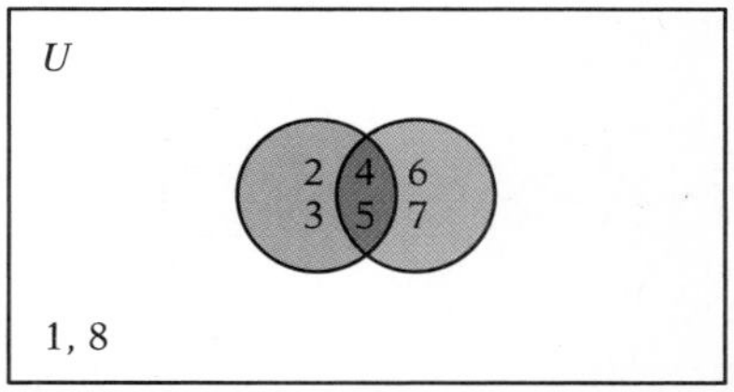

Sixty-five students at a small college enrolled in the following courses.

39 enrolled in English.
26 enrolled in mathematics.
35 enrolled in history.
19 enrolled in English and history.
11 enrolled in English and mathematics.
9 enrolled in mathematics and history.
2 enrolled in all three courses.

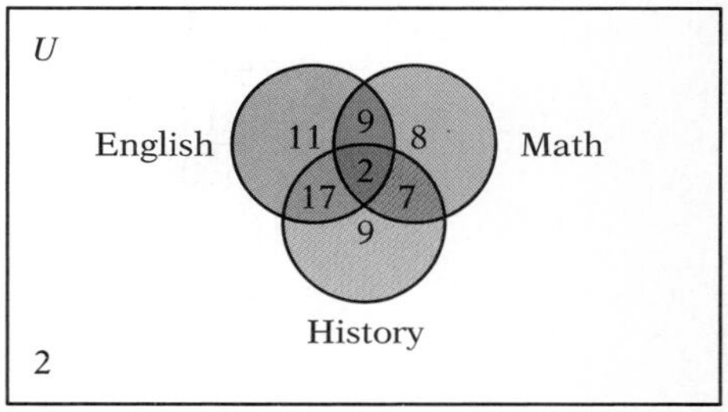

The Venn diagram is shown above. The diagram was drawn by placing the 2 students in the intersection of all three courses. Since 9 students took history and mathematics and 2 students are already in the intersection of history and mathematics, then 7 more students are placed in the intersection of history and mathematics. Continue in this manner until all information is used.

a. How many students enrolled only in English and mathematics?

b. How many students enrolled in English, but did not enroll in mathematics or history?

c. How many students did not enroll in any of the three courses?

Complete the following.

1. Sketch the Venn diagram for $U = \{1, 2, 3, 4, 5, 6, 7, 8\}$, $A = \{2, 3, 4, 5, 6\}$, $B = \{4, 5, 6, 7, 8\}$, and $C = \{4, 5\}$.
   a. What numbers are in B and not in C?
   b. What numbers are in A and C, but not in B?
   c. What numbers are in U, but not in A, B, or C?

2. A busload of 50 scouts stopped at a fast food restaurant and ordered hamburgers. The scouts could have pickles, or tomatoes, or lettuce on their hamburgers.

   31 ordered pickles.
   36 ordered tomatoes.
   31 ordered lettuce.
   21 ordered pickles and tomatoes.
   24 ordered tomatoes and lettuce.
   22 ordered pickles and lettuce.
   17 ordered all three.

   a. How many scouts ordered a hamburger with pickles only?
   b. How many scouts ordered a hamburger with lettuce and tomatoes without the pickle?
   c. How many scouts ordered a hamburger without a pickle, tomato, or lettuce?

Chapter Summary

Key Words

The *integers* are . . . , −4, −3, −2, −1, 0, 1, 2, 3, 4, ...

The *negative integers* are the integers . . . , −4, −3, −2, −1.

The *positive integers* are the integers 1, 2, 3, 4, ...

The positive integers and zero are called the *whole numbers.*

A *rational number* is a number of the form $\frac{a}{b}$, where a and b are integers, and b is not equal to zero.

An *irrational number* is a number whose decimal representation never terminates nor repeats.

The rational numbers and the irrational numbers taken together are called the *real numbers.*

The *absolute value* of a number is a measure of its distance from zero on the number line.

The expression a^n is in *exponential form*, where a is the base, and n is the exponent.

A *complex fraction* is a fraction whose numerator or denominator contains one or more fractions.

The *Order of Operations Agreement* is used to simplify numerical expressions.

A *variable expression* is an expression that contains one or more variables.

The *terms* of a variable expression are the addends of the expression.

A *variable term* is composed of a numerical coefficient and a variable part.

The *additive inverse* of a number is the opposite of the number.

The *multiplicative inverse* of a number is the reciprocal of the number.

A *set* is a collection of objects. The objects of the set are the *elements* of the set.

The *roster method* of writing a set encloses a list of the elements of the set in braces.

A *finite set* is a set in which the elements can be counted.

An *infinite set* is a set in which it is impossible to list all the elements.

The *empty set*, or *null set*, written $\emptyset$ or $\{\ \}$, is the set that contains no elements.

The *union* of two sets, written $A \cup B$, is the set that contains all the elements of A and all the elements of B. (The elements in both set A and set B are listed only once.)

The *intersection* of two sets, written $A \cap B$, is the set that contains the elements that are common to both A and B.

An *inequality* is an expression that contains the symbol $<$, $>$, $\leq$, or $\geq$.

The *solution set of an inequality* is the set of real numbers that makes the inequality true.

Essential Rules

The Commutative Property of Addition — If a and b are real numbers, then $a + b = b + a$.

The Associative Property of Addition — If a, b, and c are real numbers, then $a + (b + c) = (a + b) + c$.

The Commutative Property of Multiplication — If a and b are real numbers, then $ab = ba$.

The Associative Property of Multiplication — If a, b, and c are real numbers, then $a(bc) = (ab)c$.

The Addition Property of Zero	If a is a real number, then $a + 0 = 0 + a = a$.
The Multiplication Property of One	If a is a real number, then $a \cdot 1 = 1 \cdot a = a$.
The Inverse Property of Addition	If a is a real number, then $a + (-a) = (-a) + a = 0$.
The Inverse Property of Multiplication	If a is a nonzero real number, then $a \cdot \frac{1}{a} = \frac{1}{a} \cdot a = 1$.
The Distributive Property	If a, b, and c are real numbers, then $a(b + c) = ab + ac$.

Chapter Review

1. Find the additive inverse of 23.
2. Find the additive inverse of -15.
3. Evaluate $|-7|$.
4. Evaluate $-|-5|$.
5. Simplify: $-10 - (-3) - 8$
6. Simplify: $(-2)(-3)(-12)$
7. Simplify: $-204 \div (-17)$
8. Simplify: $|-12 - (-8)|$
9. Simplify: $|-12| - |-16|$
10. Simplify: $18 - |-12 + 8|$
11. Simplify: $-3^2(-2)^3$
12. Simplify: $(-2)^2(-3)^3$
13. Simplify: $-2 \cdot (4^2) \cdot (-3)^2$
14. Simplify: $5(-4)^3(-2)^3$
15. Simplify: $-\frac{3}{8} + \frac{3}{5} - \frac{1}{6}$
16. Simplify: $\frac{4}{15} - \frac{3}{5} - \frac{5}{6}$
17. Simplify: $\frac{3}{5}\left(\frac{10}{21}\right)\left(\frac{7}{15}\right)$
18. Simplify: $\left(-\frac{6}{7}\right)\left(\frac{4}{33}\right)\left(\frac{11}{12}\right)$
19. Simplify: $-\frac{3}{8} \div \frac{3}{5}$
20. Simplify: $-\frac{13}{24} \div \left(-\frac{3}{8}\right)$
21. Simplify: $-4.07 + 2.3 - 1.07$
22. Simplify: $(2.1)(0.6)(-3.5)$
23. Simplify: $-3.286 \div (-1.06)$
24. Simplify: $1.547 \div 0.035$
25. Simplify: $16 - 4(3 - 5)^2$
26. Simplify: $24 \div \frac{3^2 - 2^2}{3^2 + 2^2}$
27. Simplify: $\frac{3}{8} - \left(\frac{3}{5} \div \frac{9}{10}\right) + \frac{3}{4}$
28. Simplify: $\frac{3}{4} \div \left(\frac{7}{12} - \frac{3}{8}\right) + 2$
29. Evaluate $2a^2 - \frac{3b}{a}$ when $a = -3$ and $b = 2$.
30. Evaluate $\frac{3ac - b}{3b + c}$ when $a = 3$, $b = -2$, and $c = 4$.
31. Evaluate $-b|2a - b|$ when $a = -4$ and $b = 2$.
32. Evaluate $(a - 2b^2) \div (ab)$ when $a = 4$ and $b = -3$.

33. Use the Distributive Property to complete the statement.
$6x - 21y = ?(2x - 7y)$

34. Use the Commutative Property of Addition to complete the statement.
$3(x + y) = 3(? + x)$

35. Use the Commutative Property of Multiplication to complete the statement.
$(ab)14 = 14(?)$

36. Use the Associative Property of Addition to complete the statement.
$3 + (4 + y) = (3 + ?) + y$

37. Identify the property that justifies the statement.
$4 - 4 = 0$

38. Identify the property that justifies the statement.
$4 - 2(x - y) = 4 - 2x + 2y$

39. Identify the property that justifies the statement.
$2(3x) = (2 \cdot 3)x$

40. Identify the property that justifies the statement.
$3(x + y) = 3(y + x)$

41. Simplify: $3 - 3[2 - (4 - a)]$

42. Simplify: $-2(x - 3) + 4(2 - x)$

43. Simplify: $3x - 2[y - 3(2 + x)] - y$

44. Simplify: $4y - 3[x - 2(3 - 2x)] - 4y$

45. Translate and simplify "five less the product of two more than a number and two."

46. Translate and simplify "three less than twice the sum of a number and five."

47. Translate and simplify "the difference between five eighths of a number and three fourths of a number."

48. Translate and simplify "twelve minus the quotient of three more than a number and the number."

49. Find $A \cup B$ given $A = \{1, 3, 5, 7\}$ and $B = \{2, 4, 6, 8\}$.

50. Find $A \cup B$ given $A = \{-2, -1, 0, 1, 2\}$ and $B = \{0, 1, 2, 3\}$.

51. Find $A \cap B$ given $A = \{4, 6, 8, 10\}$ and $B = \{5, 7, 9, 11\}$.

52. Find $A \cap B$ given $A = \{0, 1, 2, 3\}$ and $B = \{2, 3, 4, 5\}$.

53. Graph the solution set of $x \geq -3$.

54. Graph the solution set of $x < 1$.

55. Graph the solution set of $x > 2$.

56. Graph the solution set of $x \leq -1$.

57. Graph the set $\{x|x \leq -3\} \cup \{x|x > 0\}$.

58. Graph the set $\{x|x > -2\} \cup \{x|x \leq 4\}$.

59. Graph the set $\{x|x \leq 4\} \cap \{x|x > -2\}$.

60. Graph the set $\{x|x \leq -1\} \cap \{x|x > 3\}$.

Chapter Test

1. Simplify: $8 - 4(2 - 3)^2 \div 2$

2. Simplify: $|-3 - (-5)|$

3. Simplify: $\frac{2}{3} - \frac{5}{12} + \frac{4}{9}$

4. Evaluate $\frac{b^2 - c^2}{a - 2c}$ when $a = 2$, $b = 3$, and $c = -1$.

5. Simplify: $-3^2 \cdot (-2)^2$

6. Simplify: $\left(-\frac{2}{3}\right)\left(\frac{9}{15}\right)\left(\frac{10}{27}\right)$

7. Simplify: $15 \div \frac{\frac{2}{3} - 2}{4} \cdot 6$

8. Simplify: $12 - 4\left(\frac{5^2 - 1}{3}\right) \div 16$

9. Evaluate $(a - b)^2 \div (2b + 1)$ when $a = 2$ and $b = -3$.

10. Evaluate $-2b - \left|\frac{b - a}{b^3 + 2a}\right|$ when $a = 2$ and $b = -1$.

11. Identify the property that justifies the statement.
$-2(x + y) = -2x - 2y$

12. Use the Commutative Property of Addition to complete the statement.
$(3 + 4) + 2 = (? + 3) + 2$

13. Simplify: $3x - 2(x - y) - 3(y - 4x)$

14. Simplify: $2x - 4[2 - 3(x + 4y) - 2]$

15. Simplify: $3x - [2x - 3(y - 2x) + 5y]$

16. Find $A \cup B$ given $A = \{-3, -1, 0, 1, 3\}$ and $B = \{0, 1, 3, 5\}$.

17. Find $A \cup B$ given $A = \{2, 3, 5, 8\}$ and $B = \{2, 4, 6, 8\}$.

18. Find $A \cap B$ given $A = \{-4, -2, 0, 2\}$ and $B = \{-4, 0, 4, 8\}$.

19. Graph the solution set of $x \leq 3$.

20. Graph the solution set of $x < 5$.

21. Graph the set $\{x|x \geq 2\} \cup \{x|x < -1\}$.

22. Graph the set $\{x|x < 3\} \cap \{x|x > -2\}$.

23. Translate and simplify "thirteen decreased by the product of three less than a number and nine."

24. Translate and simplify "eight minus the difference between four times a number and twelve."

25. Translate and simplify "the difference between seven eighths of a number and three fifths of the number."

2 First-Degree Equations and Inequalities

Objectives

- Solve equations using the Addition and Multiplication Properties of Equations
- Solve equations using the Distributive Property
- Application problems
- Coin and stamp problems
- Integer problems
- Value mixture problems
- Uniform motion problems
- Investment problems
- Percent mixture problems
- Solve inequalities in one variable
- Solve compound inequalities
- Application problems
- Absolute value equations
- Absolute value inequalities
- Application problems

POINT OF INTEREST

Moscow Papyrus

Most of the early history of mathematics can be traced to the Egyptians and Babylonians. This early work began around 3000 B.C. The Babylonians used clay tablets and a type of writing called *cuneiform* (wedge-shape) to record their thoughts and discoveries.

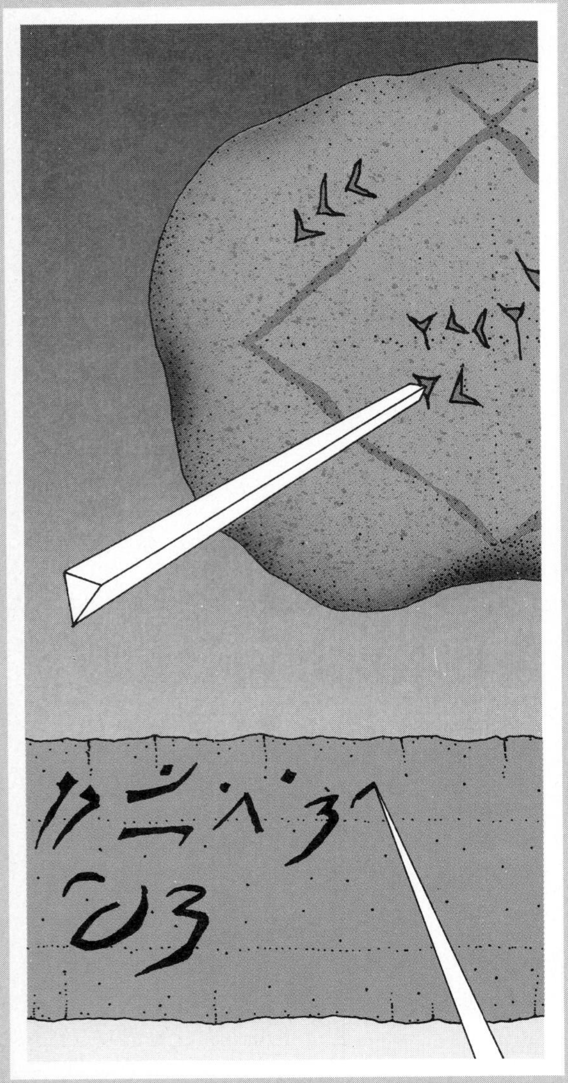

Here are the symbols for 10, 1, and subtraction.

Simple groupings of these symbols were used to represent numbers less than 60. The numbers 32 and 28 are shown below.

The Egyptians used papyrus, a plant reed dried and pounded thin, and hieratic writing, that was derived from hieroglyphics, to make records. A number of papyrus documents have been discovered over the years. One particularly famous one is called the Moscow Papyrus, which dates from 1850 B.C. It is approximately 18 feet long and three inches wide and contains 25 problems.

The Moscow Papyrus dealt with solving practical problems which were related to geometry, food preparation, and grain allotments. Here is one of the problems:

> The width of a rectangle is $\frac{3}{4}$ the length and the area is 12.
>
> Find the dimensions of the rectangle.

You might recognize this as a type of problem you have solved before. Word problems are very old indeed. This one is around 3900 years old.

SECTION **2.1**

Equations in One Variable

1 Solve equations using the Addition and Multiplication Properties of Equations

An **equation** expresses the equality of two mathematical expressions. The expressions can be either numerical or variable expressions.

$$\left.\begin{array}{l} 2 + 8 = 10 \\ x + 8 = 11 \\ x^2 + 2y = 7 \end{array}\right\} \text{Equations}$$

The equation at the right is a **conditional equation.** The equation is true if the variable is replaced by 3. The equation is false if the variable is replaced by 4.

$x + 2 = 5$	Conditional Equation
$3 + 2 = 5$	A true equation
$4 + 2 = 5$	A false equation

The replacement values of the variable that will make an equation true are called the **roots,** or **solutions,** of the equation.

The solution of the equation $x + 2 = 5$ is 3.

The equation at the right is an **identity.** Any replacement for x will result in a true equation.

$x + 2 = x + 2$ Identity

The equation at the right has **no solution** since there is no number that equals itself plus 1. Any replacement value for x will result in a false equation.

$x = x + 1$ No solution

Each of the equations at the right is a **first-degree equation in one variable.** All variables have an exponent of 1.

$$\left.\begin{array}{l} x + 2 = 12 \\ 3y - 2 = 5y \\ 3(a + 2) = 14a \end{array}\right. \quad \text{First-Degree Equations}$$

To **solve** an equation means to find a solution of the equation. The simplest equation to solve is an equation of the form

variable = constant

since the constant is the solution.

If $x = 3$, then 3 is the solution of the equation since $3 = 3$ is a true equation.

In solving an equation, the goal is to rewrite the given equation in the form variable = constant. The Addition Property of Equations can be used to rewrite an equation in this form.

The Addition Property of Equations

If $a = b$ and c is a real number, then the equations $a = b$ and $a + c = b + c$ have the same solutions.

The Addition Property of Equations states that the same number can be added to each side of an equation without changing the solution. This property is used to remove a *term* from one side of the equation by adding the opposite of that term to each side of the equation.

Solve: $x - 3 = 7$

$$x - 3 = 7$$

Add the opposite of the constant term -3 to each side of the equation and simplify. After simplifying, the equation is in the form variable = constant.

$$x - 3 + 3 = 7 + 3$$
$$x + 0 = 10$$
$$x = 10$$

To check the solution, replace the variable with 10. Simplify each side of the equation. Since $7 = 7$ is a true equation, 10 is a solution.

Check: $x - 3 = 7$

$$10 - 3 \mid 7$$
$$7 = 7$$

The solution is 10.

Because subtraction is defined in terms of addition, the Addition Property of Equations allows the same number to be subtracted from each side of an equation without changing the solution of the equation.

Solve: $x + \frac{7}{12} = \frac{1}{2}$

$$x + \frac{7}{12} = \frac{1}{2}$$

Add the opposite of the constant term $\frac{7}{12}$ to each side of the equation. This is equivalent to subtracting $\frac{7}{12}$ from each side of the equation.

$$x + \frac{7}{12} - \frac{7}{12} = \frac{1}{2} - \frac{7}{12}$$
$$x + 0 = \frac{6}{12} - \frac{7}{12}$$
$$x = -\frac{1}{12}$$

You should check this solution.

The solution is $-\frac{1}{12}$.

The Multiplication Property of Equations is used to rewrite an equation in the form variable = constant.

The Multiplication Property of Equations

If $a = b$ and c is a real number, $c \neq 0$, then the equations $a = b$ and $ac = bc$ have the same solutions.

The Multiplication Property of Equations states that each side of an equation can be multiplied by the same nonzero number without changing the solution of the equation. This property is used to remove a *coefficient* from a variable term in an equation by multiplying each side of the equation by the reciprocal of the coefficient.

Solve: $-\frac{3}{4}x = 12$

Multiply each side of the equation by $-\frac{4}{3}$, the reciprocal of $-\frac{3}{4}$. After simplifying, the equation is in the form variable = constant.

$$-\frac{3}{4}x = 12$$
$$\left(-\frac{4}{3}\right)\left(-\frac{3}{4}\right)x = \left(-\frac{4}{3}\right)12$$
$$1x = -16$$
$$x = -16$$

Check: $-\frac{3}{4}x = 12$

$-\frac{3}{4}(-16)$	12

$12 = 12$

The solution is -16.

Because division is defined in terms of multiplication, the Multiplication Property of Equations allows each side of an equation to be divided by the same nonzero number without changing the solution of the equation.

Solve: $-5x = 9$

Multiply each side of the equation by the reciprocal of -5. This is equivalent to dividing each side of the equation by -5.

$$-5x = 9$$
$$\frac{-5x}{-5} = \frac{9}{-5}$$
$$1x = -\frac{9}{5}$$
$$x = -\frac{9}{5}$$

You should check the solution.

The solution is $-\frac{9}{5}$.

When using the Multiplication Property of Equations, it is usually easier to multiply each side of the equation by the reciprocal of the coefficient when the coefficient is a fraction. Divide each side of the equation by the coefficient when the coefficient is an integer or a decimal.

In solving an equation, the application of both the Addition and the Multiplication Properties of Equations is frequently required.

Example 1 Solve: $3x - 5 = x + 2 - 7x$

Solution

$$3x - 5 = x + 2 - 7x$$

$$3x - 5 = -6x + 2$$

▶ Simplify the right side of the equation by combining like terms.

$$3x + 6x - 5 = -6x + 6x + 2$$

▶ Add $6x$ to each side of the equation.

$$9x - 5 = 2$$

$$9x - 5 + 5 = 2 + 5$$

▶ Add 5 to each side of the equation.

$$9x = 7$$

$$\frac{9x}{9} = \frac{7}{9}$$

▶ Divide each side of the equation by the coefficient 9.

$$x = \frac{7}{9}$$

The solution is $\frac{7}{9}$.

Problem 1 Solve: $6x - 5 - 3x = 14 - 5x$

Solution See page A9.

2 Solve equations using the Distributive Property

When an equation contains parentheses, one of the steps in solving the equation requires the use of the Distributive Property.

Solve: $3(x - 2) + 3 = 2(6 - x)$

$$3(x - 2) + 3 = 2(6 - x)$$

Use the Distributive Property to remove parentheses. Simplify.

$$3x - 6 + 3 = 12 - 2x$$

$$3x - 3 = 12 - 2x$$

Add $2x$ to each side of the equation.

$$5x - 3 = 12$$

Add 3 to each side of the equation.

$$5x = 15$$

Divide each side of the equation by the coefficient 5.

$$x = 3$$

Check:

$3(x - 2) + 3$	$= 2(6 - x)$
$3(3 - 2) + 3$	$2(6 - 3)$
$3(1) + 3$	$2(3)$
$3 + 3$	6
$6 =$	6

Write the solution.

The solution is 3.

Example 2 Solve: $5(2x - 7) + 2 = 3(4 - x) - 12$

Solution

$$5(2x - 7) + 2 = 3(4 - x) - 12$$

$$10x - 35 + 2 = 12 - 3x - 12$$ ▶ Use the Distributive Property.

$$10x - 33 = -3x$$ ▶ Simplify.

$$-33 = -13x$$ ▶ Subtract $10x$ from each side of the equation.

$$\frac{33}{13} = x$$ ▶ Divide each side of the equation by -13.

The solution is $\frac{33}{13}$.

Problem 2 Solve: $6(5 - x) - 12 = 2x - 3(4 + x)$

Solution See page A9.

To solve an equation containing fractions, first clear denominators by multiplying each side of the equation by the least common multiple (LCM) of the denominators.

Solve: $\frac{x}{2} - \frac{7}{9} = \frac{x}{6} + \frac{2}{3}$

$$\frac{x}{2} - \frac{7}{9} = \frac{x}{6} + \frac{2}{3}$$

Multiply each side of the equation by 18, the LCM of 2, 9, 6, and 3.

$$18\left(\frac{x}{2} - \frac{7}{9}\right) = 18\left(\frac{x}{6} + \frac{2}{3}\right)$$

Use the Distributive Property to remove parentheses.

$$\frac{18x}{2} - \frac{18 \cdot 7}{9} = \frac{18x}{6} + \frac{18 \cdot 2}{3}$$

$$9x - 14 = 3x + 12$$

Subtract $3x$ from each side of the equation.

$$6x - 14 = 12$$

Add 14 to each side of the equation.

$$6x = 26$$

Divide each side of the equation by the coefficient 6.

$$x = \frac{13}{3}$$

Check: $\frac{x}{2} - \frac{7}{9} = \frac{x}{6} + \frac{2}{3}$

$\frac{\frac{13}{3}}{2} - \frac{7}{9}$	$\frac{\frac{13}{3}}{6} + \frac{2}{3}$
$\frac{13}{6} - \frac{7}{9}$	$\frac{13}{18} + \frac{2}{3}$
$\frac{39}{18} - \frac{14}{18}$	$\frac{13}{18} + \frac{12}{18}$

$$\frac{25}{18} = \frac{25}{18}$$

Write the solution.

The solution is $\frac{13}{3}$.

Example 3 Solve: $\frac{3x-2}{12} - \frac{x}{9} = \frac{x}{2}$

Solution

$$\frac{3x-2}{12} - \frac{x}{9} = \frac{x}{2}$$ ▶ The LCM of 12, 9, and 2 is 36.

$$36\left(\frac{3x-2}{12} - \frac{x}{9}\right) = 36\left(\frac{x}{2}\right)$$ ▶ Multiply each side of the equation by the LCM.

$$\frac{36(3x-2)}{12} - \frac{36x}{9} = \frac{36x}{2}$$ ▶ Use the Distributive Property.

$$3(3x-2) - 4x = 18x$$ ▶ Simplify.

$$9x - 6 - 4x = 18x$$
$$5x - 6 = 18x$$
$$-6 = 13x$$
$$-\frac{6}{13} = x$$

The solution is $-\frac{6}{13}$.

Problem 3 Solve: $\frac{2x-7}{3} - \frac{5x+4}{5} = \frac{-x-4}{30}$

Solution See page A9.

3 Application problems

Solving application problems is primarily a skill in translating sentences into equations and then solving the equations.

An equation states that two mathematical expressions are equal. Therefore, to translate a sentence into an equation requires recognizing the words or phrases that mean *equals*. These phrases include "is," "is equal to," "amounts to," and "represents."

Once the sentence is translated into an equation, solve the equation by rewriting the equation in the form variable = constant.

An electrician works 34 h and uses \$642 of materials repairing the wiring in a home. The electrician receives an hourly wage of \$22. Find the total cost of repairing the wiring.

Strategy

- To find the total cost of repairing the wiring, write and solve an equation using C to represent the total cost.
- The total cost is the sum of the cost of labor plus the cost of materials.

Solution

$$C = (34)(22) + 642$$
$$= 748 + 642$$
$$= 1390$$

The total cost of repairing the wiring is \$1390.

Example 4 The charges for a long-distance telephone call are \$2.14 for the first three minutes and \$.47 for each additional minute or fraction of a minute. If the charges for a long-distance call were \$17.65, how many minutes did the phone call last?

Strategy
- To find the length of the phone call in minutes, write and solve an equation using n to represent the total number of minutes of the call. Then $n - 3$ is the number of additional minutes after the first three minutes of the phone call.
- The fixed charge for the three minutes plus the charge for the additional minutes is the total cost of the phone call.

Solution

$$\begin{aligned} 2.14 + 0.47(n - 3) &= 17.65 \\ 2.14 + 0.47n - 1.41 &= 17.65 \\ 0.47n + 0.73 &= 17.65 \\ 0.47n &= 16.92 \\ n &= 36 \end{aligned}$$

The phone call lasted 36 min.

Problem 4 You are making a salary of \$14,500 and receive an 8% raise for next year. Find next year's salary.

Solution See page A9.

EXERCISES 2.1

1 Solve and check.

1. $x - 2 = 7$
2. $x - 8 = 4$
3. $a + 3 = -7$
4. $-12 = x - 3$
5. $3x = 12$
6. $8x = 4$
7. $\frac{2}{7} + x = \frac{17}{21}$
8. $x + \frac{2}{3} = \frac{5}{6}$
9. $\frac{5}{8} - y = \frac{3}{4}$
10. $\frac{2}{3}y = 5$
11. $\frac{3}{5}y = 12$
12. $\frac{3t}{8} = -15$
13. $\frac{3a}{7} = -21$
14. $-\frac{5}{8}x = \frac{4}{5}$
15. $-\frac{5}{12}y = \frac{7}{16}$
16. $-\frac{3}{4}x = -\frac{4}{7}$
17. $b - 14.72 = -18.45$
18. $b + 3.87 = -2.19$

Solve and check.

19. $3x + 5x = 12$
20. $2x - 7x = 15$
21. $2x - 4 = 12$
22. $3x - 12 = 5x$
23. $4x + 2 = 4x$
24. $3m - 7 = 3m$
25. $2x + 2 = 3x + 5$
26. $7x - 9 = 3 - 4x$
27. $2 - 3t = 3t - 4$

28. $7 - 5t = 2t - 9$

29. $2a - 3a = 7 - 5a$

30. $3a - 5a = 8a + 4$

31. $\frac{5}{8}b - 3 = 12$

32. $\frac{1}{3} - 2b = 3$

33. $b + \frac{1}{5}b = 2$

34. $3x - 2x + 7 = 12 - 4x$

35. $2x - 9x + 3 = 6 - 5x$

36. $7 + 8y - 12 = 3y - 8 + 5y$

37. $2y - 4 + 8y = 7y - 8 + 3y$

38. $2x - 5 + 7x = 11 - 3x + 4x$

39. $9 + 4x - 12 = -3x + 5x + 8$

40. $3.24a + 7.14 = 5.34a$

41. $5.3y + 0.35 = 5.02y$

42. $1.2b - 3.8b = 1.6b + 4.494$

43. $6.7a - 9.2a = 6.55a - 3.91865$

2 Solve and check.

44. $2x + 2(x + 1) = 10$

45. $2x + 3(x - 5) = 15$

46. $2(a - 3) = 2(4 - 2a)$

47. $5(2 - b) = -3(b - 3)$

48. $-4(c + 2) = 3c + 6$

49. $-5(t - 4) = 10t + 10$

50. $3 - 2(y - 3) = 4y - 7$

51. $3(y - 5) - 5y = 2y + 9$

52. $4(x - 2) + 2 = 4x - 2(2 - x)$

53. $2x - 3(x - 4) = 2(3 - 2x) + 2$

54. $3(p + 2) + 4p = 6(3p - 1) + 5p$

55. $10(2x - 3) + 5x = 6(x + 1) + 3x$

56. $2(2d + 1) - 3d = 5(3d - 2) + 4d$

57. $-4(7y - 1) + 5y = -2(3y + 4) - 3y$

58. $4[3 + 5(3 - x) + 2x] = 6 - 2x$

59. $2[4 + 2(5 - x) - 2x] = 4x - 7$

60. $2[b - (4b - 5)] = 3b + 4$

61. $-3[x + 4(x + 1)] = x + 4$

62. $4[a - (3a - 5)] = a - 7$

63. $5 - 6[2t - 2(t + 3)] = 8 - t$

64. $-3(x - 2) = 2[x - 4(x - 2) + x]$

65. $3[x - (2 - x) - 2x] = 3(4 - x)$

66. $\frac{2}{9}t - \frac{5}{6} = \frac{1}{12}t$

67. $\frac{3}{4}t - \frac{7}{12}t = 1$

68. $\frac{2}{3}x - \frac{5}{6}x - 3 = \frac{1}{2}x - 5$

69. $\frac{1}{2}x - \frac{3}{4}x + \frac{5}{8} = \frac{3}{2}x - \frac{5}{2}$

70. $\frac{3x - 2}{4} - 3x = 12$

71. $\frac{2a - 9}{5} + 3 = 2a$

72. $\frac{x - 2}{4} - \frac{x + 5}{6} = \frac{5x - 2}{9}$

73. $\frac{2x - 1}{4} + \frac{3x + 4}{8} = \frac{1 - 4x}{12}$

74. $\frac{2}{3}(15 - 6a) = \frac{5}{6}(12a + 18)$

75. $\frac{1}{5}(20x + 30) = \frac{1}{3}(6x + 36)$

76. $\frac{1}{3}(x - 7) + 5 = 6x + 4$

77. $2(y - 4) + 8 = \frac{1}{2}(6y + 20)$

78. $\frac{1}{4}(2b + 50) = \frac{5}{2}\left(15 - \frac{1}{5}b\right)$

79. $\frac{1}{4}(7 - x) = \frac{2}{3}(x + 2)$

80. If $3x - 2 = x + 5$, evaluate $2(x - 3) + 4$.

81. If $3x - 4 = \frac{x + 4}{3}$, evaluate $2x^2 - 3(x - 2) + 1$.

82. If $\frac{x}{2} - \frac{3x - 2}{4} = -4$, evaluate $3(2 - 3x) + x^2$.

83. If $\frac{2x - 3}{5} + \frac{x}{2} = 3$, evaluate $\frac{1}{3}(5x - 2) - \frac{x^2}{2}$.

84. If $2[x - (3 - 2x)] = x + 4$, evaluate $2[x - 3(2x - 3)] + 2$.

85. If $\frac{2}{3}x - \frac{3}{4}x = 2$, evaluate $3(x + 20)^2 - \frac{3}{8}x$.

86. $2.3(x - 1.8) = 3.91$

87. $0.4(m + 6.5) = 3.14$

88. $-4.2(p + 3.4) = 11.13$

89. $-1.6(b - 2.35) = -11.28$

90. $0.1y + 6 = 0.12(y + 40)$

91. $0.35(n + 40) = 0.1n + 20$

3 Solve.

92. The Fahrenheit temperature is 59°. This is 32° more than $\frac{9}{5}$ the Celsius temperature. Find the Celsius temperature.

93. A total of 32% is deducted from a nurse's salary for taxes, insurance, and dues. The nurse receives a wage of $14.50 an hour and works for 40 h. Find the nurse's take-home pay.

94. A service station attendant is paid time-and-a-half for working over 40 h per week. Last week the attendant worked 47 hours and earned $530.25. Find the attendant's regular hourly rate.

95. An overnight mail service charges $3.60 for the first six ounces and $.45 for each additional ounce or fraction of an ounce. Find the number of ounces in a package that costs $7.65 to deliver.

96. At a city zoo, the admission charge for a family is $7.50 for the first person and $4.25 for each additional member of the family. How many people are in a family that is charged $28.75 for admission?

97. The charges for a long-distance telephone call are $1.42 for the first three minutes and $.65 for each additional minute or fraction of a minute. If the charges for a call were $10.52, how many minutes did the phone call last?

98. A library charges a fine for each overdue book. The fine is 25¢ the first day and 8¢ for each additional day the book is overdue. If the fine for a book is $1.21, how many days overdue is the book?

99. A library charges a fine for each overdue book. The fine is 15¢ the first day plus 7¢ a day for each additional day the book is overdue. If the fine for a book is 78¢, how many days overdue is the book?

100. At a museum, the admission charge for a family is $1.75 for the first person and $.75 for each additional member of the family. How many people are in a family that is charged $5.50 for admission?

101. The charges for a long-distance telephone call are $1.63 for the first three minutes and $.88 for each additional minute or fraction of a minute. If the charges for a call were $7.79, how many minutes did the phone call last?

SUPPLEMENTAL EXERCISES 2.1

Solve.

102. $8 \div \frac{1}{x} = -3$

103. $5 \div \frac{1}{y} = -2$

104. $\frac{1}{\frac{1}{x}} = -4$

105. $\frac{1}{\frac{1}{y}} = -9$

106. $\frac{6}{\frac{7}{a}} = -18$

107. $\frac{10}{\frac{3}{x}} - 5 = 4x$

Solve. If the equation has no solution, write "no solution."

108. $3[4(y + 2) - (y + 5)] = 3(3y + 1)$

109. $2[3(x + 4) - 2(x + 1)] = 5x + 3(1 - x)$

110. $\frac{2(3x - 4) - (x + 1)}{3} = x - 7$

111. $\frac{3(2x - 1) - (x + 2)}{5} = x - 4$

112. $\frac{4(3a + 2) - 3(a + 1)}{9} = a + 5$

113. $\frac{4[(x - 3) + 2(1 - x)]}{5} = x + 1$

114. $2584 \div x = 54\frac{46}{x}$

115. $3479 \div x = 66\frac{47}{x}$

116. $3(2x + 2) - 4(x - 3) = 2(x + 9)$

117. $2(4x - 1) + 3(2x - 2) = 7(2x - 1) - 1$

SECTION 2.2 Coin, Stamp, and Integer Problems

1 Coin and stamp problems

In solving problems dealing with coins or stamps of different values, it is necessary to represent the value of the coins or stamps in the same unit of money. The unit of money is frequently cents. For example:

The value of five 8¢ stamps is $5 \cdot 8$, or 40 cents.
The value of four 20¢ stamps is $4 \cdot 20$, or 80 cents.
The value of n 10¢ stamps is $n \cdot 10$, or $10n$ cents.

Solve: A collection of stamps consists of 5¢, 13¢, and 18¢ stamps. The number of 13¢ stamps is two more than three times the number of 5¢ stamps. The number of 18¢ stamps is five less than the number of 13¢ stamps. The total value of all the stamps is $1.68. Find the number of 18¢ stamps.

STRATEGY for solving a stamp problem

- For each denomination of stamp, write a numerical or variable expression for the number of stamps, the value of the stamp, and the total value of the stamps in cents. The results can be recorded in a table.

The number of 5¢ stamps: x
The number of 13¢ stamps: $3x + 2$
The number of 18¢ stamps: $(3x + 2) - 5 = 3x - 3$

Stamp	Number of stamps	·	Value of stamp in cents	=	Total value in cents
5¢	x	·	5	=	$5x$
13¢	$3x + 2$	·	13	=	$13(3x + 2)$
18¢	$3x - 3$	·	18	=	$18(3x - 3)$

- Determine the relationship between the total values of the stamps. Use the fact that the sum of the total values of each denomination of stamp is equal to the total value of all the stamps.

The sum of the total values of each denomination of stamp is equal to the total value of all the stamps (168 cents).

$$5x + 13(3x + 2) + 18(3x - 3) = 168$$
$$5x + 39x + 26 + 54x - 54 = 168$$
$$98x - 28 = 168$$
$$98x = 196$$
$$x = 2$$

The number of 18¢ stamps is $3x - 3$. Replace x by 2 and evaluate.

$$3x - 3 = 3(2) - 3 = 3$$

There are three 18¢ stamps in the collection.

Example 1 A coin bank contains $1.80 in nickels and dimes. In all, there are twenty-two coins in the bank. Find the number of nickels and the number of dimes in the bank.

Strategy

- Number of nickels: x
  Number of dimes: $22 - x$

Coin	Number	Value	Total Value
Nickel	x	5	$5x$
Dime	$22 - x$	10	$10(22 - x)$

- The sum of the total values of each denomination of coin equals the total value of all the coins (180 cents).

Solution

$$5x + 10(22 - x) = 180$$
$$5x + 220 - 10x = 180$$
$$-5x + 220 = 180$$
$$-5x = -40$$
$$x = 8$$

▶ There are 8 nickels in the bank.

$$22 - x = 22 - 8 = 14$$

▶ Substitute the value of x into the variable expression for the number of dimes.

The bank contains 8 nickels and 14 dimes.

Problem 1 A collection of stamps contains 3¢, 10¢, and 15¢ stamps. The number of 10¢ stamps is two more than twice the number of 3¢ stamps. There are three times as many 15¢ stamps as there are 3¢ stamps. The total value of the stamps is $1.56. Find the number of 15¢ stamps.

Solution See page A10.

2 Integer problems

Recall that an **even integer** is an integer that is divisible by 2. An **odd integer** is an integer that is not divisible by 2.

Consecutive integers are integers that follow one another in order. Examples of consecutive integers are shown at the right.

8, 9, 10
$-3, -2, -1$
$n, n + 1, n + 2$, where n is an integer

Examples of **consecutive even integers** are shown at the right.

16, 18, 20
$-6, -4, -2$
$n, n + 2, n + 4$, where n is an even integer

Examples of **consecutive odd integers** are shown at the right.

11, 13, 15
$-23, -21, -19$
$n, n + 2, n + 4$, where n is an odd integer

There is a basic strategy for solving word problems that involve integers.

Solve: The sum of three consecutive even integers is seventy-eight. Find the integers.

STRATEGY for solving an integer problem

- Let a variable represent one of the integers. Express each of the other integers in terms of that variable. Remember that for consecutive integer problems, consecutive integers will differ by 1. Consecutive even or consecutive odd integers will differ by 2.

Represent three consecutive even integers.

First even integer: n
Second even integer: $n + 2$
Third even integer: $n + 4$

- Determine the relationship among the integers.

The sum of the three even integers is 78.

$$n + (n + 2) + (n + 4) = 78$$
$$3n + 6 = 78$$
$$3n = 72$$
$$n = 24$$

$$n + 2 = 24 + 2 = 26$$
$$n + 4 = 24 + 4 = 28$$

The three consecutive even integers are 24, 26, and 28.

Example 2 One number is four more than another number. The sum of the two numbers is sixty-six. Find the two numbers.

Strategy
- The smaller number: n
  The larger number: $n + 4$
- The sum of the numbers is 66.

Solution

$$n + (n + 4) = 66$$
$$2n + 4 = 66$$
$$2n = 62$$
$$n = 31$$

▶ The smaller number is 31.

$$n + 4 = 31 + 4 = 35$$

▶ Substitute the value of n into the variable expression for the larger number.

The numbers are 31 and 35.

Problem 2 The sum of three numbers is eighty-one. The second number is twice the first number, and the third number is three less than four times the first number. Find the numbers.

Solution See page A10.

Example 3 Five times the first of three consecutive even integers is five more than the product of four and the third integer. Find the integers.

Strategy
- First even integer: n
  Second even integer: $n + 2$
  Third even integer: $n + 4$
- Five times the first integer equals five more than the product of four and the third integer.

Solution

$$5n = 4(n + 4) + 5$$
$$5n = 4n + 16 + 5$$
$$5n = 4n + 21$$
$$n = 21$$

Since 21 is not an even integer, there is no solution.

Problem 3 Find three consecutive odd integers such that three times the sum of the first two integers is ten more than the product of the third integer and four.

Solution See page A10.

EXERCISES 2.2

1 Solve.

1. A collection of 56 coins has a value of \$4.00. The collection contains only nickels and dimes. Find the number of dimes in the collection.

2. A collection of 22 coins has a value of \$4.45. The collection contains dimes and quarters. Find the number of quarters in the collection.

3. A coin bank contains 25 coins in nickels, dimes, and quarters. There are four times as many dimes as quarters. The value of the coins is \$2.05. How many dimes are in the bank?

4. A coin collection contains nickels, dimes, and quarters. There are twice as many dimes as quarters and seven more nickels than dimes. The total value of all the coins is \$3.10. How many quarters are in the collection?

5. A cashier has \$730 in twenty-dollar bills and five-dollar bills. In all, the cashier has 68 bills. How many twenty-dollar bills does the cashier have?

6. A department store uses twice as many five-dollar bills in conducting its daily business as ten-dollar bills. \$2500 was obtained in five- and ten-dollar bills for the day's business. How many five-dollar bills were obtained?

7. A stamp collector has some 15¢ stamps and some 20¢ stamps. The number of 15¢ stamps is eight less than three times the number of 20¢ stamps. The total value is \$4. Find the number of each type of stamp in the collection.

8. An office has some 20¢ stamps and some 28¢ stamps. Altogether the office has 140 stamps for a total value of \$31.20. How many of each type stamp does the office have?

9. A stamp collection consists of 3¢, 8¢, and 13¢ stamps. The number of 8¢ stamps is three less than twice the number of 3¢ stamps. The number of 13¢ stamps is twice the number of 8¢ stamps. The total value of all the stamps is \$2.53. Find the number of 3¢ stamps in the collection.

10. An account executive bought 300 stamps for \$73.80. The purchase included 15¢ stamps, 20¢ stamps, and 40¢ stamps. The number of 20¢ stamps is four times the number of 15¢ stamps. How many 40¢ stamps were purchased?

11. A stamp collector has 8¢, 11¢, and 18¢ stamps. The collector has twice as many 8¢ stamps as 18¢ stamps. There are three more 11¢ than 18¢ stamps. The total value of the stamps in the collection is \$3.48. Find the number of 18¢ stamps in the collection.

12. A stamp collection consists of 3¢, 12¢, and 15¢ stamps. The number of 3¢ stamps is five times the number of 12¢ stamps. The number of 15¢ stamps is four less than the number of 12¢ stamps. The total value of the stamps in the collection is \$3.18. Find the number of 15¢ stamps in the collection.

2 Solve.

13. What number must be added to the numerator and denominator of $\frac{5}{7}$ to produce the fraction $\frac{4}{5}$?

14. What number must be added to the numerator and denominator of $\frac{6}{11}$ to produce the fraction $\frac{2}{3}$?

15. The sum of two integers is 10. Three times the larger integer is three less than eight times the smaller integer. Find the integers.

16. The sum of two integers is thirty. Eight times the smaller integer is six more than five times the larger integer. Find the integers.

17. One integer is eight less than another integer. The sum of the two integers is fifty. Find the integers.

18. One integer is four more than another integer. The sum of the integers is twenty-six. Find the integers.

19. The sum of three numbers is one hundred twenty-three. The second number is two more than twice the first number. The third number is five less than the product of three and the first number. Find the three numbers.

20. The sum of three numbers is forty-two. The second number is twice the first number and the third number is three less than the second number. Find the three numbers.

21. The sum of three consecutive integers is negative fifty-seven. Find the integers.

22. The sum of three consecutive integers is one hundred twenty-nine. Find the integers.

23. Five times the smallest of three consecutive odd integers is ten more than twice the largest. Find the integers.

24. Find three consecutive even integers such that twice the sum of the first and third integers is twenty-one more than the second integer.

25. Find three consecutive odd integers such that three times the middle integer is seven more than the sum of the first and third integers.

26. Find three consecutive even integers such that four times the sum of the first and third integers is twenty less than six times the middle integer.

SUPPLEMENTAL EXERCISES 2.2

Solve.

27. A coin bank contains only nickels and dimes. The number of dimes in the bank is two less than twice the number of nickels. There are 52 coins in the bank. How much money is in the bank?

28. Four times the first of three consecutive odd integers is five less than the product of three and the third integer. Find the integers.

29. A collection of stamps consists of 3¢ stamps, 5¢ stamps, and 7¢ stamps. There are six more 3¢ stamps than 5¢ stamps, and two more 7¢ stamps than 3¢ stamps. The total value of the stamps is $1.94. How many 3¢ stamps are in the collection?

30. The sum of four consecutive even integers is 100. What is the sum of the smallest and the largest of the four integers?

31. The sum of four consecutive odd integers is −64. What is the sum of the smallest and the largest of the four integers?

SECTION 2.3
Value Mixture and Motion Problems

1 Value Mixture Problems

A **value mixture problem** involves combining two ingredients that have different prices into a single blend. For example, a coffee manufacturer may blend two types of coffee into a single blend.

The solution of a value mixture problem is based on the equation $AC = V$, where A is the amount of the ingredient, C is the cost per unit of the ingredient, and V is the value of the ingredient.

The value of 12 lb of coffee costing \$5.25 per pound is:

$$V = AC$$
$$V = 12(\$5.25)$$
$$V = \$63$$

Solve: How many pounds of peanuts that cost \$2.25 per pound must be mixed with 40 lb of cashews that cost \$6.00 per pound to make a mixture that costs \$3.50 per pound?

STRATEGY for solving a value mixture problem

- For each ingredient in the mixture, write a numerical or variable expression for the amount of the ingredient used, the unit cost of the ingredient, and the value of the amount used. For the mixture, write a numerical or variable expression for the amount, the unit cost of the mixture, and the value of the amount. The results can be recorded in a table.

Amount of peanuts: x
Amount of cashews: 40

	Amount, A	$\cdot$	Unit cost, C	$=$	Value, V
Peanuts	x	$\cdot$	2.25	$=$	$2.25x$
Cashews	40	$\cdot$	6.00	$=$	6.00(40)
Mixture	$40 + x$	$\cdot$	3.50	$=$	$3.50(40 + x)$

- Determine how the values of each ingredient are related. Use the fact that the sum of the values of each ingredient is equal to the value of the mixture.

The sum of the values of the peanuts and the cashews is equal to the value of the mixture.

$$2.25x + 6.00(40) = 3.50(40 + x)$$
$$2.25x + 240 = 140 + 3.50x$$
$$-1.25x + 240 = 140$$
$$-1.25x = -100$$
$$x = 80$$

The mixture must contain 80 lb of peanuts.

Example 1 How many ounces of a gold alloy that costs \$320 an ounce must be mixed with 100 oz of an alloy that costs \$100 an ounce to make a mixture that costs \$160 an ounce?

Strategy
- Ounces of the \$320 gold alloy: x
  Ounces of the \$100 gold alloy: 100

	Amount	Cost	Value
\$320 alloy	x	320	$320x$
\$100 alloy	100	100	100(100)
Mixture	$x + 100$	160	$160(x + 100)$

- The sum of the values before mixing equals the value after mixing.

Solution

$$320x + 100(100) = 160(x + 100)$$
$$320x + 10{,}000 = 160x + 16{,}000$$
$$160x + 10{,}000 = 16{,}000$$
$$160x = 6000$$
$$x = 37.5$$

The mixture must contain 37.5 oz of the \$320 gold alloy.

Problem 1 A butcher combined hamburger that cost \$3.00 per pound with hamburger that cost \$1.80 per pound. How many pounds of each were used to make a 75-lb mixture that cost \$2.20 per pound?

Solution See page A11.

2 Uniform motion problems

A car that travels constantly in a straight line at 55 mph is in uniform motion. **Uniform motion** means the speed of an object does not change.

The solution of a uniform motion problem is based on the equation $rt = d$, where r is the rate of travel, t is the time spent traveling, and d is the distance traveled.

A car travels 55 mph for 3 h.

$$d = rt$$
$$d = 55(3)$$
$$d = 165$$

The car travels a distance of 165 mi.

Solve: An executive has an appointment 785 mi from the office. The executive takes a helicopter from the office to the airport and a plane from the airport to the business appointment. The helicopter averages 70 mph, and the plane averages 500 mph. The total time spent traveling is 2 h. Find the distance from the executive's office to the airport.

STRATEGY for solving a uniform motion problem

- For each object, write a numerical or variable expression for the distance, rate, and time. The results can be recorded in a table.

The total time of travel is 2 h.

Unknown time in the helicopter: t
Time in the plane: $2 - t$

	Rate, r	·	Time, t	=	Distance, d
Helicopter	70	·	t	=	$70t$
Plane	500	·	$2 - t$	=	$500(2 - t)$

- Determine how the distances traveled by each object are related. For example, the total distance traveled by both objects may be known, or it may be known that the two objects traveled the same distance.

The total distance traveled is 785 mi.

$$70t + 500(2 - t) = 785$$
$$70t + 1000 - 500t = 785$$
$$-430t + 1000 = 785$$
$$-430t = -215$$
$$t = 0.5$$

The time spent traveling from the office to the airport in the helicopter is 0.5 h. To find the distance between these two points, substitute the values of r and t into the equation $rt = d$.

$$rt = d$$
$$70(0.5) = d$$
$$35 = d$$

The distance from the office to the airport is 35 mi.

Example 2 A long-distance runner started a course running at an average speed of 6 mph. One and one half hours later, a cyclist traveled the same course at an average speed of 12 mph. How long after the runner started did the cyclist overtake the runner?

Strategy ▬ Unknown time for the cyclist: t
Time for the runner: $t + 1.5$

	Rate	Time	Distance
Runner	6	$t + 1.5$	$6(t + 1.5)$
Cyclist	12	t	$12t$

▬ The runner and the cyclist traveled the same distance.

Solution

$$6(t + 1.5) = 12t$$
$$6t + 9 = 12t$$
$$9 = 6t$$
$$\frac{3}{2} = t$$ ▶ The cyclist traveled for 1.5 h.

$$t + 1.5 = 1.5 + 1.5 = 3$$ ▶ Substitute the value of t into the variable expression for the runner's time.

The cyclist overtook the runner 3 h after the runner started.

Problem 2 Two small planes start from the same point and fly in opposite directions. The first plane is flying 30 mph faster than the second plane. In 4 h the planes are 1160 mi apart. Find the rate of each plane.

Solution See page A11.

EXERCISES 2.3

1 Solve.

1. Forty pounds of cashews costing \$5.60 per pound were mixed with 100 lb of peanuts costing \$1.89 per pound. Find the cost per pound of the resulting mixture.

2. A coffee merchant combines coffee costing \$5.50 per pound with coffee costing \$3.00 per pound. How many pounds of each should be used to make 40 lb of a blend costing \$4.00 per pound?

3. Adult tickets for a play cost \$5.00 and children's tickets cost \$2.00. For one performance, 460 tickets were sold. Receipts for the performance were \$1880. Find the number of adult tickets sold.

4. Tickets for a school play sold for \$2.50 for each adult and \$1.00 for each child. The total receipts for 113 tickets sold were \$221. Find the number of adult tickets sold.

5. Fifty liters of pure maple syrup that cost \$9.50 per liter are mixed with imitation maple syrup that costs \$4.00 per liter. How much imitation maple syrup is needed to make a mixture that costs \$5.00 per liter?

6. To make a flour mix, a miller combined soybeans that cost \$8.50 per bushel with wheat that costs \$4.50 per bushel. How many bushels of each were used to make a mixture of 800 bushels costing \$5.50 per bushel?

7. A goldsmith combined pure gold that cost \$400 per ounce with an alloy of gold costing \$150 per ounce. How many ounces of each were used to make 50 oz of gold alloy costing \$250 per ounce?

8. A silversmith combined pure silver that cost \$5.20 an ounce with 50 oz of a silver alloy that cost \$2.80 an ounce. How many ounces of the pure silver were used to make an alloy of silver costing \$4.40 an ounce?

9. A tea mixture was made from 30 lb of tea costing \$6.00 per pound and 70 lb of tea costing \$3.20 per pound. Find the cost per pound of the tea mixture.

10. Find the cost per ounce of a face cream mixture made from 100 oz of face cream that cost \$3.46 per ounce and 60 oz of face cream that cost \$12.50 per ounce.

11. A fruitstand owner combined cranberry juice that cost \$4.20 per gallon with 50 gal of apple juice that cost \$2.10 per gallon. How much cranberry juice was used to make cranapple juice costing \$3.00 per gallon?

12. Walnuts that cost \$4.05 per kilogram were mixed with cashews that cost \$7.25 per kilogram. How many kilograms of each were used to make a 50-kilogram mixture costing \$6.25 per kilogram? Round to the nearest tenth.

2 Solve.

13. A car traveling at 56 mph overtakes a cyclist who, traveling at 14 mph, had a 1.5 h head start. How far from the starting point does the car overtake the cyclist?

14. A helicopter traveling 120 mph overtakes a speeding car traveling 90 mph. The car had a one-half hour head start. How far from the starting point did the helicopter overtake the car?

15. Two planes are 1380 mi apart and traveling toward each other. One plane is traveling 80 mph faster than the other plane. The planes meet in 1.5 h. Find the speed of each plane.

16. Two cars are 295 mi apart and traveling toward each other. One car travels 10 mph faster than the other car. The cars meet in 2.5 h. Find the speed of each car.

17. A ferry leaves a harbor and travels to a resort island at an average speed of 18 mph. On the return trip, the ferry travels at an average speed of 12 mph due to fog. The total time for the trip is 6 h. How far is the island from the harbor?

18. A commuter plane provides transportation from an international airport to the surrounding cities. One commuter plane averaged 210 mph flying to a city and 140 mph returning to the international airport. The total flying time was 4 h. Find the distance between the two airports.

19. Two planes start from the same point and fly in opposite directions. The first plane is flying 50 mph slower than the second plane. In 2.5 h the planes are 1400 mi apart. Find the rate of each plane.

20. Two hikers start from the same point and hike in opposite directions around a lake whose shoreline is 13 mi. One hiker walks 0.5 mph faster than the other hiker. How fast did each hiker walk if they meet in 2 h?

21. A student rode a bicycle to the repair shop and then walked home. The student averaged 14 mph riding to the shop and 3.5 mph walking home. The round trip took one hour. How far is it between the student's home and the bicycle shop?

22. A passenger train leaves a depot 1.5 h after a freight train leaves the same depot. The passenger train is traveling 18 mph faster than the freight train. Find the rate of each train if the passenger train overtakes the freight train in 2.5 h.

23. A plane leaves an airport at 3 P.M. At 4 P.M. another plane leaves the same airport traveling in the same direction at a speed 150 mph faster than the first plane. Four hours after the first plane takes off, the second plane is 250 mi ahead of the first plane. How far did the second plane travel?

24. A jogger and a cyclist set out at 9 A.M. from the same point headed in the same direction. The average speed of the cyclist is four times the speed of the jogger. In 2 h, the cyclist is 33 mi ahead of the jogger. How far did the cyclist ride?

SUPPLEMENTAL EXERCISES 2.3

Solve.

25. A truck leaves a depot at 10 A.M. and travels at 50 mph. At 10:30 A.M., a van leaves the same place and travels the same route at 65 mph. At what time does the van overtake the truck?

26. A grocer combines 50 gal of cranberry juice that costs \$3.50 per gallon with apple juice that costs \$2.50 per gallon. How many gallons of apple juice must be used to make cranapple juice that costs \$2.75 per gallon?

27. A commuter plane flew to a small town from a major airport at an average speed of 300 mph. The average speed on the return trip was 200 mph. What is the distance between the two airports if the total flying time was 4 h?

28. A car travels at an average speed of 30 mph for 1 mi. Is it possible to increase speed during the next mile so that the average speed for the 2 mi is 60 mph?

SECTION 2.4 Applications: Problems Involving Percent

1 Investment problems

The annual simple interest that an investment earns is given by the equation $Pr = I$, where P is the principal, or the amount invested, r is the simple interest rate, and I is the simple interest. The solution of an investment problem is based on this equation.

The annual interest rate on a \$3000 investment is 9%. The annual simple interest earned on the investment is:

$$I = Pr$$
$$I = \$3000(0.09)$$
$$I = \$270$$

Solve: You have a total of \$8000 invested in two simple interest accounts. On one account, a money market fund, the annual simple interest rate is 11.5%. On the second account, a bond fund, the annual simple interest rate is 9.75%. The total annual interest earned by the two accounts is \$823.75. How much do you have invested in each account?

STRATEGY for solving a problem involving money deposited in two simple interest accounts

- For each amount invested, use the equation $Pr = I$. Write a numerical or variable expression for the principal, the interest rate, and the interest earned. The results can be recorded in a table.

The total amount invested is \$8000

Amount invested at 11.5%: x
Amount invested at 9.75%: $8000 - x$

	Principal, P	$\cdot$	Interest rate, r	$=$	Interest earned, I
Amount at 11.5%	x	$\cdot$	0.115	$=$	$0.115x$
Amount at 9.75%	$8000 - x$	$\cdot$	0.0975	$=$	$0.0975(8000 - x)$

- Determine how the amounts of interest earned on each amount are related. For example, the total interest earned by both accounts may be known, or it may be known that the interest earned on one account is equal to the interest earned on the other account.

The total annual interest earned is \$823.75.

$$
\begin{aligned}
0.115x + 0.0975(8000 - x) &= 823.75 \\
0.115x + 780 - 0.0975x &= 823.75 \\
0.0175x + 780 &= 823.75 \\
0.0175x &= 43.75 \\
x &= 2500
\end{aligned}
$$

The amount invested at 9.75% is $8000 - x$.
Replace x by 2500 and evaluate.

$$8000 - x = 8000 - 2500 = 5500$$

The amount invested at 11.5% is \$2500.
The amount invested at 9.75% is \$5500.

Example 1 An investment of \$4000 is made at an annual simple interest rate of 10.9%. How much additional money must be invested at an annual simple interest rate of 14.5% so that the total interest earned is 12% of the total investment?

Strategy

- Additional amount to be invested at 14.5%: x

	Principal	Rate	Interest
Amount at 10.9%	4000	0.109	0.109(4000)
Amount at 14.5%	x	0.145	$0.145x$
Amount at 12%	$4000 + x$	0.12	$0.12(4000 + x)$

- The sum of the interest earned by the two investments equals the interest earned by the total investment.

Solution

$$\begin{aligned} 0.109(4000) + 0.145x &= 0.12(4000 + x) \\ 436 + 0.145x &= 480 + 0.12x \\ 436 + 0.025x &= 480 \\ 0.025x &= 44 \\ x &= 1760 \end{aligned}$$

An additional \$1760 must be invested at an annual simple interest rate of 14.5%.

Problem 1 An investment of \$3500 is made at an annual simple interest rate of 13.2%. How much additional money must be invested at an annual simple interest rate of 11.5% so that the total interest earned is \$1037?

Solution See page A12.

2 Percent mixture problems

The amount of a substance in a solution or alloy can be given as a percent of the total solution or alloy. For example, a 10% hydrogen peroxide solution means 10% of the total solution is hydrogen peroxide. The remaining 90% is water.

The solution of a percent mixture problem is based on the equation $Ar = Q$, where A is the amount of solution or alloy, r is the percent of concentration, and Q is the quantity of a substance in the solution or alloy.

The number of grams of silver in 50 g of a 40% silver alloy is:

$$\begin{aligned} Q &= Ar \\ Q &= 50(0.40) \\ Q &= 20 \end{aligned}$$

Solve: A chemist mixes an 11% acid solution with a 4% acid solution. How many milliliters of each solution should the chemist use to make a 700-ml solution that is 6% acid?

STRATEGY for solving a percent mixture problem

- For each solution, use the equation $Ar = Q$. Write a numerical or variable expression for the amount of solution, percent of concentration, and the quantity of the substance in the solution. The results can be recorded in a table.

The total amount of solution is 700 ml.

Amount of 11% solution: x
Amount of 4% solution: $700 - x$

	Amount of solution, A	$\cdot$	Percent of concentration, r	$=$	Quantity of substance, Q
11% solution	x	$\cdot$	0.11	$=$	$0.11x$
4% solution	$700 - x$	$\cdot$	0.04	$=$	$0.04(700 - x)$
6% solution	700	$\cdot$	0.06	$=$	$0.06(700)$

- Determine how the quantities of the substance in each solution are related. Use the fact that the sum of the quantities of the substances being mixed is equal to the quantity of the substance after mixing.

The sum of the quantities of the substances in the 11% solution and the 4% solution is equal to the quantity of the substances in the 6% solution.

$$0.11x + 0.04(700 - x) = 0.06(700)$$
$$0.11x + 28 - 0.04x = 42$$
$$0.07x + 28 = 42$$
$$0.07x = 14$$
$$x = 200$$

The amount of 4% solution is $700 - x$. Replace x by 200 and evaluate.

$$700 - x = 700 - 200 = 500$$

The chemist should use 200 ml of the 11% solution and 500 ml of the 4% solution.

Example 2 How many grams of pure acid must be added to 60 g of an 8% acid solution to make a 20% acid solution?

Strategy ▪ Grams of pure acid: x

	Amount	Percent	Quantity
Pure acid (100%)	x	1.00	x
8%	60	0.08	0.08(60)
20%	$60 + x$	0.20	$0.20(60 + x)$

▪ The sum of the quantities before mixing equals the quantity after mixing.

Solution

$$x + 0.08(60) = 0.20(60 + x)$$
$$x + 4.8 = 12 + 0.20x$$
$$0.8x + 4.8 = 12$$
$$0.8x = 7.2$$
$$x = 9$$

To make the 20% acid solution, 9 g of pure acid must be used.

Problem 2 A butcher has some hamburger that is 22% fat and some that is 12% fat. How many pounds of each should be mixed to make 80 lb of hamburger that is 18% fat?

Solution See page A12.

EXERCISES 2.4

1 Solve.

1. Two investments earn an annual income of $1069. One investment is in a 7.2% tax-free annual simple interest account and the other investment is in a 9.8% annual simple interest CD. The total amount invested is $12,500. How much is invested in each account?

2. Two investments earn an annual income of $765. One investment earns an annual simple interest rate of 8.5%, while the other investment earns an annual simple interest rate of 10.2%. The total amount invested is $8000. How much is invested in each account?

3. An investment club invested $5000 at an annual simple interest rate of 8.4%. How much additional money must be invested at an annual simple interest rate of 10.5% so that the total interest earned will be 9% of the total investment?

4. An investment of $4500 is made at an annual simple interest rate of 7.8%. How much additional money must be invested at an annual simple interest rate of 11% so that the total interest earned is 9% of the total investment?

5. An account executive deposited $42,000 in two simple interest accounts. On the tax-free account the annual simple interest rate is 7%, while on the money market fund the annual simple interest rate is 9.8%. How much should be invested in each account so that the interest earned by each account is the same?

6. An investment club invested $10,800 in two simple interest accounts. On one account the annual simple interest rate is 8.2%. On the other, the annual simple interest rate is 10.25%. How much should be invested in each account so that the interest earned by each account is the same?

7. A financial manager recommended an investment plan in which 30% of a client's investment be placed in a 7.5% annual simple interest tax-free account, 45% be placed in 9% high-grade bonds, and the remainder in an 11.5% high-risk investment. The total interest earned from the investments would be $2293.75. Find the total amount to be invested.

8. The manager of a trust account invests 25% of a client's account in a money market fund that earns 8% annual simple interest, 40% in bonds that earn 10.5% annual simple interest, and the remainder in trust deeds that earn 13% annual simple interest. How much should be invested in each type of investment so that the total interest earned is $4300?

9. An investment club invested $12,000 in two accounts. One investment earned 12.6% annual simple interest, while the other investment lost 5%. The total earnings from both investments were $104. Find the amount invested at 12.6%.

10. A total of $18,000 is invested in two accounts. One investment earned 11.2% annual simple interest while the other investment lost 4.7%. The total earnings from both investments were $1062. Find the amount invested at 11.2%.

2 Solve.

11. How many quarts of water must be added to 5 qt of an 80% antifreeze solution to make a 50% antifreeze solution?

12. How many milliliters of alcohol must be added to 200 ml of a 25% iodine solution to make a 10% iodine solution?

13. A goldsmith mixed 10 g of a 50% gold alloy with 40 g of a 15% gold alloy. What is the percent concentration of the resulting alloy?

14. A silversmith mixed 25 g of a 70% silver alloy with 50 g of a 15% silver alloy. What is the percent concentration of the resulting alloy?

15. How many ounces of pure water must be added to 60 oz of a 7.5% salt solution to make a 5% salt solution?

16. How many pounds of a 12% aluminum alloy must be mixed with 400 lb of a 30% aluminum alloy to make a 20% aluminum alloy?

17. A hospital staff mixed a 65% disinfectant solution with a 15% disinfectant solution. How many liters of each were used to make 50 L of a 40% disinfectant solution?

18. A butcher has some hamburger that is 20% fat and some hamburger that is 12% fat. How many pounds of each should be mixed to make 80 lb of hamburger that is 17% fat?

19. How much water must be evaporated from 8 gal of an 8% salt solution in order to obtain a 12% salt solution?

20. How much water must be evaporated from 6 qt of a 50% antifreeze solution to produce a 75% antifreeze solution?

21. A car radiator contains 12 qt of a 25% antifreeze solution. How many quarts will have to be replaced with pure antifreeze if the resulting solution is to be a 75% antifreeze solution?

22. A student mixed 50 ml of a 3% hydrogen peroxide solution with 20 ml of a 12% hydrogen peroxide solution. Find the percent concentration of the resulting mixture. Round to the nearest tenth of a percent.

23. Eighty pounds of a 54% copper alloy is mixed with 200 lb of a 22% copper alloy. Find the percent concentration of the resulting mixture. Round to the nearest tenth of a percent.

24. A druggist mixed 100 cc of a 15% alcohol solution with 50 cc of pure alcohol. Find the percent concentration of the resulting mixture. Round to the nearest tenth of a percent.

SUPPLEMENTAL EXERCISES 2.4

Solve.

25. A financial manager invested 25% of a client's money in bonds paying 9% annual simple interest, 30% in an 8% annual simple interest account, and the remainder in 9.5% corporate bonds. Find the amount invested in each if the total annual interest earned is $1785.

26. A silversmith mixed 90 g of a 40% silver alloy with 120 g of a 60% silver alloy. Find the percent concentration of the resulting alloy. Round to the nearest tenth of a percent.

27. Find the cost per pound of a tea mixture made from 50 lb of tea costing \$5.50 per pound and 75 lb of tea costing \$4.40 per pound.

28. How many kilograms of water must be evaporated from 75 kg of a 15% salt solution to produce a 20% salt solution?

29. How many grams of pure water must be added to 20 g of pure acid to make a solution that is 25% acid?

30. A radiator contains 6 L of a 25% antifreeze solution. How much should be drained and replaced with pure antifreeze to produce a 50% antifreeze solution?

SECTION 2.5
Inequalities in One Variable

1 Solve inequalities in one variable

The **solution set of an inequality** is a set of numbers, each element of which, when substituted for the variable, results in a true inequality.

The inequality at the right is true if the variable is replaced by 3, −1.98, or $\frac{2}{3}$.

$$x - 1 < 4$$
$$3 - 1 < 4$$
$$-1.98 - 1 < 4$$
$$\frac{2}{3} - 1 < 4$$

There are many values of the variable x that will make the inequality $x - 1 < 4$ true. The solution set of the inequality is any number less than 5. The solution set can be written in set builder notation as $\{x | x < 5\}$.

The graph of the solution set of $x - 1 < 4$ is shown below.

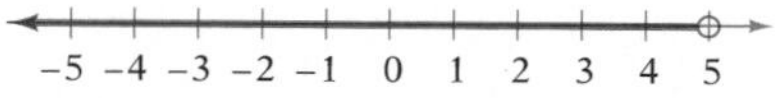

In solving an inequality, the Addition and Multiplication Properties of Inequalities are used to rewrite the inequality in the form *variable* < *constant* or *variable* > *constant*.

The Addition Property of Inequalities

If $a > b$ and c is a real number, then the inequalities $a > b$ and $a + c > b + c$ have the same solution set.

If $a < b$ and c is a real number, then the inequalities $a < b$ and $a + c < b + c$ have the same solution set.

The Addition Property of Inequalities states that the same number can be added to each side of an inequality without changing the solution set of the inequality. This property is also true for an inequality that contains the symbol $\le$ or $\ge$.

The Addition Property of Inequalities is used to remove a term from one side of an inequality by adding the additive inverse of that term to each side of the inequality. Because subtraction is defined in terms of addition, the same number can be subtracted from each side of an inequality without changing the solution set of the inequality.

Solve: $3x - 4 < 2x - 1$

	$3x - 4 < 2x - 1$
Subtract $2x$ from each side of the inequality.	$x - 4 < -1$
Add 4 to each side of the inequality.	$x < 3$
Write the solution set.	$\{x \mid x < 3\}$

The Multiplication Property of Inequalities is used to remove a coefficient from one side of an inequality so that the inequality can be written in the form variable $<$ constant or variable $>$ constant.

The Multiplication Property of Inequalities

Rule 1

If $a > b$ and $c > 0$, then the inequalities $a > b$ and $ac > bc$ have the same solution set.

If $a < b$ and $c > 0$, then the inequalities $a < b$ and $ac < bc$ have the same solution set.

Rule 2

If $a > b$ and $c < 0$, then the inequalities $a > b$ and $ac < bc$ have the same solution set.

If $a < b$ and $c < 0$, then the inequalities $a < b$ and $ac > bc$ have the same solution set.

Rule 1 states that when each side of an inequality is multiplied by a positive number, the inequality symbol remains the same. However, Rule 2 states that when each side of an inequality is multiplied by a negative number, the inequality symbol must be reversed.

Here are some examples of this property.

Rule 1		Rule 2	
$3 > 2$	$2 < 5$	$3 > 2$	$2 < 5$
$3(4) > 2(4)$	$2(4) < 5(4)$	$3(-4) < 2(-4)$	$2(-4) > 5(-4)$
$12 > 8$	$8 < 20$	$-12 < -8$	$-8 > -20$

Because division is defined in terms of multiplication, when each side of an inequality is divided by a positive number, the inequality symbol remains the same. When each side of an inequality is divided by a negative number, the inequality symbol must be reversed.

The Multiplication Property of Inequalities is true for the symbols $\leq$ or $\geq$.

Solve: $-3x > 9$

$$-3x > 9$$

Divide each side of the inequality by the coefficient -3 and reverse the inequality symbol. Simplify.

$$\frac{-3x}{-3} < \frac{9}{-3}$$

$$x < -3$$

Write the solution set.

$$\{x|x < -3\}$$

Example 1 Solve: $x + 3 > 4x + 6$

Solution

$x + 3 > 4x + 6$

$-3x + 3 > 6$ ► Subtract $4x$ from each side of the inequality.

$-3x > 3$ ► Subtract 3 from each side of the inequality.

$x < -1$ ► Divide each side of the inequality by -3 and reverse the inequality symbol.

$\{x|x < -1\}$ ► Write the solution set.

Problem 1 Solve: $2x - 1 < 6x + 7$

Solution See page A13.

When an inequality contains parentheses, the first step in solving the inequality is to use the Distributive Property to remove the parentheses.

Example 2 Solve: $5(x-2) \geq 9x - 3(2x-4)$

Solution

$5(x-2) \geq 9x - 3(2x-4)$

$5x - 10 \geq 9x - 6x + 12$ ▶ Use the Distributive Property to remove parentheses.

$5x - 10 \geq 3x + 12$ ▶ Simplify.

$2x - 10 \geq 12$ ▶ Subtract $3x$ from each side of the inequality.

$2x \geq 22$ ▶ Add 10 to each side of the inequality.

$x \geq 11$ ▶ Divide each side of the inequality by 2.

$\{x \mid x \geq 11\}$

Problem 2 Solve: $5x - 2 \leq 4 - 3(x-2)$

Solution See page A13.

2 Solve compound inequalities

A **compound inequality** is formed by joining two inequalities with a connective word such as "and" or "or." The inequalities shown below are compound inequalities.

$$2x < 4 \quad \text{and} \quad 3x - 2 > -8$$
$$2x + 3 > 5 \quad \text{or} \quad x + 2 < 5$$

The solution set of a compound inequality with the connective word *and* is the set of all elements common to the solution sets of each inequality. Therefore, it is the intersection of the solution sets of the two inequalities.

Solve: $2x < 6$ and $3x + 2 > -4$

Solve each inequality.

$2x < 6$ and $3x + 2 > -4$

$x < 3$ $\qquad 3x > -6$

$\qquad x > -2$

$\{x \mid x < 3\}$ $\qquad \{x \mid x > -2\}$

Find the intersection of the solution sets.

$\{x \mid x < 3\} \cap \{x \mid x > -2\} = \{x \mid -2 < x < 3\}$

Solve: $-3 < 2x + 1 < 5$

This inequality is equivalent to the compound inequality $-3 < 2x + 1$ and $2x + 1 < 5$.

$-3 < 2x + 1 < 5$

$-3 < 2x + 1$ and $2x + 1 < 5$

$-4 < 2x$ $\qquad 2x < 4$

$-2 < x$ $\qquad x < 2$

$\{x \mid x > -2\}$ $\qquad \{x \mid x < 2\}$

Find the intersection of the solution sets.

$\{x \mid x > -2\} \cap \{x \mid x < 2\} = \{x \mid -2 < x < 2\}$

There is an alternate method for solving the inequality in the last example.

Subtract 1 from each of the three parts of the inequality.

$$-3 < 2x + 1 < 5$$
$$-3 - 1 < 2x + 1 - 1 < 5 - 1$$
$$-4 < 2x < 4$$

Divide each of the three parts of the inequality by the coefficient 2.

$$\frac{-4}{2} < \frac{2x}{2} < \frac{4}{2}$$
$$-2 < x < 2$$

$\{x | -2 < x < 2\}$

The solution set of a compound inequality with the connective word *or* is the union of the solution sets of the two inequalities.

Solve: $2x + 3 > 7$ or $4x - 1 < 3$

Solve each inequality.

$$2x + 3 > 7 \quad \text{or} \quad 4x - 1 < 3$$
$$2x > 4 \qquad\qquad 4x < 4$$
$$x > 2 \qquad\qquad x < 1$$

$\{x | x > 2\}$ $\quad$ $\{x | x < 1\}$

Find the union of the solution sets.

$\{x | x > 2\} \cup \{x | x < 1\} =$
$\{x | x > 2 \text{ or } x < 1\}$

Example 3 Solve: $1 < 3x - 5 < 4$

Solution

$$1 < 3x - 5 < 4$$
$$1 + 5 < 3x - 5 + 5 < 4 + 5$$

▶ Add 5 to each of the three parts of the inequality.

$$6 < 3x < 9$$

▶ Simplify.

$$\frac{6}{3} < \frac{3x}{3} < \frac{9}{3}$$

▶ Divide each of the three parts of the inequality by 3.

$$2 < x < 3$$

$\{x | 2 < x < 3\}$

▶ Write the solution set.

Problem 3 Solve: $-2 \le 5x + 3 \le 13$

Solution See page A13.

Example 4 Solve: $11 - 2x > -3$ and $7 - 3x < 4$

Solution

$$11 - 2x > -3 \quad \text{and} \quad 7 - 3x < 4$$
$$-2x > -14 \qquad\qquad -3x < -3$$
$$x < 7 \qquad\qquad x > 1$$

▶ Solve each inequality.

$\{x | x < 7\}$ $\quad$ $\{x | x > 1\}$

$\{x | x < 7\} \cap \{x | x > 1\} =$
$\{x | 1 < x < 7\}$

▶ Find the intersection of the solution sets.

Problem 4 Solve: $5 - 4x > 1$ and $6 - 5x < 11$

Solution See page A13.

Example 5 Solve: $3 - 4x > 7$ or $4x + 5 < 9$

Solution

$$\begin{array}{lcl} 3 - 4x > 7 & \text{or} & 4x + 5 < 9 \\ -4x > 4 & & 4x < 4 \\ x < -1 & & x < 1 \end{array}$$

▶ Solve each inequality.

$\{x|x < -1\}$ $\{x|x < 1\}$

$\{x|x < -1\} \cup \{x|x < 1\} = \{x|x < 1\}$

▶ Find the union of the solution sets.

Problem 5 Solve: $2 - 3x > 11$ or $5 + 2x > 7$

Solution See page A13.

3 Application problems

Example 6 Company A rents cars for \$6 a day and 14¢ for every mile driven. Company B rents cars for \$12 a day and 8¢ for every mile driven. You want to rent a car for 5 days. How many miles can you drive a Company A car during the 5 days if it is to cost less than a Company B car?

Strategy

- To find the number of miles, write and solve an inequality using N to represent the number of miles.

Solution Cost of Company A car < cost of Company B car

$$\begin{aligned} 6(5) + 0.14N &< 12(5) + 0.08N \\ 30 + 0.14N &< 60 + 0.08N \\ 30 + 0.06N &< 60 \\ 0.06N &< 30 \\ N &< 500 \end{aligned}$$

It is less expensive to rent from Company A if the car is driven less than 500 mi.

Problem 6 The base of a triangle is 12 in., and the height is $(x + 2)$ in. Express as an integer the maximum height of the triangle when the area is less than 50 in.^{2}. (The formula for the area of a triangle is $A = \frac{1}{2}bh$.)

Solution See page A14.

Example 7 Find three consecutive odd integers whose sum is between 27 and 51.

Strategy ■ To find the three consecutive odd integers, write and solve a compound inequality using x to represent the first odd integer.

Solution Lower limit of the sum < sum < upper limit of the sum

$$27 < x + (x + 2) + (x + 4) < 51$$
$$27 < 3x + 6 < 51$$
$$27 - 6 < 3x + 6 - 6 < 51 - 6$$
$$21 < 3x < 45$$
$$\frac{21}{3} < \frac{3x}{3} < \frac{45}{3}$$
$$7 < x < 15$$

The three integers are 9, 11, and 13; or 11, 13, and 15; or 13, 15, and 17.

Problem 7 An average score of 80 to 89 in a history course receives a B grade. A student has grades of 72, 94, 83, and 70 on four exams. Find the range of scores on the fifth exam that will give the student a B for the course.

Solution See page A14.

EXERCISES 2.5

1 Solve.

1. $x - 3 < 2$	**2.** $x + 4 \geq 2$	**3.** $4x \leq 8$
4. $6x > 12$	**5.** $-2x > 8$	**6.** $-3x \leq -9$
7. $3x - 1 > 2x + 2$	**8.** $5x + 2 \geq 4x - 1$	**9.** $2x - 1 > 7$
10. $3x + 2 < 8$	**11.** $5x - 2 \leq 8$	**12.** $4x + 3 \leq -1$
13. $6x + 3 > 4x - 1$	**14.** $7x + 4 < 2x - 6$	**15.** $8x + 1 \geq 2x + 13$
16. $5x - 4 < 2x + 5$	**17.** $4 - 3x < 10$	**18.** $2 - 5x > 7$
19. $7 - 2x \geq 1$	**20.** $3 - 5x \leq 18$	**21.** $-3 - 4x > -11$
22. $-2 - x < 7$	**23.** $4x - 2 < x - 11$	**24.** $6x + 5 \geq x - 10$

25. $x + 7 \geq 4x - 8$

26. $3x + 1 \leq 7x - 15$

27. $6 - 2(x - 4) \leq 2x + 10$

28. $4(2x - 1) > 3x - 2(3x - 5)$

29. $2(1 - 3x) - 4 > 10 + 3(1 - x)$

30. $2 - 5(x + 1) \geq 3(x - 1) - 8$

31. $7 + 2(4 - x) < 9 - 3(6 + x)$

32. $3(4x + 3) \leq 7 - 4(x - 2)$

33. $\frac{3}{5}x - 2 < \frac{3}{10} - x$

34. $\frac{5}{6}x - \frac{1}{6} \leq x - 4$

35. $\frac{1}{3}x - \frac{3}{2} \geq \frac{7}{6} - \frac{2}{3}x$

36. $\frac{7}{12}x - \frac{3}{2} < \frac{2}{3}x + \frac{5}{6}$

37. $\frac{1}{2}x - \frac{3}{4} > \frac{7}{4}x - 2$

38. $\frac{2 - x}{4} - \frac{3}{8} \geq \frac{2}{5}x$

39. $2 - 2(7 - 2x) < 3(3 - x)$

40. $3 + 2(x + 5) \geq x + 5(x + 1) + 1$

41. $10 - 13(2 - x) < 5(3x - 2)$

42. $3 - 4(x + 2) \leq 6 + 4(2x + 1)$

2 Solve.

43. $3x < 6$ and $x + 2 > 1$

44. $x - 3 \leq 1$ and $2x \geq -4$

45. $x + 2 \geq 5$ or $3x \leq 3$

46. $2x < 6$ or $x - 4 > 1$

47. $-2x > -8$ and $-3x < 6$

48. $\frac{1}{2}x > -2$ and $5x < 10$

49. $\frac{1}{3}x < -1$ or $2x > 0$

50. $\frac{2}{3}x > 4$ or $2x < -8$

51. $x + 4 \geq 5$ and $2x \geq 6$

52. $3x < -9$ and $x - 2 < 2$

53. $-5x > 10$ and $x + 1 > 6$

54. $7x < 14$ and $1 - x < 4$

55. $2x - 3 > 1$ and $3x - 1 < 2$

56. $4x + 1 < 5$ and $4x + 7 > -1$

57. $3x + 7 < 10$ or $2x - 1 > 5$

58. $6x - 2 < -14$ or $5x + 1 > 11$

59. $-5 < 3x + 4 < 16$

60. $5 < 4x - 3 < 21$

61. $0 < 2x - 6 < 4$

62. $-2 < 3x + 7 < 1$

63. $4x - 1 > 11$ or $4x - 1 \leq -11$

64. $3x - 5 > 10$ or $3x - 5 < -10$

65. $2x + 3 \geq 5$ and $3x - 1 > 11$

66. $6x - 2 < 5$ or $7x - 5 < 16$

67. $9x - 2 < 7$ and $3x - 5 > 10$

68. $8x + 2 \leq -14$ and $4x - 2 > 10$

69. $3x - 11 < 4$ or $4x + 9 \geq 1$

70. $5x + 12 \geq 2$ or $7x - 1 \leq 13$

71. $-6 \leq 5x + 14 \leq 24$

72. $3 \leq 7x - 14 \leq 31$

73. $3 - 2x > 7$ and $5x + 2 > -18$

74. $1 - 3x < 16$ and $1 - 3x > -16$

75. $5 - 4x > 21$ or $7x - 2 > 19$

76. $6x + 5 < -1$ or $1 - 2x < 7$

77. $3 - 7x \leq 31$ and $5 - 4x > 1$

78. $9 - x \geq 7$ and $9 - 2x < 3$

79. $\frac{2}{3}x - 4 > 5$ or $x + \frac{1}{2} < 3$

80. $\frac{5}{8}x + 2 < -3$ or $2 - \frac{3}{5}x < -7$

81. $\frac{5}{7}x - 2 > 3$ and $3 - \frac{2}{3}x > -2$

82. $5 - \frac{2}{7}x < 4$ and $\frac{3}{4}x - 2 < 7$

83. $-\frac{3}{8} \leq 1 - \frac{1}{4}x \leq \frac{7}{2}$

84. $-2 \leq \frac{2}{3}x - 1 \leq 3$

3 Solve.

85. Five times the difference between a number and two is greater than the quotient of two times the number and three. Find the smallest integer that will satisfy the inequality.

86. Two times the difference between a number and eight is less than or equal to five times the sum of the number and four. Find the smallest number that will satisfy the inequality.

87. The length of a rectangle is two feet more than four times the width. Express as an integer the maximum width of the rectangle when the perimeter is less than 34 ft.

88. The length of a rectangle is 5 cm less than twice the width. Express as an integer the maximum width of the rectangle when the perimeter is less than 60 cm.

89. Find four consecutive integers whose sum is between 62 and 78.

90. Find three consecutive even integers whose sum is between 30 and 52.

91. One side of a triangle is 1 in. longer than the second side. The third side is 2 in. longer than the second side. Find the length of the second side of the triangle to the nearest whole number, if the perimeter is more than 15 in. and less than 25 in.

92. The length of a rectangle is 4 ft more than twice the width. Find the width of the rectangle to the nearest whole number, if the perimeter is more than 28 ft and less than 40 ft.

93. A cellular phone company offers its customers a rate of \$99 for up to 200 min per month of cellular phone time, or a rate of \$35 per month plus \$.40 for each minute of cellular phone time. For how many minutes per month can a customer who chooses the second option use a cellular phone before the charges exceed the first option?

94. A cellular phone company offers its customers a rate of \$36.20 per month plus \$.40 for each minute of cellular phone time, or \$20 per month plus \$.76 for each minute of cellular phone time. For how many minutes can a customer who chooses the second option use a cellular phone before the charges exceed the first option?

95. A car tested for gas mileage averages between 26 mpg and 28.5 mpg. If this test is accurate, find the range of miles that the car can travel on a full tank (13.5 gal) of gasoline.

96. A new car will average at least 22 mpg for city driving and at most 27.5 mpg for highway driving. Find the range of miles that the car can travel on a full tank (19.5 gal) of gasoline.

97. A bank offers two types of checking accounts. One account has a charge of \$7 per month plus 2¢ per check. The second account has a charge of \$2 per month and 5¢ per check. How many checks can a customer who has the second type of account write if it is to cost the customer less than the first type of account?

98. A bank offers two types of checking accounts. One account has a charge of \$2 per month plus 8¢ per check. The second account has a charge of \$8 per month and 3¢ per check. How many checks can a customer who has the first type of account write if it is to cost the customer less than the second type of account?

99. Company A rents cars for \$10 a day and 10¢ for every mile driven. Company B rents cars for \$14 per day and 6¢ for every mile driven. You want to rent a car for one week. How many miles can you drive a Company A car during the week if it is to cost you less than a Company B car?

100. Company A rents cars for \$15 a day and 5¢ for every mile driven. Company B rents cars for \$12 per day and 8¢ for every mile driven. You want to rent a car for four days. How many miles can you drive a Company B car if it is to cost you less than a Company A car?

101. An average score of 80–89 in a psychology course receives a B grade. A student has grades of 94, 88, 70, and 62 on four tests. Find the range of scores on the fifth test that will give the student a B for the course.

102. An average score of 90 or above in an English course receives an A grade. A student has grades of 85, 88, 90, and 98 on four tests. Find the range of scores on the fifth test that will give the student an A grade.

SUPPLEMENTAL EXERCISES 2.5

Use the roster method to list the set of positive integers that are solutions of the inequality.

103. $3x - 4 \leq 7$

104. $4 - 5x \geq -16$

105. $8x - 7 < 2x + 9$

106. $2x + 9 \geq 5x - 4$

107. $5 + 3(2 + x) > 8 + 4(x - 1)$

108. $6 + 4(2 - x) > 7 + 3(x + 5)$

109. $-3x < 15$ and $x + 2 < 7$

110. $3x - 2 > 1$ and $2x - 3 < 5$

111. $-4 \leq 3x + 8 < 16$

112. $5 < 7x - 3 \leq 24$

113. $2x - 9 \leq 3$ and $4x + 8 \geq 1$

114. $6x + 11 \geq 3$ and $8x - 1 \leq 12$

Solve.

115. The relationship between Celsius temperature and Fahrenheit temperature is given by the formula $F = \frac{9}{5}C + 32$. If the temperature is between 77°F and 86°F, what is the temperature range in degrees Celsius?

116. The relationship between Celsius temperature and Fahrenheit temperature is given by the formula $F = \frac{9}{5}C + 32$. If the temperature is between −22°F and −13°F, what is the temperature range in degrees Celsius?

117. The average of two positive integers is less than or equal to 40. The larger integer is 8 more than the smaller integer. Find the greatest possible value for the larger integer.

118. The average of two negative integers is less than or equal to −15. The smaller integer is 7 less than the larger integer. Find the greatest possible value for the smaller integer.

119. The charges for a long-distance telephone call are \$1.56 for the first three minutes and \$.52 for each additional minute or fraction of a minute. What is the largest whole number of minutes a call could last if it is to cost you less than \$5.40?

120. A group decides to publish a calendar to raise money. The initial cost, regardless of the number of calendars printed, is \$800. After the initial cost, each calendar costs \$1.50 to produce. What is the minimum number of calendars a group must sell at \$7.50 per calendar to make a profit of at least \$1500?

SECTION 2.6 Absolute Value Equations and Inequalities

1 Absolute value equations

The **absolute value** of a number is its distance from zero on the number line. Distance is always a positive number or zero. Therefore, the absolute value of a number is always a positive number or zero.

The distance from 0 to 3 or from 0 to -3 is 3 units.

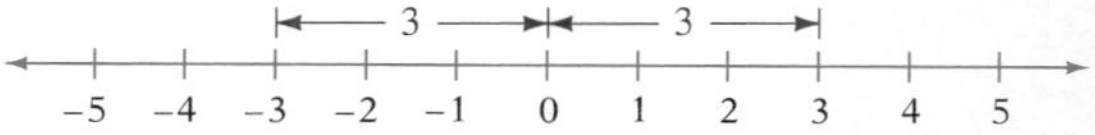

$|3| = 3$ $\qquad |-3| = 3$

Absolute value can also be used to represent the distance between any two points on the number line. The **distance between two points** on the number line is the absolute value of the difference between the coordinates of the two points.

The distance between point a and point b is given by $|b - a|$ or $|a - b|$.

The distance between -4 and 3 on the number line is 7 units. Note that the order in which the coordinates are subtracted does not affect the distance.

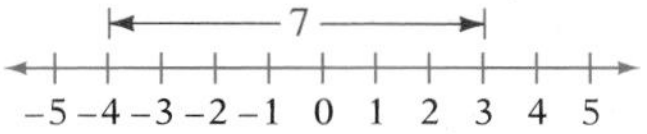

$$\text{Distance} = |3 - (-4)| = |7| = 7 \qquad \text{Distance} = |-4 - 3| = |-7| = 7$$

For any two numbers a and b, $\mathbf{|b - a| = |a - b|}$.

Find the distance from -5 to 8 on the number line.

The distance between points a and b on the number line is given by $|a - b|$. Let $a = -5$ and $b = 8$.

$$\text{Distance} = |a - b| = |-5 - 8| = |-13| = 13$$

The distance is 13.

An equation containing an absolute value symbol is called an **absolute value equation.**

$$\left.\begin{aligned} |x| &= 3 \\ |x + 2| &= 8 \\ |3x - 4| &= 5x - 9 \end{aligned}\right\} \text{Absolute Value Equations}$$

Absolute Value Equations

> If $a \geq 0$ and $|x| = a$, then $x = a$ or $x = -a$.

Given $|x| = 3$, then $x = 3$ or $x = -3$.

Solve: $|x + 2| = 8$

$|x + 2| = 8$

Remove the absolute value sign, and rewrite as two equations.

$x + 2 = 8 \qquad x + 2 = -8$

Solve each equation.

$x = 6 \qquad x = -10$

Check:

$\lvert x + 2\rvert = 8$		$\lvert x + 2\rvert = 8$	
$\lvert 6 + 2\rvert$	8	$\lvert -10 + 2\rvert$	8
$\lvert 8\rvert$		$\lvert -8\rvert$	
$8 = 8$		$8 = 8$	

Write the solution.

The solutions are 6 and -10.

Example 1 Solve.

A. $|x| = 15$ B. $|2 - x| = 12$ C. $|2x| = -4$ D. $3 - |2x - 4| = -5$

Solution

A. $|x| = 15$

$x = 15 \qquad x = -15$

► Remove the absolute value sign, and rewrite as two equations.

The solutions are 15 and -15.

B. $|2 - x| = 12$

$2 - x = 12 \qquad 2 - x = -12$

► Remove the absolute value sign, and rewrite as two equations.

$-x = 10 \qquad -x = -14$

$x = -10 \qquad x = 14$

► Solve each equation.

The solutions are -10 and 14.

C. $|2x| = -4$

► The absolute value of a number must be nonnegative.

There is no solution to the equation.

D. $3 - |2x - 4| = -5$

$-|2x - 4| = -8$

$|2x - 4| = 8$

► Solve for the absolute value. Multiply each side of the equation by -1.

$2x - 4 = 8 \qquad 2x - 4 = -8$

► Remove the absolute value sign, and rewrite as two equations.

$2x = 12 \qquad 2x = -4$

$x = 6 \qquad x = -2$

► Solve each equation.

The solutions are 6 and -2.

Problem 1 Solve.

A. $|x| = 25$ B. $|2x - 3| = 5$ C. $|x - 3| = -2$ D. $5 - |3x + 5| = 3$

Solution See page A14–A15.

2 Absolute value inequalities

Recall that absolute value represents the distance between two points. For example, the solutions of the absolute value equation $|x - 1| = 3$ are the numbers whose distance from 1 is 3. Therefore, the solutions are -2 and 4.

The solutions of the absolute value inequality $|x - 1| < 3$ are the numbers whose distance from 1 is *less than* 3. Therefore, the solutions are the numbers greater than -2 and less than 4. The solution set is $\{x \mid -2 < x < 4\}$.

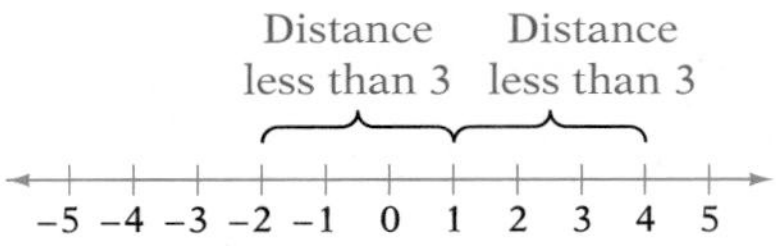

Absolute Value Inequalities of the Form $|ax + b| < c$

> To solve an absolute value inequality of the form $|ax + b| < c$, solve the equivalent compound inequality $-c < ax + b < c$.

Solve: $|3x - 1| < 5$

Solve the equivalent compound inequality.

$$|3x - 1| < 5$$
$$-5 < 3x - 1 < 5$$
$$-5 + 1 < 3x - 1 + 1 < 5 + 1$$
$$-4 < 3x < 6$$
$$\frac{-4}{3} < \frac{3x}{3} < \frac{6}{3}$$
$$-\frac{4}{3} < x < 2$$
$$\left\{x \middle| -\frac{4}{3} < x < 2\right\}$$

Example 2 Solve: $|4x - 3| < 5$

Solution

$$|4x - 3| < 5$$
$$-5 < 4x - 3 < 5$$

▶ Solve the equivalent compound inequality.

$$-5 + 3 < 4x - 3 + 3 < 5 + 3$$

▶ Add 3 to each of the three parts of the inequality.

$$-2 < 4x < 8$$
$$\frac{-2}{4} < \frac{4x}{4} < \frac{8}{4}$$

▶ Divide each of the three parts of the inequality by 4.

$$-\frac{1}{2} < x < 2$$
$$\left\{x \middle| -\frac{1}{2} < x < 2\right\}$$

▶ Write the solution set.

Problem 2 Solve: $|3x + 2| < 8$

Solution See page A15.

Example 3 Solve: $|x - 3| < 0$

Solution $|x - 3| < 0$ ▶ The absolute value of a number must be nonnegative.

$\varnothing$ ▶ The solution set is the empty set.

Problem 3 Solve: $|3x - 7| < 0$

Solution See page A15.

The solutions of the absolute value inequality $|x + 1| > 2$ are the numbers whose distance from -1 is *greater than* 2. Therefore, the solutions are the numbers less than -3 or greater than 1. The solution set is $\{x \mid x < -3 \text{ or } x > 1\}$.

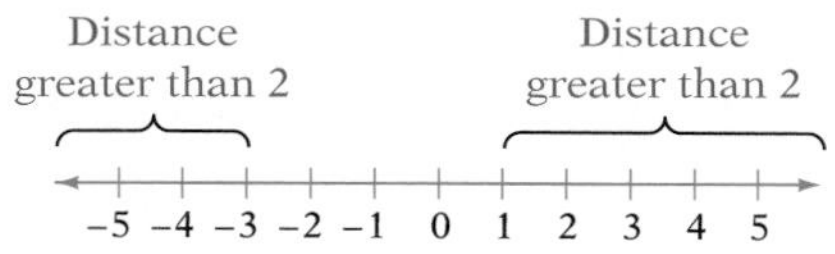

Absolute Value Inequalities of the Form $|ax + b| > c$

> To solve an absolute value inequality of the form $|ax + b| > c$, solve the equivalent compound inequality $ax + b < -c$ **or** $ax + b > c$.

Solve: $|3 - 2x| > 1$

Solve the equivalent compound inequality.

$$|3 - 2x| > 1$$

$$3 - 2x < -1 \quad \text{or} \quad 3 - 2x > 1$$

$$-2x < -4 \qquad\qquad -2x > -2$$

$$x > 2 \qquad\qquad x < 1$$

$$\{x \mid x > 2\} \qquad\qquad \{x \mid x < 1\}$$

Find the union of the solution sets.

$$\{x \mid x > 2\} \cup \{x \mid x < 1\} = \{x \mid x > 2 \text{ or } x < 1\}$$

Example 4 Solve: $|2x - 1| > 7$

Solution $|2x - 1| > 7$

$$\begin{array}{lcl} 2x - 1 < -7 & \text{or} & 2x - 1 > 7 \\ 2x < -6 & & 2x > 8 \\ x < -3 & & x > 4 \end{array}$$

▶ Solve the equivalent compound inequality.

$\{x \mid x < -3\}$ $\quad\quad$ $\{x \mid x > 4\}$

$\{x \mid x < -3\} \cup \{x \mid x > 4\} =$
$\{x \mid x < -3 \text{ or } x > 4\}$

▶ Find the union of the solution sets.

Problem 4 Solve: $|5x + 3| > 8$

Solution See page A15.

3 Application problems

The **tolerance** of a component, or part, is the acceptable amount by which the component may vary from a given measurement. For example, the diameter of a piston may vary from the given measurement of 9 cm by 0.001 cm. This is written as 9 cm ± 0.001 cm, read "9 centimeters plus or minus 0.001 centimeters." The maximum diameter, or **upper limit,** of the piston is 9 cm + 0.001 cm = 9.001 cm. The minimum diameter, or **lower limit,** is 9 cm − 0.001 cm = 8.999 cm.

The lower and upper limits of the diameter of the piston could also be found by solving the absolute value inequality $|d - 9| \leq 0.001$, where d is the diameter of the piston.

$$\begin{gathered} |d - 9| \leq 0.001 \\ -0.001 \leq d - 9 \leq 0.001 \\ -0.001 + 9 \leq d - 9 + 9 \leq 0.001 + 9 \\ 8.999 \leq d \leq 9.001 \end{gathered}$$

The lower and upper limits of the diameter of the piston are 8.999 cm and 9.001 cm.

Example 5 A doctor has prescribed 2 cc of medication for a patient. The tolerance is 0.03 cc. Find the lower and upper limits of the amount of medication to be given.

Strategy

- Let p represent the prescribed amount of medication, T the tolerance, and m the given amount of medication. Solve the absolute value inequality $|m - p| \leq T$ for m.

Solution $|m - p| \le T$
$|m - 2| \le 0.03$ ▶ Substitute the values of p and T into the inequality.
$-0.03 \le m - 2 \le 0.03$ ▶ Solve the equivalent compound inequality.
$1.97 \le m \le 2.03$

The lower and upper limits of the amount of medication to be given to the patient are 1.97 cc and 2.03 cc.

Problem 5 A machinist must make a bushing that has a tolerance of 0.003 in. The diameter of the bushing is 2.55 in. Find the lower and upper limits of the diameter of the bushing.

Solution See page A15.

EXERCISES 2.6

1 Solve.

1. Find the distance between −3 and −9 on the number line.
2. Find the distance between 8 and −4 on the number line.
3. Find the distance between −3 and 2 on the number line.
4. Find the distance between −13 and 6 on the number line.
5. $|x| = 7$
6. $|a| = 2$
7. $|-t| = 3$
8. $|-a| = 7$
9. $|-t| = -3$
10. $|-y| = -2$
11. $|x + 2| = 3$
12. $|x + 5| = 2$
13. $|y - 5| = 3$
14. $|y - 8| = 4$
15. $|a - 2| = 0$
16. $|a + 7| = 0$
17. $|x - 2| = -4$
18. $|x + 8| = -2$
19. $|2x - 5| = 4$
20. $|4 - 3x| = 4$
21. $|2 - 5x| = 2$
22. $|3 - 4x| = 9$

23. $|2 - 5x| = 3$

24. $|2x - 3| = 0$

25. $|5x + 5| = 0$

26. $|3x - 2| = -4$

27. $|2x + 5| = -2$

28. $|x - 2| - 2 = 3$

29. $|x - 9| - 3 = 2$

30. $|3a + 2| - 4 = 4$

31. $|2a + 9| + 4 = 5$

32. $|2 - y| + 3 = 4$

33. $|8 - y| - 3 = 1$

34. $|2x - 3| + 3 = 3$

35. $|4x - 7| - 5 = -5$

36. $|2x - 3| + 4 = -4$

37. $|3x - 2| + 1 = -1$

38. $|6x - 5| - 2 = 4$

39. $|4b + 3| - 2 = 7$

40. $|3t + 2| + 3 = 4$

41. $|5x - 2| + 5 = 7$

42. $3 - |x - 4| = 5$

43. $2 - |x - 5| = 4$

44. $|2x - 8| + 12 = 2$

45. $|3x - 4| + 8 = 3$

46. $2 + |3x - 4| = 5$

47. $5 + |2x + 1| = 8$

48. $5 - |2x + 1| = 5$

49. $3 - |5x + 3| = 3$

50. $6 - |2x + 4| = 3$

51. $8 - |3x - 2| = 5$

52. $8 - |1 - 3x| = -1$

2 Solve.

53. $|x| > 3$

54. $|x| < 5$

55. $|x + 1| > 2$

56. $|x - 2| > 1$

57. $|x - 5| \leq 1$

58. $|x - 4| \leq 3$

59. $|2 - x| \geq 3$

60. $|3 - x| \geq 2$

61. $|2x + 1| < 5$

62. $|3x - 2| < 4$

63. $|5x + 2| > 12$

64. $|7x - 1| > 13$

65. $|4x - 3| \leq -2$

66. $|5x + 1| \leq -4$

67. $|2x + 7| > -5$

68. $|3x - 1| > -4$

69. $|4 - 3x| \geq 5$

70. $|7 - 2x| > 9$

71. $|5 - 4x| \leq 13$

72. $|3 - 7x| < 17$

73. $|6 - 3x| \leq 0$

74. $|10 - 5x| \geq 0$

75. $|2 - 9x| > 20$

76. $|5x - 1| < 16$

3 Solve.

77. The diameter of a bushing is 1.75 in. The bushing has a tolerance of 0.008 in. Find the lower and upper limits of the diameter of the bushing.

78. A machinist must make a bushing that has a tolerance of 0.004 in. The diameter of the bushing is 3.48 in. Find the lower and upper limits of the diameter of the bushing.

79. A doctor has prescribed 2.5 cc of medication for a patient. The tolerance is 0.2 cc. Find the lower and upper limits of the amount of medication to be given.

80. A power strip is utilized on a computer to prevent the loss of programming by electrical surges. The power strip is designed to allow 110 volts plus or minus 16.5 volts. Find the lower and upper limits of voltage to the computer.

81. An electric motor is designed to run on 220 volts plus or minus 25 volts. Find the lower and upper limits of voltage on which the motor will run.

82. A piston rod for an automobile is $10\frac{3}{8}$ in. with a tolerance of $\frac{1}{32}$ in. Find the lower and upper limits of the length of the piston rod.

83. The diameter of a piston for an automobile is $3\frac{5}{16}$ in. with a tolerance of $\frac{1}{64}$ in. Find the lower and upper limits of the diameter of the piston.

The tolerance of the resistors used in electronics is given as a percent.

84. Find the lower and upper limits of a 29,000-ohm resistor with a 2% tolerance.

85. Find the lower and upper limits of a 15,000-ohm resistor with a 10% tolerance.

86. Find the lower and upper limits of a 25,000-ohm resistor with a 5% tolerance.

87. Find the lower and upper limits of a 56-ohm resistor with a 5% tolerance.

SUPPLEMENTAL EXERCISES 2.6

Solve.

88. $\left|\frac{2x-5}{3}\right| = 7$

89. $\left|\frac{4x-3}{5}\right| = 5$

90. $\left|\frac{3x-2}{4}\right| + 5 = 6$

91. $\left|\frac{5x+1}{6}\right| - 1 = 2$

92. $\left|\frac{4x-2}{3}\right| > 6$

93. $\left|\frac{2x-3}{3}\right| \geq 9$

94. $\left|\frac{2x-1}{5}\right| \leq 3$

95. $\left|\frac{3x-5}{2}\right| < 7$

For what values of the variable is the equation true? Write the answer in set builder notation.

96. $|x + 3| = x + 3$

97. $|y + 6| = y + 6$

98. $|a - 4| = 4 - a$

99. $|b - 7| = 7 - b$

100. For real numbers x and y, which of the following is always true?
a. $|x + y| = |x| + |y|$ **b.** $|x + y| \leq |x| + |y|$ **c.** $|x + y| \geq |x| + |y|$

Calculators and Computers

Solving First-Degree Equations and Absolute Value Inequalities

The program SOLVE A FIRST-DEGREE EQUATION on the Math ACE Disk will allow you to practice solving the following three types of equations:

1. $ax + b = c$
2. $ax + b = cx + d$
3. Equations with parentheses

After you select the type of equation you want to practice, a problem will be displayed on the screen. Using paper and pencil, solve the problem. When you are ready, press the RETURN key, and the complete solution will be displayed. Compare your solution with the displayed solution.

When you finish a problem, you may continue practicing the type of problem you have selected, or return to the main menu and select a different type, or quit the program.

The program ABSOLUTE VALUE INEQUALITIES on the Math ACE Disk will allow you to practice solving an absolute value inequality of the form:

$$|ax + b| < c \qquad \text{or} \qquad |ax + b| > c$$

A problem will be displayed on the screen and you, using paper and pencil, are to solve the problem. When you are ready to see the solution, press the RETURN key. The solution will be displayed.

Something Extra

Absolute Value Equations and Inequalities

The absolute value equation $|ax + b| = |cx + d|$ can be solved by using the definition of absolute value.

If both $ax + b$ and $cx + d$ are positive, then:

$$ax + b = cx + d$$

If both $ax + b$ and $cx + d$ are negative, then:

$$-(ax + b) = -(cx + d) \text{ or}$$
$$ax + b = cx + d$$

If either $ax + b$ or $cx + d$ is negative, we obtain:

$$-(ax + b) = cx + d \text{ or}$$
$$ax + b = -(cx + d)$$

From these three equations we see that if $|ax + b| = |cx + d|$ is a true equation, then either

$$ax + b = cx + d \qquad \text{or} \qquad ax + b = -(cx + d).$$

Solve: $|2x - 5| = |3 - 5x|$

The expressions are equal or the expressions are opposite.

$$2x - 5 = 3 - 5x \qquad\qquad 2x - 5 = -(3 - 5x)$$
$$7x = 8 \qquad\qquad 2x - 5 = -3 + 5x$$
$$x = \frac{8}{7} \qquad\qquad -3x = 2$$
$$x = -\frac{2}{3}$$

The solutions are $-\frac{2}{3}$ and $\frac{8}{7}$.

Solve: $|x - 4| > |3 - x|$

This problem can be solved by squaring each side of the absolute value inequality. This ensures that each side of the absolute value inequality will be positive.

$$|x - 4| > |3 - x|$$
$$(x - 4)^2 > (3 - x)^2$$
$$x^2 - 8x + 16 > 9 - 6x + x^2$$
$$-2x > -7$$
$$x < \frac{7}{2}$$

The solution is $\left\{x \middle| x < \frac{7}{2}\right\}$.

Solve.

1. $|3x - 4| = |5x - 8|$
2. $|5x - 3| = |3x - 3|$
3. $|2x - 1| = |x|$
4. $|x + 1| = |x|$
5. $\left|\frac{2x - 3}{5}\right| = |2 - x|$
6. $\left|\frac{3 - 2x}{3}\right| = \left|\frac{x - 4}{2}\right|$
7. $|2x - 3| > |2x - 5|$
8. $|x + 4| < |3 - x|$
9. $|3x - 2| < |3x + 2|$
10. $|x - 1| > |x|$

Chapter Summary

Key Words

An *equation* expresses the equality of two mathematical expressions.

A *solution* or *root* of an equation is a replacement value for the variable that will make the equation true.

To *solve* an equation means to find its solutions.

Consecutive integers are integers that follow one another in order.

The *solution set of an inequality* is a set of numbers, each element of which, when substituted for the variable, results in a true inequality.

A *compound inequality* is formed by joining two inequalities with a connective word such as "and" or "or."

An *absolute value equation* is an equation containing an absolute value symbol.

Essential Rules

The Addition Property of Equations
If $a = b$ and c is a real number, then the equations $a = b$ and $a + c = b + c$ have the same solution set.

The Multiplication Property of Equations
If $a = b$ and c is a real number, $c \neq 0$, then the equations $a = b$ and $ac = bc$ have the same solution set.

The Addition Property of Inequalities
If $a > b$ and c is a real number, then the inequalities $a > b$ and $a + c > b + c$ have the same solutions.

If $a < b$ and c is a real number, then the inequalities $a < b$ and $a + c < b + c$ have the same solutions.

The Multiplication Property of Inequalities
Rule 1
If $a > b$ and $c > 0$, then the inequalities $a > b$ and $ac > bc$ have the same solution set.

If $a < b$ and $c > 0$, then the inequalities $a < b$ and $ac < bc$ have the same solution set.

Rule 2
If $a > b$ and $c < 0$, then the inequalities $a > b$ and $ac < bc$ have the same solution set.

If $a < b$ and $c < 0$, then the inequalities $a < b$ and $ac > bc$ have the same solution set.

To solve an absolute value inequality of the form $|ax + b| < c$, solve the equivalent compound inequality

$$-c < ax + b < c.$$

To solve an absolute value inequality of the form $|ax + b| > c$, solve the equivalent compound inequality

$$ax + b < -c \text{ or } ax + b > c.$$

Value Mixture Equation Value = amount · unit cost
$$V = AC$$

Uniform Motion Equation Distance = rate · time

$d = rt$

Annual Simple Interest Equation

Simple interest = principal · simple interest rate

$I = Pr$

Percent Mixture Equation

Quantity = amount · percent of concentration

$Q = Ar$

Chapter Review

1. Solve: $x + 4 = -5$

2. Solve: $\frac{2}{3} = x + \frac{3}{4}$

3. Solve: $-3x = -21$

4. Solve: $\frac{2}{3}x = \frac{4}{9}$

5. Solve: $3y - 5 = 3 - 2y$

6. Solve: $3x - 3 + 2x = 7x - 15$

7. Solve: $\frac{3}{5}x - 3 = 2x + 5$

8. Solve: $-2x + \frac{4}{9} = x + 3$

9. Solve: $2(x - 3) = 5(4 - 3x)$

10. Solve: $2x - (3 - 2x) = 4 - 3(4 - 2x)$

11. Solve: $\frac{1}{2}x - \frac{5}{8} = \frac{3}{4}x + \frac{3}{2}$

12. Solve: $\frac{2x - 3}{3} + 2 = \frac{2 - 3x}{5}$

13. Solve: $3x - 7 > -2$

14. Solve: $2x - 9 < 8x + 15$

15. Solve: $\frac{2}{3}x - \frac{5}{8} \geq \frac{5}{4}x + 3$

16. Solve: $2 - 3(x - 4) \leq 4x - 2(1 - 3x)$

17. Solve: $-5 < 4x - 1 < 7$

18. Solve: $5x - 2 > 8$ or $3x + 2 < -4$

19. Solve: $3x < 4$ and $x + 2 > -1$

20. Solve: $3x - 2 > -4$ or $7x - 5 < 3x + 3$

21. Solve: $|2x - 3| = 8$

22. Solve: $|x - 4| = -3$

23. Solve: $|5x + 8| = 0$

24. Solve: $6 + |3x - 3| = 2$

25. Solve: $|2x - 5| \leq 3$

26. Solve: $|4x - 5| \geq 3$

27. Solve: $|5x - 4| < -2$

28. Solve: $6 - |2x - 5| > 3$

29. The diameter of a bushing is 2.75 in. The bushing has a tolerance of 0.003 in. Find the lower and upper limits of the diameter of the bushing.

30. A doctor has prescribed 2 cc of medication for a patient. The tolerance is 0.25 cc. Find the lower and upper limits of the amount of medication to be given.

31. The sum of two integers is twenty. Five times the smaller integer is two more than twice the larger integer. Find the integers.

32. Find three consecutive integers such that five times the middle integer is twice the sum of the other two integers.

33. A coin collection contains 30 coins in nickels, dimes, and quarters. There are three more dimes than nickels. The value of the coins is $3.55. Find the number of quarters in the collection.

34. A stamp collection that consists of 7¢, 13¢, and 18¢ stamps has a value of $3.35. The number of 13¢ stamps is one less than twice the number of 7¢ stamps. The number of 18¢ stamps is five more than the number of 13¢ stamps. Find the number of 13¢ stamps in the collection.

35. A silversmith combines 40 oz of pure silver that costs $8.00 per ounce with 200 oz of a silver alloy costing $3.50 per ounce. Find the cost per ounce of the mixture.

36. A grocer mixed apple juice that cost $3.20 per gallon with 40 gal of cranberry juice that cost $5.50 per gallon. How much apple juice was used to make cranapple juice costing $4.20 per gallon?

37. Two planes are 1680 mi apart and traveling toward each other. One plane is traveling 80 mph faster than the other plane. The planes meet in 1.75 h. Find the speed of each plane.

38. A cyclist traveled at a rate of 18 mph to a nearby town. The cyclist averaged 12 mph on the return trip. If the round trip took 5 h, find the distance to the nearby town.

39. Two investments earn an annual income of $635. One investment is earning 10.5% annual simple interest and the other investment is earning 6.4% annual simple interest. The total investment is $8000. Find the amount invested in each account.

40. An investment club invested $4200 at an annual simple interest rate of 8%. How much additional money must be invested at an annual simple interest rate of 9.6% so that the total interest earned is $912?

41. How many ounces of water must be added to 20 oz of a 15% salt solution to make a 6% salt solution?

42. An alloy containing 30% tin is mixed with an alloy containing 70% tin. How many pounds of each were used to make 500 lb of an alloy containing 40% tin?

43. A sales executive earns $800 per month plus a 4% commission on the amount of sales. The executive's goal is to earn at least $3000 per month. What amount of sales will enable the executive to earn $3000 or more per month?

44. An average score of 80 to 90 in a psychology class receives a B grade. A student has grades of 92, 66, 72, and 88 on four tests. Find the range of scores on the fifth test that will give the student a B for the course.

Chapter Test

1. Solve: $x - 2 = -4$

2. Solve: $x + \frac{3}{4} = \frac{5}{8}$

3. Solve: $-\frac{3}{4}y = -\frac{5}{8}$

4. Solve: $3x - 5 = 7$

5. Solve: $\frac{3}{4}y - 2 = 6$

6. Solve: $2x - 3 - 5x = 8 + 2x - 10$

7. Solve: $2[x - (2 - 3x) - 4] = x - 5$

8. Solve: $\frac{2}{3}x - \frac{5}{6}x = 4$

9. Solve: $\frac{2x + 1}{3} - \frac{3x + 4}{6} = \frac{5x - 9}{9}$

10. Solve: $2x - 5 \geq 5x + 4$

11. Solve: $4 - 3(x + 2) < 2(2x + 3) - 1$

12. Solve: $3x - 2 > 4$ or $4 - 5x < 14$

13. Solve: $4 - 3x \geq 7$ and $2x + 3 \geq 7$

14. Solve: $|3 - 5x| = 12$

15. Solve: $2 - |2x - 5| = -7$

16. Solve: $|3x - 1| \leq 2$

17. Solve: $|2x - 1| > 3$

18. Solve: $4 + |2x - 3| = 1$

19. Agency A rents cars for \$12 per day and 10¢ for every mile driven. Agency B rents cars for \$24 per day with unlimited mileage. How many miles per day can you drive an Agency A car if it is to cost you less than an Agency B car?

20. A doctor prescribed 3 cc of medication for a patient. The tolerance is 0.1 cc. Find the lower and upper limits of the amounts of medication to be given.

21. A stamp collection contains 11¢, 15¢, and 24¢ stamps. There are twice as many 11¢ stamps as 15¢ stamps. There are 30 stamps in all with a value of \$4.40. How many 24¢ stamps are in the collection?

22. A butcher combines 100 lb of hamburger that costs \$1.60 per pound with 60 lb of hamburger that costs \$3.20 per pound. Find the cost of the hamburger mixture.

23. A jogger runs a distance at a speed of 8 mph and returns the same distance running at a speed of 6 mph. Find the total distance that the jogger runs if the total time running is one hour and forty-five minutes.

24. An investment of \$12,000 is deposited into two simple interest accounts. On one account the annual simple interest rate is 7.8%. On the other, the annual simple interest rate is 9%. The total interest earned for one year is \$1020. How much was invested in each account?

25. How many ounces of pure water must be added to 60 oz of an 8% salt solution to make a 3% salt solution?

Cumulative Review

1. Simplify: $-2^2 \cdot 3^3$

2. Simplify: $4 - (2 - 5)^2 \div 3 + 2$

3. Simplify: $4 \div \dfrac{\frac{3}{8} - 1}{5} \cdot 2$

4. Evaluate $2a^2 - (b - c)^2$ when $a = 2$, $b = 3$, and $c = -1$.

5. Identify the property that justifies the statement.
$(2x + 3y) + 2 = (3y + 2x) + 2$

6. Find $A \cap B$ given $A = \{3, 5, 7, 9\}$ and $B = \{3, 6, 9\}$.

7. Simplify: $3x - 2[x - 3(2 - 3x) + 5]$

8. Simplify: $5[y - 2(3 - 2y) + 6]$

9. Solve: $4 - 3x = -2$

10. Solve: $-\frac{5}{6}b = -\frac{5}{12}$

11. Solve: $2x + 5 = 5x + 2$

12. Solve: $\frac{5}{12}x - 3 = 7$

13. Solve: $2[3 - 2(3 - 2x)] = 2(3 + x)$

14. Solve: $3[2x - 3(4 - x)] = 2(1 - 2x)$

15. Solve: $\dfrac{3x - 1}{4} - \dfrac{4x - 1}{12} = \dfrac{3 + 5x}{8}$

16. Solve: $3x - 2 \geq 6x + 7$

17. Solve: $5 - 2x \geq 6$ and $3x + 2 \geq 5$

18. Solve: $4x - 1 > 5$ or $2 - 3x < 8$

19. Solve: $|3 - 2x| = 5$

20. Solve: $3 - |2x - 3| = -8$

21. Solve: $|3x - 5| \leq 4$

22. Solve: $|4x - 3| > 5$

23. Graph the solution set of $\{x | x \geq -2\}$.

24. Graph the set $\{x | x \geq 1\} \cup \{x | x < -2\}$.

25. Translate and simplify "the sum of three times a number and six added to the product of three and the number."

26. Three times the sum of the first and third of three consecutive odd integers is fifteen more than the second integer. Find the first integer.

27. A stamp collection consists of 9¢ and 11¢ stamps. The number of 9¢ stamps is five less than twice the number of 11¢ stamps. The total value of the stamps is $1.87. Find the number of 9¢ stamps.

28. Tickets for a school play sold for $2.25 for each adult and $.75 for each child. The total receipts for 75 tickets were $128.25. Find the number of adult tickets sold.

29. Two planes are 1400 mi apart and traveling toward each other. One plane is traveling 120 mph faster than the other plane. The planes meet in 2.5 h. Find the speed of the faster plane.

30. How many liters of a 12% acid solution must be mixed with 4 L of a 5% acid solution to make an 8% acid solution?

31. An investment advisor invested $10,000 in two accounts. One investment earned 9.8% annual simple interest, while the other investment earned 12.8% annual simple interest. The amount of interest earned in one year was $1085. How much was invested in the 9.8% account?

3

Polynomials and Exponents

Objectives

- Add and subtract polynomials
- Multiply monomials
- Simplify expressions containing integer exponents
- Scientific notation
- Application problems
- Multiply a polynomial by a monomial
- Multiply two polynomials
- Multiply polynomials that have special products
- Application problems
- Divide polynomials
- Synthetic division
- Factor a monomial from a polynomial
- Factor by grouping
- Factor trinomials of the form $x^2 + bx + c$
- Factor trinomials of the form $ax^2 + bx + c$
- Factor the difference of two perfect squares and perfect square trinomials
- Factor the sum or the difference of two cubes
- Factor trinomials that are quadratic in form
- Factor completely
- Solve equations by factoring
- Application problems

POINT OF INTEREST

Origins of the Word Algebra

The word *algebra* has its origins in an Arabic book written around 825 A.D. titled *Hisab al-jabr w' almuqa-balah,* by al-Khowarizmi. The word *al-jabr,* which literally translated means reunion, was written as the word *algebra* in Latin translations of al-Khowarizmi's work and became synonymous with equations and the solutions of equations. It is interesting to note that an early meaning of the Spanish word *algebrista* was "bonesetter" or "reuniter of broken bones."

There is actually a second contribution to our language of mathematics by al-Khowarizmi. One of the translations of his work into Latin shortened his name to *Algoritmi.* A further modification of this word gives us our present word "algorithm." An algorithm is a procedure or set of instructions that is used to solve different types of problems. Computer scientists use algorithms when writing computer programs.

A further historical note is not about the word algebra, but about Omar Khayyam, a Persian who probably read al-Khowarizmi's work. Omar Khayyam is especially noted as a poet and the author of the *Rubiat.* However, he was also an excellent mathematician and astronomer and made many contributions to mathematics.

SECTION 3.1

Exponents and Operations on Polynomials

1 Add and subtract polynomials

A **monomial** is a number, a variable, or a product of a number and variables.

The examples at the right are monomials. The **degree of a monomial** is the sum of the exponents of the variables.

x	degree 1 ($x = x^1$)
$3x^2$	degree 2
$4x^2y$	degree 3
$6x^3y^4z^2$	degree 9

In this chapter, the variable n is considered a positive integer when used as an exponent.

x^n degree n

The degree of a nonzero constant term is zero.

6 degree 0

$\frac{1}{x}$ is not a monomial because a variable appears in the denominator.

A **polynomial** is a variable expression in which the terms are monomials.

A polynomial of one term is a **monomial.** $3x$

A polynomial of two terms is a **binomial.** $5x^2y + 6x$

A polynomial of three terms is a **trinomial.** $3x^2 + 9xy - 5y$

Polynomials with more than three terms do not have special names.

The **degree of a polynomial** is the greatest of the degrees of any of its terms.

$3x + 2$	degree 1
$3x^2 + 2x - 4$	degree 2
$4x^3y^2 + 6x^4 - y$	degree 5
$3x^{2n} - 5x^n + 2$	degree $2n$

The terms of a polynomial in one variable are usually arranged so that the exponents of the variable decrease from left to right. This is called **descending order.**

$2x^2 - x + 8$

$3y^5 - 3y^3 + y^2 - 12$

For a polynomial in more than one variable, descending order may refer to any one of the variables.

The polynomial at the right is shown first in descending order of the x variable and then in descending order of the y variable.

$2x^2 + 3xy + 5y^2$

$5y^2 + 3xy + 2x^2$

Polynomials can be added, using either a vertical or a horizontal format, by combining like terms.

Example 1 Simplify: $(4x^2 + 5x - 3) + (7x^3 - 7x + 1) + (4x^3 - 3x^2 + 2x + 1)$
Use a vertical format.

Solution

$$\begin{array}{rrrr} & 4x^2 & + 5x & - 3 \\ 7x^3 & & - 7x & + 1 \\ 4x^3 & - 3x^2 & + 2x & + 1 \\ \hline 11x^3 & + x^2 & & - 1 \end{array}$$

▶ Arrange the terms of each polynomial in descending order with like terms in the same column.

▶ Combine like terms in each column.

Problem 1 Simplify: $(5x^2 + 3x - 1) + (2x^2 + 4x - 6) + (x^2 - 7x + 8)$
Use a vertical format.

Solution See page A16.

Example 2 Simplify: $(3x^2 + 2x - 7) + (7x^3 - 3 + 4x^2)$
Use a horizontal format.

Solution $(3x^2 + 2x - 7) + (7x^3 - 3 + 4x^2)$

$7x^3 + (3x^2 + 4x^2) + 2x + (-7 - 3)$ ▶ Use the Commutative and Associative Properties of Addition to rearrange and group like terms.

$7x^3 + 7x^2 + 2x - 10$ ▶ Combine like terms. Write the polynomial in descending order.

Problem 2 Simplify: $(4x^2 + 3x - 5) + (6x^3 - 2 + x^2)$
Use a horizontal format.

Solution See page A16.

The **additive inverse** of the polynomial $x^2 + 5x - 4$ is $-(x^2 + 5x - 4)$.

To find the additive inverse of a polynomial, change the sign of every term of the polynomial.

$$-(x^2 + 5x - 4) = -x^2 - 5x + 4$$

Polynomials can be subtracted using either a horizontal or a vertical format. To subtract, add the additive inverse of the second polynomial to the first.

Simplify: $(6x^2 - 3x + 7) - (3x^2 - 5x + 12)$

Rewrite subtraction as the addition of the additive inverse. $(6x^2 - 3x + 7) + (-3x^2 + 5x - 12)$

Then arrange the terms in a vertical format with each polynomial in descending order with like terms in the same column. Combine the terms in each column.

$$\begin{array}{rrr} 6x^2 & - 3x & + 7 \\ -3x^2 & + 5x & - 12 \\ \hline 3x^2 & + 2x & - 5 \end{array}$$

Example 3 Simplify: $(3x^2 - 2x + 4) - (7x^2 + 3x - 12)$
Use a vertical format.

Solution $(3x^2 - 2x + 4) + (-7x^2 - 3x + 12)$ ► Rewrite subtraction as the addition of the additive inverse.

$$\begin{array}{r} 3x^2 - 2x + 4 \\ -7x^2 - 3x + 12 \\ \hline -4x^2 - 5x + 16 \end{array}$$

► Arrange the polynomials in descending order in a vertical format.

► Combine like terms in each column.

Problem 3 Simplify: $(-5x^2 + 2x - 3) - (6x^2 + 3x - 7)$
Use a vertical format.

Solution See page A16.

Example 4 Simplify: $(2x^{2n} - 3x^n + 7) - (3x^{2n} + 3x^n + 5)$
Use a horizontal format.

Solution $(2x^{2n} - 3x^n + 7) - (3x^{2n} + 3x^n + 5)$
$(2x^{2n} - 3x^n + 7) + (-3x^{2n} - 3x^n - 5)$ ► Rewrite subtraction as the addition of the additive inverse.
$-x^{2n} - 6x^n + 2$ ► Combine like terms.

Problem 4 Simplify: $(5x^{2n} - 3x^n - 7) - (-2x^{2n} - 5x^n + 8)$
Use a horizontal format.

Solution See page A16.

2 Multiply monomials

In an exponential expression, the exponent indicates the number of times the base occurs as a factor.

The product of exponential expressions with the *same* base can be simplified by adding the exponents.

$$x^3 \cdot x^4 = (x \cdot x \cdot x) \cdot (x \cdot x \cdot x \cdot x) = x^7$$
$$x^3 \cdot x^4 = x^{3+4} = x^7$$

Rule for Multiplying Exponential Expressions

If m and n are positive integers, then $x^m \cdot x^n = x^{m+n}$.

Example 5 Simplify: $(5a^2b^4)(2ab^5)$

Solution $(5a^2b^4)(2ab^5) = (5 \cdot 2)(a^2 \cdot a)(b^4 \cdot b^5)$ ▶ Use the Commutative and Associative Properties to rearrange and group factors.

$= 10a^{2+1}b^{4+5}$ ▶ Multiply variables with like bases by adding the exponents.

$= 10a^3b^9$

Problem 5 Simplify: $(7xy^3)(-5x^2y^2)(-xy^2)$

Solution See page A16.

A power of an exponential expression can be simplified by multiplying the exponents.

$(x^4)^3 = x^4 \cdot x^4 \cdot x^4 = x^{4+4+4} = x^{12}$
$(x^4)^3 = x^{4 \cdot 3} = x^{12}$

Rule for Simplifying Powers of Exponential Expressions

If m and n are positive integers, then $(x^m)^n = x^{mn}$.

Example 6 Simplify.
A. $(x^4)^5$ B. $(x^2)^n$

Solution A. $(x^4)^5 = x^{4 \cdot 5}$ ▶ Multiply the exponents.
$= x^{20}$

B. $(x^2)^n = x^{2n}$ ▶ Multiply the exponents.

Problem 6 Simplify.
A. $(y^3)^6$ B. $(x^n)^3$

Solution See page A16.

A power of the product of exponential expressions can be simplified by multiplying each exponent inside the parentheses by the exponent outside the parentheses.

$$(x^3 \cdot y^4)^2 = x^{3 \cdot 2} \cdot y^{4 \cdot 2} = x^6y^8$$

Rule for Simplifying Powers of Products

If m, n, and p are positive integers, then $(x^m \cdot y^n)^p = x^{mp}y^{np}$.

Example 7 Simplify: $(2a^3b^4)^3$

Solution $(2a^3b^4)^3 = 2^{1\cdot 3}a^{3\cdot 3}b^{4\cdot 3}$ ► Use the Rule for Simplifying Powers of Products.

$= 2^3a^9b^{12}$

$= 8a^9b^{12}$

Problem 7 Simplify: $(-2ab^3)^4$

Solution See page A16.

Example 8 Simplify: $(2ab)(3a)^2 + 5a(2a^2b)$

Solution $(2ab)(3a)^2 + 5a(2a^2b) = (2ab)(3^2a^2) + 10a^3b$

$= (2ab)(9a^2) + 10a^3b$

$= 18a^3b + 10a^3b$

$= 28a^3b$

Problem 8 Simplify: $6a(2a)^2 + 3a(2a^2)$

Solution See page A16.

3 Simplify expressions containing integer exponents

The quotient of two exponential expressions with the same base can be simplified by writing each expression in factored form, dividing by the common factors, and then writing the result with an exponent.

$$\frac{a^6}{a^2} = \frac{\overset{1}{\cancel{a}} \cdot \overset{1}{\cancel{a}} \cdot a \cdot a \cdot a \cdot a}{\underset{1}{\cancel{a}} \cdot \underset{1}{\cancel{a}}} = a^4$$

Note that subtracting the exponents gives the same result.

$$\frac{a^6}{a^2} = a^{6-2} = a^4$$

To divide two monomials with the same base, subtract the exponents of the like bases.

Simplify: $\dfrac{a^6b^2}{a^3b}$

Subtract the exponents of the like bases.

$$\frac{a^6b^2}{a^3b} = a^{6-3}b^{2-1} = a^3b$$

Recall that for any number $a \neq 0$, $\dfrac{a}{a} = 1$. This property is true for exponential expressions as well. For example, for $a \neq 0$, $\dfrac{a^4}{a^4} = 1$.

This expression also can be simplified by subtracting the exponents of the like bases.

$$\frac{a^4}{a^4} = a^{4-4} = a^0$$

Because $\dfrac{a^4}{a^4} = 1$ and $\dfrac{a^4}{a^4} = a^{4-4} = a^0$, the definition of zero as an exponent is given as follows.

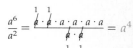

Definition of Zero as an Exponent

If $a \neq 0$, then $a^0 = 1$.

Note in this definition that $a \neq 0$. The expression 0^0 is not defined.

Simplify: $(2a - b)^0$, $2a - b \neq 0$

Any nonzero expression to the zero power is 1. $(2a - b)^0 = 1$

Simplify: $-(4a^3b^4)^0$, $a \neq 0$, $b \neq 0$

Any nonzero expression to the zero power is 1. Because the negative sign is outside the parenthesis, the answer is -1. $-(4a^3b^4)^0 = -(1) = -1$

Examine the quotient $\frac{a^4}{a^7}$.

The expression can be simplified by factoring the numerator and denominator and dividing by the common factors.

$$\frac{a^4}{a^7} = \frac{\overset{1}{\cancel{a}} \cdot \overset{1}{\cancel{a}} \cdot \overset{1}{\cancel{a}} \cdot \overset{1}{\cancel{a}}}{\underset{1}{\cancel{a}} \cdot \underset{1}{\cancel{a}} \cdot \underset{1}{\cancel{a}} \cdot \underset{1}{\cancel{a}} \cdot a \cdot a \cdot a} = \frac{1}{a^3}$$

Another way to simplify the same expression is by subtracting the exponents on the like bases.

$$\frac{a^4}{a^7} = a^{4-7} = a^{-3}$$

Since $\frac{a^4}{a^7} = \frac{1}{a^3}$ and $\frac{a^4}{a^7} = a^{-3}$, the rule for negative exponents is given as follows.

Rule of Negative Exponents

If n is a positive integer and $a \neq 0$, then

$$a^{-n} = \frac{1}{a^n} \quad \text{and} \quad a^n = \frac{1}{a^{-n}}.$$

Write 4^{-3} with a positive exponent and then evaluate.

Write the expression with a positive exponent. $4^{-3} = \frac{1}{4^3}$

Evaluate. $= \frac{1}{64}$

Now that negative exponents have been defined, the rule for dividing exponential expressions can be stated.

Rule for Dividing Exponential Expressions

If m and n are integers and $a \neq 0$, then $\frac{a^m}{a^n} = a^{m-n}$.

Write $\frac{2^{-3}}{2^2}$ with a positive exponent and then evaluate.

Use the rule for dividing exponential expressions. $\frac{2^{-3}}{2^2} = 2^{-3-2} = 2^{-5}$

Write the expression with a positive exponent. $= \frac{1}{2^5}$

Evaluate. $= \frac{1}{32}$

An exponential expression is in simplest form when it contains only positive exponents.

Simplify: $\frac{3x^6y^{-6}}{6xy^2}$

Divide variables with like bases by subtracting the exponents. $\frac{3x^6y^{-6}}{6xy^2} = \frac{x^{6-1}y^{-6-2}}{2}$

$= \frac{x^5y^{-8}}{2}$

Write the expression with positive exponents. $= \frac{x^5}{2y^8}$

Simplify: $\frac{ab^5}{a^{-4}b^6}$

Divide variables with like bases by subtracting the exponents. $\frac{ab^5}{a^{-4}b^6} = a^{1-(-4)}b^{5-6}$

$= a^5b^{-1}$

Write the expression with positive exponents. $= \frac{a^5}{b}$

The rules for multiplying and dividing exponential expressions and powers of exponential expressions are true for all integers. These rules are stated here.

Rules of Exponents

If m, n, and p are integers, then

$x^m \cdot x^n = x^{m+n}$ $\quad (x^m)^n = x^{mn}$ $\quad (x^my^n)^p = x^{mp}y^{np}$ $\quad x^{-n} = \frac{1}{x^n}, x \neq 0$

$\frac{x^m}{x^n} = x^{m-n}, x \neq 0$ $\quad x^0 = 1, x \neq 0$ $\quad \left(\frac{x^m}{y^n}\right)^p = \frac{x^{mp}}{y^{np}}, y \neq 0$ $\quad x^n = \frac{1}{x^{-n}}, x \neq 0$

Example 9 Simplify.

A. $(3x^2y^{-3})(6x^{-4}y^5)$ B. $\dfrac{x^2y^{-4}}{x^{-5}y^{-2}}$ C. $\left(\dfrac{3a^2b^{-2}c^{-1}}{27a^{-1}b^2c^{-4}}\right)^{-2}$ D. $x^{-1}y + xy^{-1}$

Solution A. $(3x^2y^{-3})(6x^{-4}y^5) = 18x^{2+(-4)}y^{-3+5}$ ▶ Use the Rule for Multiplying Exponential Expressions.

$= 18x^{-2}y^2$

$= \dfrac{18y^2}{x^2}$ ▶ Use the Rule of Negative Exponents to rewrite the expression without negative exponents.

B. $\dfrac{x^2y^{-4}}{x^{-5}y^{-2}} = x^{2-(-5)}y^{-4-(-2)}$ ▶ Use the Rule for Dividing Exponential Expressions.

$= x^7y^{-2}$

$= \dfrac{x^7}{y^2}$ ▶ Use the Rule of Negative Exponents to rewrite the expression without negative exponents.

C. $\left(\dfrac{3a^2b^{-2}c^{-1}}{27a^{-1}b^2c^{-4}}\right)^{-2} = \left(\dfrac{a^3b^{-4}c^3}{9}\right)^{-2}$ ▶ Simplify inside the parentheses by using the Rule for Dividing Exponential Expressions.

$= \dfrac{a^{-6}b^8c^{-6}}{9^{-2}}$ ▶ Multiply each exponent inside the parentheses by the exponent outside the parentheses.

$= \dfrac{9^2b^8}{a^6c^6}$ ▶ Use the Rule of Negative Exponents to rewrite the expression without negative exponents.

$= \dfrac{81b^8}{a^6c^6}$ ▶ Simplify.

D. $x^{-1}y + xy^{-1} = \dfrac{y}{x} + \dfrac{x}{y}$ ▶ Use the Rule of Negative Exponents.

$= \dfrac{y^2}{xy} + \dfrac{x^2}{xy}$ ▶ Write each fraction in terms of the LCM of the denominators.

$= \dfrac{y^2 + x^2}{xy}$ ▶ Add the two fractions.

Problem 9 Simplify.

A. $(2x^{-5}y)(5x^4y^{-3})$ B. $\dfrac{a^{-1}b^4}{a^{-2}b^{-2}}$ C. $\left(\dfrac{2^{-1}x^2y^{-3}}{4x^{-2}y^{-5}}\right)^{-2}$ D. $[(a^{-1}b)^{-2}]^3$

Solution See page A16.

4 Scientific notation

Very large and very small numbers are encountered in the fields of science and engineering. For example, the mass of the electron is 0.0000000000000000000000000009 g. Numbers such as this one are difficult to read and write, so a more convenient system for writing them has been developed. It is called **scientific notation.**

To express a number in scientific notation, write the number as the product of a number between 1 and 10 and a power of 10. The form for scientific notation is $a \times 10^n$, where $1 \le a < 10$.

For numbers greater than 10, move the decimal point to the right of the first digit. The exponent n is positive and equal to the number of places the decimal point has been moved.

$965{,}000 = 9.65 \times 10^5$

$3{,}600{,}000 = 3.6 \times 10^6$

$92{,}000{,}000{,}000 = 9.2 \times 10^{10}$

For numbers less than 1, move the decimal point to the right of the first nonzero digit. The exponent n is negative. The absolute value of the exponent is equal to the number of places the decimal point has been moved.

$0.0002 = 2 \times 10^{-4}$

$0.0000000974 = 9.74 \times 10^{-8}$

$0.000000000086 = 8.6 \times 10^{-11}$

Example 10 Write 0.000041 in scientific notation.

Solution $0.000041 = 4.1 \times 10^{-5}$

▶ The decimal point must be moved 5 digits to the right. The exponent is negative.

Problem 10 Write 942,000,000 in scientific notation.

Solution See page A16.

Converting a number written in scientific notation to decimal notation requires moving the decimal point.

When the exponent is positive, move the decimal point to the right the same number of places as the exponent.

$1.32 \times 10^4 = 13{,}200$

$1.4 \times 10^8 = 140{,}000{,}000$

When the exponent is negative, move the decimal point to the left the same number of places as the absolute value of the exponent.

$1.32 \times 10^{-2} = 0.0132$

$1.4 \times 10^{-4} = 0.00014$

Example 11 Write 3.3×10^7 in decimal notation.

Solution $3.3 \times 10^7 = 33{,}000{,}000$ ▶ Move the decimal point 7 places to the right.

Problem 11 Write 2.7×10^{-5} in decimal notation.

Solution See page A16.

Numerical calculations involving numbers that have more digits than the hand-held calculator is able to handle can be performed using scientific notation.

Simplify. $\dfrac{220{,}000 \times 0.000000092}{0.0000011}$

Write the numbers in scientific notation.

$$\frac{220{,}000 \times 0.000000092}{0.0000011} = \frac{2.2 \times 10^5 \times 9.2 \times 10^{-8}}{1.1 \times 10^{-6}}$$

Simplify.

$$= \frac{(2.2)(9.2) \times 10^{5+(-8)-(-6)}}{1.1}$$

$$= 18.4 \times 10^3$$

Write in scientific notation.

$$= 1.84 \times 10^4$$

Example 12 Simplify: $\dfrac{2{,}400{,}000{,}000 \times 0.0000063}{0.00009 \times 480}$

Solution

$$\frac{2{,}400{,}000{,}000 \times 0.0000063}{0.00009 \times 480} = \frac{2.4 \times 10^9 \times 6.3 \times 10^{-6}}{9 \times 10^{-5} \times 4.8 \times 10^2}$$

$$= \frac{(2.4)(6.3) \times 10^{9+(-6)-(-5)-2}}{(9)(4.8)}$$

$$= 0.35 \times 10^6$$

$$= 3.5 \times 10^5$$

Problem 12 Simplify: $\dfrac{5{,}600{,}000 \times 0.000000081}{900 \times 0.000000028}$

Solution See page A16.

5 Application problems

Example 13 How many miles does light travel in one day? The speed of light is 186,000 mi/s. Write the answer in scientific notation.

Strategy To find the distance traveled:

- Write the speed of light in scientific notation.
- Write the number of seconds in one day in scientific notation.
- Use the equation $d = rt$, where r is the speed of light, and t is the number of seconds in one day.

Solution $186{,}000 = 1.86 \times 10^5$

$24 \cdot 60 \cdot 60 = 86{,}400 = 8.64 \times 10^4$

$d = rt$
$d = (1.86 \times 10^5)(8.64 \times 10^4)$
$d = 1.86 \times 8.64 \times 10^9$
$d = 16.0704 \times 10^9$
$d = 1.60704 \times 10^{10}$

Light travels 1.60704×10^{10} mi in one day.

Problem 13 A computer can do an arithmetic operation in 1×10^{-7} s. How many arithmetic operations can the computer perform in one minute? Write the answer in scientific notation.

Solution See page A16.

EXERCISES 3.1

1 Simplify. Use a vertical format.

1. $(5x^2 + 2x - 7) + (x^2 - 8x + 12)$
2. $(3x^2 - 2x + 7) + (-3x^2 + 2x - 12)$
3. $(x^2 - 3xy + y^2) + (2x^2 - 3y^2)$
4. $(3x^2 + 2y^2) + (-5x^2 + 2xy - 3y^2)$
5. $(x^2 - 3x + 8) - (2x^2 - 3x + 7)$
6. $(2x^2 + 3x - 7) - (5x^2 - 8x - 2)$
7. $(x^{2n} + 7x^n - 3) + (-x^{2n} + 2x^n + 8)$
8. $(2x^{2n} - x^n - 1) + (5x^{2n} + 7x^n + 1)$
9. $(3y^3 - 7y) + (2y^2 - 8y + 2)$
10. $(-2y^2 - 4y - 12) + (5y^2 - 5y)$
11. $(2a^2 - 3a - 7) - (-5a^2 - 2a - 9)$
12. $(3a^2 - 9a) - (-5a^2 + 7a - 6)$

Simplify. Use a horizontal format.

13. $(3x^4 - 3x^3 - x^2) + (3x^3 - 7x^2 + 2x)$
14. $(3x^4 - 2x + 1) + (3x^3 - 5x - 8)$
15. $(3a^3 - 5a^2 - 6) + (-3a^3 - 6a + 2)$
16. $(6a^3 - 2a^2 - 12) + (6a^2 + 3a + 9)$
17. $(b^{2n} - b^n - 3) - (2b^{2n} - 3b^n + 4)$
18. $(x^{2n} - x^n + 2) - (3x^{2n} - x^n + 5)$
19. $(4x^2 - 3x - 9) - (-2x^3 - 3x^2 + 6x) - (5x^3 - 3x^2 + 9x + 2)$
20. $(2a^3 - 2a^2b + 2ab^2 + 3b^3) - (4a^2b - 3ab^2 - 5b^3) + (3a^3 - 2a^2b + 5ab^2 - b^3)$

21. $(3x^2 + 5x + 2) + (4x^2 - 3x - 1) + (-5x^2 + 2x - 8)$

22. $(2a^2 + 3a - 6) + (5a^2 - 7a) + (a^3 - 4a + 1)$

23. $(4x^2 - 3x - 9) - (2x^3 - 3x^2 + 6x) - (x^3 - 4x + 1)$

24. $(8b^2 + 2b - 5) - (3b^2 - 4b) - (-6b^3 - 5b^2 + 3)$

25. $(6x^4 - 5x^3 + 2x) - (4x^3 + 3x^2 - 1) + (x^4 - 2x^2 + 7x - 3)$

26. $(3x^2 - 4xy + 6y^2) + (5x^2 + 2xy - 4y^2) - (7x^2 + xy - 3y^2)$

27. $(x^{2n} - 2x^n + 3) + (4x^{2n} - 3x^n - 1) - (2x^{2n} - 6x^n - 3)$

2 Simplify.

28. $(ab^3)(a^3b)$

29. $(-2ab^4)(-3a^2b^4)$

30. $(9xy^2)(-2x^2y^2)$

31. $(x^2y)^2$

32. $(x^2y^4)^4$

33. $(-2ab^2)^3$

34. $(-3x^2y^3)^4$

35. $(2^2a^2b^3)^3$

36. $(3^3a^5b^3)^2$

37. $(xy)(x^2y)^4$

38. $(x^2y^2)(xy^3)^3$

39. $[(2x)^4]^2$

40. $[(3x)^3]^2$

41. $[(x^2y)^4]^5$

42. $[(ab)^3]^6$

43. $[(2ab)^3]^2$

44. $[(2xy)^3]^4$

45. $[(3x^2y^3)^2]^2$

46. $[(2a^4b^3)^3]^2$

47. $y^n \cdot y^{2n}$

48. $x^n \cdot x^{n+1}$

49. $y^{2n} \cdot y^{4n+1}$

50. $y^{3n} \cdot y^{3n-2}$

51. $(a^n)^{2n}$

52. $(a^{n-3})^{2n}$

53. $(y^{2n-1})^3$

54. $(x^{3n+2})^5$

55. $(b^{2n-1})^n$

56. $(2xy)(-3x^2yz)(x^2y^3z^3)$

57. $(x^2z^4)(2xyz^4)(-3x^3y^2)$

58. $(3b^5)(2ab^2)(-2ab^2c^2)$

59. $(-c^3)(-2a^2bc)(3a^2b)$

60. $(-2x^2y^3z)(3x^2yz^4)$

61. $(2a^2b)^3(-3ab^4)^2$

62. $(-3ab^3)^3(-2^2a^2b)^2$

63. $(4ab)^2(-2ab^2c^3)^3$

64. $(-2ab^2)(-3a^4b^5)^3$

3 Simplify.

65. 2^{-3}

66. $\frac{1}{3^{-5}}$

67. $\frac{1}{x^{-4}}$

68. $\frac{1}{y^{-3}}$

69. $\frac{2x^{-2}}{y^4}$

70. $\frac{a^3}{4b^{-2}}$

71. $x^{-4}x^4$

72. $x^{-3}x^{-5}$

73. $(3x^{-2})^2$ **74.** $(5x^2)^{-3}$ **75.** $\frac{x^{-3}}{x^2}$ **76.** $\frac{x^4}{x^{-5}}$

77. $a^{-2} \cdot a^4$ **78.** $a^{-5} \cdot a^7$ **79.** $(x^2y^{-4})^2$ **80.** $(x^3y^5)^{-2}$

81. $(2a^{-1})^{-2}(2a^{-1})^4$ **82.** $(3a)^{-3}(9a^{-1})^{-2}$ **83.** $(x^{-2}y)^2(xy)^{-2}$ **84.** $(x^{-1}y^2)^{-3}(x^2y^{-4})^{-3}$

85. $(2^{-1}x^2y^{-3})^2\ (2^{-2}x^{-3}y^4)$ **86.** $(x^{-1}y^{-2}z)^{-2}\ (x^2y^{-4}z)^2$

87. $(3^2x^2y^{-3})^{-2}(3^{-1}x^{-3}y^{-2})$ **88.** $(x^{-1}y^2z^{-4})^3(x^3y^{-3}z^2)^{-2}$

89. $\frac{6^2a^{-2}b^3}{3ab^4}$ **90.** $\left(\frac{x^2y^{-1}}{xy}\right)^{-4}$ **91.** $\frac{-48ab^{10}}{32a^4b^3}$ **92.** $\frac{a^2b^3c^7}{a^6bc^5}$

93. $\frac{(-4x^2y^3)^2}{(2xy^2)^3}$ **94.** $\frac{(-3a^2b^3)^2}{(-2ab^4)^3}$ **95.** $\left(\frac{x^{-3}y^{-4}}{x^{-2}y}\right)^{-2}$ **96.** $\left(\frac{a^{-2}b}{a^3b^{-4}}\right)^2$

97. $\frac{-x^{5n}}{x^{2n}}$ **98.** $\frac{y^{2n}}{-y^{8n}}$ **99.** $\frac{a^{3n-2}b^{n+1}}{a^{2n+1}b^{2n+2}}$ **100.** $\frac{x^{2n-1}y^{n-3}}{x^{n+4}y^{n+3}}$

101. $\frac{(2a^{-3}b^{-2})^3}{(a^{-4}b^{-1})^{-2}}$ **102.** $\frac{(3x^{-2}y)^{-2}}{(4xy^{-2})^{-1}}$ **103.** $\left(\frac{4^{-2}xy^{-3}}{x^{-3}y}\right)^3\left(\frac{8^{-1}x^{-2}y}{x^4y^{-1}}\right)^{-2}$ **104.** $\left(\frac{9ab^{-2}}{8a^{-2}b}\right)^{-2}\left(\frac{3a^{-2}b}{2a^2b^{-2}}\right)^3$

105. $[(xy^{-2})^3]^{-2}$ **106.** $[(x^{-2}y^{-1})^2]^{-3}$ **107.** $\left[\left(\frac{x}{y^2}\right)^{-2}\right]^3$ **108.** $\left[\left(\frac{a^2}{b}\right)^{-1}\right]^2$

109. $a + a^{-1}b$ **110.** $(a + b)^{-1}$ **111.** $x^{-1}y^{-1} + xy$ **112.** $\frac{x^{-1}}{y} + \frac{y^{-1}}{x}$

4 Write in scientific notation.

113. 0.00000467 **114.** 0.00000005 **115.** 0.0000000017

116. 4,300,000 **117.** 200,000,000,000 **118.** 9,800,000,000

Write in decimal notation.

119. 1.23×10^{-7}

120. 6.2×10^{-12}

121. 8.2×10^{15}

122. 6.34×10^{5}

123. 3.9×10^{-2}

124. 4.35×10^{9}

Simplify. Write the answer in scientific notation.

125. $(3 \times 10^{-12})(5 \times 10^{16})$

126. $(8.9 \times 10^{-5})(3.2 \times 10^{-6})$

127. $(0.0000065)(3{,}200{,}000{,}000{,}000)$

128. $(480{,}000)(0.0000000096)$

129. $\dfrac{9 \times 10^{-3}}{6 \times 10^{5}}$

130. $\dfrac{2.7 \times 10^{4}}{3 \times 10^{-6}}$

131. $\dfrac{0.0089}{500{,}000{,}000}$

132. $\dfrac{4{,}800}{0.00000024}$

133. $\dfrac{0.00056}{0.000000000004}$

134. $\dfrac{0.000000346}{0.0000005}$

135. $\dfrac{(3.2 \times 10^{-11})(2.9 \times 10^{15})}{8.1 \times 10^{-3}}$

136. $\dfrac{(6.9 \times 10^{27})(8.2 \times 10^{-13})}{4.1 \times 10^{15}}$

137. $\dfrac{(0.00000004)(84{,}000)}{(0.0003)(1{,}400{,}000)}$

138. $\dfrac{(720)(0.0000000039)}{(26{,}000{,}000{,}000)(0.018)}$

5 Solve. Write the answer in scientific notation.

139. A computer can do an arithmetic operation in 5×10^{-7} s. How many arithmetic operations can the computer perform in one hour?

140. A computer can do an arithmetic operation in 2×10^{-9} s. How many arithmetic operations can the computer perform in one minute?

141. How many meters does light travel in 8 h? The speed of light is 300,000,000 m/s.

142. How many miles does light travel in one day? The speed of light is 186,000 mi/s.

143. A high-speed centrifuge makes 4×10^{8} revolutions each minute. Find the time in seconds for the centrifuge to make one revolution.

144. The mass of an electron is 9.109×10^{-31} kg. The mass of a proton is 1.673×10^{-27} kg. How many times heavier is a proton than an electron?

145. The mass of Earth is 5.9×10^{24} kg. The mass of the sun is 2×10^{30} kg. How many times heavier is the sun than Earth?

146. One light year, an astronomical unit of distance, is the distance that light will travel in one year. Light travels 1.86×10^5 mi/s. Find the measure of one light year in miles. Use a 365-day year.

147. The sun is 3.67×10^9 mi from Pluto. How long does it take light to travel to Pluto from the sun? The speed of light is 1.86×10^5 mi/s.

148. The weight of 31 million orchid seeds is one ounce. Find the weight of one orchid seed.

149. The light from the star Alpha Centauri takes 4.3 years to reach Earth. Light travels 1.86×10^5 mi/s. How far is Alpha Centauri from Earth? Use a 365-day year.

150. The distance to Saturn is 8.86×10^8 mi. A satellite leaves Earth traveling at a constant rate of 1×10^5 mph. How long does it take for the satellite to reach Saturn?

151. The diameter of Neptune is 3×10^4 mi. Use the formula $SA = 4\pi r^2$ to find the surface area of Neptune in square miles.

152. The radius of a cell is 1.5×10^{-4} mm. Use the formula $V = \frac{4}{3}\pi r^3$ to find the volume of the cell.

153. One gram of hydrogen contains 6.023×10^{23} atoms. Find the weight of one atom of hydrogen.

154. Our galaxy is estimated to be 6×10^{17} mi across. How long would it take a space ship to cross the galaxy traveling at 25,000 mph?

SUPPLEMENTAL EXERCISES 3.1

State whether or not the expression is a polynomial.

155. $\frac{1}{3}x - 1$

156. $\frac{3}{x} - 1$

157. $5\sqrt{x} + 2$

158. $\sqrt{5}x + 2$

159. $\frac{1}{4y^2} + \frac{1}{3y}$

160. $x + \sqrt{3}$

For what value of k is the given equation an identity?

161. $(2x^3 + 3x^2 + kx + 5) - (x^3 + x^2 - 5x - 2) = x^3 + 2x^2 + 3x + 7$

162. $(6x^3 + kx^2 - 2x - 1) - (4x^3 - 3x^2 + 1) = 2x^3 - x^2 - 2x - 2$

Solve.

163. The width of a rectangle is x^n. The length of the rectangle is $3x^n$. Find the perimeter of the rectangle in terms of x^n.

164. The base of an isosceles triangle is $4x^n$. The length of each of the equal sides is $3x^n$. Find the perimeter of the triangle in terms of x^n.

165. The width of a rectangle is $4ab$. The length is $6ab$. Find the area of the rectangle in terms of ab.

166. The height of a triangle is $5xy$. The length of the base of the triangle is $8xy$. Find the area of the triangle in terms of xy.

Simplify.

167. $\frac{4m^4}{n^{-2}} + \left(\frac{n^{-1}}{m^2}\right)^{-2}$

168. $\frac{5x^3}{y^{-6}} + \left(\frac{x^{-1}}{y^2}\right)^{-3}$

169. $\left(\frac{3a^{-2}b}{a^{-4}b^{-1}}\right)^2 \div \left(\frac{a^{-1}b}{9a^2b^3}\right)^{-1}$

170. $\left(\frac{2m^3n^{-2}}{4m^4n}\right)^{-2} \div \left(\frac{mn^5}{m^{-1}n^3}\right)^3$

Solve.

171. Use the expressions $(2 + 3)^{-2}$ and $2^{-2} + 3^{-2}$ to show that $(x + y)^{-2} \neq x^{-2} + y^{-2}$.

172. Use the expressions $(2 - 3)^{-1}$ and $2^{-1} - 3^{-1}$ to show that $(x - y)^{-1} \neq x^{-1} - y^{-1}$.

173. Use the expressions $(2 + 3)^{-2}$ and $\frac{1}{2^2 + 3^2}$ to show that $(x + y)^{-2} \neq \frac{1}{x^2 + y^2}$.

174. If a and b are real nonzero numbers and $a < b$, is $a^{-1} < b^{-1}$ always a false statement?

SECTION 3.2 Multiplication of Polynomials

1 Multiply a polynomial by a monomial

To multiply a polynomial by a monomial, use the Distributive Property and the Rule for Multiplying Exponential Expressions.

Example 1 Simplify.

A. $-5x(x^2 - 2x + 3)$ B. $x^2 - x[3 - x(x - 2) + 3]$ C. $x^n(x^n - x^2 + 1)$

Solution A. $-5x(x^2 - 2x + 3)$

$-5x(x^2) - (-5x)(2x) + (-5x)(3)$ ▶ Use the Distributive Property.

$-5x^3 + 10x^2 - 15x$ ▶ Use the Rule for Multiplying Exponential Expressions.

B. $x^2 - x[3 - x(x - 2) + 3]$

$x^2 - x[3 - x^2 + 2x + 3]$ ▶ Use the Distributive Property to remove the inner grouping symbols.

$x^2 - x[6 - x^2 + 2x]$ ▶ Combine like terms.

$x^2 - 6x + x^3 - 2x^2$ ▶ Use the Distributive Property to remove the brackets.

$x^3 - x^2 - 6x$ ▶ Combine like terms, and write the polynomial in descending order.

C. $x^n(x^n - x^2 + 1)$

$x^{2n} - x^{n+2} + x^n$ ▶ Use the Distributive Property and the Rule for Multiplying Exponential Expressions.

Problem 1 Simplify.

A. $-4y(y^2 - 3y + 2)$

B. $x^2 - 2x[x - x(4x - 5) + x^2]$

C. $y^{n+3}(y^{n-2} - 3y^2 + 2)$

Solution See page A17.

2 Multiply two polynomials

The product of two polynomials is the polynomial obtained by multiplying each term of one polynomial by each term of the other polynomial and then combining like terms.

Multiply: $(2x^2 - 2x + 1)(3x + 2)$

Use the Distributive Property to multiply the trinomial by each term of the binomial.

$(2x^2 - 2x + 1)(3x + 2)$

$(2x^2 - 2x + 1)(3x) + (2x^2 - 2x + 1)(2)$

Use the Distributive Property.

$(6x^3 - 6x^2 + 3x) + (4x^2 - 4x + 2)$

Combine like terms.

$6x^3 - 2x^2 - x + 2$

A more convenient method of multiplying two polynomials is to use a vertical format similar to that used for multiplication of whole numbers.

$$\begin{array}{rrrrl} & 2x^2 & -\,2x & +\,1 & \\ & & 3x & +\,2 & \\ \hline & 4x^2 & -\,4x & +\,2 & = 2(2x^2 - 2x + 1) \\ 6x^3 & -\,6x^2 & +\,3x & & = 3x(2x^2 - 2x + 1) \\ \hline 6x^3 & -\,2x^2 & -\,x & +\,2 & \end{array}$$

Like terms are written in the same column.

Combine like terms.

Example 2 Simplify: $(4a^3 - 3a + 7)(a - 5)$

Solution

$$\begin{array}{rrrrr} & & 4a^3 & -3a & +7 \\ & & & a & -5 \\ \hline & -20a^3 & & +15a & -35 \\ 4a^4 & & -3a^2 & +7a & \\ \hline 4a^4 & -20a^3 & -3a^2 & +22a & -35 \end{array}$$

Problem 2 Simplify: $(-2b^2 + 5b - 4)(-3b + 2)$

Solution See page A17.

It is frequently necessary to find the product of two binomials. The product can be found by using a method called **FOIL,** which is based upon the Distributive Property. The letters of FOIL stand for **F**irst, **O**uter, **I**nner, and **L**ast.

Simplify: $(3x - 2)(2x + 5)$

Multiply the **F**irst terms.	$(3x - 2)(2x + 5)$	$3x \cdot 2x = 6x^2$
Multiply the **O**uter terms.	$(3x - 2)(2x + 5)$	$3x \cdot 5 = 15x$
Multiply the **I**nner terms.	$(3x - 2)(2x + 5)$	$-2 \cdot 2x = -4x$
Multiply the **L**ast terms.	$(3x - 2)(2x + 5)$	$-2 \cdot 5 = -10$

F O I L

Add the products.	$(3x - 2)(2x + 5)$	=	$6x^2 + 15x - 4x - 10$
Combine like terms.		=	$6x^2 + 11x - 10$

Example 3 Simplify: $(6x - 5)(3x - 4)$

Solution

$$\begin{aligned}(6x - 5)(3x - 4) &= 6x(3x) + 6x(-4) + (-5)(3x) + (-5)(-4) \\ &= 18x^2 - 24x - 15x + 20 \\ &= 18x^2 - 39x + 20\end{aligned}$$

Problem 3 Simplify: $(5a - 3b)(2a + 7b)$

Solution See page A17.

3 Multiply polynomials that have special products

Using FOIL, a pattern for the product of the sum and the difference of two terms and for the square of a binomial can be found.

The Sum and Difference of Two Terms

$$(a + b)(a - b) = a^2 - ab + ab - b^2$$
$$= a^2 - b^2$$

Square of the first term → a^2

Square of the second term → b^2

The Square of a Binomial

$$(a + b)^2 = (a + b)(a + b) = a^2 + ab + ab + b^2$$
$$= a^2 + 2ab + b^2$$

Square of the first term → a^2

Twice the product of the two terms → $2ab$

Square of the last term → b^2

Example 4 Simplify.

A. $(4x + 3)(4x - 3)$ B. $(2x - 3y)^2$

C. $(x^n + 5)(x^n - 5)$ D. $(x^{2n} - 2)^2$

Solution A. $(4x + 3)(4x - 3) = (4x)^2 - 3^2$
$= 16x^2 - 9$

▶ This is the sum and difference of two terms.

B. $(2x - 3y)^2 = (2x)^2 + 2(2x)(-3y) + (-3y)^2$
$= 4x^2 - 12xy + 9y^2$

▶ This is the square of a binomial.

C. $(x^n + 5)(x^n - 5) = x^{2n} - 25$

D. $(x^{2n} - 2)^2 = x^{4n} - 4x^{2n} + 4$

Problem 4 Simplify.

A. $(3x - 7)(3x + 7)$ B. $(3x - 4y)^2$

C. $(2x^n + 3)(2x^n - 3)$ D. $(2x^n - 8)^2$

Solution See page A17.

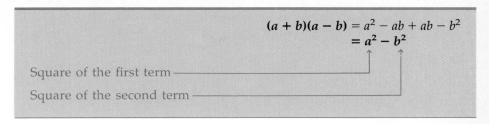

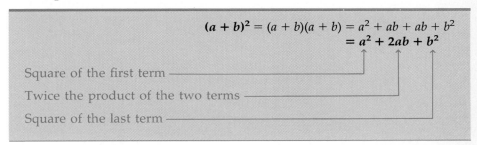

4 Application problems

Example 5 The length of a rectangle is $(2x + 3)$ ft. The width is $(x - 5)$ ft. Find the area of the rectangle in terms of the variable x.

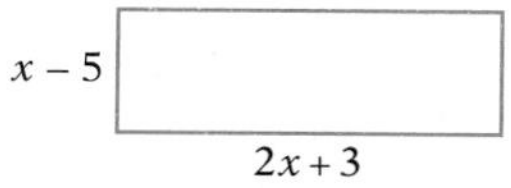

Strategy To find the area, replace the variables L and W in the equation $A = LW$ by the given values, and solve for A.

Solution

$$A = LW$$
$$A = (2x + 3)(x - 5)$$
$$A = 2x^2 - 10x + 3x - 15$$
$$A = 2x^2 - 7x - 15$$

The area is $(2x^2 - 7x - 15)$ ft^2.

Problem 5 The base of a triangle is $(2x + 6)$ ft. The height is $(x - 4)$ ft. Find the area of the triangle in terms of the variable x.

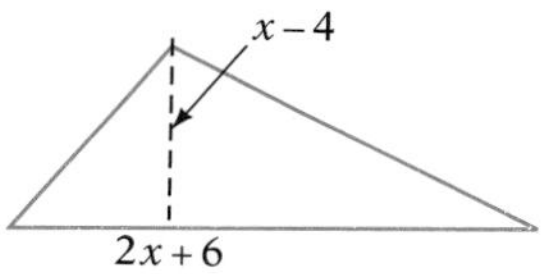

Solution See page A17.

Example 6 The corners are cut from a rectangular piece of cardboard measuring 8 in. by 12 in. The sides are folded up to make a box. Find the volume of the box in terms of the variable x, where x is the length of the side of the square cut from each corner of the rectangle.

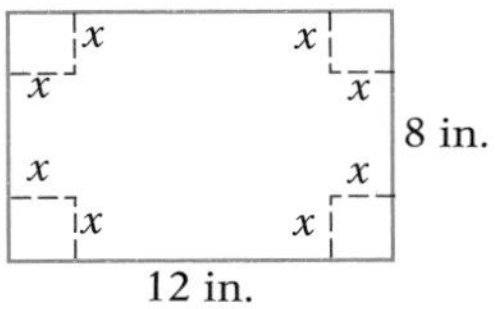

Strategy Length of the box: $12 - 2x$
Width of the box: $8 - 2x$
Height of the box: x
To find the volume, replace the variables L, W, and H in the equation $V = LWH$, and solve for V.

Solution $V = LWH$
$V = (12 - 2x)(8 - 2x)x$
$V = (96 - 24x - 16x + 4x^2)x$
$V = (96 - 40x + 4x^2)x$
$V = 96x - 40x^2 + 4x^3$
$V = 4x^3 - 40x^2 + 96x$

The volume is $(4x^3 - 40x^2 + 96x)$ in.^{3}.

Problem 6 Find the volume of the rectangular solid shown in the diagram below. All dimensions are given in feet.

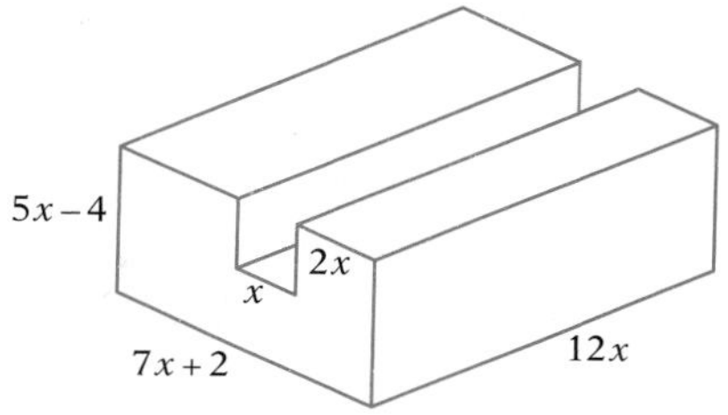

Solution See pages A17.

EXERCISES 3.2

1 Simplify.

1. $2x(x - 3)$
2. $2a(2a + 4)$
3. $3x^2(2x^2 - x)$
4. $-4y^2(4y - 6y^2)$
5. $3xy(2x - 3y)$
6. $-4ab(5a - 3b)$
7. $x^n(x + 1)$
8. $y^n(y^{2n} - 3)$
9. $x^n(x^n + y^n)$
10. $x - 2x(x - 2)$
11. $2b + 4b(2 - b)$
12. $-2y(3 - y) + 2y^2$
13. $-2a^2(3a^2 - 2a + 3)$
14. $4b(3b^3 - 12b^2 - 6)$
15. $3b(3b^4 - 3b^2 + 8)$
16. $(2x^2 - 3x - 7)(-2x^2)$
17. $(-3y^2 - 4y + 2)(y^2)$
18. $(6b^4 - 5b^2 - 3)(-2b^3)$
19. $-5x^2(4 - 3x + 3x^2 + 4x^3)$
20. $-2y^2(3 - 2y - 3y^2 + 2y^3)$
21. $-2x^2y(x^2 - 3xy + 2y^2)$
22. $3ab^2(3a^2 - 2ab + 4b^2)$
23. $x^n(x^{2n} + x^n + x)$
24. $x^{2n}(x^{2n-2} + x^{2n} + x)$

25. $a^{n+1}(a^n - 3a + 2)$

26. $a^{n+4}(a^{n-2} + 5a^2 - 3)$

27. $2y^2 - y[3 - 2(y - 4) - y]$

28. $3x^2 - x[x - 2(3x - 4)]$

29. $2y - 3[y - 2y(y - 3) + 4y]$

30. $4a^2 - 2a[3 - a(2 - a + a^2)]$

31. $7n - 4[3 + 2n(1 - 2n - 3n^2)]$

2 Simplify.

32. $(x - 2)(x + 7)$

33. $(y + 8)(y + 3)$

34. $(2y - 3)(4y + 7)$

35. $(5x - 7)(3x - 8)$

36. $(2x - 3y)(2x + 5y)$

37. $(7x - 3y)(2x - 9y)$

38. $(2a - 3b)(5a + 4b)$

39. $(3a - 5b)(a + 7b)$

40. $(5a + 2b)(3a + 7b)$

41. $(5x + 9y)(3x + 2y)$

42. $(3x - 7y)(7x + 2y)$

43. $(5x - 9y)(6x - 5y)$

44. $(xy + 4)(xy - 3)$

45. $(xy - 5)(2xy + 7)$

46. $(2x^2 - 5)(x^2 - 5)$

47. $(x^2 - 4)(x^2 - 6)$

48. $(5x^2 - 5y)(2x^2 - y)$

49. $(x^2 - 2y^2)(x^2 + 4y^2)$

50. $(x^n + 2)(x^n - 3)$

51. $(x^n - 4)(x^n - 5)$

52. $(2a^n - 3)(3a^n + 5)$

53. $(5b^n - 1)(2b^n + 4)$

54. $(2a^n - b^n)(3a^n + 2b^n)$

55. $(3x^n + b^n)(x^n + 2b^n)$

56. $(x - 2)(x^2 - 3x + 7)$

57. $(x + 3)(x^2 + 5x - 8)$

58. $(x + 5)(x^3 - 3x + 4)$

59. $(a + 2)(a^3 - 3a^2 + 7)$

60. $(2a - 3b)(5a^2 - 6ab + 4b^2)$

61. $(3a + b)(2a^2 - 5ab - 3b^2)$

62. $(2y^2 - 1)(y^3 - 5y^2 - 3)$

63. $(2b^2 - 3)(3b^2 - 3b + 6)$

64. $(2x - 5)(2x^4 - 3x^3 - 2x + 9)$

65. $(2a - 5)(3a^4 - 3a^2 + 2a - 5)$

66. $(x^2 + 2x - 3)(x^2 - 5x + 7)$

67. $(x^2 - 3x + 1)(x^2 - 2x + 7)$

68. $(a - 2)(2a - 3)(a + 7)$

69. $(b - 3)(3b - 2)(b - 1)$

70. $(x^n + 1)(x^{2n} + x^n + 1)$

71. $(a^{2n} - 3)(a^{5n} - a^{2n} + a^n)$

72. $(x^n + y^n)(x^n - 2x^ny^n + 3y^n)$

73. $(x^n - y^n)(x^{2n} - 3x^ny^n - y^{2n})$

3 Simplify.

74. $(a - 4)(a + 4)$
75. $(b - 7)(b + 7)$
76. $(3x - 2)(3x + 2)$
77. $(b - 11)(b + 11)$
78. $(3a + 5b)^2$
79. $(5x - 4y)^2$
80. $(x^2 - 3)^2$
81. $(x^2 + y^2)^2$
82. $(10 + b)(10 - b)$
83. $(2a - 3b)(2a + 3b)$
84. $(5x - 7y)(5x + 7y)$
85. $(x^2 + 1)(x^2 - 1)$
86. $(x^2 + y^2)(x^2 - y^2)$
87. $(2x^n + y^n)^2$
88. $(a^n + 5b^n)^2$
89. $(5a - 9b)(5a + 9b)$
90. $(3x + 7y)(3x - 7y)$
91. $(2x^n - 5)(2x^n + 5)$
92. $(x - 5)^2$
93. $(y + 2)^2$
94. $(2a - 3)^2$
95. $(2x - y)^2$
96. $(a - 5b)(a + 5b)$
97. $(x - yz)(x + yz)$
98. $(4y + 1)(4y - 1)$
99. $(6 - x)(6 + x)$
100. $(2x^2 - 3y^2)^2$
101. $(3a - 4b)^2$
102. $(2x^2 + 5)^2$
103. $(3x^n + 2)^2$
104. $(4b^n - 3)^2$
105. $(x^n + 3)(x^n - 3)$
106. $(x^n + y^n)(x^n - y^n)$
107. $(x^n - 1)^2$
108. $(a^n - b^n)^2$
109. $(2x^n + 5y^n)^2$

4 Solve.

110. The length of a rectangle is $(3x + 3)$ ft. The width is $(x - 4)$ ft. Find the area of the rectangle in terms of the variable x.
111. The base of a triangle is $(x + 2)$ ft. The height is $(2x - 3)$ ft. Find the area of the triangle in terms of the variable x.
112. Find the area of the figure shown below. All dimensions given are in meters.

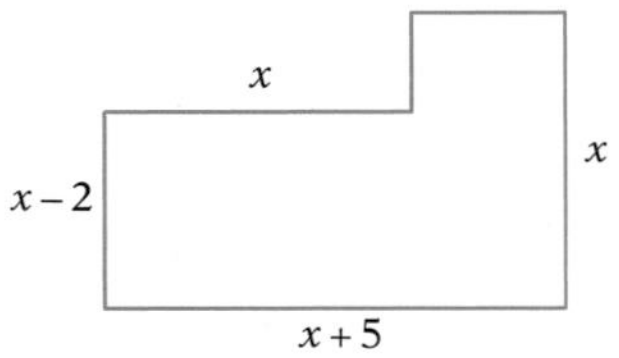

113. Find the area of the figure shown below. All dimensions given are in feet.

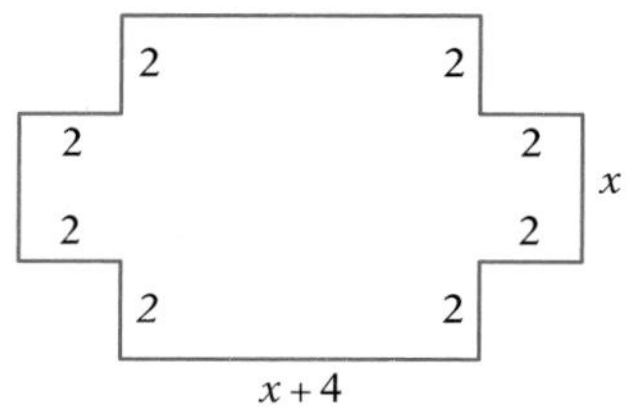

114. The length of the side of a cube is $(x - 2)$ cm. Find the volume of the cube in terms of the variable x.

115. The length of a box is $(2x + 3)$ cm, the width is $(x - 5)$ cm, and the height is x cm. Find the volume of the box in terms of the variable x.

116. Find the volume of the figure shown below. All dimensions given are in inches.

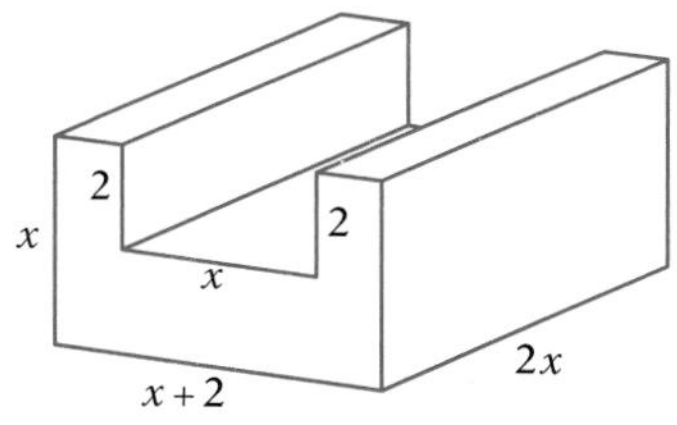

117. Find the volume of the figure shown below. All dimensions given are in centimeters.

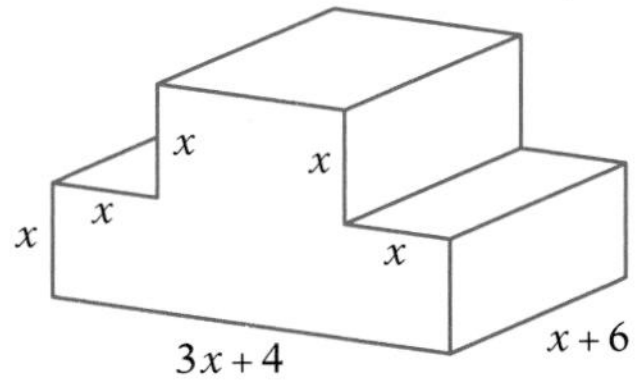

118. The radius of a circle is $(5x + 4)$ in. Find the area of the circle in terms of the variable x. Use 3.14 for π.

119. The radius of a circle is $(x - 2)$ in. Find the area of the circle in terms of the variable x. Use 3.14 for π.

SUPPLEMENTAL EXERCISES 3.2

Simplify.

120. $\frac{(2x + 1)^5}{(2x + 1)^3}$

121. $\frac{(3x - 5)^6}{(3x - 5)^4}$

122. $(a - b)^2 - (a + b)^2$

123. $(x + 2y)^2 + (x + 2y)(x - 2y)$

124. $(4y + 3)^2 - (3y + 4)^2$

125. $2x^2(3x^3 + 4x - 1) - 5x^2(x^2 - 3)$

126. $(2b + 3)(b - 4) + (3 + b)(3 - 2b)$

127. $(3x - 2y)^2 - (2x - 3y)^2$

128. $[x + (y + 1)][x - (y + 1)]$

129. $[x^2(2y - 1)]^2$

For what value of k is the given equation an identity?

130. $(5x - k)(3x + k) = 15x^2 + 4x - k^2$

131. $(2x - k)(3x - k) = 6x^2 - 25x + k^2$

132. $(kx + 1)(kx - 6) = k^2x^2 - 15x - 6$

133. $(kx - 7)(kx + 2) = k^2x^2 + 5x - 14$

Complete.

134. If $m = n + 1$, then $\frac{a^m}{a^n} =$ ______.

135. If $m = n + 2$, then $\frac{a^m}{a^n} =$ ______.

136. What polynomial when divided by $x - 4$ has a quotient of $2x + 3$?

137. What polynomial when divided by $2x - 3$ has a quotient of $x + 7$?

138. Subtract the product of $4a + b$ and $2a - b$ from $9a^2 - 2ab$.

139. Subtract the product of $5x - y$ and $x + 3y$ from $6x^2 + 12xy - 2y^2$.

140. Find $(3n^4)^3$ if $5(n - 1) = 2(3n - 2)$.

141. Find $(-2n^3)^2$ if $3(2n - 1) = 5(n - 1)$.

SECTION 3.3 Division of Polynomials

1 Divide polynomials

To divide two polynomials, use a method similar to that used for division of whole numbers. The same equation used to check division of whole numbers is used to check polynomial division.

Dividend = (Quotient × Divisor) + Remainder

Simplify: $(x^2 + 5x - 7) \div (x + 3)$

Step 1

$$\begin{array}{r} x \phantom{{}+5x-7} \\ x+3\,\overline{)\,x^2 + 5x - 7} \\ \underline{x^2 + 3x} \phantom{{}-7} \downarrow \\ 2x - 7 \end{array}$$

Think: $x\overline{)x^2} = \dfrac{x^2}{x} = x$

Multiply: $x(x + 3) = x^2 + 3x$

Subtract: $(x^2 + 5x) - (x^2 + 3x) = 2x$

Bring down the -7.

Step 2

$$\begin{array}{r} x + 2 \\ x+3\,\overline{)\,x^2 + 5x - 7} \\ \underline{x^2 + 3x} \phantom{{}-7} \\ 2x - 7 \\ \underline{2x + 6} \\ -13 \end{array}$$

Think: $x\overline{)2x} = \dfrac{2x}{x} = 2$

Multiply: $2(x + 3) = 2x + 6$

Subtract: $(2x - 7) - (2x + 6) = -13$

The remainder is -13.

Check: $(x + 2)(x + 3) + (-13) = x^2 + 3x + 2x + 6 - 13 = x^2 + 5x - 7$

$(x^2 + 5x - 7) \div (x + 3) = x + 2 - \dfrac{13}{x + 3}$

Simplify: $\dfrac{6 - 6x^2 + 4x^3}{2x + 3}$

Arrange the terms in descending order. There is no term of x in $4x^3 - 6x^2 + 6$. Insert $0x$ for the missing term so that like terms will be in columns.

$$\begin{array}{r} 2x^2 - 6x + 9 \\ 2x+3\,\overline{)\,4x^3 - 6x^2 + 0x + 6} \\ \underline{4x^3 + 6x^2} \phantom{{}+0x+6} \\ -12x^2 + 0x \phantom{{}+6} \\ \underline{-12x^2 - 18x} \phantom{{}+6} \\ 18x + 6 \\ \underline{18x + 27} \\ -21 \end{array}$$

$\dfrac{6 - 6x^2 + 4x^3}{2x + 3} = 2x^2 - 6x + 9 - \dfrac{21}{2x + 3}$

Example 1 Simplify.

A. $\dfrac{12x^2 - 11x + 10}{4x - 5}$ B. $\dfrac{x^3 + 1}{x + 1}$

Solution A.

$$\begin{array}{r} 3x + 1 \\ 4x-5\,\overline{)\,12x^2 - 11x + 10} \\ \underline{12x^2 - 15x} \phantom{{}+10} \\ 4x + 10 \\ \underline{4x - 5} \\ 15 \end{array}$$

$\dfrac{12x^2 - 11x + 10}{4x - 5} = 3x + 1 + \dfrac{15}{4x - 5}$

B.

$$\begin{array}{r} x^2 - x + 1 \\ x+1\,\overline{)\,x^3 + 0x^2 + 0x + 1} \\ \underline{x^3 + x^2} \phantom{{}+0x+1} \\ -x^2 + 0x \phantom{{}+1} \\ \underline{-x^2 - x} \phantom{{}+1} \\ x + 1 \\ \underline{x + 1} \\ 0 \end{array}$$

$\dfrac{x^3 + 1}{x + 1} = x^2 - x + 1$

Problem 1 Simplify.

A. $\dfrac{15x^2 + 17x - 20}{3x + 4}$ B. $\dfrac{3x^3 + 8x^2 - 6x + 2}{3x - 1}$

Solution See page A18.

2 Synthetic division

Synthetic division is a shorter method of dividing a polynomial by a binomial of the form $x - a$. This method of dividing uses only the coefficients of the variable terms.

Both long division and synthetic division are used below to simplify the expression $(3x^2 - 4x + 6) \div (x - 2)$.

LONG DIVISION

Compare the coefficients in this problem worked by long division with the coefficients in the same problem worked by synthetic division below.

$$\begin{array}{r} 3x\ \ + 2 \\ x - 2\overline{)3x^2 - 4x + 6} \\ \underline{3x^2 - 6x}\qquad \\ 2x + 6 \\ \underline{2x - 4} \\ 10 \end{array}$$

$$(3x^2 - 4x + 6) \div (x - 2) = 3x + 2 + \frac{10}{x - 2}$$

SYNTHETIC DIVISION

$x - a = x - 2;\ a = 2$

Value of a — Coefficients of the dividend

$$\begin{array}{r|rrr} 2 & 3 & -4 & 6 \\ \hline \end{array}$$

Bring down the 3.

$$\begin{array}{r|rrr} & 3 & & \end{array}$$

Multiply $2 \cdot 3$ and add the product (6) to -4.

$$\begin{array}{r|rrr} 2 & 3 & -4 & 6 \\ & & 6 & \\ \hline & 3 & 2 & \end{array}$$

Multiply $2 \cdot 2$ and add the product (4) to 6.

$$\begin{array}{r|rrr} 2 & 3 & -4 & 6 \\ & & 6 & 4 \\ \hline & 3 & 2 & 10 \end{array}$$

Coefficients of the quotient — Remainder

The degree of the first term of the quotient is one degree less than the degree of the first term of the dividend.

$$(3x^2 - 4x + 6) \div (x - 2) = 3x + 2 + \frac{10}{x - 2}$$

Check:

$$\begin{aligned}(3x + 2)(x - 2) + 10 &= 3x^2 - 6x + 2x - 4 + 10 \\ &= 3x^2 - 4x + 6\end{aligned}$$

Simplify: $(2x^3 + 3x^2 - 4x + 8) \div (x + 3)$

Write down the value of a and the coefficients of the dividend.
$x - a = x + 3 = x - (-3)$; $a = -3$

$$\begin{array}{r|rrrr} -3 & 2 & 3 & -4 & 8 \\ & & -6 & 9 & -15 \\ \hline & 2 & -3 & 5 & -7 \end{array}$$

Bring down the 2. Multiply $-3(2)$. Add the product to 3. Continue until all the coefficients have been used.

$\underbrace{2 \quad -3 \quad 5}_{\text{Coefficients of the quotient}}$ $\quad \underbrace{-7}_{\text{Remainder}}$

Write the quotient. The degree of the quotient is one less than the degree of the dividend.

$(2x^3 + 3x^2 - 4x + 8) \div (x + 3) =$
$2x^2 - 3x + 5 - \dfrac{7}{x+3}$

Example 2 Simplify.

A. $(5x^2 - 3x + 7) \div (x - 1)$ B. $(3x^4 - 8x^2 + 2x + 1) \div (x + 2)$

Solution A.

$$\begin{array}{r|rrr} 1 & 5 & -3 & 7 \\ & & 5 & 2 \\ \hline & 5 & 2 & 9 \end{array}$$

▶ $x - a = x - 1$; $a = 1$

$(5x^2 - 3x + 7) \div (x - 1) =$
$5x + 2 + \dfrac{9}{x-1}$

B.

$$\begin{array}{r|rrrrr} -2 & 3 & 0 & -8 & 2 & 1 \\ & & -6 & 12 & -8 & 12 \\ \hline & 3 & -6 & 4 & -6 & 13 \end{array}$$

▶ Insert a zero for the missing term.
$x - a = x + 2$; $a = -2$

$(3x^4 - 8x^2 + 2x + 1) \div (x + 2) =$
$3x^3 - 6x^2 + 4x - 6 + \dfrac{13}{x+2}$

Problem 2 Simplify.

A. $(6x^2 + 8x - 5) \div (x + 2)$ B. $(2x^4 - 3x^3 - 8x^2 - 2) \div (x - 3)$

Solution See page A18.

EXERCISES 3.3

1 Divide by using long division.

1. $(x^2 + 3x - 40) \div (x - 5)$

2. $(x^2 - 14x + 24) \div (x - 2)$

3. $(x^3 - 3x^2 + 2) \div (x - 3)$

4. $(x^3 + 4x^2 - 8) \div (x + 4)$

5. $(6x^2 + 13x + 8) \div (2x + 1)$

6. $(12x^2 + 13x - 14) \div (3x - 2)$

7. $(10x^2 + 9x - 5) \div (2x - 1)$

8. $(18x^2 - 3x + 2) \div (3x + 2)$

9. $(8x^3 - 9) \div (2x - 3)$

10. $(64x^3 + 4) \div (4x + 2)$

11. $(6x^4 - 13x^2 - 4) \div (2x^2 - 5)$

12. $(12x^4 - 11x^2 + 10) \div (3x^2 + 1)$

13. $\dfrac{3x^3 - 8x^2 - 33x - 10}{3x + 1}$

14. $\dfrac{8x^3 - 38x^2 + 49x - 10}{4x - 1}$

15. $\dfrac{4 - 7x + 5x^2 - x^3}{x - 3}$

16. $\dfrac{4 + 6x - 3x^2 + 2x^3}{2x + 1}$

17. $\dfrac{16x^2 - 13x^3 + 2x^4 - 9x + 20}{x - 5}$

18. $\dfrac{x + 3x^4 - x^2 + 5x^3 - 2}{x + 2}$

19. $\dfrac{x^3 - 4x^2 + 2x - 1}{x^2 + 1}$

20. $\dfrac{3x^3 - 2x^2 - 8}{x^2 + 5}$

21. $\dfrac{2x^3 - x + 4 - 3x^2}{x^2 - 1}$

22. $\dfrac{2 - 3x^2 + 5x^3}{x^2 + 3}$

23. $\dfrac{6x^3 + 2x^2 + x + 4}{2x^2 - 3}$

24. $\dfrac{9x^3 + 6x^2 + 2x + 1}{3x^2 + 2}$

2 Divide by using synthetic division.

25. $(2x^2 - 6x - 8) \div (x + 1)$

26. $(3x^2 + 19x + 20) \div (x + 5)$

27. $(3x^2 - 14x + 16) \div (x - 2)$

28. $(4x^2 - 23x + 28) \div (x - 4)$

29. $(3x^2 - 4) \div (x - 1)$

30. $(4x^2 - 8) \div (x - 2)$

31. $(x^2 - 9) \div (x + 4)$

32. $(x^2 - 49) \div (x + 5)$

33. $(2x^2 + 24) \div (2x + 4)$

34. $(3x^2 - 15) \div (x + 3)$

35. $(4x^2 - 8x + 3) \div (x + 1)$

36. $(3x^2 + 7x - 6) \div (x + 4)$

37. $(2x^3 - x^2 + 6x + 9) \div (x + 1)$

38. $(3x^3 + 10x^2 + 6x - 4) \div (x + 2)$

39. $(x^3 - 6x^2 + 11x - 6) \div (x - 3)$

40. $(x^3 - 4x^2 + x + 6) \div (x + 1)$

41. $(6x - 3x^2 + x^3 - 9) \div (x + 2)$

42. $(5 - 5x + 4x^2 + x^3) \div (x - 3)$

43. $(x^3 + x - 2) \div (x + 1)$

44. $(x^3 + 2x + 5) \div (x - 2)$

45. $(18 + x - 4x^3) \div (2 - x)$

46. $(12 - 3x^2 + x^3) \div (3 + x)$

47. $(2x^3 + 5x^2 - 5x + 20) \div (x + 4)$

48. $(5x^3 + 3x^2 - 17x + 6) \div (x + 2)$

49. $\dfrac{16x^2 - 13x^3 + 2x^4 - 9x + 20}{x - 5}$

50. $\dfrac{2x^3 - x^2 - 10x + 15 + x^4}{x - 2}$

51. $\dfrac{5 + 5x - 8x^2 + 4x^3 - 3x^4}{2 - x}$

52. $\dfrac{3 - 13x - 5x^2 + 9x^3 - 2x^4}{3 - x}$

53. $\dfrac{3x^4 + 3x^3 - x^2 + 3x + 2}{x + 1}$

54. $\dfrac{4x^4 + 12x^3 - x^2 - x + 2}{x + 3}$

55. $\dfrac{2x^4 - x^2 + 2}{x - 3}$

56. $\dfrac{x^4 - 3x^3 - 30}{x + 2}$

57. $\dfrac{x^3 + 125}{x + 5}$

58. $\dfrac{x^3 + 343}{x + 7}$

SUPPLEMENTAL EXERCISES 3.3

Divide by using long division.

59. $\dfrac{3x^2 - xy - 2y^2}{3x + 2y}$

60. $\dfrac{12x^2 + 11xy + 2y^2}{4x + y}$

61. $\dfrac{4a^2 - 2ab - 5b^2}{2a + b}$

62. $\dfrac{6a^2 - 5ab + 3b^2}{3a - b}$

63. $\dfrac{a^3 - b^3}{a - b}$

64. $\dfrac{a^4 + b^4}{a + b}$

65. $\dfrac{x^5 + y^5}{x + y}$

66. $\dfrac{x^6 - y^6}{x - y}$

For what value of k will the remainder be zero?

67. $(x^3 - 3x^2 - x + k) \div (x - 3)$

68. $(x^3 - 2x^2 + x + k) \div (x - 2)$

69. $(x^2 + kx - 6) \div (x - 3)$

70. $(x^3 + kx + k - 1) \div (x - 1)$

Solve.

71. When $x^2 + x + 2$ is divided by a polynomial, the quotient is $x + 4$, and the remainder is 14. Find the polynomial.

72. When $x^2 + 3x + 4$ is divided by a polynomial, the quotient is $x + 1$, and the remainder is 2. Find the polynomial.

SECTION 3.4 Factoring Polynomials

1 Factor a monomial from a polynomial

The GCF of two or more exponential expressions with the same base is the exponential expression with the smallest exponent.

$$2^5 \qquad 2^2 \qquad 2^9 \qquad \text{GCF} = 2^2 = 4$$

$$x^5 \qquad x^7 \qquad x \qquad \text{GCF} = x$$

The GCF of two or more monomials is the product of the GCF of each common factor with the smallest exponent.

$$16a^4b = 2^4 \cdot a^4 \cdot b$$
$$40a^2b^5 = 2^3 \cdot 5 \cdot a^2 \cdot b^5$$
$$\text{GCF} = 2^3 \cdot a^2 \cdot b = 8a^2b$$

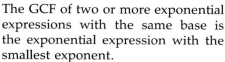

To **factor a polynomial** means to write the polynomial as a product of other polynomials.

In the example at the right, $3x$ is the GCF of the terms $3x^2$ and $6x$. $3x$ is a **common monomial factor** of the terms of the binomial. $x - 2$ is a **binomial factor** of $3x^2 - 6x$.

Multiply	
Polynomial	**Factors**
$3x^2 - 6x$	$3x(x - 2)$
Factor	

Example 1 Factor.

A. $4x^3y^2 + 12x^3y + 20xy^2$ B. $x^{2n} + x^{n+1} + x^n$

Solution A. The GCF of $4x^3y^2$, $12x^3y$, and $20xy^2$ is $4xy$.

▶ Find the GCF of the terms of the polynomial.

$4x^3y^2 = 2^2 \cdot x^3 \cdot y^2$
$12x^3y = 2^2 \cdot 3 \cdot x^3 \cdot y$
$20xy^2 = 2^2 \cdot 5 \cdot x \cdot y^2$
$\text{GCF} = 2^2 \cdot x \cdot y = 4xy$

$4x^3y^2 + 12x^3y + 20xy^2 =$
$4xy(x^2y) + 4xy(3x^2) + 4xy(5y) =$

▶ Rewrite each term of the polynomial as a product with the GCF as one of the factors.

$4xy(x^2y + 3x^2 + 5y)$

▶ Use the Distributive Property to write the polynomial as a product of factors.

B. The GCF of x^{2n}, x^{n+1}, and x^n is x^n, $n > 0$.
$x^{2n} + x^{n+1} + x^n = x^n(x^n + x + 1)$

▶ The GCF is x^n because $n < 2n$ and $n < n + 1$.

Problem 1 Factor.

A. $3x^3y - 6x^2y^2 - 3xy^3$ B. $6t^{2n} - 9t^n$

Solution See page A18.

2 Factor by grouping

In the examples at the right, the binomials in parentheses are called binomial factors.

$4x^4(2x - 3)$
$-2r^2s(5r + 2s)$

The Distributive Property is used to factor a common binomial factor from an expression.

Factor: $4a(2b + 3) - 5(2b + 3)$

The common binomial factor is $(2b + 3)$. Use the Distributive Property to write the expression as a product of factors.

$4a(2b + 3) - 5(2b + 3) =$
$(2b + 3)(4a - 5)$

Consider the binomial $y - x$. Factoring -1 from this binomial gives

$$y - x = -(x - y)$$

This equation is sometimes used to factor a common binomial from an expression.

Factor: $6r(r - s) - 7(s - r)$

Rewrite the expression as a sum of terms that have a common binomial factor.
Use $s - r = -(r - s)$.

$6r(r - s) - 7(s - r) =$
$6r(r - s) + 7(r - s) =$
$(r - s)(6r + 7)$

Some polynomials can be factored by grouping terms so that a common binomial factor is found.

Factor: $3xz - 4yz - 3xa + 4ya$

$3xz - 4yz - 3xa + 4ya =$

Group the first two terms and the last two terms. Note that $-3xa + 4ya = -(3xa - 4ya)$. $(3xz - 4yz) - (3xa - 4ya) =$

Factor the GCF from each group. $z(3x - 4y) - a(3x - 4y) =$

Write the expression as the product of factors. $(3x - 4y)(z - a)$

Factor: $8y^2 + 4y - 6ay - 3a$

$8y^2 + 4y - 6ay - 3a =$

Group the first two terms and the last two terms. Note that $-6ay - 3a = -(6ay + 3a)$. $(8y^2 + 4y) - (6ay + 3a) =$

Factor the GCF from each group. $4y(2y + 1) - 3a(2y + 1) =$

Write the expression as the product of factors. $(2y + 1)(4y - 3a)$

Example 2 Factor: $3x(y - 4) - 2(4 - y)$

Solution
$3x(y - 4) - 2(4 - y) =$
$3x(y - 4) + 2(y - 4) =$ ▶ Write the expression as a sum of terms that have a common factor. Note that $4 - y = -(y - 4)$.
$(y - 4)(3x + 2)$ ▶ Write the expression as a product of factors.

Problem 2 Factor: $6a(2b - 5) + 7(5 - 2b)$

Solution See page A18.

Example 3 Factor: $xy - 4x - 2y + 8$

Solution
$xy - 4x - 2y + 8 =$
$(xy - 4x) - (2y - 8) =$ ▶ Group the first two terms and the last two terms.
$x(y - 4) - 2(y - 4) =$ ▶ Factor out the GCF from each group.
$(y - 4)(x - 2)$

Problem 3 Factor: $3rs - 2r - 3s + 2$

Solution See page A18.

3 Factor trinomials of the form $x^2 + bx + c$

A **quadratic trinomial** is a trinomial of the form $ax^2 + bx + c$, where a, b, and c are nonzero constants. The degree of a quadratic trinomial is 2. Examples of quadratic trinomials are shown below.

$$\begin{array}{ll} 3x^2 + 4x + 7 & (a = 3,\ b = 4,\ c = 7) \\ y^2 + 2y - 9 & (a = 1,\ b = 2,\ c = -9) \\ 6x^2 - 5x + 1 & (a = 6,\ b = -5,\ c = 1) \end{array}$$

To **factor a quadratic trinomial** of the form $x^2 + bx + c$ means to express the trinomial as the product of two binomials.

The method by which factors of a trinomial are found is based on FOIL. Consider the binomial products shown below, noting the relationship between the constant terms of the binomials and the terms of the trinomials.

$$\begin{array}{ll} (x + 4)(x + 7) = x^2 + 7x + 4x + (4)(7) & = x^2 + 11x + 28 \\ (x - 3)(x - 5) = x^2 - 5x - 3x + (-3)(-5) & = x^2 - 8x + 15 \\ (x + 9)(x - 6) = x^2 - 6x + 9x + (9)(-6) & = x^2 + 3x - 54 \\ (x + 2)(x - 8) = x^2 - 8x + 2x + (2)(-8) & = x^2 - 6x - 16 \end{array}$$

The coefficient of x is the sum of the constant terms of the binomials.

The constant term is the product of the constant terms of the binomials.

Points to Remember to Factor $x^2 + bx + c$

1. In the trinomial, the coefficient of x is the sum of the constant terms of the binomials.
2. In the trinomial, the constant term is the product of the constant terms of the binomials.
3. When the constant term of the trinomial is positive, the constant terms of the binomials have the same sign as the coefficient of x in the trinomial.
4. When the constant term of the trinomial is negative, the constant terms of the binomials have opposite signs.

Use the four points listed above to factor a trinomial. For example, to factor

$$x^2 - 5x - 24,$$

find two numbers whose sum is -5 and whose product is -24 [Points 1 and 2]. Because the constant term of the trinomial is negative (-24), the numbers will have opposite signs [Point 4].

A systematic method of finding these numbers involves listing the factors of the constant term of the trinomial and the sum of those factors.

Factors of -24	Sum of the Factors
$1, -24$	$1 + (-24) = -23$
$-1, 24$	$-1 + 24 = 23$
$2, -12$	$2 + (-12) = -10$
$-2, 12$	$-2 + 12 = 10$
$3, -8$	**$3 + (-8) = -5$**
$-3, 8$	$-3 + 8 = 5$
$4, -6$	$4 + (-6) = -2$
$-4, 6$	$-4 + 6 = 2$

3 and -8 are two numbers whose sum is -5 and whose product is -24. Write the binomial factors of the trinomial.

$$x^2 - 5x - 24 = (x + 3)(x - 8)$$

Check: $(x + 3)(x - 8) = x^2 - 8x + 3x - 24 = x^2 - 5x - 24$

By the Commutative Property of Multiplication, the binomial factors can also be written as

$$x^2 - 5x - 24 = (x - 8)(x + 3)$$

Example 4 Factor.

A. $x^2 + 8x + 12$ B. $x^2 + 5x - 84$ C. $10 - 3x - x^2$

Solution A. $x^2 + 8x + 12$

Factors of 12	Sum
1, 12	13
2, 6	8
3, 4	7

▶ Try only positive factors of 12 [Point 3].

▶ Once the correct pair is found, the other factors need not be tried.

$x^2 + 8x + 12 = (x + 2)(x + 6)$

▶ Write the factors of the trinomial.

Check: $(x + 2)(x + 6) =$
$x^2 + 6x + 2x + 12 =$
$x^2 + 8x + 12$

B. $x^2 + 5x - 84$

$(-7)(12) = -84$
$-7 + 12 = 5$

▶ The factors must be of opposite signs [Point 4]. Find two factors of -84 whose sum is 5.

$x^2 + 5x - 84 = (x + 12)(x - 7)$

Check: $(x + 12)(x - 7) =$
$x^2 - 7x + 12x - 84 =$
$x^2 + 5x - 84$

C. $10 - 3x - x^2 = (5 + x)(2 - x)$

Check: $(5 + x)(2 - x) =$
$10 - 5x + 2x - x^2 =$
$10 - 3x - x^2$

Problem 4 Factor.

A. $x^2 + 13x + 42$ B. $x^2 - x - 20$ C. $x^2 + 5xy + 6y^2$

Solution See page A19.

Not all trinomials can be factored when using only integers. Consider $x^2 - 3x - 6$.

Factors of -6	Sum
1, -6	-5
-1, 6	5
2, -3	-1
-2, 3	1

Because none of the pairs of factors of -6 have a sum of -3, the trinomial is not factorable. The trinomial is said to be **nonfactorable** using integers.

Example 5 Factor: $x^2 + 2x - 4$

Solution The trinomial is nonfactorable over the integers.

▶ There are no factors of -4 whose sum is 2.

Problem 5 Factor: $x^2 + 5x - 1$

Solution See page A19.

4 Factor trinomials of the form $ax^2 + bx + c$

One method of factoring a trinomial of the form $ax^2 + bx + c$ is a trial-and-error method. Trial factors are written, using the factors of a and c to write the binomials. Then FOIL is used to check for b, the coefficient of the middle term.

Factoring polynomials of this type by trial and error may require testing many trial factors. To reduce the number of trial factors, remember the following points:

Points to Remember to Factor $ax^2 + bx + c$

1. If the terms of the trinomial do not have a common factor, then a binomial factor cannot have a common factor.
2. When the constant term of the trinomial is positive, the constant terms of the binomials have the same sign as the coefficient of x in the trinomial.
3. When the constant term of the trinomial is negative, the constant terms of the binomials have opposite signs.

Factor: $4x^2 + 31x - 8$

The terms of the trinomial do not have a common factor; therefore, the binomial factors will not have a common factor.

Because the constant term, c, of the trinomial is negative (-8), the constant terms of the binomial factors will have opposite signs.

Find the factors of a (4) and the factors of c (-8).

Factors of 4	Factors of −8
1, 4	1, −8
2, 2	−1, 8
	2, −4
	−2, 4

Using these factors, write trial factors, and use FOIL to check the middle term of the trinomial.

Remember that if the terms of the trinomial do not have a common factor, then a binomial factor cannot have a common factor (Point 1). Such trial factors need not be checked.

Trial Factors	Middle Term
$(x + 1)(4x - 8)$	Common factor
$(x - 1)(4x + 8)$	Common factor
$(x + 2)(4x - 4)$	Common factor
$(x - 2)(4x + 4)$	Common factor
$(2x + 1)(2x - 8)$	Common factor
$(2x - 1)(2x + 8)$	Common factor
$(2x + 2)(2x - 4)$	Common factor
$(2x - 2)(2x + 4)$	Common factor
$(4x + 1)(x - 8)$	$-32x + x = -31x$
$\mathbf{(4x - 1)(x + 8)}$	$\mathbf{32x - x = 31x}$

The correct factors have been found. The remaining trial factors need not be checked.

$$4x^2 + 31x - 8 = (4x - 1)(x + 8)$$

The last example illustrates that many of the trial factors may have common factors and thus need not be tried. For the remainder of this chapter, the trial factors with a common factor will not be listed.

Example 6 Factor.

A. $2x^2 - 21x + 10$ B. $6x^2 + 17x - 10$

Solution A. $2x^2 - 21x + 10$

Factors of 2	Factors of 10
1, 2	−1, −10
	−2, −5

▶ Use negative factors of 10. [Point 2]

Trial Factors	Middle Term
$(x - 2)(2x - 5)$	$-5x - 4x = -9x$
$(2x - 1)(x - 10)$	$-20x - x = -21x$

▶ Write trial factors. Use FOIL to check the middle term.

$$2x^2 - 21x + 10 = (2x - 1)(x - 10)$$

B. $6x^2 + 17x - 10$

Factors of 6	Factors of −10
1, 6	1, −10
2, 3	−1, 10
	2, −5
	−2, 5

▶ Find the factors of a (6) and the factors of c (−10).

Trial Factors	Middle Term
$(x + 2)(6x - 5)$	$-5x + 12x = 7x$
$(x - 2)(6x + 5)$	$5x - 12x = -7x$
$(2x + 1)(3x - 10)$	$-20x + 3x = -17x$
$(2x - 1)(3x + 10)$	$20x - 3x = 17x$

▶ Write trial factors. Use FOIL to check the middle term.

$$6x^2 + 17x - 10 = (2x - 1)(3x + 10)$$

Problem 6 Factor.

A. $4x^2 + 15x - 4$ B. $10x^2 + 39x + 14$

Solution See page A19.

Trinomials of the form $ax^2 + bx + c$ can also be factored by grouping. This method is an extension of the method discussed in Objective 2.

To factor $ax^2 + bx + c$, first find the factors of $a \cdot c$ whose sum is b. Use the two factors to rewrite the middle term of the trinomial as the sum of two terms. Then factor by grouping to write the factorization of the trinomial.

Factor: $3x^2 + 11x + 8$

Find two positive factors of 24 ($ac = 3 \cdot 8$) whose sum is 11, the coefficient of x.

Positive Factors of 24	Sum
1, 24	25
2, 12	14
3, 8	11

The required sum has been found. The remaining factors need not be checked.

Use the factors of 24 whose sum is 11 to write $11x$ as $3x + 8x$. Factor by grouping.

$$\begin{aligned} 3x^2 + 11x + 8 &= 3x^2 + 3x + 8x + 8 \\ &= (3x^2 + 3x) + (8x + 8) \\ &= 3x(x + 1) + 8(x + 1) \\ &= (x + 1)(3x + 8) \end{aligned}$$

Check: $(x + 1)(3x + 8) = 3x^2 + 8x + 3x + 8 = 3x^2 + 11x + 8$

Factor: $4z^2 - 17z - 21$

Find two factors of -84 [$ac = 4 \cdot (-21)$] whose sum is -17, the coefficient of z.

When the required sum is found, the remaining factors need not be checked.

Factors of -84	Sum
1, -84	-83
-1, 84	83
2, -42	-40
-2, 42	40
3, -28	-25
-3, 28	25
4, -21	-17

Use the factors of -84 whose sum is -17 to write $-17z$ as $4z - 21z$. Factor by grouping. Recall that $-21z - 21 = -(21z + 21)$.

$$\begin{aligned} 4z^2 - 17z - 21 &= 4z^2 + 4z - 21z - 21 \\ &= (4z^2 + 4z) - (21z + 21) \\ &= 4z(z + 1) - 21(z + 1) \\ &= (z + 1)(4z - 21) \end{aligned}$$

Factor: $3x^2 - 11x + 4$

Find two negative factors of 12 ($a \cdot c = 3 \cdot 4$) whose sum is -11.

Factors of 12	Sum
-1, -12	-13
-2, -6	-8
-3, -4	-7

Because no integer factors of 12 have a sum of -11, $3x^2 - 11x + 4$ is nonfactorable over the integers.

Either method of factoring discussed in this objective will always lead to a correct factorization of trinomials of the form $ax^2 + bx + c$ that are factorable.

Example 7 Factor.

A. $2x^2 - 2x + 10$ B. $10 - 17x - 6x^2$

Solution A. $2x^2 - 21x + 10$

Factors of 20	Sum
$-20, -1$	-21

▶ $a \cdot c = 2 \cdot 10 = 20$. Find two factors of 20 whose sum is -21.

$$\begin{aligned} 2x^2 - 21x + 10 &= 2x^2 - 20x - x + 10 \\ &= (2x^2 - 20x) - (x - 10) \\ &= 2x(x - 10) - (x - 10) \\ &= (x - 10)(2x - 1) \end{aligned}$$

▶ Rewrite $-21x$ as $-20x - x$.

▶ Factor by grouping.

B. $10 - 17x - 6x^2$

Factors of -60	Sum
$-60, 1$	-59
$60, -1$	59
$-30, 2$	-28
$30, -2$	28
$-20, 3$	-17

▶ $a \cdot c = -60$. Find two factors of -60 whose sum is -17.

$$\begin{aligned} 10 - 17x - 6x^2 &= 10 - 20x + 3x - 6x^2 \\ &= (10 - 20x) + (3x - 6x^2) \\ &= 10(1 - 2x) + 3x(1 - 2x) \\ &= (1 - 2x)(10 + 3x) \end{aligned}$$

▶ Rewrite $-17x$ as $-20x + 3x$.

▶ Factor by grouping.

Problem 7 Factor.

A. $6x^2 + 7x - 20$ B. $2 - x - 6x^2$

Solution See page A19.

A polynomial is factored completely when it is written as a product of factors that are nonfactorable over the integers.

Factor: $4x^3 + 12x^2 - 160x$

$4x^3 + 12x^2 - 160x =$

Factor out the GCF of the terms. The GCF of $4x^3$, $12x^2$, and $160x$ is $4x$.

$4x(x^2 + 3x - 40) =$

Factor the trinomial.

$4x(x + 8)(x - 5)$

Check: $4x(x + 8)(x - 5) =$
$(4x^2 + 32x)(x - 5) =$
$4x^3 + 12x^2 - 160x$

Example 8 Factor.

A. $30y + 2xy - 4x^2y$ B. $12x^3y^2 + 14x^2y - 6x$

Solution A. $30y + 2xy - 4x^2y =$ ▶ The GCF of $30y$, $2xy$, and $4x^2y$ is $2y$.

$2y(15 + x - 2x^2) =$ ▶ Factor out the GCF.

$2y(5 + 2x)(3 - x)$ ▶ Factor the trinomial.

B. $12x^3y^2 + 14x^2y - 6x =$ ▶ The GCF of $12x^3y^2$, $14x^2y$, and $6x$ is $2x$.

$2x(6x^2y^2 + 7xy - 3) =$

$2x(3xy - 1)(2xy + 3)$

Problem 8 Factor.

A. $3a^3b^3 + 3a^2b^2 - 60ab$ B. $40a - 10a^2 - 15a^3$

Solution See page A19.

EXERCISES 3.4

1 Factor.

1. $6a^2 - 15a$
2. $32b^2 + 12b$
3. $4x^3 - 3x^2$
4. $12a^5b^2 + 16a^4b$
5. $3a^2 - 10b^3$
6. $9x^2 + 14y^4$
7. $x^5 - x^3 - x$
8. $y^4 - 3y^2 - 2y$
9. $16x^2 - 12x + 24$
10. $2x^5 + 3x^4 - 4x^2$
11. $5b^2 - 10b^3 + 25b^4$
12. $x^2y^4 - x^2y - 4x^2$
13. $x^{2n} - x^n$
14. $a^{5n} + a^{2n}$
15. $x^{3n} - x^{2n}$
16. $y^{4n} + y^{2n}$
17. $a^{2n+2} + a^2$
18. $b^{n+5} - b^5$
19. $12x^2y^2 - 18x^3y + 24x^2y$
20. $14a^4b^4 - 42a^3b^3 + 28a^3b^2$
21. $-16a^2b^4 - 4a^2b^2 + 24a^3b^2$
22. $10x^2y + 20x^2y^2 + 30x^2y^3$
23. $y^{2n+2} + y^{n+2} - y^2$
24. $a^{2n+2} + a^{2n+1} + a^n$

2 Factor.

25. $x(a + 2) - 2(a + 2)$
26. $3(x + y) + a(x + y)$
27. $a(x - 2) - b(2 - x)$
28. $3(a - 7) - b(7 - a)$
29. $x^2 + 3x + 2x + 6$
30. $x^2 - 5x + 4x - 20$
31. $xy + 4y - 2x - 8$
32. $ab + 7b - 3a - 21$
33. $ax + bx - ay - by$
34. $2ax - 3ay - 2bx + 3by$
35. $x^2y - 3x^2 - 2y + 6$
36. $a^2b + 3a^2 + 2b + 6$

37. $6 + 2y + 3x^2 + x^2y$

38. $15 + 3b - 5a^2 - a^2b$

39. $2ax^2 + bx^2 - 4ay - 2by$

40. $4a^2x + 2a^2y - 6bx - 3by$

41. $x^ny - 5x^n + y - 5$

42. $a^nx^n + 2a^n + x^n + 2$

43. $x^3 + x^2 + 2x + 2$

44. $y^3 - y^2 + 3y - 3$

45. $2x^3 - x^2 + 4x - 2$

46. $2y^3 - y^2 + 6y - 3$

3 Factor.

47. $x^2 - 8x + 15$

48. $x^2 + 12x + 20$

49. $a^2 + 12a + 11$

50. $a^2 + a - 72$

51. $b^2 + 2b - 35$

52. $a^2 + 7a + 6$

53. $y^2 - 16y + 39$

54. $y^2 - 18y + 72$

55. $b^2 + 4b - 32$

56. $x^2 + x - 132$

57. $a^2 - 15a + 56$

58. $x^2 + 15x + 50$

59. $y^2 + 13y + 12$

60. $b^2 - 6b - 16$

61. $x^2 + 4x - 5$

62. $a^2 - 3ab + 2b^2$

63. $a^2 + 11ab + 30b^2$

64. $a^2 + 8ab - 33b^2$

65. $x^2 - 14xy + 24y^2$

66. $x^2 + 5xy + 6y^2$

67. $y^2 + 2xy - 63x^2$

68. $2 + x - x^2$

69. $21 - 4x - x^2$

70. $5 + 4x - x^2$

71. $50 + 5a - a^2$

72. $x^2 - 5x + 6$

73. $x^2 - 7x - 12$

4 Factor.

74. $2x^2 + 7x + 3$

75. $2x^2 - 11x - 40$

76. $6y^2 + 5y - 6$

77. $4y^2 - 15y + 9$

78. $6b^2 - b - 35$

79. $2a^2 + 13a + 6$

80. $3y^2 - 22y + 39$

81. $12y^2 - 13y - 72$

82. $6a^2 - 26a + 15$

83. $5x^2 + 26x + 5$

84. $4a^2 - a - 5$

85. $11x^2 - 122x + 11$

86. $10x^2 - 7x - 12$

87. $12x^2 + 16x - 3$

88. $4a^2 + 7a - 15$

89. $12a^2 - 17a - 5$

90. $2x^2 + 17x + 35$

91. $15x^2 + 19x + 6$

92. $11y^2 - 47y + 12$

93. $12x^2 - 17x + 5$

94. $12x^2 - 40x + 25$

95. $8y^2 - 18y + 9$

96. $4x^2 + 9x + 10$

97. $6a^2 - 5a - 2$

98. $10x^2 - 29x + 10$

99. $2x^2 + 5x + 12$

100. $4x^2 - 6x + 1$

101. $6x^2 + 5xy - 21y^2$

102. $6x^2 + 41xy - 7y^2$

103. $4a^2 + 43ab + 63b^2$

104. $7a^2 + 46ab - 21b^2$

105. $10x^2 - 23xy + 12y^2$

106. $18x^2 + 27xy + 10y^2$

107. $24 + 13x - 2x^2$

108. $6 - 7x - 5x^2$

109. $8 - 13x + 6x^2$

110. $30 + 17a - 20a^2$

111. $15 - 14a - 8a^2$

112. $35 - 6b - 8b^2$

113. $12 - 5a - 28a^2$

114. $24 - 2x - 15x^2$

115. $15 - 44a - 20a^2$

116. $4x^3 - 10x^2 + 6x$

117. $9a^3 + 30a^2 - 24a$

118. $12y^3 + 22y^2 - 70y$

119. $5y^4 - 29y^3 + 20y^2$

120. $30a^2 + 85ab + 60b^2$

121. $4x^3 + 10x^2y - 24xy^2$

122. $8a^4 + 37a^3b - 15a^2b^2$

123. $100 - 5x - 5x^2$

124. $50x^2 + 25x^3 - 12x^4$

125. $320x - 8x^2 - 4x^3$

126. $96y - 16xy - 2x^2y$

127. $20x^2 - 38x^3 - 30x^4$

128. $4x^2y^2 - 32xy + 60$

129. $a^4b^4 - 3a^3b^3 - 10a^2b^2$

130. $2a^2b^4 + 9ab^3 - 18b^2$

131. $90a^2b^2 + 45ab + 10$

132. $3x^3y^2 + 12x^2y - 96x$

133. $4x^4 - 45x^2 + 80$

134. $x^4 + 2x^2 + 15$

135. $2a^5 + 14a^3 + 20a$

136. $3b^6 - 9b^4 - 30b^2$

137. $3x^4y^2 - 39x^2y^2 + 120y^2$

138. $2x^4y - 7x^3y - 30x^2y$

139. $45a^2b^2 + 6ab^2 - 72b^2$

140. $16x^2y^3 + 36x^2y^2 + 20x^2y$

141. $36x^3y + 24x^2y^2 - 45xy^3$

142. $12a^3b - 70a^2b^2 - 12ab^3$

143. $48a^2b^2 - 36ab^3 - 54b^4$

144. $x^{3n} + 10x^{2n} + 16x^n$

145. $10x^{2n} + 25x^n - 60$

SUPPLEMENTAL EXERCISES 3.4

Factor.

146. $3y^2 - 2 - 5y$

147. $8b^2 - 12 - 29b$

148. $4a + 3a^3 + 13a^2$

149. $4p^2 - 3p + 4p^3$

150. $2a^3b - ab^3 - a^2b^2$

151. $3x^3y - xy^3 - 2x^2y^2$

152. $2y^3 + 2y^5 - 24y$

153. $9b^3 + 3b^5 - 30b$

154. $3(p + 1)^2 - (p + 1) - 2$

155. $4(a - 1)^2 + 7(a - 1) - 2$

156. $6(b - 2)^2 + (b - 2) - 2$

157. $12(y + 2)^2 + 5(y + 2) - 3$

158. $4y(x - 1)^2 + 5y(x - 1) - 6y$

159. $5b(a - 2)^2 - 11b(a - 2) - 12b$

Find all integers k such that the trinomial can be factored over the integers.

160. $x^2 + kx + 8$

161. $x^2 + kx - 6$

162. $x^2 + kx + 20$

163. $x^2 + kx - 30$

164. $2x^2 - kx + 3$

165. $2x^2 - kx - 5$

166. $3x^2 + kx + 5$

167. $2x^2 + kx - 3$

168. $6x^2 + kx + 1$

169. $4x^2 + kx - 1$

170. $7x^4 - kx^2 + 3$

171. $3x^4 - kx^2 - 2$

SECTION 3.5 Special Factoring

1 Factor the difference of two perfect squares and perfect square trinomials

The product of a term and itself is called a **perfect square.** The exponents on variables of perfect squares are always even numbers.

Term		Perfect Square
5	$5 \cdot 5 =$	25
x	$x \cdot x =$	x^2
$3y^4$	$3y^4 \cdot 3y^4 =$	$9y^8$
x^n	$x^n \cdot x^n =$	x^{2n}

The **square root** of a perfect square is one of the two equal factors of the perfect square. $\sqrt{}$ is the symbol for square root. To find the exponent of the square root of a perfect square variable term, divide the exponent by 2.

$\sqrt{25} = 5$
$\sqrt{x^2} = x$
$\sqrt{9y^8} = 3y^4$
$\sqrt{x^{2n}} = x^n$

The difference of two perfect squares is the product of the sum and the difference of two terms. The factors of the difference of two squares are the sum and difference of the square roots of the perfect squares.

Sum and Difference of Two Terms		**Difference of Two Perfect Squares**
$(a + b)(a - b)$	$=$	$a^2 - b^2$

The expression $4x^2 - 81y^2$ is the difference of two perfect squares.

$4x^2 - 81y^2 =$
$(2x)^2 - (9y)^2 =$

The factors are the sum and difference of the square roots of the perfect squares.

$(2x + 9y)(2x - 9y)$

The sum of two perfect squares, $a^2 + b^2$, is nonfactorable over the integers.

Example 1 Factor: $25x^2 - 1$

Solution $25x^2 - 1 = (5x)^2 - 1^2$ ► Write the binomial as the difference of two perfect squares.

$= (5x + 1)(5x - 1)$ ► The factors are the sum and difference of the square roots of the perfect squares.

Problem 1 Factor: $x^2 - 36y^4$

Solution See page A20.

A perfect square trinomial is the square of a binomial.

Square of a Binomial				**Perfect Square Trinomial**
$(a + b)^2$	$=$	$(a + b)(a + b)$	$=$	$a^2 + 2ab + b^2$
$(a - b)^2$	$=$	$(a - b)(a - b)$	$=$	$a^2 - 2ab + b^2$

In factoring a perfect square trinomial, remember that the terms of the binomial are the square roots of the perfect squares of the trinomial. The sign in the binomial is the sign of the middle term of the trinomial.

Factor: $x^2 - 14x + 49$

The trinomial is a perfect square.
Write the factors as the square of a binomial. $x^2 - 14x + 49 = (x - 7)^2$

Example 2 Factor: $4x^2 - 20x + 25$

Solution $4x^2 - 20x + 25 = (2x - 5)^2$

Problem 2 Factor: $9x^2 + 12x + 4$

Solution See page A20.

2 Factor the sum or the difference of two cubes

The product of the same three factors is called a **perfect cube.** The exponents on variables of perfect cubes are always divisible by 3.

Term		Perfect Cube
2	$2 \cdot 2 \cdot 2 =$	8
$3y$	$3y \cdot 3y \cdot 3y =$	$27y^3$
x^2	$x^2 \cdot x^2 \cdot x^2 =$	x^6
x^n	$x^n \cdot x^n \cdot x^n =$	x^{3n}

The **cube root** of a perfect cube is one of the three equal factors of the perfect cube. $\sqrt[3]{}$ is the symbol for cube root. To find the exponent of the cube root of a perfect cube variable term, divide the exponent by 3.

$$\sqrt[3]{8} = 2$$
$$\sqrt[3]{27y^3} = 3y$$
$$\sqrt[3]{x^6} = x^2$$
$$\sqrt[3]{x^{3n}} = x^n$$

The Sum or Difference of Two Cubes

The **sum or the difference of two cubes** is the product of a binomial and a trinomial.

$$a^3 + b^3 = (a + b)(a^2 - ab + b^2)$$
$$a^3 - b^3 = (a - b)(a^2 + ab + b^2)$$

Factor: $8x^3 - 27$

Write the binomial as the difference of two perfect cubes.

$$8x^3 - 27 = (2x)^3 - 3^3$$

The terms of the binomial factor are the cube roots of the perfect cubes. The sign of the binomial factor is the same sign as in the given binomial. The trinomial factor is obtained from the binomial factor.

$$= (2x - 3)(4x^2 + 6x + 9)$$

Square of the first term ($4x^2$)

Opposite of the product of the two terms ($6x$)

Square of the last term (9)

Factor: $a^3 + 64y^3$

Write the binomial as the sum of two perfect cubes. $a^3 + 64y^3 = a^3 + (4y)^3 =$

Write the binomial factor and the trinomial factor. $(a + 4y)(a^2 - 4ay + 16y^2)$

Example 3 Factor.

A. $x^3y^3 - 1$ B. $(x + y)^3 - x^3$

Solution A. $x^3y^3 - 1 = (xy)^3 - 1^3$
$= (xy - 1)(x^2y^2 + xy + 1)$

▶ Write the binomial as the difference of two perfect cubes.

B. $(x + y)^3 - x^3 =$

▶ This is the difference of two perfect cubes.

$[(x + y) - x][(x + y)^2 + x(x + y) + x^2] =$
$y(x^2 + 2xy + y^2 + x^2 + xy + x^2) =$
$y(3x^2 + 3xy + y^2)$

▶ Simplify.

Problem 3 Factor.

A. $8x^3 + y^3z^3$ B. $(x - y)^3 + (x + y)^3$

Solution See page A20.

3 Factor trinomials that are quadratic in form

Certain trinomials can be expressed as quadratic trinomials by making suitable variable substitutions. A trinomial is quadratic in form if it can be written as

$$au^2 + bu + c.$$

By letting $x^2 = u$, the trinomial at the right can be written: $x^4 + 5x^2 + 6$
$(x^2)^2 + 5(x^2) + 6$

The trinomial is quadratic in form. $u^2 + 5u + 6$

By letting $xy = u$, the trinomial at the right can be written: $2x^2y^2 + 3xy - 9$
$2(xy)^2 + 3(xy) - 9$

The trinomial is quadratic in form. $2u^2 + 3u - 9$

When a trinomial that is quadratic in form is factored, the variable part of the first term in each binomial factor will be u. For example, because $x^4 + 5x^2 + 6$ is quadratic in form when $x^2 = u$, the first term in each binomial factor will be x^2.

$$x^4 + 5x^2 + 6 = (x^2)^2 + 5(x^2) + 6$$
$$= (x^2 + 2)(x^2 + 3)$$

The trinomial $x^2y^2 - 2xy - 15$ is quadratic in form when $xy = u$. The first term in each binomial factor will be xy.

$$x^2y^2 - 2xy - 15 = (xy)^2 - 2(xy) - 15$$
$$= (xy + 3)(xy - 5)$$

Example 4 Factor.

A. $6x^2y^2 - xy - 12$ B. $2x^4 + 5x^2 - 12$

Solution A. $6x^2y^2 - xy - 12 =$ $(3xy + 4)(2xy - 3)$ ▶ The trinomial is quadratic in form when $xy = u$.

B. $2x^4 + 5x^2 - 12 =$ $(x^2 + 4)(2x^2 - 3)$ ▶ The trinomial is quadratic in form when $x^2 = u$.

Problem 4 Factor.

A. $6x^2y^2 - 19xy + 10$ B. $3x^4 + 4x^2 - 4$

Solution See page A20.

4 Factor completely

When factoring a polynomial completely, ask yourself the following questions about the polynomial:

1. Is there a common factor? If so, factor out the GCF.
2. If the polynomial is a binomial, is it the difference of two perfect squares, the sum of two cubes, or the difference of two cubes? If so, factor.
3. If the polynomial is a trinomial, is it a perfect square trinomial or the product of two binomials? If so, factor.
4. Can the polynomial be factored by grouping? If so, factor.
5. Is each factor nonfactorable over the integers? If not, factor.

Example 5 Factor.

A. $x^2y + 2x^2 - y - 2$ B. $x^{4n} - y^{4n}$

Solution A. $x^2y + 2x^2 - y - 2 =$
$(x^2y + 2x^2) - (y + 2) =$ ▶ Factor by grouping.
$x^2(y + 2) - (y + 2) =$
$(y + 2)(x^2 - 1) =$
$(y + 2)(x + 1)(x - 1)$ ▶ Factor the difference of two perfect squares.

B. $x^{4n} - y^{4n} =$
$(x^{2n})^2 - (y^{2n})^2 =$
$(x^{2n} + y^{2n})(x^{2n} - y^{2n}) =$ ▶ Factor the difference of two perfect squares.
$(x^{2n} + y^{2n})[(x^n)^2 - (y^n)^2] =$
$(x^{2n} + y^{2n})(x^n + y^n)(x^n - y^n)$ ▶ Factor the difference of two perfect squares.

Problem 5 Factor.

A. $4x - 4y - x^3 + x^2y$ B. $x^{4n} - x^{2n}y^{2n}$

Solution See page A20.

EXERCISES 3.5

1 Factor.

1. $x^2 - 16$
2. $y^2 - 49$
3. $4x^2 - 1$
4. $81x^2 - 4$
5. $b^2 - 2b + 1$
6. $a^2 + 14a + 49$
7. $16x^2 - 40x + 25$
8. $49x^2 + 28x + 4$
9. $x^2y^2 - 100$
10. $a^2b^2 - 25$
11. $x^2 + 4$
12. $a^2 + 16$
13. $x^2 + 6xy + 9y^2$
14. $4x^2y^2 + 12xy + 9$
15. $4x^2 - y^2$
16. $49a^2 - 16b^4$
17. $a^{2n} - 1$
18. $b^{2n} - 16$
19. $a^2 + 4a + 4$
20. $b^2 - 18b + 81$
21. $x^2 - 12x + 36$
22. $y^2 - 6y + 9$
23. $16x^2 - 121$
24. $49y^2 - 36$
25. $1 - 9a^2$
26. $16 - 81y^2$
27. $4a^2 + 4a - 1$
28. $9x^2 + 12x - 4$
29. $b^2 + 7b + 14$
30. $y^2 - 5y + 25$
31. $25 - a^2b^2$
32. $64 - x^2y^2$
33. $25a^2 - 40ab + 16b^2$
34. $4a^2 - 36ab + 81b^2$
35. $x^{2n} + 6x^n + 9$
36. $y^{2n} - 16y^n + 64$

2 Factor.

37. $x^3 - 27$
38. $y^3 + 125$
39. $8x^3 - 1$
40. $64a^3 + 27$

41. $x^3 - y^3$

42. $x^3 - 8y^3$

43. $m^3 + n^3$

44. $27a^3 + b^3$

45. $64x^3 + 1$

46. $1 - 125b^3$

47. $27x^3 - 8y^3$

48. $64x^3 + 27y^3$

49. $x^3y^3 + 64$

50. $8x^3y^3 + 27$

51. $16x^3 - y^3$

52. $27x^3 - 8y^3$

53. $8x^3 - 9y^3$

54. $27a^3 - 16$

55. $(a - b)^3 - b^3$

56. $a^3 + (a + b)^3$

57. $x^{6n} + y^{3n}$

58. $x^{3n} + y^{3n}$

59. $x^{3n} + 8$

60. $a^{3n} + 64$

3 Factor.

61. $x^2y^2 - 8xy + 15$

62. $x^2y^2 - 8xy - 33$

63. $x^2y^2 - 17xy + 60$

64. $a^2b^2 + 10ab + 24$

65. $x^4 - 9x^2 + 18$

66. $y^4 - 6y^2 - 16$

67. $b^4 - 13b^2 - 90$

68. $a^4 + 14a^2 + 45$

69. $x^4y^4 - 8x^2y^2 + 12$

70. $a^4b^4 + 11a^2b^2 - 26$

71. $x^{2n} + 3x^n + 2$

72. $a^{2n} - a^n - 12$

73. $3x^2y^2 - 14xy + 15$

74. $5x^2y^2 - 59xy + 44$

75. $6a^2b^2 - 23ab + 21$

76. $10a^2b^2 + 3ab - 7$

77. $2x^4 - 13x^2 - 15$

78. $3x^4 + 20x^2 + 32$

79. $2x^{2n} - 7x^n + 3$

80. $4x^{2n} + 8x^n - 5$

81. $6a^{2n} + 19a^n + 10$

4 Factor.

82. $5x^2 + 10x + 5$

83. $12x^2 - 36x + 27$

84. $3x^4 - 81x$

85. $27a^4 - a$

86. $7x^2 - 28$

87. $20x^2 - 5$

88. $y^4 - 10y^3 + 21y^2$

89. $y^5 + 6y^4 - 55y^3$

90. $x^4 - 16$

91. $16x^4 - 81$

92. $8x^5 - 98x^3$

93. $16a - 2a^4$

94. $x^3y^3 - x^3$

95. $x^3 + 2x^2 - x - 2$

96. $2x^3 - 3x^2 - 8x + 12$

97. $2x^3 + 4x^2 - 3x - 6$

98. $3x^3 - 3x^2 + 4x - 4$

99. $x^3 + x^2 - 16x - 16$

100. $4x^3 + 8x^2 - 9x - 18$

101. $a^3b^6 - b^3$

102. $x^6y^6 - x^3y^3$

103. $x^4 - 2x^3 - 35x^2$

104. $x^4 + 15x^3 - 56x^2$

105. $4x^2 + 4x - 1$

106. $8x^4 - 40x^3 + 50x^2$

107. $6x^5 + 74x^4 + 24x^3$

108. $x^4 - y^4$

109. $16a^4 - b^4$

110. $x^6 + y^6$

111. $x^4 - 5x^2 - 4$

112. $a^4 - 25a^2 - 144$

113. $3b^5 - 24b^2$

114. $16a^4 - 2a$

115. $x^4y^2 - 5x^3y^3 + 6x^2y^4$

116. $a^4b^2 - 8a^3b^3 - 48a^2b^4$

117. $16x^3y + 4x^2y^2 - 42xy^3$

118. $24a^2b^2 - 14ab^3 - 90b^4$

119. $x^3 - 2x^2 - x + 2$

120. $x^3 - 2x^2 - 4x + 8$

121. $8xb - 8x - 4b + 4$

122. $4xy + 8x + 4y + 8$

123. $4x^2y^2 - 4x^2 - 9y^2 + 9$

124. $4x^4 - x^2 - 4x^2y^2 + y^2$

125. $x^5 - 4x^3 - 8x^2 + 32$

126. $x^6y^3 + x^3 - x^3y^3 - 1$

127. $a^{2n+2} - 6a^{n+2} + 9a^2$

128. $x^{2n+1} + 2x^{n+1} + x$

129. $2x^{n+2} - 7x^{n+1} + 3x^n$

130. $3b^{n+2} + 4b^{n+1} - 4b^n$

SUPPLEMENTAL EXERCISES 3.5

Find all integers k such that the trinomial is a perfect square trinomial.

131. $x^2 + kx + 36$

132. $x^2 - kx + 81$

133. $4x^2 - kx + 25$

134. $9x^2 - kx + 1$

135. $16x^2 + kxy + y^2$

136. $49x^2 + kxy + 64y^2$

Factor.

137. $ax^3 + b - bx^3 - a$

138. $xy^2 - 2b - x + 2by^2$

139. $(p - 1)^2 - 6(p - 1) + 9$

140. $4(r - 1)^2 - 4(r - 1) + 1$

141. $(a - 3)^2 - (a + 2)^2$

142. $(b + 4)^2 - (b - 5)^2$

143. $y^{8n} - 2y^{4n} + 1$

144. $x^{6n} - 1$

145. $(y^2 - y - 6)^2 - (y^2 + y - 2)^2$

146. $(a^2 + a - 6)^2 - (a^2 + 2a - 8)^2$

Solve.

147. The product of two numbers is 63. One of the two numbers is a perfect square. The other is a prime number. Find the sum of the two numbers.

148. What is the smallest whole number by which 250 can be multiplied so that the product will be a perfect square?

149. Palindromic numbers are natural numbers that remain unchanged when their digits are written in reverse order. Find all perfect squares less than 500 that are palindromic numbers.

150. A circular cookie is cut from a square piece of dough. The diameter of the cookie is x cm. The piece of dough is x cm on a side and is 1 cm deep. In terms of x, how many cubic centimeters of dough are left over? Use 3.14 for π.

SECTION 3.6 Solving Equations by Factoring

1 Solve equations by factoring

Consider the equation $ab = 0$. If a is not zero, then b must be zero. Conversely, if b is not zero, then a must be zero. This is summarized in the Principle of Zero Products.

Principle of Zero Products

If the product of two factors is zero, then at least one of the factors must be zero.

If $ab = 0$, then $a = 0$ or $b = 0$.

The Principle of Zero Products is used to solve equations.

Solve: $(x - 4)(x + 2) = 0$

By the Principle of Zero Products, if $(x - 4)(x + 2) = 0$, then $x - 4 = 0$ or $x + 2 = 0$. Solve each equation for x.

$$\begin{aligned} x - 4 &= 0 \\ x &= 4 \end{aligned} \qquad\qquad \begin{aligned} x + 2 &= 0 \\ x &= -2 \end{aligned}$$

Check:

$$\begin{array}{r|l} (x-4)(x+2) & 0 \\ \hline (4-4)(4+2) & 0 \\ 0 \cdot 6 & 0 \end{array} \qquad \begin{array}{r|l} (x-4)(x+2) & 0 \\ \hline (-2-4)(-2+2) & 0 \\ -6 \cdot 0 & 0 \end{array}$$

$$0 = 0 \qquad\qquad 0 = 0$$

-2 and 4 check as solutions. The solutions are -2 and 4.

Quadratic Equation

An equation of the form $ax^2 + bx + c = 0$, $a \neq 0$, is a quadratic equation.

A quadratic equation is in standard form when the polynomial is in descending order and equal to zero. The quadratic equations at the right are in standard form.

$$\begin{aligned} 2x^2 + 3x + 1 &= 0 \\ 5x^2 - 2x &= 0 \\ 3x^2 - 9 &= 0 \end{aligned}$$

Some quadratic equations can be solved by factoring and then using the Principle of Zero Products.

Solve: $2x^2 - x = 1$

$$2x^2 - x = 1$$

Write the equation in standard form. $\quad 2x^2 - x - 1 = 0$

Factor the trinomial. $\quad (2x + 1)(x - 1) = 0$

Use the Principle of Zero Products. $\quad 2x + 1 = 0 \qquad x - 1 = 0$

Solve each equation for x. $\quad 2x = -1 \qquad x = 1$

$$x = -\frac{1}{2}$$

The solutions are $-\frac{1}{2}$ and 1.

The Principle of Zero Products can be extended to more than two factors. For example, if $abc = 0$, then $a = 0$, $b = 0$, or $c = 0$.

Solve: $x^3 - x^2 - 4x + 4 = 0$

Factor by grouping.

$$x^3 - x^2 - 4x + 4 = 0$$
$$(x^3 - x^2) - (4x - 4) = 0$$
$$x^2(x - 1) - 4(x - 1) = 0$$
$$(x - 1)(x^2 - 4) = 0$$
$$(x - 1)(x + 2)(x - 2) = 0$$

Use the Principle of Zero Products. $\quad x - 1 = 0 \qquad x + 2 = 0 \qquad x - 2 = 0$

Solve each equation for x. $\quad x = 1 \qquad x = -2 \qquad x = 2$

The solutions are -2, 1, and 2.

Example 1 Solve.

A. $3x^2 + 5x = 2$ B. $(x + 4)(x - 3) = 8$ C. $x^3 - x^2 - 25x + 25 = 0$

Solution A. $3x^2 + 5x = 2$

$3x^2 + 5x - 2 = 0$ ▶ Write the equation in standard form.

$(3x - 1)(x + 2) = 0$ ▶ Factor the trinomial.

$3x - 1 = 0 \qquad x + 2 = 0$ ▶ Let each factor equal zero (the Principle of Zero Products).

$3x = 1 \qquad x = -2$ ▶ Solve each equation for x.

$x = \frac{1}{3}$

The solutions are $\frac{1}{3}$ and -2. ▶ Write the solutions.

B. $(x + 4)(x - 3) = 8$

$x^2 + x - 12 = 8$ ▶ Write the equation in standard form by first multiplying the binomials.

$x^2 + x - 20 = 0$ ▶ Subtract 8 from each side of the equation.

$(x + 5)(x - 4) = 0$ ▶ Factor the trinomial.

$x + 5 = 0 \qquad x - 4 = 0$ ▶ Let each factor equal zero.

$x = -5 \qquad x = 4$ ▶ Solve each equation for x.

The solutions are -5 and 4. ▶ Write the solutions.

C. $x^3 - x^2 - 25x + 25 = 0$

$(x^3 - x^2) - (25x - 25) = 0$ ▶ Factor by grouping.

$x^2(x - 1) - 25(x - 1) = 0$

$(x^2 - 25)(x - 1) = 0$

$(x + 5)(x - 5)(x - 1) = 0$

$x + 5 = 0 \quad x - 5 = 0 \quad x - 1 = 0$ ▶ Let each factor equal zero.

$x = -5 \quad x = 5 \quad x = 1$ ▶ Solve each equation for x.

The solutions are −5, 1, and 5. ▶ Write the solutions.

Problem 1 Solve.

A. $4x^2 + 11x = 3$ B. $(x - 2)(x + 5) = 8$ C. $x^3 + 4x^2 - 9x - 36 = 0$

Solution See page A20.

2 Application problems

Example 2 The sum of the squares of two consecutive positive odd integers is 130. Find the two integers.

Strategy First positive odd integer: n

Second positive odd integer: $n + 2$

The sum of the square of the first positive odd integer and the square of the second positive odd integer is 130.

Solution

$n^2 + (n + 2)^2 = 130$

$n^2 + n^2 + 4n + 4 = 130$

$2n^2 + 4n - 126 = 0$

$2(n^2 + 2n - 63) = 0$

$2(n + 9)(n - 7) = 0$

$n + 9 = 0 \quad n - 7 = 0$

$n = -9 \quad n = 7$ ▶ −9 is not a positive odd integer. Therefore, it is not a solution.

$n + 2 = 7 + 2 = 9$ ▶ Substitute the value of n into the variable expression for the second positive odd integer and evaluate.

The two integers are 7 and 9.

Problem 2 The length of a rectangle is 5 in. longer than the width. The area of the rectangle is 66 in.^{2}. Find the length and width of the rectangle.

Solution See pages A20–A21.

EXERCISES 3.6

1 Solve.

1. $(y + 4)(y + 6) = 0$

2. $(a - 5)(a - 2) = 0$

3. $x(x - 7) = 0$

4. $b(b + 8) = 0$

5. $3z(2z + 5) = 0$

6. $4y(3y - 2) = 0$

7. $(2x + 3)(x - 7) = 0$

8. $(4a - 1)(a + 9) = 0$

9. $b^2 - 49 = 0$

10. $4z^2 - 1 = 0$

11. $9t^2 - 16 = 0$

12. $x^2 + x - 6 = 0$

13. $y^2 + 4y - 5 = 0$

14. $a^2 - 8a + 16 = 0$

15. $2b^2 - 5b - 12 = 0$

16. $t^2 - 8t = 0$

17. $x^2 - 9x = 0$

18. $2y^2 - 10y = 0$

19. $3a^2 - 12a = 0$

20. $b^2 - 4b = 32$

21. $z^2 - 3z = 28$

22. $2x^2 - 5x = 12$

23. $3t^2 + 13t = 10$

24. $4y^2 - 19y = 5$

25. $5b^2 - 17b = -6$

26. $6a^2 + a = 2$

27. $8x^2 - 10x = 3$

28. $z(z - 1) = 20$

29. $y(y - 2) = 35$

30. $t(t + 1) = 42$

31. $x(x - 12) = -27$

32. $x(2x - 5) = 12$

33. $y(3y - 2) = 8$

34. $2b^2 - 6b = b - 3$

35. $3a^2 - 4a = 20 - 15a$

36. $2t^2 + 5t = 6t + 15$

37. $(y + 5)(y - 7) = -20$

38. $(x + 2)(x - 6) = 20$

39. $(b + 5)(b + 10) = 6$

40. $(a - 9)(a - 1) = -7$

41. $(t - 3)^2 = 1$

42. $(y - 4)^2 = 4$

43. $(3 - x)^2 + x^2 = 5$

44. $(2 - b)^2 + b^2 = 10$

45. $(a - 1)^2 = 3a - 5$

46. $2x^3 + x^2 - 8x - 4 = 0$

47. $x^3 + 4x^2 - x - 4 = 0$

48. $12x^3 - 8x^2 - 3x + 2 = 0$

49. $4x^3 + 4x^2 - 9x - 9 = 0$

$x^2 + x = 132$

2 Solve.

50. The sum of a number and its square is 90. Find the number.

51. The sum of a number and its square is 132. Find the number.

52. The square of a number is 12 more than four times the number. Find the number.

53. The square of a number is 195 more than twice the number. Find the number.

54. The sum of the squares of two consecutive positive integers is equal to 145. Find the two integers.

55. The sum of the squares of two consecutive positive odd integers is equal to 290. Find the two integers.

56. The sum of the cube of a number and the product of the number and twelve is equal to seven times the square of the number. Find the number.

57. The sum of the cube of a number and the product of the number and seven is equal to eight times the square of the number. Find the number.

58. The length of a rectangle is 5 in. more than twice the width. The area is 168 in.^{2}. Find the width and length of the rectangle.

59. The width of a rectangle is 5 ft less than the length. The area of the rectangle is 300 ft^2. Find the length and width of the rectangle.

60. The length of the base of a triangle is three times the height. The area of the triangle is 24 cm^2. Find the base and height of the triangle.

61. The height of a triangle is 4 cm more than twice the length of the base. The area of the triangle is 35 cm^2. Find the height of the triangle.

62. An object is thrown downward, with an initial speed of 16 ft/s, from the top of a building 480 ft high. How many seconds later will the object hit the ground? Use the equation $d = vt + 16t^2$, where d is the distance in feet, v is the initial speed, and t is the time in seconds.

63. A stone is thrown into a well with an initial speed of 8 ft/s. The well is 624 ft deep. How many seconds later will the stone hit the bottom of the well? Use the equation $d = vt + 16t^2$, where d is the distance in feet, v is the initial speed, and t is the time in seconds.

64. The length of a rectangle is 6 cm, and the width is 3 cm. If both the length and the width are increased by equal amounts, the area of the rectangle is increased by 70 cm^2. Find the length and width of the larger rectangle.

65. The width of a rectangle is 4 cm, and the length is 8 cm. If both the width and the length are increased by equal amounts, the area of the rectangle is increased by 64 cm^2. Find the length and width of the larger rectangle.

SUPPLEMENTAL EXERCISES 3.6

Solve for x in terms of a.

66. $x^2 + 3ax - 10a^2 = 0$

67. $x^2 + 4ax - 21a^2 = 0$

68. $x^2 - 9a^2 = 0$

69. $x^2 - 16a^2 = 0$

70. $x^2 + 10ax + 25a^2 = 0$

71. $x^2 - 12ax + 36a^2 = 0$

72. $2x^2 + ax - 3a^2 = 0$

73. $3x^2 + 4ax - 4a^2 = 0$

74. $x^2 - 5ax = 15a^2 - 3ax$

75. $x^2 + 3ax = 8ax + 24a^2$

Solve.

76. Find $3n^2 + 2n - 1$ if $n(n + 6) = 16$.

77. Find $2n^3 - 3n + 1$ if $n(n - 2) = 15$.

78. The perimeter of a rectangular garden is 28 ft. The area of the garden is 40 ft^2. Find the length and width of the garden.

79. The perimeter of a rectangular garden is 44 m. The area of the garden is 120 m^2. Find the length and width of the garden.

80. A rectangular piece of cardboard is 10 in. longer than it is wide. Squares 2 in. on a side are to be cut from each corner and then the sides folded up to make an open box with a volume of 112 $in.^3$. Find the length and width of the piece of cardboard.

81. A rectangular piece of cardboard is 12 in. longer than it is wide. Squares 2 in. on a side are to be cut from each corner and then the sides folded up to make an open box with a volume of 216 $in.^3$. Find the length and width of the piece of cardboard.

82. The sides of a rectangular box have areas of 18 cm^2, 24 cm^2, and 48 cm^2. Find the volume of the box.

83. The sides of a rectangular box have areas of 16 cm^2, 20 cm^2, and 80 cm^2. Find the volume of the box.

Calculators and Computers

General Factoring

The program GENERAL FACTORING on the Math ACE Disk can be used to practice factoring. You may choose to practice polynomials of the form

$$x^2 + bx + c, \qquad ax^2 + bx + c, \quad \text{or} \quad x^3 + a^3.$$

These choices, along with the option of quitting the program, are given on a menu screen. You may practice for as long as you like with any type of problem. At the end of each problem, you may either return to the menu screen or continue practicing.

The program will present you with a polynomial to factor. When you have tried to factor the polynomial using paper and pencil, press the RETURN key on the keyboard. The correct factorization will be displayed.

Something Extra

Reverse Polish Notation

Following the Order of Operations Agreement is sometimes referred to as algebraic logic. It is a system used by many calculators. Algebraic logic is not the only system in use by calculators. Another system is called RPN logic. RPN stands for Reverse Polish Notation.

During the 1950s, Jan Lukasiewicz, a Polish logician, developed a parenthesis-free notational system for writing mathematical expressions. In this system, the operator (such as addition, multiplication, or division) follows the operands (the numbers to be added, multiplied, or divided). For RPN calculators, an [ENTER] key is used to temporarily store a number until another number and the operation can be entered. Here are some examples along with the algebraic logic equivalent.

To find	RPN Logic	Algebraic Logic
1. $3 + 4$	3 [ENTER] 4 [+]	3 [+] 4 [=]
2. $5 \times 6 \times 7$	5 [ENTER] 6 [x] 7 [x]	5 [x] 6 [x] 7 [=]
3. $4 \times (7 + 3)$	4 [ENTER] 7 [ENTER] 3 [+] [x]	4 [x] [(] 7 [+] 3 [)] [=]
4. $(3 + 4) \times (5 + 2)$	3 [ENTER] 4 [+] 5 [ENTER] 2 [+] [x]	[(] 3 [+] 4 [)] [x] [(] 5 [+] [2] [)] [=]

The Examples 3 and 4 above illustrate the concept of being "parenthesis-free." Note that the examples under RPN logic do not require parentheses, while those under algebraic logic do. If the parentheses keys are not used for Examples 3 and 4, a calculator, following algebraic logic, would use the Order of Operations Agreement. The result in Example 3 would be

$$4 \times 7 + 3 = 28 + 3 = 31 \qquad \text{instead of} \qquad 4 \times (7 + 3) = 4 \times 10 = 40.$$

This system may seem quite strange, but it is actually very efficient. A glimpse of its efficiency can be seen from Example 4. For the RPN operation, 9 keys were pushed, while 12 keys were pushed for the algebraic operation. Note also that RPN logic does not use the "equal" key. The result of an operation is displayed after the operation key is pressed.

Try the following exercises that ask you to change from algebraic logic to RPN logic or to evaluate an RPN logic expression.

Change to RPN logic.

1. $12 \div 6$

2. $7 \times 5 + 6$

3. $(9 + 3) \div 4$

4. $(5 + 7) \div (2 + 4)$

5. $6 \times 7 \times 10$

6. $(1 + 4 \times 5) \div 7$

Evaluate each of the following by using RPN logic.

7. 18 [ENTER] 2 [÷]

8. 6 [ENTER] 5 [ENTER] 3 [x] [+]

9. 3 [ENTER] 5 [x] 4 [+]

10. 7 [ENTER] 4 [ENTER] 5 [x] [+] 3 [÷]

11. 228 [ENTER] 6 [ENTER] 9 [ENTER] 12 [x] [+] [÷]

12. 1 [ENTER] 1 [ENTER] 1 [ENTER] 3 [+] [÷] [+]

Chapter Summary

Key Words

A *monomial* is a number, a variable, or a product of numbers and variables.

A *polynomial* is a variable expression in which the terms are monomials.

A polynomial of *two* terms is a *binomial.*

A polynomial of *three* terms is a *trinomial.*

The *degree of a polynomial* is the greatest of the degrees of any of its terms.

Synthetic division is a shorter method of dividing a polynomial by a binomial of the form $x - a$. (This method uses only the coefficients of the variable terms.)

A number written in *scientific notation* is a number written in the form $a \times 10^n$, where $1 \leq a < 10$.

To *factor a polynomial* means to write the polynomial as the product of other polynomials.

A *quadratic trinomial* is a polynomial of the form $ax^2 + bx + c$, where a, b, and c are nonzero constants.

A polynomial is *nonfactorable over the integers* if it does not factor using only integers.

To *factor a quadratic trinomial* of the form $ax^2 + bx + c$ means to express the trinomial as the product of two binomials.

The product of a term and itself is a *perfect square.*

The *square root* of a perfect square is one of the two equal factors of the perfect square.

The product of the same three factors is called a *perfect cube.*

The *cube root* of a perfect cube is one of the three equal factors of the perfect cube.

A *quadratic equation* is an equation of the form $ax^2 + bx + c = 0$, where $a \neq 0$. A quadratic equation is in standard form when the polynomial is in descending order and equal to zero.

Essential Rules

Rule for Multiplying Exponential Expressions	If m and n are integers, then $x^m \cdot x^n = x^{m+n}$.
Rule for Simplifying Powers of Exponential Expressions	If m and n are integers, then $(x^m)^n = x^{mn}$.
Rule for Dividing Exponential Expressions	If m and n are integers and $x \neq 0$, then $\frac{x^m}{x^n} = x^{m-n}$.
Rule for Simplifying Powers of Products	If m, n, and p are integers, then $(x^m \cdot y^n)^p = x^{mp}y^{np}$.
Rule for Simplifying Powers of Quotients	If m, n, and p are positive integers and $y \neq 0$, then $\left(\frac{x^m}{y^n}\right)^p = \frac{x^{mp}}{y^{np}}$.
Rule of Negative Exponents	If $n > 0$ and $x \neq 0$, then $x^{-n} = \frac{1}{x^n}$ and $x^n = \frac{1}{x^{-n}}$.
Zero as an Exponent	If $x \neq 0$, then $x^0 = 1$.
Difference of Two Perfect Squares	$a^2 - b^2 = (a + b)(a - b)$
Perfect Square Trinomial	$a^2 + 2ab + b^2 = (a + b)^2$
The Sum or the Difference of Two Cubes	$a^3 + b^3 = (a + b)(a^2 - ab + b^2)$ $a^3 - b^3 = (a - b)(a^2 + ab + b^2)$
The Principle of Zero Products	If $ab = 0$, then $a = 0$ or $b = 0$.

Chapter Review

1. Simplify:
$(3x^2 - 2x - 6) + (-x^2 - 3x + 4)$

2. Simplify:
$(5x^2 - 8xy + 2y^2) - (x^2 - 3y^2)$

3. Simplify:
$(3x^{2n} - 4x^n + 9) - (x^{2n} + 7x^n - 4)$

4. Simplify:
$(5x^2yz^4)(2xy^3z^{-1})(7x^{-2}y^{-2}z^3)$

5. Simplify: $(-2a^2b^4)^3(3ab^{-2})$

6. Simplify: $(2x^{-1}y^2z^5)^4(-3x^3yz^{-3})$

7. Simplify: $x^{-1}y + y^{-1}x$

8. Simplify: $\dfrac{3x^4yz^{-1}}{-12xy^3z^2}$

9. Simplify: $\dfrac{(2a^4b^{-3}c^2)^3}{(2a^3b^2c^{-1})^4}$

10. Write 93,000,000 in scientific notation.

11. Write 2.54×10^{-3} in decimal notation.

12. Simplify: $\dfrac{3 \times 10^{-3}}{15 \times 10^2}$

13. Simplify: $\dfrac{15x^2 + 2x - 2}{3x - 2}$

14. Simplify: $\dfrac{12x^2 - 16x - 7}{6x + 1}$

15. Simplify: $\dfrac{20x^2 - 23x}{5x - 7}$

16. Simplify: $\dfrac{x^2 + x + 1}{x - 3}$

17. Simplify: $\dfrac{4x^3 + 27x^2 + 10x + 2}{x + 6}$

18. Simplify: $\dfrac{x^4 - 4}{x - 4}$

19. Simplify:
$4x^2y(3x^3y^2 + 2xy - 7y^3)$

20. Simplify:
$a^{2n+3}(a^n - 5a + 2)$

21. Simplify:
$5x^2 - 4x[x - (3x + 2) + x]$

22. Simplify:
$(x^{2n} - x)(x^{n+1} - 3)$

23. Simplify:
$(x + 6)(x^3 - 3x^2 - 5x + 1)$

24. Simplify:
$(x - 4)(3x + 2)(2x - 3)$

25. Simplify: $(5a + 2b)(5a - 2b)$

26. Simplify: $(4x - 3y)^2$

27. Factor:
$18a^5b^2 - 12a^3b^3 + 30a^2b$

28. Factor:
$x^{5n} - 3x^{4n} + 12x^{3n}$

29. Factor:
$5x^{n+5} + x^{n+3} + 4x^2$

30. Factor:
$x(y - 3) + 4(3 - y)$

31. Factor: $2ax + 4bx - 3ay - 6by$

32. Factor: $x^2 + 12x + 35$

33. Factor: $12 + x - x^2$

34. Factor: $x^2 - 16x + 63$

35. Factor: $6x^2 - 31x + 18$

36. Factor: $24x^2 + 61x - 8$

37. Factor: $x^2y^2 - 9$

38. Factor: $4x^2 + 12xy + 9y^2$

39. Factor: $x^{2n} - 12x^n + 36$

40. Factor: $36 - a^{2n}$

41. Factor: $64a^3 - 27b^3$

42. Factor: $125x^{3n} + y^{3n}$

43. Factor: $(a + b)^3 + 1$

44. Factor: $8 - y^{3n}$

45. Factor: $15x^4 + x^2 - 6$

46. Factor: $36x^8 - 36x^4 + 5$

47. Factor: $21x^4y^4 + 23x^2y^2 + 6$

48. Factor: $3a^6 - 15a^4 - 18a^2$

49. Factor: $x^{4n} - 8x^{2n} + 16$

50. Factor: $3a^4b - 3ab^4$

51. Solve: $(5x - 7)(2x + 9) = 0$

52. Solve: $x^3 - x^2 - 6x = 0$

53. Solve: $6x^2 + 60 = 39x$

54. Solve: $x^3 - 16x = 0$

55. Solve: $y^3 + y^2 - 36y - 36 = 0$

56. Solve: $x^3 - 8x^2 - x + 8 = 0$

57. The most distant object visible from Earth without the aid of a telescope is the Great Galaxy of Andromeda. It takes light from the Great Galaxy of Andromeda 2.2×10^6 years to travel to Earth. Light travels about 5.9×10^{12} mph. How far from Earth is the Great Galaxy of Andromeda?

58. Light from the sun supplies Earth with 2.4×10^{14} horsepower. Earth receives only 2.2×10^{-7} of the power generated by the sun. How much power is generated by the sun?

59. The length of a rectangle is $(5x + 3)$ cm. The width is $(2x - 7)$ cm. Find the area of the rectangle in terms of the variable x.

60. The length of the side of a cube is $(3x - 1)$ ft. Find the volume of the cube in terms of the variable x.

61. Find the area of the figure shown below. All dimensions given are in inches.

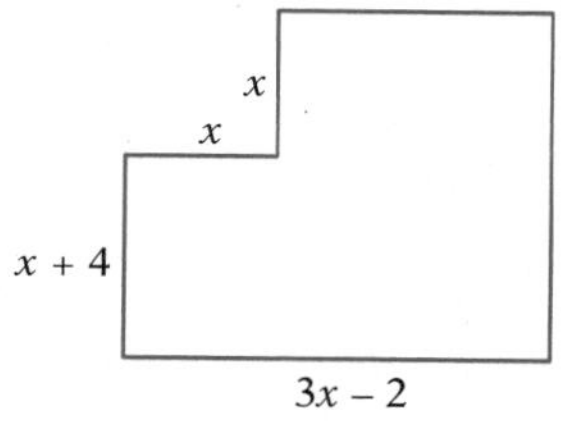

62. The sum of the squares of two consecutive even integers is 52. Find the two integers.

63. The sum of a number and its square is 56. Find the number.

64. The length of a rectangle is 2 m more than twice the width. The area of the rectangle is 60 m^2. Find the length of the rectangle.

Chapter Test

1. Simplify:
$(6x^3 - 7x^2 + 6x - 7) - (4x^3 - 3x^2 + 7)$

2. Simplify: $(-4a^2b)^3(-ab^4)$

3. Simplify: $\frac{(2a^{-4}b^2)^3}{4a^{-2}b^{-1}}$

4. Write the number 0.000000501 in scientific notation.

5. Write the number of seconds in one week in scientific notation.

6. Simplify: $(x^{-1} + y)^{-1}$

7. Simplify: $-5x[3 - 2(2x - 4) - 3x]$

8. Simplify: $(3a + 4b)(2a - 7b)$

9. Simplify: $(3t^3 - 4t^2 + 1)(2t^2 - 5)$

10. Simplify: $(3z - 5)^2$

11. Simplify: $(4x^3 + x - 15) \div (2x - 3)$

12. Simplify: $(x^3 - 5x^2 + 5x + 5) \div (x - 3)$

13. Factor: $6a^3b^2 - 4a^2b^2 + 4ab^4$

14. Factor: $12 - 17x + 6x^2$

15. Factor: $6a^4 - 13a^2 - 5$

16. Factor: $12x^3 + 12x^2 - 45x$

17. Factor: $16x^2 - 25$

18. Factor: $16t^2 + 24t + 9$

19. Factor: $27x^3 - 8$

20. Factor: $6x^2 - 4x - 3xa + 2a$

21. Factor: $3x^4 - 23x^2 - 36$

22. Solve: $6x^2 = x + 1$

23. Solve: $6x^3 + x^2 - 6x - 1 = 0$

24. The length of a rectangle is $(5x + 1)$ ft. The width is $(2x - 1)$ ft. Find the area of the rectangle in terms of the variable x.

25. A space vehicle travels 2.4×10^5 mi from Earth to the moon at an average velocity of 2×10^4 mph. How long does it take the space vehicle to reach the moon?

Cumulative Review

1. Simplify: $8 - 2[-3 - (-1)]^2 \div 4$

2. Evaluate $\frac{2a - b}{b - c}$ when $a = 4$, $b = -2$, and $c = 6$.

3. Identify the property that justifies the statement. $2x + (-2x) = 0$

4. Simplify: $2x - 4[x - 2(3 - 2x) + 4]$

5. Solve: $\frac{2}{3} - y = \frac{5}{6}$

6. Solve: $8x - 3 - x = -6 + 3x - 8$

7. Solve: $\frac{3x - 5}{3} - 8 = 2$

8. Solve: $3 - |2 - 3x| = -2$

9. Solve: $\frac{x - 2}{3} - \frac{x - 4}{5} = \frac{2x - 3}{2}$

10. Solve: $\frac{3x}{2} + 4 = \frac{7x - 2}{5}$

11. Solve: $8x + 2 < 12x + 7$

12. Solve: $2x - 3 > 2$ or $11 - 3x > 8$

13. Solve: $|2 - 3x| < 2$

14. Simplify: $3 - (3 - 3^{-1})^{-1}$

15. Simplify: $3x[8 - 2(3x - 6) + 5x] + 8$

16. Simplify: $(2x + 3)(2x^2 - 3x + 1)$

17. Simplify: $(x^n + 1)^2$

18. Factor: $8x^2 - 26x + 15$

19. Factor: $-4x^3 + 14x^2 - 12x$

20. Factor: $4x^2 - 20xy + 25y^2$

21. Factor: $a(x - y) - b(y - x)$

22. Factor: $x^4 - 16$

23. Factor: $2x^3 - 16$

24. Factor: $a^3 + 64$

25. The sum of two integers is twenty-four. The difference between four times the smaller integer and nine is three less than twice the larger integer. Find the integers.

26. How many ounces of pure gold that costs \$360 per ounce must be mixed with 80 oz of an alloy that costs \$120 per ounce to make a mixture that costs \$200 per ounce?

27. Two bicycles are 25 mi apart and traveling toward each other. One cyclist is traveling at $\frac{2}{3}$ the rate of the other cyclist. They pass in two hours. Find the rate of each cyclist.

28. If \$3000 is invested at an annual simple interest rate of 7.5%, how much additional money must be invested at an annual simple interest rate of 10% so that the total interest earned is 9% of the total investment?

29. Find three consecutive even integers whose sum is between 25 and 40.

30. The length of a rectangle is 3 in. more than the width. The area of the rectangle is 108 $in.^2$. Find the length of the rectangle.

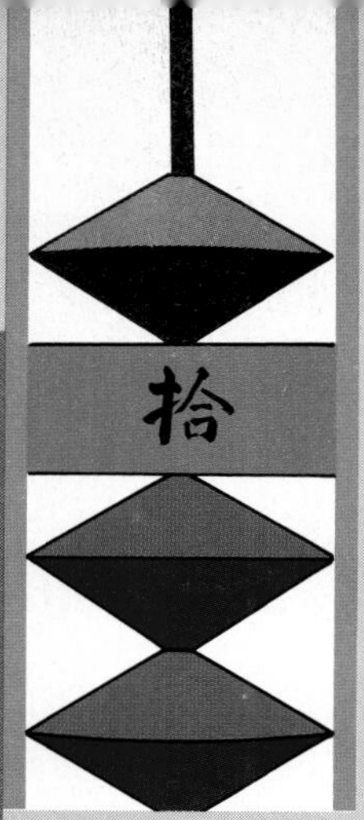

4

Rational Expressions

Objectives

- Simplify rational expressions
- Multiply and divide rational expressions
- Add and subtract rational expressions
- Simplify complex fractions
- Solve fractional equations
- Proportions
- Work problems
- Uniform motion problems
- Literal equations

POINT OF INTEREST

The Abacus

The abacus is an ancient device used to add, subtract, multiply, and divide, and to calculate square and cube roots. There are many variations of the abacus, but the one shown here has been used in China for many hundreds of years.

The abacus shown here consists of 13 rows of beads. A crossbar separates the beads. Each upper bead represents five units and each lower bead represents one unit. The place holder of each row is shown above the abacus. The first row of beads represents numbers from one to nine, the second row represents numbers from ten to ninety, etc.

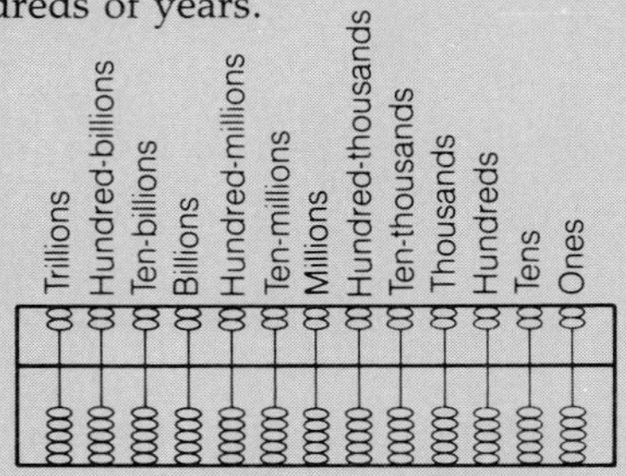

The number shown at the right is 7036.

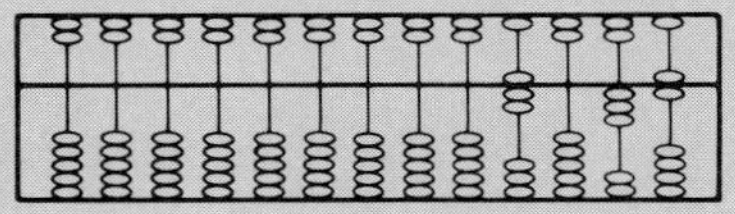

From the thousands' row—
one upper bead—5000
two lower beads—2000

From the hundreds' row—no beads
From the tens' row—three lower beads—30
From the ones' row—one upper bead—5
one lower bead—1

Add: 1247 + 2516

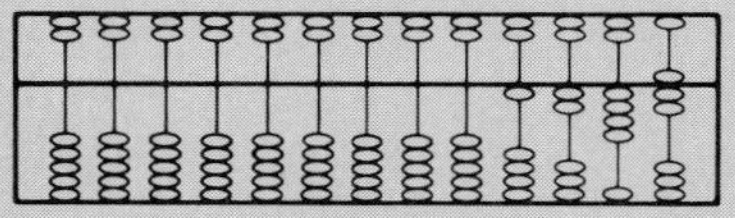

1247 is shown at the right.

Add 6 to the ones' column.

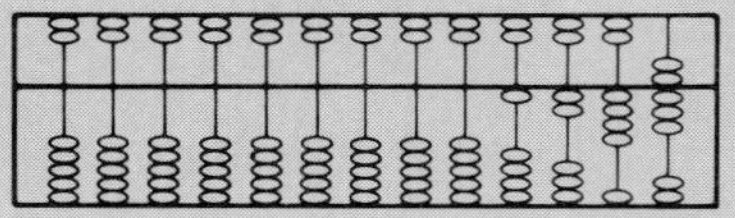

Move the 10 (the upper two beads in the ones' row) to the tens' row (one lower bead). This makes 5 beads in the lower tens' row. Remove 5 beads from the lower row and place one bead in the upper row.

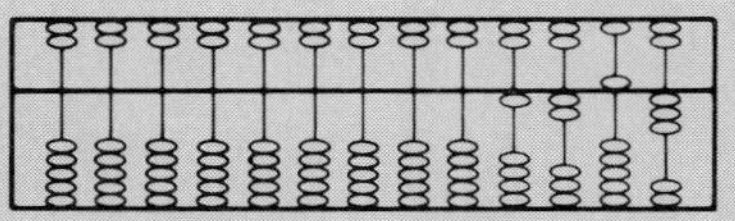

Continue adding in each row until the problem is completed. Carry if necessary.

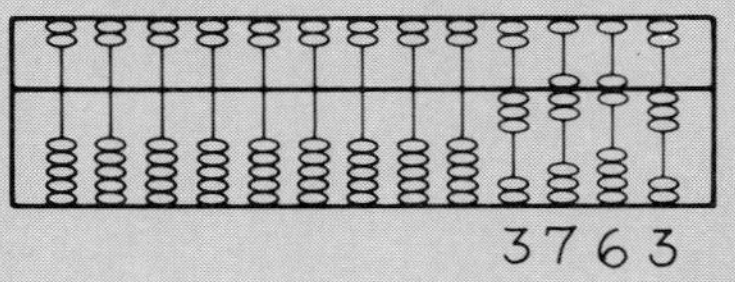

1247 + 2516 = 3763

SECTION **4.1**

Simplifying Rational Expressions

1 Simplify rational expressions

A fraction in which both the numerator and denominator are polynomials is called a **rational expression.** Examples of rational expressions are shown at the right.

$$\frac{2}{a} \qquad \frac{x^2 + y}{2x + 3} \qquad \frac{x^2 + 2x - 4}{x^4 - 2x}$$

The algebraic expression shown at the right is not a rational expression since $\sqrt{x} + 3$ is not a polynomial.

$$\frac{\sqrt{x} + 3}{x^2 + 2}$$

Rational expressions name a real number for each real number replacement of the variable or variables. Thus all properties of real numbers apply to rational expressions. Because division by zero is not defined when the variables in a rational expression are replaced with real numbers, the resulting denominator must not equal zero.

For example, the value of x cannot be 3 in the rational expression at the right.

$$\frac{2x - 5}{3x - 9}$$

$$\frac{2 \cdot 3 - 5}{3 \cdot 3 - 9} = \frac{6 - 5}{9 - 9} = \frac{1}{0} \quad \text{Not a real number}$$

A rational expression is in simplest form when the numerator and denominator have no common factors.

Simplify: $\dfrac{3x^3 - 9x^2}{6x^2 - 18x}$, $x \neq 0$, $x \neq 3$

Factor the numerator and denominator. The restrictions $x \neq 0$, $x \neq 3$ are made to ensure that the denominator is not zero.

$$\frac{3x^3 - 9x^2}{6x^2 - 18x} = \frac{3x^2(x - 3)}{6x(x - 3)}$$

Divide by the common factors.

$$= \frac{3x^2\overset{1}{\cancel{(x - 3)}}}{6x\underset{1}{\cancel{(x - 3)}}}$$

Write the answer in simplest form.

$$= \frac{x}{2}$$

This example states that $\dfrac{3x^3 - 9x^2}{6x^2 - 18x} = \dfrac{x}{2}$ as long as $x \neq 0$ and $x \neq 3$. Note that evaluating both expressions when x is replaced by 3, gives $\dfrac{0}{0} = \dfrac{3}{2}$, which is not a true statement.

If x is replaced by any number other than 0 or 3, the two expressions represent the same real number. For example, replacing x by 1 gives $\frac{-6}{-12} = \frac{1}{2}$, which is a true statement. For the remainder of the text, assume that the values of the variables in a rational expression are such that division by zero would not occur.

Example 1 Simplify.

A. $\dfrac{x^2 - 25}{x^2 + 13x + 40}$ B. $\dfrac{12 + 5x - 2x^2}{2x^2 - 3x - 20}$ C. $\dfrac{x^{2n} + x^n - 2}{x^{2n} - 1}$

Solution A. $\dfrac{x^2 - 25}{x^2 + 13x + 40} = \dfrac{(x+5)(x-5)}{(x+5)(x+8)}$ ▶ Factor the numerator and denominator.

$= \dfrac{\overset{1}{\cancel{(x+5)}}(x-5)}{\underset{1}{\cancel{(x+5)}}(x+8)}$ ▶ Divide by the common factors.

$= \dfrac{x-5}{x+8}$ ▶ Write the answer in simplest form.

B. $\dfrac{12 + 5x - 2x^2}{2x^2 - 3x - 20} = \dfrac{(4-x)(3+2x)}{(x-4)(2x+5)}$ ▶ Factor the numerator and denominator.

$= \dfrac{\overset{-1}{\cancel{(4-x)}}(3+2x)}{\underset{1}{\cancel{(x-4)}}(2x+5)}$ ▶ Divide by the common factors. Remember that $4 - x = -(x - 4)$. Therefore, $\dfrac{4-x}{x-4} = \dfrac{-(x-4)}{x-4} = \dfrac{-1}{1} = -1.$

$= -\dfrac{2x+3}{2x+5}$ ▶ Write the answer in simplest form.

C. $\dfrac{x^{2n} + x^n - 2}{x^{2n} - 1} = \dfrac{(x^n - 1)(x^n + 2)}{(x^n - 1)(x^n + 1)}$ ▶ Factor the numerator and denominator.

$= \dfrac{\overset{1}{\cancel{(x^n - 1)}}(x^n + 2)}{\underset{1}{\cancel{(x^n - 1)}}(x^n + 1)}$ ▶ Divide by the common factors.

$= \dfrac{x^n + 2}{x^n + 1}$

Problem 1 Simplify.

A. $\dfrac{6x^4 - 24x^3}{12x^3 - 48x^2}$ B. $\dfrac{20x - 15x^2}{15x^3 - 5x^2 - 20x}$ C. $\dfrac{x^{2n} + x^n - 12}{x^{2n} - 3x^n}$

Solution See page A21.

EXERCISES 4.1

1 Simplify.

1. $\dfrac{4-8x}{4}$

2. $\dfrac{8y+2}{2}$

3. $\dfrac{6x^2-2x}{2x}$

4. $\dfrac{3y-12y^2}{3y}$

5. $\dfrac{8x^2(x-3)}{4x(x-3)}$

6. $\dfrac{16y^4(y+8)}{12y^3(y+8)}$

7. $\dfrac{2x-6}{3x-x^2}$

8. $\dfrac{3a^2-6a}{12-6a}$

9. $\dfrac{6x^3-15x^2}{12x^2-30x}$

10. $\dfrac{-36a^2-48a}{18a^3+24a^2}$

11. $\dfrac{a^2+4a}{4a-16}$

12. $\dfrac{3x-6}{x^2+2x}$

13. $\dfrac{16x^3-8x^2+12x}{4x}$

14. $\dfrac{3x^3y^3-12x^2y^2+15xy}{3xy}$

15. $\dfrac{-10a^4-20a^3+30a^2}{-10a^2}$

16. $\dfrac{-7a^5-14a^4+21a^3}{-7a^3}$

17. $\dfrac{3x^{3n}-9x^{2n}}{12x^{2n}}$

18. $\dfrac{8a^n}{4a^{2n}-8a^n}$

19. $\dfrac{x^{2n}+x^ny^n}{x^{2n}-y^{2n}}$

20. $\dfrac{a^{2n}-b^{2n}}{5a^{3n}+5a^{2n}b^n}$

21. $\dfrac{x^2-7x+12}{x^2-9x+20}$

22. $\dfrac{x^2-x-20}{x^2-2x-15}$

23. $\dfrac{x^2-xy-2y^2}{x^2-3xy+2y^2}$

24. $\dfrac{2x^2+7xy-4y^2}{4x^2-4xy+y^2}$

25. $\dfrac{6-x-x^2}{3x^2-10x+8}$

26. $\dfrac{3x^2+10x-8}{8-14x+3x^2}$

27. $\dfrac{14-19x-3x^2}{3x^2-23x+14}$

28. $\dfrac{x^2+x-12}{x^2-x-12}$

29. $\dfrac{a^2-7a+10}{a^2+9a+14}$

30. $\dfrac{x^2-2x}{x^2+2x}$

31. $\dfrac{a^2-b^2}{a^3+b^3}$

32. $\dfrac{x^4-y^4}{x^2+y^2}$

33. $\dfrac{8x^3-y^3}{4x^2-y^2}$

34. $\dfrac{a^2-b^2}{a^3-b^3}$

35. $\dfrac{x^3+y^3}{3x^3-3x^2y+3xy^2}$

36. $\dfrac{3x^3+3x^2+3x}{9x^3-9}$

37. $\dfrac{3x^3-21x^2+30x}{6x^4-24x^3-30x^2}$

38. $\dfrac{3x^2y-15xy+18y}{6x^2y+6xy-36y}$

39. $\dfrac{x^3-4xy^2}{3x^3-2x^2y-8xy^2}$

40. $\dfrac{4a^2-8ab+4b^2}{4a^2-4b^2}$

41. $\dfrac{4x^3-14x^2+12x}{24x+4x^2-8x^3}$

42. $\dfrac{6x^3-15x^2-75x}{150x+30x^2-12x^3}$

43. $\dfrac{x^2-4}{a(x+2)-b(x+2)}$

44. $\dfrac{x^2(a-2)-a+2}{ax^2-ax}$

45. $\dfrac{x^4+3x^2+2}{x^4-1}$

46. $\dfrac{x^4-2x^2-3}{x^4+2x^2+1}$

47. $\dfrac{x^2y^2 + 4xy - 21}{x^2y^2 - 10xy + 21}$

48. $\dfrac{6x^2y^2 + 11xy + 4}{9x^2y^2 + 9xy - 4}$

49. $\dfrac{a^{2n} - a^n - 2}{a^{2n} + 3a^n + 2}$

50. $\dfrac{a^{2n} + a^n - 12}{a^{2n} - 2a^n - 3}$

51. $\dfrac{a^{2n} - 1}{a^{2n} - 2a^n + 1}$

52. $\dfrac{a^{2n} + 2a^nb^n + b^{2n}}{a^{2n} - b^{2n}}$

53. $\dfrac{(x - 3) - b(x - 3)}{b(x + 3) - x - 3}$

54. $\dfrac{x^2(a + b) + a + b}{x^4 - 1}$

SUPPLEMENTAL EXERCISES 4.1

For what real values of x is the rational expression undefined?

55. $\dfrac{5x}{x^2 - 9}$

56. $\dfrac{4x - 1}{x^2 - 25}$

57. $\dfrac{x - 3}{x^2 + 5x - 6}$

58. $\dfrac{2x + 3}{x^2 - 2x - 8}$

59. $\dfrac{6}{x^2 - 3x}$

60. $\dfrac{9}{x^2 - 7x}$

61. $\dfrac{4x + 5}{6x^2 + 7x - 3}$

62. $\dfrac{3x + 4}{12x^2 - 5x - 2}$

63. $\dfrac{8x - 1}{3(x + 4) - x(x + 4)}$

64. $\dfrac{x - 1}{x^3 - x^2 + x - 1}$

65. $\dfrac{2x - 4}{x^3 + x^2 - 4x - 4}$

SECTION 4.2 Operations on Rational Expressions

1 Multiply and divide rational expressions

The product of two fractions is a fraction whose numerator is the product of the numerators of the two fractions and whose denominator is the product of the denominators of the two fractions.

$$\frac{a}{b} \cdot \frac{c}{d} = \frac{ac}{bd}$$

$$\frac{5}{a + 2} \cdot \frac{b - 3}{3} = \frac{5(b - 3)}{(a + 2)3} = \frac{5b - 15}{3a + 6}$$

The product of two rational expressions can often be simplified by factoring the numerator and the denominator.

Simplify: $\dfrac{x^2 - 2x}{2x^2 + x - 15} \cdot \dfrac{2x^2 - x - 10}{x^2 - 4}$

$$\frac{x^2 - 2x}{2x^2 + x - 15} \cdot \frac{2x^2 - x - 10}{x^2 - 4} =$$

Factor the numerator and denominator of each fraction.

$$\frac{x(x-2)}{(x+3)(2x-5)} \cdot \frac{(x+2)(2x-5)}{(x+2)(x-2)} =$$

Multiply.

$$\frac{x(x-2)(x+2)(2x-5)}{(x+3)(2x-5)(x+2)(x-2)} =$$

Divide by the common factors.

$$\frac{x\overset{1}{\cancel{(x-2)}}\overset{1}{\cancel{(x+2)}}\overset{1}{\cancel{(2x-5)}}}{(x+3)\underset{1}{\cancel{(2x-5)}}\underset{1}{\cancel{(x+2)}}\underset{1}{\cancel{(x-2)}}} =$$

Write the answer in simplest form.

$$\frac{x}{x+3}$$

Example 1 Simplify.

A. $\dfrac{2x^2 - 6x}{3x - 6} \cdot \dfrac{6x - 12}{8x^3 - 12x^2}$ B. $\dfrac{6x^2 + x - 2}{6x^2 + 7x + 2} \cdot \dfrac{2x^2 + 9x + 4}{4 - 7x - 2x^2}$

Solution A. $\dfrac{2x^2 - 6x}{3x - 6} \cdot \dfrac{6x - 12}{8x^3 - 12x^2} = \dfrac{2x(x-3)}{3(x-2)} \cdot \dfrac{6(x-2)}{4x^2(2x-3)}$

$$= \frac{12x(x-3)\overset{1}{\cancel{(x-2)}}}{12x^2\underset{1}{\cancel{(x-2)}}(2x-3)} = \frac{x-3}{x(2x-3)}$$

B. $\dfrac{6x^2 + x - 2}{6x^2 + 7x + 2} \cdot \dfrac{2x^2 + 9x + 4}{4 - 7x - 2x^2} = \dfrac{(2x-1)(3x+2)}{(3x+2)(2x+1)} \cdot \dfrac{(2x+1)(x+4)}{(1-2x)(4+x)}$

$$= \frac{\overset{-1}{\cancel{(2x-1)}}\overset{1}{\cancel{(3x+2)}}\overset{1}{\cancel{(2x+1)}}\overset{1}{\cancel{(x+4)}}}{\underset{1}{\cancel{(3x+2)}}\underset{1}{\cancel{(2x+1)}}\underset{1}{\cancel{(1-2x)}}\underset{1}{\cancel{(x+4)}}} = -1$$

Problem 1 Simplify.

A. $\dfrac{12 + 5x - 3x^2}{x^2 + 2x - 15} \cdot \dfrac{2x^2 + x - 45}{3x^2 + 4x}$ B. $\dfrac{2x^2 - 13x + 20}{x^2 - 16} \cdot \dfrac{2x^2 + 9x + 4}{6x^2 - 7x - 5}$

Solution See page A21.

The **reciprocal** of a rational expression is the rational expression with the numerator and denominator interchanged.

Rational Expression			Reciprocal
	$\dfrac{a}{b}$	$\dfrac{b}{a}$	
	$\dfrac{a^2 - 2y}{4}$	$\dfrac{4}{a^2 - 2y}$	

To divide two rational expressions, multiply by the reciprocal of the divisor.

$$\frac{a}{b} \div \frac{c}{d} = \frac{a}{b} \cdot \frac{d}{c}$$

$$\frac{2}{a} \div \frac{5}{b} = \frac{2}{a} \cdot \frac{b}{5} = \frac{2b}{5a}$$

$$\frac{x+y}{2} \div \frac{x-y}{5} = \frac{x+y}{2} \cdot \frac{5}{x-y} = \frac{(x+y)5}{2(x-y)} = \frac{5x+5y}{2x-2y}$$

Example 2 Simplify.

A. $\dfrac{12x^2y^2 - 24xy^2}{5z^2} \div \dfrac{4x^3y - 8x^2y}{3z^4}$

B. $\dfrac{3y^2 - 10y + 8}{3y^2 + 8y - 16} \div \dfrac{2y^2 - 7y + 6}{2y^2 + 5y - 12}$

Solution A. $\dfrac{12x^2y^2 - 24xy^2}{5z^2} \div \dfrac{4x^3y - 8x^2y}{3z^4} = \dfrac{12x^2y^2 - 24xy^2}{5z^2} \cdot \dfrac{3z^4}{4x^3y - 8x^2y}$

$$= \frac{12xy^2(x-2)}{5z^2} \cdot \frac{3z^4}{4x^2y(x-2)}$$

$$= \frac{36xy^2z^4\overset{1}{\cancel{(x-2)}}}{20x^2yz^2\underset{1}{\cancel{(x-2)}}} = \frac{9yz^2}{5x}$$

B. $\dfrac{3y^2 - 10y + 8}{3y^2 + 8y - 16} \div \dfrac{2y^2 - 7y + 6}{2y^2 + 5y - 12} = \dfrac{3y^2 - 10y + 8}{3y^2 + 8y - 16} \cdot \dfrac{2y^2 + 5y - 12}{2y^2 - 7y + 6}$

$$= \frac{(y-2)(3y-4)}{(3y-4)(y+4)} \cdot \frac{(y+4)(2y-3)}{(y-2)(2y-3)}$$

$$= \frac{\overset{1}{\cancel{(y-2)}}\overset{1}{\cancel{(3y-4)}}\overset{1}{\cancel{(y+4)}}\overset{1}{\cancel{(2y-3)}}}{\underset{1}{\cancel{(3y-4)}}\underset{1}{\cancel{(y+4)}}\underset{1}{\cancel{(y-2)}}\underset{1}{\cancel{(2y-3)}}} = 1$$

Problem 2 Simplify.

A. $\dfrac{6x^2 - 3xy}{10ab^4} \div \dfrac{16x^2y^2 - 8xy^3}{15a^2b^2}$

B. $\dfrac{6x^2 - 7x + 2}{3x^2 + x - 2} \div \dfrac{4x^2 - 8x + 3}{5x^2 + x - 4}$

Solution See pages A21–A22.

2 Add and subtract rational expressions

When adding rational expressions in which the denominators are the same, add the numerators. The denominator of the sum is the common denominator. Write the answer in simplest form.

$$\frac{a}{c} + \frac{b}{c} = \frac{a+b}{c}$$

$$\frac{4x}{15} + \frac{8x}{15} = \frac{4x+8x}{15} = \frac{12x}{15} = \frac{4x}{5}$$

$$\frac{a}{a^2-b^2} + \frac{b}{a^2-b^2} = \frac{a+b}{a^2-b^2} = \frac{a+b}{(a-b)(a+b)} = \frac{\overset{1}{\cancel{(a+b)}}}{(a-b)\underset{1}{\cancel{(a+b)}}} = \frac{1}{a-b}$$

When subtracting rational expressions with the same denominators, subtract the numerators. The denominator of the difference is the common denominator. Write the answer in simplest form.

$$\frac{y}{y-3} - \frac{3}{y-3} = \frac{y-3}{y-3} = \frac{\overset{1}{\cancel{(y-3)}}}{\underset{1}{\cancel{(y-3)}}} = 1$$

$$\frac{7x-12}{2x^2+5x-12} - \frac{3x-6}{2x^2+5x-12} = \frac{(7x-12)-(3x-6)}{2x^2+5x-12} = \frac{4x-6}{2x^2+5x-12}$$

$$= \frac{2(2x-3)}{(2x-3)(x+4)} = \frac{2\overset{1}{\cancel{(2x-3)}}}{\underset{1}{\cancel{(2x-3)}}(x+4)} = \frac{2}{x+4}$$

Before two rational expressions with different denominators can be added or subtracted, each rational expression must be expressed in terms of a common denominator. This common denominator is the LCM of the denominators of the rational expressions.

The LCM of two or more polynomials is the simplest polynomial that contains the factors of each polynomial. To find the LCM, first factor each polynomial completely. The LCM is the product of each factor the greatest number of times it occurs in any one factorization.

To find the LCM of $3x^2 + 15x$ and $6x^4 + 24x^3 - 30x^2$, factor each polynomial.

$$3x^2 + 15x = 3x(x+5)$$
$$6x^4 + 24x^3 - 30x^2 = 6x^2(x^2 + 4x - 5) = 6x^2(x-1)(x+5)$$

The LCM is the product of the LCM of the numerical coefficients and each variable factor the greatest number of times it occurs in any one factorization.

$$\text{LCM} = 6x^2(x-1)(x+5)$$

Write the fractions $\frac{x+2}{x^2-2x}$ and $\frac{5x}{3x-6}$ in terms of the LCM of the denominators.

Find the LCM of the denominators.

$x^2 - 2x = x(x-2)$
$3x - 6 = 3(x-2)$
The LCM is $3x(x-2)$.

For each fraction, multiply the numerator and denominator by the factor whose product with the denominator is the LCM.

$$\frac{x+2}{x^2-2x} = \frac{x+2}{x(x-2)} \cdot \frac{3}{3} = \frac{3x+6}{3x(x-2)}$$

$$\frac{5x}{3x-6} = \frac{5x}{3(x-2)} \cdot \frac{x}{x} = \frac{5x^2}{3x(x-2)}$$

Simplify: $\frac{3x}{2x-3} + \frac{3x+6}{2x^2+x-6}$

The LCM of the denominators is $(2x-3)(x+2)$.

$$\frac{3x}{2x-3} + \frac{3x+6}{2x^2+x-6} =$$

Rewrite each fraction in terms of the LCM of the denominators.

$$\frac{3x}{2x-3} \cdot \frac{x+2}{x+2} + \frac{3x+6}{(2x-3)(x+2)} =$$

$$\frac{3x^2+6x}{(2x-3)(x+2)} + \frac{3x+6}{(2x-3)(x+2)} =$$

Add the fractions.

$$\frac{(3x^2+6x)+(3x+6)}{(2x-3)(x+2)} = \frac{3x^2+9x+6}{(2x-3)(x+2)} =$$

Factor the numerator to determine whether there are common factors in the numerator and denominator.

$$\frac{3(x^2+3x+2)}{(2x-3)(x+2)} = \frac{3(x+2)(x+1)}{(2x-3)(x+2)} =$$

$$\frac{3\overset{1}{\cancel{(x+2)}}(x+1)}{(2x-3)\underset{1}{\cancel{(x+2)}}} = \frac{3(x+1)}{2x-3}$$

Example 3 Simplify: $\frac{x}{2x-4} - \frac{4-x}{x^2-2x}$

Solution

$$\frac{x}{2x-4} - \frac{4-x}{x^2-2x} = \frac{x}{2(x-2)} \cdot \frac{x}{x} - \frac{4-x}{x(x-2)} \cdot \frac{2}{2} =$$

▶ Write each fraction in terms of the LCM. The LCM is $2x(x-2)$.

$$\frac{x^2}{2x(x-2)} - \frac{8-2x}{2x(x-2)} = \frac{x^2-(8-2x)}{2x(x-2)} =$$

▶ Subtract the fractions.

$$\frac{x^2+2x-8}{2x(x-2)} = \frac{(x+4)(x-2)}{2x(x-2)} = \frac{(x+4)\overset{1}{\cancel{(x-2)}}}{2x\underset{1}{\cancel{(x-2)}}} =$$

▶ Divide by the common factors.

$$\frac{x+4}{2x}$$

Problem 3 Simplify: $\frac{a-3}{a^2-5a} + \frac{a-9}{a^2-25}$

Solution See page A22.

Example 4 Simplify: $\frac{6x-23}{2x^2+x-6}+\frac{3x}{2x-3}-\frac{5}{x+2}$

Solution

$$\frac{6x-23}{2x^2+x-6}+\frac{3x}{2x-3}-\frac{5}{x+2}=$$

$$\frac{6x-23}{(2x-3)(x+2)}+\frac{3x}{2x-3}\cdot\frac{x+2}{x+2}-\frac{5}{x+2}\cdot\frac{2x-3}{2x-3}=$$

▶ Write each fraction in terms of the LCM, $(2x-3)(x+2)$.

$$\frac{6x-23}{(2x-3)(x+2)}+\frac{3x^2+6x}{(2x-3)(x+2)}-\frac{10x-15}{(2x-3)(x+2)}=$$

$$\frac{(6x-23)+(3x^2+6x)-(10x-15)}{(2x-3)(x+2)}=$$

$$\frac{6x-23+3x^2+6x-10x+15}{(2x-3)(x+2)}=$$

$$\frac{3x^2+2x-8}{(2x-3)(x+2)}=\frac{(3x-4)(x+2)}{(2x-3)(x+2)}=\frac{3x-4}{2x-3}$$

Problem 4 Simplify: $\frac{x-1}{x-2}-\frac{7-6x}{2x^2-7x+6}+\frac{4}{2x-3}$

Solution See page A22.

EXERCISES 4.2

1 Simplify.

1. $\frac{27a^2b^5}{16xy^2}\cdot\frac{20x^2y^3}{9a^2b}$

2. $\frac{15x^2y^4}{24ab^3}\cdot\frac{28a^2b^4}{35xy^4}$

3. $\frac{3x-15}{4x^2-2x}\cdot\frac{20x^2-10x}{15x-75}$

4. $\frac{2x^2+4x}{8x^2-40x}\cdot\frac{6x^3-30x^2}{3x^2+6x}$

5. $\frac{x^2y^3}{x^2-4x-5}\cdot\frac{2x^2-13x+15}{x^4y^3}$

6. $\frac{2x^2-5x+3}{x^6y^3}\cdot\frac{x^4y^4}{2x^2-x-3}$

7. $\frac{x^2-3x+2}{x^2-8x+15}\cdot\frac{x^2+x-12}{8-2x-x^2}$

8. $\frac{x^2+x-6}{12+x-x^2}\cdot\frac{x^2+x-20}{x^2-4x+4}$

9. $\frac{x^{n+1}+2x^n}{4x^2-6x}\cdot\frac{8x^2-12x}{x^{n+1}-x^n}$

10. $\frac{x^{2n}+2x^n}{x^{n+1}+2x}\cdot\frac{x^2-3x}{x^{n+1}-3x^n}$

11. $\frac{2x^2-13x-7}{3x^2-25x+28}\cdot\frac{15x^2-17x-4}{10x^2+3x-1}$

12. $\frac{4x^2-9}{6x^2+5x-6}\cdot\frac{6x^2+5x-6}{4x^2-12x+9}$

13. $\dfrac{12 + x - 6x^2}{6x^2 + 29x + 28} \cdot \dfrac{2x^2 + x - 21}{4x^2 - 9}$

14. $\dfrac{x^2 + 5x + 4}{4 + x - 3x^2} \cdot \dfrac{3x^2 + 2x - 8}{x^2 + 4x}$

15. $\dfrac{x^{2n} - x^n - 6}{x^{2n} + x^n - 2} \cdot \dfrac{x^{2n} - 5x^n - 6}{x^{2n} - 2x^n - 3}$

16. $\dfrac{x^{2n} + 3x^n + 2}{x^{2n} - x^n - 6} \cdot \dfrac{x^{2n} + x^n - 12}{x^{2n} - 1}$

17. $\dfrac{x^3 - y^3}{2x^2 + xy - 3y^2} \cdot \dfrac{2x^2 + 5xy + 3y^2}{x^2 + xy + y^2}$

18. $\dfrac{x^4 - 5x^2 + 4}{3x^2 - 4x - 4} \cdot \dfrac{3x^2 - 10x - 8}{x^2 - 4}$

19. $\dfrac{6x^2y^4}{35a^2b^5} \div \dfrac{12x^3y^3}{7a^4b^5}$

20. $\dfrac{12a^4b^7}{13x^2y^2} \div \dfrac{18a^5b^6}{26xy^3}$

21. $\dfrac{2x - 6}{6x^2 - 15x} \div \dfrac{4x^2 - 12x}{18x^3 - 45x^2}$

22. $\dfrac{4x^2 - 4y^2}{6x^2y^2} \div \dfrac{3x^2 + 3xy}{2x^2y - 2xy^2}$

23. $\dfrac{2x^2 - 2y^2}{14x^2y^4} \div \dfrac{x^2 + 2xy + y^2}{35xy^3}$

24. $\dfrac{8x^3 + 12x^2y}{4x^2 - 9y^2} \div \dfrac{16x^2y^2}{4x^2 - 12xy + 9y^2}$

25. $\dfrac{2x^2 - 5x - 3}{2x^2 + 7x + 3} \div \dfrac{2x^2 - 3x - 20}{2x^2 - x - 15}$

26. $\dfrac{3x^2 - 10x - 8}{6x^2 + 13x + 6} \div \dfrac{2x^2 - 9x + 10}{4x^2 - 4x - 15}$

27. $\dfrac{6x^2 + 23x + 20}{3x^2 - 5x - 12} \div \dfrac{6x^2 - 29x + 35}{3x^2 - 16x + 21}$

28. $\dfrac{6x^2 + 23x - 4}{6x^2 + 17x - 3} \div \dfrac{4x^2 + 20x + 25}{2x^2 + 11x + 15}$

29. $\dfrac{x^2 - 8x + 15}{x^2 + 2x - 35} \div \dfrac{15 - 2x - x^2}{x^2 + 9x + 14}$

30. $\dfrac{2x^2 + 13x + 20}{8 - 10x - 3x^2} \div \dfrac{6x^2 - 13x - 5}{9x^2 - 3x - 2}$

31. $\dfrac{x^{2n} + x^n}{2x - 2} \div \dfrac{4x^n + 4}{x^{n+1} - x^n}$

32. $\dfrac{x^{2n} - 4}{4x^n + 8} \div \dfrac{x^{n+1} - 2x}{4x^3 - 12x^2}$

33. $\dfrac{2x^2 - 13x + 21}{2x^2 + 11x + 15} \div \dfrac{2x^2 + x - 28}{3x^2 + 4x - 15}$

34. $\dfrac{2x^2 - 13x + 15}{2x^2 - 3x - 35} \div \dfrac{6x^2 + x - 12}{6x^2 + 13x - 28}$

35. $\dfrac{14 + 17x - 6x^2}{3x^2 + 14x + 8} \div \dfrac{4x^2 - 49}{2x^2 + 15x + 28}$

36. $\dfrac{16x^2 - 9}{6 - 5x - 4x^2} \div \dfrac{16x^2 + 24x + 9}{4x^2 + 11x + 6}$

37. $\dfrac{2x^{2n} - x^n - 6}{x^{2n} - x^n - 2} \div \dfrac{2x^{2n} + x^n - 3}{x^{2n} - 1}$

38. $\dfrac{x^{4n} - 1}{x^{2n} + x^n - 2} \div \dfrac{x^{2n} + 1}{x^{2n} + 3x^n + 2}$

39. $\dfrac{6x^2 + 6x}{3x + 6x^2 + 3x^3} \div \dfrac{x^2 - 1}{1 - x^3}$

40. $\dfrac{x^3 + y^3}{2x^3 + 2x^2y} \div \dfrac{3x^3 - 3x^2y + 3xy^2}{6x^2 - 6y^2}$

2 Simplify.

41. $\dfrac{3}{2xy} - \dfrac{7}{2xy} - \dfrac{9}{2xy}$

42. $-\dfrac{3}{4x^2} + \dfrac{8}{4x^2} - \dfrac{3}{4x^2}$

43. $\dfrac{x}{x^2 - 3x + 2} - \dfrac{2}{x^2 - 3x + 2}$

44. $\dfrac{3x}{3x^2 + x - 10} - \dfrac{5}{3x^2 + x - 10}$

45. $\dfrac{3}{2x^2y} - \dfrac{8}{5x} - \dfrac{9}{10xy}$

46. $\dfrac{2}{5ab} - \dfrac{3}{10a^2b} + \dfrac{4}{15ab^2}$

47. $\dfrac{2}{3x} - \dfrac{3}{2xy} + \dfrac{4}{5xy} - \dfrac{5}{6x}$

48. $\dfrac{3}{4ab} - \dfrac{2}{5a} + \dfrac{3}{10b} - \dfrac{5}{8ab}$

49. $\dfrac{2x - 1}{12x} - \dfrac{3x + 4}{9x}$

50. $\dfrac{3x - 4}{6x} - \dfrac{2x - 5}{4x}$

51. $\dfrac{3x + 2}{4x^2y} - \dfrac{y - 5}{6xy^2}$

52. $\dfrac{2y - 4}{5xy^2} + \dfrac{3 - 2x}{10x^2y}$

53. $\dfrac{2x}{x - 3} - \dfrac{3x}{x - 5}$

54. $\dfrac{3a}{a - 2} - \dfrac{5a}{a + 1}$

55. $\dfrac{3}{2a - 3} + \dfrac{2a}{3 - 2a}$

56. $\dfrac{x}{2x - 5} - \dfrac{2}{5x - 2}$

57. $\dfrac{3}{x + 5} + \dfrac{2x + 7}{x^2 - 25}$

58. $\dfrac{x}{4 - x} - \dfrac{4}{x^2 - 16}$

59. $\frac{2}{x} - 3 - \frac{10}{x - 4}$

60. $\frac{6a}{a - 3} - 5 + \frac{3}{a}$

61. $\frac{1}{2x - 3} - \frac{5}{2x} + 1$

62. $\frac{5}{x} - \frac{5x}{5 - 6x} + 2$

63. $\frac{3}{x^2 - 1} + \frac{2x}{x^2 + 2x + 1}$

64. $\frac{1}{x^2 - 6x + 9} - \frac{1}{x^2 - 9}$

65. $\frac{x}{x + 3} - \frac{3 - x}{x^2 - 9}$

66. $\frac{1}{x + 2} - \frac{3x}{x^2 + 4x + 4}$

67. $\frac{2x - 3}{x + 5} - \frac{x^2 - 4x - 19}{x^2 + 8x + 15}$

68. $\frac{-3x^2 + 8x + 2}{x^2 + 2x - 8} - \frac{2x - 5}{x + 4}$

69. $\frac{x^n}{x^{2n} - 1} - \frac{2}{x^n + 1}$

70. $\frac{2}{x^n - 1} + \frac{x^n}{x^{2n} - 1}$

71. $\frac{2}{x^n - 1} - \frac{6}{x^{2n} + x^n - 2}$

72. $\frac{2x^n - 6}{x^{2n} - x^n - 6} + \frac{x^n}{x^n + 2}$

73. $\frac{2x - 2}{4x^2 - 9} - \frac{5}{3 - 2x}$

74. $\frac{x^2 + 4}{4x^2 - 36} - \frac{13}{x + 3}$

75. $\frac{x - 2}{x + 1} - \frac{3 - 12x}{2x^2 - x - 3}$

76. $\frac{3x - 4}{4x + 1} + \frac{3x + 6}{4x^2 + 9x + 2}$

77. $\frac{x + 1}{x^2 + x - 6} - \frac{x + 2}{x^2 + 4x + 3}$

78. $\frac{x + 1}{x^2 + x - 12} - \frac{x - 3}{x^2 + 7x + 12}$

79. $\frac{x - 1}{2x^2 + 11x + 12} + \frac{2x}{2x^2 - 3x - 9}$

80. $\frac{x - 2}{4x^2 + 4x - 3} + \frac{3 - 2x}{6x^2 + x - 2}$

81. $\frac{x}{x - 3} - \frac{2}{x + 4} - \frac{14}{x^2 + x - 12}$

82. $\frac{x^2}{x^2 + x - 2} + \frac{3}{x - 1} - \frac{4}{x + 2}$

83. $\dfrac{x^2 + 6x}{x^2 + 3x - 18} - \dfrac{2x - 1}{x + 6} + \dfrac{x - 2}{3 - x}$

84. $\dfrac{2x^2 - 2x}{x^2 - 2x - 15} - \dfrac{2}{x + 3} + \dfrac{x}{5 - x}$

85. $\dfrac{4 - 20x}{6x^2 + 11x - 10} - \dfrac{4}{2 - 3x} + \dfrac{x}{2x + 5}$

86. $\dfrac{x}{4x - 1} + \dfrac{2}{2x + 1} + \dfrac{6}{8x^2 + 2x - 1}$

87. $\dfrac{7 - 4x}{2x^2 - 9x + 10} + \dfrac{x - 3}{x - 2} - \dfrac{x + 1}{2x - 5}$

88. $\dfrac{x}{3x + 4} + \dfrac{3x + 2}{x - 5} - \dfrac{7x^2 + 24x + 28}{3x^2 - 11x - 20}$

89. $\dfrac{32x - 9}{2x^2 + 7x - 15} + \dfrac{x - 2}{3 - 2x} + \dfrac{3x + 2}{x + 5}$

90. $\dfrac{x + 1}{1 - 2x} - \dfrac{x + 3}{4x - 3} + \dfrac{10x^2 + 7x - 9}{8x^2 - 10x + 3}$

91. $\dfrac{x^2}{x^3 - 8} - \dfrac{x + 2}{x^2 + 2x + 4}$

92. $\dfrac{2x}{4x^2 + 2x + 1} + \dfrac{4x + 1}{8x^3 - 1}$

93. $\dfrac{2x^2}{x^4 - 1} - \dfrac{1}{x^2 - 1} + \dfrac{1}{x^2 + 1}$

94. $\dfrac{x^2 - 12}{x^4 - 16} + \dfrac{1}{x^2 - 4} - \dfrac{1}{x^2 + 4}$

SUPPLEMENTAL EXERCISES 4.2

Simplify.

95. $\dfrac{(x + 1)^2}{1 - 2x} \cdot \dfrac{2x - 1}{x + 1}$

96. $\dfrac{2y - 3}{a - 6} \cdot \dfrac{(a - 6)^2}{3 - 2y}$

97. $\left(\dfrac{3a}{b}\right)^3 \div \left(\dfrac{a}{2b}\right)^2$

98. $\left(\dfrac{2m}{3}\right)^2 \div \left(\dfrac{m^2}{6} + \dfrac{m}{2}\right)$

99. $\left(\dfrac{y - 2}{x^2}\right)^3 \cdot \left(\dfrac{x}{2 - y}\right)^2$

100. $\dfrac{b + 3}{b - 1} \div \dfrac{b + 3}{b - 2} \cdot \dfrac{b - 1}{b + 4}$

101. $\left(\dfrac{y + 1}{y - 1}\right)^2 - 1$

102. $1 - \left(\dfrac{x - 2}{x + 2}\right)^2$

103. $\left(\dfrac{1}{3} - \dfrac{2}{a}\right) \div \left(\dfrac{3}{a} - 2 + \dfrac{a}{4}\right)$

104. $\left(\dfrac{b}{6} - \dfrac{6}{b}\right) \div \left(\dfrac{6}{b} - 4 + \dfrac{b}{2}\right)$

105. $\dfrac{3x^2 + 6x}{4x^2 - 16} \cdot \dfrac{2x + 8}{x^2 + 2x} \div \dfrac{3x - 9}{5x - 20}$

106. $\dfrac{5y^2 - 20}{3y^2 - 12y} \cdot \dfrac{9y^3 + 6y^2}{2y^2 - 4y} \div \dfrac{y^3 + 2y^2}{2y^2 - 8y}$

107. $\dfrac{a^2 + a - 6}{4 + 11a - 3a^2} \cdot \dfrac{15a^2 - a - 2}{4a^2 + 7a - 2} \div \dfrac{6a^2 - 7a - 3}{4 - 17a + 4a^2}$

108. $\dfrac{25x - x^3}{x^4 - 1} \cdot \dfrac{3 - x - 4x^2}{2x^2 + 7x - 15} \div \dfrac{4x^3 - 23x^2 + 15x}{3 - 5x + 2x^2}$

109. $\left(\dfrac{x + 1}{2x - 1} - \dfrac{x - 1}{2x + 1}\right) \cdot \left(\dfrac{2x - 1}{x} - \dfrac{2x - 1}{x^2}\right)$

110. $\left(\dfrac{y - 2}{3y + 1} - \dfrac{y + 2}{3y - 1}\right) \cdot \left(\dfrac{3y + 1}{y} - \dfrac{3y - 1}{y^2}\right)$

Solve.

111. Use $x = 3$ and $y = 5$ to show that $\dfrac{1}{x} + \dfrac{1}{y} \neq \dfrac{1}{x + y}$.

112. Use $x = 3$ and $y = 5$ to show that $\dfrac{1}{x} - \dfrac{1}{y} \neq \dfrac{1}{x - y}$.

Rewrite the expression as the sum of two fractions in simplest form.

113. $\dfrac{3x + 6y}{xy}$

114. $\dfrac{5a + 8b}{ab}$

115. $\dfrac{4a^2 + 3ab}{a^2b^2}$

116. $\dfrac{3m^2n + 2mn^2}{12m^3n^3}$

SECTION 4.3 Complex Fractions

1 Simplify complex fractions

A **complex fraction** is a fraction whose numerator or denominator contains one or more fractions. Examples of complex fractions are shown below.

$$\dfrac{5}{2 + \dfrac{1}{2}} \qquad \dfrac{5 + \dfrac{1}{y}}{5 - \dfrac{1}{y}} \qquad \dfrac{x + 4 + \dfrac{1}{x + 2}}{x - 2 + \dfrac{1}{x + 2}}$$

Simplify: $\dfrac{\frac{1}{x}+\frac{1}{y}}{\frac{1}{x}-\frac{1}{y}}$

Multiply the numerator and denominator of the complex fraction by the LCM of the denominators. The LCM of x and y is xy.

$$\frac{\frac{1}{x}+\frac{1}{y}}{\frac{1}{x}-\frac{1}{y}} = \frac{\frac{1}{x}+\frac{1}{y}}{\frac{1}{x}-\frac{1}{y}} \cdot \frac{xy}{xy} = \frac{\frac{1}{x}\cdot xy + \frac{1}{y}\cdot xy}{\frac{1}{x}\cdot xy - \frac{1}{y}\cdot xy} = \frac{y+x}{y-x}$$

Note that after multiplying the numerator and denominator of the complex fraction by the LCM of the denominators, no fraction remains in the numerator or denominator.

Example 1 Simplify.

A. $\dfrac{2-\frac{11}{x}+\frac{15}{x^2}}{3-\frac{5}{x}-\frac{12}{x^2}}$ B. $\dfrac{2x-1+\frac{7}{x+4}}{3x-8+\frac{17}{x+4}}$

Solution A.

$$\frac{2-\frac{11}{x}+\frac{15}{x^2}}{3-\frac{5}{x}-\frac{12}{x^2}} = \frac{2-\frac{11}{x}+\frac{15}{x^2}}{3-\frac{5}{x}-\frac{12}{x^2}} \cdot \frac{x^2}{x^2}$$

▶ The LCM is x^2.

$$= \frac{2\cdot x^2 - \frac{11}{x}\cdot x^2 + \frac{15}{x^2}\cdot x^2}{3\cdot x^2 - \frac{5}{x}\cdot x^2 - \frac{12}{x^2}\cdot x^2}$$

$$= \frac{2x^2-11x+15}{3x^2-5x-12}$$

$$= \frac{(2x-5)(x-3)}{(3x+4)(x-3)} = \frac{2x-5}{3x+4}$$

B.

$$\frac{2x-1+\frac{7}{x+4}}{3x-8+\frac{17}{x+4}} = \frac{2x-1+\frac{7}{x+4}}{3x-8+\frac{17}{x+4}} \cdot \frac{x+4}{x+4}$$

▶ The LCM is $x+4$.

$$= \frac{(2x-1)(x+4)+\frac{7}{x+4}(x+4)}{(3x-8)(x+4)+\frac{17}{x+4}(x+4)}$$

$$= \frac{2x^2+7x-4+7}{3x^2+4x-32+17}$$

$$= \frac{2x^2+7x+3}{3x^2+4x-15}$$

$$= \frac{(2x+1)(x+3)}{(3x-5)(x+3)} = \frac{2x+1}{3x-5}$$

Problem 1 Simplify.

A. $\dfrac{3 + \frac{16}{x} + \frac{16}{x^2}}{6 + \frac{5}{x} - \frac{4}{x^2}}$ B. $\dfrac{2x + 5 + \frac{14}{x-3}}{4x + 16 + \frac{49}{x-3}}$

Solution See page A22.

Example 2 Simplify: $2 - \dfrac{2}{2 - \frac{2}{2-x}}$

Solution $2 - \dfrac{2}{2 - \frac{2}{2-x}} = 2 - \dfrac{2}{2 - \frac{2}{2-x}} \cdot \dfrac{2-x}{2-x} =$ ▶ Simplify the term that is a complex fraction. The LCM is $2 - x$.

$2 - \dfrac{2(2-x)}{2(2-x) - 2} = 2 - \dfrac{4-2x}{4-2x-2} =$

$2 - \dfrac{4-2x}{2-2x} = 2 - \dfrac{2(2-x)}{2(1-x)} = 2 - \dfrac{2-x}{1-x} =$ ▶ The LCM of the denominators is $1 - x$.

$\dfrac{2(1-x)}{1-x} - \dfrac{2-x}{1-x} = \dfrac{2 - 2x - (2-x)}{1-x}$ ▶ Subtract.

$\dfrac{2 - 2x - 2 + x}{1-x} = \dfrac{-x}{1-x} = \dfrac{-x}{-(x-1)} = \dfrac{x}{x-1}$

Problem 2 Simplify: $3 + \dfrac{3}{3 + \frac{3}{y}}$

Solution See page A23.

EXERCISES 4.3

1 Simplify.

1. $\dfrac{2 - \frac{1}{3}}{4 + \frac{11}{3}}$

2. $\dfrac{3 + \frac{5}{2}}{8 - \frac{8}{2}}$

3. $\dfrac{3 - \frac{2}{3}}{5 + \frac{5}{6}}$

4. $\dfrac{1 + \frac{1}{x}}{1 - \frac{1}{x^2}}$

5. $\dfrac{\frac{1}{y^2} - 1}{1 + \frac{1}{y}}$

6. $\dfrac{a - 2}{\frac{4}{a} - a}$

7. $\dfrac{\dfrac{25}{a} - a}{5 + a}$

8. $\dfrac{\dfrac{1}{a^2} - \dfrac{1}{a}}{\dfrac{1}{a^2} + \dfrac{1}{a}}$

9. $\dfrac{\dfrac{1}{b} + \dfrac{1}{2}}{\dfrac{4}{b^2} - 1}$

10. $\dfrac{2 - \dfrac{4}{x+2}}{5 - \dfrac{10}{x+2}}$

11. $\dfrac{4 + \dfrac{12}{2x-3}}{5 + \dfrac{15}{2x-3}}$

12. $\dfrac{\dfrac{3}{2a-3} + 2}{\dfrac{-6}{2a-3} - 4}$

13. $\dfrac{\dfrac{-5}{b-5} - 3}{\dfrac{10}{b-5} + 6}$

14. $\dfrac{\dfrac{x}{x+1} - \dfrac{1}{x}}{\dfrac{x}{x+1} + \dfrac{1}{x}}$

15. $\dfrac{\dfrac{2a}{a-1} - \dfrac{3}{a}}{\dfrac{1}{a-1} + \dfrac{2}{a}}$

16. $\dfrac{\dfrac{3}{x}}{\dfrac{9}{x^2}}$

17. $\dfrac{\dfrac{x}{3x-2}}{\dfrac{x}{9x^2-4}}$

18. $\dfrac{\dfrac{a^2-b^2}{4a^2b}}{\dfrac{a+b}{16ab^2}}$

19. $\dfrac{1 - \dfrac{1}{x} - \dfrac{6}{x^2}}{1 - \dfrac{4}{x} + \dfrac{3}{x^2}}$

20. $\dfrac{1 - \dfrac{3}{x} - \dfrac{10}{x^2}}{1 + \dfrac{11}{x} + \dfrac{18}{x^2}}$

21. $\dfrac{1 + \dfrac{1}{x} - \dfrac{12}{x^2}}{\dfrac{9}{x^2} + \dfrac{3}{x} - 2}$

22. $\dfrac{\dfrac{15}{x^2} - \dfrac{2}{x} - 1}{\dfrac{4}{x^2} - \dfrac{5}{x} + 4}$

23. $\dfrac{6 + \dfrac{2}{x} - \dfrac{20}{x^2}}{3 - \dfrac{17}{x} + \dfrac{20}{x^2}}$

24. $\dfrac{3 + \dfrac{19}{x} + \dfrac{20}{x^2}}{6 + \dfrac{5}{x} - \dfrac{4}{x^2}}$

25. $\dfrac{1 - \dfrac{2x}{3x-4}}{x - \dfrac{32}{3x-4}}$

26. $\dfrac{1 - \dfrac{12}{3x+10}}{x - \dfrac{8}{3x+10}}$

27. $\dfrac{x - 1 + \dfrac{2}{x-4}}{x + 3 + \dfrac{6}{x-4}}$

28. $\dfrac{x - 5 - \dfrac{18}{x+2}}{x + 7 + \dfrac{6}{x+2}}$

29. $\dfrac{x - 4 + \dfrac{9}{2x+3}}{x + 3 - \dfrac{5}{2x+3}}$

30. $\dfrac{2x - 3 - \dfrac{10}{4x-5}}{3x + 2 + \dfrac{11}{4x-5}}$

31. $\dfrac{3x - 2 - \dfrac{5}{2x-1}}{x - 6 + \dfrac{9}{2x-1}}$

32. $\dfrac{x + 4 - \dfrac{7}{2x-5}}{2x + 7 - \dfrac{28}{2x-5}}$

33. $\dfrac{\dfrac{1}{a} - \dfrac{3}{a-2}}{\dfrac{2}{a} + \dfrac{5}{a-2}}$

34. $\dfrac{\dfrac{2}{b} - \dfrac{5}{b+3}}{\dfrac{3}{b} + \dfrac{3}{b+3}}$

35. $\dfrac{\dfrac{1}{y^2} - \dfrac{1}{xy} - \dfrac{2}{x^2}}{\dfrac{1}{y^2} - \dfrac{3}{xy} + \dfrac{2}{x^2}}$

36. $\dfrac{\dfrac{2}{b^2} - \dfrac{5}{ab} - \dfrac{3}{a^2}}{\dfrac{2}{b^2} + \dfrac{7}{ab} + \dfrac{3}{a^2}}$

37. $\dfrac{\dfrac{x-1}{x+1} - \dfrac{x+1}{x-1}}{\dfrac{x-1}{x+1} + \dfrac{x+1}{x-1}}$

38. $\dfrac{\dfrac{y}{y+2} - \dfrac{y}{y-2}}{\dfrac{y}{y+2} + \dfrac{y}{y-2}}$

39. $4 - \dfrac{2}{2 - \dfrac{3}{x}}$

40. $a+\dfrac{a}{a+\dfrac{1}{a}}$

41. $a-\dfrac{a}{1-\dfrac{a}{1-a}}$

42. $3-\dfrac{3}{3-\dfrac{3}{3-x}}$

43. $3-\dfrac{2}{1-\dfrac{2}{3-\dfrac{2}{x}}}$

44. $a+\dfrac{a}{2+\dfrac{1}{1-\dfrac{2}{a}}}$

45. $a-\dfrac{1}{2-\dfrac{2}{2-\dfrac{2}{a}}}$

SUPPLEMENTAL EXERCISES 4.3

Simplify.

46. $\dfrac{x^{-1}+y^{-1}}{x^{-1}-y^{-1}}$

47. $\dfrac{x^{-1}}{y^{-1}}+\dfrac{y}{x}$

48. $\dfrac{1}{(a-b)^{-1}}$

49. $\dfrac{x^{-1}+y}{x^{-1}-y}$

Find the reciprocal of the quotient. Write your answer in simplest form.

50. $\dfrac{x-\dfrac{1}{x}}{1+\dfrac{1}{x}}$

51. $\dfrac{a-\dfrac{1}{a}}{\dfrac{1}{a}+1}$

52. $1-\dfrac{1}{1-\dfrac{1}{b-2}}$

53. $2-\dfrac{2}{2-\dfrac{2}{c-1}}$

Simplify.

54. $\dfrac{\dfrac{1}{x+h}-\dfrac{1}{x}}{h}$

55. $\dfrac{\dfrac{1}{(x+h)^2}-\dfrac{1}{x^2}}{h}$

Solve. Write your answer in simplest form.

56. If $a=\dfrac{b^2+4b+4}{b^2-4}$ and $b=\dfrac{1}{c}$, express a in terms of c.

57. If $x=\dfrac{y^2-6y+9}{y^2-9}$ and $y=\dfrac{1}{2z}$, express x in terms of z.

58. If $z=\dfrac{3y^2-12y}{6y^3-96y}$ and $y=\dfrac{1}{x}$, express z in terms of x.

59. If $c=\dfrac{2b^3-8b}{6b^2+12b}$ and $b=\dfrac{1}{2a}$, express c in terms of a.

SECTION 4.4 Rational Equations

1 Solve fractional equations

To solve an equation containing fractions, **clear denominators** by multiplying each side of the equation by the LCM of the denominators. Then solve for the variable.

Solve: $\dfrac{3x}{x-5} = 5 - \dfrac{5}{x-5}$

$$\frac{3x}{x-5} = 5 - \frac{5}{x-5}$$

Multiply each side of the equation by the LCM of the denominators.

$$(x-5)\left(\frac{3x}{x-5}\right) = (x-5)\left(5 - \frac{5}{x-5}\right)$$

$$3x = (x-5)5 - (x-5)\left(\frac{5}{x-5}\right)$$

Simplify.

$$3x = 5x - 25 - 5$$
$$3x = 5x - 30$$

Solve the equation for x.

$$-2x = -30$$
$$x = 15$$

15 checks as a solution.
The solution is 15.

Occasionally, a value of the variable that appears to be a solution will make one of the denominators zero. In this case, the equation has no solution for that value of the variable.

Solve: $\dfrac{3x}{x-3} = 2 + \dfrac{9}{x-3}$

$$\frac{3x}{x-3} = 2 + \frac{9}{x-3}$$

Multiply each side of the equation by the LCM of the denominators.

$$(x-3)\left(\frac{3x}{x-3}\right) = (x-3)\left(2 + \frac{9}{x-3}\right)$$

Use the Distributive Property.

$$3x = (x-3)2 + (x-3)\left(\frac{9}{x-3}\right)$$
$$3x = 2x - 6 + 9$$
$$3x = 2x + 3$$
$$x = 3$$

Substituting 3 into the equation results in division by zero. Because division by zero is not defined, the equation has no solution.

$$\frac{3x}{x-3} = 2 + \frac{9}{x-3}$$
$$\frac{3(3)}{3-3} = 2 + \frac{9}{3-3}$$
$$\frac{9}{0} = 2 + \frac{9}{0}$$

Multiplying each side of an equation by a variable expression may produce an equation with different solutions from the original equation. Thus, anytime you multiply each side of an equation by a variable expression, you must check the resulting solution.

Example 1 Solve.

A. $\frac{3}{12} = \frac{5}{x+5}$ B. $\frac{2x}{x-2} = \frac{1}{3x-4} + 2$

Solution A.

$$\frac{3}{12} = \frac{5}{x+5}$$
$$12(x+5)\frac{3}{12} = 12(x+5)\frac{5}{x+5}$$
$$(x+5)3 = (12)5$$
$$3x + 15 = 60$$
$$3x = 45$$
$$x = 15$$

▶ Multiply each side of the equation by the LCM of the denominators.

15 checks as a solution.

The solution is 15.

B.

$$\frac{2x}{x-2} = \frac{1}{3x-4} + 2$$
$$(x-2)(3x-4)\frac{2x}{x-2} = (x-2)(3x-4)\left(\frac{1}{3x-4} + 2\right)$$
$$(3x-4)2x = (x-2)(3x-4)\left(\frac{1}{3x-4}\right) + (x-2)(3x-4)2$$
$$6x^2 - 8x = x - 2 + 6x^2 - 20x + 16$$
$$6x^2 - 8x = 6x^2 - 19x + 14$$
$$11x = 14$$
$$x = \frac{14}{11}$$

$\frac{14}{11}$ checks as a solution.

The solution is $\frac{14}{11}$.

Problem 1 Solve.

A. $\frac{5}{2x-3} = \frac{-2}{x+1}$ B. $\frac{4x+1}{2x-1} = 2 + \frac{3}{x-3}$

Solution See page A23.

2 Proportions

Quantities such as 3 feet, 5 liters, and 2 miles are number quantities written with units. In these examples, the units are feet, liters, and miles.

A **ratio** is the quotient of two quantities that have the same unit.

The weekly wages of a painter are \$425. The painter spends \$50 a week for food. The ratio of wages spent for food to the total weekly wages is written:

$\frac{\$50}{\$425} = \frac{50}{425} = \frac{2}{17}$ A ratio is in simplest form when the two numbers do not have a common factor. The units are not written.

A **rate** is the quotient of two quantities that have different units.

A car travels 120 mi on 3 gal of gas. The miles-to-gallons rate is:

$\frac{120 \text{ mi}}{3 \text{ gal}} = \frac{40 \text{ mi}}{1 \text{ gal}}$ A rate is in simplest form when the two numbers do not have a common factor. The units are written as part of the rate.

A **proportion** is an equation that states two ratios or rates are equal. For example, $\frac{90 \text{ km}}{4 \text{ L}} = \frac{45 \text{ km}}{2 \text{ L}}$ and $\frac{3}{4} = \frac{x+2}{16}$ are proportions.

Note that a proportion is a special kind of fractional equation. Many application problems can be solved by using proportions.

Solve: The sales tax on a car that costs \$4000 is \$220. To find the sales tax on a car that costs \$10,500, write a proportion using x to represent the sales tax. Solve the proportion.

$$\frac{220}{4000} = \frac{x}{10{,}500}$$

$$\frac{11}{200} = \frac{x}{10{,}500}$$

$$(200)(10{,}500)\frac{11}{200} = (200)(10{,}500)\frac{x}{10{,}500}$$

▶ Multiply each side of the equation by the LCM of the denominators.

$$(10{,}500)(11) = 200x$$

$$115{,}500 = 200x$$

$$577.50 = x$$

The sales tax on the \$10,500 car is \$577.50.

Example 2 A stock investment of 50 shares pays a dividend of \$106. At this rate, how many additional shares are required to earn a dividend of \$424?

Strategy To find the additional number of shares that are required, write and solve a proportion using x to represent the additional number of shares. Then $50 + x$ is the total number of shares of stock.

Solution

$$\frac{106}{50} = \frac{424}{50 + x}$$

$$\frac{53}{25} = \frac{424}{50 + x}$$

$$25(50 + x)\frac{53}{25} = 25(50 + x)\frac{424}{50 + x}$$

$$(50 + x)53 = (25)424$$

$$2650 + 53x = 10{,}600$$

$$53x = 7950$$

$$x = 150$$

An additional 150 shares of stock are required.

Problem 2 Two pounds of cashews cost \$3.10. At this rate, how much would 15 lb of cashews cost?

Solution See page A23.

3 Work problems

4

If a mason can build a retaining wall in 12 h, then in 1 h the mason can build $\frac{1}{12}$ of the wall. The mason's rate of work is $\frac{1}{12}$ of the wall each hour. The **rate of work** is that part of a task that is completed in one unit of time. If an apprentice can build the wall in x hours, the rate of work for the apprentice is $\frac{1}{x}$ of the wall each hour.

In solving a work problem, the goal is to determine the time it takes to complete a task. The basic equation that is used to solve work problems is

Rate of work × Time worked = Part of task completed.

For example, if a pipe can fill a tank in 5 h, then in 2 h the pipe will fill $\frac{1}{5} \times 2 = \frac{2}{5}$ of the tank. In t hours, the pipe will fill $\frac{1}{5} \times t = \frac{t}{5}$ of the tank.

Solve: A mason can build a wall in 10 h. An apprentice can build a wall in 15 h. How long will it take to build a wall when they work together?

STRATEGY for solving a work problem

■ For each person or machine, write a numerical or variable expression for the rate of work, the time worked, and the part of the task completed. The results can be recorded in a table.

Unknown time to build the wall working together: t

	Rate of work	·	Time worked	=	Part of task completed
Mason	$\frac{1}{10}$	·	t	=	$\frac{t}{10}$
Apprentice	$\frac{1}{15}$	·	t	=	$\frac{t}{15}$

■ Determine how the parts of the task completed are related. Use the fact that the sum of the parts of the task completed must equal 1, the complete task.

The sum of the part of the task completed by the mason and the part of the task completed by the apprentice is 1.

$$\frac{t}{10} + \frac{t}{15} = 1$$

$$30\left(\frac{t}{10} + \frac{t}{15}\right) = 30(1)$$

$$3t + 2t = 30$$

$$5t = 30$$

$$t = 6$$

Working together, they will build the wall in 6 h.

Example 3 An electrician requires 12 h to wire a house. The electrician's apprentice can wire a house in 16 h. After working alone on one job for 4 h, the electrician quits, and the apprentice completes the task. How long does it take the apprentice to finish wiring the house?

Strategy ■ Time required for the apprentice to finish wiring the house: t

	Rate	Time	Part
Electrician	$\frac{1}{12}$	4	$\frac{4}{12}$
Apprentice	$\frac{1}{16}$	t	$\frac{t}{16}$

- The sum of the part of the task completed by the electrician and the part of the task completed by the apprentice is 1.

Solution

$$\frac{4}{12} + \frac{t}{16} = 1$$
$$\frac{1}{3} + \frac{t}{16} = 1$$
$$48\left(\frac{1}{3} + \frac{t}{16}\right) = 48(1)$$
$$16 + 3t = 48$$
$$3t = 32$$
$$t = \frac{32}{3}$$

It will take the apprentice $10\frac{2}{3}$ h to finish wiring the house.

Problem 3 Two water pipes can fill a tank with water in 6 h. The larger pipe working alone can fill the tank in 9 h. How long will it take the smaller pipe working alone to fill the tank?

Solution See page A24.

4 Uniform motion problems

A car that travels constantly in a straight line at 55 mph is in uniform motion. **Uniform motion** means that the speed of an object does not change.

The basic equation used to solve uniform motion problems is:

$$\textbf{Distance} = \textbf{Rate} \times \textbf{Time}$$

An alternate form of this equation can be written by solving the equation for time. This form of the equation is used to solve the following problem.

$$\frac{\textbf{Distance}}{\textbf{Rate}} = \textbf{Time}$$

Solve: A motorist drove 150 mi on country roads before driving 50 mi on mountain roads. The rate of speed on the country roads was three times the rate on the mountain roads. The time spent traveling the 200 mi was 5 h. Find the rate of the motorist on the country roads.

STRATEGY for solving a uniform motion problem

- For each object, write a numerical or variable expression for the distance, rate, and time. The results can be recorded in a table.

The unknown rate of speed on the mountain roads: r
Rate of speed on the country roads: $3r$

	Distance	÷	Rate	=	Time
Country roads	150	÷	$3r$	=	$\frac{150}{3r}$
Mountain roads	50	÷	r	=	$\frac{50}{r}$

- Determine how the times traveled by each object are related. For example, it may be known that the times are equal, or the total time may be known.

The total time of the trip is 5 h.

$$\frac{150}{3r} + \frac{50}{r} = 5$$
$$\frac{50}{r} + \frac{50}{r} = 5$$
$$r\left(\frac{50}{r} + \frac{50}{r}\right) = r(5)$$
$$50 + 50 = 5r$$
$$100 = 5r$$
$$20 = r$$

The rate of speed on the country roads was $3r$. Replace r with 20 and evaluate.

$$3r = 3(20) = 60$$

The rate of speed on the country roads was 60 mph.

Example 4 A marketing executive traveled 810 mi on a corporate jet in the same amount of time that it took to travel an additional 162 mi by helicopter. The rate of the jet was 360 mph faster than the rate of the helicopter. Find the rate of the jet.

Strategy

- Rate of the helicopter: r
  Rate of the jet: $r + 360$

	Distance	Rate	Time
Jet	810	$r + 360$	$\frac{810}{r + 360}$
Helicopter	162	r	$\frac{162}{r}$

- The time traveled by jet is equal to the time traveled by helicopter.

Solution

$$\frac{810}{r + 360} = \frac{162}{r}$$

$$r(r + 360)\left(\frac{810}{r + 360}\right) = r(r + 360)\left(\frac{162}{r}\right)$$

$$810r = (r + 360)162$$

$$810r = 162r + 58{,}320$$

$$648r = 58{,}320$$

$$r = 90$$

▶ The rate of the helicopter is 90 mph.

$$r + 360 = 90 + 360 = 450$$

▶ Substitute the value of r into the variable expression for the rate of the jet.

The rate of the jet was 450 mph.

Problem 4 A plane can fly at a rate of 150 mph in calm air. Traveling with the wind, the plane flew 700 mi in the same amount of time that it flew 500 mi against the wind. Find the rate of the wind.

Solution See page A24.

EXERCISES 4.4

1 Solve.

1. $\frac{x}{2} + \frac{5}{6} = \frac{x}{3}$

2. $\frac{x}{5} - \frac{2}{9} = \frac{x}{15}$

3. $1 - \frac{3}{y} = 4$

4. $7 + \frac{6}{y} = 5$

5. $\frac{8}{2x - 1} = 2$

6. $3 = \frac{18}{3x - 4}$

7. $\frac{4}{x - 4} = \frac{2}{x - 2}$

8. $\frac{x}{3} = \frac{x + 1}{7}$

9. $\frac{x - 2}{5} = \frac{1}{x + 2}$

10. $\frac{x + 4}{10} = \frac{6}{x - 3}$

11. $\frac{3}{x - 2} = \frac{4}{x}$

12. $\frac{5}{x} = \frac{2}{x + 3}$

13. $\frac{3}{x - 4} + 2 = \frac{5}{x - 4}$

14. $\frac{5}{y + 3} - 2 = \frac{7}{y + 3}$

15. $\frac{8}{x - 5} = \frac{3}{x}$

16. $\frac{16}{2 - x} = \frac{4}{x}$

17. $5 + \frac{8}{a - 2} = \frac{4a}{a - 2}$

18. $\frac{-4}{a - 4} = 3 - \frac{a}{a - 4}$

19. $\frac{x}{2} + \frac{20}{x} = 7$

20. $3x = \frac{4}{x} - \frac{13}{2}$

21. $\frac{6}{x - 5} = \frac{1}{x}$

22. $\frac{8}{x - 2} = \frac{4}{x + 1}$

23. $\frac{x}{x+2} = \frac{6}{x+5}$

24. $\frac{x}{x-2} = \frac{3}{x-4}$

25. $-\frac{5}{x+7} + 1 = \frac{4}{x+7}$

26. $5 - \frac{2}{2x-5} = \frac{3}{2x-5}$

27. $\frac{x}{x-1} = \frac{10}{x+3}$

28. $\frac{5}{x+2} = \frac{x}{x+8}$

29. $\frac{6}{x+5} = \frac{2x}{x+1}$

30. $\frac{x}{x+2} = \frac{6}{x+5}$

31. $\frac{2}{4y^2-9} + \frac{1}{2y-3} = \frac{3}{2y+3}$

32. $\frac{5}{x-2} - \frac{2}{x+2} = \frac{3}{x^2-4}$

33. $\frac{5}{x^2-7x+12} = \frac{2}{x-3} + \frac{5}{x-4}$

34. $\frac{9}{x^2+7x+10} = \frac{5}{x+2} - \frac{3}{x+5}$

2 Solve.

35. In a wildlife preserve, 60 ducks are captured, tagged, and then released. Later, 200 ducks are examined, and three of the 200 ducks are found to have tags. Estimate the number of ducks in the preserve.

36. A pre-election survey showed that 7 out of every 12 voters would vote in an election. At this rate, how many people would be expected to vote in a city of 210,000?

37. The real estate tax for a house that cost \$110,000 is \$1375. At this rate, what is the value of a house for which the real estate tax is \$2062.50?

38. The license fee for a car that cost \$9000 was \$108. At this rate, what is the license fee for a car that cost \$12,400?

39. The scale on an architectural drawing is $\frac{1}{4}$ in. represents one foot. Find the dimensions of a room that measures $4\frac{1}{4}$ in. by $5\frac{1}{2}$ in. on the drawing.

40. A contractor estimated that 15 ft^2 of window space will be allowed for every 160 ft^2 of floor space. Using this estimate, how much window space will be allowed for 3200 ft^2 of floor space?

41. A quality control inspector found 5 defective diodes in a shipment of 4000 diodes. At this rate, how many diodes would be defective in a shipment of 3200 diodes?

42. One hundred twenty ceramic tiles are required to tile 24 ft^2 of area. At this rate, how many tiles are required to tile 300 ft^2?

43. Three-fourths of an ounce of a medication are required for a 120-lb adult. At the same rate, how many additional ounces of medication are required for a 200-lb adult?

44. A stock investment of 120 shares pays a dividend of \$288. At this rate, how many additional shares are required to earn a dividend of \$720?

45. Six ounces of an insecticide are mixed with 15 gal of water to make a spray for spraying an orange grove. At the same rate, how much additional insecticide is required to be mixed with 100 gal of water?

46. An investment of $5000 earns $425 each year. At the same rate, how much additional money must be invested to earn $765 each year?

47. A farmer estimates that 5625 bushels of corn can be harvested from 125 acres of land. Using this estimate, how many additional acres are needed to harvest 13,500 bushels of corn?

48. A mechanic's pay for 8 h of work is $120. At the same rate of pay, how much would the mechanic earn for 42 h of work?

49. A magazine pays freelance writers by the word for published articles. If the magazine pays $250 for an article that is 1200 words long, how much will be paid for an article that is 4500 words long?

50. A contractor estimated that 30 ft^3 of cement is required to make a 90-ft^2 concrete floor. Using this estimate, how many additional cubic feet of cement would be required to make a 120-ft^2 concrete floor?

51. A computer printer can print a 1000-word document in 20 s. At this rate, how many seconds are required to print a document that contains 8000 words?

52. A caterer estimates that 2 gal of fruit punch will serve 30 people. How much additional punch is necessary to serve 75 people?

53. An average jogger burns 120 calories by jogging 1 mi. How many miles would a jogger need to run in order to burn 300 calories?

54. A major league baseball player opened the season by getting 26 base hits in 95 times at bat. At the same rate, how many hits would the player get in 475 times at bat?

3 Solve.

55. One printer can print the paychecks for the employees of a company in 54 min. A second printer can print the checks in 81 min. How long would it take to print the checks with both printers operating?

56. A mason can construct a retaining wall in 18 h. The mason's apprentice can do the job in 27 h. How long would it take to construct the wall when they work together?

57. One solar heating panel can raise the temperature of water 1° in 30 min. A second solar heating panel can raise the temperature 1° in 45 min. How long would it take to raise the temperature of the water 1° when both solar panels are operating?

58. One member of a gardening team can landscape a new lawn in 36 h. The other member of the team can do the job in 45 h. How long would it take to landscape the lawn when both gardeners work together?

59. One member of a telephone crew can wire new telephone lines in 5 h, while it would take 7.5 h for the other member of the crew to do the job. How long would it take to wire new telephone lines when both members of the crew are working together?

60. A new printer can print checks three times faster than an old printer. The old printer can print the checks in 30 min. How long would it take to print the checks when both printers are operating?

61. A new machine can package transistors four times faster than an older machine. Working together, the machines can package the transistors in 8 h. How long would it take the new machine working alone to package the transistors?

62. An experienced electrician can wire a room twice as fast as an apprentice electrician. Working together, the electricians can wire a room in 5 h. How long would it take the apprentice working alone to wire a room?

63. One member of a gardening team can mow and clean up a lawn in 6 h. With both members of the team working, the job can be done in 4 h. How long would it take the second member of the team, working alone, to do the job?

64. A student can type a 60-page term paper in 4 h. With a friend's assistance, the paper can be typed in 3 h. How long would it take the friend working alone to type the paper?

65. The larger of two printers being used to print the payroll for a major corporation requires 40 min to print the payroll. After both printers have been operating for 10 min, the larger printer malfunctions. The smaller printer requires 50 more minutes to complete the payroll. How long would it take the smaller printer working alone to print the payroll?

66. An experienced bricklayer can work twice as fast as an apprentice bricklayer. After working together on a job for 8 h, the experienced bricklayer quit. The apprentice required 12 more hours to finish the job. How long would it take the experienced bricklayer working alone to do the job?

67. A roofer requires 12 h to shingle a roof. After the roofer and an apprentice work on a roof for 3 h, the roofer moves on to another job. The apprentice requires 12 more hours to finish the job. How long would it take the apprentice working alone to do the job?

68. A welder requires 25 h to do a job. After the welder and an apprentice work on a job for 10 h, the welder quits. The apprentice finishes the job in 17 h. How long would it take the apprentice working alone to do the job?

69. Three computers can print out a task in 20 min, 30 min, and 60 min, respectively. How long would it take to complete the task when all three computers are working?

70. Three machines are filling soda bottles. The machines can fill the daily quota of soda bottles in 12 h, 15 h, and 20 h, respectively. How long would it take to fill the daily quota of soda bottles when all three machines are working?

71. With both hot and cold water running, a bathtub can be filled in 10 min. The drain will empty the tub in 15 min. A child turns both faucets on and leaves the drain open. How long will it be before the bathtub starts to overflow?

72. The inlet pipe can fill a water tank in 30 min. The outlet pipe can empty the tank in 20 min. How long would it take to empty a full tank when both pipes are open?

73. An oil tank has two inlet pipes and one outlet pipe. One inlet pipe can fill the tank in 12 h, and the other inlet pipe can fill the tank in 20 h. The outlet pipe can empty the tank in 10 h. How long would it take to fill the tank when all three pipes are open?

74. Water from a tank is being used for irrigation at the same time as the tank is being filled. The two inlet pipes can fill the tank in 6 h and 12 h, respectively. The outlet pipe can empty the tank in 24 h. How long will it take to fill the tank when all three pipes are open?

4 Solve.

75. An express bus travels 320 mi in the same amount of time that a car travels 280 mi. The rate of the car is 8 mph less than the rate of the bus. Find the rate of the bus.

76. A commercial jet travels 1620 mi in the same amount of time that a corporate jet travels 1260 mi. The rate of the commercial jet is 120 mph faster than the rate of the corporate jet. Find the rate of each jet.

77. A passenger train travels 295 mi in the same amount of time that a freight train travels 225 mi. The rate of the passenger train is 14 mph faster than the rate of the freight train. Find the rate of each train.

78. The rate of a bicyclist is 7 mph faster than the rate of a long-distance runner. The bicyclist travels 30 mi in the same amount of time that the runner travels 16 mi. Find the rate of the runner.

79. A cabin cruiser travels 39 mi in the same amount of time that a power boat travels 63 mi. The rate of the cabin cruiser is 10 mph less than the rate of the power boat. Find the rate of the cabin cruiser.

80. A motorcycle travels 117 mi in the same amount of time that a car travels 99 mi. The rate of the motorcycle is 10 mph faster than the rate of the car. Find the rate of the motorcycle.

81. A cyclist rode 40 mi before having a flat tire and then walking 5 mi to a service station. The cycling rate was four times faster than the walking rate. The time spent cycling and walking was 5 h. Find the rate at which the cyclist was riding.

82. A sales executive traveled 32 mi by car and then an additional 576 mi by plane. The rate of the plane was nine times faster than the rate of the car. The total time of the trip was 3 h. Find the rate of the plane.

83. A motorist drove 72 mi before running out of gas and then walking 4 mi to a gas station. The driving rate of the motorist was twelve times the walking rate. The time spent driving and walking was 2.5 h. Find the rate at which the motorist walks.

84. An insurance representative traveled 735 mi by commercial jet and then an additional 105 mi by helicopter. The rate of the jet was four times the rate of the helicopter. The entire trip took 2.2 h. Find the rate of the jet.

85. An express train and a car leave a town at 3 P.M. and head for a town 280 mi away. The rate of the express train is twice the rate of the car. The train arrives 4 h ahead of the car. Find the rate of the train.

86. A cyclist and a jogger start from a town at the same time and head for a destination 18 mi away. The rate of the cyclist is twice the rate of the jogger. The cyclist arrives 1.5 h ahead of the jogger. Find the rate of the cyclist.

87. A single-engined plane and a commercial jet leave an airport at 10 A.M. and head for an airport 960 mi away. The rate of the jet is four times the rate of the single-engined plane. The single-engined plane arrives 4 h after the jet. Find the rate of each plane.

88. A single-engined plane and a car start from a town at 6 A.M. and head for a town 450 mi away. The rate of the plane is three times the rate of the car. The plane arrives 6 h ahead of the car. Find the rate of the plane.

89. A motorboat can travel at 18 mph in still water. Traveling with the current of a river, the boat can travel 44 mi in the same amount of time that it took to go 28 mi against the current. Find the rate of the current.

90. An account executive traveled 110 mi by car in the same amount of time that it took a plane to travel 440 mi. The rate of the plane was 165 mph faster than the rate of the car. Find the rate of the plane.

91. A plane can fly at a rate of 180 mph in calm air. Traveling with the wind, the plane flew 615 mi in the same amount of time that it flew 465 mi against the wind. Find the rate of the wind.

92. A tour boat used for river excursions can travel 7 mph in calm water. The amount of time it takes to travel 20 mi with the current is the same amount of time that it takes to travel 8 mi against the current. Find the rate of the current.

93. A canoe can travel 8 mph in still water. Rowing with the current of a river, the canoe can travel 15 mi in the same amount of time that it takes to travel 9 mi against the current. Find the rate of the current.

94. A twin-engined plane can travel 180 mph in calm air. Flying with the wind, the plane can travel 900 mi in the same amount of time that it takes to fly 500 mi against the wind. Find the rate of the wind. Round to the nearest hundredth.

95. A jet can travel 550 mph in calm air. Flying with the wind, the jet can travel 3059 mi in the same amount of time that it takes to fly 2450 mi against the wind. Find the rate of the wind. Round to the nearest hundredth.

SUPPLEMENTAL EXERCISES 4.4

Solve.

96. On a map, two cities are $3\frac{3}{8}$ in. apart. If $\frac{3}{8}$ in. on the map represents 25 mi, find the number of miles in the distance between the two cities.

97. On a map, two cities are $2\frac{5}{8}$ in. apart. If $\frac{3}{4}$ in. on the map represents 50 mi, find the number of miles in the distance between the two cities.

98. A basketball player has made 6 out of every 7 foul shots attempted. If 35 foul shots were missed by the player in one season, how many foul shots were made by the player in that season?

99. An advertisement claims that 4 out of every 5 dentists surveyed recommend a particular brand of toothpaste. If 24 of the dentists polled did not recommend this brand of toothpaste, how many dentists were included in the survey?

100. The denominator of a fraction is 4 more than the numerator. If the numerator and denominator of the fraction are increased by 3, the new fraction is $\frac{5}{6}$. Find the original fraction.

101. The numerator of a fraction is 2 less than the denominator. If the numerator and denominator of the fraction are increased by 5, the new fraction is $\frac{9}{11}$. Find the original fraction.

102. One pipe can fill a tank in 3 h, a second pipe can fill the tank in 4 h, and a third pipe can fill the tank in 6 h. How long will it take to fill the tank with all three pipes operating?

103. One printer can print a company's paychecks in 24 min, a second printer can print the checks in 16 min, and a third printer can complete the job in 12 min. How long would it take to print the checks with all three printers operating?

104. By increasing your speed by 10 mph, you can drive the 200-mi trip to your hometown in 40 min less time than it usually takes you to drive the trip. How fast do you usually drive?

105. Because of weather conditions, a bus driver reduced the usual speed along a 165-mi bus route by 5 mph. The bus arrived only 15 min later than its usual arrival time. How fast does the bus usually travel?

106. If a pump can fill a pool in A hours and a second pump can fill the pool in B hours, find a formula, in terms of A and B, for the time it takes both pumps working together to fill the pool.

107. If a parade is 1 mi long and proceeding at 3 mph, how long will it take a runner, jogging at 5 mph, to run from the beginning of the parade to the end and then back to the beginning?

SECTION 4.5 Literal equations

1 Literal equations

A **literal equation** is an equation that contains more than one variable. Examples of literal equations are shown below.

$$3x - 2y = 4$$
$$v^2 = v_0^2 + 2as$$

Formulas are used to express relationships among physical quantities. A **formula** is a literal equation that states a rule about measurement. Examples of formulas are shown below.

$s = vt - 16t^2$ (Physics)
$c^2 = a^2 + b^2$ (Geometry)
$A = P(1 + r)^t$ (Business)

The Addition and Multiplication Properties of Equations can be used to solve a literal equation for one of the variables. The goal is to rewrite the equation so that the variable being solved for is alone on one side of the equation, and all the other numbers and variables are on the other side.

Solve $C = \frac{5}{9}(F - 32)$ for F.

$$C = \frac{5}{9}(F - 32)$$

Use the Distributive Property to remove parentheses. $\quad C = \frac{5}{9}F - \frac{160}{9}$

Multiply each side of the equation by the LCM of the denominators. $\quad 9C = 5F - 160$

Add 160 to each side of the equation. $\quad 9C + 160 = 5F$

Divide each side of the equation by the coefficient 5. $\quad \frac{9C + 160}{5} = F$

Example 1 A. Solve $A = P + Prt$ for P. B. Solve $\frac{S}{S - C} = R$ for C.

Solution A.

$$A = P + Prt$$

$$A = (1 + rt)P$$ ▶ Factor P from $P + Prt$.

$$\frac{A}{1 + rt} = \frac{(1 + rt)P}{1 + rt}$$ ▶ Divide each side of the equation by $1 + rt$.

$$\frac{A}{1 + rt} = P$$

B.

$$\frac{S}{S - C} = R$$

$$(S - C)\frac{S}{S - C} = (S - C)R$$ ▶ Multiply each side of the equation by $S - C$.

$$S = SR - CR$$

$$CR + S = SR$$ ▶ Add CR to each side of the equation.

$$CR = SR - S$$ ▶ Subtract S from each side of the equation.

$$C = \frac{SR - S}{R}$$ ▶ Divide each side of the equation by R.

Problem 1 A. Solve $\frac{1}{R_1} + \frac{1}{R_2} = \frac{1}{R}$ for R. B. Solve $\frac{r}{r + 1} = t$ for r.

Solution See page A25.

EXERCISES 4.5

1

1. $P = 2L + 2W$; W (Geometry)

2. $F = \frac{9}{5}C + 32$; C (Temperature conversion)

3. $S = C - rC$; C (Business)

4. $A = P + Prt$; t (Business)

5. $PV = nRT$; R (Chemistry)

6. $A = \frac{1}{2}bh$; h (Geometry)

7. $F = \frac{Gm_1m_2}{r^2}$; m_2 (Physics)

8. $\frac{P_1V_1}{T_1} = \frac{P_2V_2}{T_2}$; P_2 (Chemistry)

9. $I = \frac{E}{R + r}$; R (Physics)

10. $S = V_0t - 16t^2$; V_0 (Physics)

11. $A = \frac{1}{2}h(b_1 + b_2)$; b_2 (Geometry)

12. $V = \frac{1}{3}\pi r^2h$; h (Geometry)

13. $\frac{1}{R} = \frac{1}{R_1} + \frac{1}{R_2}$; R_2 (Physics)

14. $\frac{1}{f} = \frac{1}{a} + \frac{1}{b}$; b (Physics)

15. $a_n = a_1 + (n - 1)d$; d (Mathematics)

16. $P = \frac{R - C}{n}$; R (Business)

17. $S = 2WH + 2WL + 2LH$; H (Geometry)

18. $S = 2\pi r^2 + 2\pi rH$; H (Geometry)

Solve for x.

19. $ax + by + c = 0$

20. $x = ax + b$

21. $ax + b = cx + d$

22. $y - y_1 = m(x - x_1)$

23. $\frac{a}{x} = \frac{b}{c}$

24. $\frac{1}{x} + \frac{1}{a} = b$

25. $\frac{1}{a} + \frac{1}{b} = \frac{1}{x}$

26. $a(a - x) = b(b - x)$

SUPPLEMENTAL EXERCISES 4.5

Solve for x.

27. $\frac{x - y}{y} = \frac{x + 5}{2y}$

28. $\frac{x - 2}{y} = \frac{x + 2}{5y}$

29. $\frac{x}{x + y} = \frac{2x}{4y}$

30. $\frac{2x}{x - 2y} = \frac{x}{2y}$

31. $\frac{x - y}{2x} = \frac{x - 3y}{5y}$

32. $\frac{x - y}{x} = \frac{2x}{9y}$

Solve for the given variable.

33. $\frac{w_1}{w_2} = \frac{f_2 - f}{f - f_1}$; f

34. $v = \frac{v_1 + v_2}{1 + \frac{v_1 v_2}{c^2}}$; v_1

Calculators and Computers

Rational Expressions

The program RATIONAL EXPRESSIONS on the Math ACE Disk will give you additional practice in multiplying and dividing rational expressions. There are three levels of difficulty, with the first level the easiest of the problems and the third level the most difficult. You may choose the level you wish to practice.

After you choose a level of difficulty, a problem will be displayed on the screen. Using paper and pencil, simplify the expression. When you are ready, press the RETURN key. The correct solution will be displayed.

After each problem, you may continue the same level of problems, return to the menu to change the level, or quit the program.

Something Extra

Errors in Algebraic Operations

Algebra involves the utilization of a basic system of properties and theorems to work problems and to prove other extended theorems. Many errors in mathematical operations occur because they "look right."

The statement $\frac{1}{a} + \frac{1}{b} = \frac{1}{a+b}$ is incorrect. Explain the strategy that should have been used and correct the statement.

This is the addition of two fractions.

Build equivalent fractions with a common denominator and place the sum of the numerators over the common denominator.

$$\frac{1}{a} + \frac{1}{b} = \frac{b}{ab} + \frac{a}{ab}$$
$$= \frac{b+a}{ab}$$

Correct the statement $(a + b)^2 = a^2 + b^2$, and identify the strategy that should have been used.

The Distributive Property was applied incorrectly. FOIL is used for squaring a binomial.

$$(a + b)^2 = a^2 + 2ab + b^2$$

The following expressions are incorrect. Correct the expressions and state the property, theorem, or operation that should have been used.

1. $(a^2)^4 = a^6$

2. $\frac{ax + by}{a + b} = x + y$

3. $x^{-1} + y^{-1} = \frac{1}{x + y}$

4. $\frac{a + b}{a} = 1 + b$

5. $2ab = 2a \cdot 2b = 4ab$

6. $x^{-1} = -x$

7. $x^2 + x^3 = x^5$

8. $(c^2x - c^2y)^2 = c^2(x - y)^2$

9. $x^2 \cdot x^{-3} = x^{-6}$

10. $x^2 - 8x - 7 = (x - 7)(x - 1)$

11. $x^3 - x = x^2$

12. $\frac{x + 4}{4} = x$

Chapter Summary

Key Words

A fraction in which both the numerator and the denominator are polynomials is a *rational expression*.

A rational expression is in *simplest form* when the numerator and denominator have no common factors.

The *least common multiple* (LCM) of two or more polynomials is the simplest polynomial that contains the factors of each polynomial.

The *reciprocal* of a rational expression is the rational expression with the numerator and denominator interchanged.

A *complex fraction* is a fraction whose numerator or denominator contains one or more fractions.

A *ratio* is the quotient of two quantities that have the same unit.

A *rate* is the quotient of two quantities that have different units.

A *proportion* is an equation that states two ratios or rates are equal.

A *literal equation* is an equation that contains more than one variable.

A *formula* is a literal equation that states a rule about measurements.

Essential Rules

To Add Fractions $\frac{a}{c} + \frac{b}{c} = \frac{a + b}{c}$

To Multiply Fractions $\frac{a}{b} \cdot \frac{c}{d} = \frac{ac}{bd}$

To Divide Fractions $\frac{a}{b} \div \frac{c}{d} = \frac{a}{b} \cdot \frac{d}{c}$

Equation for Work Problems	Rate of work × Time worked = Part of task completed
Uniform Motion Equation	Distance = Rate × Time

Chapter Review

1. Simplify: $\dfrac{6a^{5n} + 4a^{4n} - 2a^{3n}}{2a^{3n}}$

2. Simplify: $\dfrac{16 - x^2}{x^3 - 2x^2 - 8x}$

3. Simplify: $\dfrac{3x^4 + 11x^2 - 4}{3x^4 + 13x^2 + 4}$

4. Simplify: $\dfrac{x^3 - 27}{x^2 - 9}$

5. Simplify: $\dfrac{a^6b^4 + a^4b^6}{a^5b^4 - a^4b^4} \cdot \dfrac{a^2 - b^2}{a^4 - b^4}$

6. Simplify: $\dfrac{x^3 - 8}{x^3 + 2x^2 + 4x} \cdot \dfrac{x^3 + 2x^2}{x^2 - 4}$

7. Simplify: $\dfrac{x^2 + 10x + 16}{2x^2 + 13x - 24} \cdot \dfrac{6x^2 - 11x + 3}{x^2 + x - 2}$

8. Simplify: $\dfrac{16 - x^2}{6x - 6} \cdot \dfrac{x^2 + 5x + 6}{x^2 - 8x + 16}$

9. Simplify: $\dfrac{x^{2n} - 5x^n + 4}{x^{2n} - 2x^n - 8} \div \dfrac{x^{2n} - 4x^n + 3}{x^{2n} + 8x^n + 12}$

10. Simplify: $\dfrac{27x^3 - 8}{9x^3 + 6x^2 + 4x} \div \dfrac{9x^2 - 12x + 4}{9x^2 - 4}$

11. Simplify: $\dfrac{x^{n+1} + x}{x^{2n} - 1} \div \dfrac{x^{n+2} - x^2}{x^{2n} - 2x^n + 1}$

12. Simplify: $\dfrac{3 - x}{x^2 + 3x + 9} \div \dfrac{x^2 - 9}{x^3 - 27}$

13. Simplify: $\dfrac{5}{3a^2b^3} + \dfrac{7}{8ab^4}$

14. Simplify: $\dfrac{3x^2 + 2}{x^2 - 4} - \dfrac{9x - x^2}{x^2 - 4}$

15. Simplify: $\dfrac{8}{9x^2 - 4} + \dfrac{5}{3x - 2} - \dfrac{4}{3x + 2}$

16. Simplify: $\dfrac{6x}{3x^2 - 7x + 2} - \dfrac{2}{3x - 1} + \dfrac{3x}{x - 2}$

17. Simplify: $\dfrac{x}{x - 3} - 4 - \dfrac{2x - 5}{x + 2}$

18. Simplify: $\dfrac{x - 6 + \dfrac{6}{x - 1}}{x + 3 - \dfrac{12}{x - 1}}$

19. Simplify: $\dfrac{x + \dfrac{3}{x-4}}{3 + \dfrac{x}{x-4}}$

20. Simplify: $x + \dfrac{\dfrac{4}{x} - 1}{\dfrac{1}{x} - \dfrac{3}{x^2}}$

21. Simplify: $3 + \dfrac{1}{1 + \dfrac{1}{1 + \dfrac{1}{x}}}$

22. Simplify: $\dfrac{\dfrac{3x+4}{3x-4} + \dfrac{3x-4}{3x+4}}{\dfrac{3x-4}{3x+4} - \dfrac{3x+4}{3x-4}}$

23. Solve: $\dfrac{x}{45} = \dfrac{4}{15}$

24. Solve: $\dfrac{x+3}{12} = \dfrac{2}{3}$

25. Solve: $\dfrac{5x}{2x-3} + 4 = \dfrac{3}{2x-3}$

26. Solve: $-2 + \dfrac{3x}{4x+7} = \dfrac{x-2}{4x+7}$

27. Solve: $\dfrac{x}{4} + 1 = \dfrac{x}{3}$

28. Solve: $\dfrac{x}{x-3} = \dfrac{2x+5}{x+1}$

29. Solve: $\dfrac{6}{x-3} - \dfrac{1}{x+3} = \dfrac{51}{x^2-9}$

30. Solve: $\dfrac{30}{x^2+5x+4} + \dfrac{10}{x+4} = \dfrac{4}{x+1}$

31. Solve: $\dfrac{3}{3x-4} + \dfrac{2}{x+1} = \dfrac{4}{3x^2-x-4}$

32. Solve: $\dfrac{6}{2x-3} - \dfrac{5}{x+5} = \dfrac{5}{2x^2+7x-15}$

33. Solve $I = \dfrac{1}{R}V$ for R.

34. Solve $Q = \dfrac{N-S}{N}$ for N.

35. Solve $F = \dfrac{mv}{t}$ for v.

36. Solve $S = \dfrac{a}{1-r}$ for r.

37. A car uses 4 tanks of fuel to travel 1800 mi. At this rate, how many tanks of fuel would be required for a trip of 3000 mi?

38. A student reads 2 pages of text in 5 min. At this rate, how long will it take for the student to read 150 pages?

39. On a certain map, 2.5 in. represents 10 mi. How many miles would be represented by 12 in.?

40. An electrician requires 65 min to install a ceiling fan. The electrician and an apprentice working together take 40 min to install the fan. How long would it take the apprentice working alone to install the ceiling fan?

41. The inlet pipe can fill a tub in 24 min. The drain pipe can empty the tub in 15 min. How long would it take to empty a full tub when both pipes are open?

42. A gardener can mow a lawn in 42 min, while it takes an assistant 57 min to mow the same lawn. The gardener and assistant work together for 14 min and then the gardener stops. How long will it take the assistant to finish mowing the lawn?

43. Three students can paint a dormitory room in 8 h, 16 h, and 16 h respectively. How long would it take to complete the task when all three students are working?

44. A canoeist can travel 10 mph in calm water. The amount of time it takes to travel 60 mi with the current is the same amount of time as it takes to travel 40 mi against the current. Find the rate of the current.

45. A bus and a cyclist leave a school at 8 A.M. and head for a stadium 90 mi away. The rate of the bus is three times the rate of the cyclist. The cyclist arrives 4 h after the bus. Find the rate of the bus.

46. A helicopter travels 9 mi in the same amount of time as an airplane travels 10 mi. The rate of the airplane is 20 mph faster than the rate of the helicopter. Find the rate of the helicopter.

47. A tractor travels 10 mi in the same amount of time as a car travels 15 mi. The rate of the tractor is 15 mph less than the rate of the car. Find the rate of the tractor.

Chapter Test

1. Simplify: $\dfrac{v^3 - 4v}{2v^2 - 5v + 2}$

2. Simplify: $\dfrac{2a^2 - 8a + 8}{4 + 4a - 3a^2}$

3. Simplify: $\dfrac{3x^2 - 12}{5x - 15} \cdot \dfrac{2x^2 - 18}{x^2 + 5x + 6}$

4. Simplify: $\dfrac{x^2 + x - 6}{x^2 + 7x + 12} \div \dfrac{x^2 - 3x + 2}{x^2 + 6x + 8}$

5. Simplify: $\dfrac{2x^2 - x - 3}{2x^2 - 5x + 3} \div \dfrac{3x^2 - x - 4}{x^2 - 1}$

6. Simplify: $\dfrac{x^{2n} - x^n - 2}{x^{2n} + x^n} \cdot \dfrac{x^{2n} - x^n}{x^{2n} - 4}$

7. Simplify: $\dfrac{2}{x^2} + \dfrac{3}{y^2} - \dfrac{5}{2xy}$

8. Simplify: $\dfrac{3x}{x - 2} - 3 + \dfrac{4}{x + 2}$

9. Simplify: $\dfrac{2x - 1}{x + 2} - \dfrac{x}{x - 3}$

10. Simplify: $\dfrac{x + 2}{x^2 + 3x - 4} - \dfrac{2x}{x^2 - 1}$

11. Simplify: $\dfrac{1 - \dfrac{1}{x} - \dfrac{12}{x^2}}{1 + \dfrac{6}{x} + \dfrac{9}{x^2}}$

12. Simplify: $\dfrac{1 - \dfrac{1}{x + 2}}{1 - \dfrac{3}{x + 4}}$

13. Solve: $\dfrac{3}{x + 1} = \dfrac{2}{x}$

14. Solve: $\dfrac{4x}{x + 1} - x = \dfrac{2}{x + 1}$

15. Solve: $\dfrac{4x}{2x - 1} = 2 - \dfrac{1}{2x - 1}$

16. Solve $ax = bx + c$ for x.

17. The inlet pipe can fill a water tank in 48 min. The outlet pipe can empty the tank in 30 min. How long would it take to empty a full tank when both pipes are open?

18. An interior designer uses two rolls of wallpaper for every 45 ft^2 of wall space in an office. At this rate, how many rolls of wallpaper are needed for an office that has 315 ft^2 of wall space?

19. One landscaper can till the soil for a lawn in 30 min while it takes a second landscaper 15 min to do the same job. How long would it take to till the soil for the lawn with both landscapers working together?

20. A cyclist travels 20 mi in the same amount of time as a hiker walks 6 mi. The rate of the cyclist is 7 mph faster than the rate of the hiker. Find the rate of the cyclist.

Cumulative Review

1. Simplify: $8 - 4[-3 - (-2)]^2 \div 5$

2. Evaluate $3a^2 - (b^2 - c)^2$ when $a = 2$, $b = -3$, and $c = 1$.

3. Solve: $-\frac{2}{3}y = -\frac{4}{9}$

4. Solve: $\frac{2x - 3}{6} - \frac{x}{9} = \frac{x - 4}{3}$

5. Solve: $5 - |x - 4| = 2$

6. Solve: $(x - 2)(x + 1) = 4$

7. Simplify: $\frac{(2a^{-2}b^3)^{-2}}{(4a)^{-1}}$

8. Solve: $x - 3(1 - 2x) \geq 1 - 4(2 - 2x)$

9. Simplify: $(2a^2 - 3a + 1)(-2a^2)$

10. Factor: $2x^{2n} + 3x^n - 2$

11. Factor: $x^3y^3 - 27$

12. Simplify: $\frac{x^4 + x^3y - 6x^2y^2}{x^3 - 2x^2y}$

13. Simplify: $\frac{4x^3 + 2x^2 - 10x + 1}{x - 2}$

14. Simplify: $\frac{16x^2 - 9y^2}{16x^2y - 12xy^2} \div \frac{4x^2 - xy - 3y^2}{12x^2y^2}$

15. Simplify: $\frac{6x^3 - 5x^2 + 6x + 10}{3x + 2}$

16. Simplify: $\frac{5x}{3x^2 - x - 2} - \frac{2x}{x^2 - 1}$

17. Simplify: $\dfrac{x - 4 + \frac{5}{x + 2}}{x + 2 - \frac{1}{x + 2}}$

18. Solve: $\frac{2}{x - 3} = \frac{5}{2x - 3}$

19. Solve: $\frac{3}{x^2 - 36} = \frac{2}{x - 6} - \frac{5}{x + 6}$

20. Solve $I = \frac{E}{R + r}$ for r.

21. Simplify: $(1 - x^{-1})^{-1}$

22. Factor: $2xy + 8x - 3y - 12$

23. How many pounds of almonds that cost $5.40 per pound must be mixed with 50 lb of peanuts that cost $2.60 per pound to make a mixture that costs $4.00 per pound?

24. A pre-election survey showed that three out of five voters would vote in an election. At this rate, how many people would be expected to vote in a city of 125,000?

25. A new computer can work six times faster than an older computer. Working together, the computers can complete a job in 12 min. How long would it take the new computer working alone to do the job?

26. A plane can fly at a rate of 300 mph in calm air. Traveling with the wind, the plane flew 900 mi in the same amount of time as it flew 600 mi against the wind. Find the rate of the wind.

5

Rational Exponents and Radicals

Objectives

- Simplify expressions with rational exponents
- Write exponential expressions as radical expressions and radical expressions as exponential expressions
- Simplify expressions of the form $\sqrt[n]{a^n}$
- Simplify radical expressions
- Add and subtract radical expressions
- Multiply radical expressions
- Divide radical expressions
- Simplify complex numbers
- Add and subtract complex numbers
- Multiply complex numbers
- Divide complex numbers
- Solve equations containing one or more radical expressions
- Application problems

POINT OF INTEREST

Golden Ratio

The golden rectangle fascinated the early Greeks and appeared in much of their architecture. They considered this particular rectangle to be the most pleasing to the eye, and consequently, when used in the design of a building, would make the structure pleasant to see.

The golden rectangle is constructed from a square by drawing a line from the midpoint of the base of the square to the opposite vertex. Now extend the base of the square, starting from the midpoint, the length of the line. The resulting rectangle is called the golden rectangle.

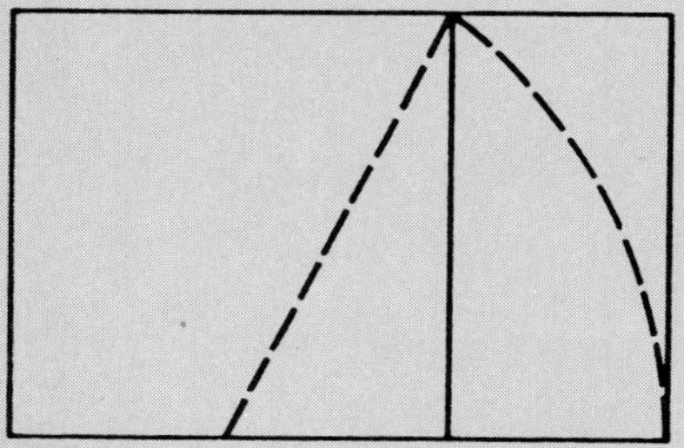

The Parthenon in Athens, Greece, is the classic example of the use of the golden rectangle in Greek architecture. A rendering of the Parthenon is shown here.

SECTION 5.1

Rational Exponents and Radical Expressions

1 Simplify expressions with rational exponents

In this section, the definition of a power is extended beyond integers so that any rational number can be used as an exponent. The definition is made so that the Rules of Exponents hold true for rational exponents.

Since the Rules of Exponents will hold true for rational exponents, the expression $\left(a^{\frac{1}{n}}\right)^n$ can be simplified by using the Rule for Simplifying Powers of Exponential Expressions.

$$\left(a^{\frac{1}{n}}\right)^n = a^{\frac{1}{n} \cdot n} = a^1 = a$$

Since $\left(a^{\frac{1}{n}}\right)^n = a$, the number $a^{\frac{1}{n}}$ is the number whose nth power is a. $a^{\frac{1}{n}}$ is called **the nth root of a.**

$$\left(a^{\frac{1}{n}}\right)^n = a \qquad \left(25^{\frac{1}{2}}\right)^2 = 25 \qquad \left(8^{\frac{1}{3}}\right)^3 = 8$$

$25^{\frac{1}{2}}$ is the number whose 2nd power is 25.

$25^{\frac{1}{2}} = 5$ because $(5)^2 = 25$.

$8^{\frac{1}{3}}$ is the number whose 3rd power is 8.

$8^{\frac{1}{3}} = 2$ because $(2)^3 = 8$.

In the expression $a^{\frac{1}{n}}$, a must be positive when n is a positive even integer.

$(-4)^{\frac{1}{2}}$ is not a real number since there is no real number whose 2nd power is -4.

As shown above, expressions that contain rational exponents do not always represent real numbers when the base of the exponential expression is a negative number For this reason, all variables in this chapter represent positive numbers unless otherwise stated.

When n is a positive odd integer, a can be a positive or a negative number.

$(-27)^{\frac{1}{3}}$ is the number whose 3rd power is -27.

$(-27)^{\frac{1}{3}} = -3$ because $(-3)^3 = -27$.

Using the definition of $a^{\frac{1}{n}}$ and the Rules of Exponents, it is possible to define any exponential expression that contains a rational exponent.

Definition of $a^{\frac{m}{n}}$

If $a^{\frac{1}{n}}$ is a real number and m and n are integers $(n \geq 0)$, then

$$a^{\frac{m}{n}} = a^{\frac{1}{n} \cdot m} = \left(a^{\frac{1}{n}}\right)^m$$

and $$a^{\frac{m}{n}} = a^{m \cdot \frac{1}{n}} = (a^m)^{\frac{1}{n}}.$$

Example 1 Simplify.

A. $27^{\frac{2}{3}}$ B. $32^{-\frac{2}{5}}$ C. $(-49)^{\frac{3}{2}}$

Solution A. $27^{\frac{2}{3}} = (3^3)^{\frac{2}{3}}$ ▶ Rewrite 27 as 3^3.

$= 3^{3\left(\frac{2}{3}\right)}$ ▶ Use the Rule for Simplifying Powers of Exponential Expressions.

$= 3^2$

$= 9$ ▶ Simplify.

B. $32^{-\frac{2}{5}} = (2^5)^{-\frac{2}{5}}$ ▶ Rewrite 32 as 2^5.

$= 2^{-2}$ ▶ Use the Rule for Simplifying Powers of Exponential Expressions.

$= \frac{1}{2^2}$ ▶ Use the Rule of Negative Exponents.

$= \frac{1}{4}$ ▶ Simplify.

C. $(-49)^{\frac{3}{2}}$ ▶ The base of the exponential expression is a negative expression, while the denominator of the exponent is a positive even number.

$(-49)^{\frac{3}{2}}$ is not a real number.

Problem 1 Simplify.

A. $64^{\frac{2}{3}}$ B. $16^{-\frac{3}{4}}$ C. $(-81)^{\frac{3}{4}}$

Solution See page A25.

Example 2 Simplify.

A. $b^{\frac{1}{2}} \cdot b^{\frac{2}{3}} \cdot b^{-\frac{1}{4}}$ B. $(x^6y^4)^{\frac{3}{2}}$ C. $\left(\frac{8a^3b^{-4}}{64a^{-9}b^2}\right)^{\frac{2}{3}}$

Solution A. $b^{\frac{1}{2}} \cdot b^{\frac{2}{3}} \cdot b^{-\frac{1}{4}} = b^{\frac{1}{2}+\frac{2}{3}-\frac{1}{4}}$ ▶ Use the Rule for Multiplying Exponential Expressions.

$= b^{\frac{6}{12}+\frac{8}{12}-\frac{3}{12}}$

$= b^{\frac{11}{12}}$

B. $(x^6y^4)^{\frac{3}{2}} = x^{6\left(\frac{3}{2}\right)}y^{4\left(\frac{3}{2}\right)}$ ▶ Use the Rule for Simplifying Powers of Products.

$= x^9y^6$

C. $\left(\dfrac{8a^3b^{-4}}{64a^{-9}b^2}\right)^{\frac{2}{3}} = \left(\dfrac{a^{12}}{8b^6}\right)^{\frac{2}{3}}$ ▶ Use the Rule for Dividing Exponential Expressions.

$= \left(\dfrac{a^{12}}{2^3b^6}\right)^{\frac{2}{3}}$ ▶ Rewrite 8 as 2^3.

$= \dfrac{a^8}{2^2b^4}$ ▶ Use the Rule for Simplifying Powers of Quotients.

$= \dfrac{a^8}{4b^4}$

Problem 2 Simplify.

A. $\dfrac{x^{\frac{1}{2}}y^{-\frac{5}{4}}}{x^{-\frac{4}{3}}y^{\frac{1}{3}}}$ B. $\left(x^{\frac{3}{4}}y^{\frac{1}{2}}z^{-\frac{2}{3}}\right)^{-\frac{4}{3}}$ C. $\left(\dfrac{16a^{-2}b^{\frac{4}{3}}}{9a^4b^{-\frac{2}{3}}}\right)^{-\frac{1}{2}}$

Solution See page A25.

2 Write exponential expressions as radical expressions and radical expressions as exponential expressions

Recall that $a^{\frac{1}{n}}$ is the nth root of a. The expression $\sqrt[n]{a}$ is another symbol for the nth root of a.

Definition of $\sqrt[n]{a}$

$$a^{\frac{1}{n}} = \sqrt[n]{a}$$

In the expression $\sqrt[n]{a}$, the symbol $\sqrt[n]{}$ is called a **radical,** n is the **index** of the radical, and a is the **radicand.** When $n = 2$, the radical expression represents a square root, and the index 2 is usually not written.

Any exponential expression with a rational exponent can be written as a radical expression.

If $a^{\frac{1}{n}}$ is a real number, then $\boldsymbol{a^{\frac{m}{n}}} = a^{m \cdot \frac{1}{n}} = (a^m)^{\frac{1}{n}} = \boldsymbol{\sqrt[n]{a^m}}$.

For $a > 0$, the expression $a^{\frac{m}{n}}$ can also be written $a^{\frac{m}{n}} = a^{\frac{1}{n} \cdot m} = (\sqrt[n]{a})^m$.

The exponential expression at the right has been written as a radical expression. $y^{\frac{2}{3}} = (y^2)^{\frac{1}{3}} = \sqrt[3]{y^2}$

The radical expressions at the right have been written as exponential expressions.

$\sqrt[5]{x^6} = (x^6)^{\frac{1}{5}} = x^{\frac{6}{5}}$

$\sqrt{17} = (17)^{\frac{1}{2}} = 17^{\frac{1}{2}}$

Example 3 Rewrite the exponential expression as a radical expression.

A. $(5x)^{\frac{2}{5}}$ B. $-2x^{\frac{2}{3}}$

Solution A. $(5x)^{\frac{2}{5}} = \sqrt[5]{(5x)^2}$
$= \sqrt[5]{25x^2}$

▶ The denominator of the rational exponent is the index of the radical. The numerator is the power of the radicand.

B. $-2x^{\frac{2}{3}} = -2(x^2)^{\frac{1}{3}}$
$= -2\sqrt[3]{x^2}$

▶ The -2 is not raised to the power.

Problem 3 Rewrite the exponential expression as a radical expression.

A. $(2x^3)^{\frac{3}{4}}$ B. $-5a^{\frac{5}{6}}$

Solution See page A25.

Example 4 Rewrite the radical expression as an exponential expression.

A. $\sqrt[5]{x^4}$ B. $\sqrt[3]{a^3 + b^3}$

Solution A. $\sqrt[5]{x^4} = (x^4)^{\frac{1}{5}} = x^{\frac{4}{5}}$

▶ The index of the radical is the denominator of the rational exponent. The power of the radicand is the numerator of the rational exponent.

B. $\sqrt[3]{a^3 + b^3} = (a^3 + b^3)^{\frac{1}{3}}$

▶ Note that $(a^3 + b^3)^{\frac{1}{3}} \neq a + b$.

Problem 4 Rewrite the radical expression as an exponential expression.

A. $\sqrt[3]{3ab}$ B. $\sqrt[4]{x^4 + y^4}$

Solution See page A25.

3 Simplify expressions of the form $\sqrt[n]{a^n}$

Every positive number has two square roots, one a positive and one a negative number. For example, because $(5)^2 = 25$ and $(-5)^2 = 25$, there are two square roots of 25, 5 and -5.

The symbol $\sqrt{}$ is used to indicate the positive or **principal square root.** To indicate the negative square root of a number, a negative sign is placed in front of the radical.

$\sqrt{25} = 5$

$-\sqrt{25} = -5$

The square root of zero is zero.

$\sqrt{0} = 0$

The square root of a negative number is not a real number since the square of a real number must be positive. | $\sqrt{-25}$ is not a real number.

The square root of a squared positive number is a positive number. | $\sqrt{5^2} = \sqrt{25} = 5$

The square root of a squared negative number is a positive number. | $\sqrt{(-5)^2} = \sqrt{25} = 5$

For any real number a, $\sqrt{a^2} = |a|$.

This says that for any real number, the square root of the number squared equals the absolute value of the number.

Every number has only one cube root.

The cube root of a positive number is positive. | $\sqrt[3]{8} = 2$, since $2^3 = 8$.

The cube root of a negative number is negative. | $\sqrt[3]{-8} = -2$, since $(-2)^3 = -8$.

For any real number a, $\sqrt[3]{a^3} = a$.

The following properties hold true for finding the nth root of a real number.

The nth Root of a^n

If n is an even integer, then $\sqrt[n]{a^n} = |a|$.

If n is an odd integer, then $\sqrt[n]{a^n} = a$.

$$\sqrt[6]{y^6} = |y| \qquad -\sqrt[12]{x^{12}} = -|x| \qquad \sqrt[5]{b^5} = b$$

Because it has been stated that all variables in this chapter represent positive numbers unless otherwise stated, the absolute signs will not be used in the examples and exercises.

To simplify $\sqrt{121}$, write the prime factorization of the radicand in exponential form. The radicand is a perfect square since its exponent is an even number. | $\sqrt{121} = \sqrt{11^2}$

Write the radical expression as an exponential expression. | $= (11^2)^{\frac{1}{2}}$

Use the Rule for Simplifying Powers of Exponential Expressions. | $= 11$

The radicand of the radical expression $\sqrt[3]{x^6y^9}$ is a perfect cube since the exponents on the variables are divisible by 3. Write the radical expression as an exponential expression.

$\sqrt[3]{x^6y^9} = (x^6y^9)^{\frac{1}{3}}$

Use the Rule for Simplifying Powers of Products.

$= x^2y^3$

Example 5 Simplify.

A. $\sqrt[5]{x^{15}}$ B. $\sqrt{49x^2y^{12}}$ C. $-\sqrt[4]{16a^4b^8}$

Solution A. $\sqrt[5]{x^{15}} = (x^{15})^{\frac{1}{5}}$

▶ The radicand is a perfect fifth power since 15 is divisible by 5. Write the radical expression as an exponential expression.

$= x^3$

▶ Use the Rule for Simplifying Powers of Exponential Expressions.

B. $\sqrt{49x^2y^{12}} = \sqrt{7^2x^2y^{12}}$

▶ Write the prime factorization of 49.

$= 7xy^6$

C. $-\sqrt[4]{16a^4b^8} = -\sqrt[4]{2^4a^4b^8}$

$= -2ab^2$

Problem 5 Simplify.

A. $-\sqrt[4]{x^{12}}$ B. $\sqrt{121x^{10}y^4}$ C. $\sqrt[3]{-125a^6b^9}$

Solution See page A25.

EXERCISES 5.1

1 Simplify.

1. $8^{\frac{1}{3}}$ **2.** $16^{\frac{1}{2}}$ **3.** $9^{\frac{3}{2}}$

4. $25^{\frac{3}{2}}$ **5.** $27^{-\frac{2}{3}}$ **6.** $64^{-\frac{1}{3}}$

7. $32^{\frac{2}{5}}$ **8.** $16^{\frac{3}{4}}$ **9.** $(-25)^{\frac{5}{2}}$

10. $(-36)^{\frac{1}{4}}$ **11.** $\left(\frac{25}{49}\right)^{-\frac{3}{2}}$ **12.** $\left(\frac{8}{27}\right)^{-\frac{2}{3}}$

13. $x^{\frac{1}{2}}x^{\frac{1}{2}}$ **14.** $a^{\frac{1}{3}}a^{\frac{5}{3}}$ **15.** $y^{-\frac{1}{4}}y^{\frac{3}{4}}$

16. $x^{\frac{2}{5}} \cdot x^{-\frac{4}{5}}$ **17.** $x^{-\frac{2}{3}} \cdot x^{\frac{3}{4}}$ **18.** $x \cdot x^{-\frac{1}{2}}$

19. $a^{\frac{1}{3}} \cdot a^{\frac{3}{4}} \cdot a^{-\frac{1}{2}}$ **20.** $y^{-\frac{1}{6}} \cdot y^{\frac{2}{3}} \cdot y^{\frac{1}{2}}$ **21.** $\dfrac{a^{\frac{1}{2}}}{a^{\frac{3}{2}}}$

22. $\dfrac{b^{\frac{1}{3}}}{b^{\frac{4}{3}}}$

23. $\dfrac{y^{-\frac{3}{4}}}{y^{\frac{1}{4}}}$

24. $\dfrac{x^{-\frac{3}{5}}}{x^{\frac{1}{5}}}$

25. $\dfrac{y^{\frac{2}{3}}}{y^{-\frac{5}{6}}}$

26. $\dfrac{b^{\frac{3}{4}}}{b^{-\frac{3}{2}}}$

27. $(x^2)^{-\frac{1}{2}}$

28. $(a^8)^{-\frac{3}{4}}$

29. $\left(x^{-\frac{2}{3}}\right)^6$

30. $\left(y^{-\frac{5}{6}}\right)^{12}$

31. $\left(a^{-\frac{1}{2}}\right)^{-2}$

32. $\left(b^{-\frac{2}{3}}\right)^{-6}$

33. $\left(x^{-\frac{3}{8}}\right)^{-\frac{4}{5}}$

34. $\left(y^{-\frac{3}{2}}\right)^{-\frac{2}{9}}$

35. $\left(a^{\frac{1}{2}} \cdot a\right)^2$

36. $\left(b^{\frac{2}{3}} \cdot b^{\frac{1}{6}}\right)^6$

37. $\left(x^{-\frac{1}{2}} \cdot x^{\frac{3}{4}}\right)^{-2}$

38. $\left(a^{\frac{1}{2}} \cdot a^{-2}\right)^3$

39. $\left(y^{-\frac{1}{2}} \cdot y^{\frac{3}{2}}\right)^{\frac{2}{3}}$

40. $\left(b^{-\frac{2}{3}} \cdot b^{\frac{1}{4}}\right)^{-\frac{4}{3}}$

41. $(x^8y^2)^{\frac{1}{2}}$

42. $(a^3b^9)^{\frac{2}{3}}$

43. $(x^4y^2z^6)^{\frac{3}{2}}$

44. $(a^8b^4c^4)^{\frac{3}{4}}$

45. $(x^{-3}y^6)^{-\frac{1}{3}}$

46. $(a^2b^{-6})^{-\frac{1}{2}}$

47. $\left(x^{-2}y^{\frac{1}{3}}\right)^{-\frac{3}{4}}$

48. $\left(a^{-\frac{2}{3}}b^{\frac{2}{3}}\right)^{\frac{3}{2}}$

49. $\left(\dfrac{x^{\frac{1}{2}}}{y^{-2}}\right)^4$

50. $\left(\dfrac{b^{-\frac{3}{4}}}{a^{-\frac{1}{2}}}\right)^8$

51. $\dfrac{x^{\frac{1}{4}} \cdot x^{-\frac{1}{2}}}{x^{\frac{2}{3}}}$

52. $\dfrac{b^{\frac{1}{2}} \cdot b^{-\frac{3}{4}}}{b^{\frac{1}{4}}}$

53. $\left(\dfrac{y^{\frac{2}{3}} \cdot y^{-\frac{5}{6}}}{y^{\frac{1}{9}}}\right)^9$

54. $\left(\dfrac{a^{\frac{1}{3}} \cdot a^{-\frac{2}{3}}}{a^{\frac{1}{2}}}\right)^4$

55. $\left(\dfrac{b^2 \cdot b^{-\frac{3}{4}}}{b^{-\frac{1}{2}}}\right)^{-\frac{1}{2}}$

56. $\dfrac{\left(x^{-\frac{5}{6}} \cdot x^3\right)^{-\frac{2}{3}}}{x^{\frac{4}{3}}}$

57. $\left(a^{\frac{2}{3}}b^2\right)^6(a^3b^3)^{\frac{1}{3}}$

58. $\left(x^3y^{-\frac{1}{2}}\right)^{-2}(x^{-3}y^2)^{\frac{1}{6}}$

59. $(16m^{-2}n^4)^{-\frac{1}{2}}\left(mn^{\frac{1}{2}}\right)$

60. $(27m^3n^{-6})^{\frac{1}{3}}\left(m^{-\frac{1}{3}}n^{\frac{5}{6}}\right)^6$

61. $\left(\dfrac{x^{\frac{1}{2}}y^{-\frac{3}{4}}}{y^{\frac{2}{3}}}\right)^{-6}$

62. $\left(\dfrac{x^{\frac{1}{2}}y^{-\frac{5}{4}}}{y^{-\frac{3}{4}}}\right)^{-4}$

63. $\left(\dfrac{2^{-6}b^{-3}}{a^{-\frac{1}{2}}}\right)^{-\frac{2}{3}}$

64. $\left(\dfrac{49c^{\frac{5}{3}}}{a^{-\frac{1}{4}}b^{\frac{5}{6}}}\right)^{-\frac{3}{2}}$

65. $\dfrac{(x^{-2}y^4)^{\frac{1}{2}}}{\left(x^{\frac{1}{2}}\right)^4}$

66. $\dfrac{(x^{-3})^{\frac{1}{3}}}{(x^9y^6)^{\frac{1}{6}}}$

67. $a^{-\frac{1}{4}}\left(a^{\frac{5}{4}} - a^{\frac{9}{4}}\right)$

68. $x^{\frac{4}{3}}\left(x^{\frac{2}{3}} + x^{-\frac{1}{3}}\right)$

69. $y^{\frac{2}{3}}\left(y^{\frac{1}{3}} + y^{-\frac{2}{3}}\right)$

70. $b^{-\frac{2}{5}}\left(b^{-\frac{3}{5}} - b^{\frac{7}{5}}\right)$

71. $a^{\frac{1}{6}}\left(a^{\frac{5}{6}} - a^{-\frac{7}{6}}\right)$

72. $\left(x^{\frac{n}{3}}\right)^{3n}$

73. $\left(a^{\frac{2}{n}}\right)^{-5n}$

74. $x^n \cdot x^{\frac{n}{2}}$

75. $a^{\frac{n}{2}} \cdot a^{-\frac{n}{3}}$

76. $\dfrac{y^{\frac{n}{2}}}{y^{-n}}$

77. $\dfrac{b^{\frac{m}{3}}}{b^m}$

78. $\left(x^{\frac{2}{n}}\right)^n$

79. $(x^{5n})^{2n}$

80. $\left(x^{\frac{n}{4}}y^{\frac{n}{8}}\right)^8$

81. $\left(x^{\frac{n}{2}}y^{\frac{n}{3}}\right)^6$

2 Rewrite the exponential expression as a radical expression.

82. $3^{\frac{1}{4}}$

83. $5^{\frac{1}{2}}$

84. $a^{\frac{3}{2}}$

85. $b^{\frac{4}{3}}$

86. $(2t)^{\frac{5}{2}}$

87. $(3x)^{\frac{2}{3}}$

88. $-2x^{\frac{2}{3}}$

89. $-3a^{\frac{2}{5}}$

90. $(a^2b)^{\frac{2}{3}}$

91. $(x^2y^3)^{\frac{3}{4}}$

92. $(a^2b^4)^{\frac{3}{5}}$

93. $(a^3b^7)^{\frac{3}{2}}$

94. $(4x + 3)^{\frac{3}{4}}$

95. $(3x - 2)^{\frac{1}{3}}$

96. $x^{-\frac{2}{3}}$

Rewrite the radical expression as an exponential expression.

97. $\sqrt{14}$

98. $\sqrt{7}$

99. $\sqrt[3]{x}$

100. $\sqrt[4]{y}$

101. $\sqrt[3]{x^4}$

102. $\sqrt[4]{a^3}$

103. $\sqrt[5]{b^3}$

104. $\sqrt[4]{b^5}$

105. $\sqrt[3]{2x^2}$

106. $\sqrt[5]{4y^7}$

107. $-\sqrt{3x^5}$

108. $-\sqrt[4]{4x^5}$

109. $3x\sqrt[3]{y^2}$

110. $2y\sqrt{x^3}$

111. $\sqrt{a^2 + 2}$

3 Simplify.

112. $\sqrt{x^{16}}$

113. $\sqrt{y^{14}}$

114. $-\sqrt{x^8}$

115. $-\sqrt{a^6}$
116. $\sqrt{x^2y^{10}}$
117. $\sqrt{a^{14}b^6}$
118. $\sqrt{25x^6}$
119. $\sqrt{121y^{12}}$
120. $\sqrt[3]{x^3y^9}$
121. $\sqrt[3]{a^6b^{12}}$
122. $-\sqrt[3]{x^{15}y^3}$
123. $-\sqrt[3]{a^9b^9}$
124. $\sqrt[3]{27a^9}$
125. $\sqrt[3]{125b^{15}}$
126. $\sqrt[3]{-8x^3}$
127. $\sqrt[3]{-a^6b^9}$
128. $\sqrt{16a^4b^{12}}$
129. $\sqrt{25x^8y^2}$
130. $\sqrt{-16x^4y^2}$
131. $\sqrt{-9a^6b^8}$
132. $\sqrt[3]{27x^9}$
133. $\sqrt[3]{8a^{21}b^6}$
134. $\sqrt[3]{-64x^9y^{12}}$
135. $\sqrt[3]{-27a^3b^{15}}$
136. $\sqrt[4]{x^{16}}$
137. $\sqrt[4]{y^{12}}$
138. $\sqrt[4]{16x^{12}}$
139. $\sqrt[4]{81a^{20}}$
140. $-\sqrt[4]{x^8y^{12}}$
141. $-\sqrt[4]{a^{16}b^4}$
142. $\sqrt[5]{x^{20}y^{10}}$
143. $\sqrt[5]{a^5b^{25}}$
144. $\sqrt[4]{81x^4y^{20}}$
145. $\sqrt[4]{16a^8b^{20}}$
146. $\sqrt[5]{32a^5b^{10}}$
147. $\sqrt[5]{-32x^{15}y^{20}}$
148. $\sqrt[5]{243x^{10}y^{40}}$
149. $\sqrt{\dfrac{16x^2}{y^{14}}}$
150. $\sqrt{\dfrac{49a^4}{b^{24}}}$
151. $\sqrt[3]{\dfrac{27b^3}{a^9}}$
152. $\sqrt[3]{\dfrac{64x^{15}}{y^6}}$
153. $\sqrt{(2x+3)^2}$
154. $\sqrt{(4x+1)^2}$
155. $\sqrt{x^2+2x+1}$
156. $\sqrt{x^2+4x+4}$

SUPPLEMENTAL EXERCISES 5.1

Which of the numbers is the largest, **a**, **b**, or **c**?

157. **a.** $16^{\frac{1}{2}}$ **b.** $16^{\frac{1}{4}}$ **c.** $16^{\frac{3}{2}}$

158. **a.** $27^{\frac{1}{3}}$ **b.** $27^{-\frac{4}{3}}$ **c.** $27^{\frac{2}{3}}$

159. **a.** $(-125)^{\frac{2}{3}}$ **b.** $(-125)^{-\frac{1}{3}}$ **c.** $(-125)^{\frac{1}{3}}$

160. **a.** $(-32)^{-\frac{1}{5}}$ **b.** $(-32)^{\frac{2}{5}}$ **c.** $(-32)^{\frac{1}{5}}$

161. **a.** $4^{\frac{1}{2}} \cdot 4^{\frac{3}{2}}$ **b.** $3^{\frac{4}{5}} \cdot 3^{\frac{6}{5}}$ **c.** $7^{\frac{1}{4}} \cdot 7^{\frac{3}{4}}$

162. **a.** $\dfrac{81^{\frac{3}{4}}}{81^{\frac{1}{4}}}$ **b.** $\dfrac{64^{\frac{2}{3}}}{64^{\frac{1}{3}}}$ **c.** $\dfrac{36^{\frac{5}{2}}}{36^{\frac{3}{2}}}$

Simplify.

163. $\sqrt[3]{\sqrt{x^6}}$
164. $\sqrt{\sqrt[3]{y^6}}$
165. $\sqrt[4]{\sqrt{a^8}}$
166. $\sqrt[5]{\sqrt[3]{b^{15}}}$
167. $\sqrt{\sqrt{16x^{12}}}$
168. $\sqrt{\sqrt{81y^8}}$
169. $\sqrt[5]{\sqrt{a^{10}b^{20}}}$
170. $\sqrt[3]{\sqrt{64x^{36}y^{30}}}$
171. $\sqrt[4]{\sqrt{256a^{16}b^{32}}}$

For what value of p is the given equation true?

172. $x^p x^{\frac{1}{4}} = x$ **173.** $y^p y^{\frac{2}{5}} = y$ **174.** $\sqrt[p]{x^8} = x^2$

175. $\sqrt[p]{y^{15}} = y^5$ **176.** $\dfrac{x^p}{x^{\frac{3}{5}}} = x^{\frac{1}{5}}$ **177.** $\dfrac{y^p}{y^{\frac{3}{4}}} = y^{\frac{1}{2}}$

178. $x^p x^{-\frac{1}{2}} = x^{\frac{1}{4}}$ **179.** $\dfrac{x^{\frac{1}{2}}}{x^p} = x^{\frac{5}{6}}$ **180.** $\dfrac{y^{\frac{1}{3}}}{y^p} = y^{\frac{5}{9}}$

Write in exponential form and then simplify.

181. $\sqrt{16^{\frac{1}{2}}}$ **182.** $\sqrt[3]{4^{\frac{3}{2}}}$ **183.** $\sqrt[4]{32^{-\frac{4}{5}}}$ **184.** $\sqrt{243^{-\frac{4}{5}}}$

SECTION 5.2 Operations on Radical Expressions

1 Simplify radical expressions

If a number is not a perfect power, its root can only be approximated, for example, $\sqrt{5}$ and $\sqrt[3]{3}$. These numbers are **irrational numbers.** Their decimal representations never terminate or repeat.

$$\sqrt{5} = 2.2360679\ldots \qquad \sqrt[3]{3} = 1.4422495\ldots$$

The approximate square roots and cube roots of the positive integers up to 105 can be found in the Appendix on page A6. The roots have been rounded to the nearest thousandth.

A radical expression is in simplest form when the radicand contains no factor that is a perfect power. The Product Property of Radicals is used to simplify radical expressions whose radicands are not perfect powers.

The Product Property of Radicals

If a and b are positive real numbers, then $\sqrt[n]{ab} = \sqrt[n]{a} \cdot \sqrt[n]{b}$.

To simplify $\sqrt{48}$, write the prime factorization of the radicand in exponential form. $\sqrt{48} = \sqrt{2^4 \cdot 3}$

Use the Product Property of Radicals to write the expression as a product. $= \sqrt{2^4}\,\sqrt{3}$

Simplify. $= 2^2\sqrt{3}$
$= 4\sqrt{3}$

To simplify $\sqrt[3]{x^7}$, write the radicand as the product of a perfect cube and a factor that does not contain a perfect cube. $\sqrt[3]{x^7} = \sqrt[3]{x^6 \cdot x}$

Use the Product Property of Radicals to write the expression as a product. $= \sqrt[3]{x^6}\,\sqrt[3]{x}$

Simplify. $= x^2\sqrt[3]{x}$

Example 1 Simplify: $\sqrt[4]{32x^7}$

Solution

$\sqrt[4]{32x^7} = \sqrt[4]{2^5x^7}$ ▶ Write the prime factorization of the coefficient of the radicand in exponential form.

$= \sqrt[4]{2^4x^4(2x^3)}$ ▶ Write the radicand as the product of a perfect fourth power and factors that do not contain a perfect fourth power.

$= \sqrt[4]{2^4x^4}\,\sqrt[4]{2x^3}$ ▶ Use the Product Property of Radicals to write the expression as a product.

$= 2x\sqrt[4]{2x^3}$ ▶ Simplify.

Problem 1 Simplify: $\sqrt[5]{x^7}$

Solution See page A26.

2 Add and subtract radical expressions

The Distributive Property is used to simplify the sum or difference of radical expressions that have the same radicand and the same index.

$$3\sqrt{5} + 8\sqrt{5} = (3 + 8)\sqrt{5} = 11\sqrt{5}$$

$$2\sqrt[3]{3x} - 9\sqrt[3]{3x} = (2 - 9)\sqrt[3]{3x} = -7\sqrt[3]{3x}$$

Radical expressions that are in simplest form and have different radicands or different indices cannot be simplified by the Distributive Property. The expressions shown below cannot be simplified by the Distributive Property.

$3\sqrt[4]{2} - 6\sqrt[4]{3}$	$2\sqrt[4]{4x} + 3\sqrt[3]{4x}$
Radicands are different	Indices are different

Simplify: $3\sqrt{32x^2} - 2x\sqrt{2} + \sqrt{128x^2}$

Simplify each term.

$3\sqrt{32x^2} - 2x\sqrt{2} + \sqrt{128x^2} =$

$3\sqrt{2^5x^2} - 2x\sqrt{2} + \sqrt{2^7x^2} =$

$3\sqrt{2^4x^2}\,\sqrt{2} - 2x\sqrt{2} + \sqrt{2^6x^2}\,\sqrt{2} =$

$3 \cdot 2^2x\sqrt{2} - 2x\sqrt{2} + 2^3x\sqrt{2} =$

$12x\sqrt{2} - 2x\sqrt{2} + 8x\sqrt{2} =$

Simplify by using the Distributive Property.

$18x\sqrt{2}$

Example 2 Simplify.

A. $5b\sqrt[4]{32a^7b^5} - 2a\sqrt[4]{162a^3b^9}$

B. $5\sqrt[5]{2x^7y^{11}} - y\sqrt[5]{64x^7y^6}$

Solution A.
$$\begin{aligned} 5b\sqrt[4]{32a^7b^5} - 2a\sqrt[4]{162a^3b^9} &= 5b\sqrt[4]{2^5a^7b^5} - 2a\sqrt[4]{3^4\cdot 2a^3b^9} \\ &= 5b\sqrt[4]{2^4a^4b^4}\,\sqrt[4]{2a^3b} - 2a\sqrt[4]{3^4b^8}\,\sqrt[4]{2a^3b} \\ &= 5b\cdot 2ab\sqrt[4]{2a^3b} - 2a\cdot 3b^2\sqrt[4]{2a^3b} \\ &= 10ab^2\sqrt[4]{2a^3b} - 6ab^2\sqrt[4]{2a^3b} \\ &= 4ab^2\sqrt[4]{2a^3b} \end{aligned}$$

B.
$$\begin{aligned} 5\sqrt[5]{2x^7y^{11}} - y\sqrt[5]{64x^7y^6} &= 5\sqrt[5]{2x^7y^{11}} - y\sqrt[5]{2^6x^7y^6} \\ &= 5\sqrt[5]{x^5y^{10}}\,\sqrt[5]{2x^2y} - y\sqrt[5]{2^5x^5y^5}\,\sqrt[5]{2x^2y} \\ &= 5\cdot xy^2\sqrt[5]{2x^2y} - y\cdot 2xy\sqrt[5]{2x^2y} \\ &= 5xy^2\sqrt[5]{2x^2y} - 2xy^2\sqrt[5]{2x^2y} \\ &= 3xy^2\sqrt[5]{2x^2y} \end{aligned}$$

Problem 2 Simplify.

A. $3xy\sqrt[3]{81x^5y} - \sqrt[3]{192x^8y^4}$

B. $4a\sqrt[3]{54a^7b^9} + a^2b\sqrt[3]{128a^4b^6}$

Solution See page A26.

3 Multiply radical expressions

The Product Property of Radicals is used to multiply radical expressions with the same index.

$$\sqrt{3x}\cdot\sqrt{5y} = \sqrt{3x\cdot 5y} = \sqrt{15xy}$$

To simplify $\sqrt[3]{2a^5b}\,\sqrt[3]{16a^2b^2}$, use the Product Property of Radicals to multiply the radicands. Then simplify.

$$\begin{aligned} \sqrt[3]{2a^5b}\,\sqrt[3]{16a^2b^2} &= \sqrt[3]{32a^7b^3} \\ &= \sqrt[3]{2^5a^7b^3} \\ &= \sqrt[3]{2^3a^6b^3}\,\sqrt[3]{2^2a} \\ &= 2a^2b\sqrt[3]{4a} \end{aligned}$$

To simplify $\sqrt{2x}(\sqrt{8x} - \sqrt{3})$, use the Distributive Property to remove parentheses. Then simplify.

$$\begin{aligned} \sqrt{2x}(\sqrt{8x} - \sqrt{3}) &= \sqrt{16x^2} - \sqrt{6x} \\ &= \sqrt{2^4x^2} - \sqrt{6x} \\ &= 2^2x - \sqrt{6x} \\ &= 4x - \sqrt{6x} \end{aligned}$$

Example 3 Simplify: $\sqrt{3x}(\sqrt{27x^2} - \sqrt{3x})$

Solution

$$\begin{aligned}\sqrt{3x}(\sqrt{27x^2} - \sqrt{3x}) &= \sqrt{81x^3} - \sqrt{9x^2}\\ &= \sqrt{3^4x^3} - \sqrt{3^2x^2}\\ &= \sqrt{3^4x^2}\sqrt{x} - \sqrt{3^2x^2}\\ &= 3^2x\sqrt{x} - 3x\\ &= 9x\sqrt{x} - 3x\end{aligned}$$

Problem 3 Simplify: $\sqrt{5b}(\sqrt{3b} - \sqrt{10})$

Solution See page A26.

To simplify $(\sqrt[3]{x} - 1)(\sqrt[3]{x} + 7)$, use the FOIL method to remove parentheses. Then simplify.

$$\begin{aligned}(\sqrt[3]{x} - 1)(\sqrt[3]{x} + 7) &= \sqrt[3]{x^2} + 7\sqrt[3]{x} - \sqrt[3]{x} - 7\\ &= \sqrt[3]{x^2} + 6\sqrt[3]{x} - 7\end{aligned}$$

The expressions $a + b$ and $a - b$, which are the sum and difference of two terms, are called **conjugates** of each other. The product of conjugates of the form $(a + b)(a - b)$ is $a^2 - b^2$.

$$(\sqrt{x} - 3)(\sqrt{x} + 3) = (\sqrt{x})^2 - 3^2 = x - 9$$

Example 4 Simplify.

A. $(2\sqrt[3]{x} - 3)(3\sqrt[3]{x} - 4)$ B. $(\sqrt{xy} - 2)(\sqrt{xy} + 2)$

Solution

A. $$\begin{aligned}(2\sqrt[3]{x} - 3)(3\sqrt[3]{x} - 4) &= 6\sqrt[3]{x^2} - 8\sqrt[3]{x} - 9\sqrt[3]{x} + 12\\ &= 6\sqrt[3]{x^2} - 17\sqrt[3]{x} + 12\end{aligned}$$

▶ Use the FOIL method.

B. $$\begin{aligned}(\sqrt{xy} - 2)(\sqrt{xy} + 2) &= (\sqrt{xy})^2 - 2^2\\ &= xy - 4\end{aligned}$$

▶ Use $(a - b)(a + b) = a^2 - b^2$.

Problem 4 Simplify.

A. $(2\sqrt[3]{2x} - 3)(\sqrt[3]{2x} - 5)$ B. $(2\sqrt{x} - 3)(2\sqrt{x} + 3)$

Solution See page A26.

4 Divide radical expressions

The Quotient Property of Radicals is used to divide radical expressions with the same index.

The Quotient Property of Radicals

> If a and b are positive real numbers, then $\sqrt[n]{\frac{a}{b}} = \frac{\sqrt[n]{a}}{\sqrt[n]{b}}$.

To simplify $\sqrt[3]{\frac{81x^5}{y^6}}$, use the Quotient Property of Radicals. Then simplify each radical expression.

$$\sqrt[3]{\frac{81x^5}{y^6}} = \frac{\sqrt[3]{81x^5}}{\sqrt[3]{y^6}} = \frac{\sqrt[3]{3^4x^5}}{\sqrt[3]{y^6}} = \frac{\sqrt[3]{3^3x^3}\sqrt[3]{3x^2}}{\sqrt[3]{y^6}} = \frac{3x\sqrt[3]{3x^2}}{y^2}$$

To simplify $\frac{\sqrt{5a^4b^7c^2}}{\sqrt{ab^3c}}$, use the Quotient Property of Radicals. Then simplify the radicand.

$$\frac{\sqrt{5a^4b^7c^2}}{\sqrt{ab^3c}} = \sqrt{\frac{5a^4b^7c^2}{ab^3c}} = \sqrt{5a^3b^4c} = \sqrt{a^2b^4}\sqrt{5ac} = ab^2\sqrt{5ac}$$

A radical expression is in simplest form when there is no fraction as part of the radicand and no radical remains in the denominator of the radical expression. The procedure used to remove a radical from the denominator is called **rationalizing the denominator.**

To simplify $\frac{2}{\sqrt{x}}$, multiply the expression by 1 in the form $\frac{\sqrt{x}}{\sqrt{x}}$. Then simplify.

$$\frac{2}{\sqrt{x}} = \frac{2}{\sqrt{x}} \cdot \frac{\sqrt{x}}{\sqrt{x}} = \frac{2\sqrt{x}}{\sqrt{x^2}} = \frac{2\sqrt{x}}{x}$$

Example 5 Simplify.

A. $\frac{5}{\sqrt{5x}}$ B. $\frac{3x}{\sqrt[3]{4x}}$

Solution A. $\frac{5}{\sqrt{5x}} = \frac{5}{\sqrt{5x}} \cdot \frac{\sqrt{5x}}{\sqrt{5x}} = \frac{5\sqrt{5x}}{\sqrt{5^2x^2}} = \frac{5\sqrt{5x}}{5x} = \frac{\sqrt{5x}}{x}$

▶ Multiply the expression by $\frac{\sqrt{5x}}{\sqrt{5x}}$.

B. $\dfrac{3x}{\sqrt[3]{4x}} = \dfrac{3x}{\sqrt[3]{2^2x}} \cdot \dfrac{\sqrt[3]{2x^2}}{\sqrt[3]{2x^2}} =$

► Multiply the expression by $\dfrac{\sqrt[3]{2x^2}}{\sqrt[3]{2x^2}}$. $\sqrt[3]{4x} = \sqrt[3]{2^2x}$. $\sqrt[3]{2^2x} \cdot \sqrt[3]{2x^2} = \sqrt[3]{2^3x^3}$, a perfect cube.

$$\frac{3x\sqrt[3]{2x^2}}{\sqrt[3]{2^3x^3}} = \frac{3x\sqrt[3]{2x^2}}{2x} = \frac{3\sqrt[3]{2x^2}}{2}$$

Problem 5 Simplify.

A. $\dfrac{y}{\sqrt{3y}}$ B. $\dfrac{3}{\sqrt[3]{3x^2}}$

Solution See page A26.

To simplify a fraction that has a binomial square root radical expression in the denominator, multiply the numerator and denominator by the conjugate of the denominator.

$$\frac{3}{5-\sqrt{7}} = \frac{3}{5-\sqrt{7}} \cdot \frac{5+\sqrt{7}}{5+\sqrt{7}}$$
$$= \frac{15+3\sqrt{7}}{25-7} = \frac{3(5+\sqrt{7})}{18}$$
$$= \frac{5+\sqrt{7}}{6}$$

$$\frac{\sqrt{x}-\sqrt{y}}{\sqrt{x}+\sqrt{y}} = \frac{\sqrt{x}-\sqrt{y}}{\sqrt{x}+\sqrt{y}} \cdot \frac{\sqrt{x}-\sqrt{y}}{\sqrt{x}-\sqrt{y}}$$
$$= \frac{\sqrt{x^2}-\sqrt{xy}-\sqrt{xy}+\sqrt{y^2}}{(\sqrt{x})^2-(\sqrt{y})^2} = \frac{x-2\sqrt{xy}+y}{x-y}$$

Example 6 Simplify.

A. $\dfrac{2-\sqrt{5}}{3+\sqrt{2}}$ B. $\dfrac{3+\sqrt{y}}{3-\sqrt{y}}$

Solution A. $\dfrac{2-\sqrt{5}}{3+\sqrt{2}} = \dfrac{2-\sqrt{5}}{3+\sqrt{2}} \cdot \dfrac{3-\sqrt{2}}{3-\sqrt{2}}$

$$= \frac{6-2\sqrt{2}-3\sqrt{5}+\sqrt{10}}{9-2}$$
$$= \frac{6-2\sqrt{2}-3\sqrt{5}+\sqrt{10}}{7}$$

B. $\dfrac{3+\sqrt{y}}{3-\sqrt{y}} = \dfrac{3+\sqrt{y}}{3-\sqrt{y}} \cdot \dfrac{3+\sqrt{y}}{3+\sqrt{y}}$

$$= \frac{9+3\sqrt{y}+3\sqrt{y}+\sqrt{y^2}}{9-(\sqrt{y})^2} = \frac{9+6\sqrt{y}+y}{9-y}$$

Problem 6 Simplify.

A. $\frac{4+\sqrt{2}}{3-\sqrt{3}}$ B. $\frac{\sqrt{2}+\sqrt{x}}{\sqrt{2}-\sqrt{x}}$

Solution See page A26.

EXERCISES 5.2

1 Simplify.

1. $\sqrt{x^4y^3z^5}$
2. $\sqrt{x^3y^6z^9}$
3. $\sqrt{8a^3b^8}$
4. $\sqrt{24a^9b^6}$
5. $\sqrt{45x^2y^3z^5}$
6. $\sqrt{60xy^7z^{12}}$
7. $\sqrt[3]{-125x^2y^4}$
8. $\sqrt[4]{16x^9y^5}$
9. $\sqrt[3]{-216x^5y^9}$
10. $\sqrt[3]{a^8b^{11}c^{15}}$
11. $\sqrt[3]{a^5b^8}$
12. $\sqrt[4]{64x^8y^{10}}$

2 Simplify.

13. $2\sqrt{x} - 8\sqrt{x}$
14. $3\sqrt{y} + 12\sqrt{y}$
15. $\sqrt{8} - \sqrt{32}$
16. $\sqrt{27} - \sqrt{75}$
17. $\sqrt{128x} - \sqrt{98x}$
18. $\sqrt{48x} + \sqrt{147x}$
19. $\sqrt{27a} - \sqrt{8a}$
20. $\sqrt{18b} + \sqrt{75b}$
21. $2\sqrt{2x^3} + 4x\sqrt{8x}$
22. $5y\sqrt{8y} + 2\sqrt{50y^3}$
23. $x\sqrt{75xy} - \sqrt{27x^3y}$
24. $3\sqrt{8x^2y^3} - 2x\sqrt{32y^3}$
25. $2\sqrt{32x^2y^3} - xy\sqrt{98y}$
26. $6y\sqrt{x^3y} - 2\sqrt{x^3y^3}$
27. $7b\sqrt{a^5b^3} - 2ab\sqrt{a^3b^3}$
28. $2a\sqrt{27ab^5} + 3b\sqrt{3a^3b}$
29. $\sqrt[3]{128} + \sqrt[3]{250}$
30. $\sqrt[3]{16} - \sqrt[3]{54}$
31. $2\sqrt[3]{3a^4} - 3a\sqrt[3]{81a}$
32. $2b\sqrt[3]{16b^2} + \sqrt[3]{128b^5}$
33. $3\sqrt[3]{x^5y^7} - 8xy\sqrt[3]{x^2y^4}$
34. $3\sqrt[4]{32a^5} - a\sqrt[4]{162a}$
35. $2a\sqrt[4]{16ab^5} + 3b\sqrt[4]{256a^5b}$
36. $2\sqrt{50} - 3\sqrt{125} + \sqrt{98}$
37. $3\sqrt{108} - 2\sqrt{18} - 3\sqrt{48}$
38. $\sqrt{9b^3} - \sqrt{25b^3} + \sqrt{49b^3}$
39. $\sqrt{4x^7y^5} + 9x^2\sqrt{x^3y^5} - 5xy\sqrt{x^5y^3}$
40. $2x\sqrt{8xy^2} - 3y\sqrt{32x^3} + \sqrt{8x^3y^2}$
41. $5a\sqrt{3a^3b} + 2a^2\sqrt{27ab} - 4\sqrt{75a^5b}$
42. $\sqrt[3]{54xy^3} - 5\sqrt[3]{2xy^3} + \sqrt[3]{128xy^3}$
43. $2\sqrt[3]{24x^3y^4} + 4x\sqrt[3]{81y^4} - 3y\sqrt[3]{24x^3y}$
44. $2a\sqrt[4]{32b^5} - 3b\sqrt[4]{162a^4b} + \sqrt[4]{2a^4b^5}$

3 Simplify.

45. $\sqrt{8}\sqrt{32}$
46. $\sqrt{14}\sqrt{35}$
47. $\sqrt[3]{4}\sqrt[3]{8}$
48. $\sqrt[3]{6}\sqrt[3]{36}$
49. $\sqrt{x^2y^5}\sqrt{xy}$
50. $\sqrt{a^3b}\sqrt{ab^4}$
51. $\sqrt{2x^2y}\sqrt{32xy}$
52. $\sqrt{5x^3y}\sqrt{10x^3y^4}$
53. $\sqrt[3]{x^2y}\sqrt[3]{16x^4y^2}$
54. $\sqrt[3]{4a^2b^3}\sqrt[3]{8ab^5}$
55. $\sqrt[4]{12ab^3}\sqrt[4]{4a^5b^2}$
56. $\sqrt[4]{36a^2b^4}\sqrt[4]{12a^5b^3}$
57. $\sqrt{3}(\sqrt{27} - \sqrt{3})$
58. $\sqrt{10}(\sqrt{10} - \sqrt{5})$
59. $\sqrt{x}(\sqrt{x} - \sqrt{2})$
60. $\sqrt{y}(\sqrt{y} - \sqrt{5})$
61. $\sqrt{2x}(\sqrt{8x} - \sqrt{32})$
62. $\sqrt{3a}(\sqrt{27a^2} - \sqrt{a})$
63. $(\sqrt{x} - 3)^2$
64. $(\sqrt{2x} + 4)^2$
65. $(4\sqrt{5} + 2)^2$
66. $2\sqrt{3x^2} \cdot 3\sqrt{12xy^3} \cdot \sqrt{6x^3y}$
67. $2\sqrt{14xy} \cdot 4\sqrt{7x^2y} \cdot 3\sqrt{8xy^2}$
68. $\sqrt[3]{8ab}\sqrt[3]{4a^2b^3}\sqrt[3]{9ab^4}$
69. $\sqrt[3]{2a^2b}\sqrt[3]{4a^3b^2}\sqrt[3]{8a^5b^6}$
70. $(\sqrt{2} - 3)(\sqrt{2} + 4)$
71. $(\sqrt{5} - 5)(2\sqrt{5} + 2)$
72. $(\sqrt{y} - 2)(\sqrt{y} + 2)$
73. $(\sqrt{x} - y)(\sqrt{x} + y)$
74. $(\sqrt{2x} - 3\sqrt{y})(\sqrt{2x} + 3\sqrt{y})$
75. $(2\sqrt{3x} - \sqrt{y})(2\sqrt{3x} + \sqrt{y})$
76. $(\sqrt{a} - 2)(\sqrt{a} - 3)$
77. $(\sqrt{x} + 4)(\sqrt{x} - 7)$
78. $(\sqrt[3]{a} + 2)(\sqrt[3]{a} + 3)$
79. $(\sqrt[3]{x} - 4)(\sqrt[3]{x} + 5)$
80. $(2\sqrt{x} - \sqrt{y})(3\sqrt{x} + \sqrt{y})$

4 Simplify.

81. $\dfrac{\sqrt{32x^2}}{\sqrt{2x}}$
82. $\dfrac{\sqrt{60y^4}}{\sqrt{12y}}$
83. $\dfrac{\sqrt{42a^3b^5}}{\sqrt{14a^2b}}$
84. $\dfrac{\sqrt{65ab^4}}{\sqrt{5ab}}$
85. $\dfrac{1}{\sqrt{5}}$
86. $\dfrac{1}{\sqrt{2}}$
87. $\dfrac{1}{\sqrt{2x}}$
88. $\dfrac{2}{\sqrt{3y}}$
89. $\dfrac{5}{\sqrt{5x}}$
90. $\dfrac{9}{\sqrt{3a}}$
91. $\sqrt{\dfrac{x}{5}}$
92. $\sqrt{\dfrac{y}{2}}$
93. $\dfrac{3}{\sqrt[3]{2}}$
94. $\dfrac{5}{\sqrt[3]{9}}$
95. $\dfrac{3}{\sqrt[3]{4x^2}}$
96. $\dfrac{5}{\sqrt[3]{3y}}$
97. $\dfrac{\sqrt{40x^3y^2}}{\sqrt{80x^2y^3}}$
98. $\dfrac{\sqrt{15a^2b^5}}{\sqrt{30a^5b^3}}$

99. $\dfrac{\sqrt{24a^2b}}{\sqrt{18ab^4}}$

100. $\dfrac{\sqrt{12x^3y}}{\sqrt{20x^4y}}$

101. $\dfrac{2}{\sqrt{5}+2}$

102. $\dfrac{5}{2-\sqrt{7}}$

103. $\dfrac{3}{\sqrt{y}-2}$

104. $\dfrac{-7}{\sqrt{x}-3}$

105. $\dfrac{\sqrt{2}-\sqrt{3}}{\sqrt{2}+\sqrt{3}}$

106. $\dfrac{\sqrt{3}+\sqrt{4}}{\sqrt{2}+\sqrt{3}}$

107. $\dfrac{4-\sqrt{2}}{2-\sqrt{3}}$

108. $\dfrac{3-\sqrt{x}}{3+\sqrt{x}}$

109. $\dfrac{\sqrt{3}-\sqrt{5}}{\sqrt{2}+\sqrt{5}}$

110. $\dfrac{\sqrt{2}+\sqrt{3}}{\sqrt{3}-\sqrt{2}}$

111. $\dfrac{3}{\sqrt[4]{8x^3}}$

112. $\dfrac{-3}{\sqrt[4]{27y^2}}$

113. $\dfrac{4}{\sqrt[5]{16a^2}}$

114. $\dfrac{a}{\sqrt[5]{81a^4}}$

115. $\dfrac{2x}{\sqrt[5]{64x^3}}$

116. $\dfrac{3y}{\sqrt[4]{32y^2}}$

117. $\dfrac{\sqrt{a}+a\sqrt{b}}{\sqrt{a}-a\sqrt{b}}$

118. $\dfrac{\sqrt{3}-3\sqrt{y}}{\sqrt{3}+3\sqrt{y}}$

119. $\dfrac{3\sqrt{xy}+2\sqrt{xy}}{\sqrt{x}-\sqrt{y}}$

120. $\dfrac{2\sqrt{x}+3\sqrt{y}}{\sqrt{x}-4\sqrt{y}}$

SUPPLEMENTAL EXERCISES 5.2

Simplify.

121. $(\sqrt{8}-\sqrt{2})^3$

122. $(\sqrt{27}-\sqrt{3})^3$

123. $(\sqrt{2}-2)^3$

124. $(\sqrt{3}-3)^3$

125. $(\sqrt{2}-3)^3$

126. $(\sqrt{5}+2)^3$

127. $\dfrac{3}{\sqrt{y+1}+1}$

128. $\dfrac{2}{\sqrt{x+4}+2}$

129. $\dfrac{\sqrt{a+4}+2}{\sqrt{a+4}-2}$

130. $\dfrac{\sqrt{b+9}-3}{\sqrt{b+9}+3}$

131. $\dfrac{3}{\sqrt{x+3}-\sqrt{x}}$

132. $\dfrac{4}{\sqrt{y+4}-\sqrt{y}}$

Rewrite as an expression with a single radical.

133. $\dfrac{\sqrt[3]{(x+y)^2}}{\sqrt{x+y}}$

134. $\dfrac{\sqrt[4]{(a+b)^3}}{\sqrt{a+b}}$

135. $\sqrt[4]{2y}\sqrt{x+3}$

136. $\sqrt[4]{2x}\sqrt{y-2}$

137. $\sqrt{a}\sqrt[3]{a+3}$

138. $\sqrt{b}\sqrt[3]{b-1}$

Rationalize the numerator and then simplify.

139. $\dfrac{\sqrt{9+h}-3}{h}$

140. $\dfrac{\sqrt{x+h}-\sqrt{x}}{h}$

Rationalize the denominator. Hint: $a + b = (\sqrt[3]{a} + \sqrt[3]{b})(\sqrt[3]{a^2} - \sqrt[3]{ab} + \sqrt[3]{b^2})$

141. $\dfrac{1}{\sqrt[3]{2}+1}$

142. $\dfrac{1}{\sqrt[3]{3}+8}$

SECTION 5.3

Complex Numbers

1 Simplify complex numbers

The radical expression $\sqrt{-4}$ is not a real number since there is no real number whose square is -4. However, the solution of an algebraic equation is sometimes the square root of a negative number.

For example, the equation $x^2 + 1 = 0$ does not have a real number solution since there is no real number whose square is a negative number.

$$x^2 + 1 = 0$$
$$x^2 = -1$$

In the seventeenth century, a new number, called an **imaginary number,** was defined so that a negative number would have a square root. The letter i was chosen to represent the number whose square is -1.

Definition of i

The number i, called the **imaginary unit,** has the property that

$$i^2 = -1.$$

An imaginary number is defined in terms of i.

Square Root of a Negative Number

If a is a positive real number, then the principal square root of negative a is the imaginary number $i\sqrt{a}$.

$$\text{For } a > 0,\ \sqrt{-a} = i\sqrt{a}.$$

This definition with $a = 1$, implies $\sqrt{-1} = i$.

An imaginary number is the product of a real number and i.

$$\sqrt{-4} = i\sqrt{4} = 2i$$
$$\sqrt{-13} = i\sqrt{13}$$
$$\sqrt{-7} = i\sqrt{7}$$

It is customary to write i in front of the radical to avoid confusing $\sqrt{a}i$ with $\sqrt{ai}$.

To simplify $\sqrt{-12}$, write $\sqrt{-12}$ as the product of a real number and i. Simplify the radical factor.

$$\sqrt{-12} = i\sqrt{12} = i\sqrt{2^2 \cdot 3} = 2i\sqrt{3}$$

Example 1 Simplify: $\sqrt{-80}$

Solution $\sqrt{-80} = i\sqrt{80} = i\sqrt{2^4 \cdot 5} = 4i\sqrt{5}$

Problem 1 Simplify: $\sqrt{-45}$

Solution See page A26.

The real numbers and the imaginary numbers make up the complex numbers.

Definition of a Complex Number

A **complex number** is a number of the form $\boldsymbol{a + bi}$, where a and b are real numbers, and $i = \sqrt{-1}$. The number a is the real part of $a + bi$, and b is the imaginary part.

Examples of complex numbers are shown below.

Real Part	Imaginary Part
a	$+ bi$
3	$+ 2i$
8	$- 10i$
$2x$	$+ 3yi$

Complex numbers $a + bi$
- Real numbers $a + 0i$
- Imaginary numbers $0 + bi$

A **real number** is a complex number in which $b = 0$.

An **imaginary number** is a complex number in which $a = 0$.

To simplify $\sqrt{20} - \sqrt{-50}$, write the complex number in the form $a + bi$.

$$\sqrt{20} - \sqrt{-50} = \sqrt{20} - i\sqrt{50}$$

Use the Product Property of Radicals to simplify each radical.

$$= \sqrt{2^2 \cdot 5} - i\sqrt{5^2 \cdot 2}$$
$$= 2\sqrt{5} - 5i\sqrt{2}$$

Example 2 Simplify: $\sqrt{25} + \sqrt{-40}$

Solution $\sqrt{25} + \sqrt{-40} = \sqrt{25} + i\sqrt{40} = \sqrt{5^2} + i\sqrt{2^2 \cdot 2 \cdot 5} = 5 + 2i\sqrt{10}$

Problem 2 Simplify: $\sqrt{98} - \sqrt{-60}$

Solution See page A26.

2 Add and subtract complex numbers

To add two complex numbers, add the real parts and add the imaginary parts.

$$\mathbf{(a + bi) + (c + di) = (a + c) + (b + d)i}$$

To subtract two complex numbers, subtract the real parts and subtract the imaginary parts.

$$\mathbf{(a + bi) - (c + di) = (a - c) + (b - d)i}$$

For example, $(3 - 7i) - (4 - 2i) = (3 - 4) + [-7 - (-2)]i = -1 - 5i$.

Example 3 Simplify: $(3 + 2i) + (6 - 5i)$

Solution $(3 + 2i) + (6 - 5i) = (3 + 6) + (2 - 5)i$ ▶ Add the real parts and add the imaginary parts.

$$= 9 - 3i$$

Problem 3 Simplify: $(-4 + 2i) - (6 - 8i)$

Solution See page A26.

To simplify $(3 + \sqrt{-12}) + (7 - \sqrt{-27})$, write each complex number in the form $a + bi$.

$$(3 + \sqrt{-12}) + (7 - \sqrt{-27}) =$$
$$(3 + i\sqrt{12}) + (7 - i\sqrt{27}) =$$

Use the Product Property of Radicals to simplify each radical.

$$(3 + i\sqrt{2^2 \cdot 3}) + (7 - i\sqrt{3^2 \cdot 3}) =$$
$$(3 + 2i\sqrt{3}) + (7 - 3i\sqrt{3}) =$$

Add the complex numbers.

$$10 - i\sqrt{3}$$

Example 4 Simplify: $(9 - \sqrt{-8}) - (5 + \sqrt{-32})$

Solution
$(9 - \sqrt{-8}) - (5 + \sqrt{-32}) =$
$(9 - i\sqrt{8}) - (5 + i\sqrt{32}) =$ ▶ Write each complex number in the form $a + bi$.
$(9 - i\sqrt{2^2 \cdot 2}) - (5 + i\sqrt{2^4 \cdot 2}) =$ ▶ Simplify each radical.
$(9 - 2i\sqrt{2}) - (5 + 4i\sqrt{2}) = 4 - 6i\sqrt{2}$

Problem 4 Simplify: $(16 - \sqrt{-45}) - (3 + \sqrt{-20})$

Solution See page A26.

3 Multiply complex numbers

When multiplying complex numbers, the term i^2 is frequently a part of the product. Recall that $i^2 = -1$.

To simplify $2i \cdot 3i$, multiply the imaginary numbers. $2i \cdot 3i = 6i^2$

Replace i^2 by -1. $= 6(-1)$

Simplify. $= -6$

When simplifying square roots of negative numbers, first rewrite the radical expressions using i.

To simplify $\sqrt{-6} \cdot \sqrt{-24}$, write each radical as the product of a real number and i. $\sqrt{-6} \cdot \sqrt{-24} = i\sqrt{6} \cdot i\sqrt{24}$

Multiply the imaginary numbers. $= i^2\sqrt{144}$

Replace i^2 by -1. $= -\sqrt{144}$

Simplify. $= -12$

Note from this example that it would have been incorrect to multiply the radicands of the two radical expressions. To illustrate,

$$\sqrt{-6} \cdot \sqrt{-24} = \sqrt{(-6)(-24)} = \sqrt{144} = 12, \textit{ not } -12.$$

To simplify $4i(3 - 2i)$, use the Distributive Property to remove parentheses. $4i(3 - 2i) = 12i - 8i^2$

Replace i^2 by -1. $= 12i - 8(-1)$

Write the answer in the form $a + bi$. $= 8 + 12i$

Example 5 Simplify: $\sqrt{-8}(\sqrt{6} - \sqrt{-2})$

Solution

$$\begin{aligned}\sqrt{-8}(\sqrt{6} - \sqrt{-2}) &= i\sqrt{8}(\sqrt{6} - i\sqrt{2}) \\ &= i\sqrt{48} - i^2\sqrt{16} \\ &= i\sqrt{2^4 \cdot 3} - (-1)\sqrt{2^4} \\ &= 4i\sqrt{3} + 4 \\ &= 4 + 4i\sqrt{3}\end{aligned}$$

▶ Write each complex number in the form $a + bi$.
▶ Use the Distributive Property.
▶ Simplify each radical. Replace i^2 by -1.
▶ Write the answer in the form $a + bi$.

Problem 5 Simplify: $\sqrt{-3}(\sqrt{27} - \sqrt{-6})$

Solution See page A26.

The product of two complex numbers can be found by using the FOIL method. For example,

$$\begin{aligned}(2 + 4i)(3 - 5i) &= 6 - 10i + 12i - 20i^2 \\ &= 6 + 2i - 20i^2 \\ &= 6 + 2i - 20(-1) \\ &= 26 + 2i\end{aligned}$$

The conjugate of $a + bi$ is $a - bi$.

The product of conjugates of the form $(a + bi)(a - bi)$ is $a^2 + b^2$.

$$(\boldsymbol{a} + \boldsymbol{bi})(\boldsymbol{a} - \boldsymbol{bi}) = a^2 - b^2i^2 = a^2 - b^2(-1) = \boldsymbol{a^2 + b^2}$$

For example, $(2 + 3i)(2 - 3i) = 2^2 + 3^2 = 4 + 9 = 13$.

Note that the product of a complex number and its conjugate is a real number.

Example 6 Simplify.

A. $(3 - 4i)(2 + 5i)$ B. $\left(\frac{9}{10} + \frac{3}{10}i\right)\left(1 - \frac{1}{3}i\right)$ C. $(4 + 5i)(4 - 5i)$

Solution

A. $(3 - 4i)(2 + 5i) =$

$6 + 15i - 8i - 20i^2 =$ ▶ Use the FOIL method.

$6 + 7i - 20i^2 =$ ▶ Combine like terms.

$6 + 7i - 20(-1) =$ ▶ Replace i^2 by -1.

$26 + 7i$ ▶ Write the answer in the form $a + bi$.

B. $\left(\frac{9}{10} + \frac{3}{10}i\right)\left(1 - \frac{1}{3}i\right) =$

$\frac{9}{10} - \frac{3}{10}i + \frac{3}{10}i - \frac{1}{10}i^2 =$ ▶ Use the FOIL method.

$\frac{9}{10} - \frac{1}{10}i^2 =$ ▶ Combine like terms.

$\frac{9}{10} - \frac{1}{10}(-1) =$ ▶ Replace i^2 by -1.

$\frac{9}{10} + \frac{1}{10} = 1$ ▶ Simplify.

C. $(4 + 5i)(4 - 5i) = 4^2 + 5^2$
$= 16 + 25$
$= 41$

▶ The product of conjugates of the form $(a + bi)(a - bi)$ is $a^2 + b^2$.

Problem 6 Simplify.

A. $(4 - 3i)(2 - i)$ B. $(3 - i)\left(\frac{3}{10} + \frac{1}{10}i\right)$ C. $(3 + 6i)(3 - 6i)$

Solution See page A26.

4 Divide complex numbers

A rational expression containing one or more complex numbers is in simplest form when no imaginary number remains in the denominator.

To simplify $\frac{2 - 3i}{2i}$, multiply the expression by 1 in the form $\frac{i}{i}$.

$$\frac{2 - 3i}{2i} = \frac{2 - 3i}{2i} \cdot \frac{i}{i} = \frac{2i - 3i^2}{2i^2}$$

Replace i^2 by -1.

$$= \frac{2i - 3(-1)}{2(-1)}$$

Simplify.

$$= \frac{3 + 2i}{-2}$$

Write the answer in the form $a + bi$.

$$= -\frac{3}{2} - i$$

Example 7 Simplify: $\frac{5 + 4i}{3i}$

Solution $\frac{5 + 4i}{3i} = \frac{5 + 4i}{3i} \cdot \frac{i}{i} = \frac{5i + 4i^2}{3i^2} = \frac{5i + 4(-1)}{3(-1)} = \frac{-4 + 5i}{-3} = \frac{4}{3} - \frac{5}{3}i$

Problem 7 Simplify: $\frac{2 - 3i}{4i}$

Solution See page A27.

To simplify a fraction that has a complex number in the denominator, multiply the numerator and denominator by the conjugate of the complex number.

$$\frac{3 + 2i}{1 + i} = \frac{(3 + 2i)}{(1 + i)} \cdot \frac{(1 - i)}{(1 - i)} = \frac{3 - 3i + 2i - 2i^2}{1^2 + 1^2}$$
$$= \frac{3 - i - 2(-1)}{2} = \frac{5 - i}{2} = \frac{5}{2} - \frac{1}{2}i$$

Example 8 Simplify: $\dfrac{5-3i}{4+2i}$

Solution $\dfrac{5-3i}{4+2i} = \dfrac{(5-3i)}{(4+2i)} \cdot \dfrac{(4-2i)}{(4-2i)} = \dfrac{20-10i-12i+6i^2}{4^2+2^2} = \dfrac{20-22i+6(-1)}{20}$

$= \dfrac{14-22i}{20} = \dfrac{7-11i}{10} = \dfrac{7}{10} - \dfrac{11}{10}i$

Problem 8 Simplify: $\dfrac{2+5i}{3-2i}$

Solution See page A27.

EXERCISES 5.3

1 Simplify.

1. $\sqrt{-4}$
2. $\sqrt{-64}$
3. $\sqrt{-98}$
4. $\sqrt{-72}$
5. $\sqrt{-27}$
6. $\sqrt{-75}$
7. $\sqrt{16} + \sqrt{-4}$
8. $\sqrt{25} + \sqrt{-9}$
9. $\sqrt{12} - \sqrt{-18}$
10. $\sqrt{60} - \sqrt{-48}$
11. $\sqrt{160} - \sqrt{-147}$
12. $\sqrt{96} - \sqrt{-125}$
13. $\sqrt{-4a^2}$
14. $\sqrt{-16b^6}$
15. $\sqrt{-49x^{12}}$
16. $\sqrt{-32x^3y^2}$
17. $\sqrt{-144a^3b^5}$
18. $\sqrt{-81a^{10}b^9}$
19. $\sqrt{4a} + \sqrt{-12a^2}$
20. $\sqrt{25b} - \sqrt{-48b^2}$
21. $\sqrt{18b^5} - \sqrt{-27b^3}$
22. $\sqrt{a^5b^2} - \sqrt{-a^5b^2}$
23. $\sqrt{-50x^3y^3} + x\sqrt{25x^4y^3}$
24. $\sqrt{-121xy} + \sqrt{60x^2y^2}$
25. $\sqrt{-49a^5b^2} - ab\sqrt{-25a^3}$
26. $\sqrt{-16x^2y} - x\sqrt{-49y}$
27. $\sqrt{12a^3} + \sqrt{-27b^3}$

2 Simplify.

28. $(2+4i) + (6-5i)$
29. $(6-9i) + (4+2i)$
30. $(-2-4i) - (6-8i)$
31. $(3-5i) + (8-2i)$

32. $(8 - \sqrt{-4}) - (2 + \sqrt{-16})$

33. $(5 - \sqrt{-25}) - (11 - \sqrt{-36})$

34. $(12 - \sqrt{-50}) + (7 - \sqrt{-8})$

35. $(5 - \sqrt{-12}) - (9 + \sqrt{-108})$

36. $(\sqrt{8} + \sqrt{-18}) + (\sqrt{32} - \sqrt{-72})$

37. $(\sqrt{40} - \sqrt{-98}) - (\sqrt{90} + \sqrt{-32})$

38. $(5 - 3i) + 2i$

39. $(6 - 8i) + 4i$

40. $(7 + 2i) + (-7 - 2i)$

41. $(8 - 3i) + (-8 + 3i)$

42. $(9 + 4i) + 6$

43. $(4 + 6i) + 7$

3 Simplify.

44. $(7i)(-9i)$

45. $(-6i)(-4i)$

46. $\sqrt{-2}\sqrt{-8}$

47. $\sqrt{-5}\sqrt{-45}$

48. $\sqrt{-3}\sqrt{-6}$

49. $\sqrt{-5}\sqrt{-10}$

50. $2i(6 + 2i)$

51. $-3i(4 - 5i)$

52. $\sqrt{-2}(\sqrt{8} + \sqrt{-2})$

53. $\sqrt{-3}(\sqrt{12} - \sqrt{-6})$

54. $(5 - 2i)(3 + i)$

55. $(2 - 4i)(2 - i)$

56. $(6 + 5i)(3 + 2i)$

57. $(4 - 7i)(2 + 3i)$

58. $(1 - i)\left(\frac{1}{2} + \frac{1}{2}i\right)$

59. $\left(\frac{4}{5} - \frac{2}{5}i\right)\left(1 + \frac{1}{2}i\right)$

60. $\left(\frac{6}{5} + \frac{3}{5}i\right)\left(\frac{2}{3} - \frac{1}{3}i\right)$

61. $(2 - i)\left(\frac{2}{5} + \frac{1}{5}i\right)$

62. $(4 - 3i)(4 + 3i)$

63. $(8 - 5i)(8 + 5i)$

64. $(3 - i)(3 + i)$

65. $(7 - i)(7 + i)$

66. $(6 - \sqrt{-2})^2$

67. $(9 - \sqrt{-1})^2$

4 Simplify.

68. $\frac{3}{i}$

69. $\frac{4}{5i}$

70. $\frac{2-3i}{-4i}$

71. $\frac{16+5i}{-3i}$

72. $\frac{4}{5+i}$

73. $\frac{6}{5+2i}$

74. $\frac{2}{2-i}$

75. $\frac{5}{4-i}$

76. $\frac{1-3i}{3+i}$

77. $\frac{2+12i}{5+i}$

78. $\frac{\sqrt{-10}}{\sqrt{8}-\sqrt{-2}}$

79. $\frac{\sqrt{-2}}{\sqrt{12}-\sqrt{-8}}$

80. $\frac{2-3i}{3+i}$

81. $\frac{3+5i}{1-i}$

SUPPLEMENTAL EXERCISES 5.3

Note the pattern when successive powers of i are simplified.

$i^1 = i$

$i^2 = -1$

$i^3 = i^2 \cdot i = -i$

$i^4 = i^2 \cdot i^2 = (-1)(-1) = 1$

$i^5 = i \cdot i^4 = i(1) = i$

$i^6 = i^2 \cdot i^4 = -1$

$i^7 = i^3 \cdot i^4 = -i$

$i^8 = i^4 \cdot i^4 = 1$

82. When the exponent on i is a multiple of 4, the power equals ______.

Use the pattern above to simplify the power of i.

83. i^6

84. i^9

85. i^{57}

86. i^{65}

87. i^{220}

88. i^{460}

89. i^0

90. i^{-2}

91. i^{-6}

92. i^{-34}

93. i^{-58}

94. i^{-180}

The property that the product of conjugates of the form $(a+bi)(a-bi) = a^2 + b^2$ can be used to factor the sum of two perfect squares over the set of complex numbers. For example, $x^2 + y^2 = (x + yi)(x - yi)$. Factor over the set of complex numbers.

95. $y^2 + 1$

96. $a^2 + 4$

97. $x^2 + 25$

98. $4b^2 + 9$

99. $49x^2 + 16$

100. $9a^2 + 64$

Solve.

101. **a.** Is $3i$ a solution of $2x^2 + 18 = 0$?
b. Is $-3i$ a solution of $2x^2 + 18 = 0$?

102. **a.** Is $7i$ a solution of $x^2 + 49 = 0$?
b. Is $-7i$ a solution of $x^2 + 49 = 0$?

103. **a.** Is $3 + i$ a solution of $x^2 - 6x + 10 = 0$?
b. Is $3 - i$ a solution of $x^2 - 6x + 10 = 0$?

104. **a.** Is $1 + 3i$ a solution of $x^2 - 2x - 10 = 0$?
b. Is $1 - 3i$ a solution of $x^2 - 2x - 10 = 0$?

105. Simplify: $2i + \cfrac{1}{2i + \cfrac{1}{2i + \cfrac{1}{i}}}$

106. Show that $\sqrt{i} = \frac{\sqrt{2}}{2} + i\frac{\sqrt{2}}{2}$ by simplifying $\left[\frac{\sqrt{2}}{2} + i\frac{\sqrt{2}}{2}\right]^2$.

107. Given $\sqrt{i} = \frac{\sqrt{2}}{2} + i\frac{\sqrt{2}}{2}$, find $\sqrt{-i}$.

108. Given $\sqrt{i} = \frac{\sqrt{2}}{2} + i\frac{\sqrt{2}}{2}$, find $i^{\frac{3}{2}}$.

SECTION 5.4

Equations Containing Radical Expressions

1 Solve equations containing one or more radical expressions

An equation that contains a variable expression in a radicand is a **radical equation.**

$$\left.\begin{array}{l}\sqrt{x+2} = \sqrt{3x-4}\\ \sqrt[3]{x-4} = 2\end{array}\right\} \text{Radical Equations}$$

The following property of equality is used to solve radical equations.

The Property of Raising Both Sides of an Equation to a Power

If a and b are real numbers and $a = b$, then $a^n = b^n$.

Solve: $\sqrt{x-2} - 6 = 0$

Rewrite the equation with the radical on one side of the equation and the constant on the other side.

$$\sqrt{x-2} - 6 = 0$$
$$\sqrt{x-2} = 6$$

Square each side of the equation.

$$(\sqrt{x-2})^2 = 6^2$$

Solve the resulting equation.

$$x - 2 = 36$$
$$x = 38$$

Check the solution.

Check:
$$\begin{array}{r|l} \sqrt{x-2} - 6 & = 0 \\ \hline \sqrt{38-2} - 6 & 0 \\ \sqrt{36} - 6 & 0 \\ 6 - 6 & 0 \\ 0 & = 0 \end{array}$$

38 checks as a solution.
The solution is 38.

Example 1 Solve.

A. $\sqrt{3x-2} - 8 = -3$ B. $\sqrt[3]{3x-1} = -4$

Solution A.
$$\sqrt{3x-2} - 8 = -3$$
$$\sqrt{3x-2} = 5$$
▶ Rewrite the equation so that the radical is alone on one side of the equation.

$$(\sqrt{3x-2})^2 = 5^2$$
▶ Square each side of the equation.

$$3x - 2 = 25$$
$$3x = 27$$
$$x = 9$$
▶ Solve the resulting equation.

Check:
$$\begin{array}{r|l} \sqrt{3x-2} - 8 & = -3 \\ \hline \sqrt{3 \cdot 9 - 2} - 8 & -3 \\ \sqrt{27-2} - 8 & -3 \\ \sqrt{25} - 8 & -3 \\ 5 - 8 & -3 \\ -3 & = -3 \end{array}$$
▶ Check the solution.

The solution is 9.

B.
$$\sqrt[3]{3x-1} = -4$$
$$(\sqrt[3]{3x-1})^3 = (-4)^3$$
▶ Cube each side of the equation.

$$3x - 1 = -64$$
$$3x = -63$$
$$x = -21$$
▶ Solve the resulting equation.

Check: $\sqrt[3]{3x-1} = -4$ ▶ Check the solution.

$\sqrt[3]{3x-1}$	-4
$\sqrt[3]{3(-21)-1}$	-4
$\sqrt[3]{-63-1}$	-4
$\sqrt[3]{-64}$	-4

$$-4 = -4$$

The solution is -21.

Problem 1 Solve.

A. $\sqrt{4x+5} - 12 = -5$ B. $\sqrt[4]{x-8} = 3$

Solution See page A27.

When raising both sides of an equation to an even power, the resulting equation may have a solution that is not a solution of the original equation. Therefore, it is necessary to check the solution of a radical equation.

Example 2 Solve.

A. $x + 2\sqrt{x-1} = 9$ B. $\sqrt{x+7} = \sqrt{x} + 1$

Solution A. $x + 2\sqrt{x-1} = 9$

$2\sqrt{x-1} = 9 - x$ ▶ Rewrite the equation with the radical on one side of the equation.

$(2\sqrt{x-1})^2 = (9-x)^2$ ▶ Square each side of the equation.

$4(x-1) = 81 - 18x + x^2$

$4x - 4 = 81 - 18x + x^2$

$0 = x^2 - 22x + 85$ ▶ Write the quadratic equation in standard form.

$0 = (x-5)(x-17)$ ▶ Factor.

$x - 5 = 0$ $x - 17 = 0$ ▶ Use the Principle of Zero Products.

$x = 5$ $x = 17$

Check:

$x + 2\sqrt{x-1} = 9$	
$5 + 2\sqrt{5-1}$	9
$5 + 2\sqrt{4}$	9
$5 + 2 \cdot 2$	9
$5 + 4$	9
$9 = 9$	

$x + 2\sqrt{x-1} = 9$	
$17 + 2\sqrt{17-1}$	9
$17 + 2\sqrt{16}$	9
$17 + 2 \cdot 4$	9
$17 + 8$	9
$25 \neq 9$	

17 does not check as a solution.

The solution is 5.

B. $\sqrt{x+7} = \sqrt{x} + 1$ ▶ A radical appears on each side of the equation.

$(\sqrt{x+7})^2 = (\sqrt{x}+1)^2$ ▶ Square each side of the equation.

$x + 7 = x + 2\sqrt{x} + 1$ ▶ Simplify the resulting equation.

$6 = 2\sqrt{x}$

$3 = \sqrt{x}$ ▶ The equation contains a radical.

$3^2 = (\sqrt{x})^2$ ▶ Square each side of the equation.

$9 = x$

Check: $\sqrt{x+7} = \sqrt{x} + 1$ ▶ Check the solution.

$\sqrt{9+7}$	$\sqrt{9}+1$
$\sqrt{16}$	$3+1$
$4 = 4$	

The solution is 9.

Problem 2 Solve.

A. $x + 3\sqrt{x+2} = 8$ B. $\sqrt{x+5} = 5 - \sqrt{x}$

Solution See page A27.

2 Application problems

A right triangle contains one 90° angle. The side opposite the 90° angle is called the **hypotenuse.** The other two sides are called **legs.**

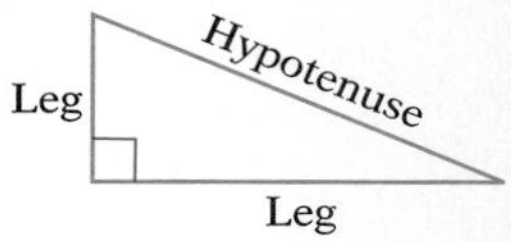

Pythagoras, a Greek mathematician, is credited with the discovery that the square of the hypotenuse of a right triangle is equal to the sum of the squares of the two legs. This is called the **Pythagorean Theorem.**

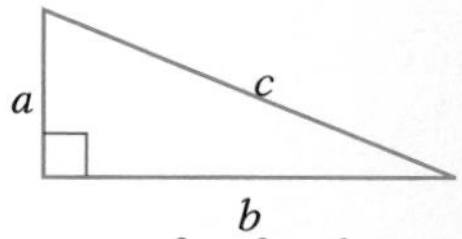

$c^2 = a^2 + b^2$

Example 3 A ladder 20 ft long is leaning against a building. How high on the building will the ladder reach when the bottom of the ladder is 8 ft from the building? Round to the nearest tenth.

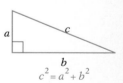

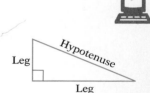

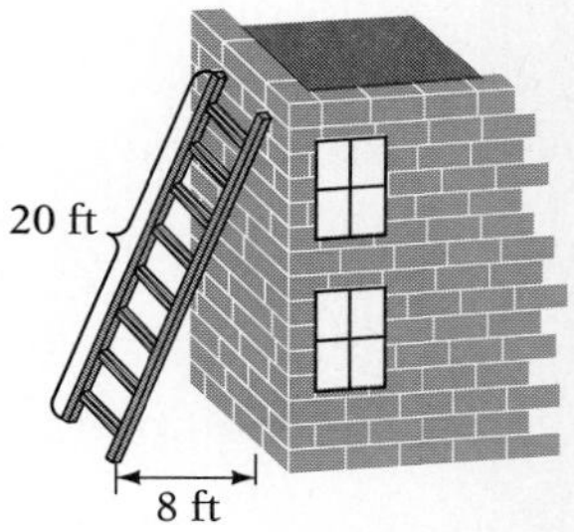

Strategy To find the distance, use the Pythagorean Theorem. The hypotenuse is the length of the ladder. One leg is the distance from the bottom of the ladder to the base of the building. The distance along the building from the ground to the top of the ladder is the unknown leg.

Solution

$$c^2 = a^2 + b^2$$
$$20^2 = 8^2 + b^2$$
$$400 = 64 + b^2$$
$$336 = b^2$$
$$(336)^{\frac{1}{2}} = (b^2)^{\frac{1}{2}}$$
$$\sqrt{336} = b$$
$$18.3 \approx b$$

The distance is 18.3 ft.

Problem 3 Find the diagonal of a rectangle that is 6 cm in length and 3 cm in width. Round to the nearest tenth.

Solution See page A28.

Example 4 An object is dropped from a high building. Find the distance the object has fallen when the speed reaches 96 ft/s. Use the equation $v = \sqrt{64d}$, where v is the speed of the object, and d is the distance.

Strategy To find the distance the object has fallen, replace v in the equation with the given value and solve for d.

Solution

$$v = \sqrt{64d}$$
$$96 = \sqrt{64d}$$
$$(96)^2 = (\sqrt{64d})^2$$
$$9216 = 64d$$
$$144 = d$$

The object has fallen 144 ft.

Problem 4 How far would a submarine periscope have to be above the water to locate a ship 5.5 mi away? The equation for the distance in miles that the lookout can see is $d = 1.4\sqrt{h}$, where h is the height in feet above the surface of the water. Round to the nearest thousandth.

Solution See page A28.

EXERCISES 5.4

1 Solve.

1. $\sqrt{x} = 5$
2. $\sqrt{y} = 2$
3. $\sqrt[3]{a} = 3$
4. $\sqrt[3]{y} = 5$
5. $\sqrt{3x} = 12$
6. $\sqrt{5x} = 10$
7. $\sqrt[3]{4x} = -2$
8. $\sqrt[3]{6x} = -3$
9. $\sqrt{2x} = -4$
10. $\sqrt{5x} = -5$
11. $\sqrt{3x - 2} = 5$
12. $\sqrt{5x - 4} = 9$
13. $\sqrt{3 - 2x} = 7$
14. $\sqrt{9 - 4x} = 4$
15. $7 = \sqrt{1 - 3x}$
16. $6 = \sqrt{8 - 7x}$
17. $\sqrt[3]{4x - 1} = 2$
18. $\sqrt[3]{5x + 2} = 3$
19. $\sqrt[3]{1 - 2x} = -3$
20. $\sqrt[3]{3 - 2x} = -2$
21. $\sqrt[3]{9x + 1} = 4$
22. $\sqrt{3x + 9} - 12 = 0$
23. $\sqrt{4x - 3} - 5 = 0$
24. $\sqrt{x - 2} = 4$
25. $\sqrt[3]{x - 3} + 5 = 0$
26. $\sqrt[3]{x - 2} = 3$
27. $\sqrt[3]{2x - 6} = 4$
28. $\sqrt{x^2 - 8x} = 3$
29. $\sqrt{x^2 + 7x + 11} = 1$
30. $\sqrt[4]{4x + 1} = 2$
31. $\sqrt[4]{2x - 9} = 3$
32. $\sqrt{2x - 3} - 2 = 1$
33. $\sqrt{3x - 5} - 5 = 3$
34. $\sqrt[3]{2x - 3} + 5 = 2$
35. $\sqrt[3]{x - 4} + 7 = 5$
36. $\sqrt{5x - 16} + 1 = 4$
37. $\sqrt{3x - 5} - 2 = 3$
38. $\sqrt{2x - 1} - 8 = -5$
39. $\sqrt{7x + 2} - 10 = -7$
40. $\sqrt[3]{4x - 3} - 2 = 3$
41. $\sqrt[3]{1 - 3x} + 5 = 3$
42. $1 - \sqrt{4x + 3} = -5$
43. $7 - \sqrt{3x + 1} = -1$
44. $\sqrt{x + 1} = 2 - \sqrt{x}$
45. $\sqrt{2x + 4} = 3 - \sqrt{2x}$
46. $\sqrt{x^2 + 3x - 2} - x = 1$
47. $\sqrt{x^2 - 4x - 1} + 3 = x$
48. $\sqrt{x^2 - 3x - 1} = 3$
49. $\sqrt{x^2 - 2x + 1} = 3$
50. $\sqrt{2x + 5} - \sqrt{3x - 2} = 1$
51. $\sqrt{4x + 1} - \sqrt{2x + 4} = 1$
52. $\sqrt{5x - 1} - \sqrt{3x - 2} = 1$
53. $\sqrt{5x + 4} - \sqrt{3x + 1} = 1$
54. $\sqrt[3]{x^2 + 2} - 3 = 0$
55. $\sqrt[3]{x^2 + 4} - 2 = 0$
56. $\sqrt[4]{x^2 + 2x + 8} - 2 = 0$
57. $\sqrt[4]{x^2 + x - 1} - 1 = 0$
58. $4\sqrt{x + 1} - x = 1$
59. $3\sqrt{x - 2} + 2 = x$
60. $x + 3\sqrt{x - 2} = 12$
61. $x + 2\sqrt{x + 1} = 7$

2 Solve.

62. Find the width of a rectangle that has a diagonal of 25 ft and a length of 24 ft.

63. A 12-ft ladder is leaning against a building. How high on the building will the ladder reach when the bottom of the ladder is 4 ft from the building? Round to the nearest tenth.

64. Find the length of a rectangle that has a diagonal of 15 m and a width of 9 m.

65. A 26-ft ladder is leaning against a building. How far is the bottom of the ladder from the wall when the ladder reaches a height of 24 ft on the building?

66. How far would a submarine periscope have to be above the water to locate a ship 3.2 mi away? The equation for the distance in miles that the lookout can see is $d = 1.4\sqrt{h}$, where h is the height in feet above the surface of the water. Round to the nearest hundredth.

67. How far would a submarine periscope have to be above the water to locate a ship 3.5 mi away? The equation for the distance in miles that the lookout can see is $d = 1.4\sqrt{h}$, where h is the height in feet above the surface of the water. Round to the nearest hundredth.

68. An object is dropped from a bridge. Find the distance the object has fallen when the speed reaches 80 ft/s. Use the equation $v = \sqrt{64d}$, where v is the speed of the object, and d is the distance.

69. An object is dropped from a high building. Find the distance the object has fallen when the speed reaches 120 ft/s. Use the equation $v = \sqrt{64d}$, where v is the speed of the object, and d is the distance.

70. Find the distance required for a car to reach a velocity of 60 m/s when the acceleration is 10 m/s^2. Use the equation $v = \sqrt{2as}$, where v is the velocity, a is the acceleration, and s is the distance.

71. Find the distance required for a car to reach a velocity of 48 ft/s when the acceleration is 12 ft/s^2. Use the equation $v = \sqrt{2as}$, where v is the velocity, a is the acceleration, and s is the distance.

72. Find the length of a pendulum that makes one swing in 3 s. The equation for the time of one swing of a pendulum is given by $T = 2\pi\sqrt{\frac{L}{32}}$, where T is the time in seconds, and L is the length in feet. Round to the nearest hundredth.

73. Find the length of a pendulum that makes one swing in 2.4 s. The equation for the time of one swing of a pendulum is given by $T = 2\pi\sqrt{\frac{L}{32}}$, where T is the time in seconds, and L is the length in feet. Round to the nearest hundredth.

SUPPLEMENTAL EXERCISES 5.4

Solve.

74. $x^{\frac{3}{4}} = 8$

75. $x^{\frac{2}{3}} = 9$

76. $x^{\frac{5}{4}} = 32$

77. $x^{\frac{3}{5}} = 27$

Solve the formula for the given variable.

78. $v = \sqrt{64d}$; d

79. $v = \sqrt{2as}$; s

80. $a^2 + b^2 = c^2$; a

81. $A = \pi r^2$; r

82. $V = \pi r^2 h$; r

83. $V = \frac{4}{3}\pi r^3$; r

Solve.

84. $\sqrt{3x - 2} = \sqrt{2x - 3} + \sqrt{x - 1}$

85. $\sqrt{2x + 3} + \sqrt{x + 2} = \sqrt{x + 5}$

86. A box has a base that measures 4 in. by 6 in. The height of the box is 3 in. Find the greatest distance between two corners. Round to the nearest hundredth.

87. Two cyclists left an intersection at the same time. The first cyclist headed due south. The second cyclist headed due east. When the first cyclist had traveled 10 mi farther than the second cyclist, the cyclists were 50 mi apart. How far had each cyclist traveled?

88. Find three odd integers, a, b, and c, such that $a^2 + b^2 = c^2$.

Calculators and Computers

The y^x Key on a Calculator

The y^x key on a calculator is used to find powers of a number. For example, to find 4^7, enter the following key strokes:

4 y^x 7 =

The number 16384 should be in the display of your calculator.

Fractional powers can also be calculated by using the [y^x] key. The memory key on your calculator is useful for this calculation. For example, to find $17^{\frac{2}{7}}$, first find the decimal equivalent for $\frac{2}{7}$. Store this result in the calculator's memory. Now the [y^x] key is used to find the power. Here are the key strokes.

2 [÷] 7 [=] [M+] 17 [y^x] [MR] [=]

The number 2.2467608 should be in the display.

The symbol M+ is used here to mean store in memory, and the symbol MR is used to mean recall from memory. These symbols may be different on your calculator.

The calculation of an installment loan payment uses the [y^x] key. Here is the formula and a sample calculation.

$$PMT = PRIN \cdot \frac{i}{1-(1+i)^{-n}}$$

In this formula, *PMT* is the monthly payment, *PRIN* is the amount borrowed, and i is the monthly interest rate as a decimal. The monthly interest rate is the annual rate divided by 12. The number of months of the loan is given by n.

A car is purchased and a loan of $8000 is secured at an annual interest rate of 8.5% for 5 years. Find the monthly payment.

First calculate $(1 + i)^{-n}$.

$(n = 5 \cdot 12 = 60;\ i = \frac{0.085}{12} = 0.0070833;\ 1 + i = 1.0070833)$

Enter: 1.0070833 [y^x] 60 [+/−] [=] [M+]

The number 0.6547513 should be displayed. This number is also stored in memory because the [M+] key was pressed.

Now complete the final calculation.

Enter: 8000 [x] 0.0070833 [÷] [( ] 1 [−] [MR] [ )] [=]

The number 164.1321 should be in the display.
The monthly payment is $164.13.

Something Extra

Diophantine Equations

Diophantus, a Greek mathematician, wrote three books. One of these books, *The Arithmetica,* is a collection of problems leading to first- and second-degree equations or to systems of equations with whole number solutions. Some of these problems have more than one solution. Such equations or systems of equations are now called Diophantine equations.

Tickets to a football game are priced at $5 for students and $8 for adults. How many tickets have been sold at each price after $360 has been collected?

Let x = number of student tickets sold.
Let y = number of adult tickets sold.

Write an equation for the total income from the tickets. $5x + 8y = 360$

Solve for x.

$$x = \frac{360}{5} - \frac{8}{5}y$$

$$x = 72 - \frac{8}{5}y$$

Since x and y must be whole numbers, y must be a multiple of 5. Let $y = 0, 5, 10, 15, 20, \ldots$

$y = 0$	$x = 72$
$y = 5$	$x = 64$
$y = 10$	$x = 56$
$y = 15$	$x = 48$
.	.
.	.
.	.
$y = 45$	$x = 0$

Thus for the information given, the following answers are possible.

(72, 0) (64, 5) (56, 10) (48, 15) (40, 20)
(32, 25) (24, 30) (16, 35) (8, 40) (0, 45)

There are 10 solutions to this problem.

A sporting goods store sells one brand of skateboard for $40 and another brand for $60. Find the number of each brand of skateboard sold from sales totaling $620.

Let x = number of $40 skateboards sold.
Let y = number of $60 skateboards sold.

Write an equation for the total income from the sale of the skateboards. $40x + 60y = 620$

Solve for x.

$$x = \frac{31}{2} - \frac{3}{2}y$$

Since x and y must be whole numbers, $y = 1, 3, 5, 7$, or 9.

$y = 1$	$x = 14$
$y = 3$	$x = 11$
$y = 5$	$x = 8$
$y = 7$	$x = 5$
$y = 9$	$x = 2$

The possible solutions are (14, 1), (11, 3), (8, 5), (5, 7), and (2, 9).

Find the possible solutions for the following problems.

1. Tickets to a football game are priced at \$10 and \$14. The total collected from ticket sales is \$410. Find the number of \$10 tickets sold.

2. Tapes cost \$8 each and compact discs cost \$13 each. Find the number of compact discs sold after \$352 has been collected.

3. A convention is planned with 45-min and 30-min sessions. The daily schedule starts at 9 A.M. and ends at 5 P.M. with a 15 min break between each session. How many sessions of each type will this schedule allow? Let x be the number of 45-min sessions and y be the number of 30-min sessions.

Chapter Summary

Key Words

The *nth root of a* is $a^{\frac{1}{n}}$. The expression $\sqrt[n]{a}$ is another symbol for the nth root of a. In the expression $\sqrt[n]{a}$, the symbol $\sqrt[n]{}$ is called a *radical*, n is the *index* of the radical, and a is the *radicand*.

If $a^{\frac{1}{n}}$ is a real number, then $a^{\frac{m}{n}} = \sqrt[n]{a^m} = (\sqrt[n]{a})^m$.

The symbol $\sqrt{}$ is used to indicate the positive or *principal square root* of a number.

The expressions $a + b$ and $a - b$ are called *conjugates* of each other. The product of conjugates of the form $(a + b)(a - b) = a^2 - b^2$.

The procedure used to remove a radical from the denominator of a radical expression is called *rationalizing the denominator.*

A *complex number* is a number of the form $a + bi$, where a and b are real numbers and $i = \sqrt{-1}$. For the complex number $a + bi$, a is the *real part* of the complex number, and b is the *imaginary part* of the complex number.

A *radical equation* is an equation that contains a variable expression in a radicand.

Essential Rules

The Product Property of Radicals	If a and b are positive real numbers, then $\sqrt[n]{ab} = \sqrt[n]{a}\sqrt[n]{b}$.
The Quotient Property of Radicals	If a and b are positive real numbers, then $\sqrt[n]{\frac{a}{b}} = \frac{\sqrt[n]{a}}{\sqrt[n]{b}}$.

Addition of Complex Numbers	If $a + bi$ and $c + di$ are complex numbers, then $(a + bi) + (c + di) = (a + c) + (b + d)i$.
Subtraction of Complex Numbers	If $a + bi$ and $c + di$ are complex numbers, then $(a + bi) - (c + di) = (a - c) + (b - d)i$.
The Property of Raising Both Sides of an Equation to a Power	If a and b are real numbers and $a = b$, then $a^n = b^n$.
The Pythagorean Theorem	The square of the hypotenuse of a right triangle is equal to the sum of the squares of the two legs. $c^2 = a^2 + b^2$

Chapter Review

1. Simplify: $81^{-\frac{1}{4}}$

2. Simplify: $\dfrac{x^{-\frac{3}{2}}}{x^{\frac{7}{2}}}$

3. Simplify: $(a^{16})^{-\frac{5}{8}}$

4. Simplify: $(16x^{-4}y^{12})(100x^6y^{-2})^{\frac{1}{2}}$

5. Rewrite $3x^{\frac{3}{4}}$ as a radical expression.

6. Rewrite $(5a + 2)^{-\frac{1}{3}}$ as a radical expression.

7. Rewrite $\sqrt{x^5}$ as an exponential expression.

8. Rewrite $7y\sqrt[3]{x^2}$ as an exponential expression.

9. Simplify: $\sqrt[4]{81a^8b^{12}}$

10. Simplify: $-\sqrt{49x^6y^{16}}$

11. Simplify: $\sqrt[3]{-8a^6b^{12}}$

12. Simplify: $\sqrt[5]{x^5y^{15}}$

13. Simplify: $\sqrt{18a^3b^6}$

14. Simplify: $\sqrt[3]{81x^6y^8}$

15. Simplify: $\sqrt[5]{-64a^8b^{12}}$

16. Simplify: $\sqrt[4]{x^6y^8z^{10}}$

17. Simplify: $\sqrt{54} + \sqrt{24}$

18. Simplify: $\sqrt{48x^5y} - x\sqrt{80x^3y}$

19. Simplify: $\sqrt{50a^4b^3} - ab\sqrt{18a^2b}$

20. Simplify: $3x\sqrt[3]{54x^8y^{10}} - 2x^2y\sqrt[3]{16x^5y^7}$

21. Simplify: $4x\sqrt{12x^2y} + \sqrt{3x^4y} - x^2\sqrt{27y}$

22. Simplify: $\sqrt{32}\sqrt{50}$

23. Simplify: $\sqrt[3]{16x^4y}\sqrt[3]{4xy^5}$

24. Simplify: $\sqrt{3x}(3 + \sqrt{3x})$

25. Simplify: $(5 - \sqrt{6})^2$

26. Simplify: $(\sqrt{3} + 8)(\sqrt{3} - 2)$

27. Simplify: $\dfrac{\sqrt{125x^6}}{\sqrt{5x^3}}$

28. Simplify: $\dfrac{8}{\sqrt{3y}}$

29. Simplify: $\dfrac{12}{\sqrt{x} - \sqrt{7}}$

30. Simplify: $\dfrac{x + 2}{\sqrt{x} + \sqrt{2}}$

31. Simplify: $\dfrac{\sqrt{x} + \sqrt{y}}{\sqrt{x} - \sqrt{y}}$

32. Simplify: $\sqrt{-36}$

33. Simplify: $\sqrt{-50}$

34. Simplify: $\sqrt{49} - \sqrt{-16}$

35. Simplify: $\sqrt{200} + \sqrt{-12}$

36. Simplify: $\sqrt{32} - \sqrt{-45}$

37. Simplify: $(5 + 2i) + (4 - 3i)$

38. Simplify: $(-8 + 3i) - (4 - 7i)$

39. Simplify: $(9 - \sqrt{-16}) + (5 + \sqrt{-36})$

40. Simplify: $(\sqrt{50} + \sqrt{-72}) - (\sqrt{162} - \sqrt{-8})$

41. Simplify: $(3 - 9i) + 7$

42. Simplify: $(8i)(2i)$

43. Simplify: $i(3 - 7i)$

44. Simplify: $\sqrt{-12}\sqrt{-6}$

45. Simplify: $(6 - 5i)(4 + 3i)$

46. Simplify: $(3 - 2i)(3 + 2i)$

47. Simplify: $\dfrac{-6}{i}$

48. Simplify: $\dfrac{5 + 2i}{3i}$

49. Simplify: $\dfrac{7}{2 - i}$

50. Simplify: $\dfrac{\sqrt{16}}{\sqrt{4} - \sqrt{-4}}$

51. Simplify: $\dfrac{5 + 9i}{1 - i}$

52. Solve: $\sqrt[3]{9x} = -6$

53. Solve: $\sqrt[3]{3x - 5} = 2$

54. Solve: $\sqrt[4]{4x + 1} = 3$

55. Solve: $\sqrt{4x + 9} + 10 = 11$

56. Solve: $\sqrt{x - 5} + \sqrt{x + 6} = 11$

57. Find the width of a rectangle that has a diagonal of 13 in. and a length of 12 in.

58. The velocity of the wind determines the amount of power generated by a windmill. A typical equation for this relationship is $v = 4.05\sqrt[3]{P}$, where v is the velocity in miles per hour and P is the power in watts. Find the amount of power generated by a 20-mph wind. Round to the nearest whole number.

59. Find the distance required for a car to reach a velocity of 88 ft/s when the acceleration is 16 ft/s^2. Use the equation $v = \sqrt{2as}$, where v is the velocity, a is the acceleration, and s is the distance.

60. A 12-ft ladder is leaning against a building. How far from the building is the bottom of the ladder when the top of the ladder touches the building 10 ft above the ground? Round to the nearest hundredth.

Chapter Test

1. Simplify: $\dfrac{r^{\frac{2}{3}}r^{-1}}{r^{-\frac{1}{2}}}$

2. Simplify: $\dfrac{\left(2x^{\frac{1}{3}}y^{-\frac{2}{3}}\right)^6}{(x^{-4}y^8)^{\frac{1}{4}}}$

3. Simplify: $\left(\dfrac{4a^4}{b^2}\right)^{-\frac{3}{2}}$

4. Rewrite $3y^{\frac{2}{5}}$ as a radical expression.

5. Rewrite $\frac{1}{2}\sqrt[4]{x^3}$ as an exponential expression.

6. Simplify: $\sqrt[3]{8x^3y^6}$

7. Simplify: $\sqrt{32x^4y^7}$

8. Simplify: $\sqrt[3]{27a^4b^3c^7}$

9. Simplify: $\sqrt{18a^3} + a\sqrt{50a}$

10. Simplify: $\sqrt[3]{54x^7y^3} - x\sqrt[3]{128x^4y^3} - x^2\sqrt[3]{2xy^3}$

11. Simplify: $\sqrt{3x}(\sqrt{x} - \sqrt{25x})$

12. Simplify: $(2\sqrt{3} + 4)(3\sqrt{3} - 1)$

13. Simplify: $(\sqrt{a} - 3\sqrt{b})(2\sqrt{a} + 5\sqrt{b})$

14. Simplify: $(2\sqrt{x} + \sqrt{y})^2$

15. Simplify: $\dfrac{\sqrt{32x^5y}}{\sqrt{2xy^3}}$

16. Simplify: $\dfrac{4 - 2\sqrt{5}}{2 - \sqrt{5}}$

17. Simplify: $\dfrac{\sqrt{x}}{\sqrt{x} - \sqrt{y}}$

18. Simplify: $(\sqrt{-8})(\sqrt{-2})$

19. Simplify: $(5 - 2i) - (8 - 4i)$

20. Simplify: $(2 + 5i)(4 - 2i)$

21. Simplify: $\dfrac{2 + 3i}{1 - 2i}$

22. Simplify: $(2 + i) + (2 - i)(3 + 2i)$

23. Solve: $\sqrt{x + 12} - \sqrt{x} = 2$

24. Solve: $\sqrt[3]{2x - 2} + 4 = 2$

25. An object is dropped from a high building. Find the distance the object has fallen when the speed reaches 192 ft/s. Use the equation $v = \sqrt{64d}$, where v is the speed of the object and d is the distance.

Cumulative Review

1. Identify the property that justifies the statement, $(a + 2)b = ab + 2b$.

2. Simplify: $2x - 3[x - 2(x - 4) + 2x]$

3. Find $A \cap B$ given $A = \{2, 4, 6\}$ and $B = \{1, 3, 5\}$.

4. Solve: $\sqrt[3]{2x - 5} + 3 = 6$

5. Solve: $5 - \frac{2}{3}x = 4$

6. Solve: $2[4 - 2(3 - 2x)] = 4(1 - x)$

7. Solve: $3x - 4 \leq 8x + 1$

8. Solve: $5 < 2x - 3 < 7$

9. Solve: $2 + |4 - 3x| = 5$

10. Solve: $|7 - 3x| > 1$

11. Factor: $64a^2 - b^2$

12. Factor: $x^5 + 2x^3 - 3x$

13. Solve: $3x^2 + 13x - 10 = 0$

14. Solve: $x^2 - 2x - 8 = 0$

15. Simplify: $\dfrac{4a^2 + 8a}{a^3 + a^2 - 2a}$

16. Simplify: $\dfrac{1 - \frac{4}{y^2}}{\frac{1}{y} + \frac{2}{y^2}}$

17. Solve $P = \dfrac{R - C}{n}$ for R.

18. Solve: $\dfrac{x - 5}{x + 2} - \dfrac{1}{2} = \dfrac{x}{x + 2}$

19. Simplify: $(3^{-1}x^3y^{-5})(3^{-1}y^{-2})^{-2}$

20. Simplify: $\left(\dfrac{x^{-\frac{1}{2}}y^{\frac{3}{4}}}{y^{-\frac{5}{4}}}\right)^4$

21. Simplify: $\sqrt{20x^3} - x\sqrt{45x}$

22. Simplify: $(\sqrt{5} - 3)(\sqrt{5} - 2)$

23. Simplify: $\dfrac{\sqrt[3]{4x^5y^4}}{\sqrt[3]{8x^2y^5}}$

24. Simplify: $(2 - \sqrt{-9})(5 + \sqrt{-16})$

25. Simplify: $\dfrac{3i}{2 - i}$

26. Graph the set $\{x|x > -1\} \cap \{x|x \leq 3\}$.

27. A collection of thirty stamps consists of 13¢ stamps and 18¢ stamps. The total value of the stamps is \$4.85. Find the number of 18¢ stamps.

28. An investment of \$2500 is made at an annual simple interest rate of 7.2%. How much additional money must be invested at an annual simple interest rate of 8.4% so that the total interest earned is \$516?

29. The width of a rectangle is 6 ft less than the length. The area of the rectangle is 72 ft^2. Find the length and width of the rectangle.

30. A sales executive traveled 25 mi by car and then an additional 625 mi by plane. The rate by plane was five times faster than the rate by car. The total time of the trip was 3 h. Find the rate of the plane.

31. How long does it take light to travel to Earth from the moon when the moon is 232,500 mi from Earth? Light travels 1.86×10^5 mi/s.

32. How far would a submarine periscope have to be above the water to locate a ship 7 mi away? The equation for the distance in miles that the lookout can see is $d = 1.4\sqrt{h}$, where h is the height in feet above the surface of the water.

6

Linear Equations and Inequalities in Two Variables

Objectives

- Points on a rectangular coordinate system
- Determine a solution of a linear equation in two variables
- Graph a linear equation in two variables
- Find the slope of a line given two points
- Find the x- and y-intercepts of a straight line
- Graph a line given a point and the slope
- Find the equation of a line given a point and the slope or given two points
- Find the equations of parallel and perpendicular lines
- Obtain data from a graph
- Graph the solution set of an inequality in two variables

POINT OF INTEREST

Brachistochrone Problem

Consider the diagram at the right. What curve should be drawn so that a ball allowed to roll along the curve will travel from *A* to *B* in the shortest time?

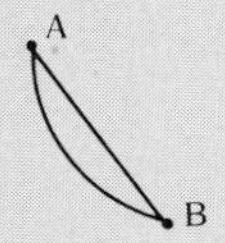

At first thought, one might conjecture that a straight line should connect the two points, since that shape is the shortest *distance* between the two points. Actually, however, the answer is half of one arch of an inverted cycloid.

A cycloid is shown below as the graph (in bold). One way to draw this curve is to think of a wheel rolling along a straight line without slipping. Then a point on the rim of the wheel traces a cycloid.

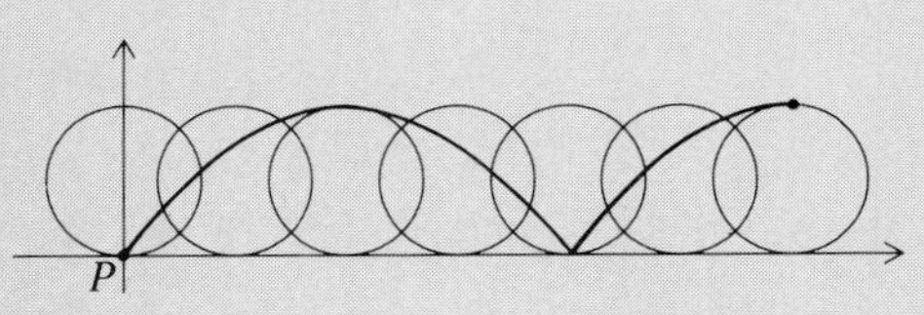

A cycloid

There are many applications of the idea of finding the shortest time between two points. As the above problem illustrates, the path of shortest time is not necessarily the path of shortest distance. Problems involving paths of shortest time are called *brachistochrone* problems.

SECTION **6.1**

The Rectangular Coordinate System

1 Points on a rectangular coordinate system

6

A **rectangular coordinate system** is formed by two number lines, one horizontal and one vertical, that intersect at the zero point of each line. The point of intersection is called the **origin.** The two lines are called the **coordinate axes,** or simply **axes.**

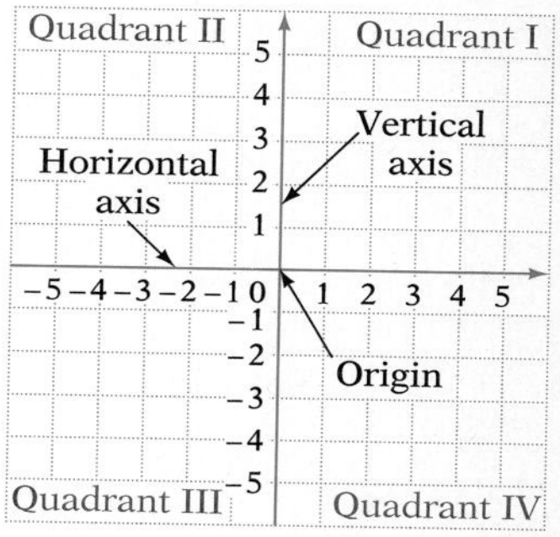

The axes determine a plane and divide the plane into four regions, called **quadrants**. The quadrants are numbered counterclockwise from I to IV.

Each point in the plane can be identified by a pair of numbers called an **ordered pair.** The first number of the pair measures a horizontal distance and is called the **abscissa.** The second number of the pair measures a vertical distance and is called the **ordinate.** The ordered pair (a, b) associated with a point is also called the **coordinates** of the point.

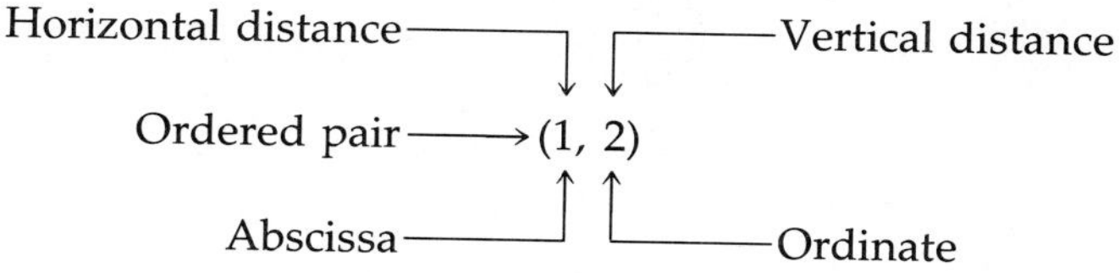

The **graph of an ordered pair** is a point in the plane. The graphs of the points whose coordinates are $(-2, 3)$ and $(3, -2)$ are shown at the right. Notice that they are different points. The order in which the numbers in an ordered pair appear *is* important.

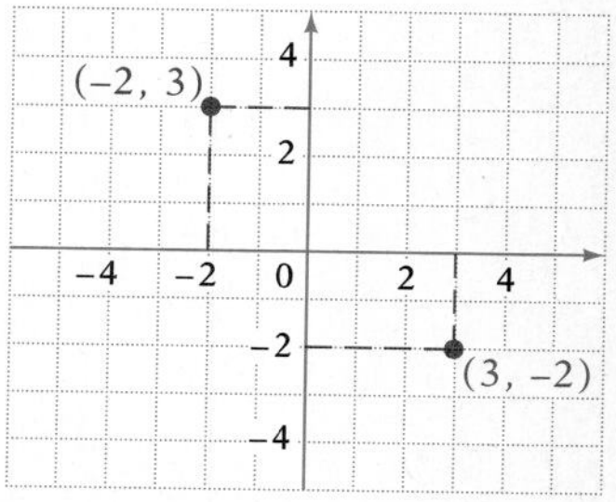

Example 1 A. Graph the points whose coordinates are (2, −1) and (−3, −4). Draw a line segment between the two points.

B. Draw a line through all points with an abscissa of 2.

Solution A.

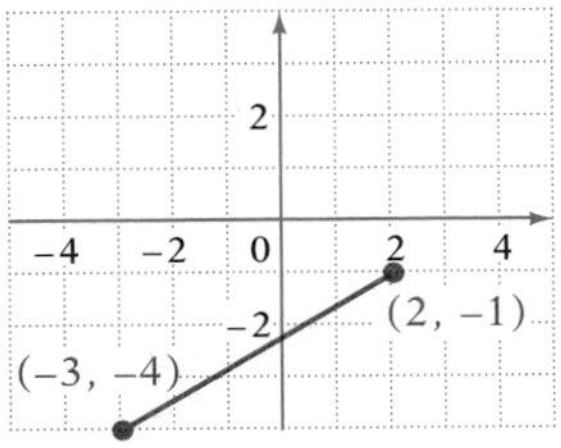

B.

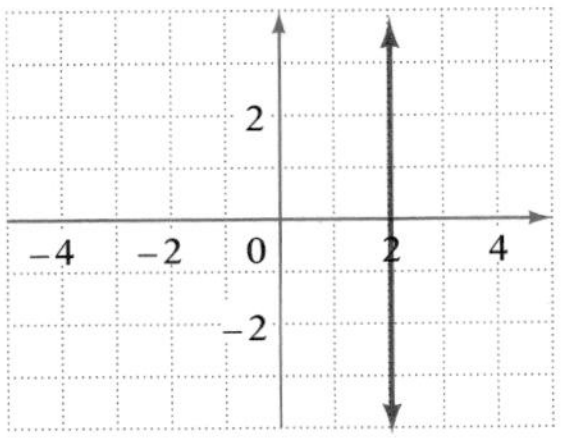

Problem 1 A. Graph the points whose coordinates are (5, −2) and (−2, 3). Draw a line segment between the two points.

B. Draw a line through all points with an ordinate of −1.

Solution See page A28.

2 Determine a solution of a linear equation in two variables

An equation in the form $y = mx + b$, where m and b are constants, is a **linear equation in two variables.** Examples of linear equations in two variables are shown below.

$$y = 5x - 7 \qquad (m = 5, \quad b = -7)$$
$$y = \frac{2}{3}x - 8 \qquad \left(m = \frac{2}{3}, \quad b = -8\right)$$
$$y = -\frac{1}{2}x + 2 \qquad \left(m = -\frac{1}{2}, \; b = 2\right)$$
$$y = x + 4 \qquad (m = 1, \quad b = 4)$$

A **solution of an equation in two variables** is an ordered pair of numbers (x, y) that makes the equation a true statement.

Is (2, 3) a solution of $y = \frac{1}{2}x + 2$?

$$y = \frac{1}{2}x + 2$$

Replace x with 2, the abscissa.

Replace y with 3, the ordinate.

$$3 \;\Big|\; \frac{1}{2}(2) + 2$$
$$\quad\Big|\; 1 + 2$$
$$3 = 3$$

Compare the results. If the results are equal, the given ordered pair is a solution. If the results are not equal, the given ordered pair is not a solution.

Yes, (2, 3) is a solution of the equation $y = \frac{1}{2}x + 2$.

Besides the ordered pair (2, 3), there are many other ordered pair solutions of the equation $y = \frac{1}{2}x + 2$. For example, the method used above can be used to show that $(-4, 0)$, $(-2, 1)$, and $(4, 4)$ are also solutions.

In general, a linear equation in two variables has an infinite number of solutions. By choosing any value for x and substituting that value into the linear equation, a corresponding value of y can be found.

Example 2 Find the ordered pair solution of $y = \frac{2}{3}x + 3$ corresponding to $x = -5$.

Solution

$$y = \frac{2}{3}x + 3$$

$$y = \frac{2}{3}(-5) + 3$$ ▶ Substitute −5 for x. Solve for y.

$$= -\frac{10}{3} + 3 = -\frac{1}{3}$$

The ordered pair solution is $\left(-5, -\frac{1}{3}\right)$.

Problem 2 Find the ordered pair solution of $y = -2x + 5$ corresponding to $x = \frac{1}{3}$.

Solution See page A28.

3 Graph a linear equation in two variables

The **graph of an equation in two variables** is a drawing of the ordered pair solutions of the equation. For a linear equation in two variables, the graph is a straight line.

To graph a linear equation, find ordered pair solutions of the equation. Do this by choosing any value of x and finding the corresponding value of y. Repeat this procedure, choosing different values for x, until you have found the number of solutions desired. Since the graph of a linear equation in two variables is a straight line, and a straight line is determined by two points, it is necessary to find only two solutions. However, it is recommended that at least three solutions be used to ensure accuracy.

To graph $y = -2x + 1$, choose any values of x, and find the corresponding values of y. It is convenient to record these solutions in a table.

x	$y = -2x + 1$	y
0	$-2(0) + 1$	1
2	$-2(2) + 1$	−3
−2	$-2(-2) + 1$	5

The horizontal axis is the x-axis. The vertical axis is the y-axis. Graph the ordered pair solutions (0, 1), (2, −3), and (−2, 5). Draw a line through the ordered pair solutions.

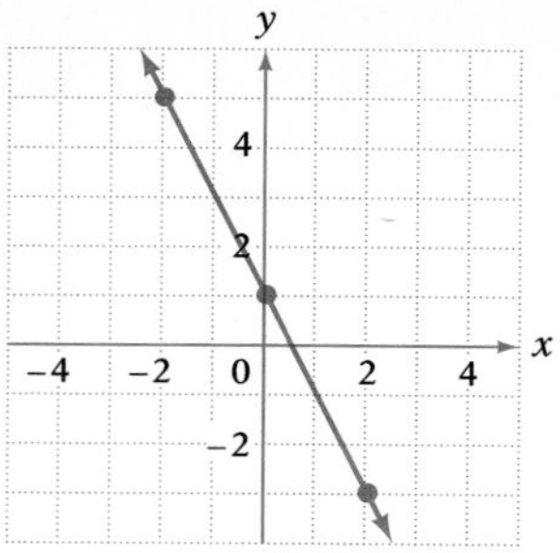

Remember that a graph is a drawing of the ordered pair solutions of the equation. Therefore, every point on the graph is a solution of the equation, and every solution of the equation is a point on the graph.

Example 3 Graph $y = -\frac{3}{2}x - 3$.

Solution

x	y
0	−3
−2	0
−4	3

▶ Find at least three solutions. When the coefficient of x is a fraction, choose values of x that will simplify the evaluations. Display the ordered pairs in a table.

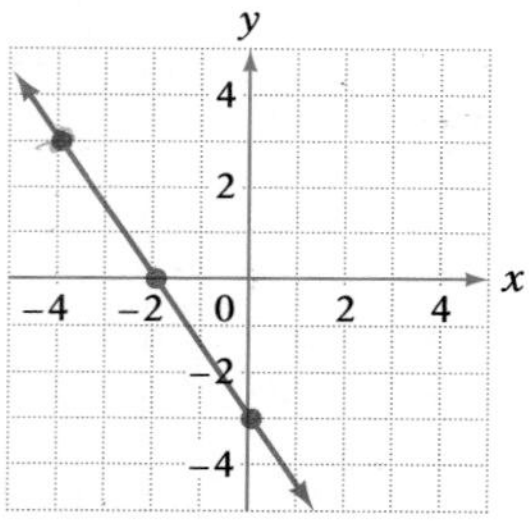

▶ Graph the ordered pairs on a rectangular coordinate system, and draw a straight line through the points.

Problem 3 Graph $y = -\frac{3}{4}x$.

Solution See page A28.

An equation of the form $Ax + By = C$, where A, B, and C are constants, is also a linear equation. Examples of equations of the form $Ax + By = C$ are shown below.

$$3x - 2y = -5 \qquad (A = 3, \quad B = -2, \; C = -5)$$
$$-\frac{1}{2}x + 2y = 4 \qquad \left(A = -\frac{1}{2}, \; B = 2, \quad C = 4\right)$$

An equation of the form $Ax + By = C$ can be written in the form $y = mx + b$.

To write the equation $3x - 2y = -5$ in the form $y = mx + b$, subtract $3x$ from each side of the equation.

$$3x - 2y = -5$$
$$-2y = -3x - 5$$

Divide each side of the equation by the coefficient -2.

$$y = \frac{3}{2}x + \frac{5}{2}$$

To graph an equation of the form $Ax + By = C$, first solve the equation for y. Then follow the same procedure used for graphing an equation of the form $y = mx + b$.

Example 4 Graph $3x + 2y = 6$.

Solution

$$3x + 2y = 6$$
$$2y = -3x + 6$$
$$y = -\frac{3}{2}x + 3$$

► Solve the equation for y.

x	y
0	3
2	0
4	−3

► Find at least three solutions.

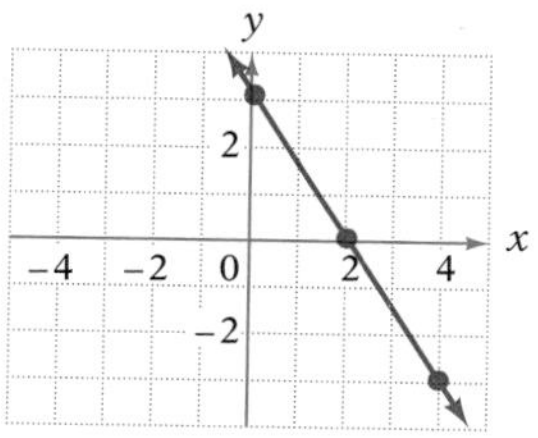

► Graph the ordered pairs on a rectangular coordinate system. Draw a straight line through the points.

Problem 4 Graph $-3x + 2y = 4$.

Solution See page A28.

The equation $y = -2$ could be written:

$$0 \cdot x + y = -2$$

No matter what value of x is chosen, y is always -2. Some solutions of the equation are $(3, -2)$, $(0, -2)$, and $(-2, -2)$. The graph is shown at the right.

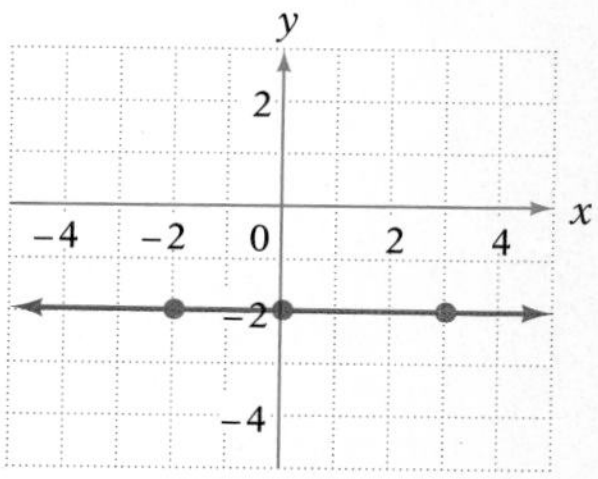

The **graph of $y = b$** is a horizontal line passing through point $(0, b)$.

The equation $x = 2$ could be written:

$$x + 0 \cdot y = 2$$

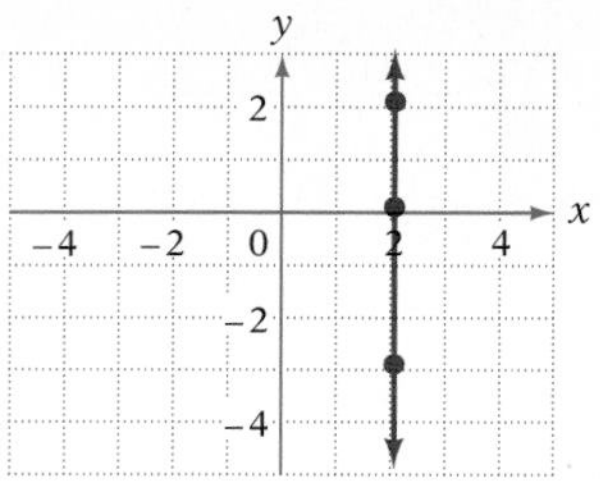

No matter what value of y is chosen, x is always 2. Some solutions of the equation are (2, 2), (2, 0), and (2, −3). The graph is shown at the right.

The **graph of $x = a$** is a vertical line passing through point $(a, 0)$.

The graph of an equation in which one of the variables is missing is either a horizontal or a vertical line.

Example 5 Graph $x = -4$.

Solution

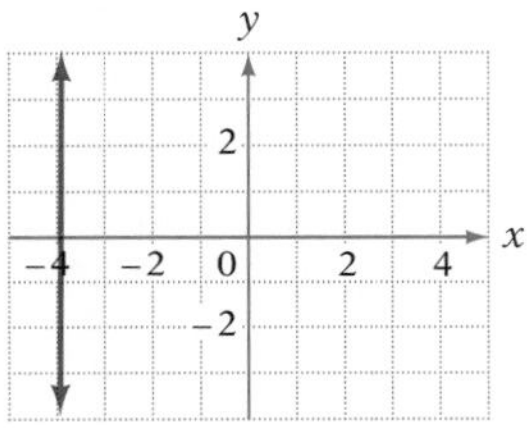

▶ The graph of an equation of the form $x = a$ is a vertical line passing through point $(a, 0)$.

Problem 5 Graph $y = 3$.

Solution See page A28.

EXERCISES 6.1

1

1. Graph the ordered pairs (3, 2) and (−1, 4). Draw a line segment between the two points.

2. Graph the ordered pairs (−1, −3) and (3, −2). Draw a line segment between the two points.

3. Graph the ordered pairs (−3, −3) and (2, −2). Draw a line segment between the two points.

4. Graph the ordered pairs (−3, 2) and (4, 2). Draw a line segment between the two points.

5. Find the coordinates of each of the points.

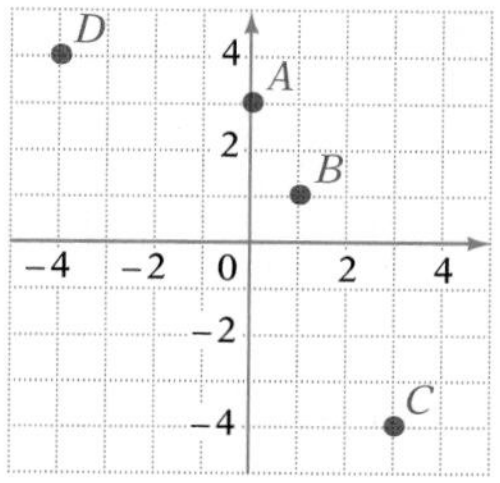

6. Find the coordinates of each of the points.

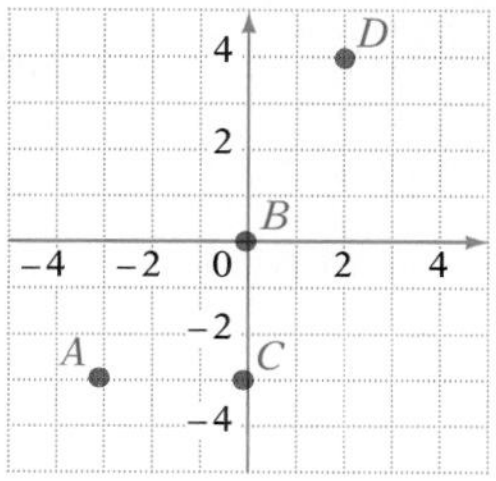

7. Find the coordinates of each of the points.

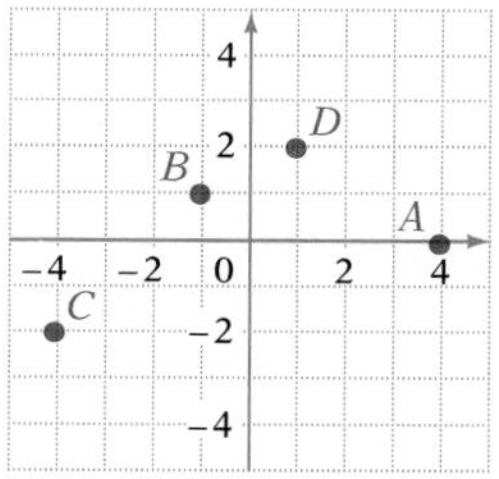

8. Find the coordinates of each of the points.

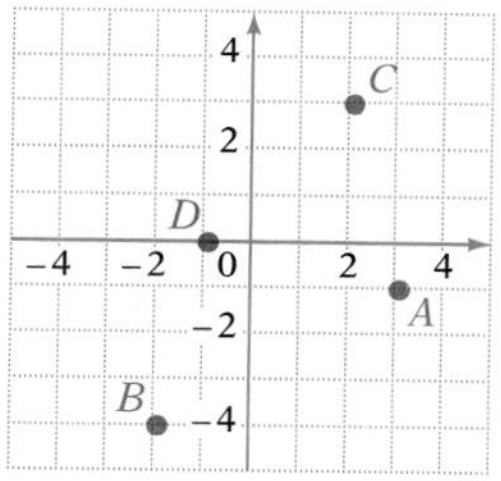

9. Draw a line through all points with an abscissa of 2.

10. Draw a line through all points with an abscissa of −3.

11. Draw a line through all points with an ordinate of −3.

12. Draw a line through all points with an ordinate of 4.

2

13. Find the ordered pair solution of $y = \frac{2}{3}x - 4$ corresponding to $x = -3$.

14. Find the ordered pair solution of $y = \frac{1}{2}x - 5$ corresponding to $x = 4$.

15. Find the ordered pair solution of $y = -\frac{2}{3}x + 4$ corresponding to $x = -6$.

16. Find the ordered pair solution of $y = -\frac{3}{4}x + 5$ corresponding to $x = -4$.

17. Find the ordered pair solution of $y = \frac{3}{2}x + 3$ corresponding to $x = -4$.

18. Find the ordered pair solution of $y = \frac{4}{3}x - 5$ corresponding to $x = 3$.

19. Find the ordered pair solution of $y = -2x - 3$ corresponding to $x = -2$.

20. Find the ordered pair solution of $y = 3x + 5$ corresponding to $x = -3$.

21. Find the ordered pair solution of $y = -\frac{4}{3}x - 3$ corresponding to $x = -2$.

22. Find the ordered pair solution of $y = \frac{3}{2}x - 4$ corresponding to $x = 3$.

3 Graph.

23. $y = 3x - 4$

24. $y = -2x + 3$

25. $y = -\frac{2}{3}x$

26. $y = \frac{3}{2}x$

27. $y = \frac{2}{3}x - 4$

28. $y = \frac{3}{4}x + 2$

29. $y = -\frac{1}{3}x + 2$

30. $y = -\frac{3}{2}x - 3$

31. $2x - y = 3$

32. $2x + y = -3$

33. $2x + 5y = 10$

34. $x - 4y = 8$

35. $y = -2$

36. $y = \frac{1}{3}x$

37. $2x - 3y = 12$

38. $3x - y = -2$

SUPPLEMENTAL EXERCISES 6.1

In each of the following exercises, three vertices of a rectangle are given. Find the coordinates of the fourth vertex.

39. (3, 5), (3, 0), (0, 0)

40. (−2, 4), (3, 4), (3, −1)

41. (−3, −2), (4, −2), (4, 5)

42. (6, −1), (6, 3), (−4, −1)

43. (−7, −2), (−1, −2), (−7, 8)

44. (5, −4), (−2, −4), (−2, −2)

For what value of k does the given point lie on the graph of the equation?

45. $3x - 2ky = 2$; (2, 1)

46. $kx + 3y = 4$; (−1, 2)

47. $x - ky = -1$; (3, 1)

48. $3x + 4ky = 2$; (−2, 1)

49. $kx + 6y = -2$; (2, −1)

50. $2kx + 3y = -3$; (1, −5)

51. A triangle has vertices whose coordinates are (−4, −3), (−4, 1), and (1, −3). Find the number of square units in the area of the triangle.

52. A triangle has vertices whose coordinates are (−2, −3), (−2, 4), and (4, −3). Find the number of square units in the area of the triangle.

53. A triangle has vertices whose coordinates are (3, 2), (6, 2), and (3, 7). Find the number of units in the perimeter of the triangle.

54. A triangle has vertices whose coordinates are (−3, −2), (−3, 3), and (9, −2). Find the number of units in the perimeter of the triangle.

55. For the linear equation $y = 3x - 2$, what is the increase in y that results when x is increased by 1?

56. For the linear equation $y = -x + 3$, what is the decrease in y that results when x is increased by 1?

SECTION 6.2 Slopes and Intercepts of Straight Lines

1 Find the slope of a line given two points

The graphs of $y = 3x + 2$ and $y = \frac{2}{3}x + 2$ are shown at the right. Each graph crosses the y-axis at the point (0, 2), but the graphs have different slants. The **slope** of a line is a measure of the slant of a line. The symbol for slope is m.

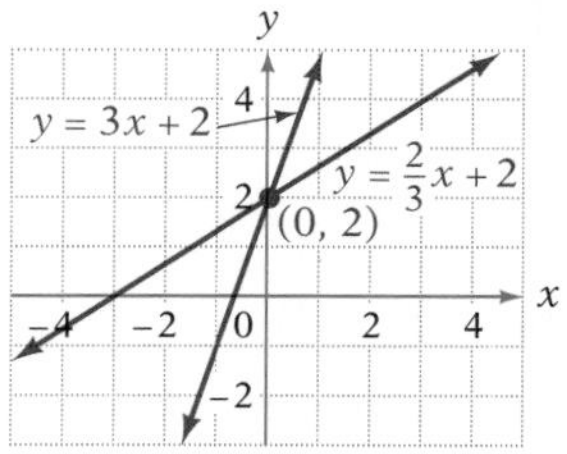

The slope of a line containing two points is the ratio of the change in the y values of the two points to the change in the x values. The line containing the points $(-1, -3)$ and $(5, 2)$ is graphed at the right.

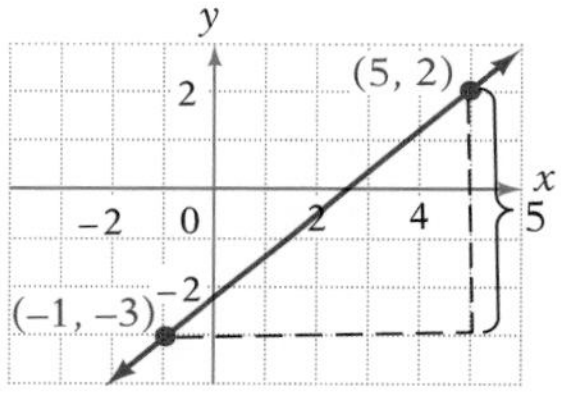

The change in the y values is the difference between the two ordinates.

Change in $y = 2 - (-3) = 5$

The change in the x values is the difference between the two abscissas.

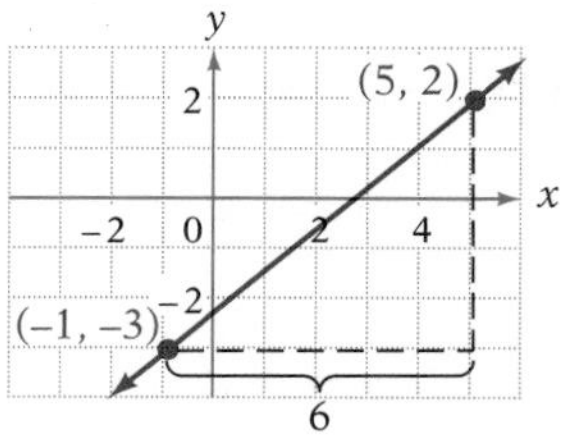

Change in $x = 5 - (-1) = 6$

$$\text{Slope} = m = \frac{\text{Change in } y}{\text{Change in } x} = \frac{5}{6}$$

Slope Formula

The slope of a line containing two points, P_1 and P_2, whose coordinates are (x_1, y_1) and (x_2, y_2), is given by:

$$\textbf{Slope} = m = \frac{y_2 - y_1}{x_2 - x_1}, \; x_1 \neq x_2.$$

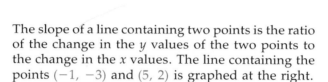

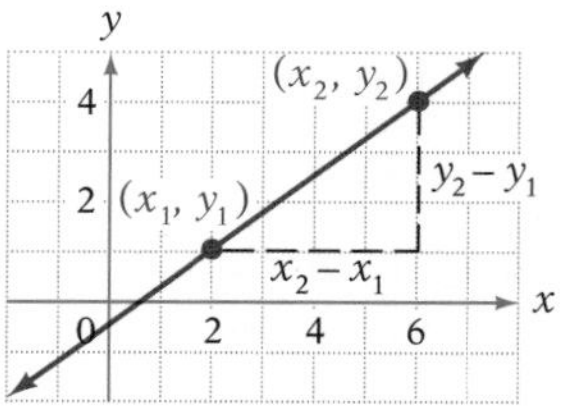

To find the slope of the line containing the points $(-2, 0)$ and $(4, 5)$, let $P_1 = (-2, 0)$ and $P_2 = (4, 5)$. It does not matter which ordered pair is named P_1 or P_2; the slope will be the same.

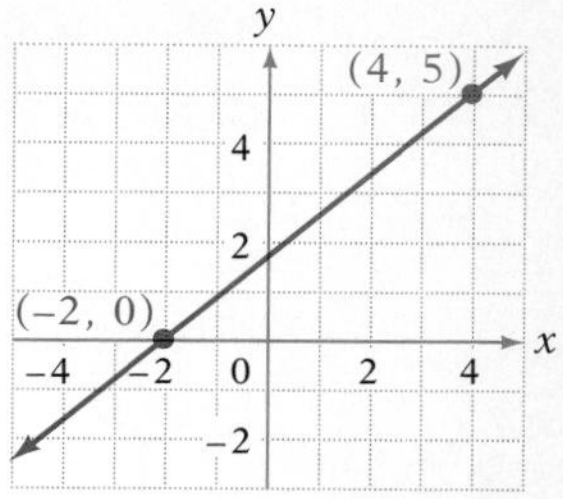

Positive slope

$$m = \frac{y_2 - y_1}{x_2 - x_1} = \frac{5 - 0}{4 - (-2)} = \frac{5}{6}$$

The slope is a positive number.

A line that slants upward to the right has a **positive slope.**

Slope is defined as $\frac{\text{change in } y}{\text{change in } x}$, a slope of $\frac{5}{6}$ means that y increases 5 units as x increases 6 units.

To find the slope of the line containing the points $(-3, 4)$ and $(4, 2)$, let $P_1 = (-3, 4)$ and $P_2 = (4, 2)$.

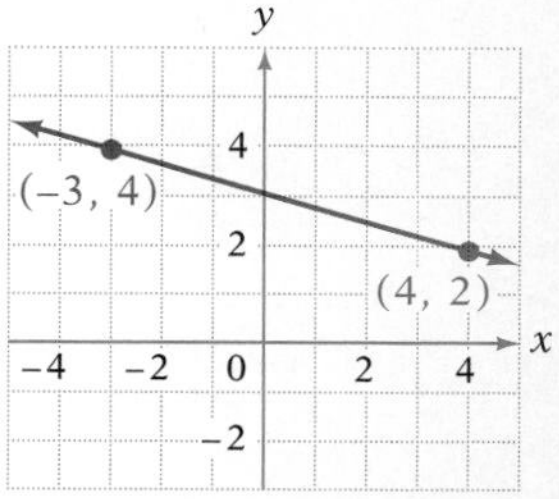

Negative slope

$$m = \frac{y_2 - y_1}{x_2 - x_1} = \frac{2 - 4}{4 - (-3)} = \frac{-2}{7} = -\frac{2}{7}$$

The slope is a negative number.

A line that slants downward to the right has a **negative slope.**

A slope of $-\frac{2}{7}$ means that y decreases 2 units as x increases 7 units.

To find the slope of the line containing the points $(-2, 2)$ and $(4, 2)$, let $P_1 = (-2, 2)$ and $P_2 = (4, 2)$.

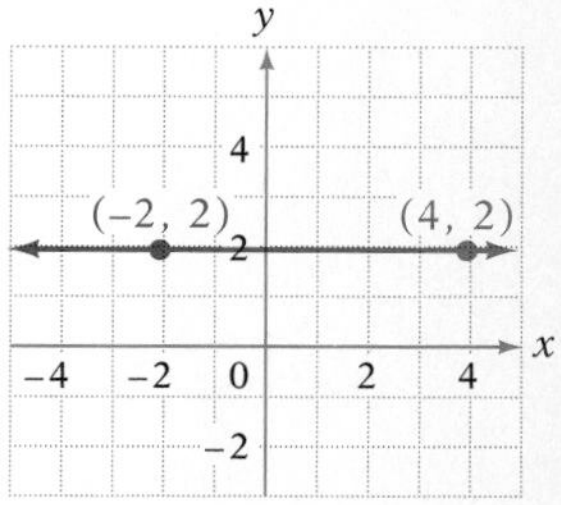

Zero slope

$$m = \frac{y_2 - y_1}{x_2 - x_1} = \frac{2 - 2}{4 - (-2)} = \frac{0}{6} = 0$$

When $y_1 = y_2$, the graph is a horizontal line.

A horizontal line has **zero slope.**

To find the slope of the line containing the points $(1, -2)$ and $(1, 3)$, let $P_1 = (1, -2)$ and $P_2 = (1, 3)$.

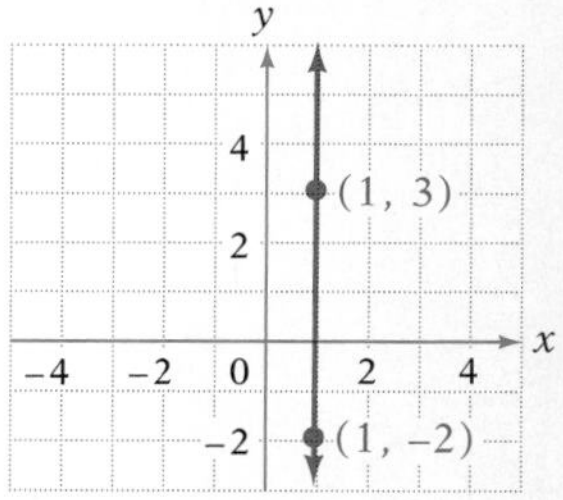

Undefined

$$m = \frac{y_2 - y_1}{x_2 - x_1} = \frac{3 - (-2)}{1 - 1} = \frac{5}{0} \quad \text{Not a real number}$$

When $x_1 = x_2$, the denominator of $\frac{y_2 - y_1}{x_2 - x_1}$ is zero and the graph is a vertical line. Because division by 0 is undefined, the slope of the line is undefined.

The slope of a vertical line is **undefined**.

Example 1 Find the slope of the line containing the points $(2, -5)$ and $(-4, 2)$.

Solution $m = \frac{y_2 - y_1}{x_2 - x_1} = \frac{2-(-5)}{-4-2} = \frac{7}{-6}$ ► Let $P_1 = (2, -5)$ and $P_2 = (-4, 2)$.

The slope is $-\frac{7}{6}$.

Problem 1 Find the slope of the line containing the points $(4, -3)$ and $(2, 7)$.

Solution See page A29.

2 Find the x- and y-intercepts of a straight line

The graph of the equation $x - 2y = 4$ is shown at the right. The graph crosses the x-axis at the point $(4, 0)$. This point is called the **x-intercept**. The graph also crosses the y-axis at the point $(0, -2)$. This point is called the **y-intercept.**

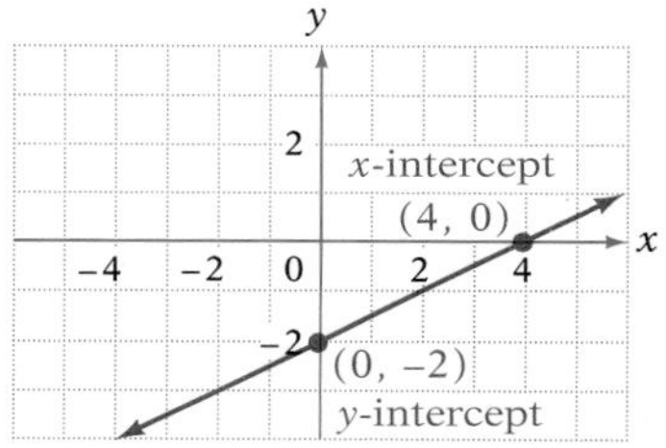

Some linear equations can be graphed by finding the x- and y-intercepts and then drawing a line through the two points. To find the x-intercept, let $y = 0$. (Any point on the x-axis has y-coordinate 0.) To find the y-intercept, let $x = 0$. (Any point on the y-axis has x-coordinate 0.)

Example 2 Graph $4x - y = 4$ by using the x- and y-intercepts.

Solution

x-intercept:
$$\begin{aligned} 4x - y &= 4 \\ 4x - 0 &= 4 \\ 4x &= 4 \\ x &= 1 \end{aligned}$$

y-intercept:
$$\begin{aligned} 4x - y &= 4 \\ 4(0) - y &= 4 \\ -y &= 4 \\ y &= -4 \end{aligned}$$

► To find the x-intercept, let $y = 0$. To find the y-intercept, let $x = 0$.

The x-intercept is $(1, 0)$. The y-intercept is $(0, -4)$.

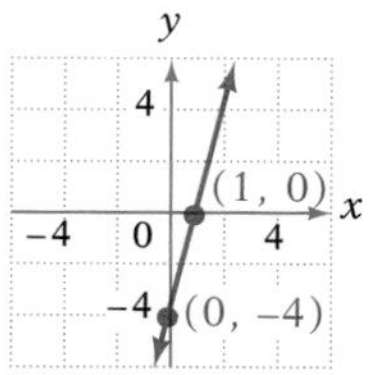

► Graph the points $(1, 0)$ and $(0, -4)$. Draw a line through the two points.

Problem 2 Graph $3x - y = 2$ by using the x- and y-intercepts.

Solution See page A29.

To find the y-intercept of $y = \frac{2}{3}x + 7$, let $x = 0$.

$$y = \frac{2}{3}x + 7$$
$$y = \frac{2}{3}(0) + 7$$
$$y = 7$$

The y-intercept is (0, 7).

For any equation of the form $y = mx + b$, the y-intercept is $(0, b)$.

Example 3 Graph $y = \frac{2}{3}x - 2$ by using the x- and y-intercepts.

Solution x-intercept:

$$y = \frac{2}{3}x - 2$$
$$0 = \frac{2}{3}x - 2$$
$$-\frac{2}{3}x = -2$$
$$x = 3$$

y-intercept: $(0, b)$

$$b = -2$$

▶ To find the x-intercept, let $y = 0$. For any equation of the form $y = mx + b$, the y-intercept is $(0, b)$.

The x-intercept is (3, 0). The y-intercept is (0, −2).

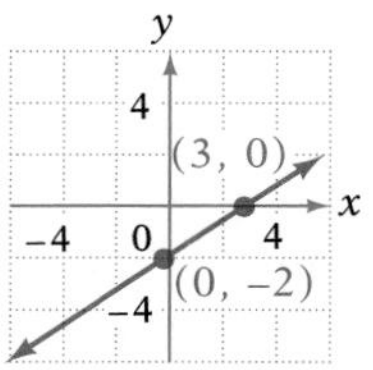

▶ Graph the points (3, 0) and (0, −2). Draw a line through the two points.

Problem 3 Graph $y = \frac{1}{4}x + 1$ by using the x- and y-intercepts.

Solution See page A29.

3 Graph a line given a point and the slope

The graph of the equation $y = -\frac{3}{4}x + 4$ is shown at the right. The points $(-4, 7)$ and $(4, 1)$ are on the graph. The slope of the line is

$$m = \frac{1-7}{4-(-4)} = \frac{-6}{8} = -\frac{3}{4}$$

Note that the slope of the line has the same value as the coefficient of x. The y-intercept is $(0, 4)$, and the constant is 4.

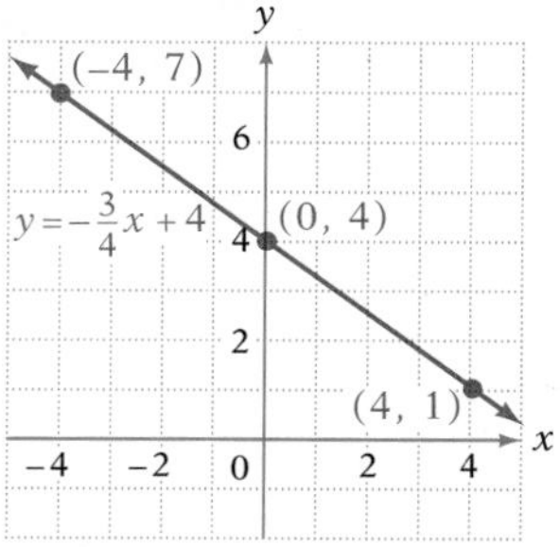

Slope-Intercept Form of a Straight Line

For an equation of the form $y = mx + b$, the slope of the line is m, the coefficient of x. The y-intercept is $(0, b)$. The equation

$$y = mx + b$$

is called the **slope-intercept form of a straight line.**

When the equation of a straight line is in the form $y = mx + b$, the graph can be drawn using the slope and y-intercept. First locate the y-intercept. Use the slope to find a second point on the line. Then draw a line through the two points.

When the equation of a straight line is in the form $Ax + By = C$, first solve the equation for y. Then follow the same procedure used for an equation in the form $y = mx + b$.

To graph $x + 2y = 4$ by using the slope and y-intercept, solve the equation for y.

$$x + 2y = 4$$
$$2y = -x + 4$$
$$y = -\frac{1}{2}x + 2$$

The y-intercept is $(0, b) = (0, 2)$.

$$m = -\frac{1}{2} = \frac{-1}{2} = \frac{\text{Change in } y}{\text{Change in } x}$$

Beginning at the y-intercept $(0, 2)$, move right 2 units (change in x) and then down 1 unit (change in y).

$(2, 1)$ is a second point on the graph.

Draw a line through the points $(0, 2)$ and $(2, 1)$.

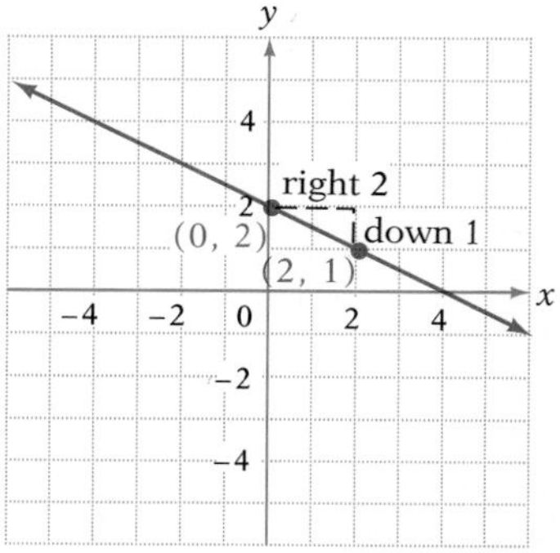

Example 4 Graph $y = -\frac{3}{2}x + 4$ by using the slope and y-intercept.

Solution y-intercept = (0, 4) ► Locate the y-intercept.

$m = -\frac{3}{2} = \frac{-3}{2}$ ► $m = \frac{\text{Change in } y}{\text{Change in } x}$

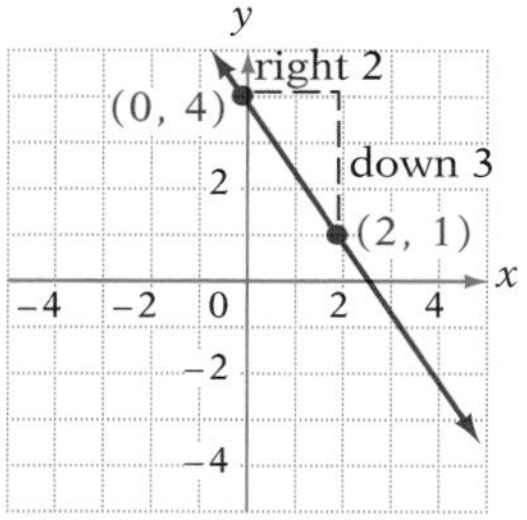

► Beginning at the y-intercept, (0, 4), move right 2 units and then down 3 units. (2, 1) is a second point on the graph. Draw a line through the points (0, 4) and (2, 1).

Problem 4 Graph $2x + 3y = 6$ by using the slope and y-intercept.

Solution See page A29.

The graph of a line can be drawn when a point on the line and the slope of the line are given.

To graph the line that passes through point (2, 1) and has slope $\frac{2}{3}$, locate the point (2, 1) on the graph.

$$m = \frac{2}{3} = \frac{\text{Change in } y}{\text{Change in } x}$$

Beginning at the point (2, 1), move right 3 units and then up 2 units.

(5, 3) is a second point on the line.

Draw a line through the points (2, 1) and (5, 3).

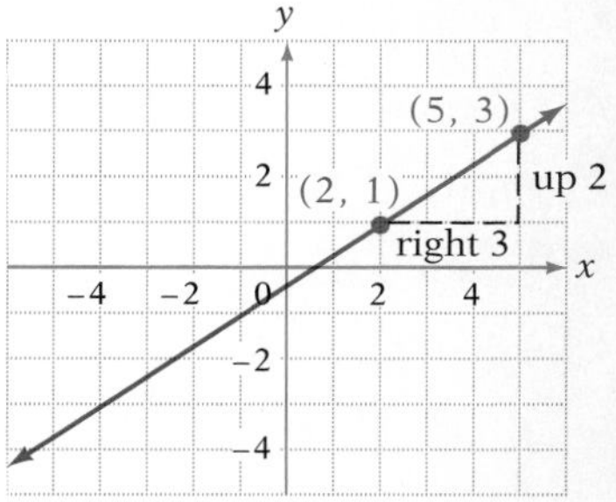

Example 5 Graph the line that passes through point (−2, 3) and has slope $-\frac{4}{3}$.

Solution $(x_1, y_1) = (-2, 3)$

$m = -\frac{4}{3} = \frac{-4}{3}$

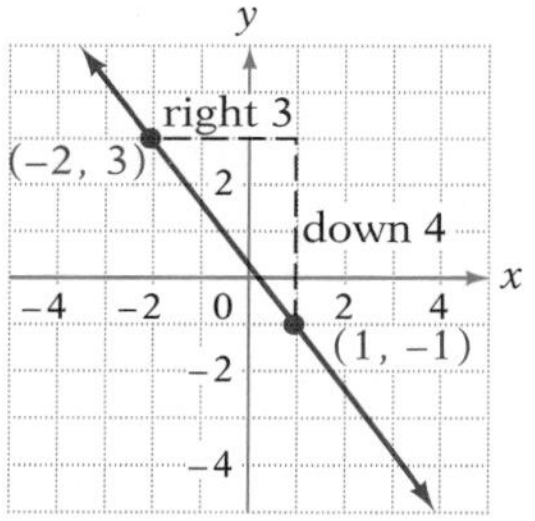

Problem 5 Graph the line that passes through point $(-3, -2)$ and has slope 3.

Solution See page A29.

EXERCISES 6.2

1 Find the slope of the line containing the points.

1. $P_1(1, 3), P_2(3, 1)$

2. $P_1(2, 3), P_2(5, 1)$

3. $P_1(-1, 4), P_2(2, 5)$

4. $P_1(3, -2), P_2(1, 4)$

5. $P_1(-1, 3), P_2(-4, 5)$

6. $P_1(-1, -2), P_2(-3, 2)$

7. $P_1(0, 3), P_2(4, 0)$

8. $P_1(-2, 0), P_2(0, 3)$

9. $P_1(2, 4), P_2(2, -2)$

10. $P_1(4, 1), P_2(4, -3)$

11. $P_1(2, 5), P_2(-3, -2)$

12. $P_1(4, 1), P_2(-1, -2)$

13. $P_1(2, 3), P_2(-1, 3)$

14. $P_1(3, 4), P_2(0, 4)$

15. $P_1(0, 4), P_2(-2, 5)$

16. $P_1(3, 0), P_2(-1, -4)$

17. $P_1(-3, 4), P_2(-2, 1)$

18. $P_1(4, -2), P_2(2, -4)$

19. $P_1(-2, 3), P_2(-2, 5)$

20. $P_1(-3, -1), P_2(-3, 4)$

21. $P_1(-2, -5), P_2(-4, -1)$

22. $P_1(-3, -2), P_2(0, -5)$

23. $P_1(3, -1), P_2(-2, -1)$

24. $P_1(0, -3), P_2(-2, -3)$

2 Find the x- and y-intercepts and graph.

25. $x - 2y = -4$

26. $3x + y = 3$

27. $4x - 2y = 5$

28. $2x - y = 4$

29. $3x + 2y = 5$

30. $4x - 3y = 8$

31. $2x - 3y = 4$

32. $3x - 5y = 9$

33. $2x - 3y = 9$

34. $3x - 4y = 4$

35. $2x + y = 3$

36. $3x + y = -5$

37. $3x + 2y = 4$

38. $3x + 4y = -12$

39. $2x - 3y = -6$

40. $4x - 3y = 6$

3 Graph by using the slope and the y-intercept.

41. $y = \frac{1}{2}x + 2$

42. $y = \frac{2}{3}x - 3$

43. $y = -\frac{2}{3}x + 4$

44. $y = -\frac{1}{2}x + 2$

45. $y = -\frac{3}{2}x$

46. $y = \frac{3}{4}x$

47. $y = \frac{2}{3}x - 1$

48. $2x - 3y = 6$

49. $3x - y = 2$

50. $4x + y = 2$

51. $3x + 2y = 8$

52. $4x - 5y = 5$

53. $3x - 2y = 6$

54. $x - 3y = 3$

55. $x + 2y = 4$

56. Graph the line that passes through point (2, 3) and has slope $\frac{1}{2}$.

57. Graph the line that passes through point (−4, 1) and has slope $\frac{2}{3}$.

58. Graph the line that passes through point (1, 4) and has slope $-\frac{2}{3}$.

59. Graph the line that passes through point (0, −2) and has slope $-\frac{1}{3}$.

60. Graph the line that passes through point (−3, 0) and has slope −3.

61. Graph the line that passes through point (2, 0) and has slope −1.

SUPPLEMENTAL EXERCISES 6.2

From the graph of the equation, write the equation of the line in the form $y = mx + b$.

62.

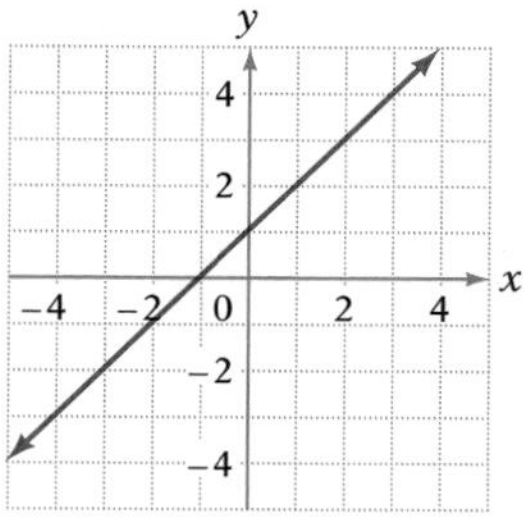

63.

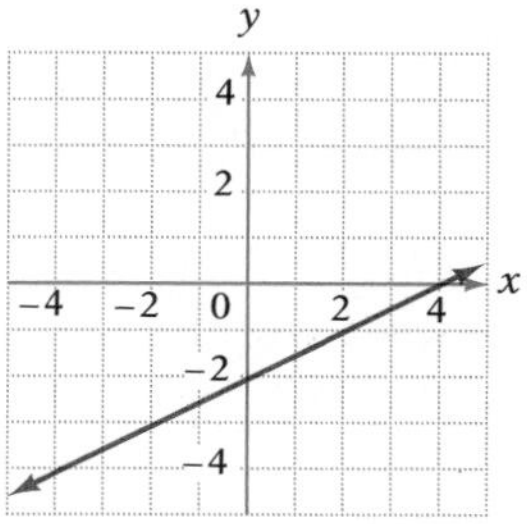

64.

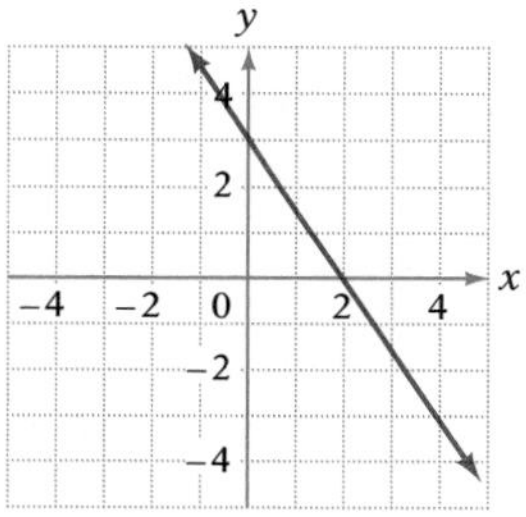

65.

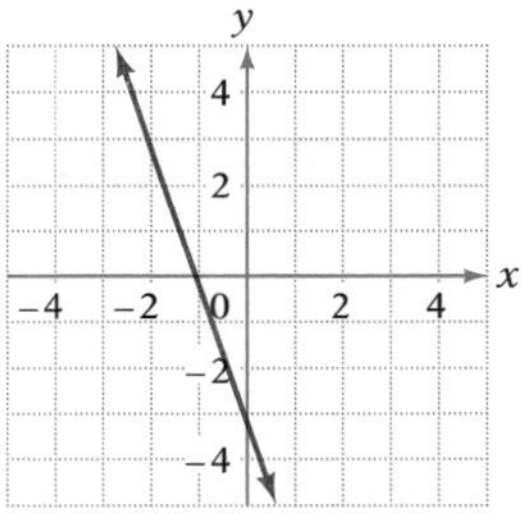

Solve.

66. **a.** Show that the equation $\frac{x}{3} + \frac{y}{4} = 1$ is a linear equation by writing it in the form $y = mx + b$.

b. Find the x- and y-intercepts.

67. **a.** Show that the equation $\frac{x}{2} - \frac{y}{5} = 1$ is a linear equation by writing it in the form $y = mx + b$.

b. Find the x- and y-intercepts.

68. **a.** Show that the equation $\frac{y}{6} - \frac{x}{4} = 1$ is a linear equation by writing it in the form $y = mx + b$.

b. Find the x- and y-intercepts.

69. **a.** Show that the equation $x - \frac{y}{8} = 1$ is a linear equation by writing it in the form $y = mx + b$.

b. Find the x- and y-intercepts.

70. Show that the x- and y-intercepts of the graph of $\frac{x}{a} + \frac{y}{b} = 1$ $(a \neq 0, b \neq 0)$ are $(a, 0)$ and $(0, b)$ respectively.

71. What effect does increasing the coefficient of x have on the graph of $y = mx + b$?

72. What effect does decreasing the coefficient of x have on the graph of $y = mx + b$?

73. What effect does increasing the constant term have on the graph of $y = mx + b$?

74. What effect does decreasing the constant term have on the graph of $y = mx + b$?

75. If $y = mx + b$, $m > 0$, then as x increases by 1, y increases by ______.

76. If $y = mx + b$, $m < 0$, then as x increases by 1, y decreases by ______.

SECTION 6.3 Finding Equations of Lines

1 Find the equation of a line given a point and the slope or given two points

When the slope of a line and a point on the line are known, the equation of the line can be determined.

To find the equation of the line whose y-intercept is $(0, 3)$ and has slope $\frac{1}{2}$, use the slope-intercept form of the equation since the given point is the y-intercept.

$$y = mx + b$$

Replace m with $\frac{1}{2}$, the given slope.

Replace b with 3, the y-intercept.

$$y = \frac{1}{2}x + 3$$

The equation of the line is $y = \frac{1}{2}x + 3$.

Example 1 Find the equation of the line whose y-intercept is $(0, -2)$ and has slope $-\frac{2}{3}$.

Solution $m = -\frac{2}{3}$ $b = -2$

$y = mx + b$

$y = -\frac{2}{3}x - 2$ ▶ Replace m with $-\frac{2}{3}$, the given slope. Replace b with -2, the y-intercept.

The equation of the line is $y = -\frac{2}{3}x - 2$.

Problem 1 Find the equation of the line whose y-intercept is (0, 3) and has slope $-\frac{5}{4}$.

Solution See page A30.

One method for finding the equation of a line, given the slope and *any* point on the line, involves use of the point-slope formula. The point-slope formula is derived from the formula for slope.

Let (x_1, y_1) be the given point on the line and (x, y) be any other point on the line. Then the slope of the line is given by $\frac{y - y_1}{x - x_1} = m$.

Formula for slope $\frac{y - y_1}{x - x_1} = m$

Multiply both sides of the equation by $(x - x_1)$. $\frac{y - y_1}{x - x_1}(x - x_1) = m(x - x_1)$

Simplify. $y - y_1 = m(x - x_1)$

Point-Slope Formula

The equation of the line with slope m and containing the point (x_1, y_1) can be found by the point-slope formula: $\boldsymbol{y - y_1 = m(x - x_1)}$.

Example 2 Find the equation of the line that contains the point (−2, 4) and has slope 2.

Solution

$y - y_1 = m(x - x_1)$ ▶ Use the point-slope formula.

$y - 4 = 2[x - (-2)]$ ▶ Substitute the slope, 2, and the coordinates of the given point, (−2, 4), into the point-slope formula.

$y - 4 = 2(x + 2)$

$y - 4 = 2x + 4$

$y = 2x + 8$

The equation of the line is $y = 2x + 8$.

Problem 2 Find the equation of the line that contains the point (4, −3) and has slope −3.

Solution See page A30.

The point-slope formula and the formula for slope are used to find the equation of a line when two points are known.

Example 3 Find the equation of the line containing the given points.

A. $P_1(-2, 5)$, $P_2(-4, -1)$ B. $P_1(2, -3)$, $P_2(2, 5)$

Solution A. $m = \frac{y_2 - y_1}{x_2 - x_1}$

$= \frac{-1 - 5}{-4 - (-2)} = \frac{-6}{-2} = 3$

▶ Find the slope. Let $(x_1, y_1) = (-2, 5)$ and $(x_2, y_2) = (-4, -1)$.

$$y - y_1 = m(x - x_1)$$
$$y - 5 = 3[x - (-2)]$$
$$y - 5 = 3(x + 2)$$
$$y - 5 = 3x + 6$$
$$y = 3x + 11$$

▶ Substitute the slope and the coordinates of either one of the known points into the point-slope formula.

The equation of the line is $y = 3x + 11$.

B. $m = \frac{y_2 - y_1}{x_2 - x_1} = \frac{5 - (-3)}{2 - 2} = \frac{8}{0}$

▶ The slope of the line is undefined. It is a vertical line passing through points $(2, -3)$ and $(2, 5)$. All points on the line have an abscissa of 2.

The equation of the line is $x = 2$.

Problem 3 Find the equation of the line containing the given points.

A. $P_1(4, -2)$, $P_2(-1, -7)$ B. $P_1(2, 3)$, $P_2(-5, 3)$

Solution See page A30.

2 Find the equations of parallel and perpendicular lines

Two lines that have the same slope do not intersect and are called **parallel lines.** Two vertical lines are parallel lines. Two horizontal lines are parallel lines.

The slope of each of the lines at the right is $\frac{2}{3}$. The lines are parallel.

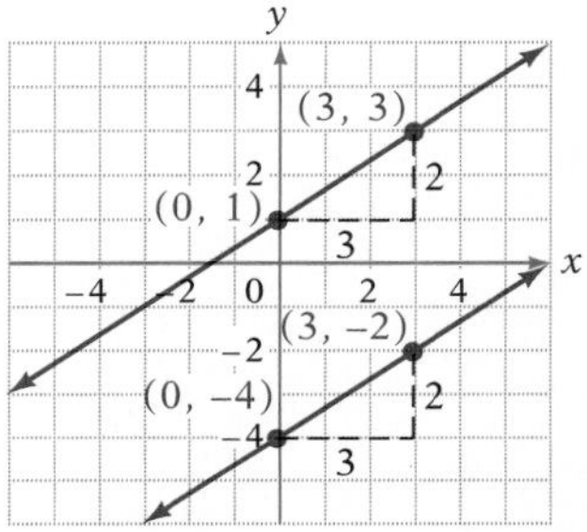

Slopes of Parallel Lines

> For two nonvertical parallel lines, one with slope m_1 and one with slope m_2, $m_1 = m_2$.

Is the line that contains the points (−2, 1) and (−5, −1) parallel to the line that contains the points (1, 0) and (4, 2)?

Find the slope of each line.

$$m_1 = \frac{-1-1}{-5-(-2)} = \frac{-2}{-3} = \frac{2}{3}$$

$$m_2 = \frac{2-0}{4-1} = \frac{2}{3}$$

$m_1 = m_2 = \frac{2}{3}$ The lines are parallel.

Are the lines $3x - 4y = 8$ and $6x - 8y = 2$ parallel?

Write each equation in slope-intercept form.

$$3x - 4y = 8 \qquad 6x - 8y = 2$$
$$-4y = -3x + 8 \qquad -8y = -6x + 2$$
$$y = \frac{3}{4}x - 2 \qquad y = \frac{3}{4}x - \frac{1}{4}$$

Find the slope of each line. $m_1 = \frac{3}{4}$ $m_2 = \frac{3}{4}$

$m_1 = m_2 = \frac{3}{4}$ The lines are parallel.

Find the equation of the line containing the point (−1, 4) and parallel to the line $2x - 3y = 2$.

Write the given equation in slope-intercept form to determine its slope.

$$2x - 3y = 2$$
$$-3y = -2x + 2$$
$$y = \frac{2}{3}x - \frac{2}{3} \qquad m = \frac{2}{3}$$

Parallel lines have the same slope.

$$y - y_1 = m(x - x_1)$$

Substitute the slope of the given line and the coordinates of the given point in the point-slope formula.

$$y - 4 = \frac{2}{3}[x - (-1)]$$
$$y - 4 = \frac{2}{3}x + \frac{2}{3}$$
$$y = \frac{2}{3}x + \frac{14}{3}$$

The equation of the line is $y = \frac{2}{3}x + \frac{14}{3}$.

Example 4 Find the equation of the line containing the point (3, −1) and parallel to the line $y = \frac{3}{2}x - 2$.

Solution

$$y - y_1 = m(x - x_1)$$
$$y - (-1) = \frac{3}{2}(x - 3)$$
$$y + 1 = \frac{3}{2}x - \frac{9}{2}$$
$$y = \frac{3}{2}x - \frac{11}{2}$$

▶ The slope of the given line is $\frac{3}{2}$. Substitute the slope of the given line and the coordinates of the given point in the point-slope formula.

The equation of the line is $y = \frac{3}{2}x - \frac{11}{2}$.

Problem 4 Are the lines $5x + 2y = 2$ and $5x + 2y = -6$ parallel?

Solution See page A30.

Two lines that intersect at right angles are **perpendicular lines.**

Any horizontal line is perpendicular to any vertical line. For example, $x = 3$ is perpendicular to $y = -2$.

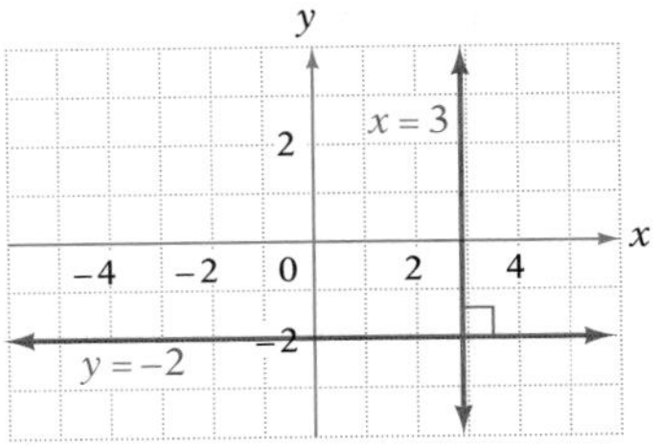

Two lines, neither of which is a vertical line, are perpendicular if the product of their slopes is -1.

Slopes of Perpendicular Lines

> For two nonvertical perpendicular lines, one with slope m_1 and one with slope m_2, $\boldsymbol{m_1 \cdot m_2 = -1}$.

Solving $m_1 \cdot m_2 = -1$ for m_2 yields $m_2 = -\frac{1}{m_1}$. This equation states that if m_1 is the slope of a nonvertical line, then the slope m_2 of a perpendicular line is $-\frac{1}{m_1}$, the **negative reciprocal** of m_1.

The line $y = \frac{1}{3}x - 2$ is perpendicular to

$y = -3x + 3$ since $\frac{1}{3}(-3) = -1$.

-3 is the negative reciprocal of $\frac{1}{3}$.

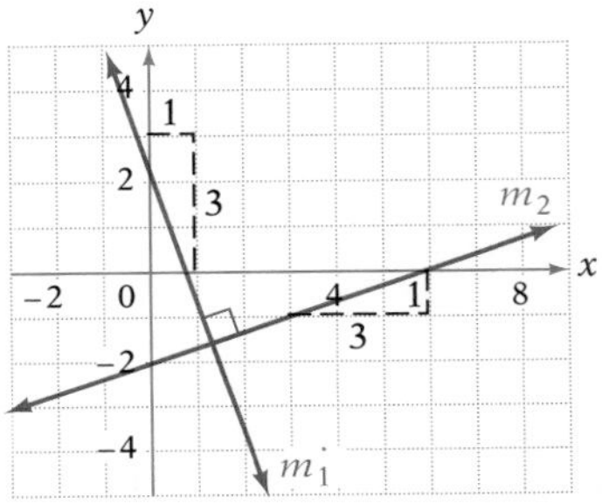

The line that contains the points (4, 2) and (−2, 5) is perpendicular to the line that contains the points (−4, 3) and (−3, 5) because the product of the slopes of the two lines is −1.

$$m_1 = \frac{5-2}{-2-4} = \frac{3}{-6} = -\frac{1}{2}$$

$$m_2 = \frac{5-3}{-3-(-4)} = \frac{2}{1} = 2$$

$$m_1 \cdot m_2 = -\frac{1}{2}(2) = -1$$

The lines $3x + 4y = -8$ and $4x - 3y = -9$ are perpendicular because the product of the slopes of the two lines is −1.

$$3x + 4y = -8 \qquad\qquad 4x - 3y = -9$$

$$4y = -3x - 8 \qquad\qquad -3y = -4x - 9$$

$$y = -\frac{3}{4}x - 2 \qquad\qquad y = \frac{4}{3}x + 3$$

$$m_1 = -\frac{3}{4} \qquad\qquad m_2 = \frac{4}{3}$$

$$m_1 \cdot m_2 = -\frac{3}{4}\left(\frac{4}{3}\right) = -1$$

Find the equation of the line containing the point (3, −4) and perpendicular to the line $2x - y = -3$.

Solve the given equation for y to determine the slope of the given line.

$$2x - y = -3$$

$$-y = -2x - 3$$

$$y = 2x + 3 \qquad m_1 = 2$$

Substitute the value of m_1 into the equation $m_1 \cdot m_2 = -1$ and solve for m_2, the slope of the perpendicular line.

$$m_1 \cdot m_2 = -1$$

$$2m_2 = -1$$

$$m_2 = -\frac{1}{2}$$

Substitute the slope, m_2, and the coordinates of the given point in the point-slope formula.

$$y - y_1 = m(x - x_1)$$

$$y - (-4) = -\frac{1}{2}(x - 3)$$

$$y + 4 = -\frac{1}{2}x + \frac{3}{2}$$

$$y = -\frac{1}{2}x - \frac{5}{2}$$

The equation of the line is $y = -\frac{1}{2}x - \frac{5}{2}$.

Example 5 Are the lines $4x - y = -2$ and $x + 4y = -12$ perpendicular?

Solution

$$4x - y = -2 \qquad\qquad x + 4y = -12$$

$$-y = -4x - 2 \qquad\qquad 4y = -x - 12$$

$$y = 4x + 2 \qquad\qquad y = -\frac{1}{4}x - 3$$

▶ Solve each equation for y.

$$m_1 = 4 \qquad\qquad m_2 = -\frac{1}{4}$$

▶ Find the slope of each line.

$$m_1 \cdot m_2 = 4\left(-\frac{1}{4}\right) = -1$$

▶ Find the product of the slopes.

The lines are perpendicular.

Problem 5 Find the equation of the line containing the point $(-2, 2)$ and perpendicular to the line $x - 4y = 3$.

Solution See page A31.

EXERCISES 6.3

1 Find the equation of the line that contains the given point and has the given slope.

1. Point $(0, 5)$, $m = 2$

2. Point $(0, 3)$, $m = 1$

3. Point $(2, 3)$, $m = \frac{1}{2}$

4. Point $(5, 1)$, $m = \frac{2}{3}$

5. Point $(-1, 4)$, $m = \frac{5}{4}$

6. Point $(-2, 1)$, $m = \frac{3}{2}$

7. Point $(3, 0)$, $m = -\frac{5}{3}$

8. Point $(-2, 0)$, $m = \frac{3}{2}$

9. Point $(2, 3)$, $m = -3$

10. Point $(1, 5)$, $m = -\frac{4}{5}$

11. Point $(-1, 7)$, $m = -3$

12. Point $(-2, 4)$, $m = -4$

13. Point $(-1, -3)$, $m = \frac{2}{3}$

14. Point $(-2, -4)$, $m = \frac{1}{4}$

15. Point $(0, 0)$, $m = \frac{1}{2}$

16. Point $(0, 0)$, $m = \frac{3}{4}$

17. Point $(2, -3)$, $m = 3$

18. Point $(4, -5)$, $m = 2$

19. Point $(3, 5)$, $m = -\frac{2}{3}$

20. Point $(5, 1)$, $m = -\frac{4}{5}$

21. Point $(-2, 0)$, $m = 0$

22. Point $(4, 0)$, $m = 0$

23. Point $(-2, -3)$, $m = \frac{3}{2}$

24. Point $(-3, -1)$, $m = \frac{4}{3}$

25. Point $(0, 2)$, undefined slope

26. Point $(0, 5)$, undefined slope

27. Point $(0, -3)$, $m = 3$

28. Point $(0, -2)$, $m = 0$

29. Point $(-5, 0)$, $m = -\frac{5}{2}$

30. Point $(-2, 0)$, $m = \frac{2}{5}$

31. Point $(0, 5)$, $m = \frac{4}{3}$

32. Point $(0, -5)$, $m = \frac{6}{5}$

33. Point $(4, -1)$, $m = -\frac{2}{5}$

34. Point $(-3, 5)$, $m = -\frac{1}{4}$

35. Point $(3, -4)$, undefined slope

36. Point $(-2, 5)$, undefined slope

37. Point $(-2, -5)$, $m = -\frac{5}{4}$

38. Point $(-3, -2)$, $m = -\frac{2}{3}$

39. Point $(-2, -3)$, $m = 0$

40. Point $(-3, -2)$, $m = 0$

Find the equation of the line containing the given points.

41. $P_1(0, 2)$, $P_2(3, 5)$

42. $P_1(0, 4)$, $P_2(1, 5)$

43. $P_1(0, -3)$, $P_2(-4, 5)$

44. $P_1(0, -2)$, $P_2(-3, 4)$

45. $P_1(2, 3)$, $P_2(5, 5)$

46. $P_1(4, 1)$, $P_2(6, 3)$

47. $P_1(-1, 3)$, $P_2(2, 4)$

48. $P_1(-1, 1)$, $P_2(4, 4)$

49. $P_1(-1, -2)$, $P_2(3, 4)$

50. $P_1(-3, -1)$, $P_2(2, 4)$

51. $P_1(0, 3)$, $P_2(2, 0)$

52. $P_1(0, 4)$, $P_2(2, 0)$

53. $P_1(-3, -1)$, $P_2(2, -1)$

54. $P_1(-3, -5)$, $P_2(4, -5)$

55. $P_1(-2, -3)$, $P_2(-1, -2)$

56. $P_1(-4, -1)$, $P_2(-5, -2)$

57. $P_1(-2, 3)$, $P_2(-1, 1)$

58. $P_1(-3, 5)$, $P_2(-2, 3)$

59. $P_1(-2, 5)$, $P_2(-2, 4)$

60. $P_1(3, 6)$, $P_2(3, -2)$

61. $P_1(3, 2)$, $P_2(-1, 5)$

62. $P_1(4, 1)$, $P_2(-2, 4)$

63. $P_1(3, -1)$, $P_2(2, -4)$

64. $P_1(4, 1)$, $P_2(3, -2)$

65. $P_1(-2, 3)$, $P_2(2, -1)$

66. $P_1(3, 1)$, $P_2(-3, -2)$

67. $P_1(-2, -3)$, $P_2(5, 0)$

68. $P_1(7, 2)$, $P_2(4, 4)$

69. $P_1(2, 0)$, $P_2(0, -1)$

70. $P_1(0, 4)$, $P_2(-2, 0)$

71. $P_1(3, -4)$, $P_2(-2, -4)$

72. $P_1(-3, 3)$, $P_2(-2, 3)$

73. $P_1(0, 0)$, $P_2(4, 3)$

74. $P_1(2, -5)$, $P_2(0, 0)$

75. $P_1(2, -1)$, $P_2(-1, 3)$

76. $P_1(3, -5)$, $P_2(-2, 1)$

77. $P_1(-2, 5)$, $P_2(-2, -5)$

78. $P_1(3, 2)$, $P_2(3, -4)$

79. $P_1(2, 1)$, $P_2(-2, -3)$

80. $P_1(-3, -2)$, $P_2(1, -4)$

2

81. Is the line $x = -2$ perpendicular to the line $y = 3$?

82. Is the line $y = \frac{1}{2}$ perpendicular to the line $y = -4$?

83. Is the line $x = -3$ parallel to the line $y = \frac{1}{3}$?

84. Is the line $x = 4$ parallel to the line $x = -4$?

85. Is the line $y = \frac{2}{3}x - 4$ parallel to the line $y = -\frac{3}{2}x - 4$?

86. Is the line $y = -2x + \frac{2}{3}$ parallel to the line $y = -2x + 3$?

87. Is the line $y = \frac{4}{3}x - 2$ perpendicular to the line $y = -\frac{3}{4}x + 2$?

88. Is the line $y = \frac{1}{2}x + \frac{3}{2}$ perpendicular to the line $y = -\frac{1}{2}x + \frac{3}{2}$?

89. Are the lines $2x + 3y = 2$ and $2x + 3y = -4$ parallel?

90. Are the lines $2x - 4y = 3$ and $2x + 4y = -3$ parallel?

91. Are the lines $x - 4y = 2$ and $4x + y = 8$ perpendicular?

92. Are the lines $4x - 3y = 2$ and $4x + 3y = -7$ perpendicular?

93. Is the line that contains the points (3, 2) and (1, 6) parallel to the line that contains the points (−1, 3) and (−1, −1)?

94. Is the line that contains the points (4, −3) and (2, 5) parallel to the line that contains the points (−2, −3) and (−4, 1)?

95. Is the line that contains the points (−3, 2) and (4, −1) perpendicular to the line that contains the points (1, 3) and (−2, −4)?

96. Is the line that contains the points (−1, 2) and (3, 4) perpendicular to the line that contains the points (−1, 3) and (−4, 1)?

97. Is the line that contains the points (−5, 0) and (0, 2) parallel to the line that contains the points (5, 1) and (0, −1)?

98. Is the line that contains the points (3, 5) and (−3, 3) perpendicular to the line that contains the points (2, −5) and (−4, 4)?

99. Find the equation of the line containing the point (−2, −4) and parallel to the line $2x - 3y = 2$.

100. Find the equation of the line containing the point (3, 2) and parallel to the line $3x + y = -3$.

101. Find the equation of the line containing the point (4, 1) and perpendicular to the line $y = -3x + 4$.

102. Find the equation of the line containing the point (2, −5) and perpendicular to the line $y = \frac{5}{2}x - 4$.

103. Find the equation of the line containing the point (−1, −3) and perpendicular to the line $3x - 5y = 2$.

104. Find the equation of the line containing the point $(-1, 3)$ and perpendicular to the line $2x + 4y = -1$.

105. Find the equation of the line containing the point $(-3, 1)$ and parallel to the line $y = \frac{2}{3}x - 1$.

106. Find the equation of the line containing the point $(4, -3)$ and parallel to the line $y = -\frac{4}{3}x + 2$.

107. Find the equation of the line containing the point $(-5, 4)$ and parallel to the line $y = -\frac{5}{3}x - 7$.

108. Find the equation of the line containing the point $(3, -4)$ and parallel to the line $y = \frac{3}{2}x$.

109. Find the equation of the line containing the point $(-4, -2)$ and parallel to the line $5x - 2y = -4$.

110. Find the equation of the line containing the point $(-2, 0)$ and parallel to the line $3x - 4y = 6$.

111. Find the equation of the line containing the point $(4, -3)$ and perpendicular to the line $y = 5x$.

112. Find the equation of the line containing the point $(-2, 5)$ and perpendicular to the line $y = \frac{2}{3}x - 5$.

113. Find the equation of the line containing the point $(-3, -3)$ and perpendicular to the line $3x - 2y = 3$.

114. Find the equation of the line containing the point $(-1, 6)$ and perpendicular to the line $4x + y = -3$.

SUPPLEMENTAL EXERCISES 6.3

Find the value of k such that the line containing P_1 and P_2 is parallel to the line containing P_3 and P_4.

115. $P_1(3, 4)$, $P_2(-2, -1)$, $P_3(4, 1)$, $P_4(0, k)$

116. $P_1(-3, 5)$, $P_2(6, -1)$, $P_3(-4, 1)$, $P_4(2, k)$

117. $P_1(6, 2)$, $P_2(-3, -1)$, $P_3(1, 5)$, $P_4(k, 4)$

118. $P_1(-4, 5)$, $P_2(2, 3)$, $P_3(5, -1)$, $P_4(k, 2)$

Find the value of k such that the line containing P_1 and P_2 is perpendicular to the line containing P_3 and P_4.

119. $P_1(2, 5)$, $P_2(6, 4)$, $P_3(-2, 1)$, $P_4(-3, k)$

120. $P_1(-3, 1)$, $P_2(3, -2)$, $P_3(-1, 5)$, $P_4(0, k)$

121. $P_1(-1, 4)$, $P_2(2, 5)$, $P_3(6, 1)$, $P_4(k, 4)$

122. $P_1(4, 2)$, $P_2(7, 6)$, $P_3(2, 5)$, $P_4(k, 8)$

Solve.

123. Find the equations of the lines that form the sides of the triangle with vertices whose coordinates are $(-3, 6)$, $(2, 0)$, and $(-2, -1)$.

124. Find the equations of the lines that form the sides of the triangle with vertices whose coordinates are $(4, 2)$, $(-1, -3)$, and $(-5, 3)$.

125. Show that the triangle with vertices whose coordinates are $(-3, -2)$, $(1, 4)$, and $(3, -6)$ is a right triangle. (*Hint:* Show that two sides of the triangle are perpendicular.)

126. Show that the triangle with vertices whose coordinates are $(2, 5)$, $(6, 3)$, and $(-1, -1)$ is a right triangle. (*Hint:* Show that two sides of the triangle are perpendicular.)

127. Show that the points whose coordinates are $(1, 6)$, $(3, 2)$, $(-1, -6)$, and $(-3, -2)$ are the vertices of a parallelogram. (*Hint:* Opposite sides of a parallelogram are parallel.)

128. Show that the points whose coordinates are $(-2, 5)$, $(8, 0)$, $(-8, 2)$, and $(2, -3)$ are the vertices of a parallelogram. (*Hint:* Opposite sides of a parallelogram are parallel.)

Is there a linear equation that contains all the given ordered pairs? If there is, find the equation.

129. $(2, 2)$, $(5, -1)$, $(3, 1)$

130. $(2, 1)$, $(-1, -5)$, $(4, 5)$

131. $(4, 3)$, $(-6, -2)$, $(8, 5)$

132. $(2, -4)$, $(-1, 5)$, $(3, 9)$

The given ordered pairs are solutions to the same linear equation. Find n.

133. $(0, 2)$, $(4, 6)$, $(-2, n)$

134. $(3, 3)$, $(-1, 4)$, $(-5, n)$

135. $(5, -2)$, $(4, 3)$, $(n, -7)$

136. $(1, -2)$, $(4, 3)$, $(n, 7)$

SECTION 6.4 Applications of Linear Equations

1 Obtain data from a graph

The rectangular coordinate system is used in business, science, and mathematics to show a relationship between two variables. One variable is represented along the horizontal axis and the other variable is represented along the vertical axis. A linear relationship between the variables is represented on the coordinate system as a straight line.

A company purchases a computer system for \$10,000. The graph at the right shows the depreciation of the computer system over a five-year period.

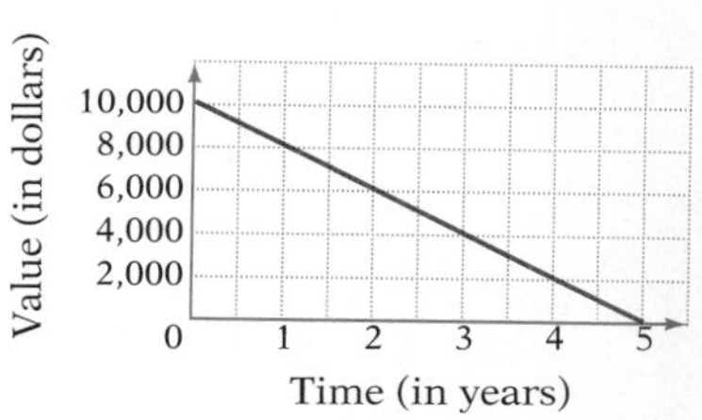

Information can be obtained from the graph. For example, the depreciated value of the computer after two years is \$6000; the depreciated value of the computer after four years is \$2000.

From the graph, an equation of the line that represents the depreciation can be written.

Use any two points shown on the graph to find the slope of the line. Note that the points (0, 10,000), (1, 8000), (2, 6000), (3, 4000), (4, 2000), and (5, 0) are all points on the graph. Points (0, 10,000) and (5, 0) are used here.

$(x_1, y_1) = (0, 10{,}000)$
$(x_2, y_2) = (5, 0)$

$$m = \frac{y_2 - y_1}{x_2 - x_1} = \frac{0 - 10{,}000}{5 - 0} = \frac{-10{,}000}{5} = -2000$$

Locate the y-intercept of the line on the graph.

The y-intercept is (0, 10,000).

Use the slope-intercept form of an equation to write the equation of the line.

$y = mx + b$
$y = -2000x + 10{,}000$

The equation of the line that represents the depreciation is $y = -2000x + 10{,}000$.

In the equation, the slope represents the annual depreciation of the computer system. The y-intercept represents the computer's value at the time of purchase.

Once the equation of the line has been found, the equation can be used to determine the depreciated value of the computer system after any given number of years.

To find the depreciated value of the computer after two and one-half years, substitute 2.5 for x in the equation and solve for y.

$$y = -2000x + 10{,}000$$
$$y = -2000(2.5) + 10{,}000$$
$$y = -5000 + 10{,}000$$
$$y = 5000$$

The depreciated value of the computer system after two and one-half years is $5000.

Example 1 The graph on the right shows the relationship between the total cost of manufacturing toasters and the number of toasters manufactured. Write the equation of the line that represents the total cost of manufacturing the toasters. Use the equation to find the total cost of manufacturing 340 toasters.

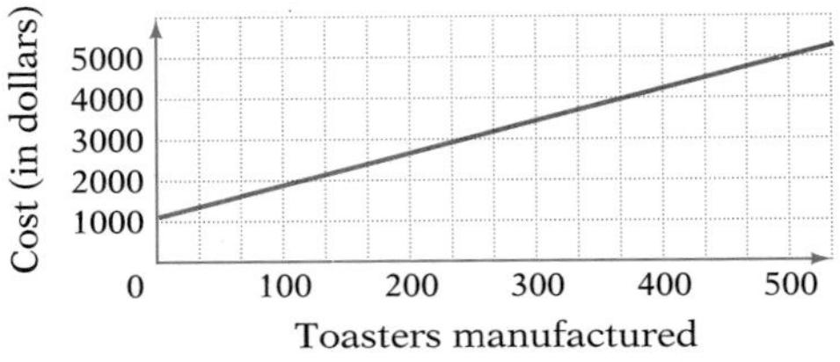

Strategy To write the equation:

- Locate two points on the graph to find the slope of the line.
- Locate the y-intercept of the line on the graph.
- Use the slope-intercept form of an equation to write the equation of the line.

To find the cost of manufacturing 340 toasters, substitute 340 for x in the equation and solve for y.

Solution $m = \dfrac{y_2 - y_1}{x_2 - x_1} = \dfrac{5000 - 1000}{500 - 0} = \dfrac{4000}{500} = 8$ ▶ Let $(x_1, y_1) = (0, 1000)$ and $(x_2, y_2) = (500, 5000)$. Find m.

The y-intercept is (0, 1000).

$y = mx + b$
$y = 8x + 1000$

▶ In the equation, the slope represents the unit cost, or the cost to manufacture one toaster. The y-intercept represents the fixed costs of operating the plant.

The equation of the line is $y = 8x + 1000$.

$y = 8x + 1000$
$y = 8(340) + 1000 = 3720$

The cost of manufacturing 340 toasters is $3720.

Problem 1 The relationship between Fahrenheit and Celsius temperature is shown on the graph. Write the equation for the Fahrenheit temperature in terms of the Celsius temperature. Use the equation to find the Fahrenheit temperature when the Celsius temperature is 40°.

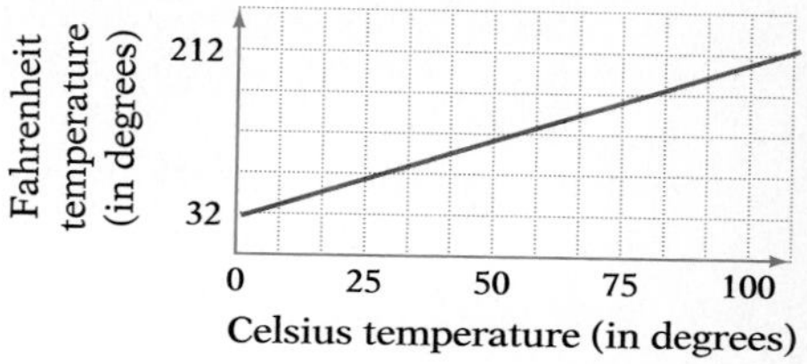

Solution See page A31.

EXERCISES 6.4

1 Solve.

1. The graph on the right represents the relationship between the amount invested and the annual income from an investment. Write the equation for the annual income in terms of the amount of the investment. Use the equation to find the annual income from a $4000 investment.

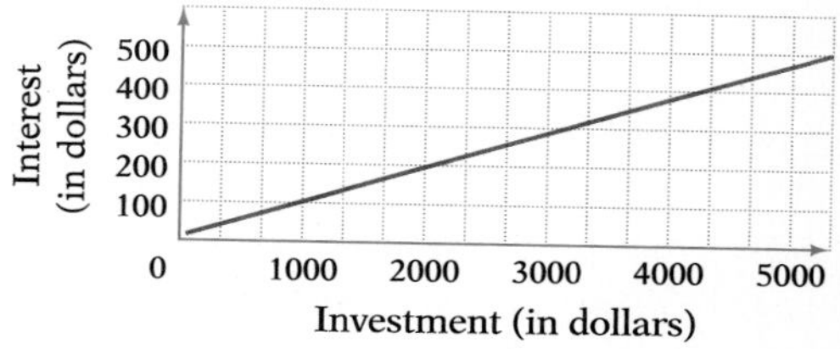

2. The graph on the right represents the relationship between the distance traveled by a motorist and the time of travel. Write the equation for the distance traveled in terms of the time of travel. Use the equation to find the distance traveled in three and one-half hours.

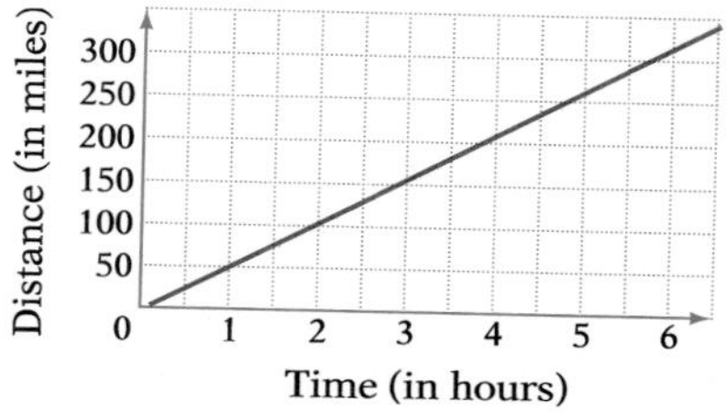

3. A milk tank has a capacity of 1000 ft^3. The graph on the right shows the relationship between the amount of milk in the tank and the time it takes to fill the tank. Write the equation for the amount of milk in terms of the time. Use the equation to find the amount of milk in the tank after one hour.

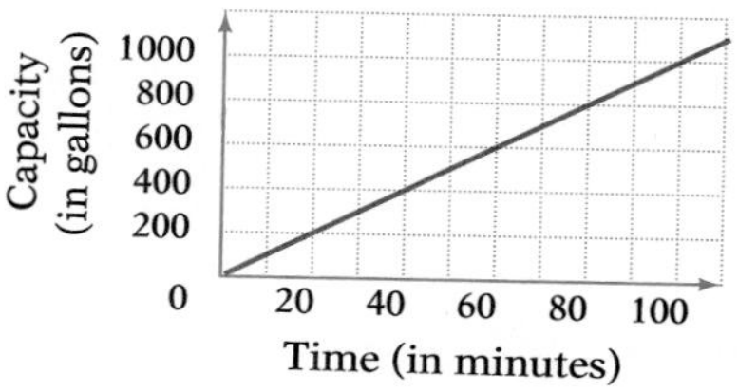

4. The graph on the right shows the relationship between the monthly income and the sales of an account executive. Write an equation for the line that represents the income of the account executive. Use the equation to find the monthly income when the monthly sales are $60,000.

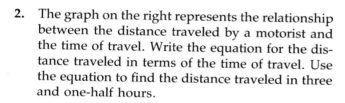

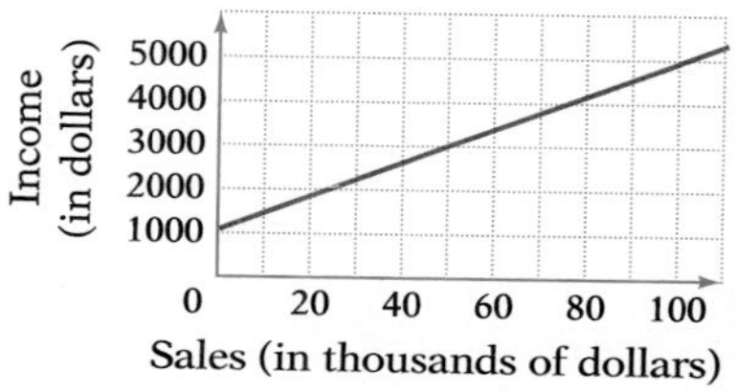

5. The relationship between the cost of a building and the depreciation allowed for income tax purposes is shown in the graph on the right. Write the equation for the line that represents the depreciated value of the building. Use the equation to find the value of the building after 12 years.

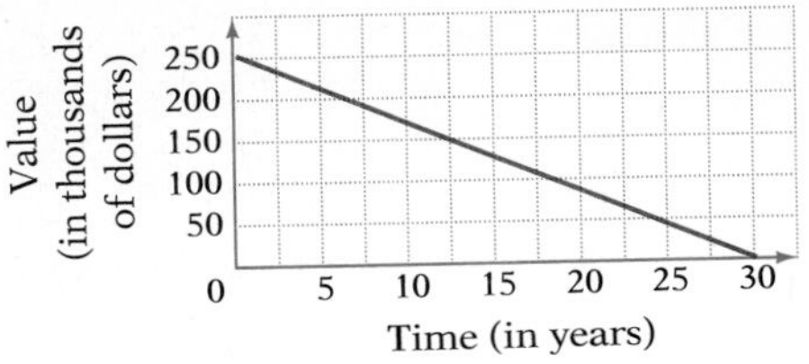

6. The graph on the right represents the relationship between a company's cost for new office equipment and the depreciation allowed for income tax purposes. Write the equation for the line that represents the depreciated value of the office equipment. Use the equation to find the depreciated value of the office equipment after two and one-half years.

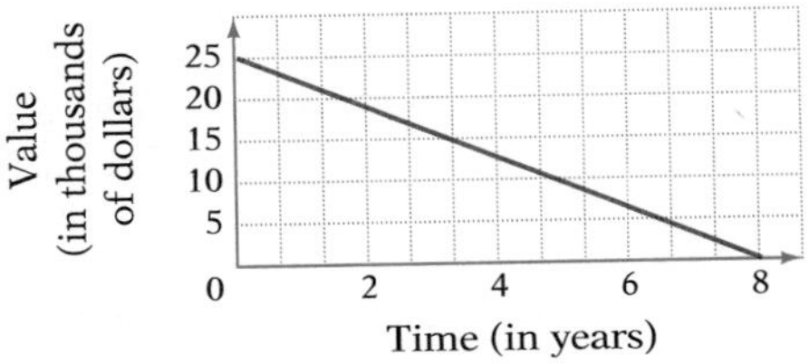

7. The relationship between the cost of manufacturing compact discs and the number of compact discs manufactured is shown in the graph on the right. Write the equation that represents the cost of manufacturing the compact discs. Use the equation to find the cost of manufacturing 35 compact discs.

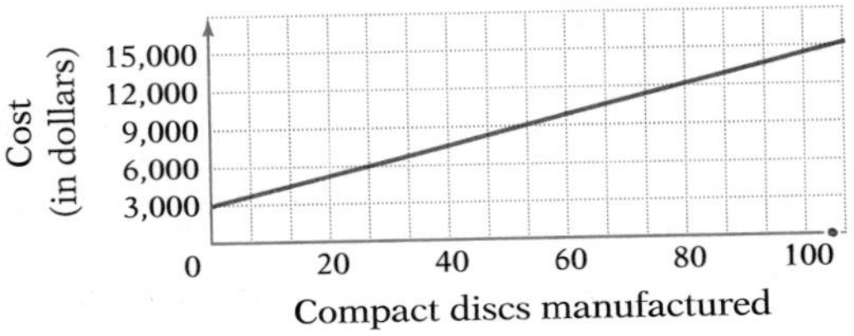

8. The graph on the right shows the relationship between the cost of a pair of skis and the number of pairs manufactured. Write the equation that represents the cost of manufacturing the skis. Use the equation to find the cost of manufacturing 150 pairs of skis.

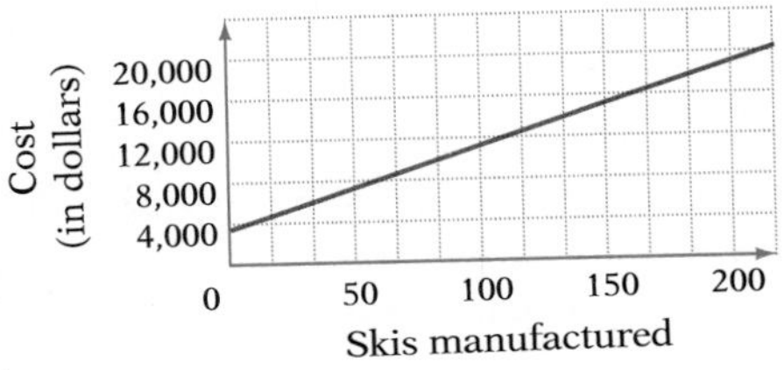

Economists frequently express the relationship between the price of a product and consumer demand for that product in a graph, called a demand curve.

9. Write the equation for the demand curve shown on the right. Use the equation to find the demand for the product when the price is \$8 per unit.

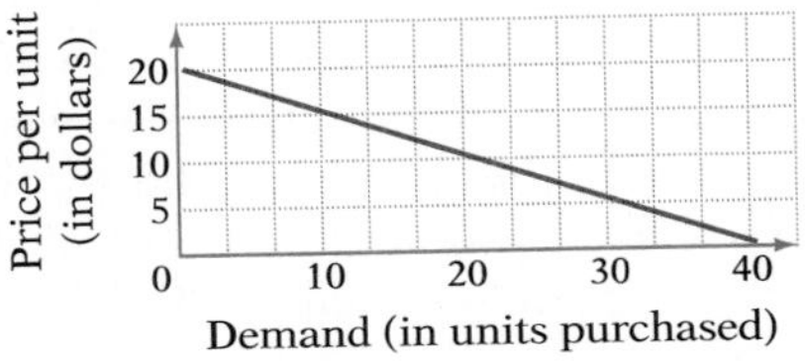

10. Write the equation for the demand curve shown on the right. Use the equation to find the demand for the product when the price is $25 per unit.

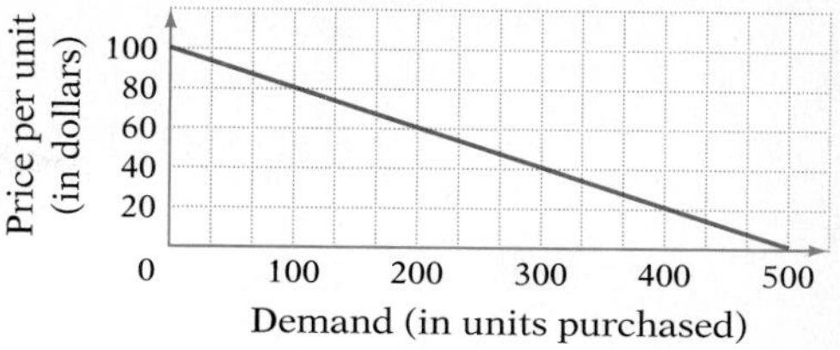

11. Write the equation for the demand curve shown on the right. Use the equation to find the demand for the product when the price is $3 per unit.

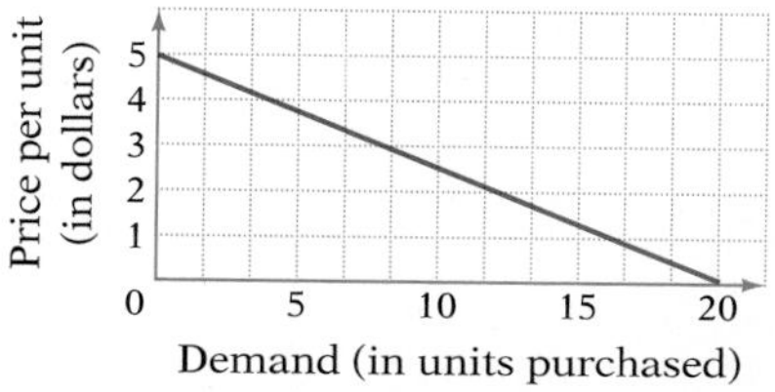

12. Write the equation for the demand curve shown on the right. Use the equation to find the demand for the product when the price is $10 per unit.

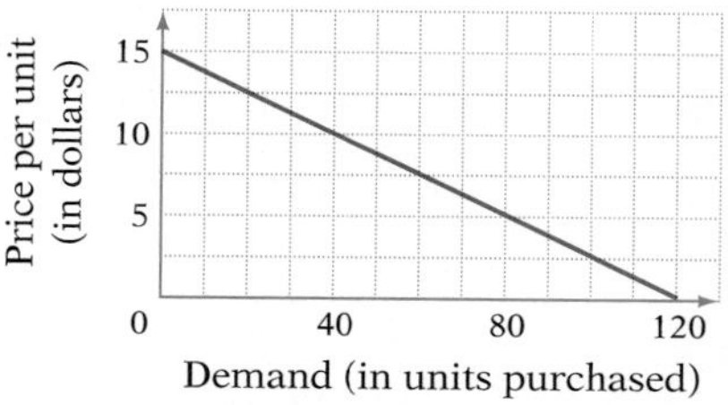

SUPPLEMENTAL EXERCISES 6.4

In 1991, the Social Security tax rate was 7.65% of gross earnings. The Social Security tax was deducted from the first $53,400 of gross earnings only.

13. Draw a graph to represent the amount of Social Security tax deducted from an employee's annual pay. Label the horizontal axis "Annual Pay." Include amounts from $0 to $50,000 on this axis.

14. Write an equation for the amount of Social Security tax deducted from an employee's annual pay.

15. For the equation written for Exercise 14, what does the slope represent? What does the y-intercept represent?

16. If the horizontal axis were extended beyond $53,400, what would the graph of the line look like beyond this point? Write an equation for this portion of the graph.

SECTION 6.5 Inequalities in Two Variables

1 Graph the solution set of an inequality in two variables

The graph of the linear equation $y = x - 1$ separates the plane into three sets:

the set of points on the line,
the set of points above the line,
the set of points below the line.

The point (2, 1) is a solution of $y = x - 1$.
The point (2, 4) is a solution of $y > x - 1$.
The point (2, −2) is a solution of $y < x - 1$.

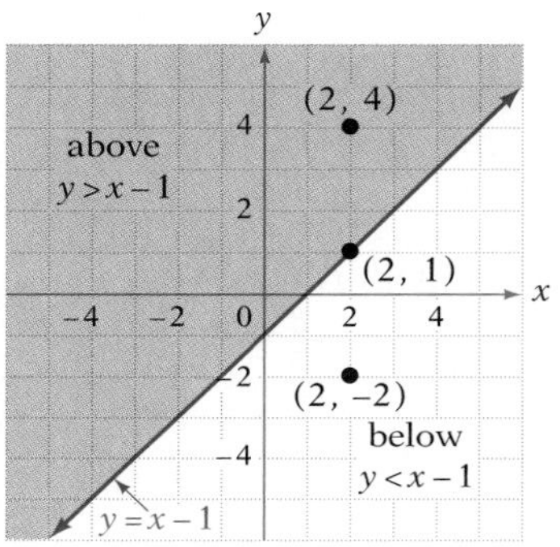

The solution set of $y = x - 1$ is all points on the line. The solution set of the linear inequality $y > x - 1$ is all points above the line. The solution set of the linear inequality $y < x - 1$ is all points below the line.

The solution set of an inequality in two variables is a **half plane**.

The following illustrates the procedure for graphing a linear inequality.

Graph the solution set of $3x - 4y < 12$.
Solve the inequality for y.

$$3x - 4y < 12$$
$$-4y < -3x + 12$$
$$y > \frac{3}{4}x - 3$$

$$y = \frac{3}{4}x - 3$$

Change the inequality to an equality, and graph the line. If the inequality is ≤ **or** ≥, the line is in the solution set and is shown by a **solid line**. If the inequality is < **or** >, the line is not part of the solution set and is shown by a **dotted line**.

If the inequality is > **or** ≥, shade the **upper half plane**. If the inequality is < **or** ≤, shade the **lower half plane**.

The line is a dotted line, and since $y > \frac{3}{4}x - 3$, the upper half plane is shaded.

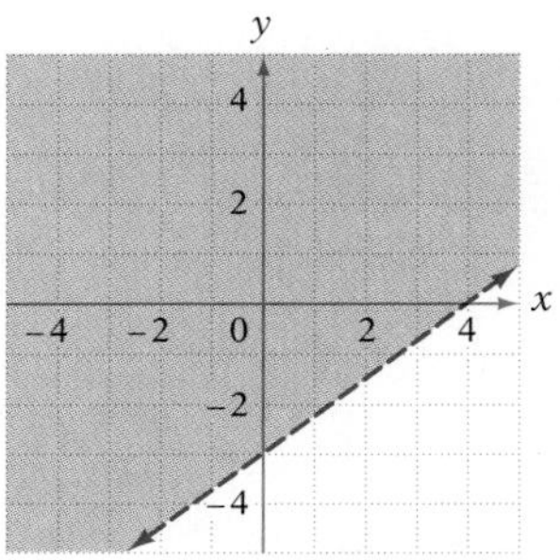

As a check, the point (0, 0) can be used to determine if the correct region of the plane has been shaded. If (0, 0) is a solution of the inequality, then (0, 0) should be in the shaded region. If (0, 0) is not a solution of the inequality, then (0, 0) should not be in the shaded region. In the above example, (0, 0) is in the shaded region, and (0, 0) is a solution of the inequality.

If the line passes through point (0, 0), another point must be used as a check, for example, (1, 0).

Example 1 Graph the solution set.

A. $x + 2y \le 4$ B. $x \ge -1$

Solution A. $x + 2y \le 4$

$2y \le -x + 4$ ▶ Solve the inequality for y.

$y \le -\frac{1}{2}x + 2$

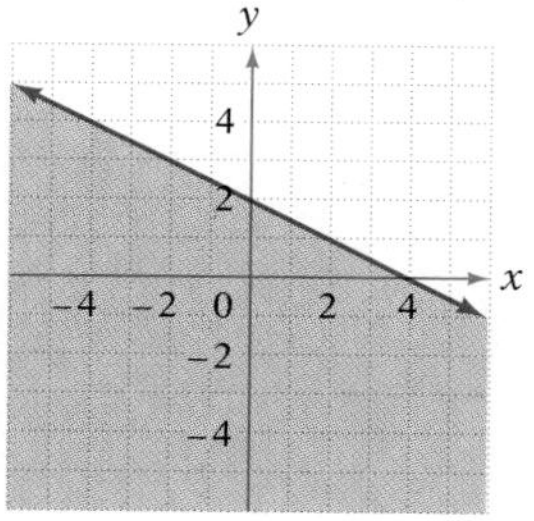

▶ Graph $y = -\frac{1}{2}x + 2$ as a solid line. Shade the lower half plane.

B. $x \ge -1$

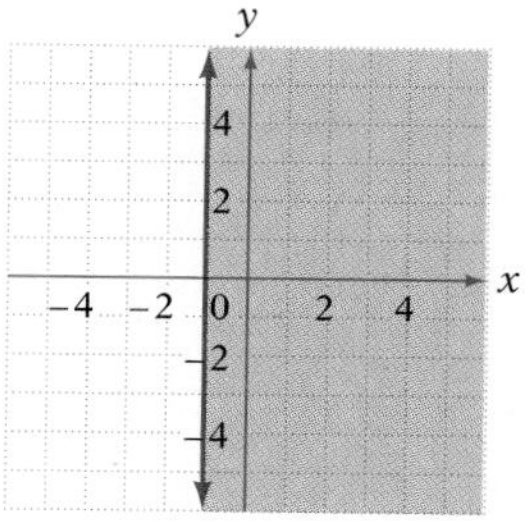

▶ Graph $x = -1$ as a solid line. The point (0, 0) satisfies the inequality.

$x \ge -1$

$0 \ge -1$

Shade the half plane to the right of the line.

Problem 1 Graph the solution set.

A. $x + 3y > 6$ B. $y < 2$

Solution See page A32.

EXERCISES 6.5

1 Graph the solution set.

1. $3x - 2y \geq 6$

2. $4x - 3y \leq 12$

3. $x - 2y < 4$

4. $x + 3y < 6$

5. $2x - 5y \leq 10$

6. $2x + 3y \geq 6$

7. $y < -\frac{2}{3}x + 2$

8. $y \leq \frac{4}{3}x - 3$

9. $y \geq -3x + 2$

10. $y \leq \frac{1}{3}x - 4$

11. $y > -\frac{5}{2}x + 4$

12. $y < 2x + 4$

13. $3y < 2x$

14. $-3y \geq x$

15. $x \leq \frac{2}{3}y$

16. $3x - 5y > 15$

17. $4x - 5y > 10$

18. $y - 4 < 0$

19. $y + 3 \geq 0$

20. $x - 3 \leq 0$

21. $x < 4$

22. $y < \frac{1}{4}x - 2$

23. $y \leq -\frac{3}{2}x + 1$

24. $y > -x - 1$

25. $x + 2 \geq 0$

26. $6x + 5y < 15$

27. $3x - 5y < 10$

SUPPLEMENTAL EXERCISES 6.5

Is the given point a solution of inequality **a**, inequality **b**, or both **a** and **b**?
a. $3x - 4y \leq 2$ **b.** $x - 2y \geq 1$

28. $(0, 2)$

29. $(4, 1)$

30. $(-4, -3)$

31. $\left(0, -\frac{1}{2}\right)$

32. $(4, -1)$

33. $(4, 4)$

34. A manufacturer makes a monochrome and color monitor. One part of the production process is the assembly of the monitors. Each monochrome monitor requires 3 h to assemble and each color monitor requires 4 h to assemble. The number of workers in the assembly division is such that there are a maximum of 300 h available to assemble monitors. Write an inequality that describes this condition. Assuming the stated conditions are true, is it possible to produce 30 monochrome monitors and 40 color monitors?

35. A food supplement, Symx A, contains 7 mg of iron per ounce. A second supplement, Symx B, contains 4 mg of iron per ounce. A nutritionist recommends at least 18 mg of iron in a daily diet. Using both food supplements, write an inequality that describes this condition. Assuming the stated conditions are true, is it possible to receive the recommended number of milligrams of iron with a diet that contains 2 oz of Symx A and 2 oz of Symx B?

Calculators and Computers

The Graphing Calculator

NOTE: For specific details on the operation of your calculator, please refer to your instruction guide.

The TI-81 calculator will be used to graph a linear equation. The x-intercept will be found as will an ordered pair solution of the equation when one of the ordered pair values is known.

The graphing calculator can be used to find the solutions to a predetermined accuracy by graphing.

Find the x-intercept of $y = -3x + 2$.

Locate the graphing keys on the calculator. They are:

[Y=] [RANGE] [ZOOM] [TRACE] [GRAPH]

First press [Y=] to display the menu. Then use the following key strokes.

[(−)] 3 [XIT] [+] 2 [GRAPH]

The graph will be displayed.

The cursor will be at the origin. Use [▶] to move the cursor to the right until the cursor is as close to the x-intercept as possible.

Use [ZOOM]. The menu for zoom will be displayed. Use [2] [ENTER].

Now use [▶], [▲], [◀], or [▼] to move the cursor to the x-intercept.

On our calculator, the intercepts read (.65789474, .03968254). Note that this is an approximation of the true solution.

Use zoom and go through the process again. The solution can now be read as (.67763158, −.0099206) and this is very close to the actual solution $\left(\frac{2}{3}, 0\right)$.

Clear the calculator. Use the key strokes: [2nd] [RESET] [2] [▼].

Find the ordered pair solution of $y = 2x - 3$ corresponding to $x = 3$.

First press [Y =] to display the menu. Then use the following key strokes.

2 [XIT] [−] 3 [GRAPH]

Move the cursor along the x-axis to 3 and then move upward until the cursor is on the graph of the equation $y = 2x - 3$.

Use the zoom feature and the cursor to locate the solution.

On our calculator, the y-coordinate corresponding to $x = 3.0004112$ is 3.000372.

Note that the solution read from the calculator is an approximation. By re-eated use of the zoom feature, a solution to a predetermined degree of accuracy can be obtained.

Solve by using a graphing calculator. The answers given are the approximate answers read from our calculator.

1. Find the x-intercept of $y = 0.2x + 1$.
2. Find the y-intercept of $y = -1.5x - 1$.
3. Find the ordered pair solution of $y = 2.4x - 4$ corresponding to $x = -1$.
4. Find the ordered pair solution of $y = -0.75x + 2$ corresponding to $y = 2$.
5. Find the ordered pair solution of $2x - 3y = 5$ corresponding to $x = -0.37$.

Something Extra

Application of Slope

The slope of a line has applications outside the realm of mathematics. In fact, one of the reasons calculus was created was to generalize the concept of slope so that it could be applied not only to lines but to curves as well.

Consider a marathon runner who runs at a constant rate of 6 mph. Some of the times and distances traveled by the runner are recorded in the table below. The graph to the right of the table is a graph of the distance traveled by the runner for all times between 0 and 4 hours.

Time	Distance
0	0
1	6
2	12
3	18
4	24

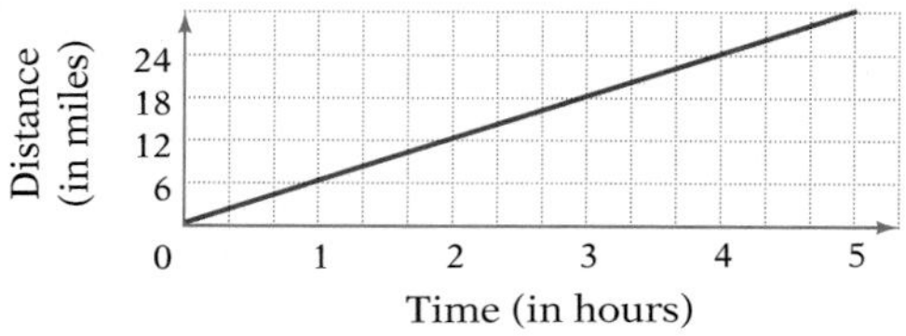

Now consider two ordered pairs taken from the table; (2, 12) and (4, 24) will be used here. The slope of the line between the two points is

$$m = \frac{24 - 12}{4 - 2} = \frac{12}{2} = 6$$

Note that the slope of the line is the same as the rate of the runner. This is not a coincidence. The rate an object moves is equal to the slope of a line.

If an object is not moving at a constant rate, the graph of the distance traveled will not be a straight line. In this case, the rate of the object is the *slope of the line tangent* to the curve. Using calculus, the slope of that line can be found.

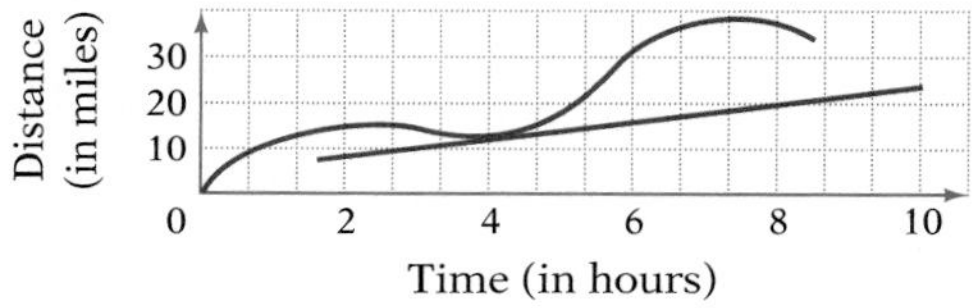

Now consider a retailer who sells, among other things, ballpoint pens for $1.50 each. The revenue to the retailer for the sale of these ballpoint pens is recorded in the table below. The graph to the right of the table shows the line through the points described in the table.

Number Sold	Revenue
0	0
1	1.50
5	7.50
12	18.00
15	22.50

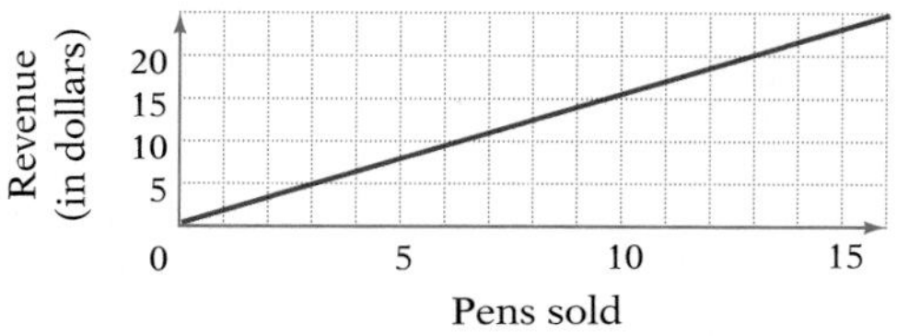

Using the ordered pairs (12, 18.00) and (15, 22.50), the slope of the line between the two points is

$$m = \frac{22.50 - 18.00}{15 - 12} = \frac{4.50}{3} = 1.50$$

In this case, the price per pen is the slope of the line. An economist refers to the $1.50 as *marginal revenue.*

One thing common to both examples given above is the word "per." In the first example it was miles **per** hour; in the second example, $1.50 **per** pen. Any quantity that can be described using the word *per* is an application of the concept of slope.

Solve.

1. An automobile manufacturer states that a certain model car averages 24 miles per gallon of gas. Make a table of the distance traveled for 0, 1, 5, 12, and 15 gallons of gas. Draw a graph of the data in the table. Using any two points in the table, show that the slope of the line is 24.

2. Name some quantities that are described using the word *per*.

Chapter Summary

Key Words

A *rectangular coordinate system* is formed by two number lines, one horizontal and one vertical, that intersect at the zero point of each line. The number lines that make up a coordinate system are called the *coordinate axes,* or simply *axes.* The *origin* is the point of intersection of the two coordinate axes.

A rectangular coordinate system divides the plane into four regions called *quadrants.*

An *ordered pair* (a, b) is used to locate a point in the plane. The first number in an ordered pair is called the *abscissa.* The second number is called the *ordinate.*

An equation of the form $y = mx + b$ or $Ax + By = C$ is a *linear equation in two variables.* The *graph* of a linear equation in two variables is a straight line.

The solution set of an inequality in two variables is a *half plane.*

The *slope* of a line is a measure of the slant or tilt of the line. The symbol for slope is m. A line that slants upward to the right has a *positive slope.* A line that slants downward to the right has a *negative slope.* A horizontal line has *zero slope.* The slope of a vertical line is *undefined.*

The point at which a graph crosses the x-axis is called the *x-intercept.* The point at which a graph crosses the y-axis is called the *y-intercept.*

Two lines that have the same slope do not intersect and are called *parallel lines*.

Two lines that intersect at right angles are called *perpendicular lines*.

Essential Rules

Slope of a straight line	Slope $= m = \frac{y_2 - y_1}{x_2 - x_1}$
Slope-intercept form of a straight line	$y = mx + b$
Point-slope formula	$y - y_1 = m(x - x_1)$
For two nonvertical parallel lines	$m_1 = m_2$
For two nonvertical perpendicular lines	$m_1 \cdot m_2 = -1$

Chapter Review

1. Find the ordered pair solution of $y = 2x + 6$ corresponding to $x = -3$.

2. Find the equation of the line that contains the point $(-5, 2)$ and has slope $\frac{2}{5}$.

3. Find the equation of the line containing the point $(-3, 4)$ and parallel to the line $2x + 3y = 9$.

4. Find the equation of the line containing the point $(-2, -3)$ and perpendicular to the line $y = -\frac{1}{2}x - 3$.

5. Find the slope of the line containing the points $(-2, 3)$ and $(4, 2)$.

6. Find the equation of the line containing the points $(3, -4)$ and $(-2, 3)$.

7. Find the equation of the line containing the points $(2, -3)$ and $(2, -5)$.

8. Find the equation of the line that contains the point $(0, 2)$ and has slope $-\frac{3}{4}$.

9. Find the x-intercept of the line $2x + 3y = 6$.

10. Find the slope of the line containing the points $(2, 5)$ and $(-2, 5)$.

11. Find the equation of the line containing the point (0, 0) and parallel to the line $y = -\frac{3}{2}x - 7$.

12. Find the ordered pair solution of $y = -\frac{3}{4}x + 2$ corresponding to $x = 3$.

13. Find the slope of the line containing the points (−2, 4) and (−2, −3).

14. Find the equation of the line containing the point (0, 0) and perpendicular to the line $2x - 3y = -2$.

15. Find the equation of the line that contains the points (3, 4) and (0, 4).

16. Find the slope of the line containing the points (3, −2) and (3, 5).

17. Find the equation of the line containing the point (3, −2) and parallel to the line $y = -3x + 4$.

18. Find the equation of the line containing the point (2, 5) and perpendicular to the line $y = -\frac{2}{3}x + 6$.

19. Find the equation of the line that contains the point (2, −4) and has slope $\frac{5}{2}$.

20. Graph the ordered pairs (3, −4) and (4, −1). Draw a line segment between the two points.

21. Graph: $y = \frac{2}{3}x - 4$

22. Graph the line that passes through point (−2, 2) and has slope 1.

23. Graph $2x - 3y = 6$ by using the x- and y-intercepts.

24. Graph the line that passes through point (−2, 3) and has slope $-\frac{3}{2}$.

25. Graph: $y = -\frac{4}{3}x + 3$

26. Graph: $3x + 2y = 1$

27. Graph: $2x + 3y = -3$

28. Graph the solution set of $2x - 3y > 9$.

29. Graph the solution set of $y > 3$.

30. Graph the solution set of $5x - 2y \leq 4$.

31. Graph the solution set of $y \geq 2x - 3$.

32. Graph the solution set of $3x - 2y < 6$.

33. The relationship between the cost of manufacturing calculators and the number of calculators manufactured is shown in the graph below. Write the equation that represents the cost of manufacturing the calculators. Use the equation to find the cost of manufacturing 125 calculators.

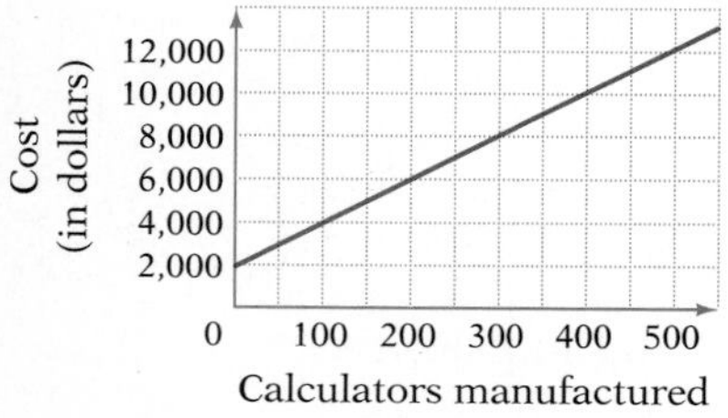

34. The graph below shows the relationship between the distance traveled by a plane and the time of travel. Write an equation for the distance traveled in terms of the time of travel. Use the equation to find the distance traveled in two and one-half hours.

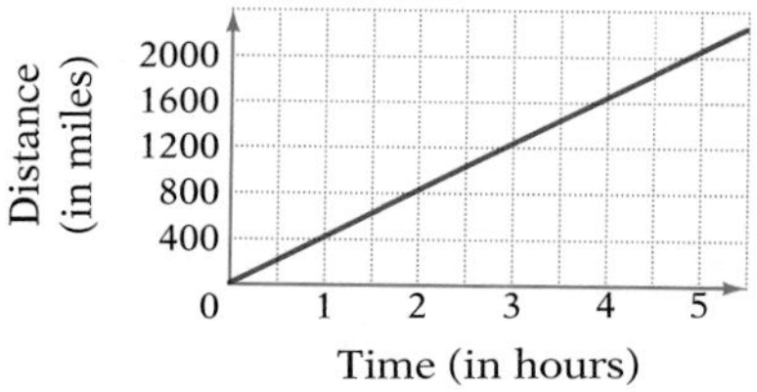

Chapter Test

1. Find the equation of the vertical line that contains the point $(-2, 3)$.

2. Find the ordered pair solution of $y = \frac{2}{3}x - 2$ corresponding to $x = -6$.

3. Find the slope of the line containing the points (2, −3) and (5, −1).

4. Find the slope of the line containing the points (4, 3) and (−4, 3).

5. Find the equation of the line that contains the point (2, −4) and has slope $-\frac{3}{2}$.

6. Find the equation of the line that contains the point (0, −3) and has slope $-\frac{4}{3}$.

7. Find the equation of the line containing the points (−2, 3) and (4, −2).

8. Find the equation of the line containing the points (−1, 2) and (−5, 3).

9. Find the equation of the line passing through (2, 3) that is perpendicular to the vertical line $x = 5$.

10. Find the equation of the line containing the point (1, 2) and parallel to the line $y = -\frac{3}{2}x - 6$.

11. Find the equation of the line containing the point (−3, 2) and perpendicular to the line $y = \frac{1}{2}x + 2$.

12. Find the equation of the horizontal line that contains the point (4, −3).

13. Find the equation of the line containing the point (5, −2) and parallel to the line $3x - 2y = 4$.

14. Graph the ordered pairs (4, 2) and (−3, 1). Draw a line between the two points.

15. Graph: $y = -\frac{3}{2}x + 2$

16. Graph: $x - 2y = 4$

17. Graph $4x + 3y = 12$ by using the x- and the y-intercepts.

18. Graph the line that passes through point $(-1, -4)$ and has slope $\frac{4}{3}$.

19. The graph below shows the relationship between the cost of a rental house and the depreciation allowed for income tax purposes. Write the equation for the depreciation.

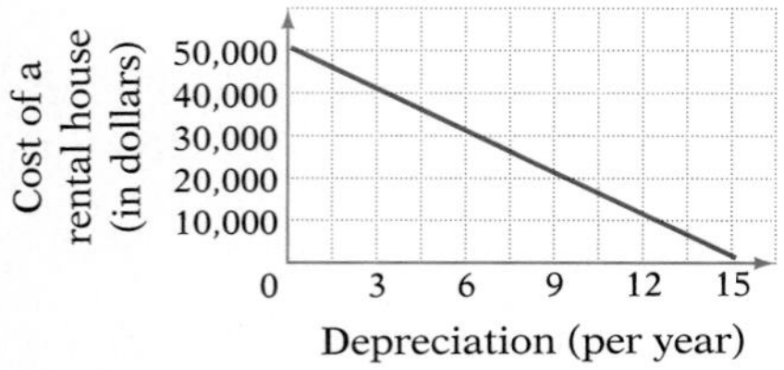

20. Graph the solution set of $3x - 4y > 8$.

Cumulative Review

1. Simplify: $-4^2 \cdot (-3)^3$

2. Simplify: $15 - 3[3 - (-2)]^2 \div 5$

3. Evaluate $\frac{-a^2 - b^2}{2}$ when $a = 2$ and $b = -3$.

4. Solve: $3 - \frac{x}{2} = \frac{3}{4}$

5. Solve: $2[y - 2(3 - y) + 4] = 4 - 3y$

6. Solve: $4 < 3x + 1 < 10$

7. Solve: $8 - |2x - 1| = 4$

8. Solve: $|4 - 5x| < 6$

9. Simplify: $\left(\frac{3ab^2}{2a}\right)^3\left(\frac{b^2}{-3ab}\right)^2$

10. Simplify: $3a^2 - a[2 - a(4 - a + a^2)]$

11. Factor: $8x^2y + 16x^2y^2 + 24xy^2$

12. Factor: $6x^2 - 9bx + 4ax - 6ab$

13. Solve: $(y - 2)^2 = 9$

14. Solve: $\sqrt{x - 2} + \sqrt{x} = 2$

15. Simplify: $\frac{2x + 3}{x + 3} + \frac{x - 2}{5 - x} - \frac{15 - 3x}{x^2 - 2x - 15}$

16. Simplify: $5b\sqrt[3]{16a^4b} - 2a\sqrt[3]{54ab^4}$

17. Simplify: $(3 - 4i)(6 - i)$

18. Solve: $\sqrt[3]{x + 3} = \sqrt[3]{4x - 9}$

19. Find the ordered pair solution of $y = -\frac{5}{2}x + 4$ corresponding to $x = 2$.

20. Find the slope of the line containing the points $(-3, 4)$ and $(-4, 2)$.

21. Find the equation of the line that contains the point $(-5, 2)$ and has slope -2.

22. Find the equation of the line containing the points $(1, -2)$ and $(-2, 4)$.

23. Find the equation of the line containing the point $(-3, 4)$ and parallel to the line $2x - 3y = 6$.

24. Find the equation of the line containing the point $(-1, 2)$ and perpendicular to the line $x - 4y = 6$.

25. Find the equation of the line containing the point $(2, 4)$ and parallel to the line $y = -\frac{3}{2}x + 2$.

26. Find the equation of the line containing the point $(4, 0)$ and perpendicular to the line $3x - 2y = 5$.

27. Graph $3x - 5y = 15$ by using the x- and y- intercepts.

28. Graph the line that passes through the point $(-3, 1)$ and has slope $-\frac{3}{2}$.

29. Graph the solution set of $3x - 2y \geq 6$.

30. A coin purse contains 17 coins with a value of \$1.60. The purse contains nickels, dimes, and quarters. There are four times as many nickels as quarters. Find the number of dimes in the purse.

31. A grocer combines coffee costing \$8.00 per pound with coffee costing \$3.00 per pound. How many pounds of each should be used to make 80 lb of a blend costing \$5.00 per pound?

32. Two planes are 1800 mi apart and traveling toward each other. One plane is traveling twice as fast as the other plane. The planes meet in 3 h. Find the speed of each plane.

33. The relationship between the cost of a truck and the depreciation allowed for income tax purposes is shown in the graph on the right. Write the equation for the line that represents the depreciated value of the truck. Use the equation to find the value of the truck after three and one half years.

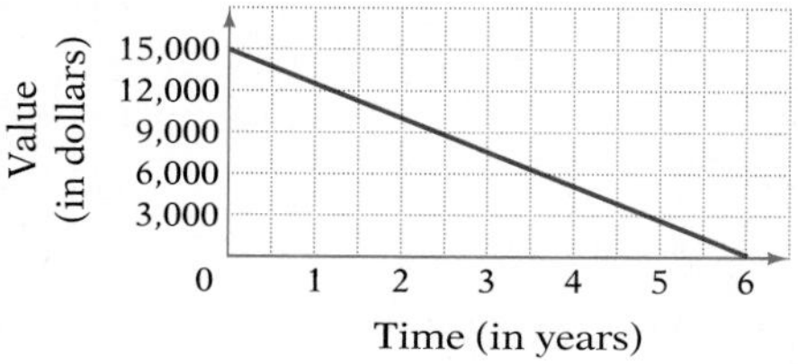

7

Quadratic Equations and Inequalities

Objectives

- Solve quadratic equations by factoring
- Write a quadratic equation given its solutions
- Solve quadratic equations by taking square roots
- Solve quadratic equations by completing the square
- Solve quadratic equations by using the quadratic formula
- Equations that are quadratic in form
- Radical equations
- Fractional equations
- Graph equations of the form $y = ax^2 + bx + c$
- Find the x-intercepts of a parabola
- Application problems
- Solve inequalities by factoring
- Solve rational inequalities

POINT OF INTEREST

Complex Numbers

Negative numbers were not universally accepted in the mathematical community until well into the 14th century. It is no wonder then that *imaginary numbers* took an even longer time to gain acceptance.

Beginning in the mid-sixteenth century, mathematicians were beginning to integrate imaginary numbers into their writings. One notation for 3i was R (0 m 3). Literally this was interpreted as $\sqrt{0 - 3}$.

By the mid-eighteenth century, the symbol i was introduced. Still later it was shown that complex numbers could be thought of as points in the plane. The complex number $3 + 4i$ was associated with the point (3, 4).

Imaginary Axis

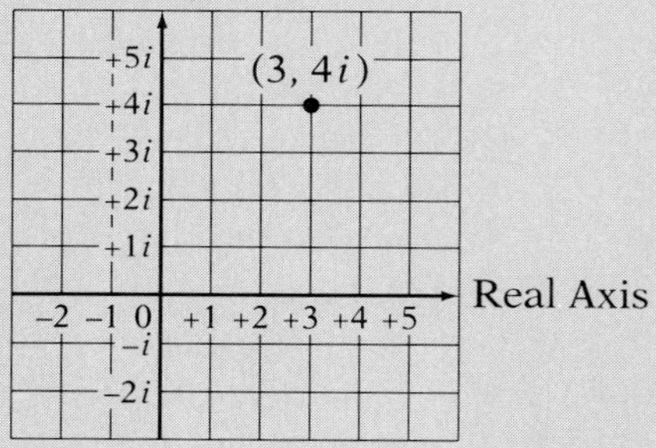

By the end of the nineteenth century, complex numbers were fully integrated into mathematics. This was due in large part to some eminent mathematicians who used complex numbers to prove theorems that had previously eluded proof.

SECTION 7.1

Solving Quadratic Equations by Factoring or by Taking Square Roots

1 Solve quadratic equations by factoring

A **quadratic equation** is an equation of the form $ax^2 + bx + c = 0$, where a, b, and c are constants and $a \neq 0$.

$$3x^2 - x + 2 = 0, \quad a = 3, \quad b = -1, \quad c = 2$$
$$-x^2 + 4 = 0, \quad a = -1, \quad b = 0, \quad c = 4$$
$$6x^2 - 5x = 0, \quad a = 6, \quad b = -5, \quad c = 0$$

A quadratic equation is in **standard form** when the polynomial is in descending order and equal to zero.

Since the degree of the polynomial $ax^2 + bx + c$ is 2, a quadratic equation is also called a **second-degree equation.**

In Chapter 3, the Principle of Zero Products was used to solve some quadratic equations. That procedure is reviewed here.

The Principle of Zero Products

If the product of two factors is zero, then at least one of the factors must be zero.

If $ab = 0$, then $a = 0$ or $b = 0$.

Solve by factoring: $x^2 - 6x = -9$

	$x^2 - 6x = -9$
Write the equation in standard form.	$x^2 - 6x + 9 = 0$
Use the Principle of Zero Products.	$(x - 3)(x - 3) = 0$
Solve each equation.	$x - 3 = 0 \qquad x - 3 = 0$ $x = 3 \qquad x = 3$
	3 checks as a solution.
Write the solutions.	The solution is 3.

When a quadratic equation has two solutions that are the same number, the solution is called a **double root** of the equation. The solution 3 is a double root of the equation $x^2 - 6x = -9$.

Example 1 Solve for x by factoring: $x^2 - 4ax - 5a^2 = 0$

Solution $x^2 - 4ax - 5a^2 = 0$ ▶ This is a literal equation. Solve for x in terms of a.

$(x + a)(x - 5a) = 0$ ▶ Factor.

$x + a = 0$ $\qquad$ $x - 5a = 0$

$x = -a$ $\qquad$ $x = 5a$

The solutions are $-a$ and $5a$.

Problem 1 Solve for x by factoring: $x^2 - 3ax - 4a^2 = 0$

Solution See page A32.

2 Write a quadratic equation given its solutions

As shown below, the solutions of the equation $(x - r_1)(x - r_2) = 0$ are r_1 and r_2.

$(x - r_1)(x - r_2) = 0$

$x - r_1 = 0$ $\qquad$ $x - r_2 = 0$

$x = r_1$ $\qquad$ $x = r_2$

Check:

$$\begin{array}{r|l} (x - r_1)(x - r_2) = 0 & \\ \hline (r_1 - r_1)(r_1 - r_2) & 0 \\ 0 \cdot (r_1 - r_2) & \\ 0 = 0 & \end{array} \qquad \begin{array}{r|l} (x - r_1)(x - r_2) = 0 & \\ \hline (r_2 - r_1)(r_2 - r_2) & 0 \\ (r_2 - r_1) \cdot 0 & \\ 0 = 0 & \end{array}$$

Using the equation $(x - r_1)(x - r_2) = 0$ and the fact that r_1 and r_2 are solutions of this equation, it is possible to write a quadratic equation given its solutions.

Write a quadratic equation that has solutions 4 and -5.

$(x - r_1)(x - r_2) = 0$

Replace r_1 by 4 and r_2 by -5. $\qquad (x - 4)[x - (-5)] = 0$

Simplify. $\qquad (x - 4)(x + 5) = 0$

Multiply. $\qquad x^2 + x - 20 = 0$

Example 2 Write a quadratic equation that has integer coefficients and has solutions $\frac{2}{3}$ and $\frac{1}{2}$.

Solution $(x - r_1)(x - r_2) = 0$

$\left(x - \frac{2}{3}\right)\left(x - \frac{1}{2}\right) = 0$ ▶ Replace r_1 by $\frac{2}{3}$ and r_2 by $\frac{1}{2}$.

$x^2 - \frac{7}{6}x + \frac{1}{3} = 0$ ▶ Multiply.

$6\left(x^2 - \frac{7}{6}x + \frac{1}{3}\right) = 6 \cdot 0$ ▶ Multiply each side of the equation by the LCM of the denominators.

$6x^2 - 7x + 2 = 0$

Problem 2 Write a quadratic equation that has integer coefficients and has solutions $-\frac{2}{3}$ and $\frac{1}{6}$.

Solution See page A32.

3 Solve quadratic equations by taking square roots

The solution of the quadratic equation $x^2 = 16$ is shown at the right.

$$x^2 = 16$$
$$x^2 - 16 = 0$$
$$(x + 4)(x - 4) = 0$$
$$x + 4 = 0 \qquad x - 4 = 0$$
$$x = -4 \qquad x = 4$$

Note that the solution is the positive or the negative square root of 16, 4 or −4.

The solution can also be found by taking the square root of each side of the equation and writing the positive and the negative square roots of the number. The notation $x = \pm 4$ means $x = 4$ or $x = -4$.

$$x^2 = 16$$
$$\sqrt{x^2} = \sqrt{16}$$
$$x = \pm\sqrt{16} = \pm 4$$

The solutions are 4 and −4.

Solve by taking square roots: $3x^2 = 54$

$$3x^2 = 54$$

Solve for x^2.

$$x^2 = 18$$

Take the square root of each side of the equation.

$$\sqrt{x^2} = \sqrt{18}$$

Simplify.

$$x = \pm\sqrt{18} = \pm 3\sqrt{2}$$

$3\sqrt{2}$ and $-3\sqrt{2}$ check as solutions.

Write the solutions.

The solutions are $3\sqrt{2}$ and $-3\sqrt{2}$.

Solving a quadratic equation by taking the square root of each side of the equation can lead to solutions that are complex numbers.

Solve by taking square roots: $2x^2 + 18 = 0$

Solve for x^2.

$$2x^2 + 18 = 0$$
$$2x^2 = -18$$
$$x^2 = -9$$

Take the square root of each side of the equation.

$$\sqrt{x^2} = \sqrt{-9}$$

Simplify. $x = \pm\sqrt{-9} = \pm 3i$

$3i$ and $-3i$ check as solutions.

Write the solutions. The solutions are $3i$ and $-3i$.

An equation containing the square of a binomial can be solved by taking square roots.

Example 3 Solve by taking square roots: $3(x-2)^2 + 12 = 0$

Solution

$$3(x-2)^2 + 12 = 0$$

$$3(x-2)^2 = -12$$ ▶ Solve for $(x-2)^2$.

$$(x-2)^2 = -4$$

$$\sqrt{(x-2)^2} = \sqrt{-4}$$ ▶ Take the square root of each side of the equation. Then simplify.

$$x - 2 = \pm\sqrt{-4} = \pm 2i$$

$x - 2 = 2i$ $\quad$ $x - 2 = -2i$ ▶ Solve for x.

$x = 2 + 2i$ $\quad$ $x = 2 - 2i$

The solutions are $2 + 2i$ and $2 - 2i$.

Problem 3 Solve by taking square roots: $2(x+1)^2 + 24 = 0$

Solution See page A32.

EXERCISES 7.1

1 Solve by factoring.

1. $x^2 - 4x = 0$
2. $y^2 + 6y = 0$
3. $t^2 - 25 = 0$
4. $p^2 - 81 = 0$
5. $s^2 - s - 6 = 0$
6. $v^2 + 4v - 5 = 0$
7. $y^2 - 6y + 9 = 0$
8. $x^2 + 10x + 25 = 0$
9. $9z^2 - 18z = 0$
10. $4y^2 + 20y = 0$
11. $r^2 - 3r = 10$
12. $p^2 + 5p = 6$
13. $v^2 + 10 = 7v$
14. $t^2 - 16 = 15t$
15. $2x^2 - 9x - 18 = 0$
16. $3y^2 - 4y - 4 = 0$
17. $4z^2 - 9z + 2 = 0$
18. $2s^2 - 9s + 9 = 0$

19. $3w^2 + 11w = 4$

20. $2r^2 + r = 6$

21. $6x^2 = 23x + 18$

22. $6x^2 = 7x - 2$

23. $4 - 15u - 4u^2 = 0$

24. $3 - 2y - 8y^2 = 0$

25. $x + 18 = x(x - 6)$

26. $t + 24 = t(t + 6)$

27. $4s(s + 3) = s - 6$

28. $3v(v - 2) = 11v + 6$

29. $u^2 - 2u + 4 = (2u - 3)(u + 2)$

30. $(3v - 2)(2v + 1) = 3v^2 - 11v - 10$

31. $(3x - 4)(x + 4) = x^2 - 3x - 28$

Solve for x by factoring.

32. $x^2 + 14ax + 48a^2 = 0$

33. $x^2 - 9bx + 14b^2 = 0$

34. $x^2 + 9xy - 36y^2 = 0$

35. $x^2 - 6cx - 7c^2 = 0$

36. $x^2 - ax - 20a^2 = 0$

37. $2x^2 + 3bx + b^2 = 0$

38. $3x^2 - 4cx + c^2 = 0$

39. $3x^2 - 14ax + 8a^2 = 0$

40. $3x^2 - 11xy + 6y^2 = 0$

41. $3x^2 - 8ax - 3a^2 = 0$

42. $3x^2 - 4bx - 4b^2 = 0$

43. $4x^2 + 8xy + 3y^2 = 0$

44. $6x^2 - 11cx + 3c^2 = 0$

45. $6x^2 + 11ax + 4a^2 = 0$

46. $12x^2 - 5xy - 2y^2 = 0$

2 Write a quadratic equation that has integer coefficients and has as solutions the given pair of numbers.

47. 2 and 5

48. 3 and 1

49. -2 and -4

50. -1 and -3

51. 6 and -1

52. -2 and 5

53. 3 and -3

54. 5 and -5

55. 4 and 4

56. 2 and 2

57. 0 and 5

58. 0 and -2

59. 0 and 3 **60.** 0 and −1 **61.** 3 and $\frac{1}{2}$

62. 2 and $\frac{2}{3}$ **63.** $-\frac{3}{4}$ and 2 **64.** $-\frac{1}{2}$ and 5

65. $-\frac{5}{3}$ and −2 **66.** $-\frac{3}{2}$ and −1 **67.** $-\frac{2}{3}$ and $\frac{2}{3}$

68. $-\frac{1}{2}$ and $\frac{1}{2}$ **69.** $\frac{1}{2}$ and $\frac{1}{3}$ **70.** $\frac{3}{4}$ and $\frac{2}{3}$

71. $\frac{6}{5}$ and $-\frac{1}{2}$ **72.** $\frac{3}{4}$ and $-\frac{3}{2}$ **73.** $-\frac{1}{4}$ and $-\frac{1}{2}$

74. $-\frac{5}{6}$ and $-\frac{2}{3}$ **75.** $\frac{3}{5}$ and $-\frac{1}{10}$ **76.** $\frac{7}{2}$ and $-\frac{1}{4}$

3 Solve by taking square roots.

77. $y^2 = 49$ **78.** $x^2 = 64$ **79.** $z^2 = -4$

80. $v^2 = -16$ **81.** $s^2 - 4 = 0$ **82.** $r^2 - 36 = 0$

83. $4x^2 - 81 = 0$ **84.** $9x^2 - 16 = 0$ **85.** $y^2 + 49 = 0$

86. $z^2 + 16 = 0$ **87.** $v^2 - 48 = 0$ **88.** $s^2 - 32 = 0$

89. $r^2 - 75 = 0$ **90.** $u^2 - 54 = 0$ **91.** $z^2 + 18 = 0$

92. $t^2 + 27 = 0$ **93.** $(x - 1)^2 = 36$ **94.** $(x + 2)^2 = 25$

95. $3(y + 3)^2 = 27$ **96.** $4(s - 2)^2 = 36$ **97.** $5(z + 2)^2 = 125$

98. $(x - 2)^2 = -4$ **99.** $(x + 5)^2 = -25$ **100.** $(x - 8)^2 = -64$

101. $3(x-4)^2 = -12$

102. $5(x+2)^2 = -125$

103. $3(x-9)^2 = -27$

104. $2(y-3)^2 = 18$

105. $\left(v - \frac{1}{2}\right)^2 = \frac{1}{4}$

106. $\left(r + \frac{2}{3}\right)^2 = \frac{1}{9}$

107. $\left(x - \frac{2}{5}\right)^2 = \frac{9}{25}$

108. $\left(y + \frac{1}{3}\right)^2 = \frac{4}{9}$

109. $\left(a + \frac{3}{4}\right)^2 = \frac{9}{16}$

110. $4\left(x - \frac{1}{2}\right)^2 = 1$

111. $3\left(x - \frac{5}{3}\right)^2 = \frac{4}{3}$

112. $2\left(x + \frac{3}{5}\right)^2 = \frac{8}{25}$

113. $(x+5)^2 - 6 = 0$

114. $(t-1)^2 - 15 = 0$

115. $(s-2)^2 - 24 = 0$

116. $(y+3)^2 - 18 = 0$

117. $(z+1)^2 + 12 = 0$

118. $(r-2)^2 + 28 = 0$

119. $(v-3)^2 + 45 = 0$

120. $(x+5)^2 + 32 = 0$

121. $\left(u + \frac{2}{3}\right)^2 - 18 = 0$

122. $\left(z - \frac{1}{2}\right)^2 - 20 = 0$

123. $\left(t - \frac{3}{4}\right)^2 - 27 = 0$

124. $\left(y + \frac{2}{5}\right)^2 - 72 = 0$

125. $\left(x + \frac{1}{2}\right)^2 + 40 = 0$

126. $\left(r - \frac{3}{2}\right)^2 + 48 = 0$

127. $\left(x - \frac{2}{3}\right)^2 + \frac{25}{9} = 0$

128. $\left(y + \frac{5}{8}\right)^2 + \frac{25}{64} = 0$

SUPPLEMENTAL EXERCISES 7.1

Write a quadratic equation that has as solutions the given pair of numbers.

129. $\sqrt{2}$ and $-\sqrt{2}$

130. $\sqrt{5}$ and $-\sqrt{5}$

131. i and $-i$

132. $2i$ and $-2i$

133. $2\sqrt{2}$ and $-2\sqrt{2}$

134. $3\sqrt{2}$ and $-3\sqrt{2}$

135. $2\sqrt{3}$ and $-2\sqrt{3}$

136. $i\sqrt{2}$ and $-i\sqrt{2}$

137. $2i\sqrt{3}$ and $-2i\sqrt{3}$

Solve for x.

138. $4a^2x^2 = 36b^2$

139. $2a^2x^2 = 32b^2$

140. $3y^2x^2 = 27z^2$

141. $5y^2x^2 = 125z^2$

142. $(x + a)^2 - 4 = 0$

143. $(x - b)^2 - 1 = 0$

144. $2(x - y)^2 - 8 = 0$

145. $(2x - 1)^2 = (2x + 3)^2$

146. $(x - 4)^2 = (x + 2)^2$

Solve.

147. Show that the solutions of the equation $ax^2 + bx = 0$ are 0 and $-\frac{b}{a}$.

148. Show that the solutions of the equation $ax^2 + c = 0$, $a > 0$, $c > 0$, are $\frac{\sqrt{ca}}{a}i$ and $-\frac{\sqrt{ca}}{a}i$.

SECTION 7.2
Solving Quadratic Equations by Completing the Square and by Using the Quadratic Formula

1 Solve quadratic equations by completing the square

Recall that a perfect square trinomial is the square of a binomial.

Perfect Square Trinomial		Square of a Binomial
$x^2 + 8x + 16$	$=$	$(x + 4)^2$
$x^2 - 10x + 25$	$=$	$(x - 5)^2$
$x^2 + 2ax + a^2$	$=$	$(x + a)^2$

For each perfect square trinomial, the square of $\frac{1}{2}$ the coefficient of x equals the constant term.

$$\left(\frac{1}{2}\textbf{ coefficient of } x\right)^2 = \textbf{Constant term}$$

$$x^2 + 8x + 16, \quad \left(\frac{1}{2} \cdot 8\right)^2 = 16$$

$$x^2 - 10x + 25, \quad \left[\frac{1}{2}(-10)\right]^2 = 25$$

$$x^2 + 2ax + a^2, \quad \left(\frac{1}{2} \cdot 2a\right)^2 = a^2$$

To complete the square on $x^2 + bx$, add $\left(\frac{1}{2}b\right)^2$ to $x^2 + bx$.

Complete the square on $x^2 - 12x$. Write the resulting perfect square trinomial as the square of a binomial.

Find the constant term. $\left[\frac{1}{2}(-12)\right]^2 = (-6)^2 = 36$

Complete the square on $x^2 - 12x$ by adding the constant term. $x^2 - 12x + 36$

Write the resulting perfect square trinomial as the square of a binomial. $x^2 - 12x + 36 = (x - 6)^2$

Complete the square on $z^2 + 3z$. Write the resulting perfect square trinomial as the square of a binomial.

Find the constant term. $\left(\frac{1}{2} \cdot 3\right)^2 = \left(\frac{3}{2}\right)^2 = \frac{9}{4}$

Complete the square on $z^2 + 3z$ by adding the constant term. $z^2 + 3z + \frac{9}{4}$

Write the resulting perfect square trinomial as the square of a binomial. $z^2 + 3z + \frac{9}{4} = \left(z + \frac{3}{2}\right)^2$

While not all quadratic equations can be solved by factoring, any quadratic equation can be solved by completing the square. Add to each side of the equation the term that completes the square. Rewrite the equation in the form $(x + a)^2 = b$. Then take the square root of each side of the equation.

Solve by completing the square: $x^2 - 4x - 14 = 0$

$$x^2 - 4x - 14 = 0$$

Add 14 to each side of the equation.

$$x^2 - 4x = 14$$

Add the constant term that completes the square on $x^2 - 4x$ to each side of the equation. $\left[\frac{1}{2}(-4)\right]^2 = 4$

$$x^2 - 4x + 4 = 14 + 4$$

Factor the perfect square trinomial.

$$(x - 2)^2 = 18$$

Take the square root of each side of the equation.

$$\sqrt{(x - 2)^2} = \sqrt{18}$$

Simplify.

$$x - 2 = \pm\sqrt{18} = \pm 3\sqrt{2}$$

Solve for x.

$$x - 2 = 3\sqrt{2} \qquad x - 2 = -3\sqrt{2}$$
$$x = 2 + 3\sqrt{2} \qquad x = 2 - 3\sqrt{2}$$

Check:

$x^2 - 4x - 14 = 0$	
$(2 + 3\sqrt{2})^2 - 4(2 + 3\sqrt{2}) - 14$	0
$4 + 12\sqrt{2} + 18 - 8 - 12\sqrt{2} - 14$	0
$0 = 0$	

$x^2 - 4x - 14 = 0$	
$(2 - 3\sqrt{2})^2 - 4(2 - 3\sqrt{2}) - 14$	0
$4 - 12\sqrt{2} + 18 - 8 + 12\sqrt{2} - 14$	0
$0 = 0$	

Write the solutions.

The solutions are $2 + 3\sqrt{2}$ and $2 - 3\sqrt{2}$.

When a, the coefficient of the x^2 term, is not 1, divide each side of the equation by a before completing the square.

Solve by completing the square: $2x^2 - x = 2$

$$2x^2 - x = 2$$

Divide each side of the equation by the coefficient of x^2.

$$\frac{2x^2 - x}{2} = \frac{2}{2}$$

The coefficient of the x^2 term is now 1.

$$x^2 - \frac{1}{2}x = 1$$

Add the term that completes the square on $x^2 - \frac{1}{2}x$ to each side of the equation.

$$x^2 - \frac{1}{2}x + \frac{1}{16} = 1 + \frac{1}{16}$$

Factor the perfect square trinomial. $\left(x - \frac{1}{4}\right)^2 = \frac{17}{16}$

Take the square root of each side of the equation. $\sqrt{\left(x - \frac{1}{4}\right)^2} = \sqrt{\frac{17}{16}}$

Simplify. $x - \frac{1}{4} = \pm\frac{\sqrt{17}}{4}$

Solve for x.

$$x - \frac{1}{4} = \frac{\sqrt{17}}{4} \qquad x - \frac{1}{4} = -\frac{\sqrt{17}}{4}$$

$$x = \frac{1}{4} + \frac{\sqrt{17}}{4} \qquad x = \frac{1}{4} - \frac{\sqrt{17}}{4}$$

$\frac{1 + \sqrt{17}}{4}$ and $\frac{1 - \sqrt{17}}{4}$ check as solutions.

Write the solutions. The solutions are $\frac{1 + \sqrt{17}}{4}$ and $\frac{1 - \sqrt{17}}{4}$.

Example 1 Solve by completing the square.

A. $4x^2 - 8x + 1 = 0$ B. $x^2 + 4x + 5 = 0$

Solution A. $4x^2 - 8x + 1 = 0$

$4x^2 - 8x = -1$ ► Subtract 1 from each side of the equation.

$\frac{4x^2 - 8x}{4} = \frac{-1}{4}$ ► The coefficient of the x^2 term must be 1. Divide each side of the equation by 4.

$x^2 - 2x = -\frac{1}{4}$

$x^2 - 2x + 1 = -\frac{1}{4} + 1$ ► Complete the square.

$(x - 1)^2 = \frac{3}{4}$ ► Factor the perfect square trinomial.

$\sqrt{(x - 1)^2} = \sqrt{\frac{3}{4}}$ ► Take the square root of each side of the equation.

$x - 1 = \pm\frac{\sqrt{3}}{2}$ ► Simplify.

$x - 1 = \frac{\sqrt{3}}{2} \qquad x - 1 = -\frac{\sqrt{3}}{2}$ ► Solve for x.

$$x = 1 + \frac{\sqrt{3}}{2} \qquad x = 1 - \frac{\sqrt{3}}{2}$$

$$x = \frac{2 + \sqrt{3}}{2} \qquad x = \frac{2 - \sqrt{3}}{2}$$

The solutions are $\frac{2 + \sqrt{3}}{2}$ and $\frac{2 - \sqrt{3}}{2}$.

B. $x^2 + 4x + 5 = 0$

$x^2 + 4x = -5$ ▶ Subtract 5 from each side of the equation.

$x^2 + 4x + 4 = -5 + 4$ ▶ Complete the square.

$(x + 2)^2 = -1$ ▶ Factor the perfect square trinomial.

$\sqrt{(x + 2)^2} = \sqrt{-1}$ ▶ Take the square root of each side of the equation.

$x + 2 = \pm i$ ▶ Simplify.

$x + 2 = i \qquad x + 2 = -i$ ▶ Solve for x.

$x = -2 + i \qquad x = -2 - i$

The solutions are $-2 + i$ and $-2 - i$.

Problem 1 Solve by completing the square.

A. $4x^2 - 4x - 1 = 0$ B. $2x^2 + x - 5 = 0$

Solution See page A33.

2 Solve quadratic equations by using the quadratic formula

A general formula known as the **quadratic formula** can be derived by applying the method of completing the square to the standard form of a quadratic equation. This formula can be used to solve any quadratic equation.

The solution of the equation $ax^2 + bx + c = 0$ by completing the square is shown below.

$$ax^2 + bx + c = 0$$

Subtract the constant term from each side of the equation.

$$ax^2 + bx + c - c = 0 - c$$
$$ax^2 + bx = -c$$

Divide each side of the equation by a, the coefficient of x^2.

$$\frac{ax^2 + bx}{a} = \frac{-c}{a}$$
$$x^2 + \frac{b}{a}x = -\frac{c}{a}$$

Complete the square by adding $\left(\frac{1}{2} \cdot \frac{b}{a}\right)^2$ to each side of the equation.

$$x^2 + \frac{b}{a}x + \left(\frac{1}{2} \cdot \frac{b}{a}\right)^2 = \left(\frac{1}{2} \cdot \frac{b}{a}\right)^2 - \frac{c}{a}$$
$$x^2 + \frac{b}{a}x + \frac{b^2}{4a^2} = \frac{b^2}{4a^2} - \frac{c}{a}$$

Simplify the right side of the equation.

$$x^2 + \frac{b}{a}x + \frac{b^2}{4a^2} = \frac{b^2}{4a^2} - \left(\frac{c}{a} \cdot \frac{4a}{4a}\right)$$

$$x^2 + \frac{b}{a}x + \frac{b^2}{4a^2} = \frac{b^2}{4a^2} - \frac{4ac}{4a^2}$$

$$x^2 + \frac{b}{a}x + \frac{b^2}{4a^2} = \frac{b^2 - 4ac}{4a^2}$$

Factor the perfect square trinomial on the left side of the equation.

$$\left(x + \frac{b}{2a}\right)^2 = \frac{b^2 - 4ac}{4a^2}$$

Take the square root of each side of the equation.

$$\sqrt{\left(x + \frac{b}{2a}\right)^2} = \sqrt{\frac{b^2 - 4ac}{4a^2}}$$

$$x + \frac{b}{2a} = \pm\frac{\sqrt{b^2 - 4ac}}{2a}$$

Solve for x.

$$x + \frac{b}{2a} = \frac{\sqrt{b^2 - 4ac}}{2a} \qquad x + \frac{b}{2a} = -\frac{\sqrt{b^2 - 4ac}}{2a}$$

$$x = -\frac{b}{2a} + \frac{\sqrt{b^2 - 4ac}}{2a} \qquad x = -\frac{b}{2a} - \frac{\sqrt{b^2 - 4ac}}{2a}$$

$$= \frac{-b + \sqrt{b^2 - 4ac}}{2a} \qquad = \frac{-b - \sqrt{b^2 - 4ac}}{2a}$$

The Quadratic Formula

The solutions of $ax^2 + bx + c = 0$, $a \neq 0$, are

$$\frac{-b + \sqrt{b^2 - 4ac}}{2a} \quad \text{and} \quad \frac{-b - \sqrt{b^2 - 4ac}}{2a}.$$

The quadratic formula is frequently written in the form

$$\boldsymbol{x = \frac{-b \pm \sqrt{b^2 - 4ac}}{2a}}.$$

Solve by using the quadratic formula: $4x^2 = 8x - 13$

$$4x^2 = 8x - 13$$

Write the equation in standard form.

$$4x^2 - 8x + 13 = 0$$

$$a = 4,\ b = -8,\ c = 13$$

Replace a, b, and c in the quadratic formula by their values.

$$x = \frac{-b \pm \sqrt{b^2 - 4ac}}{2a}$$

$$= \frac{-(-8) \pm \sqrt{(-8)^2 - 4 \cdot 4 \cdot 13}}{2 \cdot 4}$$

Simplify.

$$= \frac{8 \pm \sqrt{64 - 208}}{8} = \frac{8 \pm \sqrt{-144}}{8}$$

$$= \frac{8 \pm 12i}{8} = \frac{2 \pm 3i}{2} = 1 \pm \frac{3}{2}i$$

Check:

$4x^2 = 8x - 13$	
$4\left(1 + \frac{3}{2}i\right)^2$	$8\left(1 + \frac{3}{2}i\right) - 13$
$4\left(1 + 3i - \frac{9}{4}\right)$	$8 + 12i - 13$
$4\left(-\frac{5}{4} + 3i\right)$	$-5 + 12i$

$-5 + 12i = -5 + 12i$

$4x^2 = 8x - 13$	
$4\left(1 - \frac{3}{2}i\right)^2$	$8\left(1 - \frac{3}{2}i\right) - 13$
$4\left(1 - 3i - \frac{9}{4}\right)$	$8 - 12i - 13$
$4\left(-\frac{5}{4} - 3i\right)$	$-5 - 12i$

$-5 - 12i = -5 - 12i$

The solutions are $1 + \frac{3}{2}i$ and $1 - \frac{3}{2}i$.

Example 2 Solve by using the quadratic formula.

A. $4x^2 + 12x + 9 = 0$ B. $2x^2 - x + 5 = 0$

Solution A. $4x^2 + 12x + 9 = 0$ ▶ $a = 4,\ b = 12,\ c = 9$

$$x = \frac{-b \pm \sqrt{b^2 - 4ac}}{2a}$$

▶ Replace a, b, and c in the quadratic formula by their values. Then simplify.

$$= \frac{-12 \pm \sqrt{12^2 - 4 \cdot 4 \cdot 9}}{2 \cdot 4}$$

$$= \frac{-12 \pm \sqrt{0}}{8} = \frac{-12}{8} = -\frac{3}{2}$$

▶ The equation has a double root.

The solution is $-\frac{3}{2}$.

B. $2x^2 - x + 5 = 0$ ▶ $a = 2,\ b = -1,\ c = 5$

$$x = \frac{-b \pm \sqrt{b^2 - 4ac}}{2a}$$

▶ Replace a, b, and c in the quadratic formula by their values. Then simplify.

$$= \frac{-(-1) \pm \sqrt{(-1)^2 - 4 \cdot 2 \cdot 5}}{2 \cdot 2}$$

$$= \frac{1 \pm \sqrt{1 - 40}}{4} = \frac{1 \pm \sqrt{-39}}{4}$$

$$= \frac{1 \pm i\sqrt{39}}{4}$$

The solutions are $\frac{1}{4} + \frac{\sqrt{39}}{4}i$ and $\frac{1}{4} - \frac{\sqrt{39}}{4}i$.

Problem 2 Solve by using the quadratic formula.

A. $x^2 + 6x - 9 = 0$ B. $4x^2 = 4x - 1$

Solution See page A33.

In Example 2A, the solution of the equation is a double root, and in Example 2B, the solutions are complex numbers.

In the quadratic formula, the quantity $b^2 - 4ac$ is called the **discriminant.** When a, b, and c are real numbers, the discriminant determines whether a quadratic equation will have a double root, two real number solutions that are not equal, or two complex number solutions.

The Effect of the Discriminant on the Solutions of a Quadratic Equation

1. If $b^2 - 4ac = 0$, the equation has one real number solution, a double root.
2. If $b^2 - 4ac > 0$, the equation has two real number solutions that are not equal.
3. If $b^2 - 4ac < 0$, the equation has two complex number solutions.

The equation $x^2 - 4x - 5 = 0$ has two real number solutions because the discriminant is greater than zero.

$a = 1,\ b = -4,\ c = -5$
$b^2 - 4ac$
$(-4)^2 - 4(1)(-5) = 16 + 20 = 36$
$36 > 0$

Example 3 Use the discriminant to determine whether $4x^2 - 2x + 5 = 0$ has one real number solution, two real number solutions, or two complex number solutions.

Solution $b^2 - 4ac$ ▶ $a = 4,\ b = -2,\ c = 5$
$(-2)^2 - 4(4)(5) = 4 - 80 = -76$
$-76 < 0$ ▶ The discriminant is less than 0.

The equation has two complex number solutions.

Problem 3 Use the discriminant to determine whether $3x^2 - x - 1 = 0$ has one real number solution, two real number solutions, or two complex number solutions.

Solution See page A34.

EXERCISES 7.2

1 Solve by completing the square.

1. $x^2 - 4x - 5 = 0$
2. $y^2 + 6y + 5 = 0$
3. $v^2 + 8v - 9 = 0$
4. $w^2 - 2w - 24 = 0$
5. $z^2 - 6z + 9 = 0$
6. $u^2 + 10u + 25 = 0$

7. $r^2 + 4r - 7 = 0$
8. $s^2 + 6s - 1 = 0$
9. $x^2 - 6x + 7 = 0$
10. $y^2 + 8y + 13 = 0$
11. $z^2 - 2z + 2 = 0$
12. $t^2 - 4t + 8 = 0$
13. $s^2 - 5s - 24 = 0$
14. $v^2 + 7v - 44 = 0$
15. $x^2 + 5x - 36 = 0$
16. $y^2 - 9y + 20 = 0$
17. $p^2 - 3p + 1 = 0$
18. $r^2 - 5r - 2 = 0$
19. $t^2 - t - 1 = 0$
20. $u^2 - u - 7 = 0$
21. $y^2 - 6y = 4$
22. $w^2 + 4w = 2$
23. $x^2 = 8x - 15$
24. $z^2 = 4z - 3$
25. $v^2 = 4v - 13$
26. $x^2 = 2x - 17$
27. $p^2 + 6p = -13$
28. $x^2 + 4x = -20$
29. $y^2 - 2y = 17$
30. $x^2 + 10x = 7$
31. $z^2 = z + 4$
32. $r^2 = 3r - 1$
33. $x^2 + 13 = 2x$
34. $x^2 + 27 = 6x$
35. $2y^2 + 3y + 1 = 0$
36. $2t^2 + 5t - 3 = 0$
37. $4r^2 - 8r = -3$
38. $4u^2 - 20u = -9$
39. $6y^2 - 5y = 4$
40. $6v^2 - 7v = 3$
41. $4x^2 - 4x + 5 = 0$
42. $4t^2 - 4t + 17 = 0$
43. $9x^2 - 6x + 2 = 0$
44. $9y^2 - 12y + 13 = 0$
45. $2s^2 = 4s + 5$
46. $3u^2 = 6u + 1$
47. $2r^2 = 3 - r$
48. $2x^2 = 12 - 5x$
49. $y - 2 = (y - 3)(y + 2)$
50. $8s - 11 = (s - 4)(s - 2)$
51. $6t - 2 = (2t - 3)(t - 1)$

52. $2z + 9 = (2z + 3)(z + 2)$ **53.** $(x - 4)(x + 1) = x - 3$ **54.** $(y - 3)^2 = 2y + 10$

Solve by completing the square. Approximate the solutions to the nearest thousandth.

55. $z^2 + 2z = 4$ **56.** $t^2 - 4t = 7$ **57.** $2x^2 = 4x - 1$

58. $3y^2 = 5y - 1$ **59.** $4z^2 + 2z - 1 = 0$ **60.** $4w^2 - 8w = 3$

2 Solve by using the quadratic formula.

61. $x^2 - 3x - 10 = 0$ **62.** $z^2 - 4z - 8 = 0$ **63.** $y^2 + 5y - 36 = 0$

64. $z^2 - 3z - 40 = 0$ **65.** $w^2 = 8w + 72$ **66.** $t^2 = 2t + 35$

67. $v^2 = 24 - 5v$ **68.** $x^2 = 18 - 7x$ **69.** $2y^2 + 5y - 3 = 0$

70. $4p^2 - 7p + 3 = 0$ **71.** $8s^2 = 10s + 3$ **72.** $12t^2 = 5t + 2$

73. $v^2 - 2v - 7 = 0$ **74.** $t^2 - 2t - 11 = 0$ **75.** $y^2 - 8y - 20 = 0$

76. $x^2 = 14x - 24$ **77.** $v^2 = 12v - 24$ **78.** $2z^2 - 2z - 1 = 0$

79. $4x^2 - 4x - 7 = 0$ **80.** $2p^2 - 8p + 5 = 0$ **81.** $2s^2 - 3s + 1 = 0$

82. $4w^2 - 4w - 1 = 0$ **83.** $3x^2 + 10x + 6 = 0$ **84.** $3v^2 = 6v - 2$

85. $6w^2 = 19w - 10$ **86.** $z^2 + 2z + 2 = 0$ **87.** $p^2 - 4p + 5 = 0$

88. $y^2 - 2y + 5 = 0$ **89.** $x^2 + 6x + 13 = 0$ **90.** $s^2 - 4s + 13 = 0$

91. $t^2 - 6t + 10 = 0$ **92.** $2w^2 - 2w + 5 = 0$ **93.** $4v^2 + 8v + 3 = 0$

94. $2x^2 + 6x + 5 = 0$ **95.** $2y^2 + 2y + 13 = 0$ **96.** $4t^2 - 6t + 9 = 0$

97. $3v^2 + 6v + 1 = 0$ **98.** $2r^2 = 4r - 11$ **99.** $3y^2 = 6y - 5$

100. $2x(x - 2) = x + 12$ **101.** $10y(y + 4) = 15y - 15$

102. $(3s - 2)(s + 1) = 2$ **103.** $(2t + 1)(t - 3) = 9$

Use the discriminant to determine whether the quadratic equation has one real number solution, two real number solutions, or two complex number solutions.

104. $2z^2 - z + 5 = 0$ **105.** $3y^2 + y + 1 = 0$ **106.** $9x^2 - 12x + 4 = 0$

107. $4x^2 + 20x + 25 = 0$ **108.** $2v^2 - 3v - 1 = 0$ **109.** $3w^2 + 3w - 2 = 0$

110. $2p^2 + 5p + 1 = 0$ **111.** $2t^2 + 9t + 3 = 0$ **112.** $5z^2 + 2 = 0$

Solve by using the quadratic formula. Approximate the solutions to the nearest thousandth.

113. $x^2 + 6x - 6 = 0$ **114.** $p^2 - 8p + 3 = 0$ **115.** $r^2 - 2r - 4 = 0$

116. $w^2 + 4w - 1 = 0$ **117.** $3t^2 = 7t + 1$ **118.** $2y^2 = y + 5$

SUPPLEMENTAL EXERCISES 7.2

Solve.

119. $\sqrt{2}y^2 + 3y - 2\sqrt{2} = 0$ **120.** $\sqrt{3}z^2 + 10z - 3\sqrt{3} = 0$

121. $\sqrt{2}x^2 + 5x - 3\sqrt{2} = 0$ **122.** $\sqrt{3}w^2 + w - 2\sqrt{3} = 0$

123. $t^2 - t\sqrt{3} + 1 = 0$ **124.** $y^2 + y\sqrt{7} + 2 = 0$

Solve for x.

125. $x^2 - ax - 2a^2 = 0$

126. $x^2 - ax - 6a^2 = 0$

127. $x^2 + 3ax - 4a^2 = 0$

128. $x^2 + 3ax - 10a^2 = 0$

129. $x^2 - 6ax + 8a^2 = 0$

130. $x^2 + 9ax + 18a^2 = 0$

131. $2x^2 + 3ax - 2a^2 = 0$

132. $2x^2 - 7ax + 3a^2 = 0$

133. $x^2 - 2x - y = 0$

134. $x^2 - 4xy - 4 = 0$

For what values of p does the quadratic equation have two real number solutions that are not equal? Write the answer in set builder notation.

135. $x^2 - 6x + p = 0$

136. $x^2 + 10x + p = 0$

For what values of p does the quadratic equation have two complex number solutions? Write the answer in set builder notation.

137. $x^2 - 2x + p = 0$

138. $x^2 + 4x + p = 0$

Solve.

139. Show that the equation $x^2 + bx - 1 = 0$ always has real number solutions regardless of the value of b.

140. Show that the equation $2x^2 + bx - 2 = 0$ always has real number solutions regardless of the value of b.

SECTION 7.3 Equations That Are Reducible to Quadratic Equations

1 Equations that are quadratic in form

Certain equations that are not quadratic equations can be expressed in quadratic form by making suitable substitutions. An equation is **quadratic in form** if it can be written as $au^2 + bu + c = 0$.

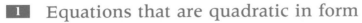

The equation at the right is quadratic in form.

Let $x^2 = u$. Replace x^2 by u. The equation is quadratic in form.

$$x^4 - 4x^2 - 5 = 0$$
$$(x^2)^2 - 4(x^2) - 5 = 0$$
$$u^2 - 4u - 5 = 0$$

The equation at the right is quadratic in form.

Let $y^{\frac{1}{2}} = u$. Replace $y^{\frac{1}{2}}$ by u. The equation is quadratic in form.

$$y - y^{\frac{1}{2}} - 6 = 0$$
$$(y^{\frac{1}{2}})^2 - (y^{\frac{1}{2}}) - 6 = 0$$
$$u^2 - u - 6 = 0$$

The key to recognizing equations that are quadratic in form: when the equation is written in standard form, the exponent on one variable term is $\frac{1}{2}$ the exponent on the other variable term.

Solve: $z + 7z^{\frac{1}{2}} - 18 = 0$

The equation $z + 7z^{\frac{1}{2}} - 18 = 0$ is quadratic in form.

$$z + 7z^{\frac{1}{2}} - 18 = 0$$
$$(z^{\frac{1}{2}})^2 + 7(z^{\frac{1}{2}}) - 18 = 0$$

To solve this equation, let $z^{\frac{1}{2}} = u$.

$$u^2 + 7u - 18 = 0$$

Solve for u by factoring.

$$(u - 2)(u + 9) = 0$$

$$u - 2 = 0 \qquad u + 9 = 0$$
$$u = 2 \qquad u = -9$$

Replace u by $z^{\frac{1}{2}}$.

$$z^{\frac{1}{2}} = 2 \qquad z^{\frac{1}{2}} = -9$$

Solve for z by squaring each side of the equation.

$$(z^{\frac{1}{2}})^2 = 2^2 \qquad (z^{\frac{1}{2}})^2 = (-9)^2$$
$$z = 4 \qquad z = 81$$

Check the solution. When squaring each side of an equation, the resulting equation may have a solution that is not a solution of the original equation.

Check:

$z + 7z^{\frac{1}{2}} - 18 = 0$	
$4 + 7(4)^{\frac{1}{2}} - 18$	0
$4 + 7 \cdot 2 - 18$	
$4 + 14 - 18$	
$0 = 0$	

$z + 7z^{\frac{1}{2}} - 18 = 0$	
$81 + 7(81)^{\frac{1}{2}} - 18$	0
$81 + 7 \cdot 9 - 18$	
$81 + 63 - 18$	
$126 \neq 0$	

4 checks as a solution, but 81 does not check as a solution.

Write the solution.

The solution is 4.

Example 1 Solve.

A. $x^4 + x^2 - 12 = 0$ B. $x^{\frac{2}{3}} - 2x^{\frac{1}{3}} - 3 = 0$

Solution A.

$$x^4 + x^2 - 12 = 0$$
$$(x^2)^2 + (x^2) - 12 = 0$$

▶ The equation is quadratic in form.

$$u^2 + u - 12 = 0$$

▶ Let $x^2 = u$.

$$(u - 3)(u + 4) = 0$$

▶ Solve for u by factoring.

$$u - 3 = 0 \qquad u + 4 = 0$$
$$u = 3 \qquad u = -4$$

$$x^2 = 3 \qquad x^2 = -4$$

▶ Replace u by x^2.

$$\sqrt{x^2} = \sqrt{3} \qquad \sqrt{x^2} = \sqrt{-4}$$

▶ Solve for x by taking square roots.

$$x = \pm\sqrt{3} \qquad x = \pm 2i$$

The solutions are $\sqrt{3}$, $-\sqrt{3}$, $2i$, and $-2i$.

B.

$$x^{\frac{2}{3}} - 2x^{\frac{1}{3}} - 3 = 0$$
$$(x^{\frac{1}{3}})^2 - 2(x^{\frac{1}{3}}) - 3 = 0$$

▶ The equation is quadratic in form.

$$u^2 - 2u - 3 = 0$$

▶ Let $x^{\frac{1}{3}} = u$.

$$(u - 3)(u + 1) = 0$$

▶ Solve for u by factoring.

$$u - 3 = 0 \qquad u + 1 = 0$$
$$u = 3 \qquad u = -1$$

$$x^{\frac{1}{3}} = 3 \qquad x^{\frac{1}{3}} = -1$$

▶ Replace u by $x^{\frac{1}{3}}$.

$$(x^{\frac{1}{3}})^3 = 3^3 \qquad (x^{\frac{1}{3}})^3 = (-1)^3$$

▶ Solve for x by cubing both sides of the equation.

$$x = 27 \qquad x = -1$$

The solutions are 27 and -1.

Problem 1 Solve.

A. $x - 5x^{\frac{1}{2}} + 6 = 0$ B. $4x^4 + 35x^2 - 9 = 0$

Solution See page A34.

2 Radical equations

Certain equations containing a radical can be solved by first solving the equation for the radical expression and then squaring each side of the equation.

Remember that when squaring each side of an equation, the resulting equation may have a solution that is not a solution of the original equation. Therefore, the solutions of a radical equation must be checked.

Solve: $\sqrt{x+2}+4=x$

$$\sqrt{x+2}+4=x$$

Solve for the radical expression. $\qquad \sqrt{x+2}=x-4$

Square each side of the equation. $\qquad (\sqrt{x+2})^2=(x-4)^2$

Simplify. $\qquad x+2=x^2-8x+16$

Write the equation in standard form. $\qquad 0=x^2-9x+14$

Solve for x by factoring. $\qquad 0=(x-7)(x-2)$

$$x-7=0 \qquad x-2=0$$
$$x=7 \qquad x=2$$

Check the solution.

Check:

$\sqrt{x+2}+4=x$		$\sqrt{x+2}+4=x$	
$\sqrt{7+2}+4$	7	$\sqrt{2+2}+4$	2
$\sqrt{9}+4$		$\sqrt{4}+4$	
$3+4$		$2+4$	
$7=7$		$6\neq 2$	

7 checks as a solution, but 2 does not check as a solution.

Write the solution. The solution is 7.

Example 2 Solve: $\sqrt{7y-3}+3=2y$

Solution

$$\sqrt{7y-3}+3=2y$$
$$\sqrt{7y-3}=2y-3$$ ▶ Solve for the radical expression.
$$(\sqrt{7y-3})^2=(2y-3)^2$$ ▶ Square each side of the equation.
$$7y-3=4y^2-12y+9$$
$$0=4y^2-19y+12$$ ▶ Write the equation in standard form.
$$0=(4y-3)(y-4)$$ ▶ Solve by y by factoring.

$$4y-3=0 \qquad y-4=0$$
$$4y=3 \qquad y=4$$
$$y=\frac{3}{4}$$

▶ 4 checks as a solution. $\frac{3}{4}$ does not check as a solution.

The solution is 4.

Problem 2 Solve: $\sqrt{2x+1}+x=7$

Solution See page A34.

If an equation contains more than one radical, the procedure of solving for the radical expression and squaring each side of the equation may have to be repeated.

Example 3 Solve: $\sqrt{2y+1} - \sqrt{y} = 1$

Solution

$$\begin{aligned} \sqrt{2y+1} - \sqrt{y} &= 1 \\ \sqrt{2y+1} &= \sqrt{y} + 1 && \blacktriangleright \text{ Solve for one of the radical expressions.} \\ (\sqrt{2y+1})^2 &= (\sqrt{y}+1)^2 && \blacktriangleright \text{ Square each side of the equation.} \\ 2y+1 &= y + 2\sqrt{y} + 1 \\ y &= 2\sqrt{y} && \blacktriangleright \text{ Solve for the radical expression.} \\ y^2 &= (2\sqrt{y})^2 && \blacktriangleright \text{ Square each side of the equation.} \\ y^2 &= 4y \\ y^2 - 4y &= 0 \\ y(y-4) &= 0 \end{aligned}$$

$$y = 0 \qquad y - 4 = 0$$
$$y = 4 \qquad \blacktriangleright \text{ 0 and 4 check as solutions.}$$

The solutions are 0 and 4.

Problem 3 Solve: $\sqrt{2x-1} + \sqrt{x} = 2$

Solution See page A34.

3 Fractional equations

After each side of a fractional equation has been multiplied by the LCM of the denominators, the resulting equation is sometimes a quadratic equation. The solutions to the resulting equation must be checked because multiplying each side of an equation by a variable expression may produce an equation that has a solution that is not a solution of the original equation.

Solve: $\dfrac{1}{r} + \dfrac{1}{r+1} = \dfrac{3}{2}$

$$\frac{1}{r} + \frac{1}{r+1} = \frac{3}{2}$$

Multiply each side of the equation by the LCM of the denominators.

$$\begin{aligned} 2r(r+1)\left(\frac{1}{r} + \frac{1}{r+1}\right) &= 2r(r+1) \cdot \frac{3}{2} \\ 2(r+1) + 2r &= r(r+1) \cdot 3 \\ 2r + 2 + 2r &= 3r(r+1) \\ 4r + 2 &= 3r^2 + 3r \end{aligned}$$

Write the equation in standard form.

$$0 = 3r^2 - r - 2$$

Solve for r by factoring.

$$0 = (3r + 2)(r - 1)$$

$$3r + 2 = 0 \qquad r - 1 = 0$$
$$3r = -2 \qquad r = 1$$
$$r = -\frac{2}{3}$$

$-\frac{2}{3}$ and 1 check as solutions.

Write the solutions.

The solutions are $-\frac{2}{3}$ and 1.

Example 4 Solve.

A. $\frac{9}{x-3} = 2x + 1$ B. $\frac{18}{2a-1} + 3a = 17$

Solution A.

$$\frac{9}{x-3} = 2x + 1$$

► The LCM of the denominators is $x - 3$.

$$(x-3)\frac{9}{x-3} = (x-3)(2x+1)$$
$$9 = 2x^2 - 5x - 3$$
$$0 = 2x^2 - 5x - 12$$

► Write the equation in standard form.

$$0 = (2x+3)(x-4)$$

► Solve for x by factoring.

$$2x + 3 = 0 \qquad x - 4 = 0$$
$$2x = -3 \qquad x = 4$$
$$x = -\frac{3}{2}$$

The solutions are $-\frac{3}{2}$ and 4.

B.

$$\frac{18}{2a-1} + 3a = 17$$
$$(2a-1)\left(\frac{18}{2a-1} + 3a\right) = (2a-1)17$$
$$(2a-1)\frac{18}{2a-1} + (2a-1)(3a) = (2a-1)17$$
$$18 + 6a^2 - 3a = 34a - 17$$
$$6a^2 - 37a + 35 = 0$$
$$(6a-7)(a-5) = 0$$

$$6a - 7 = 0 \qquad a - 5 = 0$$
$$6a = 7 \qquad a = 5$$
$$a = \frac{7}{6}$$

The solutions are $\frac{7}{6}$ and 5.

Problem 4 Solve.

A. $3y + \frac{25}{3y - 2} = -8$ B. $\frac{5}{x + 2} = 2x - 5$

Solution See page A35.

EXERCISES 7.3

1 Solve.

1. $x^4 - 13x^2 + 36 = 0$

2. $y^4 - 5y^2 + 4 = 0$

3. $z^4 - 6z^2 + 8 = 0$

4. $t^4 - 12t^2 + 27 = 0$

5. $p - 3p^{\frac{1}{2}} + 2 = 0$

6. $v - 7v^{\frac{1}{2}} + 12 = 0$

7. $x - x^{\frac{1}{2}} - 12 = 0$

8. $w - 2w^{\frac{1}{2}} - 15 = 0$

9. $z^4 + 3z^2 - 4 = 0$

10. $y^4 + 5y^2 - 36 = 0$

11. $x^4 + 12x^2 - 64 = 0$

12. $x^4 - 81 = 0$

13. $p + 2p^{\frac{1}{2}} - 24 = 0$

14. $v + 3v^{\frac{1}{2}} - 4 = 0$

15. $y^{\frac{2}{3}} - 9y^{\frac{1}{3}} + 8 = 0$

16. $z^{\frac{2}{3}} - z^{\frac{1}{3}} - 6 = 0$

17. $x^6 - 9x^3 + 8 = 0$

18. $y^6 + 9y^3 + 8 = 0$

19. $z^8 - 17z^4 + 16 = 0$

20. $v^4 - 15v^2 - 16 = 0$

21. $p^{\frac{2}{3}} + 2p^{\frac{1}{3}} - 8 = 0$

22. $w^{\frac{2}{3}} + 3w^{\frac{1}{3}} - 10 = 0$

23. $2x - 3x^{\frac{1}{2}} + 1 = 0$

24. $3y - 5y^{\frac{1}{2}} - 2 = 0$

2 Solve.

25. $\sqrt{x+1} + x = 5$

26. $\sqrt{x-4} + x = 6$

27. $x = \sqrt{x} + 6$

28. $\sqrt{2y-1} = y - 2$

29. $\sqrt{3w+3} = w + 1$

30. $\sqrt{2s+1} = s - 1$

31. $\sqrt{4y+1} - y = 1$

32. $\sqrt{3s+4} + 2s = 12$

33. $\sqrt{10x+5} - 2x = 1$

34. $\sqrt{t+8} = 2t + 1$

35. $\sqrt{p+11} = 1 - p$

36. $x - 7 = \sqrt{x-5}$

37. $\sqrt{x-1} - \sqrt{x} = -1$

38. $\sqrt{y} + 1 = \sqrt{y+5}$

39. $\sqrt{2x-1} = 1 - \sqrt{x-1}$

40. $\sqrt{x+6} + \sqrt{x+2} = 2$

41. $\sqrt{t+3} + \sqrt{2t+7} = 1$

42. $\sqrt{5-2x} = \sqrt{2-x} + 1$

3 Solve.

43. $x = \dfrac{10}{x-9}$

44. $z = \dfrac{5}{z-4}$

45. $\dfrac{t}{t+1} = \dfrac{-2}{t-1}$

46. $\dfrac{2v}{v-1} = \dfrac{5}{v+2}$

47. $\dfrac{y-1}{y+2} + y = 1$

48. $\dfrac{2p-1}{p-2} + p = 8$

49. $\dfrac{3r+2}{r+2} - 2r = 1$

50. $\dfrac{2v+3}{v+4} + 3v = 4$

51. $\dfrac{2}{2x+1} + \dfrac{1}{x} = 3$

52. $\dfrac{3}{s} - \dfrac{2}{2s-1} = 1$

53. $\dfrac{16}{z-2} + \dfrac{16}{z+2} = 6$

54. $\dfrac{2}{y+1} + \dfrac{1}{y-1} = 1$

55. $\dfrac{t}{t-2} + \dfrac{2}{t-1} = 4$

56. $\dfrac{4t+1}{t+4} + \dfrac{3t-1}{t+1} = 2$

57. $\dfrac{5}{2p-1} + \dfrac{4}{p+1} = 2$

58. $\dfrac{3w}{2w+3} + \dfrac{2}{w+2} = 1$

59. $\dfrac{2v}{v+2} + \dfrac{3}{v+4} = 1$

60. $\dfrac{x+3}{x+1} - \dfrac{x-2}{x+3} = 5$

SUPPLEMENTAL EXERCISES 7.3

Solve.

61. $\frac{x^2}{4} + \frac{x}{2} = 6$

62. $\frac{x^2}{9} + \frac{2x}{3} = 3$

63. $3\left(\frac{x+1}{2}\right)^2 = 54$

64. $2\left(\frac{x-2}{3}\right)^2 = 24$

65. $\frac{x+2}{3} + \frac{2}{x-2} = 3$

66. $\frac{x-1}{2} + \frac{4}{x+1} = 2$

67. $\frac{x^4}{4} + 1 = \frac{5x^2}{4}$

68. $\frac{x^4}{4} + 2 = \frac{9x^2}{4}$

69. $\frac{x^4}{3} - \frac{8x^2}{3} = 3$

70. $\frac{x^4}{6} + \frac{x^2}{6} = 2$

71. $\frac{x^2}{4} + \frac{x}{2} + \frac{1}{8} = 0$

72. $\frac{x^2}{2} + \frac{x}{3} + \frac{1}{6} = 0$

73. $\frac{x^4}{8} + \frac{x^2}{4} = 3$

74. $\frac{x^4}{8} - \frac{x^2}{2} = 4$

75. $\sqrt{x^4 - 2} = x$

76. $\sqrt{x^4 + 4} = 2x$

77. $(\sqrt{x} - 2)^2 - 5\sqrt{x} + 14 = 0$ *Hint:* Let $u = \sqrt{x} - 2$.

78. $(\sqrt{x} - 1)^2 - 7\sqrt{x} + 19 = 0$ *Hint:* Let $u = \sqrt{x} - 1$.

79. $(\sqrt{x} + 3)^2 - 4\sqrt{x} - 17 = 0$ *Hint:* Let $u = \sqrt{x} + 3$.

SECTION 7.4

Graphing Quadratic Equations in Two Variables

1 Graph equations of the form $y = ax^2 + bx + c$

An equation of the form $y = ax^2 + bx + c$, $a \neq 0$, is a **quadratic equation in two variables.** Examples of quadratic equations in two variables are shown below.

$$y = 2x^2 - x - 3$$
$$y = -x^2 + 4$$
$$y = 3x^2 - 2x$$

The graph of a quadratic equation in two variables is a **parabola.** The graph is "cup" shaped and opens either up or down. The coefficient of x^2 determines whether the parabola opens up or down. When a is **positive,** the parabola **opens up.** When a is **negative,** the parabola **opens down.** The graphs of two parabolas are shown below.

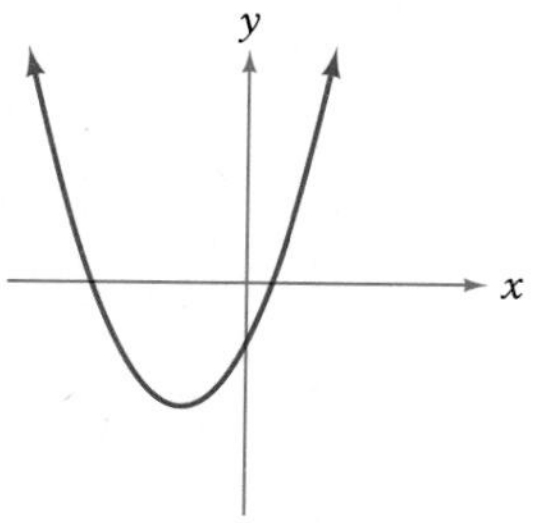

Parabola that opens up
$y = ax^2 + bx + c,\ a > 0$

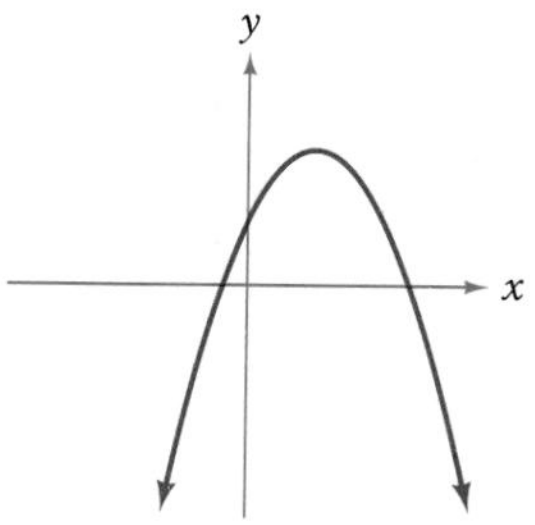

Parabola that opens down
$y = ax^2 + bx + c,\ a < 0$

Since the graph of a parabola is cup shaped, drawing its graph requires finding enough ordered pair solutions of the equation so that its cup shape can be determined. Remember that when a is positive, the parabola will open up, and when a is negative, the parabola will open down.

For example, the graph of $y = x^2 - x - 2$ will open up because a is positive ($a = 1$).

To graph $y = x^2 - x - 2$, find enough ordered pair solutions to determine the cup shape. These ordered pairs can be recorded in a table.

x	y
0	−2
1	−2
−1	0
2	0
−2	4
3	4

Graph the ordered pair solutions on a rectangular coordinate system.

Draw a parabola through the points.

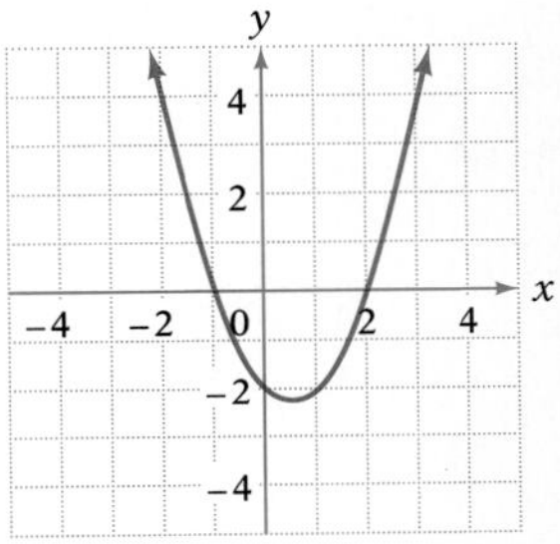

Example 1 Graph.

A. $y = -2x^2 + x + 4$ B. $y = -2x^2 - 4x$

Solution A.

x	y
0	4
1	3
−1	1
2	−2
−2	−6

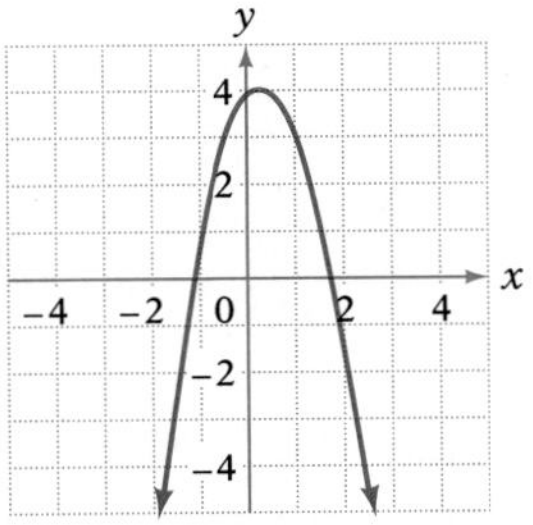

▶ Since a is negative ($a = -2$), the parabola will open down.

B.

x	y
0	0
1	−6
−1	2
−2	0
−3	−6

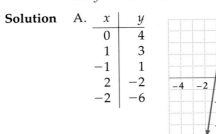

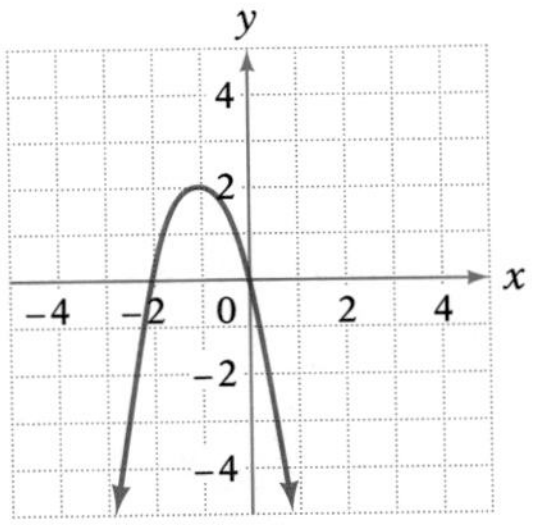

Problem 1 Graph.

A. $y = \frac{1}{2}x^2 - 2x - 1$ B. $y = -x^2 + 3x - 4$

Solution See page A35.

Every parabola has an axis of symmetry and a vertex that is on the axis of symmetry. To understand the axis of symmetry, think of folding the paper along that axis. The two halves of the graph will match up.

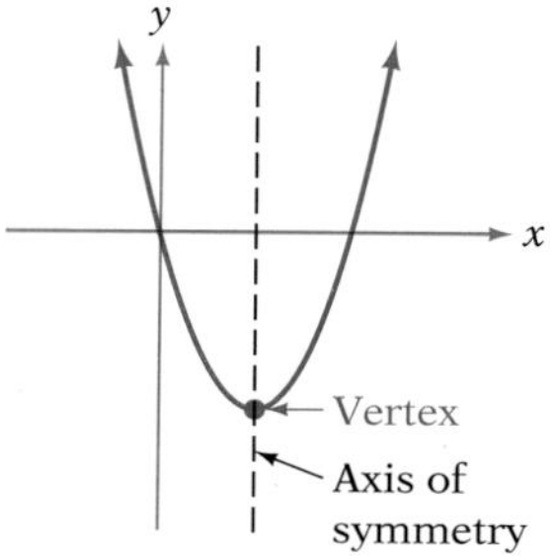

The coordinates of the vertex and the axis of symmetry of a parabola can be found by completing the square.

To find the vertex of the parabola whose equation is $y = x^2 - 4x + 5$, group the variable terms.

$$y = x^2 - 4x + 5$$
$$y = (x^2 - 4x) + 5$$

Complete the square on $x^2 - 4x$. Note that 4 is added and subtracted. Since $4 - 4 = 0$, the equation is not changed.

$$y = (x^2 - 4x + 4) - 4 + 5$$

Factor the trinomial and combine like terms.

$$y = (x - 2)^2 + 1$$

Since the coefficient of x^2 is positive, the parabola opens up. The vertex is the lowest point on the parabola, or that point which has the least y-coordinate.

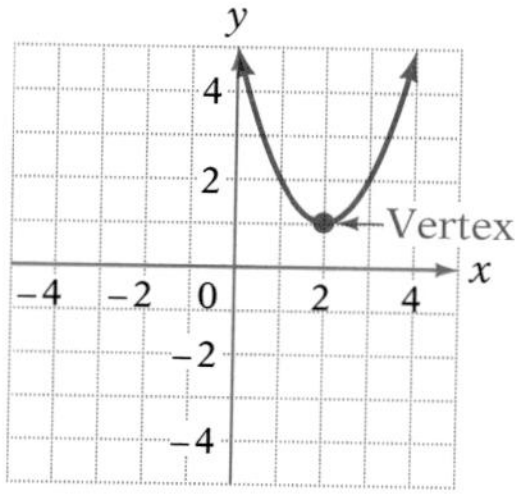

Since $(x - 2)^2 \geq 0$ for any x, the least y-coordinate occurs when $(x - 2)^2 = 0$. $(x - 2)^2 = 0$ when $x = 2$. The x-coordinate of the vertex is 2.

To find the y-coordinate of the vertex, replace x by 2, and solve for y.

$$\begin{aligned} y &= (x - 2)^2 + 1 \\ &= (2 - 2)^2 + 1 \\ &= 1 \end{aligned}$$

The vertex is (2, 1).

Since the axis of symmetry is parallel to the y-axis and passes through the vertex, the equation of the axis of symmetry is $x = 2$.

By following the procedure of this example and completing the square on the equation $y = ax^2 + bx + c$, the **x-coordinate of the vertex** is $-\frac{b}{2a}$. The y-coordinate of the vertex can then be determined by substituting this value of x into $y = ax^2 + bx + c$ and solving for y. Since the axis of symmetry is parallel to the y-axis and passes through the vertex, the equation of the **axis of symmetry** is $x = -\frac{b}{2a}$.

The coordinates of the vertex and the axis of symmetry can be used to graph a parabola. These concepts are used below in graphing the parabola given by the equation $y = x^2 + 2x - 3$.

Find the x-coordinate of the vertex. $a = 1$, $b = 2$.

$$x = -\frac{b}{2a} = -\frac{2}{2(1)} = -1$$

Find the y-coordinate of the vertex by replacing x with -1 and solving for y.

$$\begin{aligned} y &= x^2 + 2x - 3 \\ &= (-1)^2 + 2(-1) - 3 \\ &= 1 - 2 - 3 \\ &= -4 \end{aligned}$$

The vertex is $(-1, -4)$.

The axis of symmetry is the line $x = -1$.

Find some ordered pair solutions of the equation and record these in a table. Because the graph is symmetric to the line $x = -1$, choose values of x greater than -1.

x	y
0	-3
1	0
2	5

Graph the ordered pair solutions on a rectangular coordinate system. Use symmetry to locate points of the graph on the other side on the axis of symmetry. Remember that corresponding points on the graph are the same distance from the axis of symmetry.

Draw a parabola through the points.

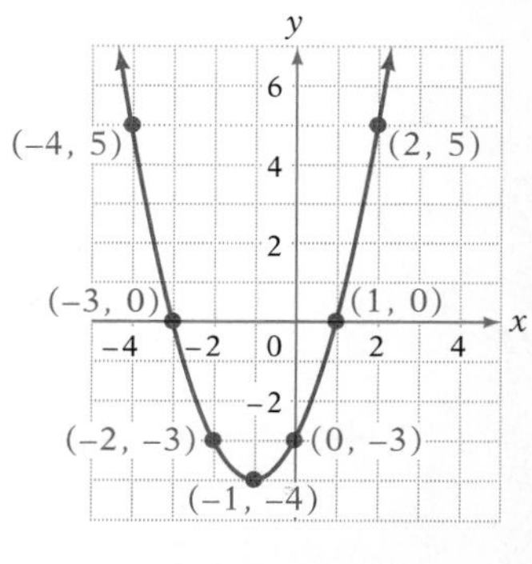

Example 2 Find the vertex and the axis of symmetry of the parabola whose equation is $y = -3x^2 + 6x + 1$. Then sketch its graph.

Solution $-\frac{b}{2a} = -\frac{6}{2(-3)} = 1$

▶ Find the x-coordinate of the vertex. $a = -3$, $b = 6$

$$y = -3x^2 + 6x + 1$$
$$= -3(1)^2 + 6(1) + 1$$
$$= 4$$

▶ Find the y-coordinate of the vertex by replacing x by 1 and solving for y.

The vertex is (1, 4).

The axis of symmetry is the line $x = 1$.

▶ The axis of symmetry is the line $x = -\frac{b}{2a}$.

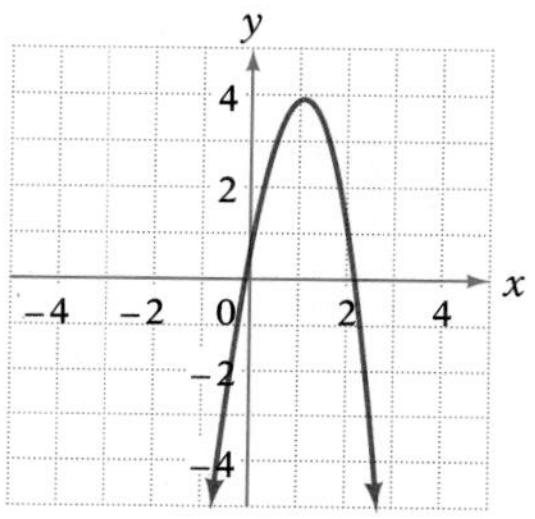

▶ Since a is negative, the parabola opens down. Find a few ordered pairs and use symmetry to sketch the graph.

Problem 2 Find the vertex and the axis of symmetry of the parabola whose equation is $y = x^2 - 2$. Then sketch its graph.

Solution See page A35.

7

2 Find the x-intercepts of a parabola

The points at which a graph crosses or touches a coordinate axis are called the **intercepts** of the graph.

When the graph of a parabola crosses the x-axis, the y-coordinate is zero.

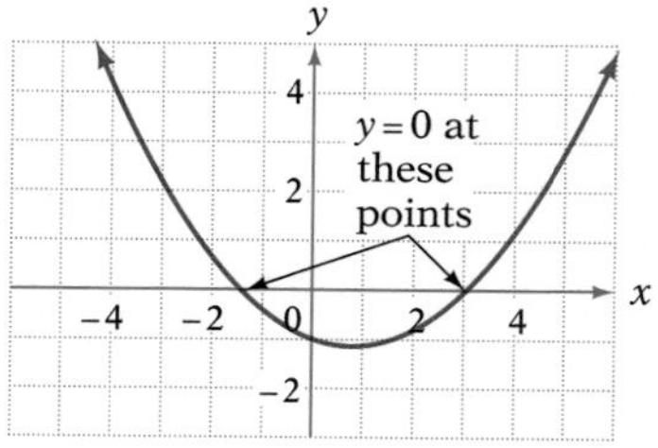

To find the x-intercepts for the parabola $y = ax^2 + bx + c$, let $y = 0$. Then solve for x.

Example 3 Find the x-intercepts of the parabola given by each equation.

A. $y = x^2 - 2x - 1$ B. $y = 4x^2 - 4x + 1$

Solution A.

$$y = x^2 - 2x - 1$$
$$0 = x^2 - 2x - 1$$

▶ Let $y = 0$.

$$x = \frac{-b \pm \sqrt{b^2 - 4ac}}{2a}$$
$$= \frac{-(-2) \pm \sqrt{(-2)^2 - 4(1)(-1)}}{2 \cdot 1}$$
$$= \frac{2 \pm \sqrt{4 + 4}}{2} = \frac{2 \pm \sqrt{8}}{2}$$
$$= \frac{2 \pm 2\sqrt{2}}{2} = 1 \pm \sqrt{2}$$

▶ The equation is nonfactorable over the integers. Use the quadratic formula to solve for x.

The x-intercepts are $(1 + \sqrt{2}, 0)$ and $(1 - \sqrt{2}, 0)$.

B.

$$y = 4x^2 - 4x + 1$$
$$0 = 4x^2 - 4x + 1$$
$$0 = (2x - 1)(2x - 1)$$

▶ Let $y = 0$.
▶ Solve for x by factoring.

$$2x - 1 = 0 \qquad 2x - 1 = 0$$
$$2x = 1 \qquad 2x = 1$$
$$x = \frac{1}{2} \qquad x = \frac{1}{2}$$

▶ The equation has a double root.

The x-intercept is $\left(\frac{1}{2}, 0\right)$.

Problem 3 Find the x-intercepts of the parabola given by each equation.

A. $y = 2x^2 - 5x + 2$ B. $y = x^2 + 4x + 4$

Solution See page A36.

In Example 3B, the parabola only touches the x-axis. In this case, the parabola is said to be **tangent** to the x-axis at $x = \frac{1}{2}$.

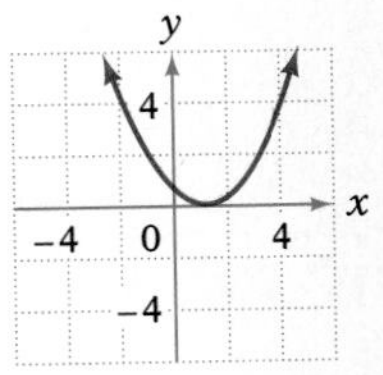

A parabola can have one x-intercept, as in Example 3B; two x-intercepts, as in the graph at the bottom of the previous page; or no x-intercepts, as in the graph at the right.

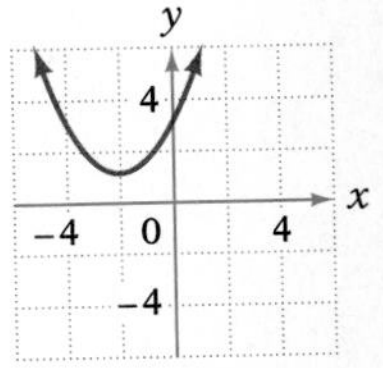

The parabola whose equation is $y = ax^2 + bx + c$ will have one x-intercept if the equation $0 = ax^2 + bx + c$ has one real number solution, two x-intercepts if the equation has two real number solutions, or no x-intercepts if the equation has no real number solutions.

Since the discriminant determines whether there are one, two, or no real number solutions of the equation $0 = ax^2 + bx + c$, it can also be used to determine whether there are one, two, or no x-intercepts of a parabola.

The Effect of the Discriminant on the Number of x-Intercepts of a Parabola

1. If $b^2 - 4ac = 0$, the parabola has one x-intercept.
2. If $b^2 - 4ac > 0$, the parabola has two x-intercepts.
3. If $b^2 - 4ac < 0$, the parabola has no x-intercepts.

The parabola whose equation is $y = 2x^2 - x + 2$ has no x-intercepts since the discriminant is less than zero.

$a = 2,\ b = -1,\ c = 2$
$b^2 - 4ac$
$(-1)^2 - 4(2)(2) = 1 - 16 = -15$
$-15 < 0$

Example 4 Use the discriminant to determine the number of x-intercepts of the parabola whose equation is $y = x^2 - 6x + 9$.

Solution $b^2 - 4ac$ ▶ $a = 1,\ b = -6,\ c = 9$

$(-6)^2 - 4(1)(9) = 36 - 36 = 0$ ▶ The discriminant is equal to zero.

The parabola has one x-intercept.

Problem 4 Use the discriminant to determine the number of x-intercepts of the parabola whose equation is $y = x^2 - x - 6$.

Solution See page A36.

EXERCISES 7.4

1 Find the vertex and axis of symmetry of the parabola given by each equation. Then sketch its graph.

1. $y = x^2$

2. $y = -x^2$

3. $y = x^2 - 2$

4. $y = x^2 + 2$

5. $y = -x^2 + 3$

6. $y = -x^2 - 1$

7. $y = \frac{1}{2}x^2$

8. $y = 2x^2$

9. $y = 2x^2 - 1$

10. $y = -\frac{1}{2}x^2 + 2$

11. $y = x^2 - 2x$

12. $y = x^2 + 2x$

13. $y = -2x^2 + 4x$

14. $y = \frac{1}{2}x^2 - x$

15. $y = x^2 - x - 2$

16. $y = x^2 - 3x + 2$

17. $y = 2x^2 - x - 5$

18. $y = 2x^2 - x - 3$

19. $y = -2x^2 - 3x + 2$

20. $y = 2x^2 - 7x + 3$

21. $y = x^2 - 4x + 4$

22. $y = -x^2 + 6x - 9$

23. $y = x^2 + 4x + 5$

24. $y = -x^2 - 2x - 1$

2 Find the x-intercepts of the parabola given by each equation.

25. $y = x^2 - 4$

26. $y = x^2 - 9$

27. $y = 2x^2 - 4x$

28. $y = 3x^2 + 6x$

29. $y = x^2 - x - 2$

30. $y = x^2 - 2x - 8$

31. $y = 2x^2 - 5x - 3$

32. $y = 4x^2 + 11x + 6$

33. $y = 3x^2 - 19x - 14$

34. $y = 6x^2 + 7x + 2$

35. $y = 3x^2 - 19x + 20$

36. $y = 3x^2 + 19x + 28$

37. $y = 9x^2 - 12x + 4$

38. $y = x^2 - 2$

39. $y = 9x^2 - 2$

40. $y = 2x^2 - x - 1$

41. $y = 2x^2 - 5x - 3$

42. $y = x^2 + 2x - 1$

43. $y = x^2 + 4x - 3$

44. $y = x^2 + 6x + 10$

45. $y = -x^2 - 4x - 5$

46. $y = x^2 - 2x - 2$

47. $y = -x^2 - 2x + 1$

48. $y = -x^2 + 4x + 1$

49. $y = x^2 + 4x + 2$

50. $y = x^2 + 2x - 4$

51. $y = x^2 + 8x + 14$

52. $y = x^2 - 6x + 7$

53. $y = x^2 - 2x - 4$

54. $y = x^2 - 2x + 2$

Use the discriminant to determine the number of x-intercepts of the graph of the parabola.

55. $y = 2x^2 + x + 1$

56. $y = 2x^2 + 2x - 1$

57. $y = -x^2 - x + 3$

58. $y = -2x^2 + x + 1$

59. $y = x^2 - 8x + 16$

60. $y = x^2 - 10x + 25$

61. $y = -3x^2 - x - 2$

62. $y = -2x^2 + x - 1$

63. $y = 4x^2 - x - 2$

64. $y = 2x^2 + x + 4$

65. $y = -2x^2 - x - 5$

66. $y = -3x^2 + 4x - 5$

67. $y = x^2 + 8x + 16$

68. $y = x^2 - 12x + 36$

69. $y = x^2 + x - 3$

70. $y = x^2 + 7x - 2$

71. $y = x^2 + 5x - 4$

72. $y = x^2 - 5x + 8$

73. $y = 3x^2 - 2x + 4$

74. $y = 4x^2 - 9x + 1$

75. $y = 5x^2 + 2x + 3$

76. $y = 4x^2 - x + 1$

77. $y = 2x^2 + 3x + 4$

78. $y = -5x^2 + x - 1$

79. $y = -10x^2 + x + 1$

80. $y = -5x^2 - 7x - 3$

81. $y = -7x^2 + 3x + 7$

SUPPLEMENTAL EXERCISES 7.4

Find the value of k such that the graph of the equation contains the given point.

82. $y = x^2 - 3x + k$; $(2, 5)$

83. $y = x^2 + 2x + k$; $(-3, 1)$

84. $y = 2x^2 + 3x + k$; $(-4, 8)$

85. $y = 3x^2 - 4x + k$; $(2, -6)$

86. $y = x^2 + kx - 4$; $(-1, 3)$

87. $y = x^2 + kx + 2$; $(3, -1)$

88. $y = 2x^2 + kx - 3$; $(4, -3)$

89. $y = 3x^2 + kx - 6$; $(-2, 4)$

Solve.

90. The point (x_1, y_1) lies in Quadrant I and is a solution of the equation $y = 3x^2 - 2x - 1$. Given $y_1 = 5$, find x_1.

91. The point (x_1, y_1) lies in Quadrant II and is a solution of the equation $y = 2x^2 + 5x - 3$. Given $y_1 = 9$, find x_1.

92. The point (x_1, y_1) lies in Quadrant II and is a solution of the equation $y = 2x^2 + 11x + 4$. Given $y_1 = 10$, find x_1.

93. The point (x_1, y_1) lies in Quadrant II and is a solution of the equation $y = 3x^2 + 7x + 2$. Given $y_1 = 8$, find x_1.

94. What effect does increasing the coefficient of x^2 have on the graph of $y = ax^2 + bx + c$, $a > 0$?

95. What effect does decreasing the coefficient of x^2 have on the graph of $y = ax^2 + bx + c$, $a > 0$?

96. What effect does increasing the constant term have on the graph of $y = ax^2 + bx + c$?

97. What effect does decreasing the constant term have on the graph of $y = ax^2 + bx + c$?

An equation of the form $y = ax^2 + bx + c$ can be written in the form $y = a(x - h)^2 + k$, where (h, k) are the coordinates of the vertex of the parabola. Use the process of completing the square to rewrite the equation in the form $y = a(x - h)^2 + k$. Find the vertex. (*Hint:* Review the example at the bottom of page 346.)

98. $y = x^2 - 4x + 7$

99. $y = x^2 - 2x - 2$

100. $y = x^2 - 6x + 3$

101. $y = x^2 + 4x - 1$

102. $y = x^2 + x + 2$

103. $y = x^2 - x - 3$

Using $y = a(x - h)^2 + k$ as the equation of a parabola with the vertex at (h, k), find the equation of the parabola satisfying the given information. Write the final equation in the form $y = ax^2 + bx + c$.

104. Vertex (1, 2); Graph passes through $P(2, 5)$

105. Vertex (−2, 2); Graph passes through $P(4, 6)$

106. Vertex $(3, -1)$; Graph passes through $P(4, -3)$

107. Vertex $(0, -3)$; Graph passes through $P(3, -2)$

SECTION 7.5 Applications of Quadratic Equations

1 Application problems

The application problems in this section are similar to those problems that were solved earlier in the text. Each of the strategies for the problems in this section will result in a quadratic equation.

Solve: A small pipe takes 16 min longer to empty a tank than does a larger pipe. Working together, the pipes can empty the tank in 6 min. How long would it take each pipe working alone to empty the tank?

STRATEGY for solving an application problem

- Determine the type of problem. Is it a uniform motion problem, a geometry problem, an integer problem, or a work problem?

The problem is a work problem.

- Choose a variable to represent the unknown quantity. Write numerical or variable expressions for all the remaining quantities. These results can be recorded in a table.

The unknown time of the larger pipe: t
The unknown time of the smaller pipe: $t + 16$

	Rate of work	·	Time worked	=	Part of task completed
Larger pipe	$\frac{1}{t}$	·	6	=	$\frac{6}{t}$
Smaller pipe	$\frac{1}{t+16}$	·	6	=	$\frac{6}{t+16}$

- Determine how the quantities are related. If necessary, review the strategies presented in Chapter 2.

The sum of the parts of the task completed must equal 1.

$$\frac{6}{t} + \frac{6}{t+16} = 1$$

$$t(t+16)\left(\frac{6}{t} + \frac{6}{t+16}\right) = t(t+16) \cdot 1$$

$$(t+16)6 + 6t = t^2 + 16t$$

$$6t + 96 + 6t = t^2 + 16t$$

$$0 = t^2 + 4t - 96$$

$$0 = (t+12)(t-8)$$

$$t + 12 = 0 \qquad t - 8 = 0$$

$$t = -12 \qquad t = 8$$

The solution $t = -12$ is not possible since time cannot be a negative number.

The time for the smaller pipe is $t + 16$. Replace t by 8 and evaluate.

$$t + 16 = 8 + 16 = 24$$

The larger pipe requires 8 min to empty the tank.
The smaller pipe requires 24 min to empty the tank.

Example 1 In 8 h, two campers rowed 15 mi down a river and then rowed back to their campsite. The rate of the river's current was 1 mph. Find the rate at which the campers row in calm water.

Strategy

- This is a uniform motion problem.
- Unknown rowing rate of the campers: r

	Distance	÷	Rate	=	Time
Down river	15	÷	$r + 1$	=	$\frac{15}{r+1}$
Up river	15	÷	$r - 1$	=	$\frac{15}{r-1}$

- The total time of the trip was 8 h.

Solution

$$\frac{15}{r+1} + \frac{15}{r-1} = 8$$

$$(r+1)(r-1)\left(\frac{15}{r+1} + \frac{15}{r-1}\right) = (r+1)(r-1)8$$

$$(r-1)15 + (r+1)15 = (r^2-1)8$$

$$15r - 15 + 15r + 15 = 8r^2 - 8$$

$$30r = 8r^2 - 8$$

$$0 = 8r^2 - 30r - 8$$

$$0 = 2(4r^2 - 15r - 4)$$

$$0 = 2(4r+1)(r-4)$$

$$4r + 1 = 0 \qquad r - 4 = 0$$

$$4r = -1 \qquad r = 4$$

$$r = -\frac{1}{4}$$

▶ The solution $r = -\frac{1}{4}$ is not possible because the rate cannot be a negative number.

The rowing rate is 4 mph.

Problem 1 The length of a rectangle is 3 m more than the width. The area is 54 m^2. Find the length of the rectangle.

Solution See page A36.

EXERCISES 7.5

1 Solve.

1. The base of a triangle is one less than five times the height of the triangle. The area of the triangle is 21 cm^2. Find the height and the length of the base of the triangle.

2. The height of a triangle is 3 in. less than the length of the base of a triangle. The area of the triangle is 90 $in.^2$. Find the height and the length of the base of the triangle.

3. The length of a rectangle is two feet less than three times the width of the rectangle. The area of the rectangle is 65 ft^2. Find the length and width of the rectangle.

4. The length of a rectangle is 2 cm less than twice the width. The area of the rectangle is 180 cm^2. Find the length and width of the rectangle.

5. The sum of the squares of two consecutive odd integers is thirty-four. Find the two integers.

6. The sum of the squares of three consecutive odd integers is eighty-three. Find the three integers.

7. Five times an integer plus the square of the integer is twenty-four. Find the integer.

8. The sum of five times an integer and twice the square of the integer is three. Find the integer.

9. The height of a projectile fired upward is given by the formula $s = v_0 t - 16t^2$, where s is the height, v_0 is the initial velocity, and t is the time. Find the time for a projectile to return to Earth if it has an initial velocity of 200 ft/s.

10. The height of a projectile fired upward is given by the formula $s = v_0 t - 16t^2$, where s is the height, v_0 is the initial velocity, and t is the time. Find the time for a projectile to reach a height of 64 ft if it has an initial velocity of 128 ft/s. Round to the nearest hundredth of a second.

11. A chemistry experiment requires that a vacuum be created in a chamber. A small vacuum pump requires 15 s longer than does a second, larger pump to evacuate the chamber. Working together, the pumps can evacuate the chamber in 4 s. Find the time required for the larger vacuum pump working alone to evacuate the chamber.

12. An old computer requires 3 h longer to print the payroll than does a new computer. With both computers running, the payroll can be completed in 2 h. Find the time required for the new computer running alone to complete the payroll.

13. A small air conditioner requires 16 min longer to cool a room 3° than does a larger air conditioner. Working together, the two air conditioners can cool the room 3° in 6 min. How long would it take each air conditioner working alone to cool the room 3°?

14. A small pipe can fill a tank in 6 min more time than it takes a larger pipe to fill the same tank. Working together, both pipes can fill the tank in 4 min. How long would it take each pipe working alone to fill the tank?

15. An old mechanical sorter takes 21 min longer to sort a batch of mail than does a second, newer model. With both sorters working, a batch of mail can be sorted in 10 min. How long would it take each sorter working alone to sort the batch of mail?

16. A small heating unit takes 8 h longer to melt a piece of iron than does a larger unit. Working together, the heating units can melt the iron in 3 h. How long would it take each heating unit working alone to melt the iron?

17. A cruise ship made a trip of 100 mi in 8 h. The ship traveled the first 40 mi at a constant rate before increasing its speed by 5 mph. Another 60 mi was traveled at the increased speed. Find the rate of the cruise ship for the first 40 mi.

18. A cyclist traveled 60 mi at a constant rate before reducing the speed by 2 mph. Another 40 mi was traveled at the reduced speed. The total time for the 100-mile trip was 9 h. Find the rate during the first 60 mi.

19. The rate of a single-engined plane in calm air is 100 mph. Flying with the wind, the plane can fly 240 mi in one hour less time than is required to make the return trip of 240 mi. Find the rate of the wind.

20. A car travels 120 mi. A second car, traveling 10 mph faster than the first car, makes the same trip in 1 h less time. Find the speed of each car.

21. The rate of a river's current is 2 mph. A rowing crew can row 16 mi down this river and back in 6 h. Find the rowing rate of the crew in calm water.

22. A boat traveled 30 mi down a river and then returned. The total time for the round trip was 4 h, and the rate of the river's current was 4 mph. Find the rate of the boat in still water.

SUPPLEMENTAL EXERCISES 7.5

Solve.

23. The sum of a number and twice its reciprocal is $\frac{33}{4}$. Find the number.

24. The difference between a number and twice its reciprocal is $\frac{49}{5}$. Find the number.

25. The numerator of a fraction is 2 less than the denominator. The sum of the fraction and three times its reciprocal is $\frac{13}{2}$. Find the fraction.

26. The numerator of a fraction is 3 less than the denominator. The sum of the fraction and four times its reciprocal is $\frac{17}{2}$. Find the fraction.

27. Find two consecutive integers whose cubes differ by 127.

28. Find two consecutive even integers whose cubes differ by 488.

29. An open box is formed from a rectangular piece of cardboard whose length is 8 cm more than its width by cutting squares 2 cm in length from each corner and then folding up the sides. Find the dimensions of the box if its volume is 256 cm^3.

30. An open box is formed from a rectangular piece of cardboard whose length is 6 cm more than its width by cutting squares 4 cm in length from each corner and then folding up the sides. Find the dimensions of the box if its volume is 288 cm^3.

31. The volumes of two spheres differ by 684π cm^3. The radius of the larger sphere is 3 cm more than the radius of the smaller sphere. Find the radius of the larger sphere. (*Hint:* The formula for the volume of a sphere is $V = \frac{4}{3}\pi r^3$.)

32. The volumes of two spheres differ by 372π cm^3. The radius of the larger sphere is 3 cm more than the radius of the smaller sphere. Find the radius of the larger sphere. (*Hint:* The formula for the volume of a sphere is $V = \frac{4}{3}\pi r^3$.)

SECTION 7.6 Non-linear Inequalities

1 Solve inequalities by factoring

A **quadratic inequality in one variable** is one that can be written in the form $ax^2 + bx + c < 0$ or $ax^2 + bx + c > 0$, where $a \neq 0$. The symbols $\leq$ and $\geq$ can also be used.

Quadratic inequalities can be solved by algebraic means. However, it is often easier to use a graphical method to solve these inequalities. The graphical method is used here.

Solve and graph the solution set of $x^2 - x - 6 < 0$.

Factor the trinomial.

$$x^2 - x - 6 < 0$$
$$(x - 3)(x + 2) < 0$$

On a number line, draw vertical lines at the numbers that make each factor equal to zero.

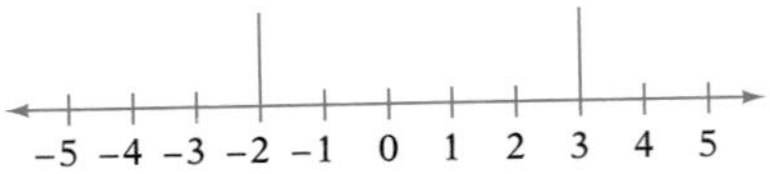

$x - 3 = 0$ $\quad$ $x + 2 = 0$

$x = 3$ $\quad$ $x = -2$

For each factor, place plus signs above the number line for those regions where the factor is positive and negative signs where the factor is negative. $x - 3$ is positive for $x > 3$ and $x + 2$ is positive for $x > -2$.

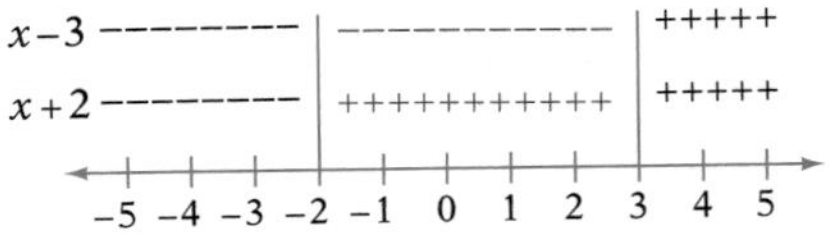

Since $x^2 - x - 6 < 0$, the solution set will be the regions where one factor is positive and the other factor is negative.

Write the solution set. $\{x \mid -2 < x < 3\}$

The graph of the solution set of $x^2 - x - 6 < 0$:

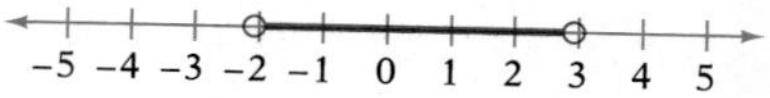

This method of solving quadratic inequalities can be used on any polynomial that can be factored into linear factors.

Solve and graph the solution set of $x^3 - 4x^2 - 4x + 16 > 0$.

Factor the polynomial by grouping.

$$x^3 - 4x^2 - 4x + 16 > 0$$
$$x^2(x - 4) - 4(x - 4) > 0$$
$$(x^2 - 4)(x - 4) > 0$$
$$(x - 2)(x + 2)(x - 4) > 0$$

On a number line, identify for each factor the regions where the factor is positive and where the factor is negative.

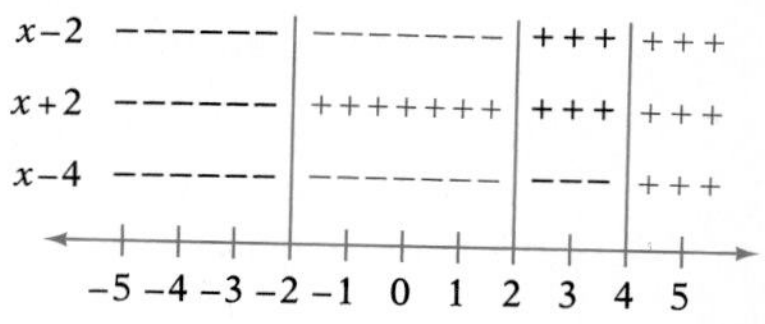

There are two regions where the product of the three factors is positive.

Write the solution set. $\{x \mid -2 < x < 2 \text{ or } x > 4\}$

The graph of the solution set of $x^3 - 4x^2 - 4x + 16 > 0$:

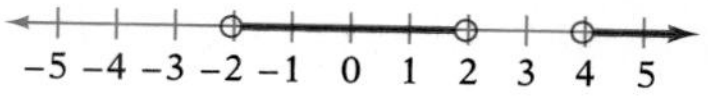

Example 1 Solve and graph the solution set of $2x^2 - x - 3 \geq 0$.

Solution

$$2x^2 - x - 3 \geq 0$$
$$(2x - 3)(x + 1) \geq 0$$

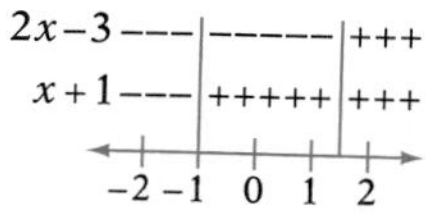

$\left\{x \mid x \leq -1 \text{ or } x \geq \frac{3}{2}\right\}$

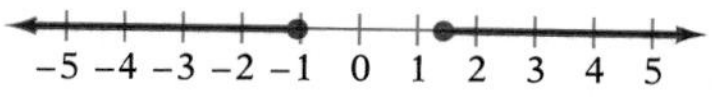

Problem 1 Solve and graph the solution set of $2x^2 - x - 10 \leq 0$.

Solution See page A37.

2 Solve rational inequalities

The graphical method used in the last objective can be used to solve rational inequalities.

For example, to solve the rational inequality $\frac{2x-5}{x-4} \le 1$, rewrite the inequality so that 0 appears on the right side of the inequality.

$$\frac{2x-5}{x-4} \le 1$$

Then simplify.

$$\frac{2x-5}{x-4} - 1 \le 0$$

$$\frac{2x-5}{x-4} - \frac{x-4}{x-4} \le 0$$

$$\frac{x-1}{x-4} \le 0$$

On a number line, identify for each factor of the numerator and each factor of the denominator the regions where the factor is positive and where the factor is negative.

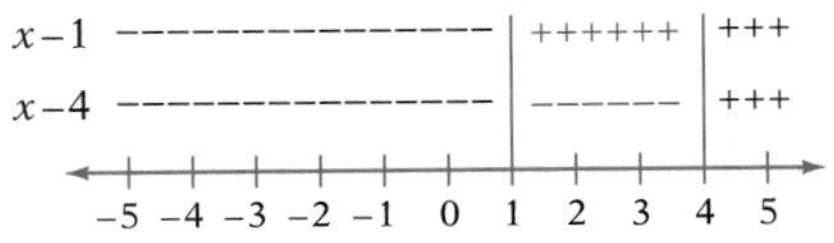

The region where the quotient of the two factors is negative is between 1 and 4.

Write the solution set. $\{x|1 \le x < 4\}$

Note that 1 is part of the solution set, but 4 is not part of the solution set since the denominator of the rational expression is zero when $x = 4$.

Example 2 Solve and graph the solution set of $\frac{x+4}{x-3} \ge 0$.

Solution $\frac{x+4}{x-3} \ge 0$

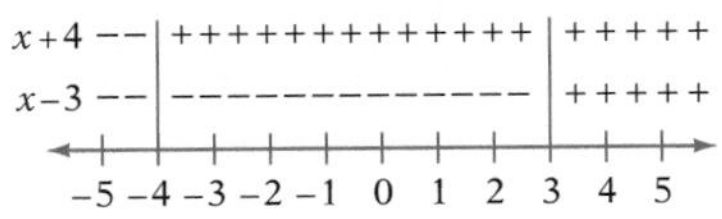

$\{x|x > 3 \text{ or } x \le -4\}$

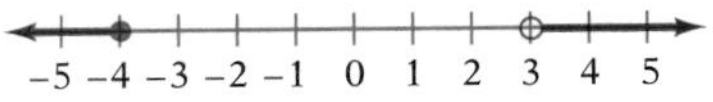

Problem 2 Solve and graph the solution set of $\frac{x}{x-2} \le 0$.

Solution See page A37.

EXERCISES 7.6

1 Solve and graph the solution set.

1. $(x - 4)(x + 2) > 0$

2. $(x + 1)(x - 3) > 0$

3. $x^2 - 3x + 2 \geq 0$

4. $x^2 + 5x + 6 > 0$

5. $x^2 - x - 12 < 0$

6. $x^2 + x - 20 < 0$

7. $(x - 1)(x + 2)(x - 3) < 0$

8. $(x + 4)(x - 2)(x + 1) > 0$

9. $(x + 4)(x - 2)(x - 1) \geq 0$

10. $(x - 1)(x + 5)(x - 2) \leq 0$

Solve.

11. $x^2 - 16 > 0$

12. $x^2 - 4 \geq 0$

13. $x^2 - 4x + 4 > 0$

14. $x^2 + 6 + 9 > 0$

15. $x^2 - 9x \leq 36$

16. $x^2 + 4x > 21$

17. $2x^2 - 5x + 2 \geq 0$

18. $4x^2 - 9x + 2 < 0$

19. $4x^2 - 8x + 3 < 0$

20. $2x^2 + 11x + 12 \geq 0$

21. $(x - 6)(x + 3)(x - 2) \leq 0$

22. $(x + 5)(x - 2)(x - 3) > 0$

23. $(2x - 1)(x - 4)(2x + 3) > 0$

24. $(x - 2)(3x - 1)(x + 2) \leq 0$

25. $(x - 5)(2x - 7)(x + 1) < 0$

26. $(x + 4)(2x + 7)(x - 2) > 0$

27. $x^3 + 3x^2 - x - 3 \leq 0$

28. $x^3 + x^2 - 9x - 9 < 0$

29. $x^3 - x^2 - 4x + 4 \geq 0$

30. $2x^3 + 3x^2 - 8x - 12 \geq 0$

2 Solve and graph the solution set.

31. $\frac{x - 4}{x + 2} > 0$

32. $\frac{x + 2}{x - 3} > 0$

33. $\frac{x - 3}{x + 1} \leq 0$

34. $\frac{x - 1}{x} > 0$

35. $\frac{(x - 1)(x + 2)}{x - 3} \leq 0$

36. $\frac{(x + 3)(x - 1)}{x - 2} \geq 0$

Solve.

37. $\frac{3x}{x - 2} > 1$

38. $\frac{2x}{x + 1} < 1$

39. $\frac{2}{x + 1} \geq 2$

40. $\frac{3}{x - 1} < 2$

41. $\frac{x}{(x - 1)(x + 2)} \geq 0$

42. $\frac{x - 2}{(x + 1)(x - 1)} \leq 0$

43. $\frac{1}{x} < 2$

44. $\frac{x}{2x - 1} \geq 1$

SUPPLEMENTAL EXERCISES 7.6

Graph the solution set.

45. $(x + 2)(x - 3)(x + 1)(x + 4) > 0$

46. $(x - 1)(x + 3)(x - 2)(x - 4) \geq 0$

47. $(x^2 + 2x - 8)(x^2 - 2x - 3) < 0$

48. $(x^2 + 2x - 3)(x^2 + 3x + 2) \geq 0$

49. $(x^2 + 1)(x^2 - 3x + 2) > 0$

50. $(x^2 - 9)(x^2 + 5x + 6) \leq 0$

51. $\dfrac{x^2(3 - x)(2x + 1)}{(x + 4)(x + 2)} \geq 0$

52. $x < x^2$

53. $x^3 > x$

54. $\dfrac{1}{x} + x > 2$

55. $3x - \dfrac{1}{x} \leq 2$

56. $x^2 - x < \dfrac{1 - x}{x}$

Calculators and Computers

Solving Quadratic Equations

The solutions of all quadratic equations can be found by using the quadratic formula. Although this formula is available, sometimes the coefficients of the equation make the use of the formula difficult because the computations are very tedious. Consider trying to solve the equation

$$2.984x^2 + 9834.1x - 509.0023 = 0$$

by using the quadratic formula. The computations would take a long time even with a calculator.

The program QUADRATIC EQUATIONS on the Math Ace Disk will solve any quadratic equation with real number coefficients, including those whose solutions are complex numbers. You can use the program to test your ability to solve equations by comparing your answer to that of the program.

Something Extra

Trajectories

When an object is projected upward with a velocity v_0, the distance s above the ground after time t is given by $s = -16t^2 + v_0t + h$, where v_0 is the initial velocity and h is the starting height.

At the end of the thrust (or burn out), a rocket is 3000 ft high with a velocity of 640 ft/s.

a. Find the maximum height.
b. Find the time necessary before reaching the maximum height.
c. Find the time from burn out until the rocket hits the ground.

Substitute 640 for v_0 and 3000 for h in the equation.

$$s = -16t^2 + v_0t + h$$
$$s = -16t^2 + 640t + 3000$$

This is the equation of a parabola. The t-coordinate of the vertex is the time to reach the maximum height, and the s-coordinate is the maximum height.

$$t = -\frac{b}{2a} = -\frac{640}{2(-16)} = 20$$
$$\begin{aligned} s &= -16(20)^2 + 640(20) + 3000 \\ &= -6400 + 12{,}800 + 3000 \\ &= 9400 \end{aligned}$$

The time necessary for reaching the maximum height is 20 s and the maximum height is 9400 ft.

When the rocket hits the ground, the distance above the ground equals 0. Solve for t.

$$0 = -16t^2 + 640t + 3000$$
$$t = \frac{-640 \pm \sqrt{640^2 - 4(-16)3000}}{2(-16)}$$
$$\approx -4.24 \text{ or } 44.24$$

Since time must be positive, the answer -4.24 is not possible. The rocket will hit the ground after 44.24 s.

When an object is projected in a direction 60 degrees from the horizontal, the distance above the ground (s) is given by $s = \frac{-64x^2}{v_0^2} + 2x + h$, where x is the horizontal distance that the object travels.

Solve.

1. A ball is thrown upward with a velocity of 45 ft/s from a building 80 ft high. Find the maximum height of the ball, the time for the ball to reach the maximum height, and the time for the ball to reach the ground. Graph the height as a function of time.

2. A cannon is fired at an angle of 60 degrees with the horizontal. The muzzle velocity of the projectile is 320 ft/s. Find the maximum height of the projectile. Graph the height as a function of the distance.

Chapter Summary

Key Words

A *quadratic equation* is an equation of the form $ax^2 + bx + c = 0$, where a, b, and c are constants and $a \neq 0$. A quadratic equation is also called a *second-degree equation.* A quadratic equation is in *standard form* when the polynomial is in descending order and equal to zero.

Adding to a binomial the constant term that makes the binomial a perfect square trinomial is called *completing the square.*

A *quadratic inequality in one variable* is one that can be written in the form $ax^2 + bx + c < 0$ or $ax^2 + bx + c > 0$, where $a \neq 0$. The symbols $\leq$ and $\geq$ can also be used.

An equation is quadratic in form if it can be written as $au^2 + bu + c = 0$.

Essential Rules

The Principle of Zero Products If $ab = 0$, then $a = 0$ or $b = 0$.

The Quadratic Formula $x = \dfrac{-b \pm \sqrt{b^2 - 4ac}}{2a}$

Equation of a Parabola $y = ax^2 + bx + c$
When $a > 0$, the parabola opens up.
When $a < 0$, the parabola opens down.

The x-coordinate of the vertex is $-\dfrac{b}{2a}$.

The axis of symmetry is the line $x = -\dfrac{b}{2a}$.

Chapter Review

1. Solve: $2x^2 - 3x = 0$

2. Solve for x: $6x^2 + 9xc = 6c^2$

3. Solve: $x^2 = 48$

4. Solve: $\left(x + \frac{1}{2}\right)^2 + 4 = 0$

5. Solve: $6x^2 - 5x - 6 = 0$

6. Solve: $3(x - 2)^2 - 24 = 0$

7. Solve: $3x^2 - 6x = 2$

8. Solve: $x^2 + 4x + 12 = 0$

9. Write a quadratic equation that has integer coefficients and has solutions $\frac{1}{3}$ and -3.

10. Write a quadratic equation that has integer coefficients and has solutions $\frac{1}{2}$ and -4.

11. Solve: $2x^2 + 9x = 5$

12. Solve: $2(x + 1)^2 - 36 = 0$

13. Solve: $x^2 + 6x + 10 = 0$

14. Solve: $\frac{2}{x - 4} + 3 = \frac{x}{2x - 3}$

15. Solve: $x^4 - 6x^2 + 8 = 0$

16. Solve: $x^2 - 4x - 6 = 0$

17. Solve: $\sqrt{2x - 1} + \sqrt{2x} = 3$

18. Solve: $2x^{\frac{2}{3}} + 3x^{\frac{1}{3}} - 2 = 0$

19. Solve: $\sqrt{3x - 2} + 4 = 3x$

20. Solve: $x^4 - 4x^2 + 3 = 0$

21. Solve: $3x^2 + 10x = 8$

22. Solve: $x^2 - 6x - 2 = 0$

23. Solve: $\frac{2x}{x - 4} + \frac{6}{x + 1} = 11$

24. Solve: $2x^2 - 2x = 1$

25. Solve: $x^{\frac{2}{3}} + x^{\frac{1}{3}} - 12 = 0$

26. Solve: $2x = 4 - 3\sqrt{x - 1}$

27. Solve: $x = \sqrt{x} + 2$

28. Solve: $2x = \sqrt{5x + 24} + 3$

29. Solve: $3x = \dfrac{9}{x - 2}$

30. Solve: $\dfrac{3x + 7}{x + 2} + x = 3$

31. Solve: $\dfrac{x - 2}{2x + 3} - \dfrac{x - 4}{x} = 2$

32. Solve: $1 - \dfrac{x + 4}{2 - x} = \dfrac{x - 3}{x + 2}$

33. Find the axis of symmetry of the parabola whose equation is $y = -x^2 + 6x - 5$.

34. Find the vertex of the parabola whose equation is $y = -x^2 + 3x - 2$.

35. Use the discriminant to determine the number of x-intercepts of the parabola whose equation is $y = -2x^2 + 2x - 3$.

36. Use the discriminant to determine the number of x-intercepts of the parabola whose equation is $y = 3x^2 - 2x - 4$.

37. Find the x-intercepts of the parabola whose equation is $y = 4x^2 + 12x + 4$.

38. Find the x-intercepts of the parabola whose equation is $y = -2x^2 - 3x + 2$.

39. Use the discriminant to determine the number of x-intercepts of the parabola whose equation is $y = -3x^2 + 4x + 6$.

40. Use the discriminant to determine the number of x-intercepts of the parabola whose equation is $y = 2x^2 - 2x + 5$.

41. Find the x-intercepts of the parabola whose equation is $y = 3x^2 + 9x$.

42. Find the x-intercepts of the parabola whose equation is $y = 2x^2 + 7x - 12$.

43. Solve: $(x + 3)(2x - 5) < 0$

44. Solve: $(x - 2)(x + 4)(2x + 3) \le 0$

45. Solve and graph the solution set of $\dfrac{x - 2}{2x - 3} \ge 0$.

46. Solve and graph the solution set of $\dfrac{(2x - 1)(x + 3)}{x - 4} \le 0$.

47. Graph: $y = -x^2 - 2x + 3$

48. Graph: $y = x^2 - 5x + 5$

49. The length of a rectangle is 2 cm more than twice the width. The area of the rectangle is 60 cm^2. Find the length and width of the rectangle.

50. The sum of the squares of three consecutive even integers is fifty-six. Find the three integers.

51. An older computer requires 12 min longer to print the payroll than does a newer computer. Together the computers can print the payroll in 8 min. Find the time for the new computer working alone to complete the payroll.

52. A car travels 200 mi. A second car, traveling 10 mph faster than the first car, makes the same trip in 1 h less time. Find the speed of each car.

Chapter Test

1. Solve: $2x^2 + x = 6$

2. Solve: $12x^2 + 7x - 12 = 0$

3. Write a quadratic equation that has integer coefficients and has solutions 3 and -3.

4. Write a quadratic equation that has integer coefficients and has solutions $-\frac{1}{3}$ and 3.

5. Solve: $2(x + 3)^2 - 36 = 0$

6. Solve: $x^2 + 4x - 1 = 0$

7. Solve: $x^2 + 2x + 8 = 0$

8. Solve: $3x^2 - x + 8 = 0$

9. Solve: $\frac{2x}{x - 1} + \frac{3}{x + 2} = 1$

10. Solve: $2x + 7x^{\frac{1}{2}} - 4 = 0$

11. Solve: $x^4 - 11x^2 + 18 = 0$

12. Solve: $\sqrt{2x + 1} + 5 = 2x$

13. Solve: $\sqrt{x - 2} = \sqrt{x} - 2$

14. Solve: $\frac{2x}{x - 3} + \frac{5}{x - 1} = 1$

15. Find the x-intercepts of the parabola whose equation is $y = 2x^2 + 5x - 12$.

16. Find the axis of symmetry of the parabola whose equation is $y = 2x^2 + 6x + 3$.

17. Graph $y = \frac{1}{2}x^2 + x - 4$.

18. Solve and graph the solution set of $\frac{2x - 3}{x + 4} \le 0$.

19. The base of a triangle is 3 ft more than three times the height. The area of the triangle is 30 ft^2. Find the base and height of the triangle.

20. The rate of a river's current is 2 mph. A canoe was rowed 6 mi down the river and back in 4 h. Find the rowing rate in calm water.

Cumulative Review

1. Evaluate $2a^2 - b^2 \div c^2$ when $a = 3$, $b = -4$, and $c = -2$.

2. Solve: $\frac{2x - 3}{4} - \frac{x + 4}{6} = \frac{3x - 2}{8}$

3. Find the slope of the line containing the points $(3, -4)$ and $(-1, 2)$.

4. Find the equation of the line containing the point $(1, 2)$ and parallel to the line $x - y = 1$.

5. Factor: $-3x^3y + 6x^2y^2 - 9xy^3$

6. Factor: $6x^2 - 7x - 20$

7. Factor: $a^nx + a^ny - 2x - 2y$

8. Divide: $(3x^3 - 13x^2 + 10) \div (3x - 4)$

9. Simplify: $\frac{x^2 + 2x + 1}{8x^2 + 8x} \cdot \frac{4x^3 - 4x^2}{x^2 - 1}$

10. Solve: $\frac{x}{2x + 3} - \frac{3}{4x^2 - 9} = \frac{x}{2x - 3}$

11. Solve $S = \frac{n}{2}(a + b)$ for b.

12. Simplify: $-2i(7 - 4i)$

13. Simplify: $a^{-\frac{1}{2}}(a^{\frac{1}{2}} - a^{\frac{3}{2}})$

14. Simplify: $\dfrac{\sqrt[3]{8x^4y^5}}{\sqrt[3]{16xy^6}}$

15. Solve: $3x^2 + 7x = 6$

16. Solve: $x^2 + 6x + 10 = 0$

17. Solve: $x^4 - 6x^2 + 8 = 0$

18. Solve: $\sqrt{3x + 1} - 1 = x$

19. Solve: $\dfrac{x}{x+2} - \dfrac{4x}{x+3} = 1$

20. Find the x- and y-intercepts of the graph of $6x - 5y = 15$.

21. Graph $y = -(x - 2)^2$.

22. Solve and graph the solution set of $\dfrac{(x-1)(x-5)}{x+3} \geq 0$.

23. A piston rod for an automobile is $9\frac{3}{8}$ in. with a tolerance of $\frac{1}{64}$ in. Find the lower and upper limits of the length of the piston rod.

24. The base of a triangle is $(x + 8)$ ft. The height is $(2x - 4)$ ft. Find the area of the triangle in terms of the variable x.

25. How many quarts of pure antifreeze should be mixed with 6 qt of a 25% solution of antifreeze to make a 60% antifreeze solution?

26. The rate of a plane in calm air is 450 mph. Flying with the wind, the plane can fly 2000 mi in one hour less time than is required to make the return trip of 2000 mi. Find the rate of the wind.

8

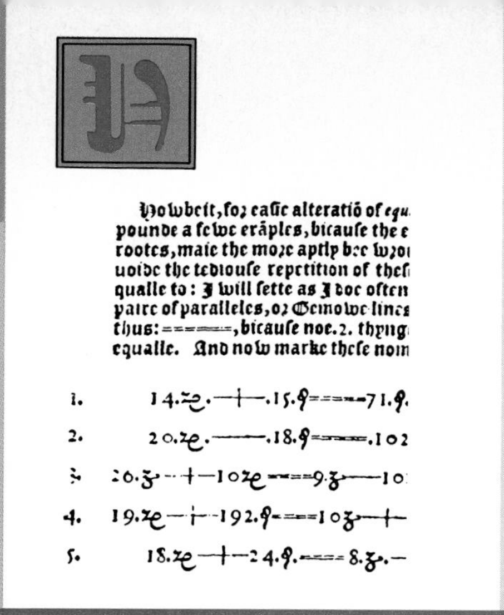
Howbeit, for easie alteratiō of equ
pounde a fewe exāples, bicause the e
rootes, maie the more aptly bee wro
uoide the tediouse repetition of thes
qualle to : I will sette as I doe often
paire of paralleles, or Gemowe lines
thus: ═══════, bicause noe.2. thyng
equalle. And now marke these nom

1. 14.ze.—+—.15.ϑ═══71.ϑ.
2. 20.ze.——.18.ϑ═══.102
3. 26.ȝ.—+—10ze═══9.ȝ.——10
4. 19.ze—+—192.ϑ═══10ȝ.—+—
5. 18.ze—+—24.ϑ.═══8.ȝ.—

Functions and Relations

Objectives

- Evaluate functions
- Find the domain and range of a function
- Graph functions
- Determine whether or not a relation is a function
- Determine from a graph whether a function is one-to-one
- Composite functions
- Inverse functions
- Variation problems
- Minimum and maximum problems

POINT OF INTEREST

History of Equal Signs

A portion of a page of the first book that used an equal sign, =, is shown below. This book was written in 1557 by Robert Recorde and was titled *The Whetstone of Witte.*

Notice in the illustration the words "bicause noe 2 thyngs can be more equalle." Recorde decided that two things could not be more equal than two parallel lines of the same length. Therefore, it made sense to use this symbol to show equality.

This page also illustrates the use of the plus sign, +, and the minus sign, −. These symbols had been widely used only for about 100 years when this book was written.

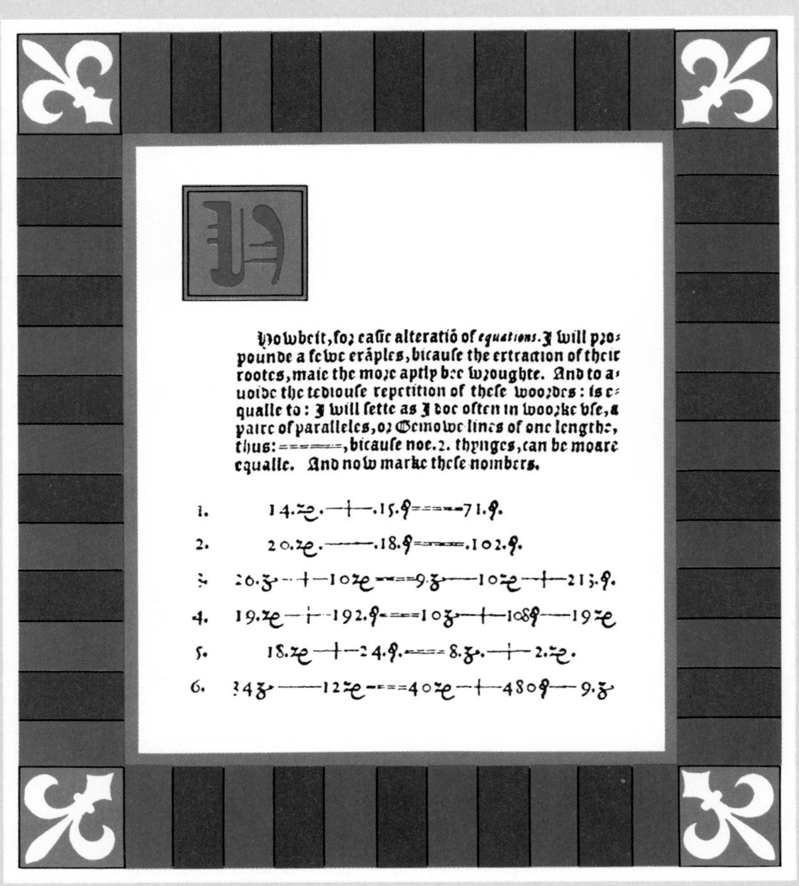

Howbeit, for easie alteratiō of *equations*. I will propounde a fewe exāples, bicause the extraction of their rootes, maie the more aptly bee wroughte. And to auoide the tediouse repetition of these woordes: is equalle to: I will sette as I doe often in woorke use, a paire of paralleles, or Gemowe lines of one lengthe, thus: ======, bicause noe .2. thynges, can be moare equalle. And now marke these nombers.

1. 14.ze. —+— .15.ϑ ====== 71.ϑ.
2. 20.ze. —— .18.ϑ ====== .102.ϑ.
3. 26.ʒ —+— 10ze ====== 9.ʒ —— 10ze —+— 213.ϑ.
4. 19.ze —+— 192.ϑ ====== 10ʒ —+— 108ϑ —— 19ze
5. 18.ze —+— 24.ϑ. ====== 8.ʒ. —+— 2.ze.
6. 34ʒ —— 12ze ====== 40ze —+— 480ϑ —— 9.ʒ

SECTION 8.1
Functions and Relations

1 Evaluate functions

There are many situations in science, business, and mathematics where a correspondence exists between two quantities. The correspondence can be described by a formula, in a table, or in a graph, and then recorded as ordered pairs.

The formula $d = 16t^2$ describes a correspondence between the distance (d) a rock will fall and the time (t) of its fall. For each value of t, the formula assigns only one value for the distance. The ordered pair (2, 64) indicates that in 2 seconds a rock will fall 64 feet. Some of the other ordered pairs determined by the correspondence are shown at the right.

Time in seconds ↓   Distance in feet ↓

(1, 16)
(3, 144)
(4, 256)
(5, 400)

The table at the right describes a grading scale that defines a correspondence between a percent score and a letter grade. For any percent score, the table assigns only one letter grade. The ordered pair (86, B) indicates that a score of 86% receives a letter grade of B.

Score	Grade
90–100	A
80–89	B
70–79	C
60–69	D
0–59	F

The graph at the right defines a correspondence between the depth of a scuba diver and the pressure on the diver. For each depth, the graph assigns only one pressure. The ordered pair (10, 20) indicates that when at a depth of 10 feet, the pressure is 20 pounds per square inch.

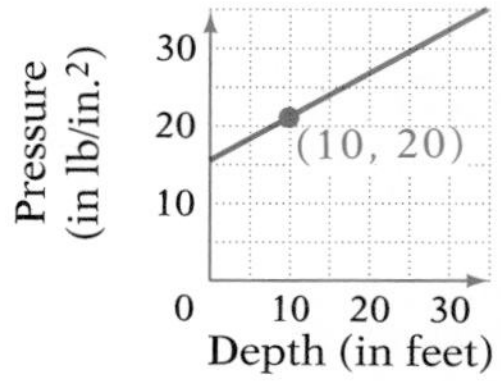

In each of these examples, a correspondence or a rule determines a set of ordered pairs. The set of ordered pairs determined by the correspondence is called a function.

Definition of a Function

A **function** is a set of ordered pairs in which no two ordered pairs that have the same first component have different second components.

For example, the ordered pair (2, 64) was determined by the formula $d = 16t^2$. The first component, 2, can only have a second component of 64; no other second component can be paired with this first component. There is exactly one value of d that can be paired with a value of t.

The ordered pairs of a function can have different first components paired with the same second component. For example, the ordered pairs determined by the grading scale shown on the previous page include (80, B), (83, B), and (87, B). The fact that 80, or any number between 80 and 89, cannot be paired with any letter other than B is the condition that makes this correspondence a function.

Not every correspondence between two sets is a function. Consider the correspondence that assigns to a positive real number a square root of that number. This is not a function because each positive real number can be paired with the positive or the negative square root of that number. For example, 9 can be paired with 3 or with -3. Thus the set of ordered pairs would contain (9, 3) and (9, -3). But the definition of function requires that no two ordered pairs with the same first component have *different* second components.

A **relation** assigns to each element of a first set one or more elements of a second set.

The correspondence that pairs a positive real number with a square root of that number is a relation. In general, a relation is *any* set of ordered pairs. A function is a special type of relation.

Although a function can always be described in terms of ordered pairs, frequently functions are described by an equation. The letter f is commonly used to represent a function, but any letter can be used.

The "square" function assigns to each real number its square. The square function is described by the equation

$$f(x) = x^2.$$ Read $f(x)$ as "f of x" or "the value of f at x."

$f(x)$ is the symbol for the number that is paired with x. In terms of ordered pairs, this is written $(x, f(x))$.

It is important to remember that $f(x)$ does not mean f times x. The letter f stands for the function, and $f(x)$ is the number that is paired with x.

To **evaluate a function** means to find the number that is paired with a given number. To evaluate $f(x) = x^2$ at 4 means to find the number that is paired with 4.

The notation $f(4)$ is used to indicate the number that is paired with 4. To evaluate $f(x) = x^2$ at 4, replace x by 4 and simplify.

$$f(x) = x^2$$
$$f(4) = 4^2 = 16$$

The **value** of the function at 4 is 16.

Example 1 Evaluate $f(x) = 2x + 1$ at $x = 3$.

Solution
$f(x) = 2x + 1$ ▶ $f(3)$ is the number that is paired with 3. Replace x by 3 and simplify.
$f(3) = 2(3) + 1$
$f(3) = 7$

Problem 1 Evaluate $s(t) = 2t^2 + 3t - 4$ at $t = -3$.

Solution See page A37.

Example 2 Evaluate $f(x) = x^2$ at $x = a + h$.

Solution
$f(x) = x^2$
$f(a + h) = (a + h)^2$ ▶ Replace x by its value.
$f(a + h) = a^2 + 2ah + h^2$ ▶ Simplify.

Problem 2 Evaluate $f(x) = x^2$ at $x = a + 1$.

Solution See page A37.

The basic operations of addition and subtraction can be performed on functions.

Example 3 For $f(x) = x^2$ and $g(x) = 3x$, evaluate $f(x) + g(x)$ at $x = 5$.

Solution
$f(x) + g(x) = x^2 + 3x$
$f(5) + g(5) = 5^2 + 3(5)$ ▶ Replace x by its value.
$f(5) + g(5) = 25 + 15$ ▶ Simplify.
$f(5) + g(5) = 40$

Problem 3 For $f(x) = 2x^2$ and $g(x) = 4x - 1$, evaluate $f(x) - g(x)$ at $x = -1$.

Solution See page A37.

2 Find the domain and range of a function

The basic concept of a function is a set of ordered pairs. When a function is given by a set of ordered pairs, the **domain** of the function is the set of the first components of each ordered pair. The **range** of the function is the set of the second components of each ordered pair.

For the function defined by the ordered pairs

$$\{(1, 2), (4, 3), (7, 9), (8, 11)\},$$

the domain is {1, 4, 7, 8} and the range is {2, 3, 9, 11}.

When the ordered pairs are formed by using a correspondence or rule that pairs a member of a first set with only one member of a second set, the first set is the domain of the function. The second set is the range of the function.

In the grading scale shown at the right, the correspondence pairs a percent score with one of the letters A, B, C, D, or F.

Score	Grade
90–100	A
80–89	B
70–79	C
60–69	D
0–59	F

The domain of the grading scale is the percent scores 0 to 100.

The range of the grading scale is the set of letters A, B, C, D, and F.

Example 4 Find the domain and range of the function {(1, 0), (2, 3), (3, 8), (4, 15)}.

Solution The domain is {1, 2, 3, 4}. ▶ The domain of the function is the set of the first components in the ordered pairs.

The range is {0, 3, 8, 15}. ▶ The range of the function is the set of the second components in the ordered pairs.

Problem 4 Find the domain and range of the function {(0, 1), (1, 3), (2, 5), (3, 7), (4, 9)}.

Solution See page A37.

When a function is described by an equation and the domain is specified, the range of a function can be found by evaluating the function at each point of the domain.

Example 5 Find the range of the function given by the equation $f(x) = 2x - 3$ if the domain is $\{0, 1, 2, 3\}$.

Solution

$$f(x) = 2x - 3$$
$$f(0) = 2(0) - 3 = -3$$
$$f(1) = 2(1) - 3 = -1$$
$$f(2) = 2(2) - 3 = 1$$
$$f(3) = 2(3) - 3 = 3$$

▶ Replace x by each member of the domain. The range includes the values of $f(0)$, $f(1)$, $f(2)$, and $f(3)$.

The range is $\{-3, -1, 1, 3\}$.

Problem 5 Find the range of the function given by the equation $f(x) = x^2 - 2x + 1$ if the domain is $\{-2, -1, 0, 1, 2\}$.

Solution See page A37.

Given an element a in the range of a function, it is possible to find an element in the domain that corresponds to a.

The number 3 is in the range of the function given by the equation $f(x) = x^2 - 1$. Find an element in the domain that corresponds to 3 and write an ordered pair that belongs to the function.

Because 3 is in the range, $f(x) = 3$. $\quad f(x) = 3$

Replace $f(x)$ by $x^2 - 1$. $\quad x^2 - 1 = 3$

Solve for x. $\quad x^2 = 4$

$$x = \pm 2$$

In this case, there are two values in the domain that can be paired with the range element 3. The two values are 2 and -2. Two ordered pairs that belong to the function are $(2, 3)$ and $(-2, 3)$. *Remember:* A function can have different first elements paired with the same second element. A function cannot have the same first element paired with different second elements.

Example 6 The number -2 is in the range of the function given by the equation $f(x) = 3x + 1$. Find an element in the domain that corresponds to -2, and write an ordered pair that belongs to the function.

Solution

$f(x) = -2$ ▶ Because -2 is in the range, $f(x) = -2$.

$3x + 1 = -2$ ▶ Replace $f(x)$ by $3x + 1$.

$3x = -3$ ▶ Solve for x.

$x = -1$ ▶ The element in the domain that corresponds to -2 is -1.

$(-1, -2)$ ▶ The ordered pair $(-1, -2)$ belongs to the function.

Problem 6 The number -4 is in the range of the function given by the equation $f(x) = 2x - 1$. Find an element in the domain that corresponds to -4, and write an ordered pair that belongs to the function.

Solution See page A38.

When an equation is used to define a function and the domain is not stated, the domain is the set of real numbers for which the equation produces real numbers. For example:

For all real numbers x, $f(x) = x^2 + 5$ is a real number. The domain of f is the set of real numbers.

The domain of the function given by the equation $g(x) = \frac{1}{x-4}$ is all real numbers except 4. When $x = 4$, $g(4)$ is undefined.

The domain of the function given by the equation $h(x) = \sqrt{x+3}$ is all real numbers greater than or equal to -3. When x is less than -3, $h(x)$ is not a real number. For example, $h(-5) = \sqrt{-5+3} = \sqrt{-2}$, which is not a real number.

Domain of a Function

Unless otherwise stated, the domain of a function is all real numbers except:
(a) those numbers for which the denominator of the function is zero; or
(b) those numbers for which the value of the function is not a real number.

Example 7 What numbers are excluded from the domain of the function given by the equation $f(x) = \frac{5}{x^2 - 6x}$?

Solution

$x^2 - 6x = 0$ ► Find all values of x for which the denominator equals zero.

$x(x - 6) = 0$

$x = 0 \qquad x - 6 = 0$

$x = 6$

0 and 6 are excluded from the domain of the function.

Problem 7 What numbers are excluded from the domain of the function given by the equation $g(x) = \sqrt{2x - 5}$?

Solution See page A38.

EXERCISES 8.1

1 For $f(x) = 3x^2$, find:

1. $f(2)$ **2.** $f(1)$ **3.** $f(-1)$

4. $f(-2)$ **5.** $f(a)$ **6.** $f(w)$

For $g(x) = x^2 - x + 1$, find:

7. $g(0)$ **8.** $g(1)$ **9.** $g(-2)$

10. $g(-1)$ **11.** $g(t)$ **12.** $g(s)$

For $f(x) = 4x - 3$, find:

13. $f(-2)$ **14.** $f(3)$ **15.** $f(2 + h)$

16. $f(3 + h)$ **17.** $f(1 + h) - f(1)$ **18.** $f(-1 + h) - f(-1)$

For $g(x) = x^2 - 1$, find:

19. $g(2)$ **20.** $g(-3)$ **21.** $g(1 + h)$

22. $g(2 + h)$ **23.** $g(3 + h) - g(3)$ **24.** $g(-1 + h) - g(-1)$

For $f(x) = 2x^2 - 3$ and $g(x) = -2x + 3$, find:

25. $f(2) - g(2)$ **26.** $f(3) - g(3)$ **27.** $f(0) + g(0)$

28. $f(1) + g(1)$ **29.** $f(1 + h) - f(1)$ **30.** $f(2 + h) - f(2)$

31. $\dfrac{g(1 + h) - g(1)}{h}$ **32.** $\dfrac{g(-2 + h) - g(-2)}{h}$ **33.** $\dfrac{f(3 + h) - f(3)}{h}$

34. $\dfrac{f(-1 + h) - f(-1)}{h}$ **35.** $\dfrac{g(a + h) - g(a)}{h}$ **36.** $\dfrac{f(a + h) - f(a)}{h}$

2 Find the domain and range of the function.

37. {(1, 1), (2, 4), (3, 7), (4, 10), (5, 13)}
38. {(2, 6), (4, 18), (6, 38), (8, 66), (10, 102)}
39. {(0, 1), (2, 2), (4, 3), (6, 4)}
40. {(0, 1), (1, 2), (4, 3), (9, 4)}
41. {(1, 0), (3, 0), (5, 0), (7, 0), (9, 0)}
42. {(−2, −4), (2, 4), (−1, 1), (1, 1), (−3, 9), (3, 9)}
43. {(0, 0), (1, 1), (−1, 1), (2, 2), (−2, 2)}
44. {(0, −5), (5, 0), (10, 5), (15, 10)}

Find the range of the function defined by each equation.

45. $f(x) = 4x - 3$; domain = {0, 1, 2, 3, 4}
46. $g(x) = x^2 + 2x - 1$; domain = {−2, −1, 0, 1, 2}
47. $h(x) = \frac{x}{2} + 3$; domain = {−4, −2, 0, 2, 4}
48. $F(x) = \sqrt{x + 1}$; domain = {−1, 0, 3, 8, 15}
49. $G(x) = \frac{2}{x + 3}$; domain = {−2, −1, 0, 1, 2}
50. $f(x) = |5x - 2|$; domain = {−10, −5, 0, 5, 10}
51. $g(a) = \frac{a^2 + 1}{3a - 1}$; domain = {−1, 0, 1, 2}
52. $h(a) = (a^2 + 3a)^2$; domain = {−2, −1, 0, 1, 2}

In the exercises below, a number in the range of a function is given. Find an element in the domain that corresponds to the number, and write an ordered pair that belongs to the function.

53. $f(x) = x + 5$; −3

54. $g(x) = x - 4$; 6

55. $h(a) = 3a + 2$; −1

56. $f(a) = 2a - 5$; 0

57. $g(x) = \frac{2}{3}x + \frac{1}{3}$; 1

58. $h(x) = \frac{3}{2}x - 1$; −4

59. $h(x) = x^2 + 3$; 7

60. $g(x) = x^2 - 3$; −2

61. $f(x) = \frac{x+1}{5}$; 7

62. $g(a) = \frac{a-3}{4}$; 5

What values of x are excluded from the domain of the function defined by each equation?

63. $f(x) = x^2 + 3x - 4$

64. $g(x) = 2x^2 - x + 5$

65. $h(x) = \frac{2x}{3x^2 - x}$

66. $F(x) = \frac{6}{5x^2 - 2x}$

67. $G(x) = \frac{x+1}{x^2 + x - 6}$

68. $f(x) = \frac{x-2}{x^2 - 3x - 4}$

69. $g(x) = \sqrt{6x - 2}$

70. $h(x) = \sqrt{3x - 9}$

71. $F(x) = |3x - 7|$

72. $G(x) = \frac{x}{|x|}$

SUPPLEMENTAL EXERCISES 8.1

Solve.

73. If $f(x) = \sqrt{x - 2}$ and $f(a) = 4$, find a.

74. If $f(x) = \sqrt{x + 5}$, and $f(a) = 3$, find a.

75. $f(a, b) =$ the sum of a and b
$g(a, b) =$ the product of a and b
Find $f(2, 5) + g(2, 5)$.

76. $f(a, b) =$ the greatest common divisor of a and b
$g(a, b) =$ the least common multiple of a and b
Find $f(14, 35) + g(14, 35)$.

77. The sale price of an item is a function, s, of the original price, p, where $s(p) = 0.80p$. If an item's original price is \$200, what is the sale price of the item?

78. The markup on an item is a function, m, of its cost, c, where $m(c) = 0.25c$. If the cost of an item is \$150, what is the markup on the item?

79. Let $f(x)$ be the digit in the xth decimal place of the repeating digit $0.\overline{387}$. For example, $f(3) = 7$ because 7 is the digit in the third decimal place. Find $f(14)$.

80. Let $f(x)$ be the digit in the xth decimal place of the repeating digit $0.\overline{018}$. For example, $f(1) = 0$ because 0 is the digit in the first decimal place. Find $f(21)$.

81. Given $f(x) = (x + 1)(x - 1)$, for what values of x is $f(x)$ negative? Write your answer in set builder notation.

82. Given $f(x) = (x + 2)(x - 2)$, for what values of x is $f(x)$ negative? Write your answer in set builder notation.

83. Given $f(x) = -|x + 3|$, for what value of x is $f(x)$ greatest?

84. Given $f(x) = |2x - 2|$, for what value of x is $f(x)$ smallest?

SECTION 8.2 Graphs of Functions

1 Graph functions

A function is a set of ordered pairs in which no two ordered pairs that have the same first component have different second components. The ordered pairs of a function can be written as $(x, f(x))$. However, often the value of the function, $f(x)$, is labeled y, and the ordered pairs are written (x, y), where $y = f(x)$. The **graph of a function** is a graph of the ordered pairs (x, y) of the function.

Since the graph of the equation $y = mx + b$ is a straight line, a function defined by the equation $f(x) = mx + b$ is a **linear function.**

To graph $f(x) = 2x + 1$, think of the function as the equation $y = 2x + 1$.

This is the equation of a straight line. The slope is 2, and the y-intercept is 1.

Graph the straight line.

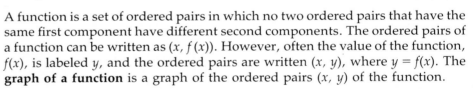

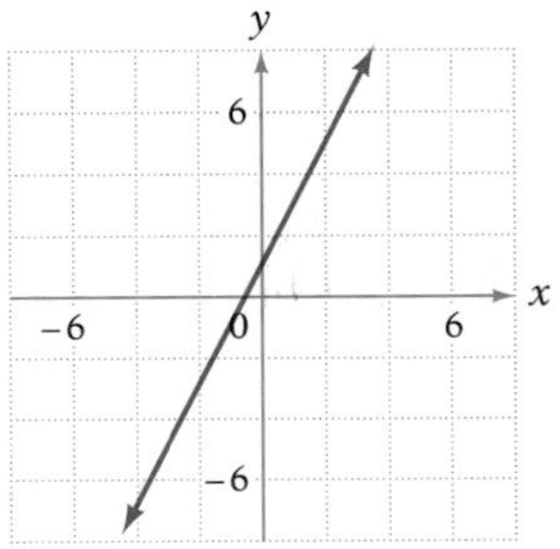

In the equation $y = 2x + 1$, the value of the variable y *depends* on the value of x. Thus y is called the **dependent variable,** and x is called the **independent variable.** For $f(x) = 2x + 1$, $f(x)$ is a symbol for the dependent variable.

In graphing functions, it is important to remember that $f(x)$ is the y-coordinate of an ordered pair.

The graph of $f(x) = x + 2$ is shown at the right.

When $x = -6$, $y = f(-6) = -4$.

When $x = 5$, $y = f(5) = 7$.

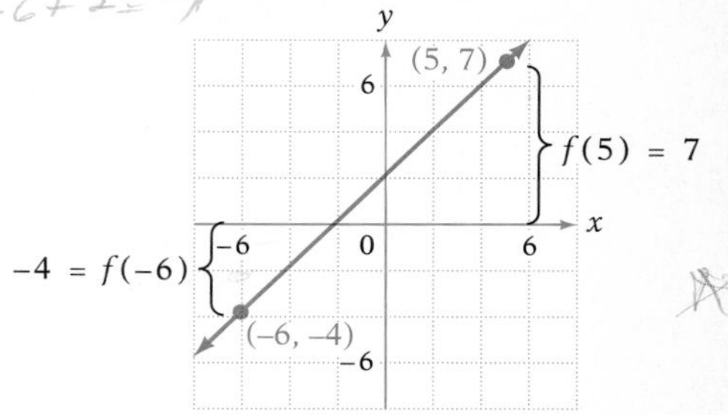

By associating the horizontal axis with the domain and the vertical axis with the range of a function, the graph of a function can be used to help determine its domain and range. For example, because the graph of $f(x) = x + 2$ extends infinitely in both directions, the domain is the real numbers, and the range is the real numbers.

In general, for any linear function defined by the equation $f(x) = mx + b$, $m \neq 0$, the domain is the real numbers, and the range is the real numbers.

To graph $f(x) = 2$, think of the function as the equation $y = 2$.

The graph of an equation of the form $y = b$ is a horizontal line passing through the point $(0, b)$.

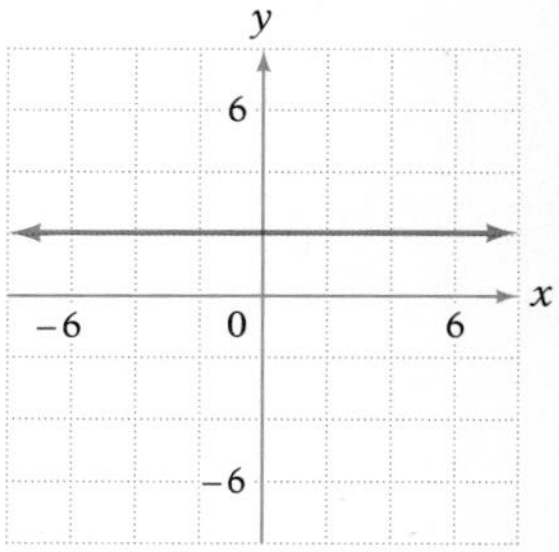

This function is an example of a **constant function;** for any value in the domain, the corresponding value of $f(x)$ is the constant b. The domain of the function $f(x) = 2$ is all real numbers. The range is $\{2\}$.

In general, a function defined by the equation $f(x) = b$ is a constant function and its graph is a horizontal line.

A function of the form $f(x) = ax^2 + bx + c$, $a \neq 0$, is a **quadratic function.**

To graph $f(x) = x^2 - 1$, think of the function as the equation $y = x^2 - 1$.

This is a quadratic equation in two variables; the graph is a parabola.

The coefficient of x^2 is positive; the graph will open upward.

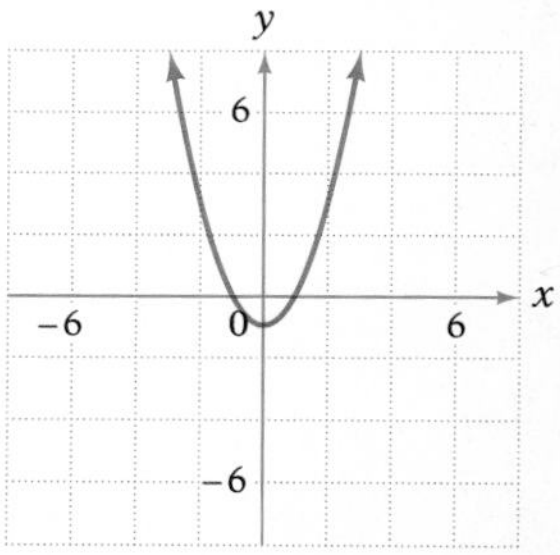

Because $f(x) = x^2 - 1$ is a real number for all values of x, the domain of the function is the real numbers. But note from the graph of this function that no portion of the parabola is below a y value of -1. For any value of x, $y \geq -1$. The range of this function is $\{y|y \geq -1\}$.

To graph $f(x) = -x^2 - 2x + 2$, think of the function as the equation $y = -x^2 - 2x + 2$.

This is a quadratic equation in two variables; the graph is a parabola.

The coefficient of x^2 is negative; the graph will open downward.

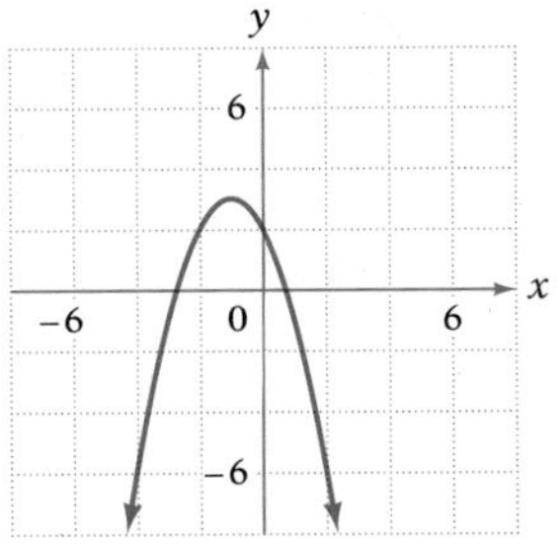

Because $f(x) = -x^2 - 2x + 2$ is a real number for all values of x, the domain of the function f is the real numbers.

To find the range of this function, complete the square on $-x^2 - 2x + 2$.

$$\begin{aligned} -x^2 - 2x + 2 &= -(x^2 + 2x) + 2 \\ &= -(x^2 + 2x + 1) + 1 + 2 \\ &= -(x + 1)^2 + 3 \end{aligned}$$

Because $-(x + 1)^2 \leq 0$ for all x, $-(x + 1)^2 + 3 \leq 3$ for all values of x. The range of the function is $\{y|y \leq 3\}$.

On page 384, the graph of a linear function, such as $f(x) = x + 2$, was shown to be a straight line. On page 385, the graph of a quadratic function, such as $f(x) = x^2 - 1$, was shown to be a parabola. Since $x + 2$ and $x^2 - 1$ are polynomials, the functions given by the equations $f(x) = x + 2$ and $f(x) = x^2 - 1$ are called **polynomial functions.**

A polynomial function of degree 3 is called a **cubic function.** The polynomial function given by the equation $f(x) = x^3 + 2x^2 - 4x + 1$ is an example of a cubic function.

To graph the function given by the equation $f(x) = x^3$, think of the function as the equation $y = x^3$.

Find enough ordered pairs to draw the graph.

x	y
-2	-8
-1	-1
$-\frac{1}{2}$	$-\frac{1}{8}$
0	0
$\frac{1}{2}$	$\frac{1}{8}$
1	1
2	8

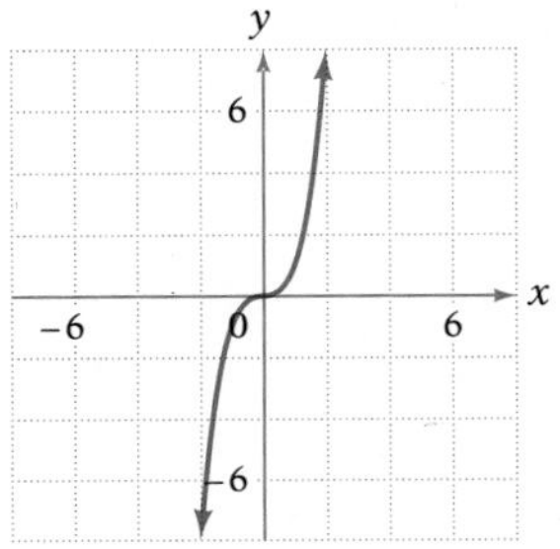

Because the function given by the equation $f(x) = x^3$ is a real number for all values of x, the domain of the function f is the real numbers. The range of the function f is the real numbers.

To graph $f(x) = x^3 + 1$, think of the function as the equation $y = x^3 + 1$.

Find enough ordered pairs to draw the graph.

x	y
-2	-7
-1	0
0	1
1	2
2	9

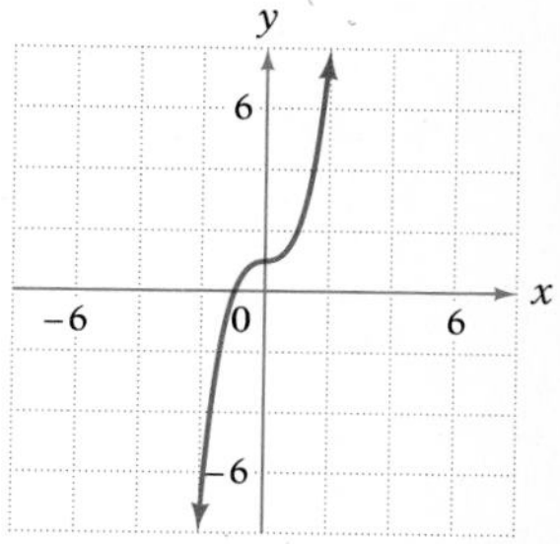

The domain of $f(x) = x^3 + 1$ is all real numbers. The range of the function is all real numbers.

To graph $f(x) = 2x^3 - 5x$, think of the function as the equation $y = 2x^3 - 5x$.

Find enough ordered pairs to draw the graph.

x	y
-2	-6
-1	3
0	0
1	-3
2	6

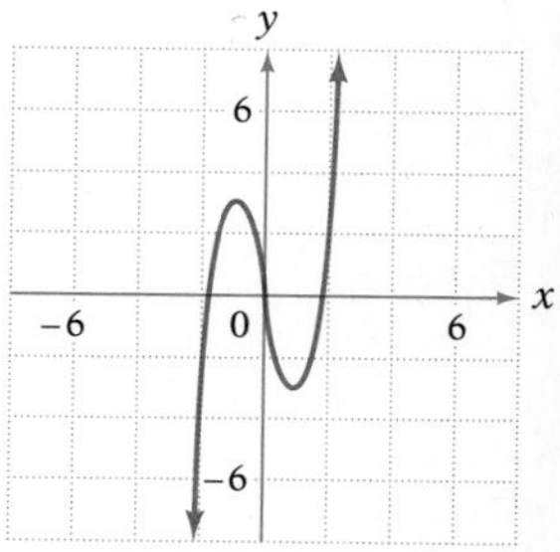

The domain of $f(x) = 2x^3 - 5x$ is all real numbers. The range of the function is all real numbers.

To graph $f(x) = -x^3 + 4x - 1$, think of the function as the equation $y = -x^3 + 4x - 1$.

Find enough ordered pairs to draw the graph.

x	y
-3	14
-2	-1
-1	-4
0	-1
1	2
2	-1
3	-16

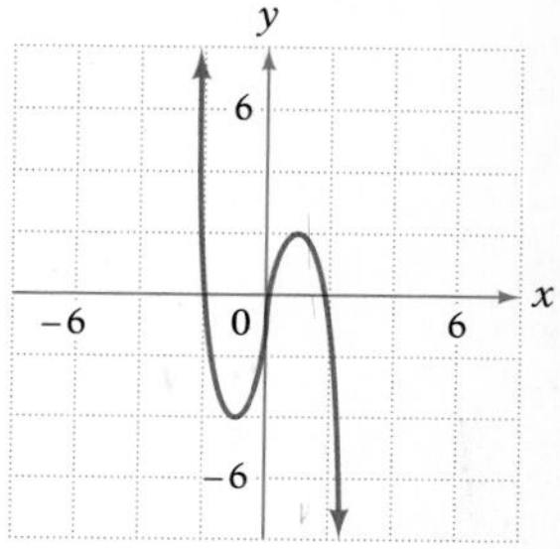

The domain of $f(x) = -x^3 + 4x - 1$ is all real numbers. The range of the function is all real numbers.

The function given by the equation $f(x) = |x|$ is an example of an **absolute value function.** The graph of this function is V shaped.

To graph $f(x) = |x|$, think of the function as the equation $y = |x|$.

Find enough ordered pairs to draw the graph.

x	y
−3	3
−2	2
−1	1
0	0
1	1
2	2
3	3

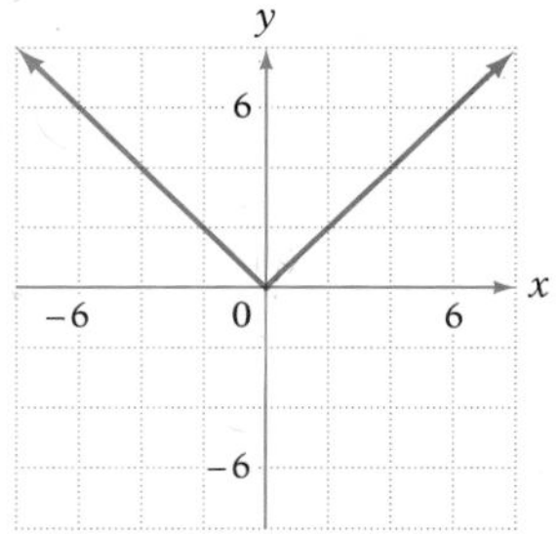

The domain of $f(x) = |x|$ is all real numbers. The range is $\{y|y \geq 0\}$.

To graph $f(x) = |x - 2|$, think of the function as the equation $y = |x - 2|$.

Find enough ordered pairs to draw the graph.

x	y
−1	3
0	2
1	1
2	0
3	1
4	2

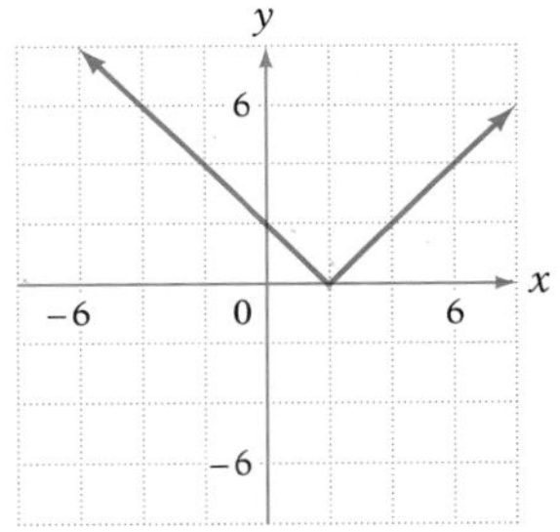

The domain of $f(x) = |x - 2|$ is all real numbers. The range is $\{y|y \geq 0\}$.

To graph $f(x) = 3|x| - 1$, think of the function as the equation $y = 3|x| - 1$.

Find enough ordered pairs to draw the graph.

x	y
−2	5
−1	2
0	−1
1	2
2	5

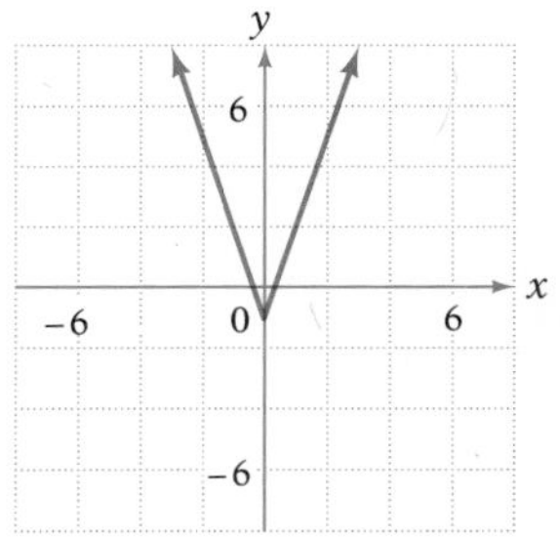

The domain of $f(x) = 3|x| - 1$ is all real numbers. The range is $\{y|y \geq -1\}$.

Example 1 Graph. State the domain and range of the function defined by each equation.

A. $f(x) = |x| + 2$ B. $f(x) = -x^3 - 2x^2 + 2$

Solution A.

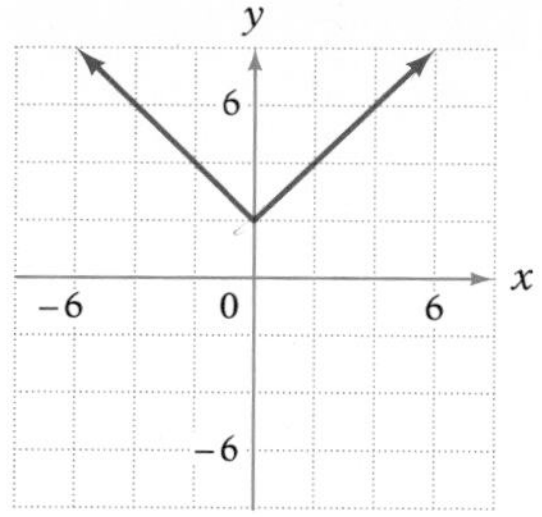

▶ This is an absolute value function. The graph is V shaped.

The domain is all real numbers. The range is $\{y|y \geq 2\}$.

B.

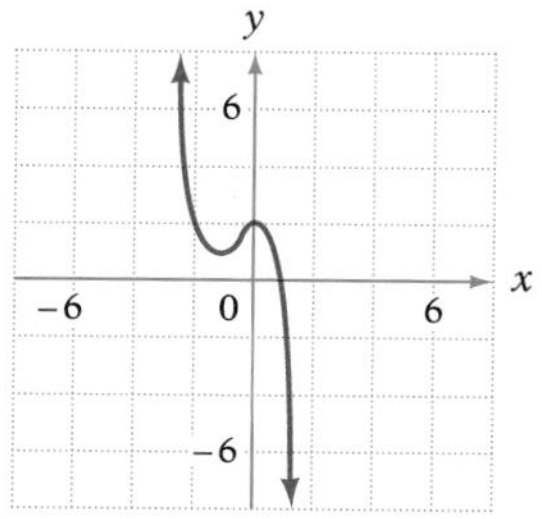

▶ This is a cubic function. Some ordered pairs are $(-2, 2)$, $(-1, 1)$, $(0, 2)$, and $(1, -1)$.

The domain is all real numbers. The range is all real numbers.

Problem 1 Graph. State the domain and range of the function defined by each equation.

A. $f(x) = |x + 2|$ B. $f(x) = x^3 + 2x + 1$

Solution See page A38.

2 Determine whether or not a relation is a function

As stated previously, not every correspondence between two sets is a function. A function assigns to each element of the domain one and only one element of the range. A relation assigns to each element of the domain one or more elements of the range.

For example, the correspondence that pairs a positive real number with a square root is not a function because each positive real number can be paired with a positive or negative square root of that number. For example, since 16 can be paired with 4 or -4, the ordered pairs (16, 4) and (16, -4) would belong to the correspondence. But by the definition of a function, no two ordered pairs with the same first component can have different second components.

The ordered pairs (1, 1), (1, -1), (4, 2), (4, -2), (9, 3), and (9, -3) belong to the relation that pairs a positive real number with its square root. The graph of the relation is shown below.

Because each positive real number x has two square roots, a vertical line intersects the graph of the relation more than once.

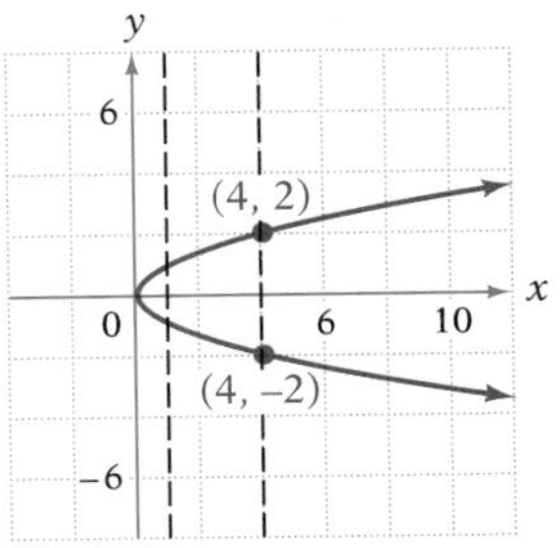

The ordered pairs (-2, -8), (-1, -1), (0, 0), (1, 1), (2, 8) belong to the function that pairs a number with its cube. The graph of the function is shown below.

Because every real number has only one cube, there is only one y-coordinate for each x. A vertical line intersects the graph of the function no more than once.

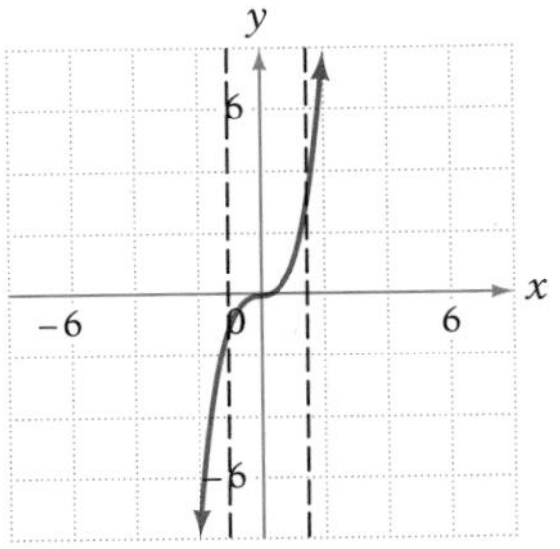

With these two graphs in mind, a test can be stated which allows you to determine whether a graph is the graph of a function.

Vertical Line Test

A graph is the graph of a function if any vertical line intersects the graph at no more than one point.

For example, the graph of a circle is not the graph of a function because a vertical line can intersect the graph more than once.

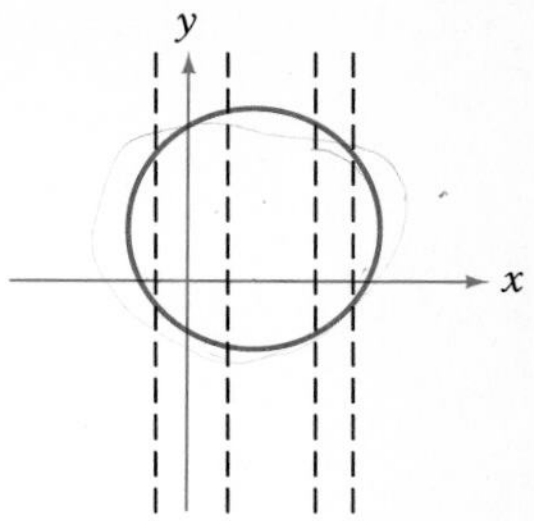

The graph of a quadratic function is the graph of a function. Any vertical line intersects the graph at most once.

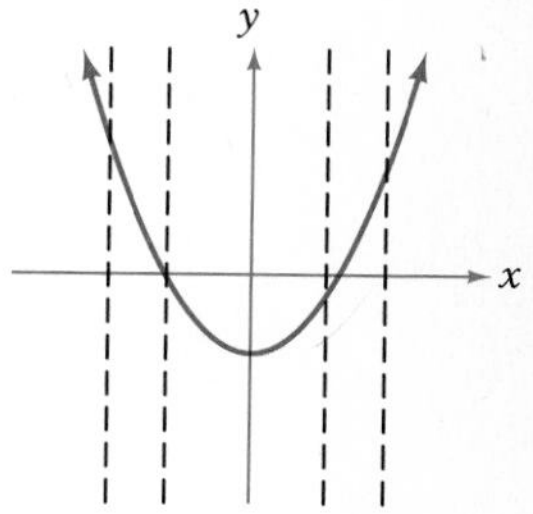

Example 2 Use the vertical line test to determine if the graph shown at the right is the graph of a function.

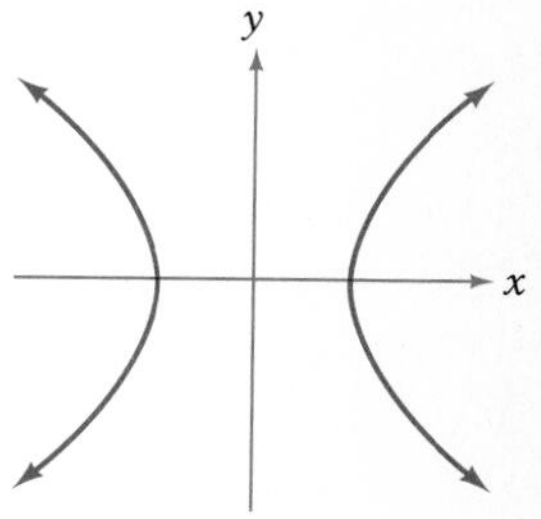

Solution

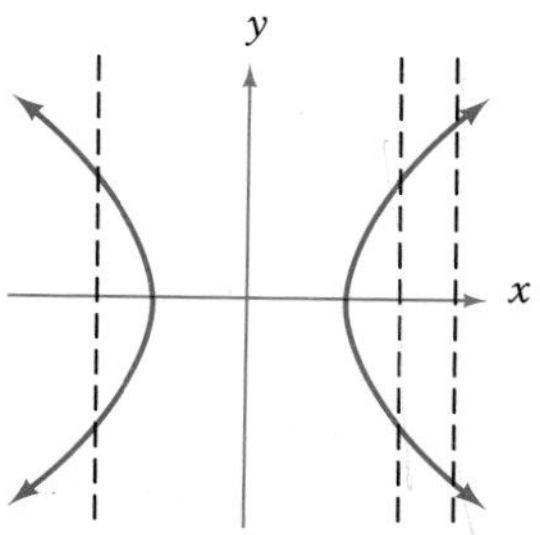

▶ A vertical line can intersect the graph at more than one point. Therefore, the relation includes ordered pairs with the same first component and different second components.

The graph is not the graph of a function.

Problem 2 Use the vertical line test to determine if the graph shown at the right is the graph of a function.

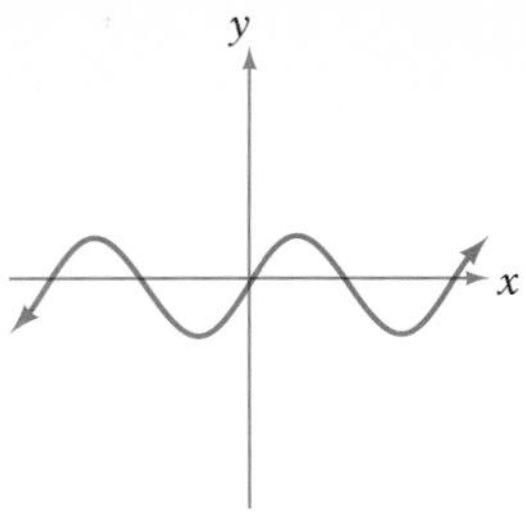

Solution See page A38.

3 Determine from a graph whether a function is one-to-one

Recall that a function is a set of ordered pairs in which no two ordered pairs that have the same first component have different second components. This means that given any x there is only one y that can be paired with that x. A **one-to-one function** satisfies the additional condition that given any y, there is only one x that can be paired with the given y. One-to-one functions are commonly expressed by writing 1–1.

The function given by the equation $y = |x|$ is not a 1–1 function since, given $y = 2$, there are two possible values of x, 2 and -2, which can be paired with the given y-value. The graph at the right illustrates that a horizontal line intersects the graph more than once.

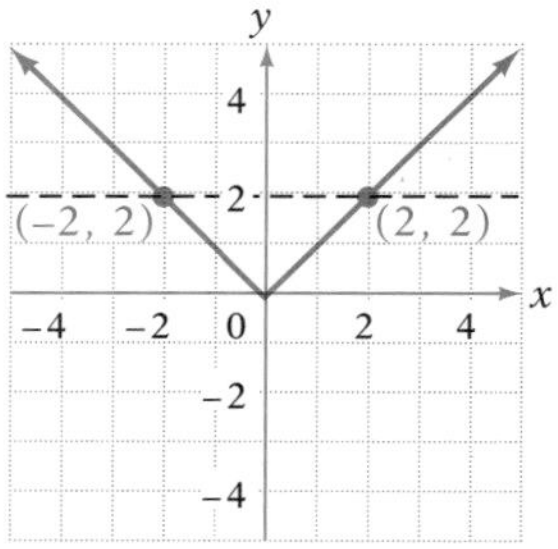

Just as the vertical line test can be used to determine if a graph represents a function, a **horizonal line test** can be used to determine if the graph of a function represents a 1–1 function.

Horizontal Line Test

A graph of a function is the graph of a 1–1 function if any horizontal line intersects the graph at no more than one point.

The graph of a quadratic function is shown at the right. Note that a horizontal line can intersect the graph at more than one point. Therefore, this function is not a 1–1 function. In general, $f(x) = ax^2 + bx + c$, $a \neq 0$, is not a 1–1 function.

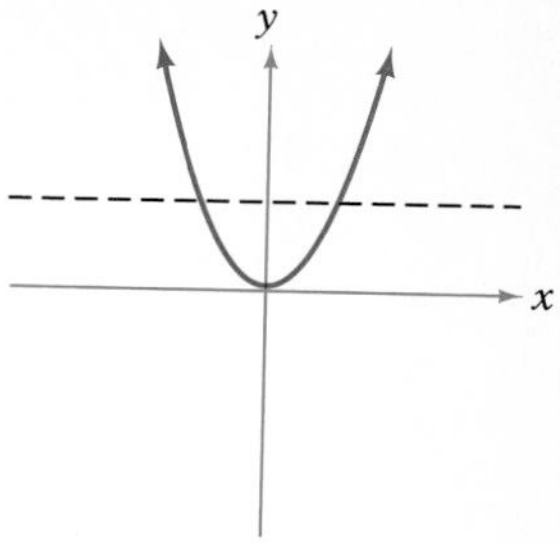

Since any vertical line will intersect the graph at the right at no more than one point, the graph is the graph of a function. Since any horizontal line will intersect the graph at no more than one point, the graph is the graph of a 1–1 function.

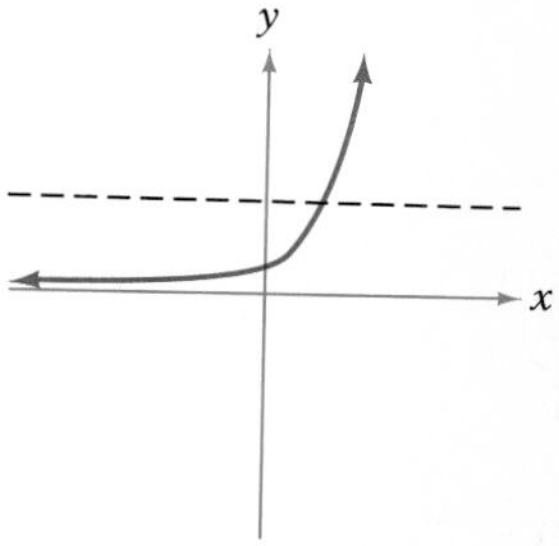

Example 3 Determine if the graph represents the graph of a 1–1 function.

A.

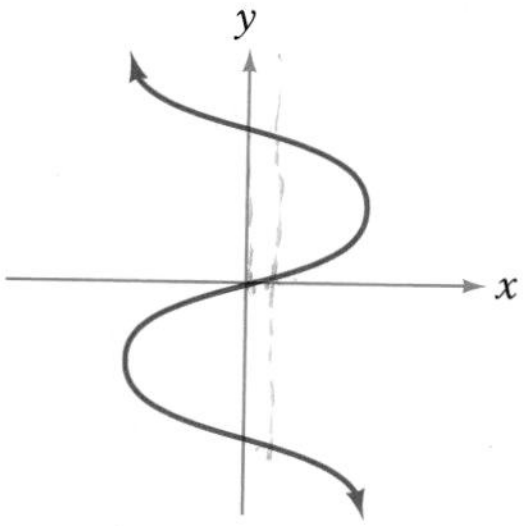

B.

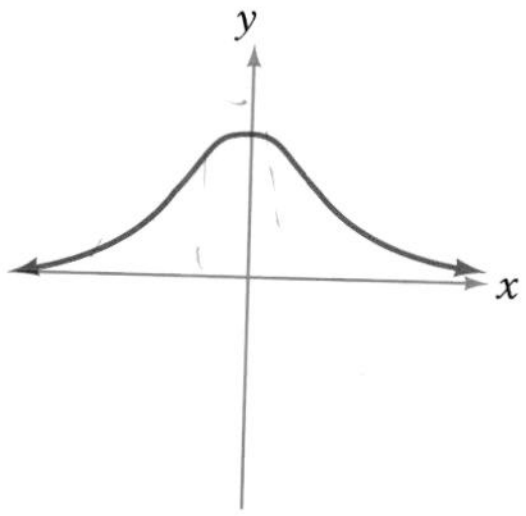

Solution A.

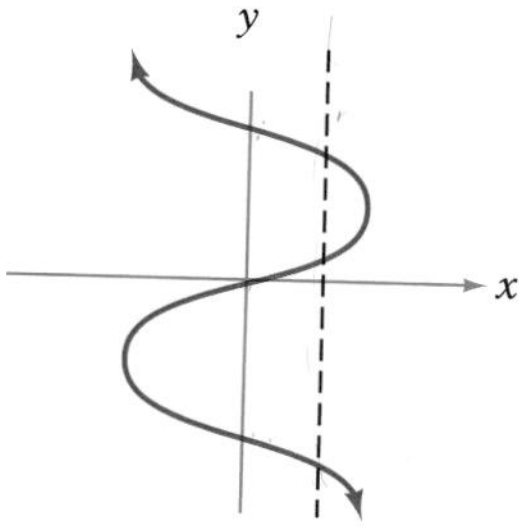

▶ A vertical line can intersect the graph at more than one point.
The graph does not represent a function.

It is not the graph of a 1–1 function.

B.

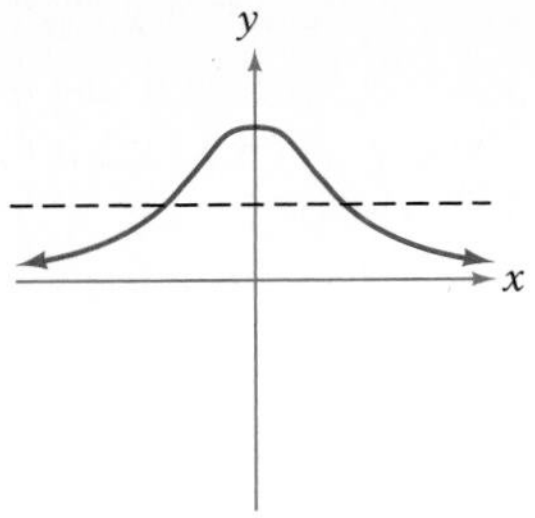

▶ A horizontal line can intersect the curve at more than one point.

It is not the graph of a 1–1 function.

Problem 3 Determine if the graph represents the graph of a 1–1 function.

A.

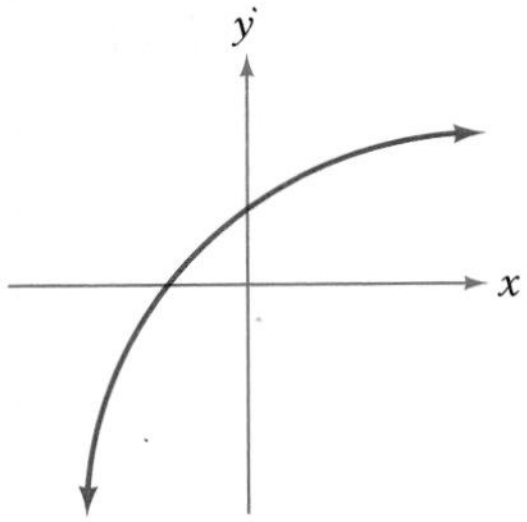

B.

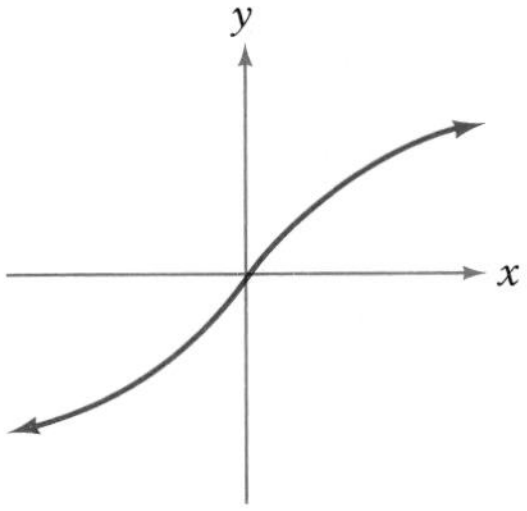

Solution See page A38.

EXERCISES 8.2

1 Graph the function defined by each equation. State the domain and range of the function.

1. $f(x) = 2x - 1$

2. $f(x) = 4$

3. $f(x) = 2x^2 - 1$

4. $f(x) = 2x^2 - 3$

5. $f(x) = |x - 1|$

6. $f(x) = |x - 3|$

7. $f(x) = x^3 - 1$

8. $f(x) = -x^3 + 1$

9. $f(x) = -3x - 1$

10. $f(x) = -\frac{1}{2}x + 1$

11. $f(x) = \frac{1}{2}x^2$

12. $f(x) = \frac{1}{3}x^2$

13. $f(x) = |x| + 1$

14. $f(x) = |x| - 1$

15. $f(x) = x^3 + 2x^2$

16. $f(x) = x^3 - 3x^2$

17. $f(x) = \frac{2}{3}x + 4$

18. $f(x) = \frac{3}{4}x + 1$

19. $f(x) = 2x^2 - 4x - 3$

20. $f(x) = -1$

21. $f(x) = 2|x| - 1$

22. $f(x) = 2|x| + 2$

23. $f(x) = 2x^3 + 3x$

24. $f(x) = -2x^3 + 4x$

25. $f(x) = -\frac{1}{2}x - 2$

26. $f(x) = -3$

27. $f(x) = x^2 + 2x - 4$

28. $f(x) = -x^2 + 2x - 1$

29. $f(x) = -|x|$

30. $f(x) = -|x + 2|$

31. $f(x) = x^3 - x^2 + x - 1$

32. $f(x) = x^3 + x^2 - x + 1$

2. Is the set of ordered pairs a function?

33. $\{(0, 0), (1, 0), (2, 0), (3, 0)\}$

34. $\{(1, 1), (1, 2), (1, 3), (1, 4)\}$

35. $\{(-1, 1), (0, 1), (1, 1), (2, 1)\}$

36. $\{(2, 2), (4, 4), (6, 6), (8, 8)\}$

37. $\{(-3, 3), (-2, 2), (-1, 1), (-2, -2)\}$

38. $\{(-5, -5), (5, 5), (-5, 5), (5, -5)\}$

39. $\{(1, 4), (2, 3), (3, 2), (4, 1)\}$

40. $\{(-4, -1), (-3, -2), (-2, -3), (-1, -4)\}$

Use the vertical line test to determine if the graph is the graph of a function.

41.

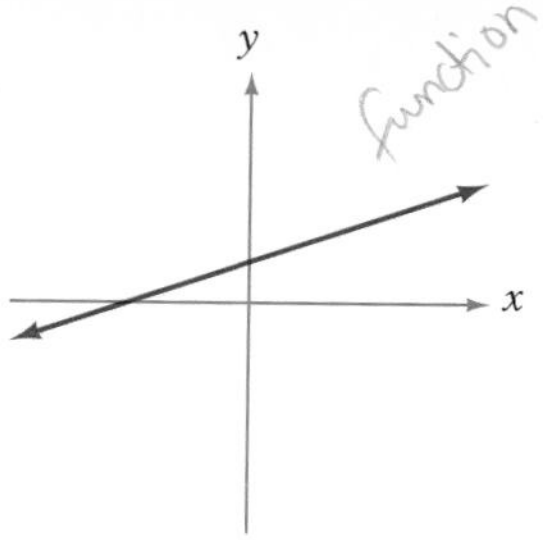

42.

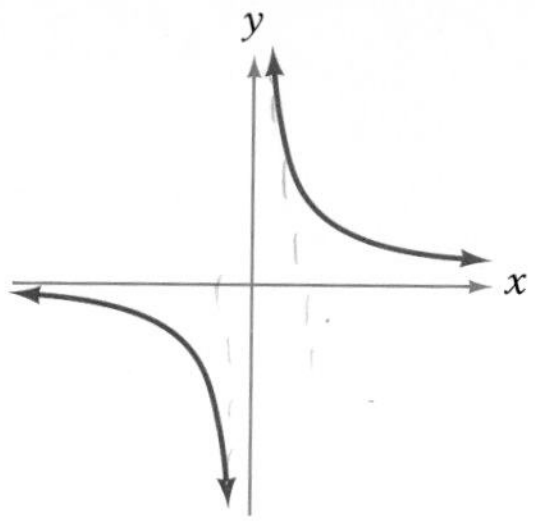

43.

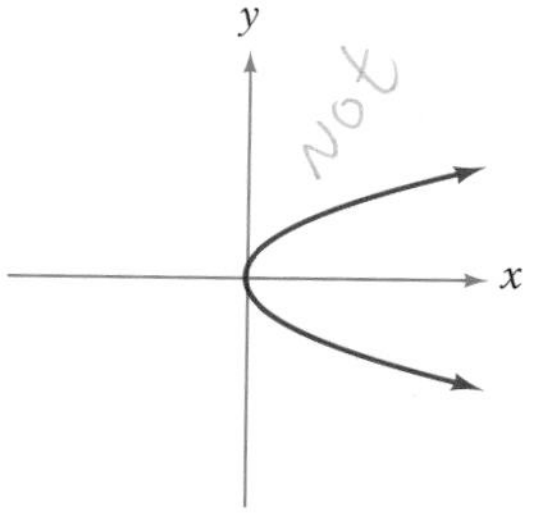

44.

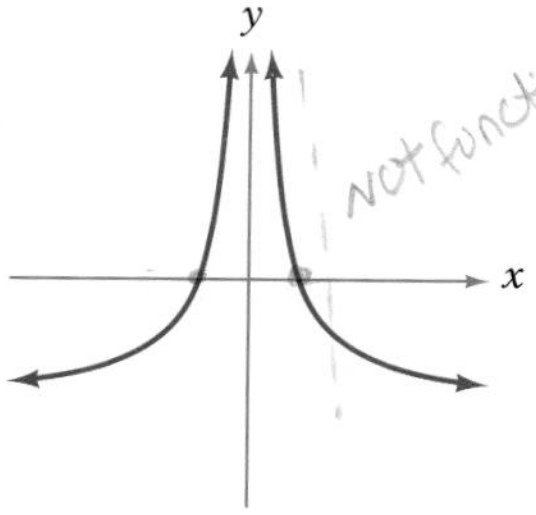

45.

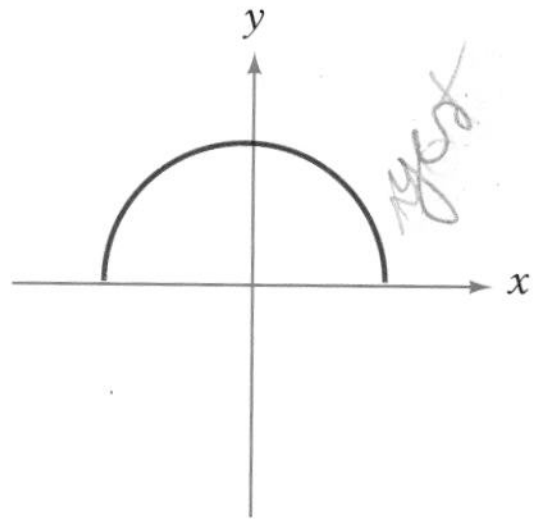

46.

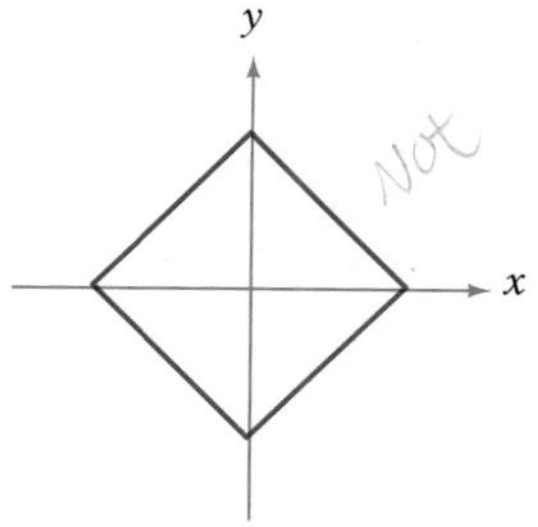

3 Determine if the graph represents the graph of a 1–1 function.

47.

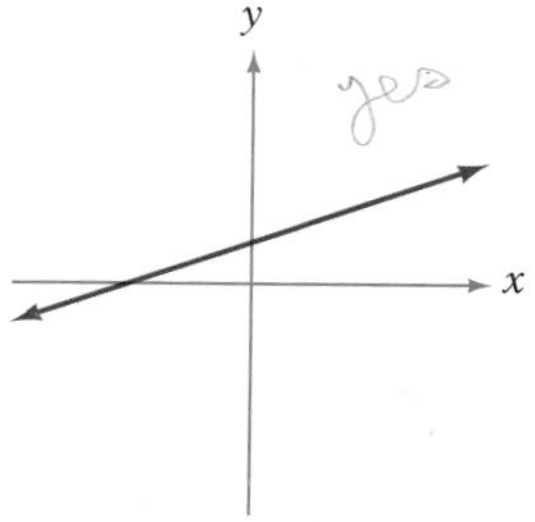

48.

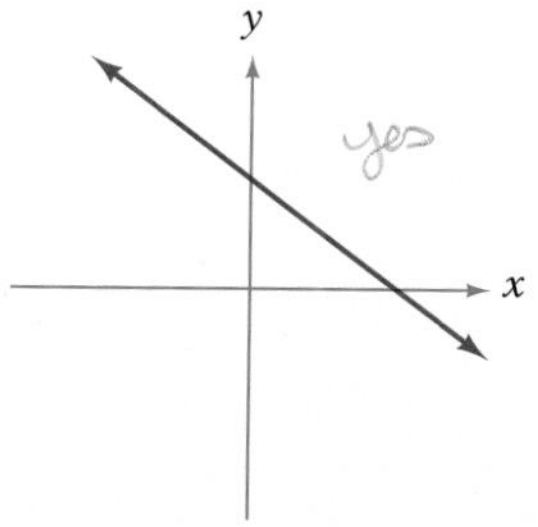

49.

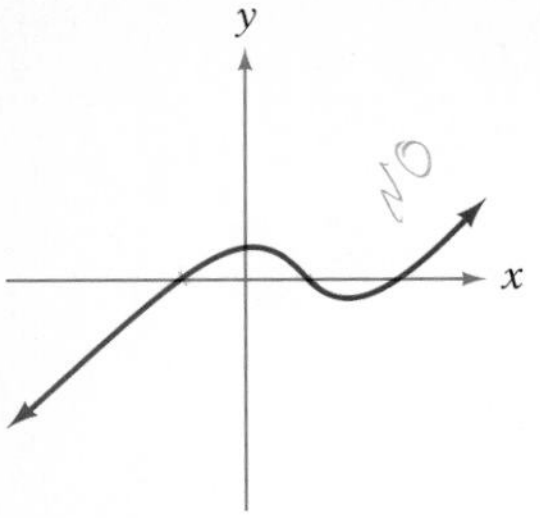

50.

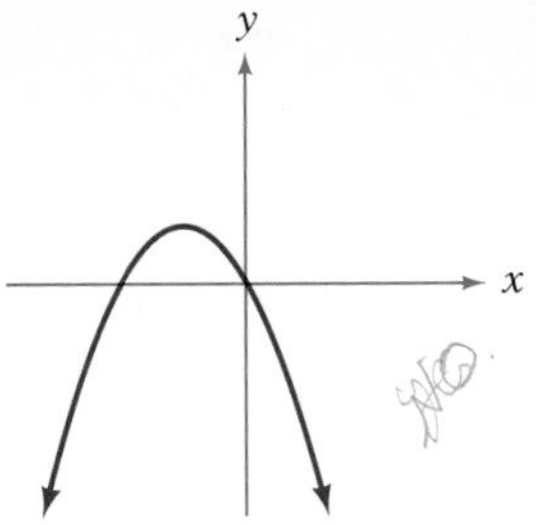

51.

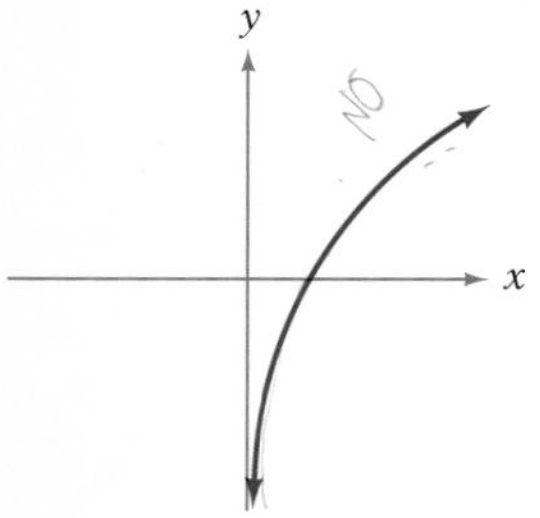

52.

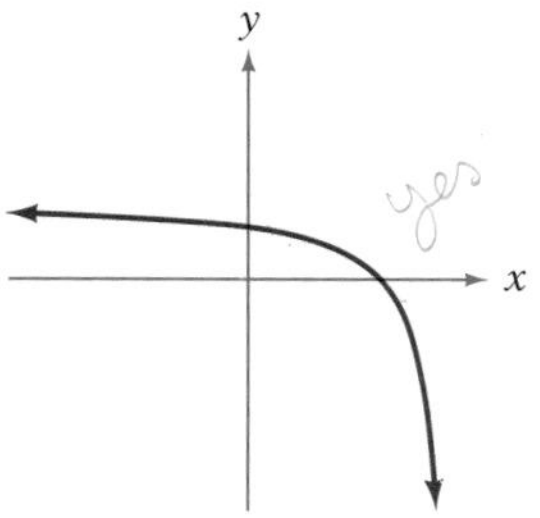

53.

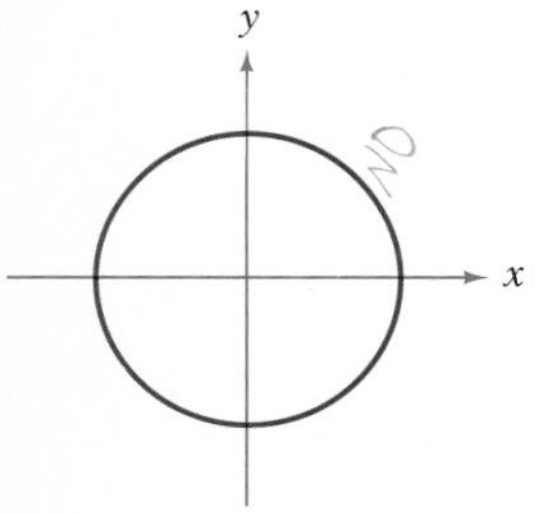

54.

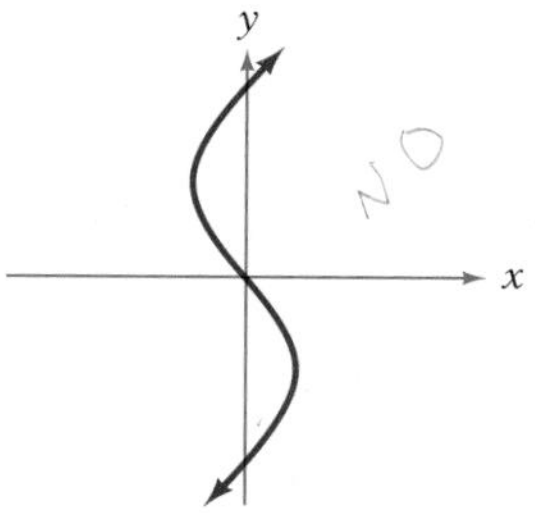

SUPPLEMENTAL EXERCISES 8.2

Which of the following relations are not functions?

55. **a.** $f(x) = x$ **b.** $f(x) = \left|\frac{x}{2}\right|$ **c.** $\{(3, 1), (1, 3), (3, 0), (0, 3)\}$

56. **a.** $f(x) = -x$ **b.** $f(x) = \frac{2}{\sqrt{x}}$ **c.** $\{(1, 4), (4, 1), (1, -4), (-4, 1)\}$

Solve.

57. A linear function f includes the ordered pairs (2, 4) and (4, 10). Find $f(-1)$.

58. A linear function f includes the ordered pairs (2, 3) and (4, 1). Find $f(7)$.

59. f is a linear function with $f(2) = 0$ and $f(-1) = 0$. Find $f(1)$.

60. f is a linear function with $f(1) = 2$ and $f(-1) = 8$. Find $f(2)$.

61. f is a constant function with $f(3) = 5$. Find $f(-1)$.

62. f is a constant function with $f(1) = 4$. Find $f(6)$.

63. For every increase of one unit in the value of x, the value of the linear function $f(x)$ increases by 4 units. Given $f(3) = 8$, find $f(0)$.

64. For every increase of one unit in the value of x, the value of the linear function $f(x)$ increases by 6 units. Given $f(4) = 21$, find $f(0)$.

65. a. For what real value of x does $x^3 = 0$?
    b. What is the x-intercept of the graph of $f(x) = x^3$?

66. a. For what real value of x does $x^3 - 1 = 0$?
    b. What is the x-intercept of the graph of $f(x) = x^3 - 1$?

SECTION 8.3 Composite Functions and Inverse Functions

8

1 Composite functions

Recall from Section 8.1 that to evaluate a function, replace the variable by the given value and then simplify. For example, to evaluate $f(x) = x^2 - 3x$ at 5, replace x by 5 and then simplify.

$$f(x) = x^2 - 3x$$
$$f(5) = 5^2 - 3(5) = 25 - 15 = 10$$

The value of the function at 5 is 10.

A function can be evaluated at the value of another function. Consider the two functions f and g defined by the equations $f(x) = x^2 - 3x$ and $g(x) = 2x - 5$. Then the symbol $f(g(-1))$ means the *value of the function f at* $g(-1)$. Using the equations for g and f,

$$g(-1) = 2(-1) - 5 = -2 - 5 = -7$$
$$f(g(-1)) = f(-7) = (-7)^2 - 3(-7) = 49 + 21 = 70$$

Now evaluate $g(f(-1))$. Again using the equations for f and g,

$$f(-1) = (-1)^2 - 3(-1) = 1 + 3 = 4$$
$$g(f(-1)) = g(4) = 2(4) - 5 = 8 - 5 = 3$$

Note from evaluating these functions that $f(g(-1)) \neq g(f(-1))$. The order of evaluating the functions is important and can give different values.

Evaluate $K(v(2))$ given $K(x) = 2x^2 + 1$ and $v(x) = \frac{3}{2x - 3}$.

Use the equations for $K(x)$ and $v(x)$.

$$v(2) = \frac{3}{2(2) - 3} = \frac{3}{4 - 3} = 3$$
$$K(v(2)) = K(3) = 2(3)^2 + 1 = 2(9) + 1 = 19$$

In the previous examples, specific values of x were used to evaluate the functions. It is possible to evaluate the functions using a variable rather than a number.

Let $f(x) = x^2 + 2x$ and $g(x) = 2x - 1$. Using the equations for g and f, find $f(g(x))$.

$$g(x) = 2x - 1$$

Replace $g(x)$ with $2x - 1$.
Evaluate $f(2x - 1)$.
Simplify.

$$\begin{aligned} f(g(x)) &= f(2x - 1) \\ &= (2x - 1)^2 + 2(2x - 1) \\ &= 4x^2 - 4x + 1 + 4x - 2 \\ &= 4x^2 - 1 \end{aligned}$$

The function produced in this manner is called a **composite function** and is denoted by $f \circ g$ (read "f circle g").

For functions f and g, the composite function $f \circ g$ is given by $(f \circ g)(x) = f(g(x))$. The domain of the composite function is all x in the domain of g such that $g(x)$ is in the domain of f.

The composite function is also called the composition function.

For $v(x) = 1 - 3x$ and $w(x) = x^2 - x$, find $(v \circ w)(x)$ and $(w \circ v)(x)$.

Evaluate $(v \circ w)(x)$.
Replace $w(x)$ by $x^2 - x$.

$$\begin{aligned} (v \circ w)(x) &= v(w(x)) \\ &= v(x^2 - x) \\ &= 1 - 3(x^2 - x) \\ &= 1 - 3x^2 + 3x \\ &= -3x^2 + 3x + 1 \end{aligned}$$

Evaluate $(w \circ v)(x)$.
Replace $v(x)$ by $1 - 3x$.

$$\begin{aligned} (w \circ v)(x) &= w(v(x)) \\ &= w(1 - 3x) \\ &= (1 - 3x)^2 - (1 - 3x) \\ &= 1 - 6x + 9x^2 - 1 + 3x \\ &= 9x^2 - 3x \end{aligned}$$

Note again that $(v \circ w)(x) \neq (w \circ v)(x)$.

Example 1 Given $f(x) = x^2 - 1$ and $g(x) = 3x + 4$, evaluate each of the following.

A. $f(g(0))$ B. $g(f(x))$

Solution A. $g(x) = 3x + 4$

$g(0) = 3(0) + 4 = 4$ ▶ To evaluate $f(g(0))$, first evaluate $g(0)$.

$f(x) = x^2 - 1$

$f(4) = 4^2 - 1 = 15$ ▶ Substitute the value of $g(0)$ for x in $f(x)$. $g(0) = 4$

$f(g(0)) = 15$

B. $g(f(x)) = g(x^2 - 1)$ ▶ $f(x) = x^2 - 1$

$= 3(x^2 - 1) + 4$ ▶ Substitute $x^2 - 1$ for x in the function $g(x)$. $g(x) = 3x + 4$

$= 3x^2 - 3 + 4$

$= 3x^2 + 1$

Problem 1 Evaluate each of the following given $g(x) = 3x - 2$ and $h(x) = x^2 + 1$.

A. $g(h(0))$ B. $h(g(x))$

Solution See page A38.

2 Inverse functions

The **inverse of a function** is a function in which the components of each ordered pair are reversed.

For example, some of the ordered pairs of the function defined by the equation $f(x) = 2x$ are $(0, 0)$, $(-1, -2)$, $(3, 6)$, and $\left(\frac{1}{2}, 1\right)$.

The inverse of this function would contain the ordered pairs $(0, 0)$, $(-2, -1)$, $(6, 3)$, and $\left(1, \frac{1}{2}\right)$.

Now consider the function defined by the equation $g(x) = x^2$. Some of the ordered pairs of this function are $(0, 0)$, $(-1, 1)$, $(1, 1)$, $(-3, 9)$, and $(3, 9)$. Reversing the ordered pairs gives $(0, 0)$, $(1, -1)$, $(1, 1)$, $(9, -3)$, and $(9, 3)$. These ordered pairs do not satisfy the definition of a function because there are ordered pairs with the same first components and different second components. This example illustrates that not all functions have an inverse function.

The graphs of $f(x) = 2x$ and $g(x) = x^2$ are shown below.

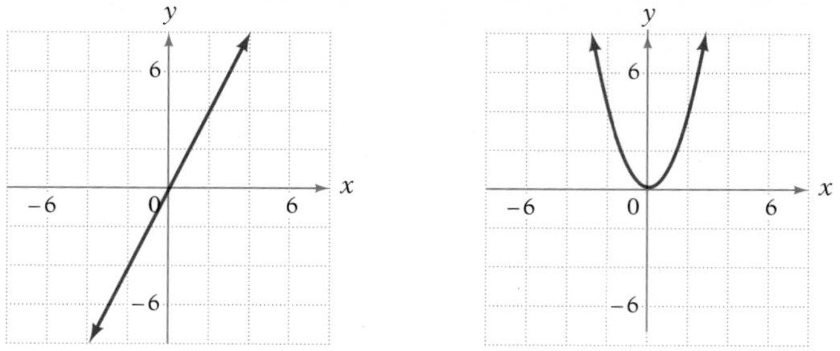

By the horizontal line test, the function f is a 1–1 function while the function g is not a 1–1 function.

Condition for an Inverse Function

A function has an inverse function if and only if it is a 1–1 function.

The symbol f^{-1} is used to denote the inverse of a function. $f^{-1}(x)$ is read "f inverse of x." Note that this is not the reciprocal of $f(x)$ but is the notation used for the inverse of a 1–1 function.

The domain of the inverse of the function f is the range of f, and the range of f^{-1} is the domain of f.

To find the inverse of a function, interchange x and y. Then solve for y.

To find the inverse function of $f(x) = 3x + 6$, think of the function as the equation $y = 3x + 6$.

$$f(x) = 3x + 6$$
$$y = 3x + 6$$

Interchange x and y.

$$x = 3y + 6$$

Solve for y.

$$3y = x - 6$$
$$y = \frac{1}{3}x - 2$$

Replace y with $f^{-1}(x)$.

$$f^{-1}(x) = \frac{1}{3}x - 2$$

The inverse function of $f(x) = 3x + 6$ is $f^{-1}(x) = \frac{1}{3}x - 2$.

The inverse of a linear function with $m \neq 0$ will always be another linear function.

Example 2 Find the inverse of the function defined by the equation $f(x) = 2x - 4$.

Solution

$$f(x) = 2x - 4$$

$y = 2x - 4$ ▶ Think of the function as the equation $y = 2x - 4$.

$x = 2y - 4$ ▶ Interchange x and y.

$2y = x + 4$ ▶ Solve for y.

$$y = \frac{1}{2}x + 2$$

$$f^{-1}(x) = \frac{1}{2}x + 2$$

Problem 2 Find the inverse of the function defined by the equation $f(x) = 4x + 2$.

Solution See page A38.

The graphs of $f(x) = 2x - 4$ and $f^{-1}(x) = \frac{1}{2}x + 2$, found in Example 2, are shown at the right.

The inverse function f^{-1} is the mirror image of f with respect to the line $y = x$.

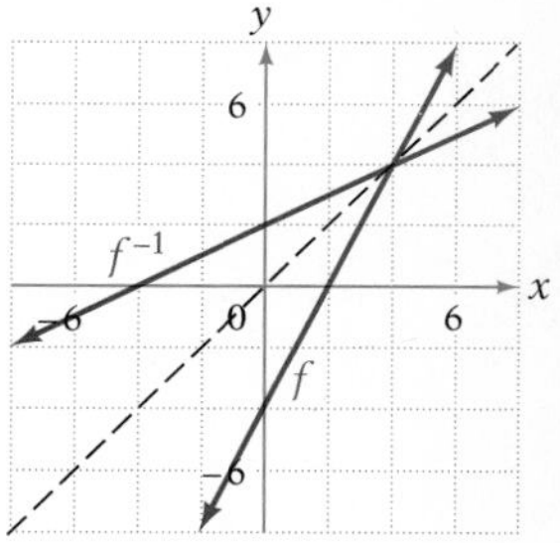

If two functions are inverse functions of one another, then their graphs are mirror images of each other with respect to the line $y = x$.

The composite functions $f(f^{-1}(x))$ and $f^{-1}(f(x))$ have the following property:

$$\mathbf{f(f^{-1}(x)) = f^{-1}(f(x)) = x}$$

For the functions in Example 2,

$$\begin{aligned} f(f^{-1}(x)) &= f\left(\frac{1}{2}x + 2\right) \\ &= 2\left(\frac{1}{2}x + 2\right) - 4 \\ &= x + 4 - 4 \\ &= x \end{aligned} \qquad \begin{aligned} f^{-1}(f(x)) &= f^{-1}(2x - 4) \\ &= \frac{1}{2}(2x - 4) + 2 \\ &= x - 2 + 2 \\ &= x \end{aligned}$$

This concept of inverse functions is similar to the additive inverse and multiplicative inverse used in arithmetic operations. For example, adding the number a and its additive inverse $(-a)$ to an expression results in the original expression.

$$x = x + a + (-a) = x$$

Inverse functions operate in a similar manner; one undoes the other.

For the functions $f(x) = 2x - 4$ and $f^{-1}(x) = \frac{1}{2}x + 2,$

$$f^{-1}(f(2)) = f^{-1}(0) = 2 \qquad f(f^{-1}(2)) = f(3) = 2$$

Example 3 Are the functions defined by the equations $f(x) = -2x + 3$ and $g(x) = -\frac{1}{2}x + \frac{3}{2}$ inverses of each other?

Solution

$$\begin{aligned} f(g(x)) &= f\left(-\frac{1}{2}x + \frac{3}{2}\right) \\ &= -2\left(-\frac{1}{2}x + \frac{3}{2}\right) + 3 \\ &= x - 3 + 3 \\ &= x \end{aligned}$$

▶ Use the property that for inverses $f(f^{-1}(x)) = f^{-1}(f(x)) = x$.

▶ $f(g(x)) = x$

$$\begin{aligned} g(f(x)) &= g(-2x + 3) \\ &= -\frac{1}{2}(-2x + 3) + \frac{3}{2} \\ &= x - \frac{3}{2} + \frac{3}{2} \\ &= x \end{aligned}$$

▶ $g(f(x)) = x$

The functions are inverses of each other.

Problem 3 Are the functions defined by the equations $h(x) = 4x + 2$ and $g(x) = \frac{1}{4}x - \frac{1}{2}$ inverses of each other?

Solution See page A39.

The function given by the equation $f(x) = \frac{1}{2}x^2$ does not have an inverse that is a function. Two of the ordered pair solutions of this function are (4, 8) and (−4, 8).

The graph of $f(x) = \frac{1}{2}x^2$ is shown at the right. This graph does not pass the horizontal line test for the graph of a 1–1 function. The mirror image of the graph with respect to the line $y = x$ is also shown. This graph does not pass the vertical line test for the graph of a function.

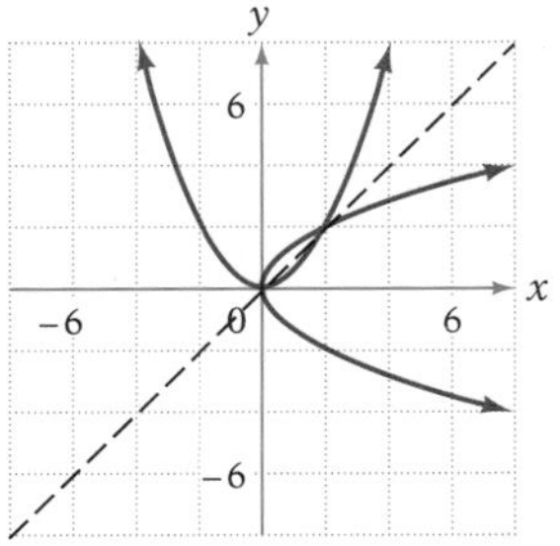

A quadratic function with domain the real numbers does not have an inverse function.

EXERCISES 8.3

1 Given $f(x) = 2x - 3$ and $g(x) = 4x - 1$, evaluate the composite function.

1. $f(g(0))$ **2.** $g(f(0))$ **3.** $f(g(2))$

4. $g(f(-2))$ **5.** $f(g(x))$ **6.** $g(f(x))$

Given $h(x) = 2x + 4$ and $f(x) = \frac{1}{2}x + 2$, evaluate the composite function.

7. $h(f(0))$ **8.** $f(h(0))$ **9.** $h(f(2))$

10. $f(h(-1))$ **11.** $h(f(x))$ **12.** $f(h(x))$

Given $g(x) = x^2 + 3$ and $h(x) = x - 2$, evaluate the composite function.

13. $g(h(0))$ **14.** $h(g(0))$ **15.** $g(h(4))$

16. $h(g(-2))$ **17.** $g(h(x))$ **18.** $h(g(x))$

Given $f(x) = x^2 + x + 1$ and $h(x) = 3x + 2$, evaluate the composite function.

19. $f(h(0))$ **20.** $h(f(0))$ **21.** $f(h(-1))$

22. $h(f(-2))$ **23.** $f(h(x))$ **24.** $h(f(x))$

2 Find the inverse of the function. If the function does not have an inverse, write "no inverse."

25. $\{(1, 0), (2, 3), (3, 8), (4, 15)\}$ **26.** $\{(1, 0), (2, 1), (-1, 0), (-2, 1)\}$

27. $\{(3, 5), (-3, -5), (2, 5), (-2, -5)\}$ **28.** $\{(-5, -5), (-3, -1), (-1, 3), (1, 7)\}$

29. $f(x) = 4x - 8$ **30.** $f(x) = 3x + 6$ **31.** $f(x) = x^2 - 1$

32. $f(x) = 2x + 4$ **33.** $f(x) = x - 5$ **34.** $f(x) = \frac{1}{2}x - 1$

35. $f(x) = \frac{1}{3}x + 2$ **36.** $f(x) = -2x + 2$ **37.** $f(x) = -3x - 9$

38. $f(x) = 2x^2 + 2$

39. $f(x) = \frac{2}{3}x + 4$

40. $f(x) = \frac{3}{4}x - 4$

41. $f(x) = -\frac{1}{3}x + 1$

42. $f(x) = -\frac{1}{2}x + 2$

43. $f(x) = 2x - 5$

44. $f(x) = 3x + 4$

45. $f(x) = x^2 + 3$

46. $f(x) = 5x - 2$

47. $f(x) = 4x - 2$

48. $f(x) = 6x - 3$

49. $f(x) = -8x + 4$

50. $f(x) = -6x + 2$

51. $f(x) = 8x + 6$

52. $f(x) = \frac{1}{2}x^2 - 4$

Are the functions inverses of each other?

53. $f(x) = 4x$; $g(x) = \frac{x}{4}$

54. $g(x) = x + 5$; $h(x) = x - 5$

55. $f(x) = 3x$; $h(x) = \frac{1}{3x}$

56. $h(x) = x + 2$; $g(x) = 2 - x$

57. $g(x) = 3x + 2$; $f(x) = \frac{1}{3}x - \frac{2}{3}$

58. $h(x) = 4x - 1$; $f(x) = \frac{1}{4}x + \frac{1}{4}$

59. $f(x) = \frac{1}{2}x - \frac{3}{2}$; $g(x) = 2x + 3$

60. $g(x) = -\frac{1}{2}x - \frac{1}{2}$; $h(x) = -2x + 1$

Complete.

61. The domain of the inverse function f^{-1} is the ______ of f.

62. The range of the inverse function f^{-1} is the ______ of f.

63. For any function f and its inverse f^{-1}, $f(f^{-1}(3)) =$ ______.

64. For any function f and its inverse f^{-1}, $f^{-1}(f(-4)) =$ ______.

SUPPLEMENTAL EXERCISES 8.3

If f is a 1–1 function and $f(0) = 5$, $f(1) = 7$, and $f(2) = 9$, find:

65. $f^{-1}(5)$

66. $f^{-1}(7)$

67. $f^{-1}(9)$

If f is a 1–1 function and $f(1) = 2$, $f(2) = 5$, and $f(3) = 8$, find:

68. $f^{-1}(2)$ **69.** $f^{-1}(5)$ **70.** $f^{-1}(8)$

Given $f(x) = -x + 4$, find:

71. $f^{-1}(2)$ **72.** $f^{-1}(4)$ **73.** $f^{-1}(6)$

Given $f(x) = 3x - 5$, find:

74. $f^{-1}(0)$ **75.** $f^{-1}(2)$ **76.** $f^{-1}(4)$

The graphs of the functions f and g are shown at the right. Use the graphs to determine the values of the composite functions.

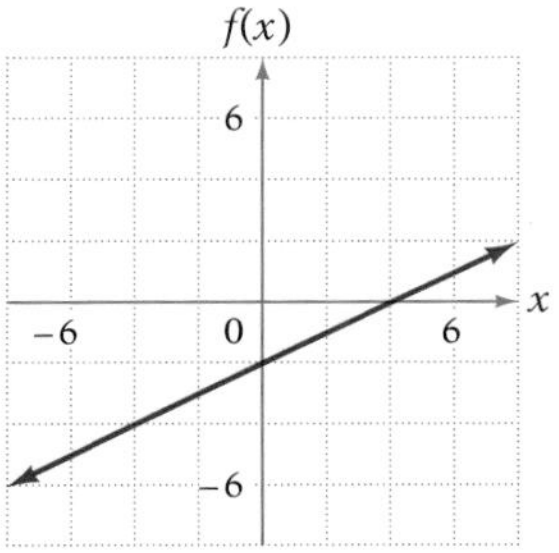

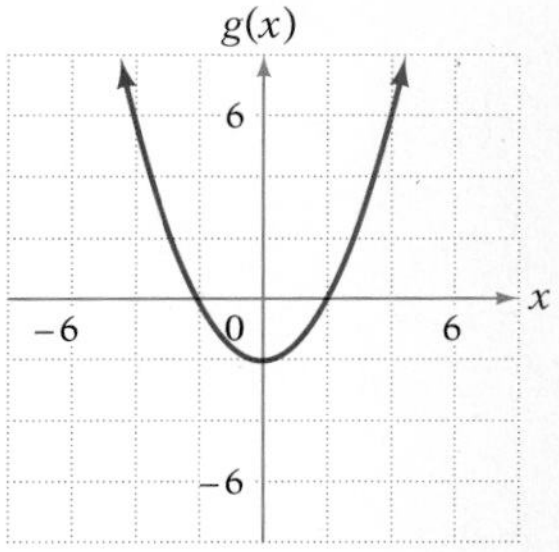

77. $f(g(0))$ **78.** $f(g(2))$

79. $f(g(4))$ **80.** $f(g(-2))$

81. $f(g(-4))$ **82.** $g(f(0))$

83. $g(f(4))$ **84.** $g(f(-4))$

Graph the function defined by each equation. Then use the property that the inverse function is the mirror image of f with respect to the line $y = x$ to graph the inverse function on the same coordinate axes. (*Hint:* Interchange the coordinates in the ordered pair solutions of f to find ordered pair solutions of f^{-1}.)

85. $f(x) = 3x - 4$ **86.** $f(x) = 3x + 2$ **87.** $f(x) = \frac{1}{4}x + 2$ **88.** $f(x) = \frac{1}{3}x - 1$

89. $f(x) = x^3 + 1$ **90.** $f(x) = x^3 - 1$ **91.** $f(x) = 2x^3 - 2$ **92.** $f(x) = 2x^3 + 3$

For $f(x) = x^2 + 1$, $g(x) = 2x - 1$, and $h(x) = 1 - x$, find each of the following.

93. $g(f(h(4)))$ **94.** $f(g(h(-1)))$ **95.** $g(h(f(x)))$ **96.** $h(g(f(x)))$

SECTION 8.4 Application Problems

1 Variation problems

Direct variation is a special function that can be expressed as the equation $y = kx$, where k is a constant. The equation $y = kx$ is read "y varies directly as x" or "y is proportional to x." The constant k is called the **constant of variation** or the **constant of proportionality**.

The circumference (C) of a circle varies directly as the diameter (d). The direct variation equation is written $C = \pi d$. The constant of variation is π.

A nurse makes $18 per hour. The nurse's total wage (w) is directly proportional to the number of hours (h) worked. The equation of variation is $w = 18h$. The constant of proportionality is 18.

A direct variation equation can be written in the form $y = kx^n$, where n is a positive number. For example, the equation $y = kx^2$ is read "y varies directly as the square of x."

The area (A) of a circle varies directly as the square of the radius (r) of the circle. The direct variation equation is $A = \pi r^2$.

Given that V varies directly as r and that $V = 20$ when $r = 4$, the constant of variation can be found by writing the basic direct variation equation, replacing V and r by the given values and solving for the constant of variation.

$$V = kr$$
$$20 = k \cdot 4$$
$$5 = k$$

The direct variation equation can then be written by substituting the value of k into the basic direct variation equation.

$$V = 5r$$

Example 1 The amount (A) of medication prescribed for a person is directly related to the person's weight (W). For a 50-kg person, 2 ml of medication are prescribed. How many milliliters of medication are required for a person who weighs 75 kg?

Strategy To find the required amount of medication:

- Write the basic direct variation equation, replace the variables by the given values, and solve for k.
- Write the direct variation equation, replacing k by its value. Substitute 75 for W, and solve for A.

Solution

$$A = kW$$
$$2 = k \cdot 50$$
$$\frac{1}{25} = k$$
$$A = \frac{1}{25}W$$
$$= \frac{1}{25} \cdot 75 = 3$$

The required amount of medication is 3 ml.

Problem 1 The distance (s) a body falls from rest varies directly as the square of the time (t) of the fall. An object falls 64 ft in 2 s. How far will it fall in 5 s?

Solution See page A39.

Joint variation is a variation in which a variable varies directly as the product of two or more other variables. A joint variation can be expressed as the equation $z = kxy$, where k is a constant. The equation $z = kxy$ is read "z varies jointly as x and y."

The area (A) of a triangle varies jointly as the base (b) and the height (h). The joint variation equation is written $A = \frac{1}{2}bh$. The constant of variation is $\frac{1}{2}$.

Inverse variation is a function that can be expressed as the equation $y = \frac{k}{x}$, where k is a constant. The equation $y = \frac{k}{x}$ is read "y varies inversely as x" or "y is inversely proportional to x."

In general, an inverse variation equation can be written $y = \frac{k}{x^n}$, where n is a positive number. For example, the equation $y = \frac{k}{x^2}$ is read "y varies inversely as the square of x."

Given that P varies inversely as the square of x and that $P = 5$ when $x = 2$, the variation constant can be found by writing the basic inverse variation equation, replacing P and x by the given values, and solving for the constant of variation.

$$P = \frac{k}{x^2}$$
$$5 = \frac{k}{2^2}$$
$$5 = \frac{k}{4}$$
$$20 = k$$

The inverse variation equation can then be found by substituting the value of k into the basic inverse variation equation.

$$P = \frac{20}{x^2}$$

Example 2 A company that produces personal computers has determined that the number of computers it can sell (s) is inversely proportional to the price (P) of the computer. Two thousand computers can be sold when the price is $2500. How many computers can be sold if the price of a computer is $2000?

Strategy To find the number of computers:

- Write the basic inverse variation equation, replace the variables by the given values, and solve for k.
- Write the inverse variation equation, replacing k by its value. Substitute 2000 for P, and solve for s.

Solution

$$s = \frac{k}{P}$$

$$2000 = \frac{k}{2500}$$

$$5{,}000{,}000 = k$$

$$s = \frac{5{,}000{,}000}{P}$$

$$= \frac{5{,}000{,}000}{2000} = 2500$$

At a price of $2000, 2500 computers can be sold.

Problem 2 The resistance (R) to the flow of electric current in a wire of fixed length is inversely proportional to the square of the diameter (d) of a wire. If a wire of diameter 0.01 cm has a resistance of 0.5 ohms, what is the resistance in a wire that is 0.02 cm in diameter?

Solution See page A39.

A **combined variation** is a variation in which two or more types of variation occur at the same time. For example, in physics, the volume (V) of a gas varies directly as the temperature (T) and inversely as the pressure (P). This combined variation is written $V = \frac{kT}{P}$.

Example 3 The pressure (P) of a gas varies directly as the temperature (T) and inversely as the volume (V). When $T = 50°$ and $V = 275$ in.^{3}, $P = 20$ lb/in.^{2}. Find the pressure of a gas when $T = 60°$ and $V = 250$ in.^{3}.

Strategy To find the pressure:

- Write the basic combined variation equation, replace the variables by the given values, and solve for k.
- Write the combined variation equation, replacing k by its value. Substitute 60 for T and 250 for V, and solve for P.

Solution

$$P = \frac{kT}{V}$$

$$20 = \frac{k(50)}{275}$$

$$110 = k$$

$$P = \frac{110T}{V}$$

$$= \frac{110(60)}{250} = 26.4$$

The pressure is 26.4 lb/in.^{2}.

Problem 3 The strength (s) of a rectangular beam varies directly as its width (w) and inversely as the square of its depth (d). If the strength of a beam 2 in. wide and 12 in. deep is 1200 lb, find the strength of a beam 4 in. wide and 8 in. deep.

Solution See page A40.

2 Minimum and maximum problems

In many applications it is important to be able to determine the maximum or minimum of a function.

The graph of $f(x) = x^2 - 2x + 3$ is shown at the right. Since a is positive, the parabola opens upward. The vertex of the parabola is the lowest point on the parabola. It is the point that has the minimum y-coordinate. The minimum y-coordinate is the minimum value of the function.

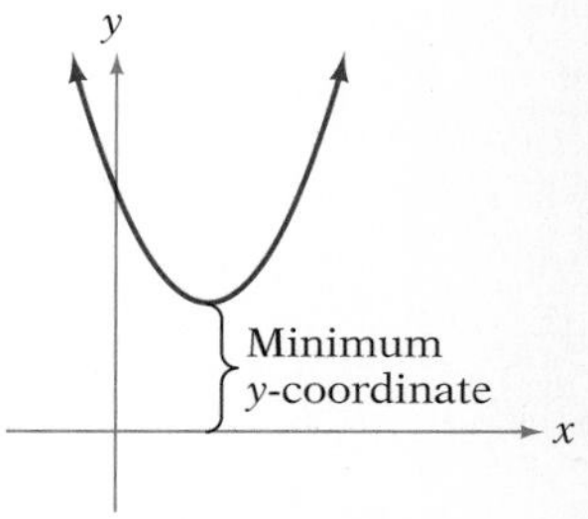

The graph of $f(x) = -x^2 + 2x + 1$ is shown at the right. Since a is negative, the parabola opens downward. The vertex of the parabola is the highest point on the parabola. It is the point that has the maximum y-coordinate. The maximum y-coordinate is the maximum value of the function.

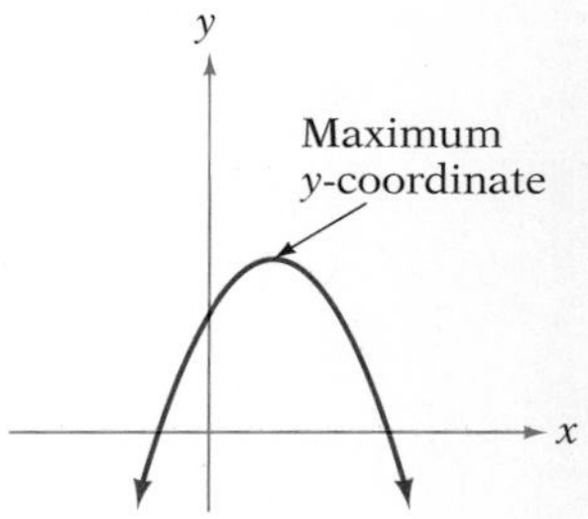

To find the minimum or maximum value of a quadratic function, first find the x-coordinate of the vertex. Then evaluate the function at that value.

Example 4 Find the maximum value of $f(x) = -2x^2 + 4x + 3$.

Solution $x = -\frac{b}{2a} = -\frac{4}{2(-2)} = 1$ ▶ Find the x-coordinate of the vertex. $a = -2$, $b = 4$

$f(x) = -2x^2 + 4x + 3$ ▶ Evaluate the function at $x = 1$.

$f(1) = -2(1)^2 + 4(1) + 3$
$= 5$

The maximum value of the function is 5.

Problem 4 Find the minimum value of $f(x) = 2x^2 - 3x + 1$.

Solution See page A40.

Example 5 A mining company has determined that the cost in dollars (c) per ton of mining a mineral is given by $c(x) = 0.2x^2 - 2x + 12$, where x is the number of tons of the mineral that is mined. Find the number of tons of the mineral that should be mined to minimize the cost. What is the minimum cost?

Strategy
- To find the number of tons that will minimize the cost, find the x-coordinate of the vertex.
- To find the minimum cost, evaluate the function at the x-coordinate of the vertex.

Solution $x = -\frac{b}{2a} = -\frac{-2}{2(0.2)} = 5$

To minimize the cost, 5 tons should be mined.

$c(x) = 0.2x^2 - 2x + 12$
$c(5) = 0.2(5)^2 - 2(5) + 12 = 5 - 10 + 12 = 7$

The minimum cost per ton is \$7.

Problem 5 The height in feet (s) of a ball thrown straight up is given by $s(t) = -16t^2 + 64t$, where t is the time in seconds. Find the time it takes the ball to reach its maximum height. What is the maximum height?

Solution See page A40.

Example 6 Find two numbers whose difference is 10 and whose product is a minimum.

Strategy Let x represent one number. Since the difference between the two numbers is 10, $x + 10$ represents the other number. Then their product is represented by $x^2 + 10x$.
- To find the first of the two numbers, find the x-coordinate of the vertex of $f(x) = x^2 + 10x$.
- To find the other number, replace x in $x + 10$ by the x-coordinate of the vertex and evaluate.

Solution $x = -\frac{b}{2a} = -\frac{10}{2(1)} = -5$

$x + 10 = -5 + 10 = 5$

The numbers are -5 and 5.

Problem 6 A mason is forming a rectangular floor for a storage shed. The perimeter of the rectangle is 44 ft. What dimensions would give the floor a maximum area?

Solution See pages A40–41.

EXERCISES 8.4

1 Solve.

1. The profit (P) realized by a company varies directly as the number of products it sells (s). If a company makes a profit of \$2500 on the sale of 20 products, what is the profit when the company sells 300 products?

2. The number of bushels of wheat (b) produced by a farm is directly proportional to the number of acres (A) planted in wheat. If a 25-acre farm yields 1125 bushels of wheat, what is the yield of a farm that has 220 acres of wheat?

3. The pressure (p) on a diver in the water varies directly as the depth (d). If the pressure is 3.6 lb/in.^{2} when the depth is 8 ft, what is the pressure when the depth is 30 ft?

4. The distance (d) a spring will stretch varies directly as the force (f) applied to the spring. If a force of 5 lb is required to stretch a spring 2 in., what force is required to stretch the spring 5 in.?

5. The distance (d) a person can see to the horizon from a point above the surface of Earth varies directly as the square root of the height (H). If, for a height of 500 ft, the horizon is 19 mi away, how far is the horizon from a point that is 800 ft high? Round to the nearest hundredth.

6. The period (p) of a pendulum, or the time it takes for a pendulum to make one complete swing, varies directly as the square root of the length (L) of the pendulum. If the period of a pendulum is 1.5 s when the length is 2 ft, find the period when the length is 4.5 ft. Round to the nearest hundredth.

7. The distance (s) a ball will roll down an inclined plane is directly proportional to the square of the time (t). If the ball rolls 5 ft in one second, how far will it roll in 4 s?

8. The stopping distance (s) of a car varies directly as the square of its speed (v). If a car traveling 30 mph requires 60 ft to stop, find the stopping distance for a car traveling 55 mph.

9. The length (L) of a rectangle of fixed area varies inversely as the width (w). If the length of a rectangle is 10 ft when the width is 4 ft, find the length of the rectangle when the width is 5 ft.

10. The number of items (n) that can be purchased for a given amount of money is inversely proportional to the cost (C) of an item. If 50 items can be purchased when the cost per item is \$.30, how many items can be purchased when the cost per item is \$.25?

11. For a constant temperature, the pressure (P) of a gas varies inversely as the volume (V). If the pressure is 25 lb/in.^{2} when the volume is 400 ft^3, find the pressure when the volume is 150 ft^3.

12. The speed (v) of a gear varies inversely as the number of teeth (t). If a gear that has 48 teeth makes 20 revolutions per minute, how many revolutions per minute will a gear that has 30 teeth make?

13. The pressure (p) in a liquid varies directly as the product of the depth (d) and the density (D) of the liquid. If the pressure is 37.5 lb/in.^{2} when the depth is 100 in. and the density is 1.2, find the pressure when the density remains the same and the depth is 60 in.

14. The current (I) in a wire varies directly as the voltage (v) and inversely as the resistance (r). If the current is 27.5 amps when the voltage is 110 volts and the resistance is 4 ohms, find the current when the voltage is 195 volts and the resistance is 12 ohms.

15. The repulsive force (f) between the north poles of two magnets is inversely proportional to the square of the distance (d) between them. If the repulsive force is 18 lb when the distance is 3 in., find the repulsive force when the distance is 1.2 in.

16. The intensity (I) of a light source is inversely proportional to the square of the distance (d) from the source. If the intensity is 8 lumens at a distance of 6 m, what is the intensity when the distance is 4 m?

17. The resistance (R) of a wire varies directly as the length (L) of the wire and inversely as the square of the diameter (d). If the resistance is 9 ohms in 50 ft of wire that has a diameter of 0.05 in., find the resistance in 50 ft of a similar wire that has a diameter of 0.02 in.

18. The frequency of vibration (f) of a string varies directly as the square root of the tension (T) and inversely as the length (L) of the string. If the frequency is 40 vibrations per second when the tension is 25 lb and the length of the string is 3 ft, find the frequency when the tension is 36 lb and the string is 4 ft.

19. The wind force (w) on a vertical surface varies directly as the product of the area (A) of the surface and the square of the wind velocity (v). When the wind is blowing at 30 mph, the force on a 10-square-foot area is 45 lb. Find the force on this area when the wind is blowing at 60 mph.

20. The power (P) in an electric circuit is directly proportional to the product of the current (I) and the square of the resistance (R). If the power is 100 watts when the current is 4 amps and the resistance is 5 ohms, find the power when the current is 2 amps and the resistance is 10 ohms.

2 Find the minimum or maximum value of the quadratic function defined by the equation.

21. $f(x) = x^2 - 2x + 3$

22. $f(x) = x^2 + 3x - 4$

23. $f(x) = -2x^2 + 4x - 3$

24. $f(x) = -x^2 - x + 2$

25. $f(x) = 3x^2 + 3x - 2$

26. $f(x) = x^2 - 5x + 3$

27. $f(x) = -3x^2 + 4x - 2$

28. $f(x) = -2x^2 - 5x + 1$

Solve.

29. The height in feet (s) of a rock thrown upward at an initial speed of 64 ft/s from a cliff 50 ft high is given by $s(t) = -16t^2 + 64t + 50$, where t is the time in seconds. Find the maximum height above the ground that the rock will attain.

30. The height in feet (s) of a ball thrown upward at an initial speed of 80 ft/s from a platform 50 ft high is given by $s(t) = -16t^2 + 80t + 50$, where t is the time in seconds. Find the maximum height above the ground that the ball will attain.

31. A pool is treated with a chemical to reduce the amount of algae. The amount of algae in the pool t days after the treatment can be approximated by $A(t) = 40t^2 - 400t + 500$. How many days after treatment will the pool have the least amount of algae?

32. The suspension cable that supports a small footbridge hangs in the shape of a parabola. The height in feet of the cable above the bridge is given by $h(x) = 0.25x^2 - 0.8x + 25$, where x is the distance from one end of the bridge. What is the minimum height of the cable above the bridge?

33. A manufacturer of microwave ovens believes that the revenue the company receives is related to the price (p) of an oven by $R(p) = 125p - \frac{1}{4}p^2$. What price will give the maximum revenue?

34. A manufacturer of camera lenses estimated that the average monthly cost of producing lenses can be given by $C(x) = 0.1x^2 - 20x + 2000$, where x is the number of lenses produced each month. Find the number of lenses to produce in order to minimize the average cost.

35. Find two numbers whose sum is 20 and whose product is a maximum.

36. Find two numbers whose sum is 50 and whose product is a maximum.

37. Find two numbers whose difference is 24 and whose product is a minimum.

38. Find two numbers whose difference is 14 and whose product is a minimum.

SUPPLEMENTAL EXERCISES 8.4

Solve.

39. **a.** Graph $y = kx$ when $k = 2$.
b. What kind of function does the graph represent?

40. **a.** Graph $y = kx$ when $k = \frac{1}{2}$.
b. What kind of function does the graph represent?

41. **a.** Graph $y = kx^2$ when $k = 2$.
b. What kind of function does the graph represent?

42. **a.** Graph $y = kx^2$ when $k = \frac{1}{2}$.
b. What kind of function does the graph represent?

43. **a.** Graph $y = \frac{k}{x}$ when $k = 2$ and $x > 0$.

b. Is this the graph of a function?

44. **a.** Graph $y = \frac{k}{x^2}$ when $k = 2$ and $x > 0$.

b. Is this the graph of a function?

45. In the inverse variation equation $y = \frac{k}{x}$, what is the effect on x if y doubles?

46. In the direct variation equation $y = kx$, what is the effect on y when x doubles?

Complete using the word *directly* or *inversely*.

47. If a varies directly as b and inversely as c, then c varies ______ as b and ______ as a.

48. If a varies ______ as b and c, then abc is constant.

49. If the length of a rectangle is held constant, the area of the rectangle varies ______ as the width.

50. If the area of a rectangle is held constant, the length of the rectangle varies ______ as the width.

Calculators and Computers

Using the Graphing Calculator to Graph Functions

The Texas Instruments TI-81 calculator can be used to graph different types of functions.

A previous calculator exercise utilized the graphing calculator to graph straight lines (linear functions). The TI-81 can be used to graph higher-degree polynomial functions and absolute value functions.

Graph $f(x) = 2x^2 - 1$ and $f(x) = -2x^2 + 1$ on the same set of axes. First press [Y=] to display the menu. Then use the following key strokes.

2 [XIT] [^] 2 [−] 1 [ENTER]

[(−)] 2 [XIT] [^] 2 [+] 1 [ENTER] [GRAPH]

The graphs are shown below.

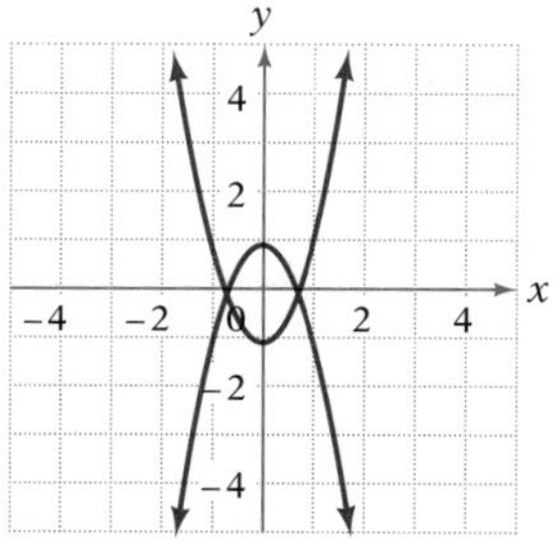

Graph each of the following on a graphing calculator.

1. Graph $f(x) = x^3 + 2$ and $f(x) = -x^3 + 2$ on the same set of axes.

2. Graph $f(x) = x^3 + 2$ and $f(x) = x^3 - 2$ on the same set of axes.

3. Graph $f(x) = x^3 + 2x^2 - 2x - 1$. How many x-intercepts does the graph have? Find the x-intercepts of the graph to two decimal places.

4. Graph $f(x) = x^4 - 1$ and $f(x) = -x^4 + 1$ on the same set of axes.

5. Graph $f(x) = x^4 - 2x^2 + 4$. How many x-intercepts does the graph have? Does the function have any real roots?

6. Graph $f(x) = 2x^4 - 5x^3 - 11x^2 + 20x + 12$. How many x-intercepts does the graph have? Find the x-intercepts of the graph to two decimal places.

7. Graph $f(x) = |x|$, $g(x) = |x - 2|$, and $h(x) = |x + 2|$ on the same set of axes. Make a conjecture about the effect of the number inside the absolute value symbol.

8. Graph $f(x) = |x|$, $g(x) = |x| - 2$, and $h(x) = |x| + 2$ on the same set of axes. Make a conjecture about the number outside the absolute value symbol.

9. Graph $f(x) = |x|$, $g(x) = |x - 2| + 2$, and $h(x) = |x - 2| - 2$ on the same set of axes. Note that the number inside the absolute value symbol moves the function right and left. The number outside the absolute value symbol moves the function up or down.

Something Extra

RSA Public-Key Cryptography

Cryptography is the study of techniques of concealing the meaning of a message. For example, the message

THE IDES OF MARCH,

called **plaintext,** could be concealed as

WKH LGHV RI PDUFK,

called **ciphertext.** Unless a person knows the procedure for writing the plaintext message as ciphertext, it is difficult to understand the ciphertext message. The procedure to change plaintext to ciphertext is called **encryption.**

For the example given, each letter of the plaintext message was replaced by the third letter after it in the alphabet. This method of encryption is sometimes called the "Caesar Cipher" after Julius Caesar, who supposedly used this method to send messages to his troops.

For the message above, knowing how the message was encrypted would enable you to rewrite the ciphertext as plaintext. Writing the ciphertext message as plaintext is called **decrypting** the message.

Not all encryption schemes are quite so easy, however. Ronald Rivest, Adi Shamir, and Leonard Adleman at the Massachusetts Institute of Technology developed a system of encrypting a message for which knowing how the message was encrypted does not help to decrypt the message. The method is called RSA Public-Key Cryptography. The method is based on large prime numbers and a function named *modulo n*, which is abbreviated *mod n*. The domain for a modulo n function is the integers.

To evaluate the modulo n function at b, divide b by n and write the remainder a. This is written $a \equiv b \pmod{n}$. For example,

$1 \equiv 17 \pmod{8}$ because the remainder is 1 when 17 is divided by 8.
$0 \equiv 22 \pmod{2}$ because the remainder is 0 when 22 is divided by 2.
$7 \equiv 7 \pmod{12}$ because the remainder is 7 when 7 is divided by 12.

The RSA Public-Key Cryptography system uses two very large prime numbers, p and q. The number n is the product of p and q ($n = pq$). For our example, let $p = 3$, $q = 11$, then $n = 33$. In actual practice, p and q would be prime numbers with at least 100 digits.

Before a message is encrypted, each letter of the alphabet is designated by a two-digit number. The table below shows one method, but any scheme will do.

A-01	B-02	C-03	D-04	E-05	F-06
G-07	H-08	I-09	J-10	K-11	L-12
M-13	N-14	O-15	P-16	Q-17	R-18
S-19	T-20	U-21	V-22	W-23	X-24
Y-25	Z-26				

Here is how the word MATH would be encrypted using RSA Public-Key Cryptography. First replace each letter by its numerical equivalent.

M A T H
13 01 20 08

Now choose a number to use as an exponent; 7 will be used here. (The RSA Public-Key Cryptography method shows how to choose the exponent.) To encrypt the first digit, evaluate $13^7 \pmod{33}$. Using a calculator, $07 \equiv 13^7 \pmod{33}$. (The zero is written to keep each number as a two-digit number.) The second letter is encrypted similarly ($01 \equiv 01^7 \pmod{33}$). The remaining letters, T and H, are encrypted as 26 and 02. The encrypted message is

07 01 26 02.

To decrypt this message, the modulo 33 function is again used, but this time a different exponent is chosen; 3 will be used here. (Again, the RSA Public-Key

Cryptography System shows how to arrive at this exponent.) Evaluate 07^3 (mod 33), 01^3 (mod 33), 26^3 (mod 33), and 02^3 (mod 33). The result is 13 01 20 08, the original message.

It is important to note here that the method of retrieving the original message is different from the way the message was encrypted. In one case, an exponent of 7 was used and in the second case an exponent of 3 was used. Thus knowing how the message was encrypted does not help to decrypt it. This is the major point of the RSA Public-Key Cryptography. The encryption key can be public (anyone can know it) and yet that does not help to decrypt a message.

In practice, this method requires using a computer to evaluate the modulo n function because n is very large, often over 200 digits, and the exponents are also very large. For more information on RSA Public-Key Cryptography see: *BYTE,* January 1983, pages 198–218; or *Mathematics Teacher,* January 1991, pages 54–61.

Evaluate.

1. 6 mod 8

2. 42 mod 5

3. 62 mod 6

4. 3^3 mod 4

5. 6^2 mod 10

6. $3 \cdot 5$ mod 3

7. $(8 \cdot 9)$ mod 5

8. $(16 + 3)$ mod 3

9. $(45 - 22)$ mod 4

Chapter Summary

Key Words

A *relation* is a set of ordered pairs and assigns to each member of a first set one or more members of a second set.

A *function* is a relation in which each element of the first set is assigned to one and only one member of the second set. No two ordered pairs that have the same first component have different second components.

The *domain* of a function is the set of first components of the ordered pairs of the function. The *range* of a function is the set of second components of the ordered pairs of the function.

A function given by the equation $f(x) = mx + b$, $m \neq 0$, is a *linear function.*

A function given by the equation $f(x) = b$ is a *constant function.*

A function given by the equation $f(x) = |x|$ is an *absolute value function.*

A function given by the equation $f(x) = ax^2 + bx + c$, $a \neq 0$, is a *quadratic function.*

A polynomial function of degree 3 is called a *cubic function.*

A *one-to-one function* is a function that satisfies the additional condition that given any y, there is only one x that can be paired with the given y.

The *vertical line test* is used to determine whether or not a graph is the graph of a function. The *horizontal line test* is used to determine whether or not the graph of a function is the graph of a 1–1 function.

The value of the *composition* of the functions f and g at a is $f(g(a))$.

The *inverse of a one-to-one function* is a function in which the components of each ordered pair are reversed.

Direct variation is a special function that can be expressed as the equation $y = kx^n$, where k is a constant called the *constant of variation* or the *constant of proportionality*.

Joint variation is a variation in which a variable varies directly as the product of two or more variables. A joint variation can be expressed as the equation $z = kxy$, where k is a constant.

Inverse variation is a function that can be expressed as the equation $y = \frac{k}{x^n}$, where k is a constant.

Combined variation is a variation in which two or more types of variation occur at the same time.

The y-coordinate of the vertex of the parabola whose equation is $f(x) = ax^2 + bx + c$, $a \neq 0$, is either a minimum or maximum value of the function.

Essential Rule

For the function f and its inverse f^{-1}, $f(f^{-1}(x)) = f^{-1}(f(x)) = x$.

Chapter Review

1. For $f(x) = x^3 - 8$, find $f(-2)$.

2. For $f(x) = x^2 + x + 2$, find $f(4 + h) - f(4)$.

3. For $f(x) = 3x^2 + 8$, find $\frac{f(2 + h) - f(2)}{h}$.

4. Find the domain and range of the function $\{(3, 8), (5, 8), (7, 8)\}$.

5. Find the range of $f(x) = 3x$ if the domain is $\{0, 2, 4, 6\}$.

6. The number 8 is in the range of $f(x) = x^2 + 4$. Find an element in the domain that corresponds to 8.

7. What values of x are excluded from the domain of $f(x) = \frac{3x + 1}{x^2 - 2x - 3}$?

8. What values of x are excluded from the domain of $f(x) = \sqrt{4x - 12}$?

9. Is the set of ordered pairs a function? $\{(3, 2), (4, 8), (4, 7), (1, 1)\}$

10. Is the set of ordered pairs a 1–1 function? $\{(1, 2), (3, 8), (8, 3), (2, 1)\}$

11. Given $f(x) = 3x^2 - 4$ and $g(x) = 2x + 1$, find $f(g(x))$.

12. Given $f(x) = x^2 + 4$ and $g(x) = 4x - 1$, find $f(g(0))$.

13. Given $f(x) = 6x + 8$ and $g(x) = 4x + 2$, find $g(f(-1))$.

14. Given $f(x) = 2x^2 + x - 5$ and $g(x) = 3x - 1$, find $g(f(x))$.

15. Find the inverse of the function defined by the equation $f(x) = \frac{1}{2}x + 8$.

16. Find the inverse of the function defined by the equation $f(x) = -6x + 4$.

17. Find the minimum value of the function defined by the equation $f(x) = 2x^2 - 6x + 1$.

18. Are the functions defined by the equations $f(x) = -\frac{1}{4}x + \frac{5}{4}$ and $g(x) = -4x + 5$ inverses of each other?

19. Determine if the graph represents the graph of a 1–1 function.

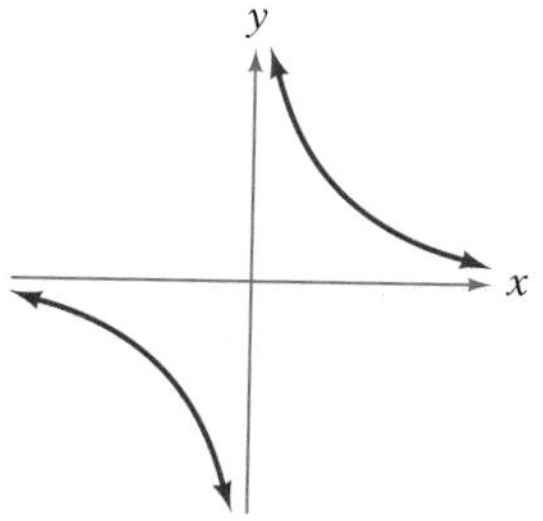

20. Determine if the graph represents the graph of a 1–1 function.

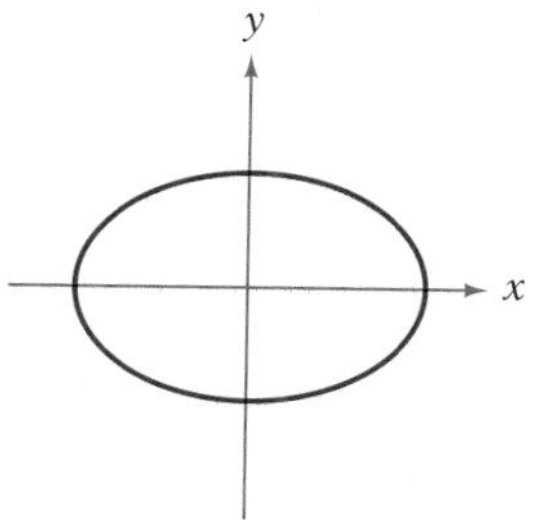

21. Determine if the graph represents the graph of a 1–1 function.

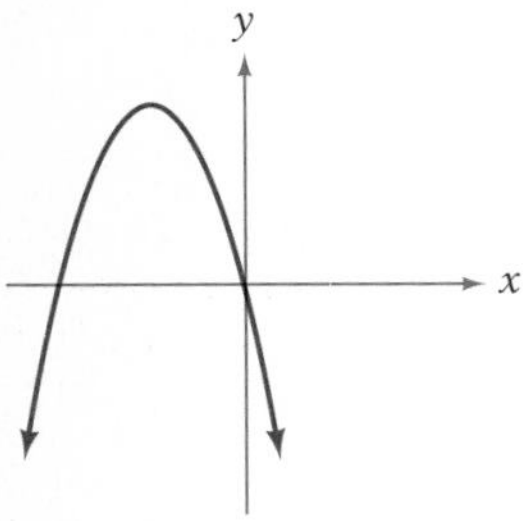

22. Determine if the graph represents the graph of a function.

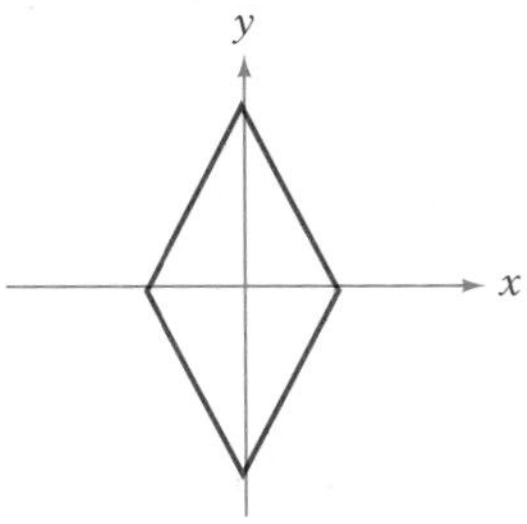

23. Graph $f(x) = -3x + 6$. State the domain and range.

24. Graph $x^2 + 2x - 4$. State the domain and range.

25. Graph $f(x) = |x| - 3$. State the domain and range.

26. Determine if the graph represents the graph of a function.

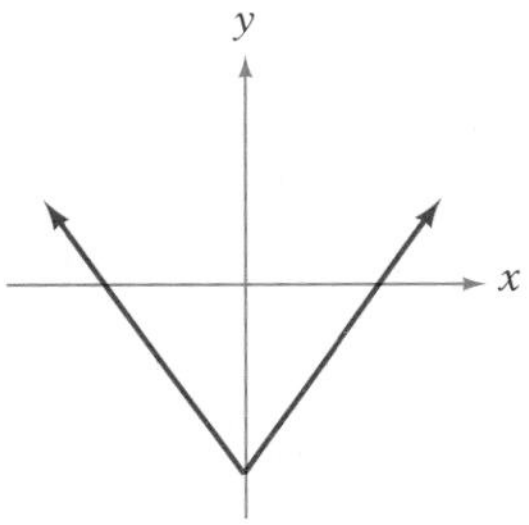

27. The pressure (p) of wind on a flat surface varies jointly as the area (A) of the surface and the square of the wind's velocity (v). If the pressure on 22 ft^2 is 10 lb when the wind's velocity is 10 mph, find the pressure on the same surface when the wind's velocity is 20 mph.

28. The illumination (I) produced by a light varies inversely as the square of the distance (d) from the light. If the illumination produced 10 ft from a light is 12 lumens, find the illumination 2 ft from the light.

29. The electrical resistance (r) of a cable varies directly as its length (l) and inversely as the square of its diameter (d). If a cable 16,000 ft long and $\frac{1}{4}$ in. in diameter has a resistance of 3.2 ohms, what is the resistance of a cable that is 8000 ft long and $\frac{1}{2}$ in. in diameter?

30. The profit function for a certain business is given by the equation $p(x) = -2x^2 + 360x - 600$. How many items should be sold to maximize the profit?

Chapter Test

1. For $f(x) = x^2 - 1$, find $f(4)$.

2. Evaluate $g(x) = -x^2 + 3x - 2$ at $x = -2$.

3. Is the set of ordered pairs a function? $\{(1, 3), (3, 3), (2, 4), (4, 4)\}$

4. What values of x are excluded from the domain of $f(x) = \frac{3}{x^2 - 8x}$?

5. Find the range of $f(x) = 4x - 3$ if the domain is $\{-2, -1, 0, 1, 2\}$.

6. Find the domain of the function $\{(3, -1), (5, -3), (7, -5), (9, -7)\}$.

7. For $g(x) = -2x + 3$, find $g(1 + h)$.

8. What values of x are excluded from the domain of $f(x) = \sqrt{x - 7}$?

9. Find the maximum value of the function whose equation is $f(x) = -3x^2 + 6x + 1$.

10. For $g(x) = 2x^2 + 1$ and $h(x) = 3x + 4$, find $g(0) + h(0)$.

11. Given $f(x) = 2x^2 - 3$ and $g(x) = 4x + 1$, evaluate $f(g(0))$.

12. Find the inverse of the function defined by the equation $f(x) = \frac{1}{2}x - 1$.

13. Given $g(x) = -3x + 2$ and $h(x) = x - 4$, evaluate $g(h(x))$.

14. Are the functions defined by the equations $f(x) = 3x + 6$ and $g(x) = \frac{1}{3}x + 2$ inverses of each other?

15. Determine if the graph represents the graph of a function.

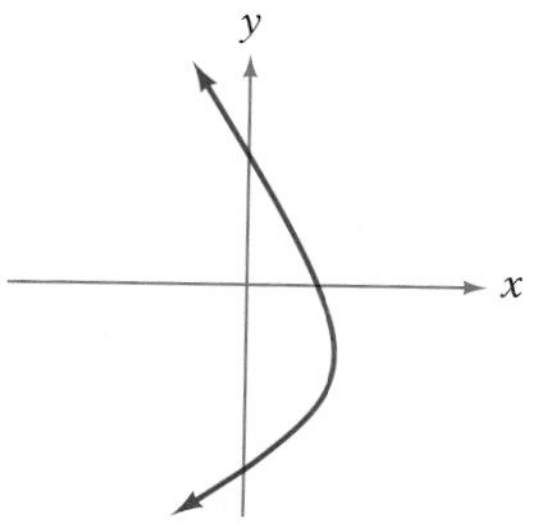

16. Determine if the graph represents the graph of a 1–1 function.

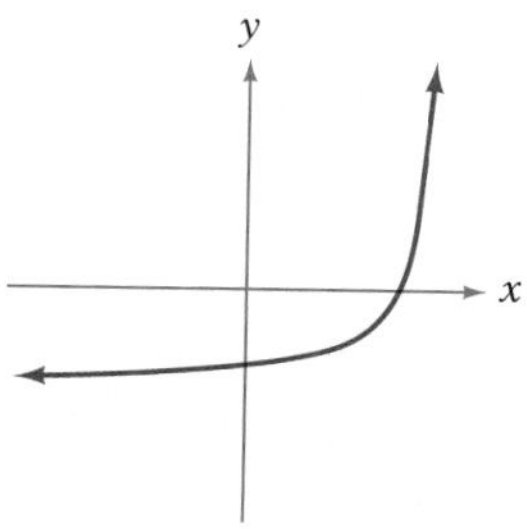

17. Graph $f(x) = |x| - 2$. State the domain and range of the function.

18. Graph $f(x) = x^3 + 2$. State the domain and range of the function.

19. Graph $f(x) = \frac{1}{2}x - 2$.

20. Graph $f(x) = -x^2 + 1$.

21. Graph $f(x) = |x + 1|$. State the domain and range of the function.

22. Graph $f(x) = -x^2 + 2x$. State the domain and range of the function.

23. The stopping distance (s) of a car varies directly as the square of the speed (v) of the car. For a car traveling at 50 mph, the stopping distance is 170 ft. Find the stopping distance of a car that is traveling at 30 mph.

24. The current (I) in an electric circuit varies inversely as the resistance (R). If the current in the circuit is 4 amps when the resistance is 50 ohms, find the current in the circuit when the resistance is 100 ohms.

25. The perimeter of a rectangle is 48 cm. What dimensions would give the rectangle a maximum area?

Cumulative Review

1. Evaluate $-3a + \left|\frac{3b - ab}{3b - c}\right|$ when $a = 2$, $b = 2$, and $c = -2$.

2. Divide: $x^3 - 3x^2 + 2x + 1 \div (x - 2)$

3. Solve: $\frac{3x-1}{6} - \frac{5-x}{4} = \frac{5}{6}$

4. Solve: $4x - 2 < -10$ or $3x - 1 > 8$

5. Solve: $|8 - 2x| \geq 0$

6. Simplify: $\left(\frac{3a^3b}{2a}\right)^2\left(\frac{a^2}{-3b^2}\right)^3$

7. Simplify: $(x - 4)(2x^2 + 4x - 1)$

8. Factor: $a^4 - 2a^2 - 8$

9. Factor: $x^3y + x^2y^2 - 6xy^3$

10. Solve: $(b + 2)(b - 5) = 2b + 14$

11. Solve: $x^2 - 2x > 15$

12. Simplify: $\frac{x^2 + 4x - 5}{2x^2 - 3x + 1} - \frac{x}{2x - 1}$

13. Solve: $\frac{5}{x^2 + 7x + 12} = \frac{9}{x + 4} - \frac{2}{x + 3}$

14. Simplify: $\frac{4 - 6i}{2i}$

15. Find the equation of the line containing the points $(-3, 4)$ and $(2, -6)$.

16. Find the equation of the line containing the point $(-3, 1)$ and perpendicular to the line $2x - 3y = 6$.

17. Solve: $3x^2 = 3x - 1$

18. Solve: $\sqrt{8x + 1} = 2x - 1$

19. Evaluate $f(x) = 2x^2 - 3$ at $x = -2$.

20. Find the range of $f(x) = |3x - 4|$ if the domain is $\{0, 1, 2, 3\}$.

21. Is the set of ordered pairs a function? $\{(-3, 0), (-2, 0), (-1, 1), (0, 1)\}$

22. Solve: $\sqrt[3]{5x - 2} = 2$

23. Find the inverse of the function given by the equation $f(x) = -3x + 9$.

24. Graph the set $\{x|x < -3\} \cap \{x|x > -4\}$.

25. Graph $f(x) = \frac{1}{4}x^2$.

26. Graph the solution set of $3x - 4y \geq 8$.

27. Find the cost per pound of a tea mixture made from 30 lb of tea costing $4.50 per pound and 45 lb of tea costing $3.60 per pound.

28. How many pounds of an 80% copper alloy must be mixed with 50 lb of a 20% copper alloy to make an alloy that is 40% copper?

29. Six ounces of an insecticide are mixed with 16 gal of water to make a spray for spraying an orange grove. How much additional insecticide is required to be mixed with 28 gal of water?

30. A large pipe can fill a tank in 8 min less time than it takes a smaller pipe to fill the same tank. Working together, both pipes can fill the tank in 3 min. How long would it take the larger pipe working alone to fill the tank?

31. The distance (d) a spring stretches varies directly as the force (f) used to stretch the spring. If a force of 50 lb can stretch the spring 30 in., how far will a force of 40 lb stretch the spring?

32. The frequency of vibration (f) in an open pipe organ varies inversely as the length (L) of the pipe. If the air in a pipe 2 m long vibrates 60 times per minute, find the frequency in a pipe that is 1.5 m long.

9

Conic Sections

Objectives

- Graph parabolas
- Find the distance between two points in the plane
- Find the midpoint of a line segment
- Find the equation of a circle and then graph the circle
- Write the equation of a circle in standard form and then graph the circle
- Graph an ellipse with center at the origin
- Graph a hyperbola with center at the origin
- Graph the solution set of a quadratic inequality in two variables

POINT OF INTEREST

Conic Sections

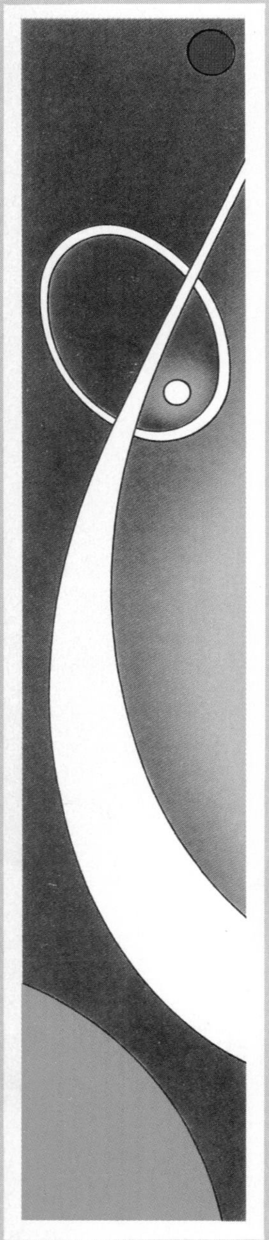

The graphs of three curves—the ellipse, the parabola, and the hyperbola—are discussed in this chapter. These curves were studied by the Greeks and were known prior to 400 B.C. The names of these curves were first used by Apollonius around 250 B.C. in *Conic Sections,* the most authoritative Greek discussion of these curves. Apollonius borrowed the names from a school founded by Pythagoras.

The diagram at the right shows the path of a planet around the sun. The curve traced out by the planet is an ellipse. The **aphelion** is the position of the planet when it is farthest from the sun. The **perihelion** is the position when the planet is nearest to the sun.

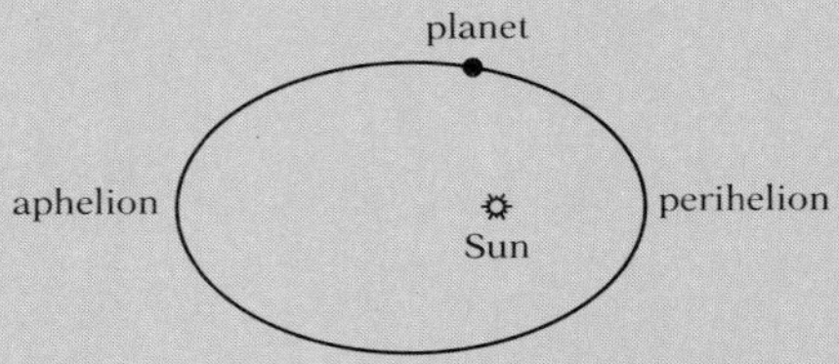

A telescope, like the one at the Palomar Observatory, has a cross section that is in the shape of a parabola. A parabolic mirror has the unusual property that all light rays parallel to the axis of symmetry that hit the mirror are reflected to the same point. This point is called the **focus of the parabola.**

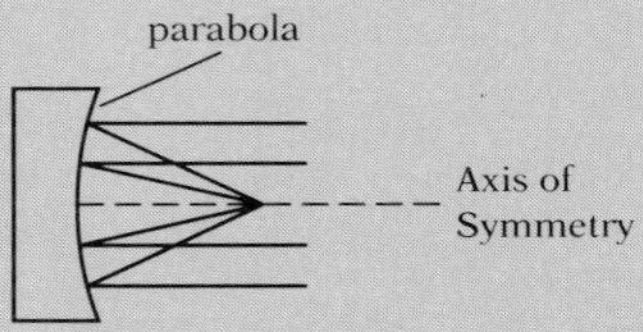

Some comets, unlike Halley's Comet, travel with such speed that they are not captured by the sun's gravitational field. The path of the comet as it comes around the sun is in the shape of a hyperbola.

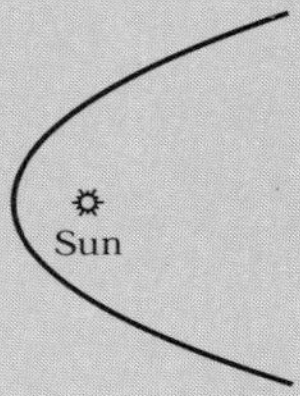

SECTION 9.1
The Parabola

1 Graph parabolas

A parabola is one of a number of curves called conic sections. The graph of a **conic section** can be represented by the intersection of a plane and a cone.

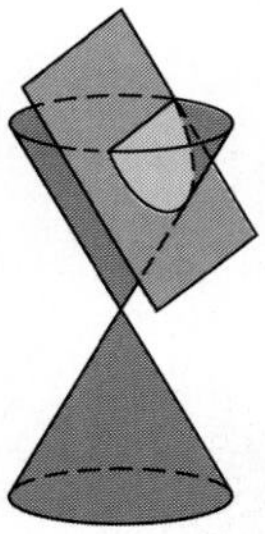

Recall from Chapter 7 that the graph of the equation $y = ax^2 + bx + c$, $a \neq 0$, is a parabola with the axis of symmetry parallel to the y-axis.

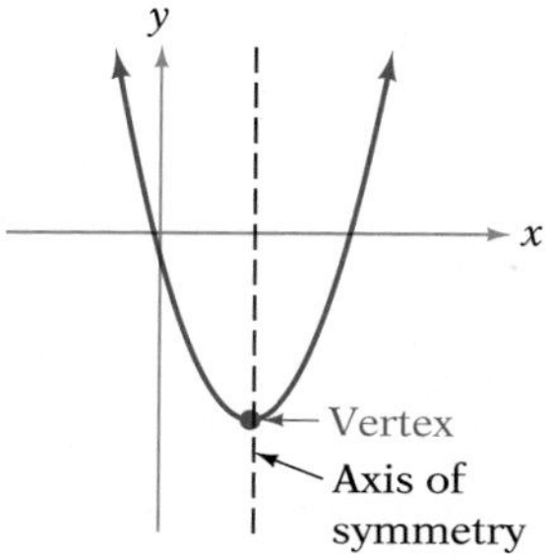

A review of finding the axis of symmetry and the coordinates of the vertex is contained in the example below.

Example 1 Find the vertex and the axis of symmetry of the parabola given by the equation $y = x^2 + 2x - 3$. Then sketch the graph of the parabola.

Solution $-\frac{b}{2a} = -\frac{2}{2(1)} = -1$ ▶ The x-coordinate of the vertex is $-\frac{b}{2a}$.

$y = x^2 + 2x - 3$
$= (-1)^2 + 2(-1) - 3$
$= -4$

▶ Find the y-coordinate of the vertex by replacing x by -1 and solving for y.

The vertex is $(-1, -4)$.

The axis of symmetry is the line $x = -1$. ▶ The axis of symmetry is the line $x = -\frac{b}{2a}$.

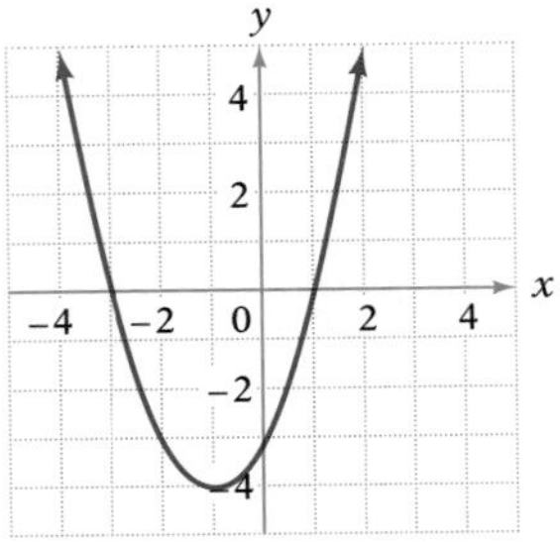

▶ Since a is positive, the parabola opens up. Use the vertex and axis of symmetry to sketch the graph.

Problem 1 Find the vertex and the axis of symmetry of the parabola given by the equation $y = -x^2 + x + 3$. Then sketch the graph of the parabola.

Solution See page A41.

The graph of an equation of the form $x = ay^2 + by + c$, $a \neq 0$, is also a parabola. In this case, the parabola opens to the right when a is positive and opens to the left when a is negative.

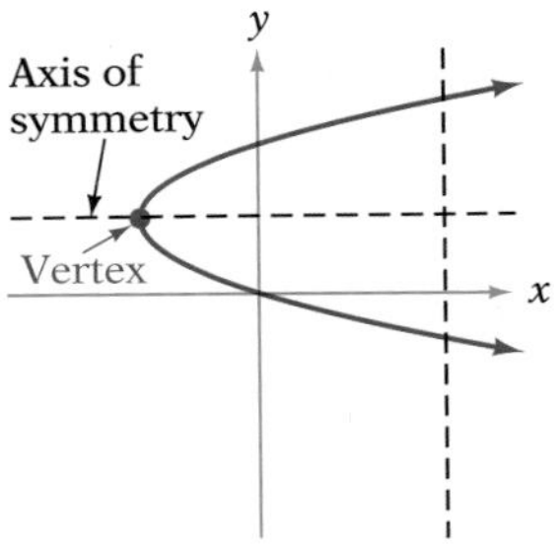

For a parabola of this form, the ***y*-coordinate of the vertex** is $-\frac{b}{2a}$. The **axis of symmetry** is the line $y = -\frac{b}{2a}$.

Using the vertical line test, the graph of a parabola of this form is not the graph of a function. The graph of $x = ay^2 + by + c$ is the graph of a relation.

Example 2 Find the vertex and the axis of symmetry of the parabola whose equation is $x = 2y^2 - 8y + 5$. Then sketch its graph.

Solution $-\frac{b}{2a} = -\frac{-8}{2(2)} = 2$

▶ Find the y-coordinate of the vertex. $a = 2$, $b = -8$.

$$x = 2y^2 - 8y + 5$$
$$= 2(2)^2 - 8(2) + 5$$
$$= -3$$

▶ Find the x-coordinate of the vertex by replacing y by 2 and solving for x.

The vertex is $(-3, 2)$.

The axis of symmetry is the line $y = 2$.

▶ The axis of symmetry is the line $y = -\frac{b}{2a}$.

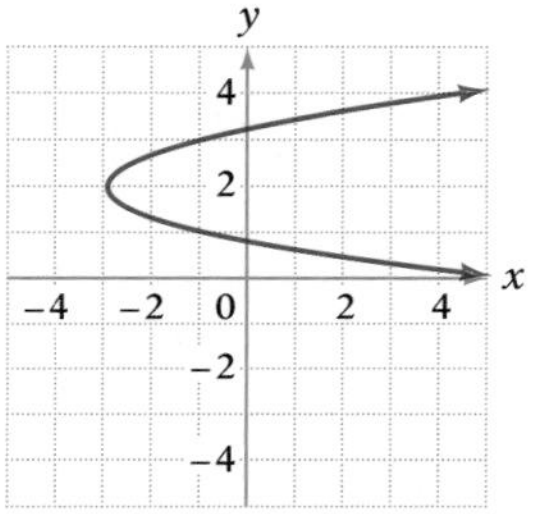

▶ Since a is positive, the parabola opens to the right. Use the vertex and axis of symmetry to sketch the graph.

Problem 2 Find the axis of symmetry and the vertex of the parabola whose equation is $x = -2y^2 - 4y - 3$. Then sketch its graph.

Solution See page A41.

EXERCISES 9.1

1 Find the vertex and axis of symmetry of the parabola given by the equation. Then sketch its graph.

1. $y = x^2 - 2x - 4$

2. $y = x^2 + 4x - 4$

3. $y = -x^2 + 2x - 3$

4. $y = -x^2 + 4x - 5$

5. $x = y^2 + 6y + 5$

6. $x = y^2 - y - 6$

7. $y = 2x^2 - 4x + 1$

8. $y = 2x^2 + 4x - 5$

9. $y = x^2 - 5x + 4$

10. $y = x^2 + 5x + 6$

11. $x = y^2 - 2y - 5$

12. $x = y^2 - 3y - 4$

13. $y = -3x^2 - 9x$

14. $y = -2x^2 + 6x$

15. $x = -\frac{1}{2}y^2 + 4$

16. $x = -\frac{1}{4}y^2 - 1$

17. $x = \frac{1}{2}y^2 - y + 1$

18. $x = -\frac{1}{2}y^2 + 2y - 3$

19. $y = \frac{1}{2}x^2 + 2x - 6$

20. $y = -\frac{1}{2}x^2 + x - 3$

SUPPLEMENTAL EXERCISES 9.1

Use the vertex and the direction in which the parabola opens to determine the domain and range of the relation.

21. $y = x^2 - 4x - 2$

22. $y = x^2 - 6x + 1$

23. $y = -x^2 + 2x - 3$

24. $y = -x^2 - 2x + 4$

25. $x = y^2 + 6y - 5$

26. $x = y^2 + 4y - 3$

27. $x = -y^2 - 2y + 6$

28. $x = -y^2 - 6y + 2$

Recall from the Point of Interest feature at the beginning of this chapter that an application of parabolas as mirrors for telescopes was mentioned. The light from a source strikes the mirror and is reflected to a point called the **focus** of the parabola. The focus is $\frac{1}{4a}$ units from the vertex of the parabola on the axis of symmetry in the direction the parabola opens. In the expression $\frac{1}{4a}$, a is the coefficient of the second-degree term. In each of the following, find the coordinates of the focus of each parabola.

29. $y = 2x^2 - 4x + 1$

30. $y = -\frac{1}{4}x^2 + 2$

31. $x = \frac{1}{2}y^2 + y - 2$

32. $x = -y^2 - 4y + 1$

SECTION 9.2 The Circle

1 Find the distance between two points in the plane

The distance between two points on a horizontal or vertical number line is the absolute value of the difference between the coordinates of the two points.

Find the distance between the points −2 and 4 on the vertical number line shown at the right.

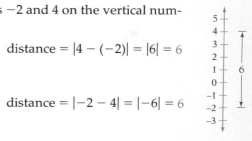

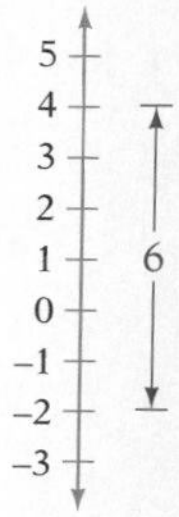

The distance is the absolute value of the difference between the coordinates.

$\text{distance} = |4 - (-2)| = |6| = 6$

Absolute value is used so that the coordinates can be subtracted in either order.

$\text{distance} = |-2 - 4| = |-6| = 6$

Now consider the points and the right triangle in the coordinate plane shown at the right. The vertical distance between $P_1(x_1, y_1)$ and $P_2(x_2, y_2)$, is $|y_2 - y_1|$.

The horizontal distance between the two points $P_1(x_1, y_1)$ and $P_2(x_2, y_2)$ is $|x_2 - x_1|$.

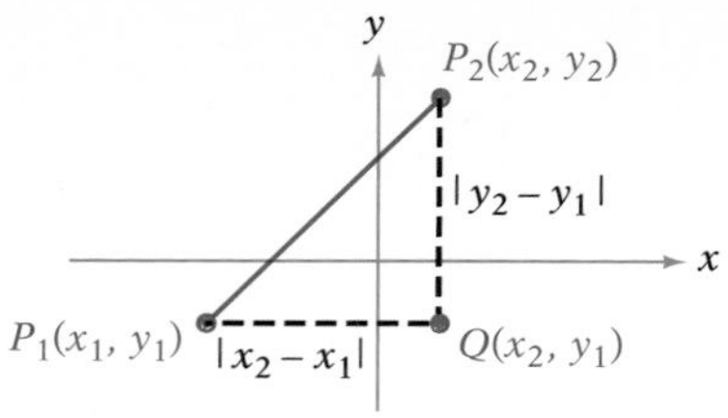

The Pythagorean Theorem is used to find the distance between $P_1(x_1, y_1)$ and $P_2(x_2, y_2)$.

$$d^2 = |x_2 - x_1|^2 + |y_2 - y_1|^2$$

Since the square of a number is always nonnegative, the absolute value signs are not necessary.

$$d^2 = (x_2 - x_1)^2 + (y_2 - y_1)^2$$

Taking the square root of each side of the equation gives the distance formula.

$$d = \sqrt{(x_2 - x_1)^2 + (y_2 - y_1)^2}$$

The Distance Formula

If $P_1(x_1, y_1)$ and $P_2(x_2, y_2)$ are two points in the plane, then the distance between the two points is given by $d = \sqrt{(x_2 - x_1)^2 + (y_2 - y_1)^2}$.

Example 1 Find the distance between the points whose coordinates are $(-4, 3)$ and $(2, -1)$.

Solution $(x_1, y_1) = (-4, 3) \qquad (x_2, y_2) = (2, -1)$

$$d = \sqrt{(x_2 - x_1)^2 + (y_2 - y_1)^2} = \sqrt{[2 - (-4)]^2 + (-1 - 3)^2}$$
$$= \sqrt{6^2 + (-4)^2} = \sqrt{36 + 16} = \sqrt{52} = 2\sqrt{13}$$

Problem 1 Find the distance between the points whose coordinates are $(-1, -5)$ and $(3, -2)$.

Solution See page A41.

2 Find the midpoint of a line segment

The midpoint of a line segment is equidistant from its endpoints. The midpoint of the line segment P_1P_2 is (x_m, y_m). The intersection of the horizontal line segment through P_1 and the vertical line through P_2 is Q with coordinates (x_2, y_1).

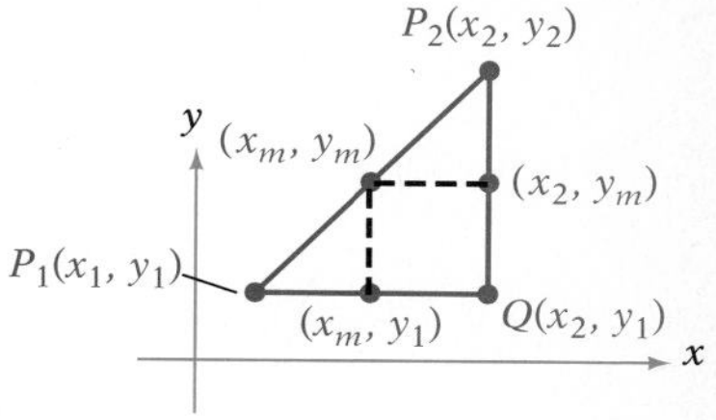

The x-coordinate x_m of the midpoint of the line segment P_1P_2 is the same as the x-coordinate of the midpoint of the line segment P_1Q. It is the average of the x-coordinates of the points P_1 and P_2, or

$$x_m = \frac{x_1 + x_2}{2}.$$

In like manner, the y-coordinate of the midpoint of the line segment P_1P_2 is

$$y_m = \frac{y_1 + y_2}{2}.$$

The Midpoint Formula

> If $P_1(x_1, y_1)$ and $P_2(x_2, y_2)$ are the endpoints of a line segment, then the coordinates of the midpoint (x_m, y_m) of the line segment are given by
>
> $$x_m = \frac{x_1 + x_2}{2} \quad \text{and} \quad y_m = \frac{y_1 + y_2}{2}.$$

Find the coordinates of the midpoint of the line segment with endpoints $(-2, 5)$ and $(6, -8)$.

Substitute the coordinates of the endpoints in the midpoint formulas and simplify.

$$x_m = \frac{x_1 + x_2}{2} \qquad y_m = \frac{y_1 + y_2}{2}$$

$$x_m = \frac{-2 + 6}{2} \qquad y_m = \frac{5 + (-8)}{2}$$

$$x_m = 2 \qquad y_m = -\frac{3}{2}$$

The coordinates of the midpoint of the line segment are $\left(2, -\frac{3}{2}\right)$.

Example 2 Find the coordinates of the midpoint of the line segment with endpoints $(-3, -6)$ and $(0, 7)$.

Solution $x_m = \frac{x_1 + x_2}{2}$ $\qquad$ $y_m = \frac{y_1 + y_2}{2}$

$= \frac{-3 + 0}{2}$ $\qquad$ $= \frac{-6 + 7}{2}$

$= -\frac{3}{2}$ $\qquad$ $= \frac{1}{2}$

The coordinates of the midpoint of the line segment are $\left(-\frac{3}{2}, \frac{1}{2}\right)$.

Problem 2 Find the coordinates of the midpoint of the line segment with endpoints (−8, 4) and (−4, 5).

Solution See page A41.

3 Find the equation of a circle and then graph the circle

A **circle** is a conic section that is formed by the intersection of a cone by a plane parallel to the base of the cone.

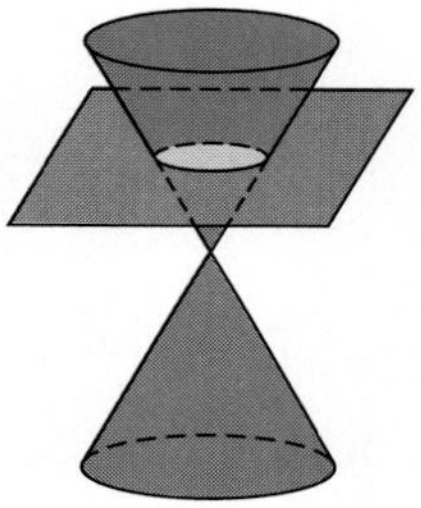

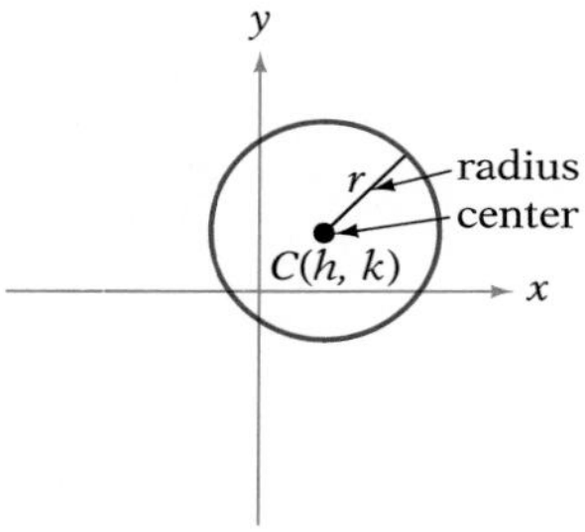

A **circle** can be defined as the set of all points $P(x, y)$ in the plane that are a fixed distance from a given point $C(h, k)$ called the **center.** The fixed distance is the **radius** of the circle.

Using the distance formula, the equation of a circle can be determined.

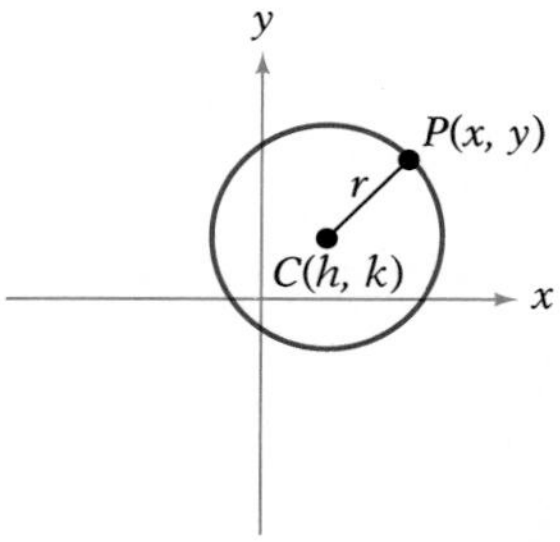

Let (h, k) be the coordinates of the center of the circle, r the radius, and $P(x, y)$ any point on the circle.

Then $r = \sqrt{(x - h)^2 + (y - k)^2}$.

Squaring each side of the equation gives the equation of a circle.

The Standard Form of the Equation of a Circle

Let r be the radius of a circle and $C(h, k)$ the center of the circle. Then the equation of the circle is given by

$$(x - h)^2 + (y - k)^2 = r^2.$$

Recall that the graph of a circle is not the graph of a function. The graph of a circle is the graph of a relation.

To find the equation of the circle with radius 4 and center $C(-1, 2)$, use the standard form of the equation of a circle.

$$(x - h)^2 + (y - k)^2 = r^2$$

Replace r by 4, h by -1, and k by 2.

$$[x - (-1)]^2 + (y - 2)^2 = 4^2$$
$$(x + 1)^2 + (y - 2)^2 = 16$$

To sketch the graph of this circle, draw a circle with center $C(-1, 2)$ and radius 4.

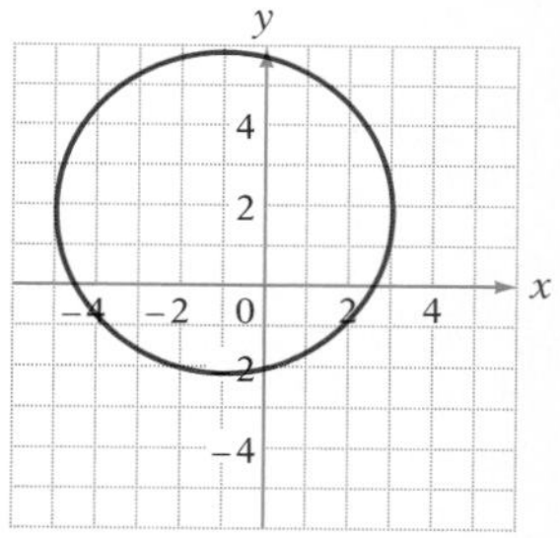

Example 3 Find the equation of the circle with radius 5 and center $C(-1, 3)$. Then sketch its graph.

Solution

$$(x - h)^2 + (y - k)^2 = r^2$$
$$[x - (-1)]^2 + (y - 3)^2 = 5^2$$
$$(x + 1)^2 + (y - 3)^2 = 25$$

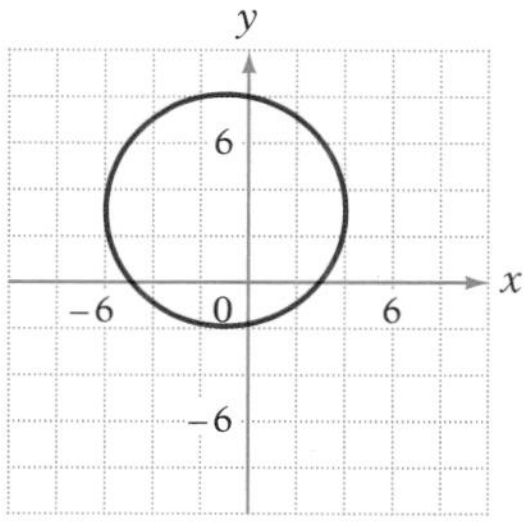

Problem 3 Find the equation of the circle with radius 4 and center $C(2, -3)$. Then sketch its graph.

Solution See page A42.

A circle passes through point $P(2, 1)$ and has as its center the point $C(3, -4)$. Because the center of the circle and a point on the circle are known, use the distance formula to find the radius of the circle.

$$\sqrt{(x_2 - x_1)^2 + (y_2 - y_1)^2} = r$$
$$\sqrt{(3 - 2)^2 + (-4 - 1)^2} = r$$
$$\sqrt{1^2 + (-5)^2} = r$$
$$\sqrt{1 + 25} = r$$
$$\sqrt{26} = r$$

The radius is $\sqrt{26}$. The equation of the circle is $(x - 3)^2 + (y + 4)^2 = 26$.

Example 4 Find the equation of the circle in which a diameter has endpoints $P_1(-4, -1)$ and $P_2(2, 3)$.

Solution

$$x_m = \frac{x_1 + x_2}{2} \qquad y_m = \frac{y_1 + y_2}{2}$$
$$x_m = \frac{-4 + 2}{2} \qquad y_m = \frac{-1 + 3}{2}$$
$$x_m = -1 \qquad y_m = 1$$
$$(x_m, y_m) = (-1, 1)$$

▶ Let $(x_1, y_1) = (-4, -1)$ and $(x_2, y_2) = (2, 3)$. Find the center of the circle by finding the midpoint of the diameter.

$$r = \sqrt{(x_1 - x_m)^2 + (y_1 - y_m)^2}$$
$$r = \sqrt{(-4 - [-1])^2 + (-1 - 1)^2}$$
$$= \sqrt{9 + 4}$$
$$= \sqrt{13}$$

▶ Find the radius of the circle. Use either point on the circle and the coordinates of the center of the circle. P_1 is used here.

$$(x + 1)^2 + (y - 1)^2 = 13$$

▶ Write the equation of the circle with center $C(-1, 1)$ and radius $\sqrt{13}$.

Problem 4 Find the equation of the circle for which a diameter has endpoints $P_1(-2, 1)$ and $P_2(4, -1)$.

Solution See page A42.

4 Write the equation of a circle in standard form and then graph the circle

The equation of a circle can also be expressed as the equation $\mathbf{x^2 + y^2 + ax + by + c = 0}$. To rewrite this equation in standard form, it is necessary to complete the square on the x and y terms.

Write the equation of the circle $x^2 + y^2 + 4x + 2y + 1 = 0$ in standard form.

$$x^2 + y^2 + 4x + 2y + 1 = 0$$

Subtract the constant term from each side of the equation.

$$x^2 + y^2 + 4x + 2y = -1$$

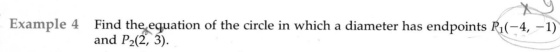

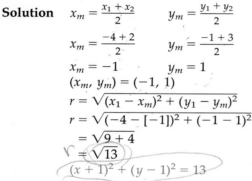

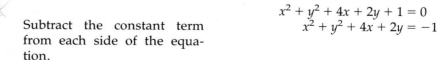

Rewrite the equation by grouping terms involving x and terms involving y. $(x^2 + 4x) + (y^2 + 2y) = -1$

Complete the square on $x^2 + 4x$ and $y^2 + 2y$.
$(x^2 + 4x + 4) + (y^2 + 2y + 1) = -1 + 4 + 1$
$(x^2 + 4x + 4) + (y^2 + 2y + 1) = 4$

Factor each trinomial. $(x + 2)^2 + (y + 1)^2 = 4$

Example 5 Write the equation of the circle $x^2 + y^2 + 3x - 2y = 1$ in standard form. Then sketch its graph.

Solution

$$x^2 + y^2 + 3x - 2y = 1$$

$$(x^2 + 3x) + (y^2 - 2y) = 1$$

▶ Group terms involving x and terms involving y.

$$\left(x^2 + 3x + \frac{9}{4}\right) + (y^2 - 2y + 1) = 1 + \frac{9}{4} + 1$$

▶ Complete the square on $x^2 + 3x$ and $y^2 - 2y$.

$$\left(x + \frac{3}{2}\right)^2 + (y - 1)^2 = \frac{17}{4}$$

▶ Factor each trinomial.

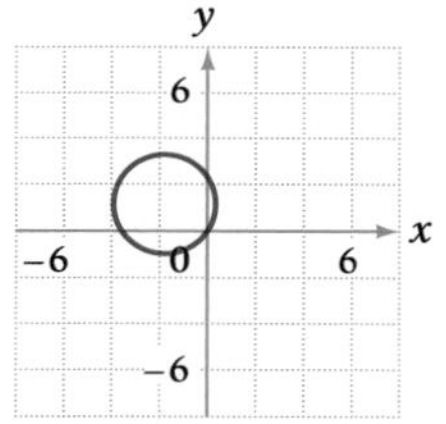

▶ Draw a circle with center $\left(-\frac{3}{2}, 1\right)$ and radius $\sqrt{\frac{17}{4}} = \frac{\sqrt{17}}{2}$.
$\frac{\sqrt{17}}{2} \approx 2.1$

Problem 5 Write the equation of the circle $x^2 + y^2 - 4x + 8y + 15 = 0$ in standard form. Then sketch its graph.

Solution See page A42.

EXERCISES 9.2

1 Find the distance between the two points.

1. $P_1(3, 2); P_2(4, 5)$
2. $P_1(4, 5); P_2(1, 1)$
3. $P_1(-4, 2); P_2(1, -1)$
4. $P_1(3, -2); P_2(-1, 4)$
5. $P_1(-1, -2); P_2(1, 5)$
6. $P_1(-2, -3); P_2(3, 1)$
7. $P_1(-1, -1); P_2(-4, -5)$
8. $P_1(-4, 1); P_2(3, -2)$

2 Find the midpoint of the line segment joining the points.

9. $P_1(0, 5)$; $P_2(3, -1)$

10. $P_1(6, -8)$; $P_2(7, -4)$

11. $P_1(6, 4)$; $P_2(-3, -2)$

12. $P_1(-2, 2)$; $P_2(4, -5)$

13. $P_1(0, -6)$; $P_2(0, -4)$

14. $P_1(0, 0)$; $P_2(0, -2)$

15. $P_1(-3, -3)$; $P_2(-4, 4)$

16. $P_1(-2, 5)$; $P_2(-5, 3)$

3 Sketch a graph of the equation of the circle.

17. $(x - 2)^2 + (y + 2)^2 = 9$

18. $(x + 2)^2 + (y - 3)^2 = 16$

19. $(x + 3)^2 + (y - 1)^2 = 25$

20. $(x - 2)^2 + (y + 3)^2 = 4$

21. $(x - 4)^2 + (y + 2)^2 = 1$

22. $(x - 3)^2 + (y - 2)^2 = 16$

23. $(x + 5)^2 + (y + 2)^2 = 4$

24. $(x + 1)^2 + (y - 1)^2 = 9$

25. Find the equation of the circle with radius 2 and center $C(2, -1)$. Then sketch its graph.

26. Find the equation of the circle with radius 3 and center $C(-1, -2)$. Then sketch its graph.

27. Find the equation of the circle that passes through point $P(1, 2)$ and whose center is $C(-1, 1)$. Then sketch its graph.

28. Find the equation of the circle that passes through point $P(-1, 3)$ and whose center is $C(-2, 1)$. Then sketch its graph.

29. Find the equation of the circle for which the diameter has endpoints $P_1(-1, 4)$ and $P_2(-5, 8)$.

30. Find the equation of the circle for which the diameter has endpoints $P_1(2, 3)$ and $P_2(5, -2)$.

31. Find the equation of the circle for which the diameter has endpoints $P_1(-4, 2)$ and $P_2(0, 0)$.

32. Find the equation of the circle for which the diameter has endpoints $P_1(-8, -3)$ and $P_2(0, -4)$.

4 Write the equation of the circle in standard form. Then sketch its graph.

33. $x^2 + y^2 - 2x + 4y - 20 = 0$

34. $x^2 + y^2 - 4x + 8y + 4 = 0$

35. $x^2 + y^2 + 6x + 8y + 9 = 0$

36. $x^2 + y^2 - 6x + 10y + 25 = 0$

37. $x^2 + y^2 - x + 4y + \frac{13}{4} = 0$

38. $x^2 + y^2 + 4x + y + \frac{1}{4} = 0$

39. $x^2 + y^2 - 6x + 4y + 4 = 0$

40. $x^2 + y^2 - 10x + 8y + 40 = 0$

SUPPLEMENTAL EXERCISES 9.2

The vertices of a triangle are the given points. Use the Pythagorean Theorem to determine whether the triangle is a right triangle.

41. (0, 2), (3, 5), (6, 2)

42. (−3, 4), (−1, 4), (7, −2)

43. (−1, 0), (7, −8), (0, −7)

44. (−3, 6), (6, −6), (−6, 0)

Write the equation of the circle in standard form.

45. The circle has center $C(3, 0)$ and passes through the origin.

46. The circle has center $C(-2, 0)$ and passes through the origin.

47. The diameter of the circle has endpoints $P_1(-1, 3)$ and $P_2(5, 5)$.

48. The diameter of the circle has endpoints $P_1(-2, 4)$ and $P_2(2, -2)$.

49. The circle has radius 1, is tangent to both the x- and y-axes, and lies in Quadrant II.

50. The circle has radius 1, is tangent to both the x- and y-axes, and lies in Quadrant IV.

SECTION 9.3 The Ellipse and the Hyperbola

1 Graph an ellipse with center at the origin

The orbits of the planets around the sun are oval shaped. This oval shape can be described as an **ellipse,** which is another of the conic sections.

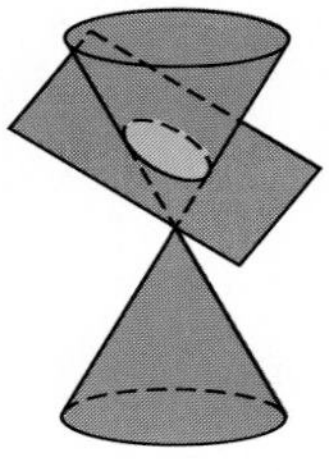

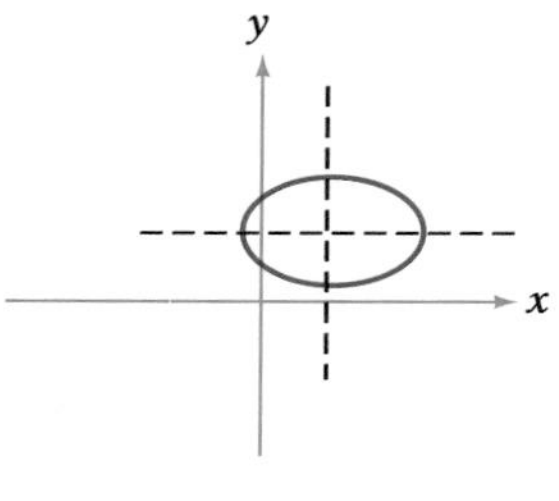

There are two **axes of symmetry** for an ellipse. The intersection of the two axes is the **center** of the ellipse.

An ellipse with center at the origin is shown at the right. Note that there are two x-intercepts and two y-intercepts.

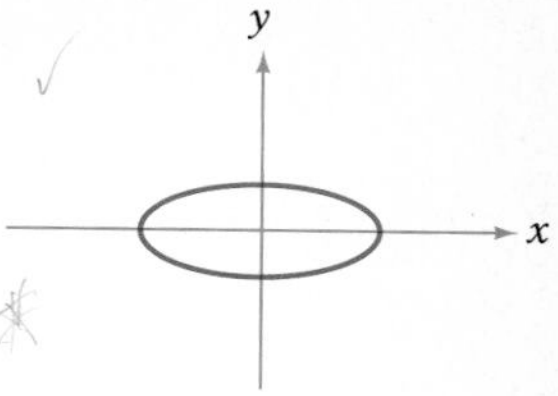

Using the vertical line test, the graph of an ellipse is not the graph of a function. The graph of an ellipse is the graph of a relation.

The Standard Form of the Equation of an Ellipse with Center at the Origin

The equation of an ellipse with center at the origin is $\frac{x^2}{a^2} + \frac{y^2}{b^2} = 1$.

The x-intercepts are $(a, 0)$ and $(-a, 0)$.
The y-intercepts are $(0, b)$ and $(0, -b)$.

By finding the x- and y-intercepts for an ellipse and using the fact that the ellipse is oval shaped, a graph of an ellipse can be sketched.

Example 1 Sketch a graph of the ellipse given by the equation.

A. $\frac{x^2}{9} + \frac{y^2}{4} = 1$ B. $\frac{x^2}{16} + \frac{y^2}{12} = 1$

Solution A. $\frac{x^2}{9} + \frac{y^2}{4} = 1$

▶ $a^2 = 9$, $b^2 = 4$

x-intercepts: $(3, 0)$ and $(-3, 0)$

▶ The x-intercepts are $(a, 0)$ and $(-a, 0)$.

y-intercepts: $(0, 2)$ and $(0, -2)$

▶ The y-intercepts are $(0, b)$ and $(0, -b)$.

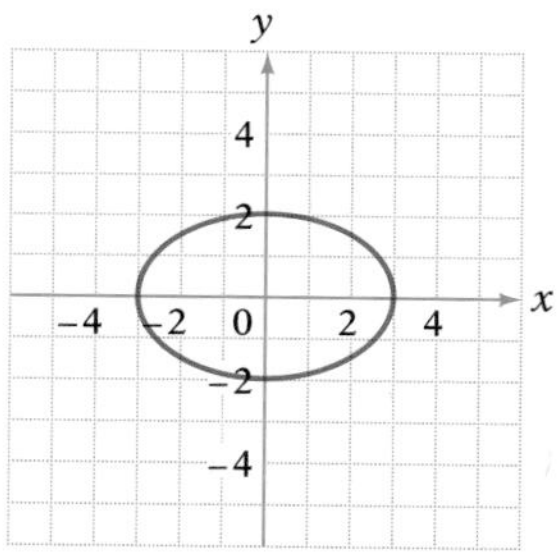

▶ Use the intercepts and symmetry to sketch the graph of the ellipse.

B. $\frac{x^2}{16} + \frac{y^2}{12} = 1$

▶ $a^2 = 16$, $b^2 = 12$

x-intercepts: (4, 0) and (−4, 0)

▶ The x-intercepts are $(a, 0)$ and $(-a, 0)$.

y-intercepts: $(0, 2\sqrt{3})$ and $(0, -2\sqrt{3})$

▶ The y-intercepts are $(0, b)$ and $(0, -b)$.

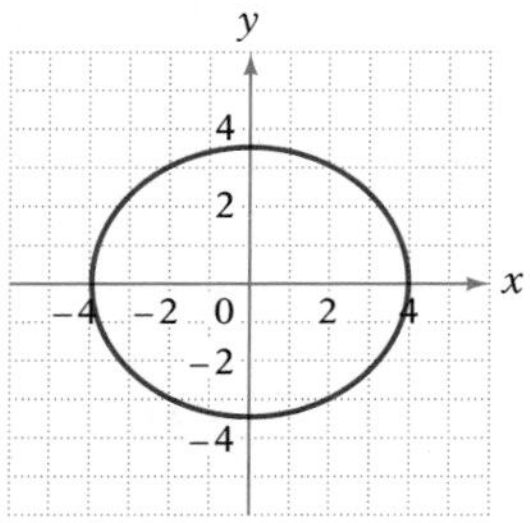

▶ Use the intercepts and symmetry to sketch the graph of the ellipse. $2\sqrt{3} \approx 3.5$

Problem 1 Sketch a graph of the ellipse given by the equation.

A. $\frac{x^2}{4} + \frac{y^2}{25} = 1$ B. $\frac{x^2}{18} + \frac{y^2}{9} = 1$

Solution See page A42.

2 Graph a hyperbola with center at the origin

A **hyperbola** is a conic section that is formed by the intersection of a cone by a plane perpendicular to the base of the cone.

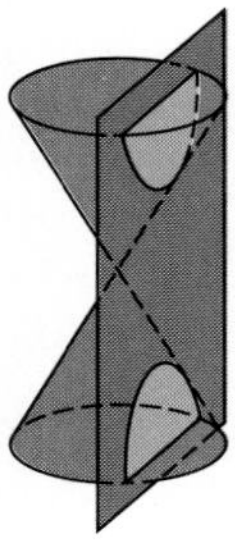

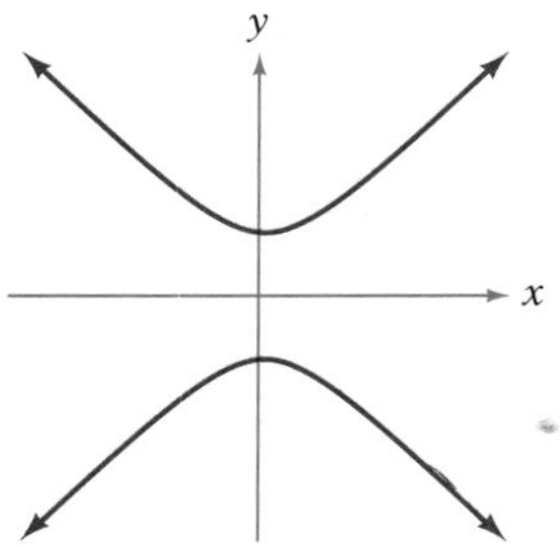

The hyperbola has two **vertices** and an **axis of symmetry** that passes through the vertices. The **center** of a hyperbola is the midpoint between the two vertices.

The graphs at the top of the next page show two possible graphs of a hyperbola with center at the origin. In the first graph, the vertices are x-intercepts. In the second graph, the vertices are y-intercepts.

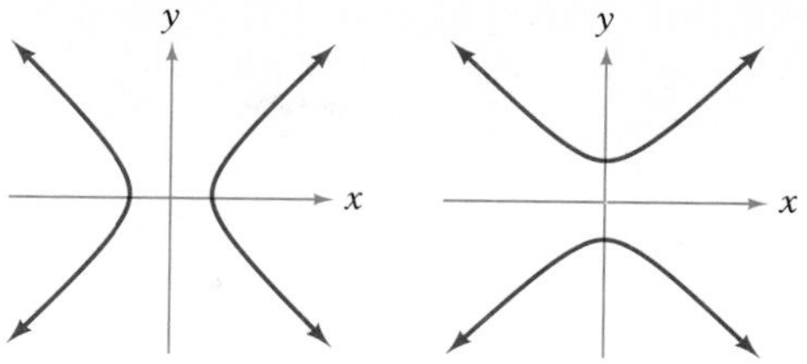

In either case, the graph of a hyperbola is not the graph of a function. The graph of a hyperbola is the graph of a relation.

The Standard Form of the Equation of a Hyperbola with Center at the Origin

The equation of a hyperbola for which the axis of symmetry is the x-axis is $\frac{x^2}{a^2} - \frac{y^2}{b^2} = 1$. The vertices are $(a, 0)$ and $(-a, 0)$.

The equation of a hyperbola for which the axis of symmetry is the y-axis is $\frac{y^2}{a^2} - \frac{x^2}{b^2} = 1$. The vertices are $(0, a)$ and $(0, -a)$.

To sketch a hyperbola, it is helpful to draw two lines that are "approached" by the hyperbola. These two lines are called **asymptotes.** As the hyperbola gets farther from the origin, the hyperbola "gets closer to" the asymptotes.

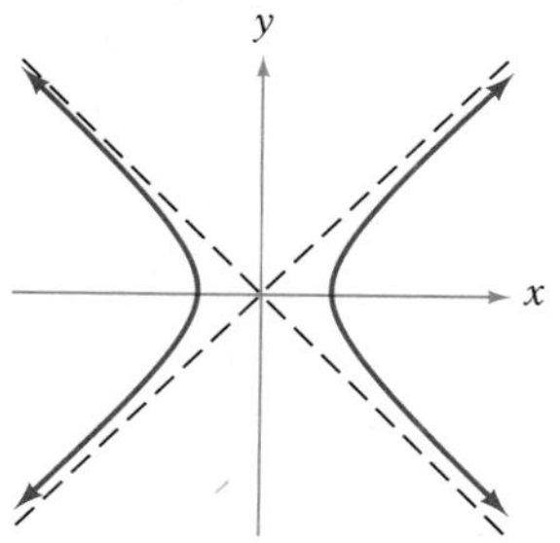

Since the asymptotes are straight lines, their equations are linear equations.

Asymptotes of a Hyperbola with Center at the Origin

The equations of the asymptotes for the hyperbola $\frac{x^2}{a^2} - \frac{y^2}{b^2} = 1$ are $y = \frac{b}{a}x$ and $y = -\frac{b}{a}x$.

The equations of the asymptotes for the hyperbola $\frac{y^2}{a^2} - \frac{x^2}{b^2} = 1$ are $y = \frac{a}{b}x$ and $y = -\frac{a}{b}x$.

Example 2 Sketch a graph of the hyperbola given by the equation.

A. $\frac{x^2}{16} - \frac{y^2}{4} = 1$ B. $\frac{y^2}{16} - \frac{x^2}{25} = 1$

Solution A. $\frac{x^2}{16} - \frac{y^2}{4} = 1$

▶ $a^2 = 16$, $b^2 = 4$

Axis of symmetry: x-axis

Vertices: (4, 0) and (−4, 0)

▶ The vertices are $(a, 0)$ and $(-a, 0)$.

Asymptotes: $y = \frac{1}{2}x$ and $y = -\frac{1}{2}x$

▶ The asymptotes are $y = \frac{b}{a}x$ and $y = -\frac{b}{a}x$.

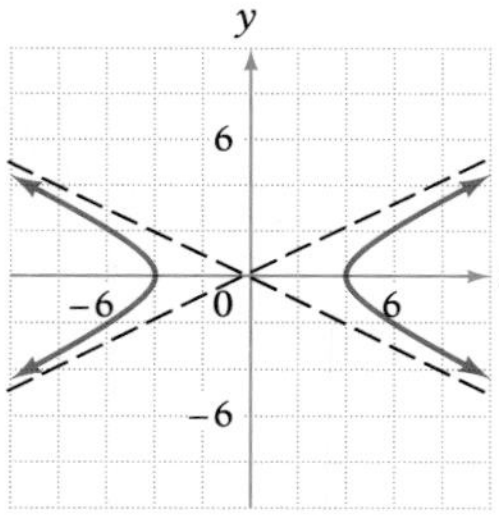

▶ Sketch the asymptotes. Use symmetry and the fact that the hyperbola will approach the asymptotes to sketch its graph.

B. $\frac{y^2}{16} - \frac{x^2}{25} = 1$

▶ $a^2 = 16$, $b^2 = 25$

Axis of symmetry: y-axis

Vertices: (0, 4) and (0, −4)

▶ The vertices are $(0, a)$ and $(0, -a)$.

Asymptotes: $y = \frac{4}{5}x$ and $y = -\frac{4}{5}x$

▶ The asymptotes are $y = \frac{a}{b}x$ and $y = -\frac{a}{b}x$.

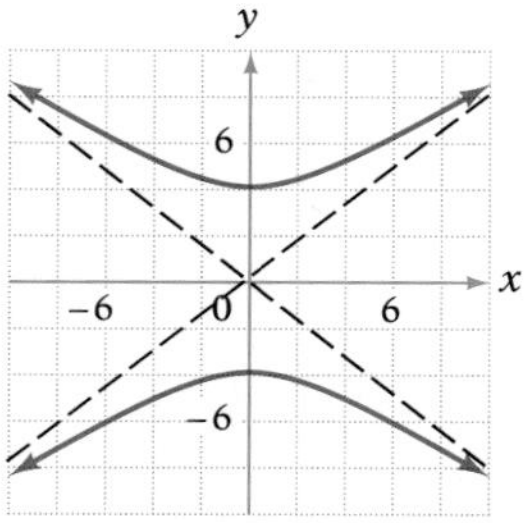

▶ Sketch the asymptotes. Use symmetry and the fact that the hyperbola will approach the asymptotes to sketch its graph.

Problem 2 Sketch a graph of the hyperbola given by the equation.

A. $\frac{x^2}{9} - \frac{y^2}{25} = 1$ B. $\frac{y^2}{9} - \frac{x^2}{9} = 1$

Solution See page A43.

EXERCISES 9.3

1 Sketch a graph of the ellipse given by the equation.

1. $\frac{x^2}{4} + \frac{y^2}{9} = 1$

2. $\frac{x^2}{25} + \frac{y^2}{16} = 1$

3. $\frac{x^2}{25} + \frac{y^2}{9} = 1$

4. $\frac{x^2}{16} + \frac{y^2}{9} = 1$

5. $\frac{x^2}{36} + \frac{y^2}{16} = 1$

6. $\frac{x^2}{49} + \frac{y^2}{64} = 1$

7. $\frac{x^2}{9} + \frac{y^2}{25} = 1$

8. $\frac{x^2}{8} + \frac{y^2}{25} = 1$

9. $\frac{x^2}{12} + \frac{y^2}{4} = 1$

10. $\frac{x^2}{16} + \frac{y^2}{36} = 1$

11. $\frac{x^2}{36} + \frac{y^2}{9} = 1$

12. $\frac{x^2}{4} + \frac{y^2}{16} = 1$

2 Sketch a graph of the hyperbola given by the equation.

13. $\frac{x^2}{9} - \frac{y^2}{16} = 1$

14. $\frac{x^2}{25} - \frac{y^2}{4} = 1$

15. $\frac{y^2}{16} - \frac{x^2}{9} = 1$

16. $\frac{y^2}{4} - \frac{x^2}{9} = 1$

17. $\frac{x^2}{4} - \frac{y^2}{25} = 1$

18. $\frac{x^2}{9} - \frac{y^2}{49} = 1$

19. $\frac{y^2}{25} - \frac{x^2}{9} = 1$

20. $\frac{y^2}{4} - \frac{x^2}{16} = 1$

21. $\frac{x^2}{25} - \frac{y^2}{16} = 1$

22. $\frac{x^2}{9} - \frac{y^2}{9} = 1$

23. $\frac{y^2}{16} - \frac{x^2}{4} = 1$

24. $\frac{y^2}{9} - \frac{x^2}{36} = 1$

25. $\frac{x^2}{25} - \frac{y^2}{9} = 1$

26. $\frac{x^2}{16} - \frac{y^2}{25} = 1$

SUPPLEMENTAL EXERCISES 9.3

Sketch a graph of the conic section given by the equation.

27. $4x^2 + y^2 = 16$

Hint: Divide each term by 16.

28. $x^2 - y^2 = 9$

29. $y^2 - 4x^2 = 16$

30. $9x^2 + 4y^2 = 144$

31. $9x^2 - 25y^2 = 225$

32. $4y^2 - x^2 = 36$

33. $x^2 + 4y^2 = 36$

34. $25x^2 - 16y^2 = 400$

Just as a parabola has a focus, an ellipse and a hyperbola have foci (plural of focus). The foci of an ellipse have an application in "whispering galleries." The foci of hyperbolas are used in navigation systems. The foci of the ellipse whose equation is $\frac{x^2}{a^2} + \frac{y^2}{b^2} = 1$ $(a > b)$ are $F_1(c, 0)$ and $F_2(-c, 0)$, where $c = \sqrt{a^2 - b^2}$. The foci of the hyperbola whose equation is $\frac{x^2}{a^2} - \frac{y^2}{b^2} = 1$ are $F_1(c, 0)$ and $F_2(-c, 0)$, where $c = \sqrt{a^2 + b^2}$. Find the foci for each of the following.

35. $\frac{x^2}{16} + \frac{y^2}{7} = 1$

36. $\frac{x^2}{25} + \frac{y^2}{9} = 1$

37. $\frac{x^2}{9} - \frac{y^2}{16} = 1$

38. $\frac{x^2}{4} - \frac{y^2}{9} = 1$

SECTION 9.4
Quadratic Inequalities

1 Graph the solution set of a quadratic inequality in two variables

The **graph of a quadratic inequality in two variables** is a region of the plane that is bounded by one of the conic sections (parabola, circle, ellipse, or hyperbola). When graphing an inequality of this type, use the point (0, 0) to determine which portion of the plane to shade.

To graph the solution set of $x^2 + y^2 > 9$, change the inequality to an equality.

$$x^2 + y^2 > 9$$
$$x^2 + y^2 = 9$$

This is the equation of a circle with center (0, 0) and radius 3.

Since the inequality is $>$, the graph is drawn as a dotted circle.

Substitute the point (0, 0) in the inequality. Because $0^2 + 0^2 > 9$ is not true, the point (0, 0) should not be in the shaded region.

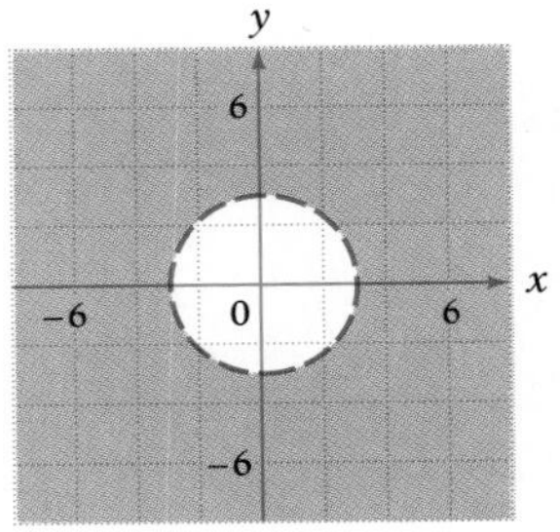

Example 1 Graph the solution set.

A. $y \leq x^2 + 2x + 2$ B. $\frac{y^2}{9} - \frac{x^2}{4} \geq 1$

Solution A. $y \leq x^2 + 2x + 2$

$y = x^2 + 2x + 2$

▶ Change the inequality to an equality.
This is the equation of a parabola that opens up.
The vertex is $(-1, 1)$.
The axis of symmetry is the line $x = -1$.

▶ Because the inequality is $\leq$, the graph is drawn as a solid line.

▶ Substitute the point (0, 0) into the inequality.
Because $0 < 0^2 + 2(0) + 2$ is true, the point (0, 0) should be in the shaded region.

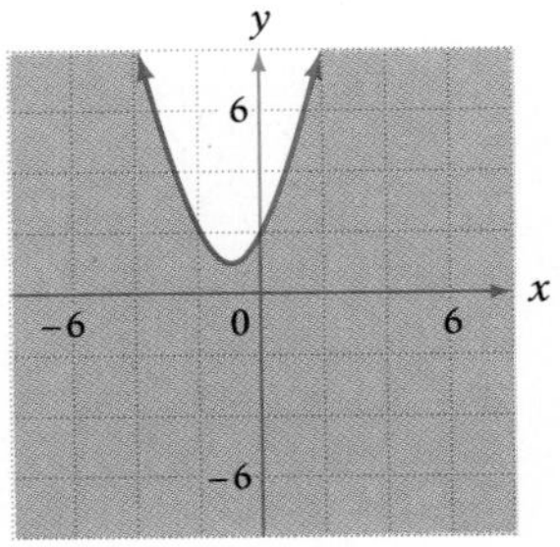

B. $\frac{y^2}{9} - \frac{x^2}{4} \geq 1$

$\frac{y^2}{9} - \frac{x^2}{4} = 1$

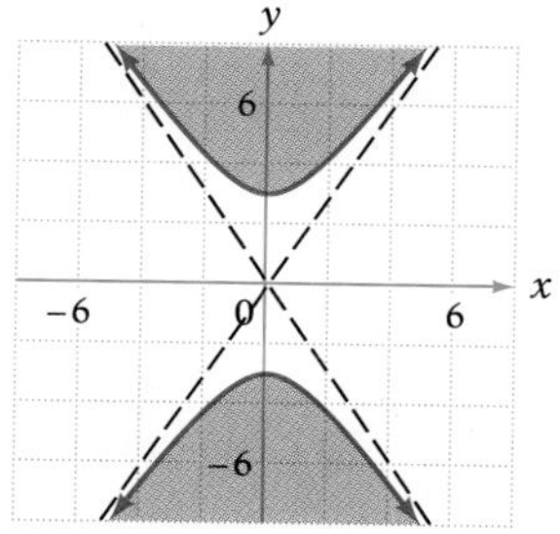

▶ Change the inequality to an equality. This is the equation of a hyperbola. The vertices are $(0, -3)$ and $(0, 3)$. The equations of the asymptotes are $y = \frac{3}{2}x$ and $y = -\frac{3}{2}x$.

▶ Because the inequality is $\geq$, the graph is drawn as a solid line.

▶ Substitute the point $(0, 0)$ into the inequality. Because $\frac{0^2}{9} - \frac{0^2}{4} \geq 1$ is not true, the point $(0, 0)$ should not be in the shaded region.

Problem 1 Graph the solution set.

A. $\frac{x^2}{9} + \frac{y^2}{16} \leq 1$ B. $\frac{x^2}{9} - \frac{y^2}{4} \leq 1$

Solution See page A43.

EXERCISES 9.4

1 Graph the solution set.

1. $y < x^2 - 4x + 3$

2. $y < x^2 - 2x - 3$

3. $(x - 1)^2 + (y + 2)^2 \leq 9$

4. $(x + 2)^2 + (y - 3)^2 > 4$

5. $(x + 3)^2 + (y - 2)^2 \geq 9$

6. $(x - 2)^2 + (y + 1)^2 \leq 16$

7. $\frac{x^2}{16} + \frac{y^2}{25} < 1$

8. $\frac{x^2}{9} + \frac{y^2}{4} \geq 1$

9. $\frac{x^2}{25} - \frac{y^2}{9} \leq 1$

10. $\frac{y^2}{25} - \frac{x^2}{36} > 1$

11. $\frac{x^2}{4} + \frac{y^2}{16} \geq 1$

12. $\frac{x^2}{4} - \frac{y^2}{16} \leq 1$

SUPPLEMENTAL EXERCISES 9.4

Graph the solution set.

13. $x < y^2 - 6y + 1$

14. $x \leq 2y^2 - 8y + 7$

15. $x^2 + y^2 + 2x + 4y + 1 > 0$

16. $x^2 + y^2 + 6x - 7 \leq 0$

17. $2x^2 + y^2 < 8$

18. $9x^2 + 8y^2 > 144$

19. $x^2 + y^2 + 4x - 6y + 14 \geq 0$

20. $x^2 + y^2 - 8x + 4y + 25 < 0$

Calculators and Computers

Graph of an Ellipse or Hyperbola

The program GRAPH OF AN ELLIPSE OR HYPERBOLA on the Math Ace Disk will graph an ellipse or hyperbola with the center at the origin. The program asks you to select from one of the following:

1) $\frac{x^2}{a^2} + \frac{y^2}{b^2} = 1$ 2) $\frac{x^2}{a^2} - \frac{y^2}{b^2} = 1$ 3) $\frac{y^2}{a^2} - \frac{x^2}{b^2} = 1$

You provide the values for a and b, and the program will draw the graph. The values of a and b should be kept between 1 and 8. After the graph has been drawn, you may draw another graph without erasing the first graph, erase the graph and then draw another graph, or quit the program.

By choosing various values for a and b and viewing the resulting graphs, you can better understand the effect these numbers have on the graph.

Something Extra

The Difference Quotient

Consider the graph of a straight line shown at the right. $P(x, f(x))$ and $Q(x + h, f(x + h))$ are two points on the line. The change in the y value over the change in the x value is given by

$$\frac{\text{rise}}{\text{run}} = \frac{f(x + h) - f(x)}{h}.$$

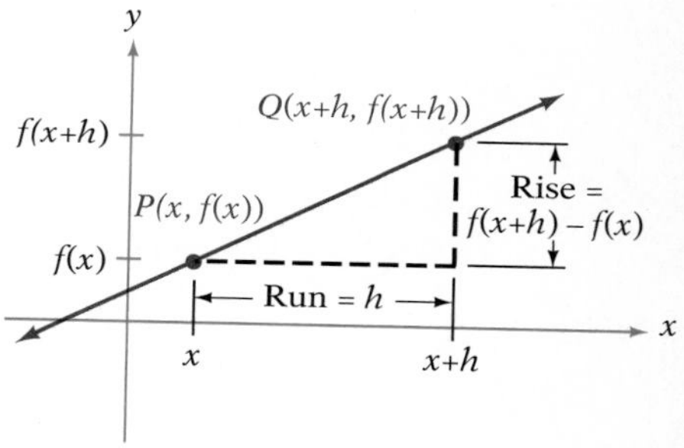

The expression $\frac{f(x + h) - f(x)}{h}$ is called the difference quotient of f. The difference quotient is used in calculus.

Find the difference quotient for $f(x) = 2x + 3$.

Substitute in the difference quotient.

$$\begin{aligned}\frac{f(x + h) - f(x)}{h} &= \frac{[2(x + h) + 3] - (2x + 3)}{h} \\ &= \frac{2x + 2h + 3 - 2x - 3}{h} \\ &= \frac{2h}{h} = 2\end{aligned}$$

Find the difference quotient for $f(x) = x^2 - 3$.

$$\begin{aligned}\frac{f(x + h) - f(x)}{h} &= \frac{[(x + h)^2 - 3] - (x^2 - 3)}{h} \\ &= \frac{x^2 + 2xh + h^2 - 3 - x^2 + 3}{h} \\ &= \frac{2xh + h^2}{h} = 2x + h\end{aligned}$$

Solve.

1. Find the difference quotient for $f(x) = -3x + 2$. Compare the slope of the function with the difference quotient.

2. Find the difference quotient for $f(x) = \frac{1}{2}x - 2$. Compare the slope of the function with the difference quotient. From Exercises 1 and 2, make a tentative conclusion relating the difference quotient and the slope of a linear equation.

3. Find the difference quotient for $f(x) = x^2 + 2$.

4. Find the difference quotient for $f(x) = x^2 - 3x$.

5. Find the difference quotient for $f(x) = 2x^2 - 2x - 3$.

Chapter Summary

Key Words

The graph of a *conic section* can be represented by the intersection of a plane and a cone.

A *circle* is the set of all points $P(x, y)$ in the plane that are a fixed distance from a given point $C(h, k)$ called the center. The fixed distance is the *radius* of the circle.

The *asymptotes* of a hyperbola are two straight lines that are "approached" by the hyperbola. As the hyperbola gets farther from the origin, the hyperbola gets "closer to" the asymptotes.

The *graph of a quadratic inequality in two variables* is a region of the plane that is bounded by one of the conic sections.

Essential Rules

Equation of a Parabola $y = ax^2 + bx + c$

When $a > 0$, the parabola opens up.

When $a < 0$, the parabola opens down.

The x-coordinate of the vertex is $-\frac{b}{2a}$.

The axis of symmetry is the line $x = -\frac{b}{2a}$.

$x = ay^2 + by + c$

When $a > 0$, the parabola opens to the right.
When $a < 0$, the parabola opens to the left.

The y-coordinate of the vertex is $-\frac{b}{2a}$.

The axis of symmetry is the line $y = -\frac{b}{2a}$.

Distance Formula

If $P_1(x_1, y_1)$ and $P_2(x_2, y_2)$ are two points in the plane, then the distance between the two points is given by $d = \sqrt{(x_2 - x_1)^2 + (y_2 - y_1)^2}$.

Midpoint Formula

If $P_1(x_1, y_1)$ and $P_2(x_2, y_2)$ are the endpoints of a line segment, then the midpoint (x_m, y_m) of the line segment is given by $x_m = \frac{x_1 + x_2}{2}$ and $y_m = \frac{y_1 + y_2}{2}$.

Equation of a Circle

$(x - h)^2 + (y - k)^2 = r^2$

The center is (h, k), and the radius is r.

Equation of an Ellipse

$\frac{x^2}{a^2} + \frac{y^2}{b^2} = 1$

The x-intercepts are $(a, 0)$ and $(-a, 0)$.
The y-intercepts are $(0, b)$ and $(0, -b)$.

Equation of a Hyperbola

$\frac{x^2}{a^2} - \frac{y^2}{b^2} = 1$

The axis of symmetry is the x-axis.
The vertices are $(a, 0)$ and $(-a, 0)$.
The equations of the asymptotes are $y = \frac{b}{a}x$ and $y = -\frac{b}{a}x$.

$\frac{y^2}{a^2} - \frac{x^2}{b^2} = 1$

The axis of symmetry is the y-axis.
The vertices are $(0, a)$ and $(0, -a)$.
The equations of the asymptotes are $y = \frac{a}{b}x$ and $y = -\frac{a}{b}x$.

Chapter Review

1. Find the axis of symmetry of the parabola whose equation is $x = -2y^2 + 4y - 3$.

2. Find the x-intercepts of the parabola whose equation is $y = 2x^2 + 3x - 1$.

3. Find the vertex of the parabola whose equation is $x = y^2 - 3y + 5$.

4. Find the equation of the circle with center at the point $C(-3, 7)$ and whose radius is 2.

5. Find the axis of symmetry of the parabola whose equation is $y = 2x^2 - 3x + 2$.

6. Find the equation of the circle that passes through $P(5, 4)$ and whose center is $C(2, -1)$.

7. Find the distance between the points $P_1(3, -4)$ and $P_2(0, 3)$.

8. Find the midpoint of a line segment with endpoints $P_1(0, -5)$ and $P_2(8, -4)$.

9. Find the midpoint of a line segment with endpoints $P_1(2, -3)$ and $P_2(-6, 9)$.

10. Find the equation of the circle in which a diameter has endpoints $P_1(2, -3)$ and $P_2(4, 7)$.

11. Find the midpoint of the line segment with endpoints $P_1(8, -3)$ and $P_2(0, 5)$.

12. Find the equation of the circle in which a diameter has endpoints $P_1(-1, -1)$ and $P_2(1, 7)$.

13. Find the equation of the circle with radius 3 and center $C(-2, 4)$.

14. Find the equation of the circle that passes through the point $P(2, 5)$ and whose center is $C(-2, 1)$.

15. Find the midpoint of the line segment with endpoints $P_1(5, -4)$ and $P_2(3, 0)$.

16. Graph $(x - 2)^2 + (y + 1)^2 = 9$.

17. Write the equation $x^2 + y^2 - 4x + 2y + 1 = 0$ in standard form, then sketch its graph.

18. Graph $\frac{y^2}{25} - \frac{x^2}{16} = 1$.

19. Graph $\frac{x^2}{9} - \frac{y^2}{4} = 1$.

20. Graph $\frac{x^2}{16} + \frac{y^2}{4} = 1$.

21. Graph $y = \frac{1}{2}x^2 - 2x + 3$.

22. Graph $x = y^2 - y - 2$.

23. Graph the solution set of $(x + 3)^2 + (y - 1)^2 \geq 16$.

24. Graph the solution set of $\frac{x^2}{16} - \frac{y^2}{25} < 1$.

25. Graph the solution set of $(x - 2)^2 + (y + 1)^2 \leq 16$.

26. Graph the solution set of $\frac{x^2}{16} + \frac{y^2}{4} > 1$.

27. Graph the solution set of $x < -y^2 + 6y - 9$.

28. Graph the solution set of $y \geq -x^2 - 2x + 3$.

29. Graph $\frac{x^2}{36} + \frac{y^2}{4} = 1$.

30. Graph the solution set of $x^2 - 4x + y^2 - 4y + 4 < 0$.

Chapter Test

1. Find the x-intercepts of the parabola whose equation is $y = 4x^2 - 4x - 3$.

2. Find the equation of the circle that passes through the point $P(1, -2)$ and whose center is $C(4, 1)$.

3. Find the midpoint of the line segment with endpoints $P_1(-3, 2)$ and $P_2(-5, -6)$.

4. Find the vertex of the parabola whose equation is $x = 2y^2 - 3y + 2$.

5. Find the axis of symmetry of the parabola whose equation is $y = 3x^2 - 2x + 5$.

6. Find the vertex of the parabola whose equation is $x = y^2 - 4y + 1$.

7. Find the axis of symmetry of the parabola whose equation is $x = 2y^2 + 3y - 1$.

8. Write the equation $x^2 + y^2 - 6x - 2y + 1 = 0$ in standard form.

9. Find the midpoint of a line segment with endpoints $P_1(-3, -7)$ and $P_2(4, 2)$.

10. Find the distance between the points $P_1(-5, 1)$ and $P_2(2, 0)$.

11. Find the equation of the circle for which a diameter has endpoints $P_1(-1, -3)$ and $P_2(5, -1)$.

12. Write the equation $x^2 + y^2 - 4x + 2y - 4 = 0$ in standard form. Then sketch its graph.

13. Graph $y = -2x^2 - 3x + 2$.

14. Graph $\frac{x^2}{4} + \frac{y^2}{36} = 1$.

15. Graph the solution set of $(x + 2)^2 + (y - 1)^2 \geq 25$.

16. Graph $x = -y^2 - y + 2$.

17. Graph $\frac{y^2}{4} - \frac{x^2}{16} = 1$.

18. Graph the solution set of $\frac{x^2}{16} - \frac{y^2}{25} \geq 1$.

19. Graph $y = x^2 - 2x + 3$.

20. Graph the solution set of $\frac{x^2}{9} + \frac{y^2}{16} \le 1$.

Cumulative Review

1. Solve: $\frac{5x - 2}{3} - \frac{1 - x}{5} = \frac{x + 4}{10}$

2. Solve: $\frac{6x}{2x - 3} - \frac{1}{2x - 3} = 7$

3. Solve: $4 + |3x + 2| < 6$

4. Find the equation of the line that contains the point $(2, -3)$ and has slope $-\frac{3}{2}$.

5. Find the equation of the line containing the point $(4, -2)$ and perpendicular to the line $y = -x + 5$.

6. Simplify: $x^{2n}(x^{2n} + 2x^n - 3x)$

7. Factor: $(x - 1)^3 - y^3$

8. Solve: $\frac{3x - 2}{x + 4} \le 1$

9. Simplify: $\frac{ax - bx}{ax + ay - bx - by}$

10. Simplify: $\frac{x - 4}{3x - 2} - \frac{1 + x}{3x^2 + x - 2}$

11. Simplify: $\left(\frac{12a^2b^{-2}}{a^{-3}b^{-4}}\right)^{-1}\left(\frac{ab}{4^{-1}a^{-2}b^4}\right)^2$

12. Write $2\sqrt[4]{x^3}$ as an exponential expression.

13. Simplify: $\sqrt{18} - \sqrt{-25}$

14. Solve: $2x^2 + 2x - 3 = 0$

15. Solve: $a^{\frac{2}{3}} - 2a^{\frac{1}{3}} - 3 = 0$

16. Solve: $x - \sqrt{2x - 3} = 3$

17. Evaluate $f(x) = -x^2 + 3x - 2$ at $x = -3$.

18. For $f(x) = 4x + 8$, find the equation of the inverse function.

19. Find the maximum value of the function given by the equation $f(x) = -2x^2 + 4x - 2$.

20. Find the maximum product of two numbers whose sum is 40.

21. Find the distance between the points $P_1(2, 4)$ and $P_2(-1, 0)$.

22. Find the equation of the circle that passes through $P(3, 1)$ and whose center is $C(-1, 2)$.

23. Graph $\{x|x < 4\} \cap \{x|x > 2\}$.

24. Graph the solution set of $5x + 2y > 10$.

25. Graph $x = y^2 - 2y + 3$.

26. Graph $\frac{x^2}{25} + \frac{y^2}{4} = 1$.

27. Graph $\frac{y^2}{4} - \frac{x^2}{25} = 1$.

28. Graph $\frac{x^2}{9} - \frac{y^2}{36} = 1$.

29. Tickets for a school play sold for $4.00 for each adult and $1.50 for each child. The total receipts for the 192 tickets sold were $493. Find the number of adult tickets sold.

30. A motorcycle travels 180 mi in the same amount of time as a car travels 144 mi. The rate of the motorcycle is 12 mph faster than the rate of the car. Find the rate of the motorcycle.

31. The rate of a river's current is 1.5 mph. A rowing crew can row 12 mi down this river and 12 mi back in 6 h. Find the rowing rate of the crew in calm water.

32. The speed (v) of a gear varies inversely as the number of teeth (t). If a gear that has 36 teeth makes 30 revolutions per minute, how many revolutions per minute will a gear that has 60 teeth make?

10

Systems of Equations and Inequalities

Objectives

- Solve systems of linear equations by graphing
- Solve systems of linear equations by the substitution method
- Solve systems of two linear equations in two variables by the addition method
- Solve systems of three linear equations in three variables by the addition method
- Evaluate determinants
- Solve systems of equations by using Cramer's Rule
- Solve systems of equations by using matrices
- Rate-of-wind and water current problems
- Application problems
- Solve nonlinear systems of equations
- Graph the solution set of a system of inequalities

POINT OF INTEREST

Analytic Geometry

Euclid's Geometry (plane geometry is still taught in high school) was developed around 300 B.C. Euclid depends on a synthetic proof—a proof based on pure logic.

Algebra is the science of solving equations such as $ax^2 + bx + c = 0$. As early as the third century, equations were solved by trial and error. However, the appropriate terminology and symbols of operation were not fully developed until the seventeenth century.

Analytic geometry is the combining of algebraic methods with geometric concepts. The development of analytic geometry in the seventeenth century is attributed to Descartes.

The development of analytic geometry depended on the invention of a system of coordinates in which any point in the plane could be represented by an ordered pair of numbers. This invention allowed lines and curves (such as the conic sections) to be represented and described by algebraic equations.

Some equations are quite simple and have a simple graph. For example, the graph of the equation $y = 2x - 3$ is a line.

Other graphs are quite complicated and have complicated equations. The graph of the Mandelbrot set is very complicated. The graph is called a fractal.

SECTION **10.1**

Solving Systems of Linear Equations by Graphing and by the Substitution Method

1 Solve systems of linear equations by graphing

A **system of equations** is two or more equations considered together. The system shown below is a system of two linear equations in two variables.

$$3x + 4y = 7$$
$$2x - 3y = 6$$

The graphs of the equations are straight lines.

A **solution of a system of equations in two variables** is an ordered pair that is a solution of each equation of the system.

Is $(3, -2)$ a solution of the system
$2x - 3y = 12$
$5x + 2y = 11$?

$2x - 3y = 12$	
$2(3) - 3(-2)$	12
$6 - (-6)$	
$12 = 12$	

$5x + 2y = 11$	
$5(3) + 2(-2)$	11
$15 + (-4)$	
$11 = 11$	

Yes, since $(3, -2)$ is a solution of each equation, it is a solution of the system of equations.

A solution of a system of linear equations can be found by graphing the two equations on the same coordinate axes. Three possible conditions result.

A. The lines can intersect at one point. The point of intersection of the lines is the ordered pair that is a solution of each equation of the system. It is the solution of the system of equations. The system of equations is **independent.**

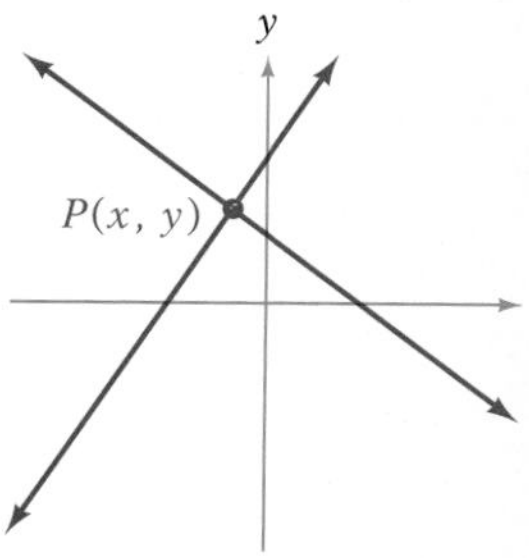

B. The lines can be parallel and not intersect. The system of equations is **inconsistent** and has no solution. The slopes of the lines in an inconsistent system of equations are equal.

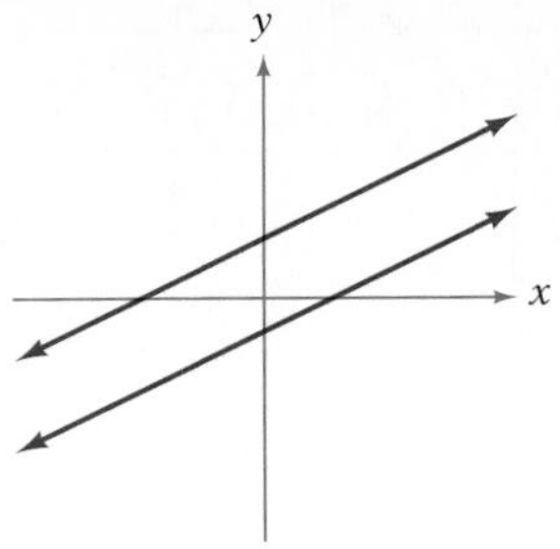

C. The lines can represent the same line. The lines intersect at infinitely many points; therefore, there are infinitely many solutions. The system of equations is **dependent.** The solutions are the ordered pairs that are solutions of either one of the two equations in the system.

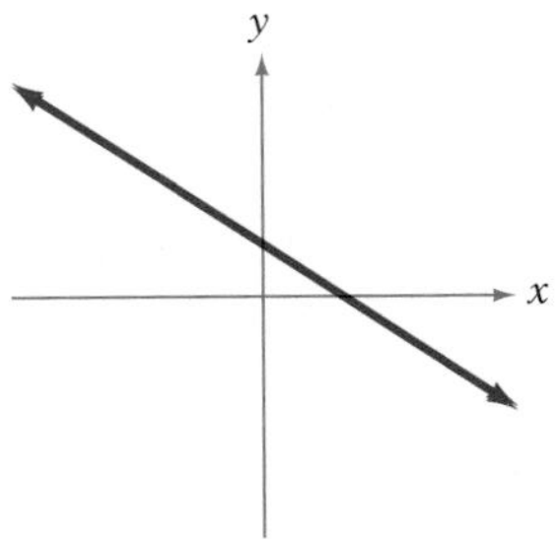

Solve by graphing: $x + 2y = 4$
$2x + y = -1$

Graph each line.

The equations intersect at one point.
The system of equations is independent.

Find the point of intersection.

The solution is $(-2, 3)$.

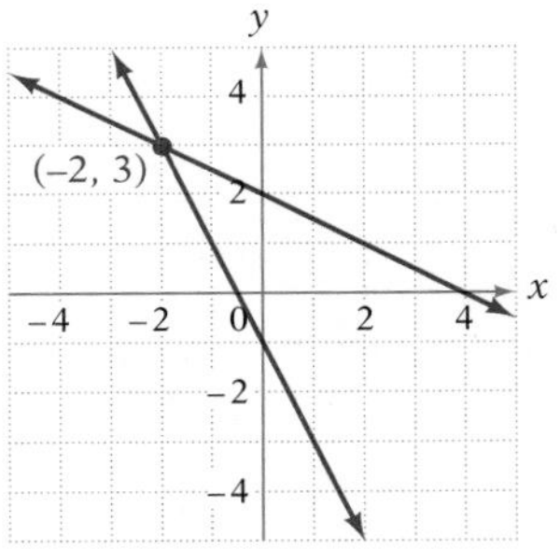

Solve by graphing: $2x + 3y = 6$
$4x + 6y = -12$

Graph each line.

The lines are parallel and therefore do not intersect.

The system of equations is inconsistent and has no solution.

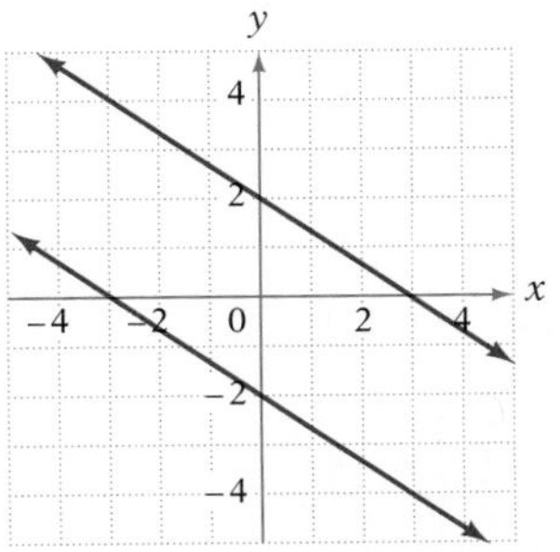

Solve by graphing: $x - 2y = 4$
$2x - 4y = 8$

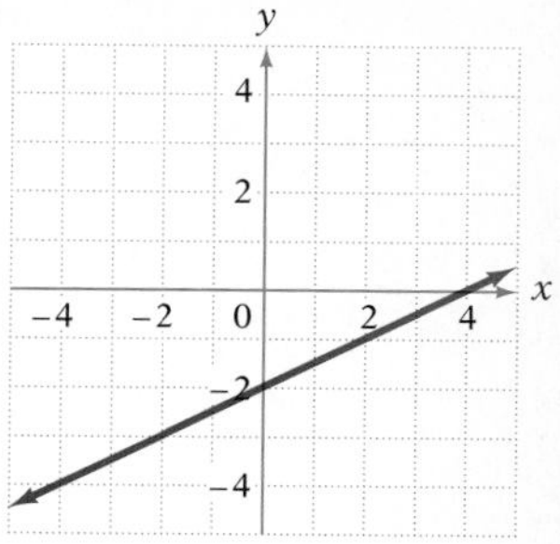

Graph each line.

The two equations represent the same line. The system of equations is dependent and, therefore, has an infinite number of solutions.

The solutions are the ordered pairs that are solutions of the equation $x - 2y = 4$.

Example 1 Solve by graphing.

A. $2x - y = 3$
$3x + y = 2$

B. $2x + 3y = 6$
$y = -\frac{2}{3}x + 1$

Solution A.

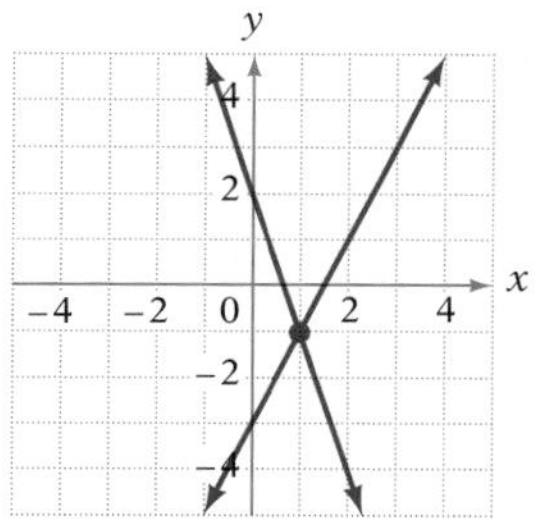

The solution is $(1, -1)$.

B.

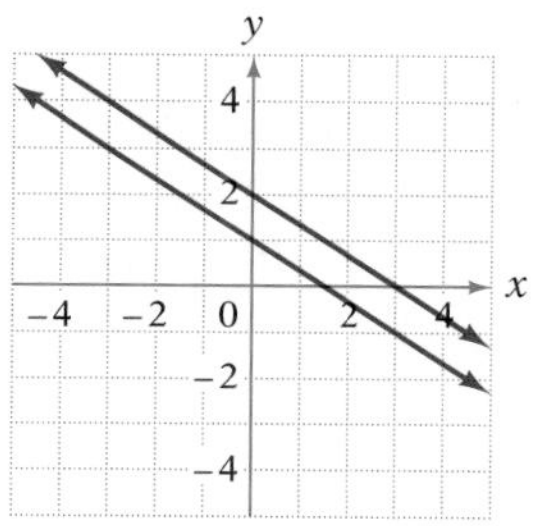

The system of equations is inconsistent and has no solution.

Problem 1 Solve by graphing.

A. $x + y = 1$
$2x + y = 0$

B. $3x - 4y = 12$
$y = \frac{3}{4}x - 3$

Solution See page A43.

2 Solve systems of linear equations by the substitution method

Graphical methods of solving systems of equations are often unsatisfactory since it can be difficult to read exact values of a point of intersection. For example, the point $\left(\frac{2}{3}, -\frac{4}{7}\right)$ would be difficult to read from a graph.

An algebraic method, called the **substitution method,** can be used to find an exact solution of a system of equations.

In the system of equations at the right, equation (2) states that $y = 3x - 4$.

$$\begin{aligned} (1)\quad 2x - 3y &= 5 \\ (2)\quad y &= 3x - 4 \end{aligned}$$

Substitute $3x - 4$ for y in equation (1).

$$2x - 3(3x - 4) = 5$$

Solve for x.

$$\begin{aligned} 2x - 9x + 12 &= 5 \\ -7x + 12 &= 5 \\ -7x &= -7 \\ x &= 1 \end{aligned}$$

Substitute the value of x into equation (2), and solve for y.

$$\begin{aligned} y &= 3x - 4 \\ y &= 3(1) - 4 \\ y &= 3 - 4 \\ y &= -1 \end{aligned}$$

The solution is $(1, -1)$.

Solve by substitution: $3x - y = 5$
$2x + 5y = 9$

$$\begin{aligned} (1)\quad 3x - y &= 5 \\ (2)\quad 2x + 5y &= 9 \end{aligned}$$

Solve equation (1) for y. Equation (1) is chosen because it is the easier equation to solve for one variable in terms of the other.

$$\begin{aligned} 3x - y &= 5 \\ -y &= -3x + 5 \\ y &= 3x - 5 \end{aligned}$$

Substitute $3x - 5$ for y in equation (2).

$$\begin{aligned} 2x + 5y &= 9 \\ 2x + 5(3x - 5) &= 9 \\ 2x + 15x - 25 &= 9 \\ 17x - 25 &= 9 \\ 17x &= 34 \\ x &= 2 \end{aligned}$$

Substitute the value of x into the equation $y = 3x - 5$ and solve for y.

$$\begin{aligned} y &= 3x - 5 \\ y &= 3(2) - 5 \\ y &= 6 - 5 \\ y &= 1 \end{aligned}$$

The solution is $(2, 1)$.

Example 2 Solve by substitution.

A. $3x - 3y = 2$
$y = x + 2$

B. $9x + 3y = 12$
$y = -3x + 4$

Solution A. $3x - 3y = 2$ (1)
$y = x + 2$ (2)

► Equation (2) states that $y = x + 2$.

$$\begin{aligned} 3x - 3(x + 2) &= 2 \\ 3x - 3x - 6 &= 2 \\ -6 &= 2 \end{aligned}$$

► Substitute $x + 2$ for y in equation (1).

This is not a true equation. The system of equations is inconsistent. The system does not have a solution.

B. $9x + 3y = 12$ (1)
$y = -3x + 4$ (2) ▶ Equation (2) states that $y = -3x + 4$.

$9x + 3(-3x + 4) = 12$ ▶ Substitute $-3x + 4$ for y in equation (1).
$9x - 9x + 12 = 12$
$12 = 12$

This is a true equation. The system of equations is dependent. The solutions are the ordered pairs that are solutions of the equation $y = -3x + 4$.

Problem 2 Solve by substitution.

A. $3x - y = 3$
$6x + 3y = -4$

B. $6x - 3y = 6$
$2x - y = 2$

Solution See page A44.

EXERCISES 10.1

1 Solve by graphing.

1. $x + y = 2$
$x - y = 4$

2. $x + y = 1$
$3x - y = -5$

3. $x - y = -2$
$x + 2y = 10$

4. $2x - y = 5$
$3x + y = 5$

5. $3x - 2y = 6$
$y = 3$

6. $x = 4$
$3x - 2y = 4$

7. $x = 4$
$y = -1$

8. $x + 2 = 0$
$y - 1 = 0$

9. $y = x - 5$
$2x + y = 4$

10. $2x - 5y = 4$
$y = x + 1$

11. $y = \frac{1}{2}x - 2$
$x - 2y = 8$

12. $2x + 3y = 6$
$y = -\frac{2}{3}x + 1$

13. $2x - 5y = 10$
$y = \frac{2}{5}x - 2$

14. $3x - 2y = 6$
$y = \frac{3}{2}x - 3$

15. $3x - 4y = 12$
$5x + 4y = -12$

16. $2x - 3y = 6$
$2x - 5y = 10$

2 Solve by substitution.

17. $3x - 2y = 4$
$x = 2$

18. $y = -2$
$2x + 3y = 4$

19. $y = 2x - 1$
$x + 2y = 3$

20. $y = -x + 1$
$2x - y = 5$

21. $x = 3y + 1$
$x - 2y = 6$

22. $x = 2y - 3$
$3x + y = 5$

23. $4x - 3y = 5$
$y = 2x - 3$

24. $3x + 5y = -1$
$y = 2x - 8$

25. $5x - 2y = 9$
$y = 3x - 4$

26. $4x - 3y = 2$
$y = 3x + 1$

27. $x = 2y + 4$
$4x + 3y = -17$

28. $3x - 2y = -11$
$x = 2y - 9$

29. $5x + 4y = -1$
$y = 2 - 2x$

30. $3x + 2y = 4$
$y = 1 - 2x$

31. $2x - 5y = -9$
$y = 9 - 2x$

32. $5x + 2y = 15$
$x = 6 - y$

33. $7x - 3y = 3$
$x = 2y + 2$

34. $3x - 4y = 6$
$x = 3y + 2$

35. $2x + 2y = 7$
$y = 4x + 1$

36. $3x + 7y = -5$
$y = 6x - 5$

37. $3x + y = 5$
$2x + 3y = 8$

38. $4x + y = 9$
$3x - 4y = 2$

39. $x + 3y = 5$
$2x + 3y = 4$

40. $x - 4y = 2$
$2x - 5y = 1$

41. $3x - y = 10$
$5x + 2y = 2$

42. $2x - y = 5$
$3x - 2y = 9$

43. $3x + 4y = 14$
$2x + y = 1$

44. $5x + 3y = 8$
$3x + y = 8$

45. $3x + 5y = 0$
$x - 4y = 0$

46. $2x - 7y = 0$
$3x + y = 0$

47. $5x - 3y = -2$
$-x + 2y = -8$

48. $2x + 7y = 1$
$-x + 4y = 7$

49. $y = 3x + 2$
$y = 2x + 3$

50. $y = 3x - 7$
$y = 2x - 5$

51. $y = 3x + 1$
$y = 6x - 1$

52. $y = 2x - 3$
$y = 4x - 4$

53. $x = 2y + 1$
$x = 3y - 1$

54. $x = 4y + 1$
$x = -2y - 5$

55. $y = 5x - 1$
$y = 5 - x$

56. $y = 3 - 2x$
$y = 2 - 3x$

57. $-x + 2y = 13$
$5x + 3y = 13$

58. $3x - y = 8$
$4x - 7y = 5$

SUPPLEMENTAL EXERCISES 10.1

For what values of k will the system of equations be inconsistent?

59. $2x - 2y = 5$
$kx - 2y = 3$

60. $6x - 3y = 4$
$3x - ky = 1$

61. $x = 6y + 6$
$kx - 3y = 6$

62. $x = 2y + 2$
$kx - 8y = 2$

Solve using two variables.

63. The sum of two numbers is 44. One number is 8 less than the other number. Find the numbers.

64. The sum of two numbers is 76. One number is 12 less than the other number. Find the numbers.

65. The sum of two numbers is 128. One number is 16 more than the other number. Find the numbers.

66. The sum of two numbers is 116. One number is 14 more than the other number. Find the numbers.

67. The sum of two numbers is 19. Five less than twice the first number equals the second number. Find the numbers.

68. The sum of two numbers is 22. Three times the first plus 2 equals the second number. Find the numbers.

Solve. (*Hint:* These equations are not linear equations. First rewrite the equations as linear equations by substituting x for $\frac{1}{a}$ and y for $\frac{1}{b}$.)

69. $\frac{2}{a} + \frac{3}{b} = 4$
$\frac{4}{a} + \frac{1}{b} = 3$

70. $\frac{2}{a} + \frac{1}{b} = 1$
$\frac{8}{a} - \frac{2}{b} = 0$

71. $\frac{1}{a} + \frac{3}{b} = 2$
$\frac{4}{a} - \frac{1}{b} = 3$

72. $\frac{3}{a} + \frac{4}{b} = -1$
$\frac{1}{a} + \frac{6}{b} = 2$

73. $\frac{6}{a} + \frac{1}{b} = -1$
$\frac{3}{a} - \frac{2}{b} = -3$

74. $\frac{1}{a} - \frac{4}{b} = 1$
$\frac{5}{a} - \frac{2}{b} = -3$

SECTION 10.2 Solving Systems of Linear Equations by the Addition Method

1 Solve systems of two linear equations in two variables by the addition method

The **addition method** is an alternative method for solving a system of equations. This method is based on the Addition Property of Equations. It is appropriate when it is not convenient to solve one equation for one variable in terms of another variable.

Note, for the system of equations at the right, the effect of adding equation (2) to equation (1). Since $-3y$ and $3y$ are additive inverses, adding the equations results in an equation with only one variable.

$$\begin{aligned} (1)\quad 5x - 3y &= 14 \\ (2)\quad 2x + 3y &= -7 \\ 7x + 0y &= 7 \\ 7x &= 7 \end{aligned}$$

The solution of the resulting equation is the first component of the ordered pair solution of the system.

$$\begin{aligned} 7x &= 7 \\ x &= 1 \end{aligned}$$

The second component is found by substituting the value of x into equation (1) or (2) and then solving for y. Equation (1) is used here.

$$\begin{aligned} (1)\quad 5x - 3y &= 14 \\ 5(1) - 3y &= 14 \\ 5 - 3y &= 14 \\ -3y &= 9 \\ y &= -3 \end{aligned}$$

The solution is $(1, -3)$.

Sometimes adding the two equations does not eliminate one of the variables. In this case, use the Multiplication Property of Equations to rewrite one or both of the equations so that when the equations are added, one of the variables is eliminated. To do this, first choose which variable to eliminate. The coefficients of that variable must be additive inverses. Multiply each equation by a constant that will produce coefficients that are additive inverses.

Solve by the addition method: $3x + 4y = 2$
$2x + 5y = -1$

$$\begin{aligned} (1)\quad 3x + 4y &= 2 \\ (2)\quad 2x + 5y &= -1 \end{aligned}$$

Eliminate x. Multiply equation (1) by 2 and equation (2) by -3. Note how the constants are selected. The negative sign is used so that the coefficients will be additive inverses.

$$\begin{aligned} 2(3x + 4y) &= 2(2) \\ -3(2x + 5y) &= -3(-1) \end{aligned}$$

The coefficients of the x terms are additive inverses.

$$\begin{aligned} 6x + 8y &= 4 \\ -6x - 15y &= 3 \end{aligned}$$

Add the equations.
Solve for y.

$$\begin{aligned} -7y &= 7 \\ y &= -1 \end{aligned}$$

Substitute the value of y into one of the equations, and solve for x. Equation (1) is used here.

$$\begin{aligned} (1)\quad 3x + 4y &= 2 \\ 3x + 4(-1) &= 2 \\ 3x - 4 &= 2 \\ 3x &= 6 \\ x &= 2 \end{aligned}$$

The solution is $(2, -1)$.

Solve by the addition method: $\frac{2}{3}x + \frac{1}{2}y = 4$
$\frac{1}{4}x - \frac{3}{8}y = -\frac{3}{4}$

$$\begin{aligned} (1)\quad \tfrac{2}{3}x + \tfrac{1}{2}y &= 4 \\ (2)\quad \tfrac{1}{4}x - \tfrac{3}{8}y &= -\tfrac{3}{4} \end{aligned}$$

Clear the fractions. Multiply each equation by the LCM of the denominators.

$$\begin{aligned} 6\left(\tfrac{2}{3}x + \tfrac{1}{2}y\right) &= 6(4) \\ 8\left(\tfrac{1}{4}x - \tfrac{3}{8}y\right) &= 8\left(-\tfrac{3}{4}\right) \end{aligned}$$

$$\begin{aligned} 4x + 3y &= 24 \\ 2x - 3y &= -6 \end{aligned}$$

Eliminate y. Add the equations.
Solve for x.

$$\begin{aligned} 6x &= 18 \\ x &= 3 \end{aligned}$$

Substitute the value of x into equation (1), and solve for y.

$$\frac{2}{3}x + \frac{1}{2}y = 4$$
$$\frac{2}{3}(3) + \frac{1}{2}y = 4$$
$$2 + \frac{1}{2}y = 4$$
$$\frac{1}{2}y = 2$$
$$y = 4$$

The solution is (3, 4).

To solve the system of equations at the right, eliminate x. Multiply equation (1) by -2 and add to equation (2).

(1) $3x - 2y = 5$
(2) $6x - 4y = 1$

$$-6x + 4y = -10$$
$$6x - 4y = 1$$
$$0 = -9$$

This is not a true equation. The system of equations is inconsistent. The system does not have a solution.

The graph of the equations of the system is shown at the right. Note that the lines are parallel and therefore do not intersect.

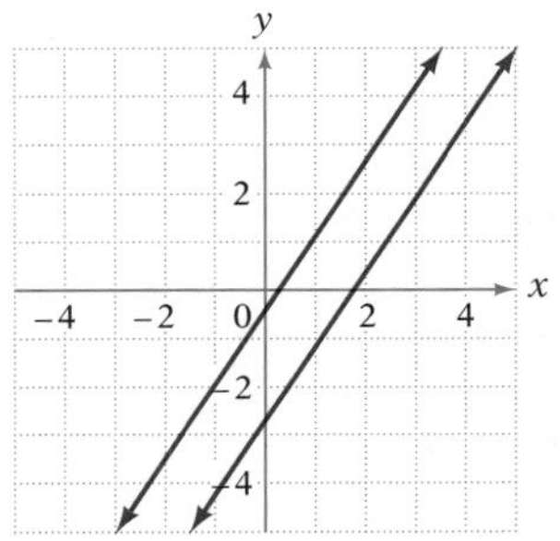

Example 1 Solve by the addition method.

A. $3x - 2y = 2x + 5$
$\quad 2x + 3y = -4$

B. $4x - 8y = 36$
$\quad 3x - 6y = 27$

Solution A. $3x - 2y = 2x + 5$ (1)
$\quad 2x + 3y = -4$ (2)

$x - 2y = 5$
$2x + 3y = -4$
▶ Write equation (1) in the form $Ax + By = C$.

$-2(x - 2y) = -2(5)$
$2x + 3y = -4$
▶ To eliminate x, multiply each side of equation (1) by -2.

$-2x + 4y = -10$
$2x + 3y = -4$
$7y = -14$
$y = -2$
▶ Add the equations.

$$2x + 3y = -4$$
$$2x + 3(-2) = -4$$ ▶ Replace y in equation (2) by its value.
$$2x - 6 = -4$$
$$2x = 2$$
$$x = 1$$

The solution is (1, −2).

B. $4x - 8y = 36$ (1)
$3x - 6y = 27$ (2)

$$3(4x - 8y) = 3(36)$$
$$-4(3x - 6y) = -4(27)$$
▶ To eliminate x, multiply each side of equation (1) by 3 and each side of equation (2) by −4.

$$12x - 24y = 108$$
$$-12x + 24y = -108$$
$$0 = 0$$ ▶ Add the equations.

This is a true equation. The system of equations is dependent. The solutions are the ordered pairs that are solutions of the equation $4x - 8y = 36$.

Problem 1 Solve by the addition method.

A. $2x + 5y = 6$
$3x - 2y = 6x + 2$

B. $2x + y = 5$
$4x + 2y = 6$

Solution See pages A44–A45.

2 Solve systems of three linear equations in three variables by the addition method

An equation of the form $Ax + By + Cz = D$, where A, B, C, and D are constants, is a **linear equation in three variables.** Examples of linear equations in three variables are shown below.

$$2x + 4y - 3z = 7 \qquad x - 6y + z = -8$$

Just as a solution of an equation in two variables is an ordered pair (x, y), a solution of an equation in three variables is an ordered triple (x, y, z). For example, (2, 1, −3) and (3, 1, −2) are solutions of the equation

$$2x - y - 2z = 9.$$

The ordered triples (1, 3, 2) and (2, −1, 3) are not solutions of this equation.

Graphing an equation in three variables requires a third coordinate axis perpendicular to the *xy*-plane. The third axis is commonly called the *z*-axis. The result is a three-dimensional coordinate system called the *xyz*-coordinate system. To help visualize a three-dimensional coordinate system, think of a corner of a room: the floor is the *xy*-plane, one wall is the *yz*-plane, and the other wall is the *xz*-plane. A three-dimensional coordinate system is shown at the right.

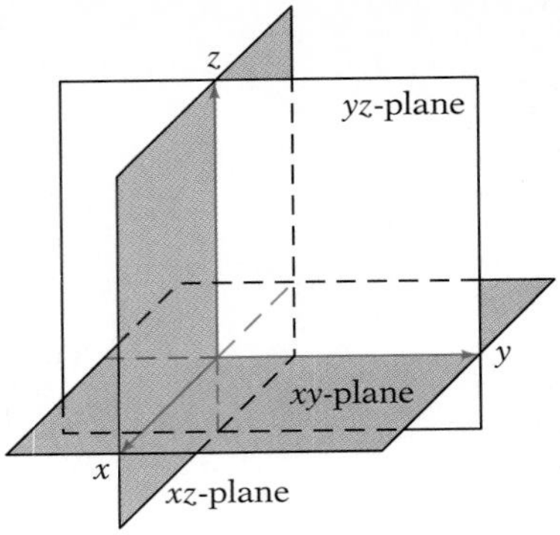

Graphing an ordered triple requires three moves, the first along the *x*-axis, the second along the *y*-axis, and the third along the *z*-axis. The graph of the points (−4, 2, 3) and (3, 4, −2) is shown at the right.

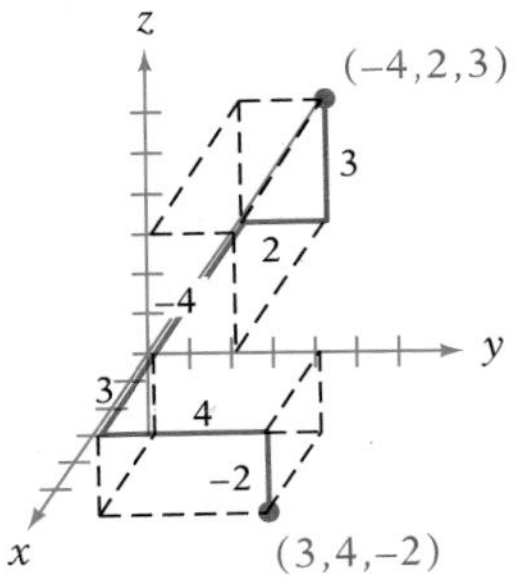

The graph of a linear equation in three variables is a plane. That is, if all the solutions of a linear equation in three variables were plotted in an *xyz*-coordinate system, the graph would look like a large piece of paper extending infinitely. The graph of $x + y + z = 3$ is shown at the right.

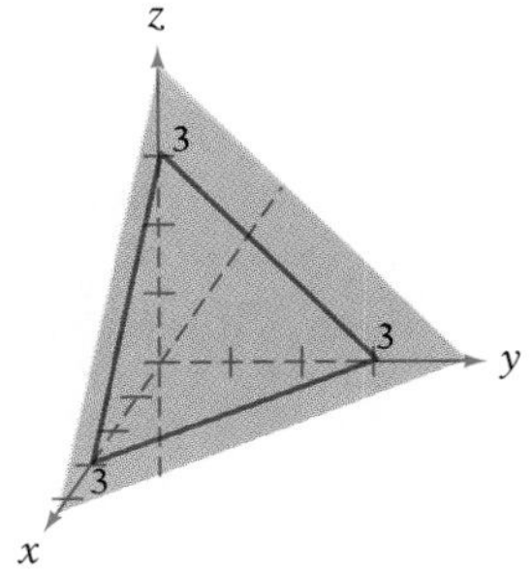

There are different ways three planes can be oriented in an *xyz*-coordinate system. The systems of equations represented by the planes below are inconsistent.

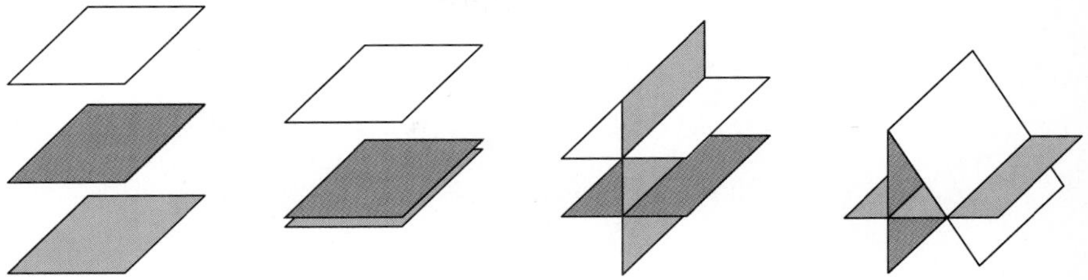

A B C D

Graphs of Inconsistent Systems of Equations

For a system of three equations in three variables to have a solution, the graphs of the planes must intersect at a single point, along a common line, or all equations must have a graph that is the same plane. These situations are shown in the figures below.

The three planes shown in Figure E intersect at a point. The system of equations represented by planes that intersect at a point are independent.

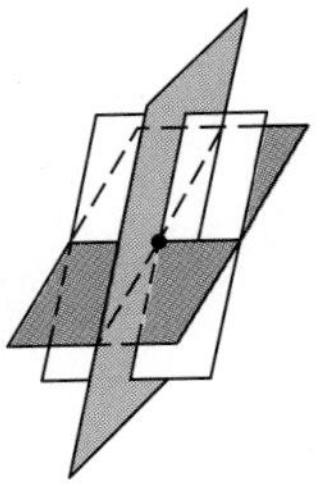

E

The planes shown in Figures F and G intersect along a common line. The system of equations represented by the planes in Figure H have a graph that is the same plane. The systems of equations represented by the planes below are dependent.

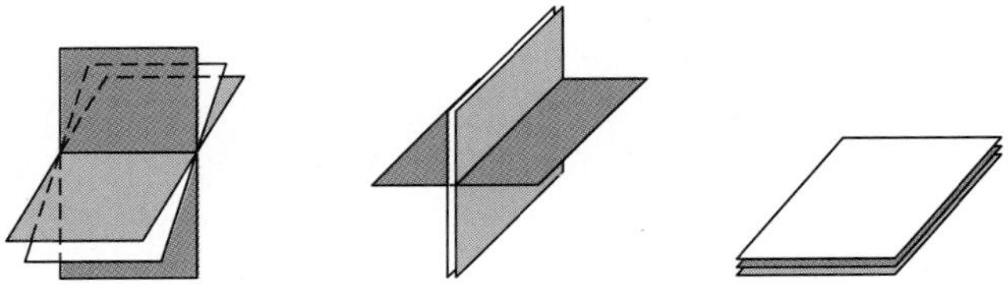

F G H

A system of linear equations in three variables can be solved by using the addition method. First, eliminate one variable from any two of the given equations. Then eliminate the same variable from any other two equations. The result will be a system of two equations in two variables. Solve this system by the addition method.

Solve: $x + 4y - z = 10$, $3x + 2y + z = 4$, $2x - 3y + 2z = -7$

$$\begin{aligned} (1)\quad x + 4y - z &= 10 \\ (2)\quad 3x + 2y + z &= 4 \\ (3)\quad 2x - 3y + 2z &= -7 \end{aligned}$$

Eliminate z from equations (1) and (2) by adding the two equations.

$$\begin{aligned} x + 4y - z &= 10 \\ 3x + 2y + z &= 4 \\ (4)\quad 4x + 6y &= 14 \end{aligned}$$

Eliminate z from equations (1) and (3). Multiply equation (1) by 2 and add to equation (3).

$$\begin{aligned} 2x + 8y - 2z &= 20 \\ 2x - 3y + 2z &= -7 \\ (5)\quad 4x + 5y &= 13 \end{aligned}$$

Solve the system of two equations in two variables.

$$\begin{aligned} (4)\quad 4x + 6y &= 14 \\ (5)\quad 4x + 5y &= 13 \end{aligned}$$

Eliminate x. Multiply equation (5) by -1 and add to equation (4).

$$\begin{aligned} 4x + 6y &= 14 \\ -4x - 5y &= -13 \\ y &= 1 \end{aligned}$$

Substitute the value of y into equation (4) or (5), and solve for x. Equation (4) is used here.

$$\begin{aligned} 4x + 6y &= 14 \\ 4x + 6(1) &= 14 \\ 4x + 6 &= 14 \\ 4x &= 8 \\ x &= 2 \end{aligned}$$

Substitute the value of y and the value of x into one of the equations in the original system. Equation (2) is used here.

$$\begin{aligned} 3x + 2y + z &= 4 \\ 3(2) + 2(1) + z &= 4 \\ 6 + 2 + z &= 4 \\ 8 + z &= 4 \\ z &= -4 \end{aligned}$$

The solution is $(2, 1, -4)$.

Solve: $2x - 3y - z = 1$, $x + 4y + 3z = 2$, $4x - 6y - 2z = 5$

$$\begin{aligned} (1)\quad 2x - 3y - z &= 1 \\ (2)\quad x + 4y + 3z &= 2 \\ (3)\quad 4x - 6y - 2z &= 5 \end{aligned}$$

Eliminate x from equations (1) and (2). Multiply equation (2) by -2 and add to equation (1).

$$\begin{aligned} 2x - 3y - z &= 1 \\ -2x - 8y - 6z &= -4 \\ -11y - 7z &= -3 \end{aligned}$$

Eliminate x from equations (1) and (3). Multiply equation (1) by -2 and add to equation (3).

$$\begin{aligned} -4x + 6y + 2z &= -2 \\ 4x - 6y - 2z &= 5 \\ 0 &= 3 \end{aligned}$$

This is not a true equation.

The system is inconsistent.

Example 2 Solve:
$$\begin{aligned} 3x - y + 2z &= 1 \\ 2x + 3y + 3z &= 4 \\ x + y - 4z &= -9 \end{aligned}$$

Solution
$$\begin{aligned} 3x - y + 2z &= 1 && (1) \\ 2x + 3y + 3z &= 4 && (2) \\ x + y - 4z &= -9 && (3) \end{aligned}$$

$$\begin{aligned} 3x - y + 2z &= 1 \\ x + y - 4z &= -9 \\ 4x - 2z &= -8 \\ 2x - z &= -4 && (4) \end{aligned}$$

▶ Eliminate y. Add equations (1) and (3).

▶ Multiply each side of the equation by $\frac{1}{2}$.

$$\begin{aligned} 9x - 3y + 6z &= 3 \\ 2x + 3y + 3z &= 4 \\ 11x + 9z &= 7 && (5) \end{aligned}$$

▶ Multiply equation (1) by 3 and add to equation (2).

$$\begin{aligned} (4) \quad 2x - z &= -4 \\ (5) \quad 11x + 9z &= 7 \end{aligned}$$

▶ Solve the system of two equations.

$$\begin{aligned} 18x - 9z &= -36 \\ 11x + 9z &= 7 \\ 29x &= -29 \\ x &= -1 \end{aligned}$$

▶ Multiply equation (4) by 9 and add to equation (5).

$$\begin{aligned} 2x - z &= -4 \\ 2(-1) - z &= -4 \\ -2 - z &= -4 \\ -z &= -2 \\ z &= 2 \end{aligned}$$

▶ Replace x by -1 in equation (4).

$$\begin{aligned} x + y - 4z &= -9 \\ -1 + y - 4(2) &= -9 \\ -1 + y - 8 &= -9 \\ -9 + y &= -9 \\ y &= 0 \end{aligned}$$

▶ Replace x by -1 and z by 2 in equation (3).

The solution is $(-1, 0, 2)$.

Problem 2
$$\begin{aligned} x - y + z &= 6 \\ 2x + 3y - z &= 1 \\ x + 2y + 2z &= 5 \end{aligned}$$

Solution See page A45.

EXERCISES 10.2

1 Solve by the addition method.

1. $x - y = 5$
   $x + y = 7$

2. $x + y = 1$
   $2x - y = 5$

3. $3x + y = 4$
   $x + y = 2$

4. $x - 3y = 4$
   $x + 5y = -4$

5. $3x + y = 7$
   $x + 2y = 4$

6. $x - 2y = 7$
   $3x - 2y = 9$

7. $2x + 3y = -1$
   $x + 5y = 3$

8. $x + 5y = 7$
   $2x + 7y = 8$

9. $3x - y = 4$
   $6x - 2y = 8$

10. $x - 2y = -3$
    $-2x + 4y = 6$

11. $2x + 5y = 9$
    $4x - 7y = -16$

12. $8x - 3y = 21$
    $4x + 5y = -9$

13. $4x - 6y = 5$
    $2x - 3y = 7$

14. $3x + 6y = 7$
    $2x + 4y = 5$

15. $3x - 5y = 7$
    $x - 2y = 3$

16. $3x + 4y = 25$
    $2x + y = 10$

17. $x + 3y = 7$
    $-2x + 3y = 22$

18. $2x - 3y = 14$
    $5x - 6y = 32$

19. $3x + 2y = 16$
    $2x - 3y = -11$

20. $2x - 5y = 13$
    $5x + 3y = 17$

21. $4x + 4y = 5$
    $2x - 8y = -5$

22. $3x + 7y = 16$
    $4x - 3y = 9$

23. $5x + 4y = 0$
    $3x + 7y = 0$

24. $3x - 4y = 0$
    $4x - 7y = 0$

25. $5x + 2y = 1$
    $2x + 3y = 7$

26. $3x + 5y = 16$
    $5x - 7y = -4$

27. $3x - 6y = 6$
    $9x - 3y = 8$

28. $4x - 8y = 5$
    $8x + 2y = 1$

29. $5x + 2y = 2x + 1$
    $2x - 3y = 3x + 2$

30. $3x + 3y = y + 1$
    $x + 3y = 9 - x$

31. $\frac{2}{3}x - \frac{1}{2}y = 3$
$\frac{1}{3}x - \frac{1}{4}y = \frac{3}{2}$

32. $\frac{3}{4}x + \frac{1}{3}y = -\frac{1}{2}$
$\frac{1}{2}x - \frac{5}{6}y = -\frac{7}{2}$

33. $\frac{2}{5}x - \frac{1}{3}y = 1$
$\frac{3}{5}x + \frac{2}{3}y = 5$

34. $\frac{5}{6}x + \frac{1}{3}y = \frac{4}{3}$
$\frac{2}{3}x - \frac{1}{2}y = \frac{11}{6}$

35. $\frac{3}{4}x + \frac{2}{5}y = -\frac{3}{20}$
$\frac{3}{2}x - \frac{1}{4}y = \frac{3}{4}$

36. $\frac{2}{5}x - \frac{1}{2}y = \frac{13}{2}$
$\frac{3}{4}x - \frac{1}{5}y = \frac{17}{2}$

37. $4x - 5y = 3y + 4$
$2x + 3y = 2x + 1$

38. $5x - 2y = 8x - 1$
$2x + 7y = 4y + 9$

39. $2x + 5y = 5x + 1$
$3x - 2y = 3y + 3$

2 Solve by the addition method.

40. $x + 2y - z = 1$
$2x - y + z = 6$
$x + 3y - z = 2$

41. $x + 3y + z = 6$
$3x + y - z = -2$
$2x + 2y - z = 1$

42. $2x - y + 2z = 7$
$x + y + z = 2$
$3x - y + z = 6$

43. $x - 2y + z = 6$
$x + 3y + z = 16$
$3x - y - z = 12$

44. $3x + y = 5$
$3y - z = 2$
$x + z = 5$

45. $2y + z = 7$
$2x - z = 3$
$x - y = 3$

46. $x - y + z = 1$
$2x + 3y - z = 3$
$-x + 2y - 4z = 4$

47. $2x + y - 3z = 7$
$x - 2y + 3z = 1$
$3x + 4y - 3z = 13$

48. $2x + 3z = 5$
$3y + 2z = 3$
$3x + 4y = -10$

49. $3x + 4z = 5$
$2y + 3z = 2$
$2x - 5y = 8$

50. $2x + 4y - 2z = 3$
$x + 3y + 4z = 1$
$x + 2y - z = 4$

51. $x - 3y + 2z = 1$
$x - 2y + 3z = 5$
$2x - 6y + 4z = 3$

52. $2x + y - z = 5$
$x + 3y + z = 14$
$3x - y + 2z = 1$

53. $3x - y - 2z = 11$
$2x + y - 2z = 11$
$x + 3y - z = 8$

54. $3x + y - 2z = 2$
$x + 2y + 3z = 13$
$2x - 2y + 5z = 6$

55. $4x + 5y + z = 6$
$2x - y + 2z = 11$
$x + 2y + 2z = 6$

56. $2x - y + z = 6$
$3x + 2y + z = 4$
$x - 2y + 3z = 12$

57. $3x + 2y - 3z = 8$
$2x + 3y + 2z = 10$
$x + y - z = 2$

58. $3x - 2y + 3z = -4$
$2x + y - 3z = 2$
$3x + 4y + 5z = 8$

59. $3x - 3y + 4z = 6$
$4x - 5y + 2z = 10$
$x - 2y + 3z = 4$

60. $3x - y + 2z = 2$
$4x + 2y - 7z = 0$
$2x + 3y - 5z = 7$

61. $2x + 2y + 3z = 13$
$-3x + 4y - z = 5$
$5x - 3y + z = 2$

62. $2x - 3y + 7z = 0$
$x + 4y - 4z = -2$
$3x + 2y + 5z = 1$

63. $5x + 3y - z = 5$
$3x - 2y + 4z = 13$
$4x + 3y + 5z = 22$

SUPPLEMENTAL EXERCISES 10.2

Solve. (*Hint:* Multiply both sides of each equation in the system by a multiple of 10 so that the coefficients and constants are integers.)

64. $0.2x - 0.3y = 0.5$
$0.3x - 0.2y = 0.5$

65. $0.4x - 0.9y = -0.1$
$0.3x + 0.2y = 0.8$

66. $0.5x - 0.4y = 1.8$
$0.3x - 0.5y = 1.6$

67. $0.1x + 0.6y = -1.7$
$0.4x + 0.5y = -1.1$

68. $1.25x - 0.25y = -1.5$
$1.5x + 2.5y = 1$

69. $2.25x + 1.5y = 3$
$1.75x + 2.25y = 1.25$

70. $1.5x + 2.5y + 1.5z = 8$
$0.5x - 2y - 1.5z = -1$
$2.5x - 1.5y + 2z = 2.5$

71. $1.6x - 0.9y + 0.3z = 2.9$
$1.6x + 0.5y - 0.1z = 3.3$
$0.8x - 0.7y + 0.1z = 1.5$

72. $0.6x + 0.5y - 0.2z = 0.6$
$0.3x + 0.4y + 0.2z = 0.3$
$0.5x + 0.6y - 0.4z = 1.1$

73. $1.2x - 1.4y + 1.1z = 3.8$
$1.6x + 1.5y + 1.2z = 1.7$
$1.2x + 1.2y + 0.5z = 1.2$

Solve.

74. The point of intersection of the graphs of the equations $Ax + 5y = 7$ and $2x + By = 8$ is $(2, -1)$. Find A and B.

75. The point of intersection of the graphs of the equations $Ax + 3y = 6$ and $2x + By = -4$ is $(3, -2)$. Find A and B.

76. The point of intersection of the graphs of the equations $Ax + 2y - 3z = 13$, $3x + By + z = 11$, and $2x - 3y + Cz = 0$ is $(2, 3, -1)$. Find A, B, and C.

77. The point of intersection of the graphs of the equations $Ax + 3y + 2z = 8$, $2x + By - 3z = -12$, and $3x - 2y + Cz = 1$ is $(3, -2, 4)$. Find A, B, and C.

78. Find the equation in standard form of the circle that passes through $(4, 1)$, $(2, -3)$, and $(-2, 1)$.

79. Find a, b, and c such that the graph of $y = ax^2 + bx + c$ passes through $(0, 1)$, $(1, 2)$, and $(-1, 4)$.

80. The distance between a point and a line is the perpendicular distance from the point to the line. Find the distance between the point $(3, 1)$ and the line $y = x$.

81. A coin bank contains only nickels, dimes, and quarters. There is a total of 30 coins in the bank. The value of all the coins is \$3.25. Find the number of nickels, dimes, and quarters in the bank. *Hint:* There is more than one solution.

SECTION 10.3 Solving Systems of Equations by Using Determinants and by Using Matrices

1 Evaluate determinants

A **matrix** is a rectangular array of numbers. Each number in the matrix is called an **element** of the matrix. The matrix at the right, with three rows and four columns, is called a 3×4 (read 3 by 4) matrix.

$$A = \begin{bmatrix} 1 & -3 & 2 & 4 \\ 0 & 4 & -3 & 2 \\ 6 & -5 & 4 & -1 \end{bmatrix}$$

A matrix of m rows and n columns is said to be of **order** $\mathbf{m \times n}$. The matrix A, on the previous page has order 3×4. The notation a_{ij} refers to the element of a matrix in the ith row and jth column. For matrix A, $a_{23} = -3$, $a_{31} = 6$, and $a_{13} = 2$.

A **square matrix** is one that has the same number of rows as columns. An example of a 2×2 and a 3×3 matrix is shown at the right.

$$\begin{bmatrix} -1 & 3 \\ 5 & 2 \end{bmatrix} \qquad \begin{bmatrix} 4 & 0 & 1 \\ 5 & -3 & 7 \\ 2 & 1 & 4 \end{bmatrix}$$

Associated with every square matrix is a number called its **determinant.**

Determinant of a 2 × 2 Matrix

The determinant of a 2×2 matrix $\begin{bmatrix} a_{11} & a_{12} \\ a_{21} & a_{22} \end{bmatrix}$ is written $\begin{vmatrix} a_{11} & a_{12} \\ a_{21} & a_{22} \end{vmatrix}$. The value of this determinant is given by the formula

$$\begin{vmatrix} a_{11} & a_{12} \\ a_{21} & a_{22} \end{vmatrix} = a_{11}a_{22} - a_{21}a_{12}.$$

Note that vertical bars are used to represent the determinant and brackets are used to represent the matrix.

Find the value of the determinant $\begin{vmatrix} 3 & 4 \\ -1 & 2 \end{vmatrix}$.

Use the formula.

$$\begin{vmatrix} 3 & 4 \\ -1 & 2 \end{vmatrix} = 3 \cdot 2 - (-1) \cdot 4 = 6 - (-4) = 10$$

The value of the determinant is 10.

To evaluate the determinant of a square matrix of order greater than 2, the *minor* and *cofactor* of a matrix are used.

Definition of a Minor of a Matrix

The **minor** M_{ij} of a square matrix A of order greater than 2 is the determinant of A after row i and column j have been removed from A.

Consider the matrix $A = \begin{bmatrix} 2 & -3 & 4 \\ 0 & 4 & 8 \\ -1 & 3 & 6 \end{bmatrix}$. The minor M_{12} of A is the determinant of matrix A after row 1 and column 2 have been removed from A.

$$\begin{bmatrix} 2 & 3 & 4 \\ 0 & 4 & 8 \\ -1 & 3 & 6 \end{bmatrix}$$

The minor $M_{12} = \begin{vmatrix} 0 & 8 \\ -1 & 6 \end{vmatrix}$.

Find the value of M_{33} for the matrix $A = \begin{bmatrix} -5 & -3 & 4 \\ 2 & 1 & -4 \\ 3 & 2 & 5 \end{bmatrix}$.

M_{33} is the determinant after eliminating row 3 and column 3.

$$\begin{bmatrix} -5 & -3 & 4 \\ 2 & 1 & -4 \\ 3 & 2 & -5 \end{bmatrix}$$

$$M_{33} = \begin{vmatrix} -5 & -3 \\ 2 & 1 \end{vmatrix}$$

Find the value of the determinant.

$$= -5(1) - 2(-3) = 1$$

The value of M_{33} is 1.

Definition of the Cofactor of a Matrix

> The cofactor C_{ij} of a matrix A is the product $(-1)^{i+j}M_{ij}$, where M_{ij} is the minor of A.
>
> $$C_{ij} = (-1)^{i+j}M_{ij}$$

Find the value of (a) C_{23} and (b) C_{31} for $A = \begin{bmatrix} 2 & -3 & 4 \\ 0 & 4 & 8 \\ -1 & 3 & 6 \end{bmatrix}$.

(a) $C_{23} = (-1)^{2+3}M_{23} = -M_{23}$.

$$C_{23} = -\begin{vmatrix} 2 & -3 \\ -1 & 3 \end{vmatrix}$$

Find the value of the determinant.

$$= -[2(3) - (-1)(-3)] = -3$$

The value of C_{23} is -3.

(b) $C_{31} = (-1)^{3+1}M_{31} = M_{31}$.

$$C_{31} = \begin{vmatrix} -3 & 4 \\ 4 & 8 \end{vmatrix}$$

Find the value of the determinant.

$$= -3(8) - 4(4) = -40$$

The value of C_{31} is -40.

If $i + j$ is an even integer, then $(-1)^{i+j} = 1$. If $i + j$ is an odd integer, then $(-1)^{i+j} = -1$. Thus the cofactor of a matrix is M_{ij} or $-M_{ij}$ depending on whether $i + j$ is an even or odd integer. Another way to remember the sign associated with the cofactor is to consider the following sign pattern.

$$\begin{bmatrix} + & - & + \\ - & + & - \\ + & - & + \end{bmatrix}$$

Use the sign pattern to find the value of (a) C_{22} and (b) C_{32} for

$$A = \begin{vmatrix} 2 & -3 & 0 \\ 1 & 5 & -3 \\ -1 & 2 & 5 \end{vmatrix}.$$

(a) From the sign pattern, the cofactor is $+M_{22}$.

$$C_{22} = +M_{22} = +\begin{vmatrix} 2 & 0 \\ -1 & 5 \end{vmatrix}$$

$$= 2(5) - (-1)0 = 10$$

The value of C_{22} is 10.

(b) From the sign pattern, the cofactor is $-M_{32}$.

$$C_{32} = -M_{32} = -\begin{vmatrix} 2 & 0 \\ 1 & -3 \end{vmatrix}$$

$$= -[2(-3) - 1(0)] = 6$$

The value of C_{32} is 6.

Cofactor Expansion of a Determinant by the First Row

The determinant $|A|$ of matrix A of order 3 is the sum of the products of each entry in the first row of A and its cofactor.

$$\begin{vmatrix} a_{11} & a_{12} & a_{13} \\ a_{21} & a_{22} & a_{23} \\ a_{31} & a_{32} & a_{33} \end{vmatrix} = a_{11}C_{11} + a_{12}C_{12} + a_{13}C_{13}$$

$$= a_{11}\begin{vmatrix} a_{22} & a_{23} \\ a_{32} & a_{33} \end{vmatrix} - a_{12}\begin{vmatrix} a_{21} & a_{23} \\ a_{31} & a_{33} \end{vmatrix} + a_{13}\begin{vmatrix} a_{21} & a_{22} \\ a_{31} & a_{32} \end{vmatrix}$$

To find the value of the determinant $\begin{vmatrix} 2 & -1 & -3 \\ 1 & 2 & 0 \\ 3 & -1 & 2 \end{vmatrix}$, expand by the cofactors of the first row.

$$\begin{vmatrix} 2 & -1 & -3 \\ 1 & 2 & 0 \\ 3 & -1 & 2 \end{vmatrix} = 2 \cdot C_{11} + (-1) \cdot C_{12} + (-3)C_{13}$$

$$= 2\begin{vmatrix} 2 & 0 \\ -1 & 2 \end{vmatrix} - (-1)\begin{vmatrix} 1 & 0 \\ 3 & 2 \end{vmatrix} + (-3)\begin{vmatrix} 1 & 2 \\ 3 & -1 \end{vmatrix}$$

$$= 2(4 - 0) - (-1)(2 - 0) + (-3)(-1 - 6)$$
$$= 2(4) - (-1)(2) + (-3)(-7)$$
$$= 8 + 2 + 21$$
$$= 31$$

The value of the determinant is 31.

For the previous example, when the value of the determinant is found by expanding about the elements of the third column, the answer is the same.

$$\begin{vmatrix} 2 & -1 & -3 \\ 1 & 2 & 0 \\ 3 & -1 & 2 \end{vmatrix} = -3\begin{vmatrix} 1 & 2 \\ 3 & -1 \end{vmatrix} - 0\begin{vmatrix} 2 & -1 \\ 3 & -1 \end{vmatrix} + 2\begin{vmatrix} 2 & -1 \\ 1 & 2 \end{vmatrix}$$
$$= -3(-1 - 6) - 0 + 2(4 + 1)$$
$$= -3(-7) + 2(5)$$
$$= 21 + 10$$
$$= 31$$

Cofactor Expansion of a Determinant

The determinant $|A|$ of matrix A of order 3 is the cofactor expansion by any row or column of A.

To find the value of a 3×3 determinant, expand by the cofactors of the elements of any row or column. However, note that by choosing a row or column that contains a zero, the computation is simplified.

Example 1 Evaluate the determinant.

A. $\begin{vmatrix} 3 & -2 \\ 6 & -4 \end{vmatrix}$ B. $\begin{vmatrix} -2 & 3 & 1 \\ 4 & -2 & 0 \\ 1 & -2 & 3 \end{vmatrix}$

Solution A. $\begin{vmatrix} 3 & -2 \\ 6 & -4 \end{vmatrix} = 3(-4) - (6)(-2)$

$$= -12 + 12 = 0$$

The value of the determinant is 0.

B. $$\begin{vmatrix} -2 & 3 & 1 \\ 4 & -2 & 0 \\ 1 & -2 & 3 \end{vmatrix} = -4\begin{vmatrix} 3 & 1 \\ -2 & 3 \end{vmatrix} + (-2)\begin{vmatrix} -2 & 1 \\ 1 & 3 \end{vmatrix} + 0$$
$$= -4(9 + 2) - 2(-6 - 1)$$
$$= -4(11) - 2(-7)$$
$$= -44 + 14$$
$$= -30$$

▶ Because there is a zero in the second row, expand by cofactors of the second row.

The value of the determinant is -30.

Problem 1 Evaluate the determinant.

A. $\begin{vmatrix} -1 & -4 \\ 3 & -5 \end{vmatrix}$ B. $\begin{vmatrix} 1 & 4 & -2 \\ 3 & 1 & 1 \\ 0 & -2 & 2 \end{vmatrix}$

Solution See page A46.

2 Solve systems of equations by using Cramer's Rule

The connection between determinants and systems of equations can be understood by solving a general system of linear equations.

Solve: $a_{11}x + a_{12}y = b_1$
$a_{21}x + a_{22}y = b_2$

$$\begin{aligned} (1)\quad a_{11}x + a_{12}y &= b_1 \\ (2)\quad a_{21}x + a_{22}y &= b_2 \end{aligned}$$

Eliminate y. Multiply equation (1) by a_{22} and equation (2) by $-a_{12}$.

$$\begin{aligned} a_{11}a_{22}x + a_{12}a_{22}y &= b_1a_{22} \\ -a_{21}a_{12}x - a_{12}a_{22}y &= -b_2a_{12} \end{aligned}$$

Add the equations.

$$(a_{11}a_{22} - a_{21}a_{12})x = b_1a_{22} - b_2a_{12}$$

Assuming $a_{11}a_{22} - a_{21}a_{12} \neq 0$, solve for x.

$$x = \frac{b_1a_{22} - b_2a_{12}}{a_{11}a_{22} - a_{21}a_{12}}$$

The denominator $a_{11}a_{22} - a_{21}a_{12}$ is the determinant of the coefficients of x and y. This is called the **coefficient determinant.**

$$a_{11}a_{22} - a_{21}a_{12} = \begin{vmatrix} a_{11} & a_{12} \\ a_{21} & a_{22} \end{vmatrix}$$

Coefficients of x (first column)

Coefficients of y (second column)

The numerator for x, $b_1a_{22} - b_2a_{12}$, is the determinant obtained by replacing the first column in the coefficient determinant by the constants b_1 and b_2. This is called the **numerator determinant.**

$$b_1a_{22} - b_2a_{12} = \begin{vmatrix} b_1 & a_{12} \\ b_2 & a_{22} \end{vmatrix}$$

Constants of the equations (first column)

Following a similar procedure and eliminating x, the y-component of the solution can also be expressed in determinant form. These results are summarized in Cramer's Rule.

Cramer's Rule

The solution of the system of equations $\begin{aligned} a_{11}x + a_{12}y &= b_1 \\ a_{21}x + a_{22}y &= b_2 \end{aligned}$ is given by $x = \dfrac{D_x}{D}$ and $y = \dfrac{D_y}{D}$, where

$$D = \begin{vmatrix} a_{11} & a_{12} \\ a_{21} & a_{22} \end{vmatrix},\ D_x = \begin{vmatrix} b_1 & a_{12} \\ b_2 & a_{22} \end{vmatrix},\ D_y = \begin{vmatrix} a_{11} & b_1 \\ a_{21} & b_2 \end{vmatrix},\ \text{and } D \neq 0.$$

Example 2 Solve by using Cramer's Rule: $2x - 3y = 8$
$5x + 6y = 11$

Solution $D = \begin{vmatrix} 2 & -3 \\ 5 & 6 \end{vmatrix} = 27$

▶ Find the value of the coefficient determinant.

$D_x = \begin{vmatrix} 8 & -3 \\ 11 & 6 \end{vmatrix} = 81 \qquad D_y = \begin{vmatrix} 2 & 8 \\ 5 & 11 \end{vmatrix} = -18$

▶ Find the value of each of the numerator determinants.

$x = \frac{D_x}{D} = \frac{81}{27} = 3 \qquad y = \frac{D_y}{D} = \frac{-18}{27} = \frac{-2}{3}$

▶ Use Cramer's Rule to write the solutions.

The solution is $\left(3, -\frac{2}{3}\right)$.

Problem 2 Solve by using Cramer's Rule: $6x - 6y = 5$
$2x - 10y = -1$

Solution See page A46.

For the system shown at the right, $D = 0$. Therefore, $\frac{D_x}{D}$ and $\frac{D_y}{D}$ are undefined.

$6x - 9y = 5$
$4x - 6y = 4$

When $D = 0$, the system of equations is dependent if both D_x and D_y are zero. The system of equations is inconsistent if either D_x or D_y is not zero.

$D = \begin{vmatrix} 6 & -9 \\ 4 & -6 \end{vmatrix} = 0$

A procedure similar to that followed for two equations in two variables can be used to extend Cramer's Rule to three equations in three variables.

Cramer's Rule for Three Equations in Three Variables

The solution of the system of equations
$$\begin{aligned} a_{11}x + a_{12}y + a_{13}z &= b_1 \\ a_{21}x + a_{22}y + a_{23}z &= b_2 \\ a_{31}x + a_{32}y + a_{33}z &= b_3 \end{aligned}$$
is given by $x = \frac{D_x}{D}$, $y = \frac{D_y}{D}$, and $z = \frac{D_z}{D}$, where

$$D = \begin{vmatrix} a_{11} & a_{12} & a_{13} \\ a_{21} & a_{22} & a_{23} \\ a_{31} & a_{32} & a_{33} \end{vmatrix},\ D_x = \begin{vmatrix} b_1 & a_{12} & a_{13} \\ b_2 & a_{22} & a_{23} \\ b_3 & a_{32} & a_{33} \end{vmatrix},\ D_y = \begin{vmatrix} a_{11} & b_1 & a_{13} \\ a_{21} & b_2 & a_{23} \\ a_{31} & b_3 & a_{33} \end{vmatrix},$$

$$D_z = \begin{vmatrix} a_{11} & a_{12} & b_1 \\ a_{21} & a_{22} & b_2 \\ a_{31} & a_{32} & b_3 \end{vmatrix}, \text{ and } D \neq 0.$$

Example 3 Solve by using Cramer's Rule:

$$\begin{aligned} 3x - y + z &= 5 \\ x + 2y - 2z &= -3 \\ 2x + 3y + z &= 4 \end{aligned}$$

Solution

$$D = \begin{vmatrix} 3 & -1 & 1 \\ 1 & 2 & -2 \\ 2 & 3 & 1 \end{vmatrix} = 28$$

▶ Find the value of the coefficient determinant.

$$D_x = \begin{vmatrix} 5 & -1 & 1 \\ -3 & 2 & -2 \\ 4 & 3 & 1 \end{vmatrix} = 28$$

▶ Find the value of each of the numerator determinants.

$$D_y = \begin{vmatrix} 3 & 5 & 1 \\ 1 & -3 & -2 \\ 2 & 4 & 1 \end{vmatrix} = 0$$

$$D_z = \begin{vmatrix} 3 & -1 & 5 \\ 1 & 2 & -3 \\ 2 & 3 & 4 \end{vmatrix} = 56$$

$$x = \frac{D_x}{D} = \frac{28}{28} = 1$$

▶ Use Cramer's Rule to write the solution.

$$y = \frac{D_y}{D} = \frac{0}{28} = 0$$

$$z = \frac{D_z}{D} = \frac{56}{28} = 2$$

The solution is (1, 0, 2).

Problem 3 Solve by using Cramer's Rule:

$$\begin{aligned} 2x - y + z &= -1 \\ 3x + 2y - z &= 3 \\ x + 3y + z &= -2 \end{aligned}$$

Solution See page A46.

3 Solve systems of equations by using matrices

As previously stated in this section, a **matrix** is a rectangular array of numbers. Each number in the matrix is called an **element** of the matrix. The matrix shown below, with 3 rows and 4 columns, is a 3×4 matrix.

$$\begin{bmatrix} 1 & 4 & -3 & 6 \\ -2 & 5 & 2 & 0 \\ -1 & 3 & 7 & -4 \end{bmatrix}$$

The notation a_{ij} refers to the element of a matrix in the ith row and jth column. For the matrix shown on the previous page, $a_{23} = 2$ (2nd row, 3rd column), $a_{31} = -1$ (3rd row, 1st column), and $a_{13} = -3$ (1st row, 3rd column).

The elements $a_{11}, a_{22}, a_{33}, \ldots, a_{nn}$ form the **main diagonal** of a matrix. The elements 1, 5, and 7 form the main diagonal of the matrix shown on the previous page.

By considering only the coefficients and constants for the system of equations below, the corresponding 3×4 **augmented matrix** can be formed.

System of Equations	**Augmented Matrix**
$\begin{aligned} 3x - 2y + z &= 2 \\ x \quad\quad - 3z &= -2 \\ 2x - y + 4z &= 5 \end{aligned}$	$\begin{bmatrix} 3 & -2 & 1 & 2 \\ 1 & 0 & -3 & -2 \\ 2 & -1 & 4 & 5 \end{bmatrix}$

Note that when a term is missing from one of the equations of the system, the coefficient of that term is 0, and a 0 is entered in the matrix.

A system of equations can be written from an augmented matrix.

$$\begin{bmatrix} 2 & -1 & 4 & 1 \\ 1 & 1 & 0 & 3 \\ 3 & -2 & -1 & 5 \end{bmatrix} \qquad \begin{aligned} 2x - y + 4z &= 1 \\ x + y \quad\quad &= 3 \\ 3x - 2y - z &= 5 \end{aligned}$$

A system of equations can be solved by writing the system in matrix form and then performing operations on the matrix similar to those performed on the equations of the system. These operations are called **elementary row operations.**

Elementary Row Operations

1. Interchange two rows.
2. Multiply all the elements in a row by the same nonzero number.
3. Replace a row by the sum of that row and a multiple of any other row.

The goal is to use the elementary row operations to rewrite the matrix with 1's down the main diagonal and 0's to the left of the 1's in all rows except the first. This is called the **echelon form** of the matrix. Examples of echelon form are shown below.

$$\begin{bmatrix} 1 & 2 & 2 \\ 0 & 1 & -1 \end{bmatrix} \qquad \begin{bmatrix} 1 & -3 & -1 & -6 \\ 0 & 1 & -2 & 7 \\ 0 & 0 & 1 & -2 \end{bmatrix} \qquad \begin{bmatrix} 1 & \frac{1}{2} & -\frac{1}{3} & 2 \\ 0 & 1 & -\frac{2}{3} & -2 \\ 0 & 0 & 1 & 6 \end{bmatrix}$$

A system of equations can be solved by using elementary row operations to rewrite the augmented matrix of a system of equations in echelon form.

The system of equations at the right can be solved by first writing the system in matrix form. Then use elementary row operations to write the matrix in echelon form.

$$2x + 5y = 8 \qquad \begin{bmatrix} 2 & 5 & 8 \\ 3 & 4 & 5 \end{bmatrix}$$
$$3x + 4y = 5$$

Element a_{11} must be a 1. Multiply row 1 by $\frac{1}{2}$.

[Elementary row operation (2)]

$$\begin{bmatrix} 1 & \frac{5}{2} & 4 \\ 3 & 4 & 5 \end{bmatrix}$$

Element a_{21} must be a 0. Multiply row 1 by -3, and add it to row 2. Replace row 2 by the sum.

[Elementary row operation (3)]

$$\begin{bmatrix} 1 & \frac{5}{2} & 4 \\ 0 & -\frac{7}{2} & -7 \end{bmatrix}$$

Element a_{22} must be a 1. Multiply row 2 by $-\frac{2}{7}$.

[Elementary row operation (2)]

$$\begin{bmatrix} 1 & \frac{5}{2} & 4 \\ 0 & 1 & 2 \end{bmatrix}$$

The matrix is now in echelon form.

Write the system of equations represented by the matrix.

$$(1) \quad x + \frac{5}{2}y = 4$$

$$(2) \quad y = 2$$

Substitute the value of y into equation (1), and solve for x.

$$x + \frac{5}{2}(2) = 4$$
$$x + 5 = 4$$
$$x = -1$$

The solution is $(-1, 2)$.

The order in which the elements in the 2×3 matrix were changed is important.

$$\begin{bmatrix} a_{11} & a_{12} & a_{13} \\ a_{21} & a_{22} & a_{23} \end{bmatrix}$$

1. Change a_{11} to a 1.
2. Change a_{21} to a 0.
3. Change a_{22} to a 1.

Example 4 Solve by using a matrix: $3x + 2y = 3$
$2x - 3y = 15$

Solution

$\begin{bmatrix} 3 & 2 & 3 \\ 2 & -3 & 15 \end{bmatrix}$ $\begin{bmatrix} 1 & \frac{2}{3} & 1 \\ 2 & -3 & 15 \end{bmatrix}$ ▶ Write the system in matrix form. Multiply row 1 by $\frac{1}{3}$.

$\begin{bmatrix} 1 & \frac{2}{3} & 1 \\ 0 & -\frac{13}{3} & 13 \end{bmatrix}$ $\begin{bmatrix} 1 & \frac{2}{3} & 1 \\ 0 & 1 & -3 \end{bmatrix}$ ▶ Multiply row 1 by -2 and add to row 2. ▶ Multiply row 3 by $-\frac{3}{13}$.

(1) $x + \frac{2}{3}y = 1$ $x + \frac{2}{3}(-3) = 1$ ▶ Write the system of equations represented by the matrix.

(2) $y = -3$ $x - 2 = 1$ Substitute the value of y into equation

$x = 3$ (1), and solve for x.

The solution is $(3, -3)$.

Problem 4 Solve by using a matrix: $3x - 5y = -12$
$4x - 3y = -5$

Solution See pages A46–A47.

The matrix method of solving systems of equations can be extended to larger systems of equations. A system of three equations in three unknowns is written as a 3×4 augmented matrix.

The order in which the elements in a 3×4 matrix are changed is

$$\begin{bmatrix} a_{11} & a_{12} & a_{13} & a_{14} \\ a_{21} & a_{22} & a_{23} & a_{24} \\ a_{31} & a_{32} & a_{33} & a_{34} \end{bmatrix}$$

1. Change a_{11} to a 1.
2. Change a_{21} and a_{31} to 0's.
3. Change a_{22} to a 1.
4. Change a_{32} to a 0.
5. Change a_{33} to a 1.

To solve the system shown at the right, write the system in matrix form.

$$\begin{aligned} 2x + 3y + 3z &= -2 \\ x + 2y - 3z &= 9 \\ 3x - 2y - 4z &= 1 \end{aligned} \qquad \begin{bmatrix} 2 & 3 & 3 & -2 \\ 1 & 2 & -3 & 9 \\ 3 & -2 & -4 & 1 \end{bmatrix}$$

Element a_{11} must be a 1. Interchange rows 1 and 2.

$$\begin{bmatrix} 1 & 2 & -3 & 9 \\ 2 & 3 & 3 & -2 \\ 3 & -2 & -4 & 1 \end{bmatrix}$$

Element a_{21} must be a 0. Multiply row 1 by -2 and add to row 2. Replace row 2 by the sum.
Element a_{31} must be a 0. Multiply row 1 by -3 and add to row 3. Replace row 3 by the sum.

$$\begin{bmatrix} 1 & 2 & -3 & 9 \\ 0 & -1 & 9 & -20 \\ 0 & -8 & 5 & -26 \end{bmatrix}$$

Element a_{22} must be a 1. Multiply row 2 by -1.

$$\begin{bmatrix} 1 & 2 & -3 & 9 \\ 0 & 1 & -9 & 20 \\ 0 & -8 & 5 & -26 \end{bmatrix}$$

Element a_{32} must be a 0. Multiply row 2 by 8 and add to row 3. Replace row 3 by the sum.

$$\begin{bmatrix} 1 & 2 & -3 & 9 \\ 0 & 1 & -9 & 20 \\ 0 & 0 & -67 & 134 \end{bmatrix}$$

Element a_{33} must be a 1. Multiply row 3 by $-\frac{1}{67}$.

$$\begin{bmatrix} 1 & 2 & -3 & 9 \\ 0 & 1 & -9 & 20 \\ 0 & 0 & 1 & -2 \end{bmatrix}$$

Write the system represented by the matrix.

$$\begin{aligned} &(1) \quad x + 2y - 3z = 9 \\ &(2) \qquad\quad\; y - 9z = 20 \\ &(3) \qquad\qquad\quad\; z = -2 \end{aligned}$$

Substitute the value of z into equation (2), and solve for y.

$$\begin{aligned} y - 9z &= 20 \\ y - 9(-2) &= 20 \\ y + 18 &= 20 \\ y &= 2 \end{aligned}$$

Substitute the values of y and z into equation (1), and solve for x.

$$\begin{aligned} x + 2y - 3z &= 9 \\ x + 2(2) - 3(-2) &= 9 \\ x + 4 + 6 &= 9 \\ x &= -1 \end{aligned}$$

The solution is $(-1, 2, -2)$.

Solve by using matrices:

$$\begin{aligned} x - y + z &= 2 \\ x + 2y - z &= 3 \\ 3x + 3y - z &= 6 \end{aligned}$$

Write the system in matrix form.

$$\begin{bmatrix} 1 & -1 & 1 & 2 \\ 1 & 2 & -1 & 3 \\ 3 & 3 & -1 & 6 \end{bmatrix}$$

Element a_{11} is a 1. Element a_{21} must be a 0. Multiply row 1 by -1 and add to row 2. Replace row 2 by the sum.
Element a_{31} must be a 0. Multiply row 1 by -3 and add to row 3. Replace row 3 by the sum.

$$\begin{bmatrix} 1 & -1 & 1 & 2 \\ 0 & 3 & -2 & 1 \\ 0 & 6 & -4 & 0 \end{bmatrix}$$

Element a_{22} must be a 1. Multiply row 2 by $\frac{1}{3}$.

$$\begin{bmatrix} 1 & -1 & 1 & 2 \\ 0 & 1 & -\frac{2}{3} & \frac{1}{3} \\ 0 & 6 & -4 & 0 \end{bmatrix}$$

Element a_{32} must be a 0. Multiply row 2 by −6 and add to row 3. Replace row 3 by the sum.

$$\begin{bmatrix} 1 & -1 & 1 & 2 \\ 0 & 1 & -\frac{2}{3} & \frac{1}{3} \\ 0 & 0 & 0 & -2 \end{bmatrix}$$

Write the system represented by the matrix. The equation $0 = -2$ is not a true equation. The system of equations is inconsistent.

$$\begin{aligned} x - y + z &= 2 \\ y - \tfrac{2}{3}z &= \tfrac{1}{3} \\ 0 &= -2 \end{aligned}$$

The system of equations has no solution.

Example 5 Solve by using a matrix:

$$\begin{aligned} 3x + 2y + 3z &= 2 \\ 2x - 3y + 4z &= 5 \\ x + 4y + 2z &= 8 \end{aligned}$$

Solution

$$\begin{bmatrix} 3 & 2 & 3 & 2 \\ 2 & -3 & 4 & 5 \\ 1 & 4 & 2 & 8 \end{bmatrix}$$

▶ Write the system in matrix form.

$$\begin{bmatrix} 1 & 4 & 2 & 8 \\ 2 & -3 & 4 & 5 \\ 3 & 2 & 3 & 2 \end{bmatrix}$$

▶ Interchange rows 1 and 3.

$$\begin{bmatrix} 1 & 4 & 2 & 8 \\ 0 & -11 & 0 & -11 \\ 0 & -10 & -3 & -22 \end{bmatrix}$$

▶ Multiply row 1 by −2 and add to row 2. Multiply row 1 by −3 and add to row 3.

$$\begin{bmatrix} 1 & 4 & 2 & 8 \\ 0 & 1 & 0 & 1 \\ 0 & -10 & -3 & -22 \end{bmatrix}$$

▶ Multiply row 2 by $-\frac{1}{11}$.

$$\begin{bmatrix} 1 & 4 & 2 & 8 \\ 0 & 1 & 0 & 1 \\ 0 & 0 & -3 & -12 \end{bmatrix}$$

▶ Multiply row 2 by 10 and add to row 3.

$$\begin{bmatrix} 1 & 4 & 2 & 8 \\ 0 & 1 & 0 & 1 \\ 0 & 0 & 1 & 4 \end{bmatrix}$$

▶ Multiply row 3 by $-\frac{1}{3}$.

$$\begin{aligned} (1)\quad x + 4y + 2z &= 8 \\ (2)\quad y &= 1 \\ (3)\quad z &= 4 \end{aligned}$$

▶ Write the system of equations represented by the matrix.

$$\begin{aligned} x + 4y + 2z &= 8 \\ x + 4(1) + 2(4) &= 8 \\ x + 4 + 8 &= 8 \\ x &= -4 \end{aligned}$$

▶ Substitute the values of y and z into equation (1), and solve for x.

The solution is (−4, 1, 4).

Problem 5 Solve by using a matrix:
$$\begin{aligned} 3x - 2y - 3z &= 5 \\ x + 3y - 2z &= -4 \\ 2x + 6y + 3z &= 6 \end{aligned}$$

Solution See page A47.

EXERCISES 10.3

1 Evaluate the determinant.

1. $\begin{vmatrix} 2 & -1 \\ 3 & 4 \end{vmatrix}$

2. $\begin{vmatrix} 5 & 1 \\ -1 & 2 \end{vmatrix}$

3. $\begin{vmatrix} 6 & -2 \\ -3 & 4 \end{vmatrix}$

4. $\begin{vmatrix} -3 & 5 \\ 1 & 7 \end{vmatrix}$

5. $\begin{vmatrix} 3 & 6 \\ 2 & 4 \end{vmatrix}$

6. $\begin{vmatrix} 5 & -10 \\ 1 & -2 \end{vmatrix}$

7. $\begin{vmatrix} 1 & -1 & 2 \\ 3 & 2 & 1 \\ 1 & 0 & 4 \end{vmatrix}$

8. $\begin{vmatrix} 4 & 1 & 3 \\ 2 & -2 & 1 \\ 3 & 1 & 2 \end{vmatrix}$

9. $\begin{vmatrix} 3 & -1 & 2 \\ 0 & 1 & 2 \\ 3 & 2 & -2 \end{vmatrix}$

10. $\begin{vmatrix} 4 & 5 & -2 \\ 3 & -1 & 5 \\ 2 & 1 & 4 \end{vmatrix}$

11. $\begin{vmatrix} 4 & 2 & 6 \\ -2 & 1 & 1 \\ 2 & 1 & 3 \end{vmatrix}$

12. $\begin{vmatrix} 3 & 6 & -3 \\ 4 & -1 & 6 \\ -1 & -2 & 3 \end{vmatrix}$

2 Solve by using Cramer's Rule.

13. $\begin{aligned} 2x - 5y &= 26 \\ 5x + 3y &= 3 \end{aligned}$

14. $\begin{aligned} 3x + 7y &= 15 \\ 2x + 5y &= 11 \end{aligned}$

15. $\begin{aligned} x - 4y &= 8 \\ 3x + 7y &= 5 \end{aligned}$

16. $\begin{aligned} 5x + 2y &= -5 \\ 3x + 4y &= 11 \end{aligned}$

17. $\begin{aligned} 2x + 3y &= 4 \\ 6x - 12y &= -5 \end{aligned}$

18. $\begin{aligned} 5x + 4y &= 3 \\ 15x - 8y &= -21 \end{aligned}$

19. $\begin{aligned} 2x + 5y &= 6 \\ 6x - 2y &= 1 \end{aligned}$

20. $\begin{aligned} 7x + 3y &= 4 \\ 5x - 4y &= 9 \end{aligned}$

21. $\begin{aligned} -2x + 3y &= 7 \\ 4x - 6y &= 9 \end{aligned}$

22. $\begin{aligned} 9x + 6y &= 7 \\ 3x + 2y &= 4 \end{aligned}$

23. $\begin{aligned} 2x - 5y &= -2 \\ 3x - 7y &= -3 \end{aligned}$

24. $\begin{aligned} 8x + 7y &= -3 \\ 2x + 2y &= 5 \end{aligned}$

25. $\begin{aligned} 2x - y + 3z &= 9 \\ x + 4y + 4z &= 5 \\ 3x + 2y + 2z &= 5 \end{aligned}$

26. $\begin{aligned} 3x - 2y + z &= 2 \\ 2x + 3y + 2z &= -6 \\ 3x - y + z &= 0 \end{aligned}$

27. $\begin{aligned} 3x - y + z &= 11 \\ x + 4y - 2z &= -12 \\ 2x + 2y - z &= -3 \end{aligned}$

28. $x + 2y + 3z = 8$
$2x - 3y + z = 5$
$3x - 4y + 2z = 9$

29. $4x - 2y + 6z = 1$
$3x + 4y + 2z = 1$
$2x - y + 3z = 2$

30. $x - 3y + 2z = 1$
$2x + y - 2z = 3$
$3x - 9y + 6z = -3$

31. $5x - 4y + 2z = 4$
$3x - 5y + 3z = -4$
$3x + y - 5z = 12$

32. $2x + 4y + z = 7$
$x + 3y - z = 1$
$3x + 2y - 2z = 5$

3 Solve by using matrices.

33. $3x + y = 6$
$2x - y = -1$

34. $2x + y = 3$
$x - 4y = 6$

35. $x - 3y = 8$
$3x - y = 0$

36. $2x + 3y = 16$
$x - 4y = -14$

37. $y = 4x - 10$
$2y = 5x - 11$

38. $2y = 4 - 3x$
$y = 1 - 2x$

39. $2x - y = -4$
$y = 2x - 8$

40. $3x - 2y = -8$
$y = \frac{3}{2}x - 2$

41. $4x - 3y = -14$
$3x + 4y = 2$

42. $5x + 2y = 3$
$3x + 4y = 13$

43. $5x + 4y + 3z = -9$
$x - 2y + 2z = -6$
$x - y - z = 3$

44. $x - y - z = 0$
$3x - y + 5z = -10$
$x + y - 4z = 12$

45. $5x - 5y + 2z = 8$
$2x + 3y - z = 0$
$x + 2y - z = 0$

46. $2x + y - 5z = 3$
$3x + 2y + z = 15$
$5x - y - z = 5$

47. $2x + 3y + z = 5$
$3x + 3y + 3z = 10$
$4x + 6y + 2z = 5$

48. $x - 2y + 3z = 2$
$2x + y + 2z = 5$
$2x - 4y + 6z = -4$

49. $3x + 2y + 3z = 2$
$6x - 2y + z = 1$
$3x + 4y + 2z = 3$

50. $2x + 3y - 3z = -1$
$2x + 3y + 3z = 3$
$4x - 4y + 3z = 4$

51. $5x - 5y - 5z = 2$
$5x + 5y - 5z = 6$
$10x + 10y + 5z = 3$

52. $3x - 2y + 2z = 5$
$6x + 3y - 4z = -1$
$3x - y + 2z = 4$

53. $4x + 4y - 3z = 3$
$8x + 2y + 3z = 0$
$4x - 4y + 6z = -3$

SUPPLEMENTAL EXERCISES 10.3

Solve for x.

54. $\begin{vmatrix} 3 & 2 \\ 4 & x \end{vmatrix} = -11$

55. $\begin{vmatrix} -1 & 4 \\ 2 & x \end{vmatrix} = -11$

56. $\begin{vmatrix} -2 & 3 \\ 5 & x \end{vmatrix} = -3$

57. $\begin{vmatrix} 1 & 0 & 2 \\ 4 & 3 & -1 \\ 0 & 2 & x \end{vmatrix} = -24$

58. $\begin{vmatrix} -2 & 1 & 3 \\ 0 & x & 4 \\ -1 & 2 & -3 \end{vmatrix} = -24$

59. $\begin{vmatrix} 3 & -2 & 1 \\ -1 & 0 & x \\ 2 & 4 & 0 \end{vmatrix} = 44$

Complete.

60. If all the elements in one row or one column of a 2×2 matrix are zeros, the value of the determinant of the matrix is ______.

61. If all the elements in one row or one column of a 3×3 matrix are zeros, the value of the determinant of the matrix is ______.

62. a. The value of the determinant $\begin{vmatrix} x & x & a \\ y & y & b \\ z & z & c \end{vmatrix}$ is ______.

b. If two columns of a 3×3 matrix contain identical elements, the value of the determinant is ______.

SECTION 10.4 Application Problems in Two Variables

1 Rate-of-wind and water current problems

Motion problems that involve an object moving with or against a wind or current normally require two variables to solve.

Solve: A motorboat traveling with the current can go 24 mi in 2 h. Against the current it takes 3 h to go the same distance. Find the rate of the motorboat in calm water and the rate of the current.

STRATEGY for solving rate-of-wind or water current problems

- Choose one variable to represent the rate of the object in calm conditions and a second variable to represent the rate of the wind or current. Using these variables, express the rate of the object with and against the wind or current. Use the equation $rt = d$ to write expressions for the distance traveled by the object. The results can be recorded in a table.

Rate of the boat in calm water: x
Rate of the current: y

	Rate	·	Time	=	Distance
With current	$x + y$	·	2	=	$2(x + y)$
Against current	$x - y$	·	3	=	$3(x - y)$

- Determine how the expressions for distance are related.

The distance traveled with the current is 24 mi.
The distance traveled against the current is 24 mi.

$$2(x + y) = 24$$
$$3(x - y) = 24$$

Solve the system of equations.

$$2(x + y) = 24 \quad \Rightarrow \quad \frac{1}{2} \cdot 2(x + y) = \frac{1}{2} \cdot 24 \quad \Rightarrow \quad x + y = 12$$

$$3(x - y) = 24 \quad \Rightarrow \quad \frac{1}{3} \cdot 3(x - y) = \frac{1}{3} \cdot 24 \quad \Rightarrow \quad x - y = 8$$

$$2x = 20$$
$$x = 10$$

Replace x by 10 in the equation $x + y = 12$. Solve for y.

$$x + y = 12$$
$$10 + y = 12$$
$$y = 2$$

The rate of the boat in calm water is 10 mph.
The rate of the current is 2 mph.

Example 1 Flying with the wind, a plane flew 1000 mi in 5 h. Flying against the wind, the plane could fly only 500 mi in the same amount of time. Find the rate of the plane in calm air and the rate of the wind.

Strategy
- Rate of the plane in still air: p
  Rate of the wind: w

	Rate	Time	Distance
With wind	$p + w$	5	$5(p + w)$
Against wind	$p - w$	5	$5(p - w)$

- The distance traveled with the wind is 1000 mi.
  The distance traveled against the wind is 500 mi.

Solution

$$5(p + w) = 1000$$
$$5(p - w) = 500$$

$$p + w = 200$$
$$p - w = 100$$
▶ Multiply each side of the equations by $\frac{1}{5}$.

$$2p = 300$$
$$p = 150$$
▶ Add the equations.

$$p + w = 200$$
$$150 + w = 200$$
$$w = 50$$
▶ Substitute the value of p into one of the equations.

The rate of the plane in calm air is 150 mph.
The rate of the wind is 50 mph.

Problem 1 A rowing team rowing with the current traveled 18 mi in 2 h. Against the current, the team rowed 10 mi in 2 h. Find the rate of the rowing team in calm water and the rate of the current.

Solution See page A48.

2 Application problems

The application problems in this section are varieties of those problems solved earlier in the text. Each of the strategies for the problems in this section will result in a system of equations.

Solve: A store owner purchased twenty 60-watt light bulbs and 30 fluorescent lights for a total cost of \$40. A second purchase, at the same prices, included thirty 60-watt bulbs and 10 fluorescent lights for a total cost of \$25. Find the cost of a 60-watt bulb and a fluorescent light.

STRATEGY for solving an application problem in two variables

- Choose one variable to represent one of the unknown quantities and a second variable to represent the other unknown quantity. Write numerical or variable expressions for all the remaining quantities. These results can be recorded in two tables, one for each of the conditions.

Cost of a 60-watt bulb: b
Cost of a fluorescent light: f

First purchase

	Amount	·	Unit cost	=	Value
60-watt	20	·	b	=	$20b$
Fluorescent	30	·	f	=	$30f$

Second purchase

	Amount	·	Unit cost	=	Value
60-watt	30	·	b	=	$30b$
Fluorescent	10	·	f	=	$10f$

- Determine a system of equations. The strategies presented in Chapter 2 can be used to determine the relationships between the expressions in the tables. Each table will give one equation of the system.

The total of the first purchase was \$40.
The total of the second purchase was \$25.

$$20b + 30f = 40$$
$$30b + 10f = 25$$

Solve the system of equations.

$$\begin{aligned} 20b + 30f &= 40 \\ 30b + 10f &= 25 \end{aligned} \quad \Rightarrow \quad \begin{aligned} 3(20b + 30f) &= 3 \cdot 40 \\ -2(30b + 10f) &= -2 \cdot 25 \end{aligned} \quad \Rightarrow \quad \begin{aligned} 60b + 90f &= 120 \\ -60b - 20f &= -50 \\ 70f &= 70 \\ f &= 1 \end{aligned}$$

Replace f by 1 in the equation $20b + 30f = 40$. Solve for b.

$$
\begin{aligned}
20b + 30f &= 40 \\
20b + 30(1) &= 40 \\
20b + 30 &= 40 \\
20b &= 10 \\
b &= 0.5
\end{aligned}
$$

The cost of a 60-watt bulb was \$.50.
The cost of a fluorescent light was \$1.00.

Example 2 The total value of the nickels and dimes in a coin bank is \$2.50. If the nickels were dimes and the dimes were nickels, the total value of the coins would be \$3.05. Find the number of dimes and the number of nickels in the bank.

Strategy

- Number of nickels in the bank: n
  Number of dimes in the bank: d

Coins in the bank now:

Coin	Number	Value	Total Value
Nickels	n	5	$5n$
Dimes	d	10	$10d$

Coins in the bank if the nickels were dimes and the dimes were nickels:

Coin	Number	Value	Total Value
Nickels	d	5	$5d$
Dimes	n	10	$10n$

- The value of the nickels and dimes in the bank is \$2.50.
  The value of the nickels and dimes in the bank would be \$3.05.

Solution

$$
\begin{aligned}
5n + 10d &= 250 \quad (1) \\
10n + 5d &= 305 \quad (2)
\end{aligned}
$$

$$
\begin{aligned}
10n + 20d &= 500 \\
-10n - 5d &= -305
\end{aligned}
$$

▶ Multiply equation (1) by 2 and equation (2) by −1.

$$
\begin{aligned}
15d &= 195 \\
d &= 13
\end{aligned}
$$

▶ Add the equations.

$$
\begin{aligned}
5n + 10d &= 250 \\
5n + 10(13) &= 250 \\
5n + 130 &= 250 \\
5n &= 120 \\
n &= 24
\end{aligned}
$$

▶ Substitute the value of d into one of the equations.

There are 24 nickels and 13 dimes in the bank.

Problem 2 A citrus fruit grower purchased 25 orange trees and 20 grapefruit trees for \$290. The next week, at the same prices, the grower bought 20 orange trees and 30 grapefruit trees for \$330. Find the cost of an orange tree and the cost of a grapefruit tree.

Solution See pages A48–A49.

EXERCISES 10.4

1 Solve.

1. Flying with the wind, a small plane flew 320 mi in 2 h. Against the wind, the plane could fly only 280 mi in the same amount of time. Find the rate of the plane in calm air and the rate of the wind.

2. A jet plane flying with the wind went 2100 mi in 4 h. Against the wind, the plane could fly only 1760 mi in the same amount of time. Find the rate of the plane in calm air and the rate of the wind.

3. A cabin cruiser traveling with the current went 48 mi in 3 h. Against the current, it took 4 h to travel the same distance. Find the rate of the cabin cruiser in calm water and the rate of the current.

4. A motorboat traveling with the current went 48 mi in 2 h. Against the current, it took 3 h to travel the same distance. Find the rate of the boat in calm water and the rate of the current.

5. Flying with the wind, a pilot flew 450 mi between two cities in 2.5 h. The return trip against the wind took 3 h. Find the rate of the plane in calm air and the rate of the wind.

6. A turbo-prop plane flying with the wind flew 600 mi in 2 h. Flying against the wind, the plane required 3 h to travel the same distance. Find the rate of the wind and the rate of the plane in calm air.

7. A motorboat traveling with the current went 88 km in 4 h. Against the current, the boat could go only 64 km in the same amount of time. Find the rate of the boat in calm water and the rate of the current.

8. A rowing team rowing with the current traveled 18 km in 2 h. Rowing against the current, the team rowed 12 km in the same amount of time. Find the rate of the team in calm water and the rate of the current.

9. A plane flying with a tailwind flew 360 mi in 3 h. Against the wind, the plane required 4 h to fly the same distance. Find the rate of the plane in calm air and the rate of the wind.

10. Flying with the wind, a plane flew 1000 mi in 4 h. Against the wind, the plane required 5 h to fly the same distance. Find the rate of the plane in calm air and the rate of the wind.

11. A motorboat traveling with the current went 54 mi in 3 h. Against the current, it took 3.6 h to travel the same distance. Find the rate of the boat in calm water and the rate of the current.

12. A plane traveling with the wind flew 3625 mi in 6.25 h. Against the wind, the plane required 7.25 h to fly the same distance. Find the rate of the plane in calm air and the rate of the wind.

13. A cabin cruiser traveling with the current went 41 mi in 2 h. Against the current, the boat could go only 31 mi in the same amount of time. Find the rate of the cabin cruiser and the rate of the current.

14. Flying with the wind, a plane flew 786 mi in 3 h. Against the wind, the plane could fly only 654 mi in the same amount of time. Find the rate of the plane in calm air and the rate of the wind.

2 Solve.

15. A carpenter purchased 50 ft of redwood and 90 ft of pine for a total cost of $31.20. A second purchase, at the same prices, included 200 ft of redwood and 100 ft of pine for a total cost of $78. Find the cost per foot of redwood and pine.

16. A merchant mixed 10 lb of a cinnamon tea with 5 lb of spice tea. The 15-lb mixture cost $40. A second mixture included 12 lb of the cinnamon tea and 8 lb of the spice tea. The 20-lb mixture cost $54. Find the cost per pound of the cinnamon tea and the spice tea.

17. During one month, a homeowner used 400 units of electricity and 120 units of gas for a total cost of $73.60. The next month, 350 units of electricity and 200 units of gas were used for a total cost of $72. Find the cost per unit of gas.

18. A contractor buys 20 yd of nylon carpet and 28 yd of wool carpet for $1360. A second purchase, at the same prices, includes 15 yd of nylon carpet and 20 yd of wool carpet for $990. Find the cost per yard of the wool carpet.

19. The total value of the quarters and dimes in a coin bank is $6.90. If the quarters were dimes and the dimes were quarters, the total value of the coins would be $7.80. Find the number of quarters in the bank.

20. A coin bank contains only nickels and dimes. The total value of the coins in the bank is $2.50. If the nickels were dimes and the dimes were nickels, the total value of the coins would be $3.20. Find the number of nickels in the bank.

21. A company manufactures both color and black-and-white television sets. The cost of materials for a black-and-white TV is \$25, while the cost of materials for a color TV is \$75. The cost of labor to manufacture a black-and-white TV is \$40, while the cost of labor to manufacture a color TV is \$65. During a week when the company has budgeted \$4800 for materials and \$4380 for labor, how many color TV's does the company plan to manufacture?

22. A company manufactures both 10-speed and standard model bicycles. The cost of materials for a 10-speed bicycle is \$40, while the cost of materials for a standard bicycle is \$30. The cost of labor to manufacture a 10-speed bicycle is \$50, while the cost of labor to manufacture a standard bicycle is \$25. During a week when the company has budgeted \$1740 for materials and \$1950 for labor, how many 10-speed bicycles does the company plan to manufacture?

23. A pharmacist has two vitamin-supplement powders. The first powder is 25% vitamin B_1 and 15% vitamin B_2. The second is 15% vitamin B_1 and 20% vitamin B_2. How many milligrams of each of the two powders should the pharmacist use to make a mixture that contains 117.5 mg of vitamin B_1 and 120 mg of vitamin B_2?

24. A chemist has two alloys, one of which is 10% gold and 15% lead, the other of which is 30% gold and 40% lead. How many grams of each of the two alloys should be used to make an alloy that contains 60 g of gold and 88 g of lead?

SUPPLEMENTAL EXERCISES 10.4

Solve.

25. Two angles are complementary. The larger angle is 9° more than eight times the measure of the smaller angle. Find the measure of the two angles. (Complementary angles are two angles whose sum is 90°.)

26. Two angles are supplementary. The larger angle is 40° more than three times the measure of the smaller angle. Find the measure of the two angles. (Supplementary angles are two angles whose sum is 180°.)

27. The sum of the ages of a gold coin and a silver coin is 75 years. The age of the gold coin ten years from now is five years less than the age of the silver coin ten years ago. Find the present ages of the two coins.

28. The difference between the ages of an oil painting and a watercolor is 35 years. The age of the oil painting five years from now is twice the age of the watercolor five years ago. Find the present ages of each.

29. The total value of the nickels, dimes, and quarters in a coin bank is \$3.50. If the nickels were dimes and the dimes were nickels, the total value of the coins would be \$4.25. If the quarters were dimes and the dimes were quarters, the total value of the coins would be \$4.25. Find the number of nickels, dimes, and quarters in the bank.

30. The total value of the nickels, dimes, and quarters in a coin bank is \$4.00. If the nickels were dimes and the dimes were nickels, the total value of the coins would be \$3.75. If the quarters were dimes and the dimes were quarters, the total value of the coins would be \$6.25. Find the number of nickels, dimes, and quarters in the bank.

31. The sum of the digits of a two-digit number equals $\frac{1}{7}$ of the number. If the digits of the number are reversed, the new number is equal to 36 less than the original number. Find the original number.

32. The sum of the digits of a two-digit number equals $\frac{1}{5}$ of the number. If the digits of the number are reversed, the new number is equal to 9 more than the original number. Find the original number.

SECTION 10.5 Solving Systems of Quadratic Equations and Systems of Inequalities

1 Solve nonlinear systems of equations

10

A nonlinear system of equations is one in which one or more equations of the system is not a linear equation. Listed below are some examples of nonlinear systems and their graphs.

$$x^2 + y^2 = 4$$
$$y = x^2 + 2$$

The graphs intersect at one point.
The system of equations has one solution.

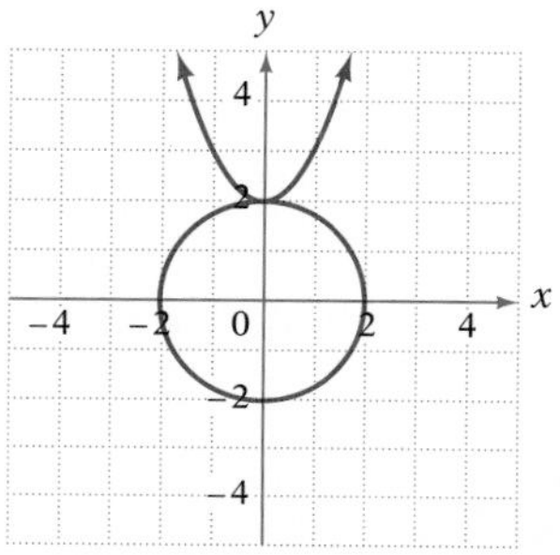

$y = x^2$
$y = -x + 2$

The graphs intersect at two points.
The system of equations has two solutions.

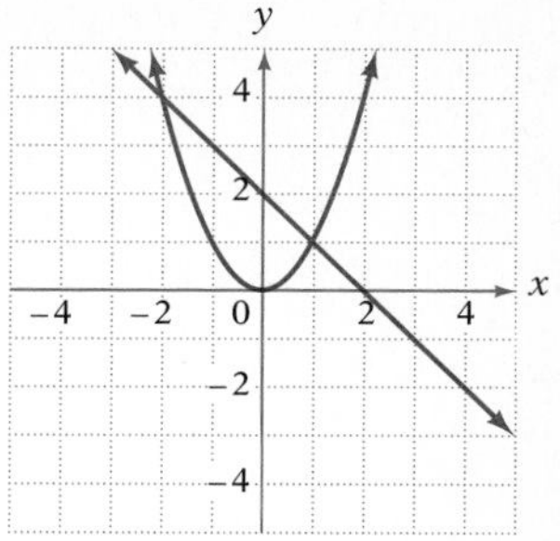

$(x + 2)^2 + (y - 2)^2 = 4$
$x = y^2$

The graphs do not intersect.
The system of equations has no solutions.

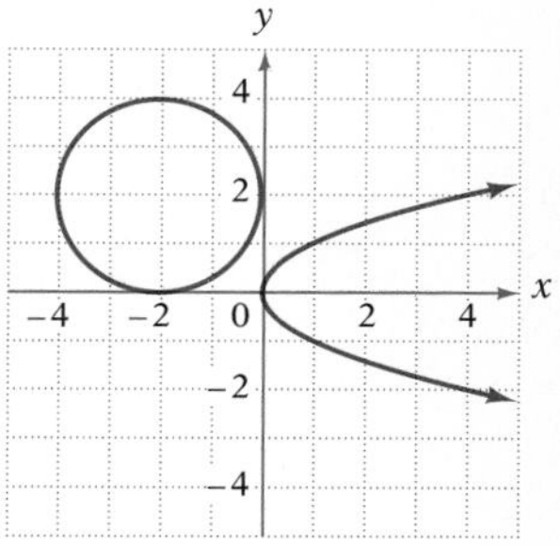

Nonlinear systems of equations can be solved by using either a substitution method or an addition method.

The system of equations at the right contains a linear equation and a quadratic equation. When a system contains both a linear and a quadratic equation, the substitution method is used.

(1) $2x - y = 4$
(2) $y^2 = 4x$

Solve equation (1) for y.

$$2x - y = 4$$
$$-y = -2x + 4$$
$$y = 2x - 4$$

Substitute $2x - 4$ for y into equation (2).

$$y^2 = 4x$$
$$(2x - 4)^2 = 4x$$

Write the equation in standard form.

$$4x^2 - 16x + 16 = 4x$$
$$4x^2 - 20x + 16 = 0$$

Solve for x by factoring.

$$4(x^2 - 5x + 4) = 0$$
$$4(x - 4)(x - 1) = 0$$

$$x - 4 = 0 \qquad x - 1 = 0$$
$$x = 4 \qquad x = 1$$

Substitute the values of x into the equation $y = 2x - 4$, and solve for y.

$$y = 2x - 4 \qquad y = 2x - 4$$
$$y = 2(4) - 4 \qquad y = 2(1) - 4$$
$$y = 4 \qquad y = -2$$

The solutions are (4, 4) and (1, −2).

The graph of the system that was solved above is shown at the right. Note that the line intersects the parabola at two points. These points correspond to the solutions.

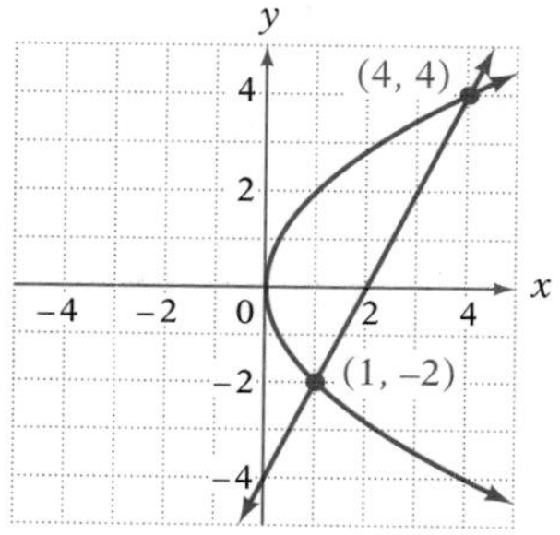

The system of equations at the right contains a linear equation. The substitution method is used to solve the system.

(1) $x^2 + y^2 = 4$
(2) $y = x + 4$

Substitute the expression for y into equation (1).

$$x^2 + y^2 = 4$$
$$x^2 + (x + 4)^2 = 4$$

Write the equation in standard form.

$$x^2 + x^2 + 8x + 16 = 4$$
$$2x^2 + 8x + 16 = 4$$
$$2x^2 + 8x + 12 = 0$$

Because the discriminant of the quadratic equation is less than zero, the equation has two complex number solutions. Therefore, the system of equations has no real number solutions.

$$b^2 - 4ac = 8^2 - 4(2)(12)$$
$$= 64 - 96$$
$$= -32$$

The graph of the system of equations is shown at the right. Note that the two graphs do not intersect.

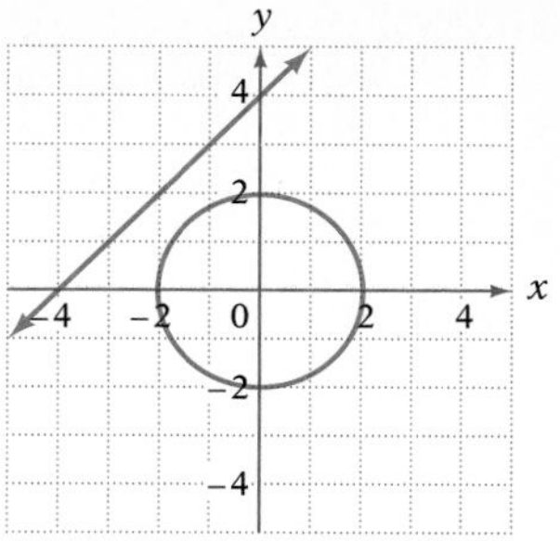

The addition method is used to solve the system of equations shown at the right.

(1) $4x^2 + y^2 = 16$
(2) $x^2 + y^2 = 4$

Multiply equation (2) by -1 and add to equation (1).

$$4x^2 + y^2 = 16$$
$$-x^2 - y^2 = -4$$
$$3x^2 = 12$$

Solve for x.

$$x^2 = 4$$
$$x = \pm 2$$

Substitute the values of x into equation (2), and solve for y.

$$x^2 + y^2 = 4 \qquad x^2 + y^2 = 4$$
$$2^2 + y^2 = 4 \qquad (-2)^2 + y^2 = 4$$
$$y^2 = 0 \qquad y^2 = 0$$
$$y = 0 \qquad y = 0$$

The solutions are (2, 0) and (−2, 0).

The graphs of the equations in this system are shown at the right. Note that the graphs intersect at two points.

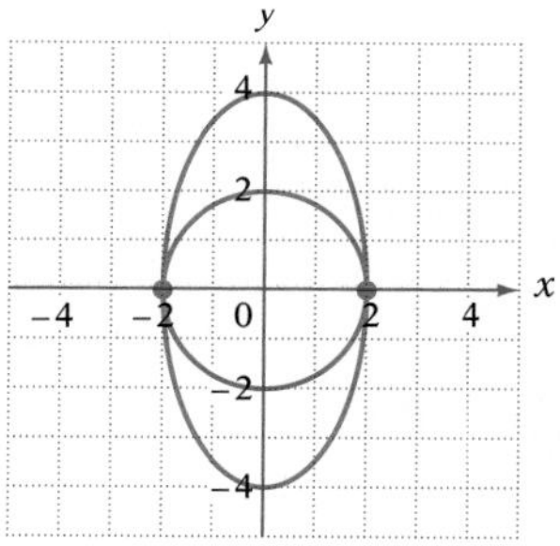

Example 1 Solve.

A. $y = 2x^2 - 3x - 1$
$y = x^2 - 2x + 5$

B. $3x^2 - 2y^2 = 26$
$x^2 - y^2 = 5$

Solution A. (1) $y = 2x^2 - 3x - 1$
(2) $y = x^2 - 2x + 5$

$$2x^2 - 3x - 1 = x^2 - 2x + 5$$
$$x^2 - x - 6 = 0$$
$$(x - 3)(x + 2) = 0$$

▶ Use the substitution method.

$x - 3 = 0 \qquad x + 2 = 0$
$x = 3 \qquad x = -2$

$y = 2x^2 - 3x - 1$	$y = 2x^2 - 3x - 1$
$y = 2(3)^2 - 3(3) - 1$	$y = 2(-2)^2 - 3(-2) - 1$
$y = 18 - 9 - 1$	$y = 8 + 6 - 1$
$y = 8$	$y = 13$

▶ Substitute the values of x into equation (1).

The solutions are (3, 8) and (−2, 13).

B. (1) $3x^2 - 2y^2 = 26$
(2) $x^2 - y^2 = 5$

$$3x^2 - 2y^2 = 26$$
$$-2x^2 + 2y^2 = -10$$
$$x^2 = 16$$
$$x = \pm 4$$

▶ Use the addition method. Multiply equation (2) by −2.

$x^2 - y^2 = 5$	$x^2 - y^2 = 5$
$4^2 - y^2 = 5$	$(-4)^2 - y^2 = 5$
$16 - y^2 = 5$	$16 - y^2 = 5$
$-y^2 = -11$	$-y^2 = -11$
$y^2 = 11$	$y^2 = 11$
$y = \pm\sqrt{11}$	$y = \pm\sqrt{11}$

▶ Substitute the values of x into equation (2).

The solutions are $(4, \sqrt{11})$, $(4, -\sqrt{11})$, $(-4, \sqrt{11})$, and $(-4, -\sqrt{11})$.

Problem 1 Solve.

A. $y = 2x^2 + x - 3$
$y = 2x^2 - 2x + 9$

B. $x^2 - y^2 = 10$
$x^2 + y^2 = 8$

Solution See page A49.

2 Graph the solution set of a system of inequalities

The **solution set of a system of inequalities** is the intersection of the solution sets of each inequality. To graph the solution set of a system of inequalities, first graph the solution set for each inequality. The solution set of the system of inequalities is the region of the plane represented by the intersection of the two shaded regions.

To graph the solution set of $2x - y \le 3$
$3x + 2y \ge 8$,
graph the solution set of each inequality.

The solution set is the region of the plane represented by the intersection of the solution sets of each inequality.

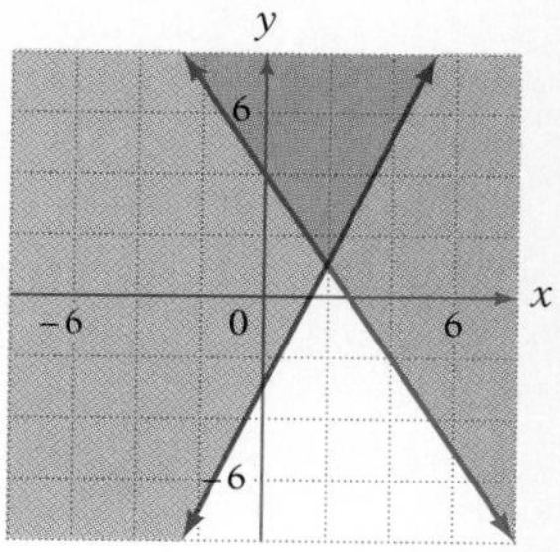

Example 2 Graph the solution set.

A. $\dfrac{x^2}{9} + \dfrac{y^2}{4} \ge 1$
$\dfrac{x^2}{4} - \dfrac{y^2}{9} > 1$

B. $y > x^2$
$y < x + 2$

Solution A.

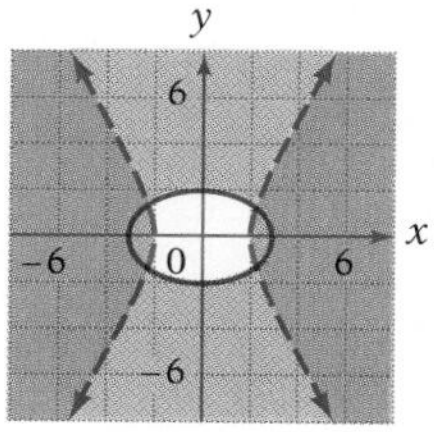

B.

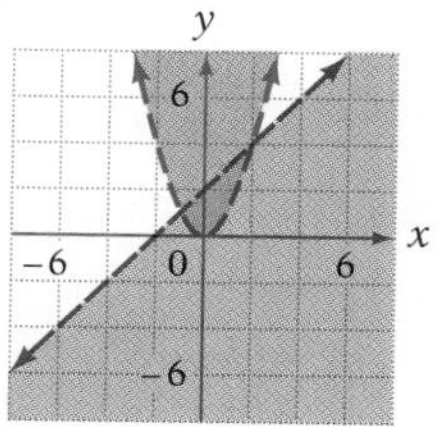

Problem 2 Graph the solution set.

A. $x^2 + y^2 < 16$
$y^2 > x$

B. $y \ge x - 1$
$y < -2x$

Solution See page A49.

EXERCISES 10.5

1 Solve.

1. $y = x^2 - x - 1$
$y = 2x + 9$

2. $y = x^2 - 3x + 1$
$y = x + 6$

3. $y^2 = -x + 3$
$x - y = 1$

4. $y^2 = 4x$
$x - y = -1$

5. $y^2 = 2x$
$x + 2y = -2$

6. $y^2 = 2x$
$x - y = 4$

7. $x^2 + 2y^2 = 12$
$2x - y = 2$

8. $x^2 + 4y^2 = 37$
$x - y = -4$

9. $x^2 + y^2 = 13$
$x + y = 5$

10. $x^2 + y^2 = 16$
$x - 2y = -4$

11. $4x^2 + y^2 = 12$
$y = 4x^2$

12. $2x^2 + y^2 = 6$
$y = 2x^2$

13. $y = x^2 - 2x - 3$
$y = x - 6$

14. $y = x^2 + 4x + 5$
$y = -x - 3$

15. $3x^2 - y^2 = -1$
$x^2 + 4y^2 = 17$

16. $x^2 + y^2 = 10$
$x^2 + 9y^2 = 18$

17. $2x^2 + 3y^2 = 30$
$x^2 + y^2 = 13$

18. $x^2 + y^2 = 61$
$x^2 - y^2 = 11$

19. $y = 2x^2 - x + 1$
$y = x^2 - x + 5$

20. $y = -x^2 + x - 1$
$y = x^2 + 2x - 2$

21. $2x^2 + 3y^2 = 24$
$x^2 - y^2 = 7$

22. $2x^2 + 3y^2 = 21$
$x^2 + 2y^2 = 12$

23. $x^2 + y^2 = 36$
$4x^2 + 9y^2 = 36$

24. $2x^2 + 3y^2 = 12$
$x^2 - y^2 = 25$

25. $11x^2 - 2y^2 = 4$
$3x^2 + y^2 = 15$

26. $x^2 + 4y^2 = 25$
$x^2 - y^2 = 5$

27. $2x^2 - y^2 = 7$
$2x - y = 5$

28. $3x^2 + 4y^2 = 7$
$x - 2y = -3$

29. $y = 3x^2 + x - 4$
$y = 3x^2 - 8x + 5$

30. $y = 2x^2 + 3x + 1$
$y = 2x^2 + 9x + 7$

2 Graph the solution set.

31. $2x + y \geq 1$
$3x + 2y < 6$

32. $3x - 4y < 12$
$x + 2y < 6$

33. $x - 2y \leq 6$
$2x + 3y \leq 6$

34. $x - 3y > 6$
$2x + 3y > 9$

35. $2x + 3y \leq 15$
$3x - y \leq 6$
$y \geq 0$

36. $x + y \leq 6$
$x - y \leq 2$
$x \geq 0$

37. $x^2 + y^2 < 16$
$y > x + 1$

38. $y > x^2 - 4$
$y < x - 2$

39. $x^2 + y^2 < 25$
$\frac{x^2}{9} + \frac{y^2}{36} < 1$

40. $\frac{x^2}{9} - \frac{y^2}{4} < 1$
$\frac{x^2}{25} + \frac{y^2}{9} < 1$

41. $x^2 + y^2 > 4$
$x^2 + y^2 < 25$

42. $\frac{x^2}{25} + \frac{y^2}{16} \leq 1$
$\frac{x^2}{4} + \frac{y^2}{4} \geq 1$

SUPPLEMENTAL EXERCISES 10.5

Graph the solution set.

43. $x - 3y \le 6$
$5x - 2y \ge 4$
$y \ge 0$

44. $2x - y \le 4$
$3x + y \le 1$
$y \le 0$

45. $y > x^2 - 3$
$y < x + 3$
$x \ge 0$

46. $x^2 + y^2 \le 25$
$y > x + 1$
$x \ge 0$

47. $x^2 + y^2 < 3$
$x > y^2 - 1$
$y \ge 0$

48. $\frac{x^2}{4} - \frac{y^2}{25} \le 1$
$\frac{x^2}{25} + \frac{y^2}{4} \le 1$
$y \ge 0$

49. $\frac{x^2}{16} + \frac{y^2}{4} \le 1$
$x^2 + y^2 \le 4$
$x \ge 0$
$y \le 0$

50. $\frac{x^2}{4} + \frac{y^2}{25} \le 1$
$x > y^2 - 4$
$x \le 0$
$y \ge 0$

Solve.

51. Is it possible for two ellipses with centers at the origin to intersect in exactly three points?

52. Is it possible for two circles with centers at the origin to intersect in exactly two points?

53. Graph $xy > 1$ and $y > \frac{1}{x}$ on the same coordinate grid. Dividing each side of $xy > 1$ by x yields $y > \frac{1}{x}$, but the graphs are not the same. Explain.

Calculators and Computers

Solving Simultaneous Equations or Inequalities by Using the TI-81 Texas Instruments Graphing Calculator

Solve: $3x - 2y = 4$
$x + 4y = 9$

Solve each equation for y.
$$y = \frac{3}{2}x - 2 = 1.5x - 2$$
$$y = -\frac{1}{4}x + \frac{9}{4} = -0.25x + 2.25$$

First press [Y=] to display the menu. Then use the following key strokes.

1.5 [XIT] [−] 2 [ENTER]

[(−)] .25 [XIT] [+] 2.25 [ENTER] [GRAPH]

You will now see the graphs on the screen. Move the cursor to the intersection. On our calculator the solution read $x \approx 2.501662$, $y \approx 1.6232048$. Your answer may be slightly different.

Now use [ZOOM] to obtain a closer approximation. From the zoom menu, use [1]. Place the cursor at the upper left of the intersection. Use [ENTER]. Now move the cursor to the lower right of the solution. Use [ENTER]. The box you formed now fills the screen and a better approximation of the solution can be found. Use the cursor to find the solution. On our calculator the solution read as $x \approx 2.434349$, $y \approx 1.6553288$. This process can be repeated to a predetermined degree of accuracy. The actual answer is $\left(\frac{17}{7}, \frac{23}{14}\right)$.

The same procedure can be used to find the graphical solution of nonlinear systems of equations.

Solve: $y = 0.5x^2 + 1$
$y = x^3 + 4x^2 - 1$

Enter the equations into the graphing calculator. You will notice that there are three points of intersection. You may have to move the cursor to different points of the graph and use the [ZOOM] feature to find the coordinates of the three intersections.

The TI-81 calculator can be used to graph a system of inequalities.

Solve: $y > 3x - 3$
$y < x^2 + x - 4$

Enter the inequalities. Consult the instruction manual for specific keystrokes.

[2nd] [DRAW-7] [SHADE (2x − 3, x² + x − 4)] [ENTER]

The [DRAW] key is one of the advanced functions keys and [DRAW-7] means to use the number 7 from the menu. After using number 7 from the menu, [SHADE] will appear on the screen. At this time the two functions will be entered. Then when [ENTER] is used, the shaded portion above the first entry and below the second entry will be shown on the screen.

Find the approximate solutions of the following systems by using a graphing calculator.

1. $3x - 4y = 6$
$x + 2y = 4$

2. $2x + y = 4$
$3x - 4y = 8$

3. $y = 0.35x - 0.8$
$y = 1.22x - 1.2$

4. $y = -0.55x + 1.33$
$y = 2x - 3.2$

5. $y = x^2 - 7$
$y = -x^2 + 3$

6. $y = x^4 - 3x^2 + x$
$y = 0.5x - 1$

7. $y > 3x - 2$
$y < -0.5x + 4$

8. $y > 3x - 4$
$y < -4x + 2$

9. $y > 2x - 3$
$y < -x^2 + 3$

10. $y > 2x^2 - 4$
$y < -x^3 + 1$

Something Extra

Linear Programming

During World War II, George Dantzig was asked by the Air Force to determine the most cost-effective way to supply squadrons deployed in countries all over the world. His solution involved solving a system of inequalities. Through his work and the work of others, a branch of mathematics called linear programming was born.

The goal of a linear programming problem is to maximize or minimize a set of conditions. For example, the Air Force was trying to minimize the cost of supplying squadrons. On the other hand, a manufacturer may be interested in a method of maximizing profit. Linear programming can be helpful in each case. The following problem contains the basic qualities of any linear programming problem. Key parts of the problem are highlighted.

A manufacturer produces two types of computers: a table model and a laptop. Past sales experience shows that (1) *at least twice as many table models are sold as laptops*. The size of the manufacturing facility limits (2) *the number of computer systems produced per day to a maximum of 12*. The demand for the computers is such that the manufacturer can make (3) *a profit of \$250 for each table model sold and a profit of \$300 for each laptop sold*. How many of each type of computer system should the manufacturer produce to maximize profit?

Let x represent the number of table models sold and y the number of laptops sold. Then from (1), $x \geq 2y$. From (2), $x + y \leq 12$. Because the manufacturer cannot produce less than zero computers, $x \geq 0$ and $y \geq 0$. These four inequalities are called *constraints*. The constraints are conditions the manufacturer must satisfy when making different computer models. For example, the manufacturer cannot make 5 laptops because that would require producing at least 10 table models (condition 1) and $5 + 10 = 15$, which is not less than 12 (condition 2).

The constraints result in a system of inequalities. This system of inequalities is shown below along with the graph of the solution set.

constraints $\begin{cases} x \geq 2y \\ x + y \leq 12 \\ x \geq 0,\ y \geq 0 \end{cases}$

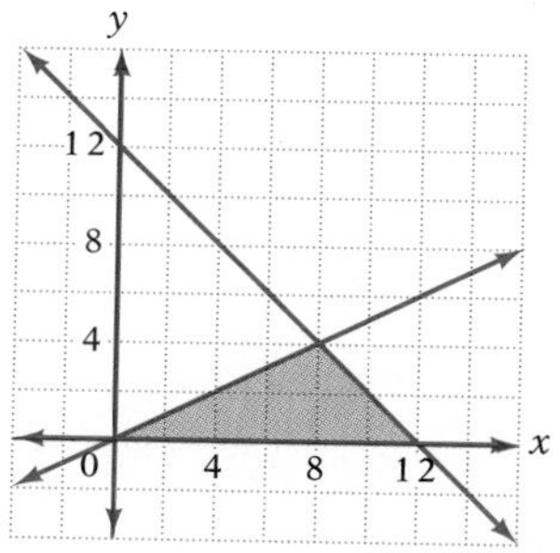

The constraints are one element of a linear programming problem. The second element is called the *objective function.* The goal or objective of a linear programming problem is to maximize or minimize this function. From (3), the total profit from selling computer systems is Profit = $250x + 300y$.

The shaded region in the graph on the previous page is called the *set of feasible solutions* for the linear programming problem. Any ordered pair in the shaded region satisfies the manufacturer's constraints. The ordered pair (5, 2) belongs to the set of feasible solutions and indicates selling 5 table models and 2 laptops. In this case, Profit = 250(5) + 300(2) = \$1850. Another ordered pair, (8, 3), indicates selling 8 table models and 3 laptop models. For this ordered pair, Profit = 250(8) + 300(3) = \$2900. Notice that the profit for the second ordered pair is greater than for the first ordered pair.

The goal of this linear programming problem is to maximize profit. A restatement of this goal is: find the ordered pair that yields the maximum profit when substituted into the profit function. One way to do this would be to continue choosing ordered pairs in the set of feasible solutions until the maximum profit was obtained. This could be quite time consuming.

Fortunately, a theorem from linear programming states that the maximum or minimum will be one of the "corner points" of the set of feasible solutions. The corner points are the points at which two lines intersect. For this problem, the corner points are (0, 0), (8, 4), and (12, 0). Substituting each of these ordered pairs into the objective function gives a maximum profit for the ordered pair (8, 4). Profit = 250(8) + 300(4) = \$3200. The solution to the linear programming problem is now complete. The manufacturer should produce 8 table models and 4 laptops. The total profit is \$3200.

Changing either the constraints or the objective function can change the outcome of a linear programming problem. For example, if the manufacturer made a profit of \$300 for each table top model and \$250 for each laptop, the objective function would be Profit = $300x + 250y$. Substituting the corner points into this objective function gives (12, 0) as the ordered pair that maximizes profit. In this case, the manufacturer would produce 12 table models and no laptops.

1. How can you algebraically determine the coordinates of the corner point (8, 4)?

2. Verify that (12, 0) maximizes profit for the objective function Profit = $300x + 250y$.

3. Find the maximum profit if the objective function is Profit = $150x + 300y$.

Chapter Summary

Key Words

Equations considered together are called a *system of equations.*

A *solution of a system of equations in two variables* is an ordered pair that is a solution of each equation of the system.

When the graphs of a system of equations intersect at only one point, the equations are called *independent equations.* When a system of equations has no solution, it is called an *inconsistent system of equations.* When the graphs of a system coincide, the equations are called *dependent equations.*

An equation of the form $Ax + By + Cz = D$, where A, B, C, and D are constants, is called a *linear equation in three variables.* A *solution of a system of equations in three variables* is an ordered triple that is a solution of each equation of the system.

A *matrix* is a rectangular array of numbers. Each number in the matrix is an *element* of the matrix. A *square matrix* has the same number of rows as columns.

A *determinant* is a number associated with a square matrix. The *minor of an element* in a 3×3 determinant is the 2×2 determinant that is obtained by eliminating the row and column that contain that element.

The *cofactor* C_{ij} of an element of a matrix A is the product $(-1)^{i+j}M_{ij}$ where M_{ij} is the minor of A.

The *solution set of a system of inequalities* is the intersection of the solution sets of each inequality.

Essential Rules

Cramer's Rule is a method of solving a system of equations by using determinants.

Elementary row operations include:

1. Interchange two rows.
2. Multiply all the elements in a row by the same nonzero number.
3. Replace a row by the sum of that row and a multiple of any other row.

Chapter Review

1. Solve by substitution: $2x - 6y = 15$
$x = 3y + 8$

2. Solve by substitution: $3x + 12y = 18$
$x + 4y = 6$

3. Solve by the addition method:
$3x + 2y = 2$
$x + y = 3$

4. Solve by the addition method:
$5x - 15y = 30$
$x - 3y = 6$

5. Solve by the addition method:
$\frac{3}{x} - \frac{6}{y} = 14$
$\frac{1}{x} + \frac{9}{y} = 1$

6. Solve by the addition method:
$x + y + z = 4$
$2x - y + z = -1$
$2x + y + z = 3$

7. Solve by the addition method:
$3x + y = 13$
$2y + 3z = 5$
$x + 2z = 11$

8. Solve by the addition method:
$3x - 4y - 2z = 17$
$4x - 3y + 5z = 5$
$5x - 5y + 3z = 14$

9. Evaluate the determinant: $\begin{vmatrix} 6 & 1 \\ 2 & 5 \end{vmatrix}$

10. Evaluate the determinant: $\begin{vmatrix} 1 & 5 & -2 \\ -2 & 1 & 4 \\ 4 & 3 & -8 \end{vmatrix}$

11. Solve by using Cramer's Rule:
$2x - y = 7$
$3x + 2y = 7$

12. Solve by using Cramer's Rule:
$3x - 4y = 10$
$2x + 5y = 15$

13. Solve by using Cramer's Rule:
$x + y + z = 0$
$x + 2y + 3z = 5$
$2x + y + 2z = 3$

14. Solve by using Cramer's Rule:
$x + 3y + z = 6$
$2x + y - z = 12$
$x + 2y - z = 13$

15. Solve: $y = x^2 + 5x - 6$
$y = x - 10$

16. Solve: $x^2 + y^2 = 20$
$x^2 - y^2 = 12$

17. Solve: $x^2 - y^2 = 24$
$2x^2 + 5y^2 = 55$

18. Solve: $2x^2 + y^2 = 19$
$3x^2 - y^2 = 6$

19. Evaluate the determinant: $\begin{vmatrix} 5 & -2 \\ 3 & 4 \end{vmatrix}$

20. Solve by using a matrix: $2x + 3y = 0$
$3x - 2y = 13$

21. Solve by using Cramer's Rule:
$2x - 2y + z = -5$
$4x + 3y + 6z = 8$
$x + y + 2z = 3$

22. Solve: $3x^2 - y^2 = 2$
$x^2 + 2y^2 = 3$

23. Solve by the addition method:
$5x + 2y = 5$
$6x + 3y = 4$

24. Solve by the addition method:
$3x - 4y = 8$
$5x + 3y = -6$

25. Solve: $2x + y = 9$
$4y^2 = x$

26. Solve by substitution: $2x - 5y = 23$
$y = 2x - 3$

27. Solve by the addition method:
$x - 2y + z = 7$
$3x - z = -1$
$3y + z = 1$

28. Solve by using Cramer's Rule:
$3x - 2y = 2$
$-2x + 3y = 1$

29. Solve by using a matrix:
$2x - 2y - 6z = 1$
$4x + 2y + 3z = 1$
$2x - 3y - 3z = 3$

30. Evaluate the determinant:
$\begin{vmatrix} 3 & -2 & 5 \\ 4 & 6 & 3 \\ 1 & 2 & 1 \end{vmatrix}$

31. Solve by using Cramer's Rule:
$4x - 3y = 17$
$3x - 2y = 12$

32. Solve by using a matrix:
$3x + 2y - z = -1$
$x + 2y + 3z = -1$
$3x + 4y + 6z = 0$

33. Solve by graphing: $x + y = 3$
$3x - 2y = -6$

34. Solve by graphing: $2x - y = 4$
$y = 2x - 4$

35. Graph the solution set:
$4x - 3y > 6$
$x + 2y < 4$

36. Graph the solution set:
$2x - 5y \leq 10$
$x + 2y > 2$

37. Graph the solution set:
$x + 3y \leq 6$
$2x - y \geq 4$

38. Graph the solution set:
$x^2 + y^2 < 36$
$x + y > 4$

39. Graph the solution set:
$y \geq x^2 - 4x + 2$
$y \leq \frac{1}{3}x - 1$

40. Graph the solution set:
$\frac{x^2}{25} + \frac{y^2}{16} \leq 1$
$\frac{y^2}{4} - \frac{x^2}{4} \geq 1$

41. A cabin cruiser traveling with the current went 60 mi in 3 h. Against the current it took 5 h to travel the same distance. Find the rate of the cabin cruiser in calm water and the rate of the current.

42. A plane flying with the wind flew 600 mi in 3 h. Flying against the wind, the plane required 4 h to travel the same distance. Find the rate of the plane in calm air and the rate of the wind.

43. At a movie theater, admission tickets are $5 for children and $8 for adults. The receipts for one Friday evening were $2500. The next day there were three times as many children as the preceding evening but only half the number of adults as the night before, yet the receipts were still $2500. Find the number of children who attended the movie Friday evening.

44. A confectioner mixed 3 lb of milk chocolate candy with 3 lb of semi-sweet chocolate candy. The 6-lb mixture costs $30. A second mixture included 6 lb of the milk chocolate candy and 2 lb of the semi-sweet chocolate candy. The 8-lb mixture costs $42. Find the cost per pound of the milk chocolate candy and the semi-sweet chocolate candy.

45. Either 3 pencils and 6 pens or 21 pencils and 2 pens can be bought for $6. Find the price per pencil.

46. A wallet contains $44 in one-dollar bills and five-dollar bills. If the one-dollar bills were five-dollar bills and the five-dollar bills were ten-dollar bills, the wallet would contain $130. Find the number of one-dollar bills.

Chapter Test

1. Solve by substitution: $3x + 2y = 4$
$x = 2y - 1$

2. Solve by substitution: $5x + 2y = -23$
$2x + y = -10$

3. Solve by substitution: $y = 3x - 7$
$y = -2x + 3$

4. Solve by using a matrix:
$3x + 4y = -2$
$2x + 5y = 1$

5. Solve by the addition method:
$4x - 6y = 5$
$6x - 9y = 4$

6. Solve by the addition method:
$3x - y = 2x + y - 1$
$5x + 2y = y + 6$

7. Solve by the addition method:
$2x + 4y - z = 3$
$x + 2y + z = 5$
$4x + 8y - 2z = 7$

8. Solve by using a matrix:
$x - y - z = 5$
$2x + z = 2$
$3y - 2z = 1$

9. Evaluate the determinant: $\begin{vmatrix} 3 & -1 \\ -2 & 4 \end{vmatrix}$

10. Evaluate the determinant: $\begin{vmatrix} 1 & -2 & 3 \\ 3 & 1 & 1 \\ 2 & -1 & -2 \end{vmatrix}$

11. Solve by using Cramer's Rule:
$5x + 2y = 9$
$3x + 5y = -7$

12. Solve by using Cramer's Rule:
$3x + 2y + 2z = 2$
$x - 2y - z = 1$
$2x - 3y - 3z = -3$

13. Solve: $x - 2y = 6$
$y = x^2 - 17$

14. Solve: $2x^2 + 3y^2 = 20$
$3x^2 - y^2 = 8$

15. Solve by graphing: $2x - 3y = -6$
$2x - y = 2$

16. Solve by graphing: $x - 2y = -5$
$3x + 4y = -15$

17. Graph the solution set: $2x - y < 3$
$4x + 3y < 11$

18. Graph the solution set: $(x - 1)^2 + y^2 \le 25$
$y^2 < x$

19. A plane flying with the wind went 350 mi in 2 h. The return trip, flying against the wind, took 2.8 h. Find the rate of the plane in calm air and the rate of the wind.

20. A clothing manufacturer purchased 60 yd of cotton and 90 yd of wool for a total cost of $900. Another purchase, at the same prices, included 80 yd of cotton and 20 yd of wool for a total cost of $500. Find the cost per yard of the cotton and the wool.

Cumulative Review

1. Solve: $\frac{3}{2}x - \frac{3}{8} + \frac{1}{4}x = \frac{7}{12}x - \frac{5}{6}$

2. Solve: $3x + 2 \le 5$ and $x + 5 \ge 1$

3. Find the equation of the line containing the points $(2, -1)$ and $(3, 4)$.

4. Factor: $6x^2 - 19x + 10$

5. Simplify: $\frac{2x}{x^2 - 5x + 6} - \frac{3}{x^2 - 2x - 3}$

6. Solve: $\frac{3}{x^2 - 5x + 6} - \frac{x}{x - 3} = \frac{2}{x - 2}$

7. Simplify: $a^{-1} + a^{-1}b$

8. Simplify: $\sqrt[3]{-4ab^4}\,\sqrt[3]{2a^3b^4}$

9. Solve: $\sqrt{3x + 10} = 1$

10. Solve: $2x^2 + 9x = 5$

11. Solve: $3x^2 = 2x + 2$

12. Solve: $x^{\frac{2}{3}} - 5x^{\frac{1}{3}} + 6 = 0$

13. Use the discriminant to determine the number of x-intercepts of the parabola whose equation is $y = -2x^2 - x - 2$.

14. Solve by graphing:
$$5x - 2y = 10$$
$$3x + 2y = 6$$

15. Find the inverse of the function given by the equation $f(x) = \frac{2}{3}x - 1$.

16. Find the distance between the points $P_1(-4, 0)$ and $P_2(2, 2)$.

17. Solve by substitution:
$$3x - 2y = 7$$
$$y = 2x - 1$$

18. Solve by the addition method:
$$3x + 2z = 1$$
$$2y - z = 1$$
$$x + 2y = 1$$

19. Evaluate the determinant: $\begin{vmatrix} 2 & -5 & 1 \\ 3 & 1 & 2 \\ 6 & -1 & 4 \end{vmatrix}$

20. Solve by using Cramer's Rule:
$$3x - 3y = 2$$
$$6x - 4y = 5$$

21. Solve by using a matrix:
$$3x + 3y + 2z = 1$$
$$x + 2y + z = 1$$
$$2x - 2y + z = 0$$

22. Solve:
$$x^2 + y^2 = 5$$
$$y^2 = 4x$$

23. Write the equation $x^2 + y^2 - 4x + 2y - 4 = 0$ in standard form. Then sketch its graph.

24. Graph: $f(x) = |x| - 2$

25. A coin purse contains 40 coins in nickels, dimes, and quarters. There are three times as many dimes as quarters. The total value of the coins is \$4.10. How many nickels are in the coin purse?

26. The distance (d) a spring stretches varies directly as the force (f) used to stretch the spring. If a force of 40 lb can stretch the spring 24 in., how far will a force of 60 lb stretch the spring?

27. The height of a triangle is twice the length of the base. The area of the triangle is 36 m^2. Find the length of the base.

28. How many milliliters of pure water must be added to 100 ml of a 4% salt solution to make a 2.5% salt solution?

29. An average score of 75–84 in a chemistry class receives a C grade. A student has grades of 70, 79, 76, and 73 on four chemistry tests. Find the range of scores on the fifth test that will give the student a C grade for the course.

30. A rowboat traveling with the current went 22.5 mi in 3 h. Against the current, it took 5 h to travel the same distance. Find the rate of the rowboat in calm water and the rate of the current.

11

Exponential and Logarithmic Functions

Objectives

- Evaluate exponential functions
- Graph exponential functions
- Write equivalent exponential and logarithmic equations
- Graph $f(x) = \log_b x$
- The Properties of Logarithms
- Find common logarithms
- Find common antilogarithms
- Solve exponential equations
- Solve logarithmic equations
- Find the logarithm of a number other than base 10
- Application problems

POINT OF INTEREST

Napier Rods

The labor that is involved in calculating the products of large numbers has led many people to devise ways to short-cut the procedure. One such way was first described in the early 1600's by John Napier and is based on Napier Rods.

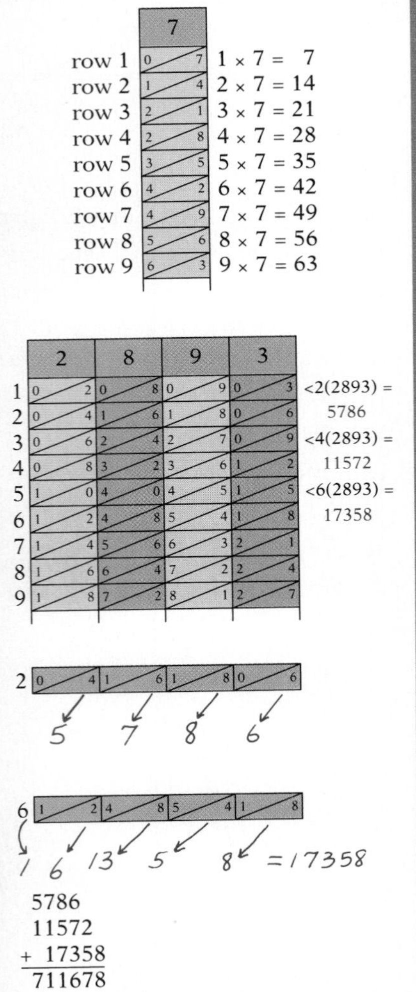

A Napier Rod consists of placing a number and the first 9 multiples of that number on a rectangular piece of paper (Napier's Rod). This is done for the first 9 positive integers. It is necessary to have more than one rod for each number. The rod for 7 is shown at the left.

To illustrate how these rods were used to multiply, an example will be used.

Multiply: $\begin{array}{r} 2893 \\ \times\ 246 \\ \hline \end{array}$

Place the rods for 2, 8, 9, and 3 next to one another. The products for 2, 4, and 6 are found by using the numbers along the 2nd, 4th, and 6th rows.

Each number is found by adding the digits diagonally downward on the diagonal and carrying to the next diagonal when necessary. The products for 2 and 6 are shown at the left.

Notice that the sum of the 3rd diagonal is 13, thus carrying to the next diagonal is necessary.

The final product is then found by addition.

Before the invention of the electronic calculator, logarithms were used to ease the drudgery of lengthy calculations. John Napier is also credited with the invention of logarithms.

SECTION 11.1

The Exponential and Logarithmic Functions

1 Evaluate exponential functions

A function frequently encountered in applications of mathematics is an **exponential function.**

Definition of Exponential Functions

> The exponential function f with base b is defined by
>
> $$f(x) = b^x$$
>
> where $b > 0$, $b \neq 1$, and x is any real number.

Examples of exponential functions are $f(x) = 2^x$, $g(x) = \left(\frac{2}{3}\right)^x$, and $h(x) = \pi^x$.

The value of $f(x) = 2^x$ at $x = 3$ is $f(3) = 2^3 = 8$.

The value of $f(x) = 3^x$ at $x = -2$ is $f(-2) = 3^{-2} = \frac{1}{3^2} = \frac{1}{9}$.

If the base of an exponential function were allowed to be a negative number, the value of the function could be a complex number. For example, the value of $f(x) = (-4)^x$ at $x = \frac{1}{2}$ is $f\left(\frac{1}{2}\right) = (-4)^{\frac{1}{2}} = \sqrt{-4} = 2i$.

For this reason, the base of an exponential function is required to be a positive real number.

Using a calculator, the value of $f(x) = 4^x$ at $x = \sqrt{2}$ can be found to the desired degree of accuracy by using approximations for $\sqrt{2}$.

$$f(\sqrt{2}) \approx f(1.41) = 4^{1.41} \approx 7.06$$
$$f(\sqrt{2}) \approx f(1.414) = 4^{1.414} \approx 7.101$$
$$f(\sqrt{2}) \approx f(1.4142) = 4^{1.4142} \approx 7.1029$$
$$f(\sqrt{2}) \approx f(1.41421) = 4^{1.41421} \approx 7.10296$$

Example 1 Evaluate $f(x) = \left(\frac{1}{2}\right)^x$ at $x = 2$ and $x = -3$.

Solution $f(x) = \left(\frac{1}{2}\right)^x$

$f(2) = \left(\frac{1}{2}\right)^2 = \frac{1}{4}$ $\qquad$ $f(-3) = \left(\frac{1}{2}\right)^{-3} = 2^3 = 8$

Problem 1 Evaluate $f(x) = \left(\frac{2}{3}\right)^x$ at $x = 3$ and $x = -2$.

Solution See page A50.

Example 2 Evaluate $f(x) = 2^{3x-1}$ at $x = 1$ and $x = -1$.

Solution $f(x) = 2^{3x-1}$

$f(1) = 2^{3(1)-1} = 2^2 = 4$ $\qquad$ $f(-1) = 2^{3(-1)-1} = 2^{-4} = \frac{1}{2^4} = \frac{1}{16}$

Problem 2 Evaluate $f(x) = 2^{2x+1}$ at $x = 0$ and $x = -2$.

Solution See page A50.

$F(0)2^{2\cdot 0+1} = 2^3 = 8$ $\qquad$ $F(-2) = 2^{2(-2)+1} = 2^5 = 64$

The exponential key $\boxed{y^x}$ on a scientific calculator can be used to approximate the value of an exponential function. The following examples illustrate the keystrokes used on a scientific calculator.

Power	Key Sequence	Calculator Display
$3^{2.5}$	3 $\boxed{y^x}$ 2.5 $\boxed{=}$	15.58845727
$2.4^{\frac{11}{3}}$	2.4 $\boxed{y^x}$ $\boxed{(}$ 11 $\boxed{\div}$ 3 $\boxed{)}$ $\boxed{=}$	24.78037568
$\pi^{-4.16}$	π $\boxed{y^x}$ 4.16 $\boxed{+/-}$ $\boxed{=}$	0.008547843

The examples above retain the number as shown in the calculator display. The directions in many problems will require rounding the answer to a given place value.

Example 3 Evaluate $f(x) = (\sqrt{5})^x$ at $x = 4$, $x = -2.1$, and $x = \pi$. Round to the nearest ten-thousandth.

Solution $f(x) = (\sqrt{5})^x$

$f(4) = (\sqrt{5})^4 = 25$ $\qquad$ $f(-2.1) = (\sqrt{5})^{-2.1} \approx 0.1845$ $\qquad$ $f(\pi) = (\sqrt{5})^{\pi} \approx 12.5297$

Problem 3 Evaluate $f(x) = \pi^x$ at $x = 3$, $x = -2$, and $x = \pi$. Round to the nearest ten-thousandth.

Solution See page A50.

A base that is frequently used in applications of exponential functions is an irrational number designated by e. The number e is approximately equal to 2.718281828.

Example 4 Evaluate $f(x) = e^x$ at $x = 2$, $x = -3$, and $x = \pi$. Round to the nearest ten-thousandth.

Solution $f(x) = e^x$

$f(2) = e^2 \approx 7.3891$ $\quad f(-3) = e^{-3} \approx 0.0498$ $\quad f(\pi) = e^{\pi} \approx 23.1407$

Problem 4 Evaluate $f(x) = e^x$ at $x = 1.2$, $x = -2.5$, and $x = e$. Round to the nearest ten-thousandth.

Solution See page A50.

2 Graph exponential functions

Some of the properties of an exponential function can be seen by considering its graph.

To graph $f(x) = 2^x$, think of the function as the equation $y = 2^x$.

Choose values of x, and find the corresponding values of y. The results can be recorded in a table.

x	$f(x) = y$
-2	$2^{-2} = \frac{1}{4}$
-1	$2^{-1} = \frac{1}{2}$
0	$2^0 = 1$
1	$2^1 = 2$
2	$2^2 = 4$
3	$2^3 = 8$

Graph the ordered pairs on a rectangular coordinate system.

Connect the points with a smooth curve.

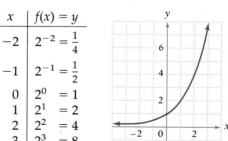

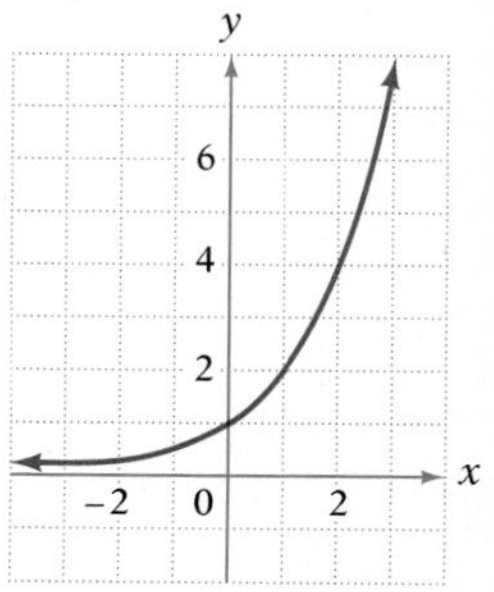

Note that a vertical line would intersect the graph at only one point. Therefore, by the vertical line test, $f(x) = 2^x$ is the graph of a function. Also notice that a horizontal line would intersect the graph at only one point. Therefore, $f(x) = 2^x$ is the graph of a one-to-one function.

To graph $f(x) = \left(\frac{1}{2}\right)^x$, think of the function as the equation $y = \left(\frac{1}{2}\right)^x$.

Choose values of x, and find the corresponding values of y.

Graph the ordered pairs on a rectangular coordinate system.

Connect the points with a smooth curve.

x	$f(x) = y$
-3	$\left(\frac{1}{2}\right)^{-3} = 8$
-2	$\left(\frac{1}{2}\right)^{-2} = 4$
-1	$\left(\frac{1}{2}\right)^{-1} = 2$
0	$\left(\frac{1}{2}\right)^{0} = 1$
1	$\left(\frac{1}{2}\right)^{1} = \frac{1}{2}$
2	$\left(\frac{1}{2}\right)^{2} = \frac{1}{4}$

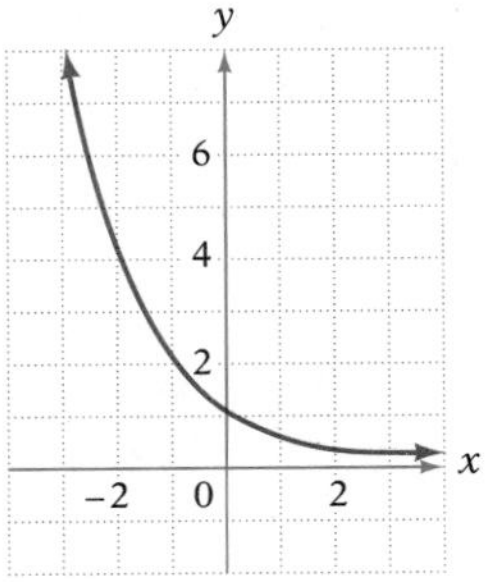

Applying the vertical and horizontal line tests, $f(x) = \left(\frac{1}{2}\right)^x$ is also the graph of a one-to-one function.

The graph of $f(x) = 2^{-x}$ is shown at the right.

Note that since $2^{-x} = (2^{-1})^x = \left(\frac{1}{2}\right)^x$, the graphs of $f(x) = 2^{-x}$ and $f(x) = \left(\frac{1}{2}\right)^x$ are the same.

x	y
-3	8
-2	4
-1	2
0	1
1	$\frac{1}{2}$
2	$\frac{1}{4}$

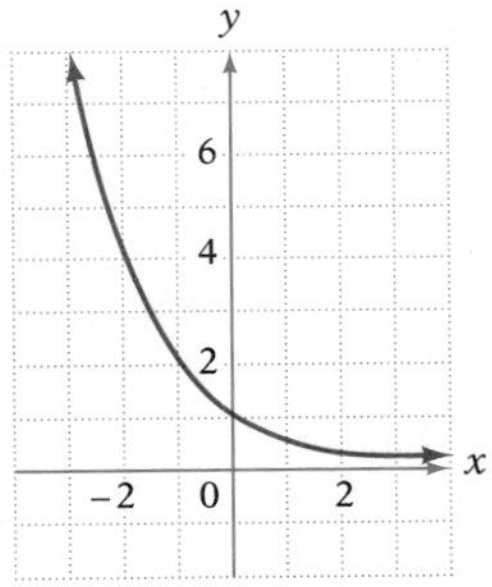

Example 5 Graph.

A. $f(x) = 3^{\frac{1}{2}x-1}$

B. $f(x) = 2^x - 1$

Solution A.

x	y
-2	$\frac{1}{9}$
0	$\frac{1}{3}$
2	1
4	3

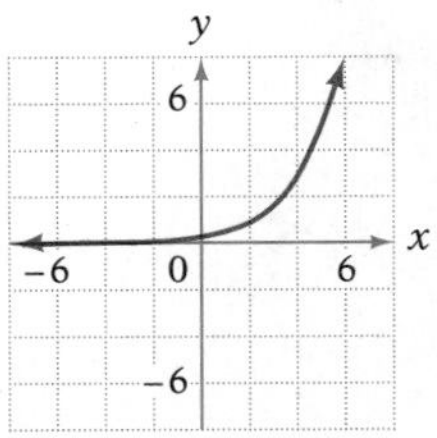

B.

x	y
-2	$-\frac{3}{4}$
-1	$-\frac{1}{2}$
0	0
1	1
2	3
3	7

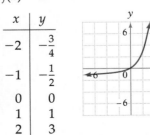

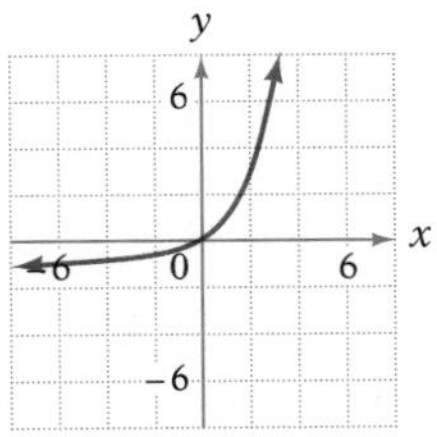

Problem 5 Graph.

A. $f(x) = 2^{-\frac{1}{2}x}$ B. $f(x) = 2^x + 1$

Solution See page A50.

3 Write equivalent exponential and logarithmic equations

Because the exponential function is a 1–1 function, it has an inverse function. Recall that the inverse of a function is formed by interchanging x and y.

For $y = 2^x$, the inverse function is given by the equation $x = 2^y$, and y is called the **logarithm** of x with base 2, written $y = \log_2 x$, where the abbreviation log is used for logarithm.

Definition of a Logarithm

For $b > 0$, $b \neq 1$, $y = \log_b x$ is equivalent to $x = b^y$.

The notation $\log_b x$ is read "the logarithm of x with base b." The definition of logarithms indicates that a logarithm is an exponent.

The graph of $x = 2^y$ or $y = \log_2 x$ is shown at the right. Using this graph, it is possible to estimate the logarithm of a number.

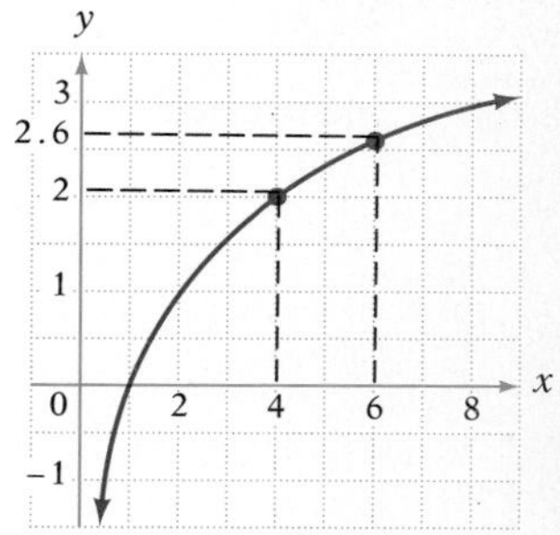

When $x = 4$, $y = 2$. This can be written $\log_2 4 = 2$. Read $\log_2 4$ as "the logarithm of 4 with base 2" or "the log base 2 of 4."

When $x = 6$, $y \approx 2.6$. Therefore, $\log_2 6 \approx 2.6$.

The table below shows equivalent equations written in both exponential and logarithmic form.

Exponential Form	Logarithmic Form
$2^4 = 16$	$\log_2 16 = 4$
$\left(\frac{2}{3}\right)^2 = \frac{4}{9}$	$\log_{\frac{2}{3}}\left(\frac{4}{9}\right) = 2$
$10^{-1} = 0.1$	$\log_{10} 0.1 = -1$

Example 6 Write $3^4 = 81$ in logarithmic form.

Solution $3^4 = 81$ is equivalent to $\log_3 81 = 4$.

Problem 6 Write $7^3 = 343$ in logarithmic form.

Solution See page A50.

Example 7 Write $\log_5 125 = 3$ in exponential form.

Solution $\log_5 125 = 3$ is equivalent to $5^3 = 125$.

Problem 7 Write $\log_{\frac{1}{2}}\left(\frac{1}{8}\right) = 3$ in exponential form.

Solution See page A50.

Logarithms with base 10 are called **common logarithms.** Usually the base, 10, is omitted when writing the common logarithm of a number. Therefore, $\log_{10}x$ is written $\log x$.

Example 8 Write $\log 100 = 2$ in exponential form.

Solution $\log 100 = 2$ is equivalent to $10^2 = 100$.

Problem 8 Write $\log 0.1 = -1$ in exponential form.

Solution See page A50.

When e is used as the base of a logarithm, the logarithm is referred to as the **natural logarithm** and is abbreviated $\ln x$. This is read "el en x."

Example 9 Write $\ln 0.368 = -1$ in exponential form.

Solution $\ln 0.368 = -1$ is equivalent to $e^{-1} = 0.368$.

Problem 9 Write $\ln 7.389 = 2$ in exponential form.

Solution See page A50.

Some exponential and logarithmic equations can be solved by using the following property.

Equality of Exponents Property

If $b^x = b^y$, then $x = y$.

Evaluate: $\log_2 8$

Write an equation. $\log_2 8 = x$

Write the equation in its equivalent exponential form. $8 = 2^x$

Write 8 in exponential form using 2 as the base. $2^3 = 2^x$

Use the Equality of Exponents Property. $3 = x$

$\log_2 8 = 3$

Example 10 Evaluate: $\log_3\left(\frac{1}{9}\right)$

Solution

$\log_3\left(\frac{1}{9}\right) = x$ ▶ Write an equation.

$\frac{1}{9} = 3^x$ ▶ Write the equation in its equivalent exponential form.

$3^{-2} = 3^x$ ▶ Write $\frac{1}{9}$ in exponential form using 3 as the base.

$-2 = x$ ▶ Solve for x using the Equality of Exponents Property.

$\log_3\left(\frac{1}{9}\right) = -2$

Problem 10 Evaluate: $\log_4 64$

Solution See page A50.

To solve $\log_4 x = -2$ for x, write the equation in its equivalent exponential form.

$\log_4 x = -2$

$4^{-2} = x$

Solve for x.

$\frac{1}{16} = x$

The solution is $\frac{1}{16}$.

Example 11 Solve for x: $\log_5 x = 2$

Solution

$\log_5 x = 2$

$5^2 = x$ ▶ Write the equation in its equivalent exponential form.

$25 = x$

The solution is 25.

Problem 11 Solve for x: $\log_2 x = -4$

Solution See page A50.

4 Graph $f(x) = \log_b x$

The graph of a logarithmic function can be drawn by using the relationship between the exponential and logarithmic functions.

To graph $f(x) = \log_2 x$, think of the function as the equation $y = \log_2 x$.

$$f(x) = \log_2 x$$
$$y = \log_2 x$$

Write the equivalent exponential equation.

$$x = 2^y$$

Since the equation is solved for x in terms of y, it is easier to choose values of y and find the corresponding values of x. The results can be recorded in a table.

x	y
$\frac{1}{4}$	-2
$\frac{1}{2}$	-1
1	0
2	1
4	2

Graph the ordered pairs on a rectangular coordinate system.

Connect the points with a smooth curve.

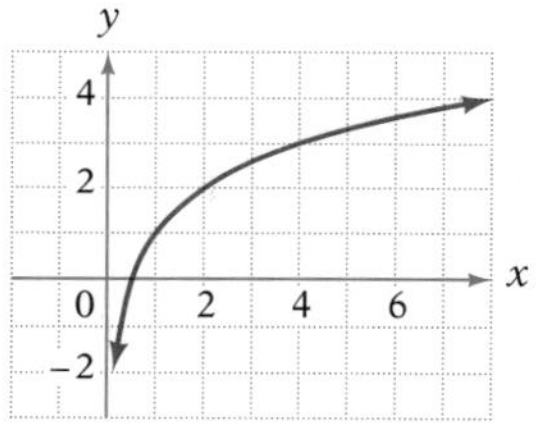

Applying the vertical and horizontal line tests, $f(x) = \log_2 x$ is the graph of a one-to-one function.

Recall that the graph of the inverse of a function f is the mirror image of f with respect to the line $y = x$. The graph of $f(x) = 2^x$ is shown on page 531. Since $g(x) = \log_2 x$ is the inverse of $f(x) = 2^x$, the graphs of these functions are mirror images of each other with respect to the line $y = x$.

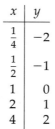

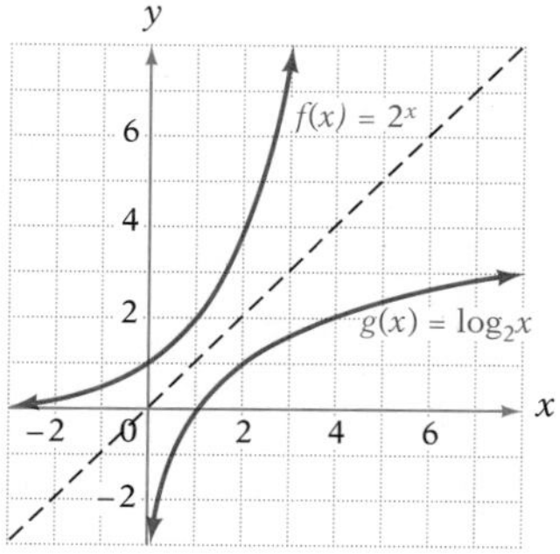

To graph $f(x) = \log_2 x + 1$, think of the function as the equation $y = \log_2 x + 1$.

$$f(x) = \log_2 x + 1$$
$$y = \log_2 x + 1$$

Solve the equation for $\log_2 x$.

$$y - 1 = \log_2 x$$

Write the equivalent exponential equation.

$$2^{y-1} = x$$

Choose values of y, and find the corresponding values of x.

x	y
$\frac{1}{4}$	-1
$\frac{1}{2}$	0
1	1
2	2
4	3

Graph the ordered pairs on a rectangular coordinate system.

Connect the points with a smooth curve.

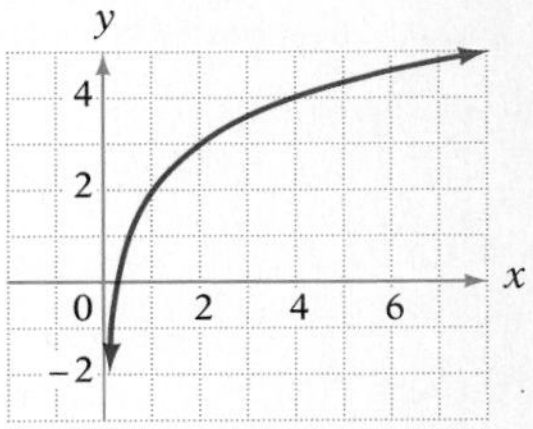

Example 12 Graph.

A. $f(x) = \log_3 x$ B. $f(x) = 2\log_3 x$

Solution A. $f(x) = \log_3 x$

$y = \log_3 x$ ▶ Substitute y for $f(x)$.

$x = 3^y$ ▶ Write the equivalent exponential equation.

▶ Choose values of y, and find the corresponding values of x. Graph the ordered pairs on a rectangular coordinate system. Connect the points with a smooth curve.

x	y
$\frac{1}{9}$	-2
$\frac{1}{3}$	-1
1	0
3	1

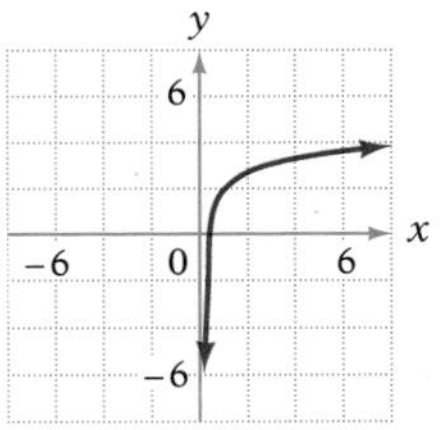

B. $f(x) = 2\log_3 x$

$y = 2\log_3 x$ ▶ Substitute y for $f(x)$.

$\frac{y}{2} = \log_3 x$ ▶ Solve the equation for $\log_3 x$.

$x = 3^{\frac{y}{2}}$ ▶ Write the equivalent exponential equation.

▶ Choose values of y, and find the corresponding values of x. Graph the ordered pairs on a rectangular coordinate system. Connect the points with a smooth curve.

x	y
$\frac{1}{9}$	-4
$\frac{1}{3}$	-2
1	0
3	2

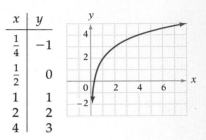

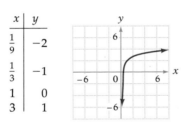

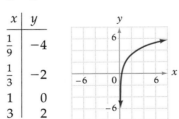

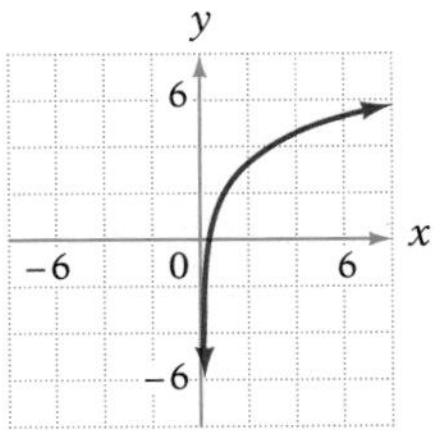

Problem 12 Graph.

A. $f(x) = \log_2(x - 1)$

B. $f(x) = \log_3 2x$

Solution See page A51.

EXERCISES 11.1

1 Given $f(x) = 3^x$, find:

1. $f(2)$ **2.** $f(3)$ **3.** $f(-2)$

4. $f(-1)$ **5.** $f(0)$ **6.** $f(1)$

Given $f(x) = 2^{x+1}$, find:

7. $f(3)$ **8.** $f(1)$ **9.** $f(-3)$

10. $f(-4)$ **11.** $f(-1)$ **12.** $f(0)$

Given $f(x) = \left(\frac{1}{2}\right)^{2x}$, find:

13. $f(0)$ **14.** $f(1)$ **15.** $f(-2)$

16. $f(-1)$ **17.** $f\left(\frac{3}{2}\right)$ **18.** $f\left(-\frac{1}{2}\right)$

Given $f(x) = \left(\frac{1}{3}\right)^{x-1}$, find:

19. $f(2)$ **20.** $f(3)$ **21.** $f(-1)$

22. $f(-2)$ **23.** $f(0)$ **24.** $f(1)$

Given $f(x) = 2^{x^2}$, find:

25. $f(1)$ **26.** $f(0)$ **27.** $f(2)$

28. $f(3)$ **29.** $f(-1)$ **30.** $f(-2)$

Solve. Round to the nearest ten-thousandth.

31. Given $f(x) = (\sqrt{2})^{2x}$, find $f(1)$.

32. Given $f(x) = 3(\sqrt{2})^x$, find $f(2)$.

33. Given $f(x) = (\sqrt{2})^{-x}$, find $f(-1)$.

34. Given $f(x) = (\sqrt{2})^{-x}$, find $f(-2)$.

35. Given $f(x) = e^{2x}$, find $f\left(\frac{1}{2}\right)$.

36. Given $f(x) = e^{3x}$, find $f(1)$.

37. Given $f(x) = 3e^{x-1}$, find $f(2)$.

38. Given $f(x) = e^{-2x}$, find $f(-1)$.

2 Graph.

39. $f(x) = 3^x$

40. $f(x) = 3^{-x}$

41. $f(x) = 2^{x+1}$

42. $f(x) = 2^{x-1}$

43. $f(x) = \left(\frac{1}{3}\right)^x$

44. $f(x) = \left(\frac{2}{3}\right)^x$

45. $f(x) = 2^{-x} + 1$

46. $f(x) = 2^x - 3$

47. $f(x) = \left(\frac{1}{2}\right)^{2x}$

48. $f(x) = 2^{\frac{1}{2}x}$

3 Write the exponential equation in logarithmic form.

49. $2^5 = 32$

50. $3^4 = 81$

51. $5^2 = 25$

52. $10^3 = 1000$

53. $4^{-2} = \frac{1}{16}$

54. $3^{-3} = \frac{1}{27}$

55. $\left(\frac{1}{2}\right)^2 = \frac{1}{4}$

56. $\left(\frac{1}{3}\right)^4 = \frac{1}{81}$

57. $3^0 = 1$

58. $10^0 = 1$

59. $a^x = w$

60. $b^y = c$

Write the logarithmic equation in exponential form.

61. $\log_3 9 = 2$ **62.** $\log_2 32 = 5$ **63.** $\log_4 4 = 1$ **64.** $\log_7 7 = 1$

65. $\log 1 = 0$ **66.** $\log_8 1 = 0$ **67.** $\log 0.01 = -2$ **68.** $\log_5 \frac{1}{5} = -1$

69. $\log_{\frac{1}{3}}\left(\frac{1}{9}\right) = 2$ **70.** $\log_{\frac{1}{4}}\left(\frac{1}{16}\right) = 2$ **71.** $\log_b u = v$ **72.** $\log_c x = y$

73. $\log 1000 = 3$ **74.** $\ln 20.086 = 3$ **75.** $\ln 1 = 0$ **76.** $\ln 0.135 = -2$

Evaluate.

77. $\log_4 16$ **78.** $\log_3 27$ **79.** $\log_2 32$ **80.** $\log 1000$

81. $\log 100$ **82.** $\log_5 125$ **83.** $\log_6 216$ **84.** $\log_7 1$

Solve for x.

85. $\log_3 x = 2$ **86.** $\log_5 x = 1$ **87.** $\log_4 x = 3$ **88.** $\log_2 x = 6$

89. $\log_7 x = -1$ **90.** $\log_8 x = -2$ **91.** $\log_6 x = 3$ **92.** $\log_4 x = 0$

4 Graph.

93. $f(x) = \log_4 x$ **94.** $f(x) = \log_2(x + 1)$

95. $f(x) = \log_3(2x - 1)$ **96.** $f(x) = \log_2\left(\frac{1}{2}x\right)$

97. $f(x) = 3 \log_2 x$ **98.** $f(x) = \frac{1}{2} \log_2 x$

99. $f(x) = -\log_2 x$

100. $f(x) = -\log_3 x$

101. $f(x) = e^x$

102. $f(x) = \ln x$

Graph the functions on the same rectangular coordinate system.

103. $f(x) = 3^x$; $g(x) = \log_3 x$

104. $f(x) = 4^x$; $g(x) = \log_4 x$

105. $f(x) = 10^x$; $g(x) = \log_{10} x$

106. $f(x) = \left(\frac{1}{2}\right)^x$; $g(x) = \log_{\frac{1}{2}} x$

SUPPLEMENTAL EXERCISES 11.1

Evaluate. Round to the nearest hundredth.

107. $f(x) = 2^x$ at $x = \sqrt{2}$

108. $f(x) = 2^x$ at $x = \sqrt{3}$

109. $f(x) = 3^x$ at $x = \sqrt{2}$

110. $f(x) = 3^x$ at $x = \sqrt{3}$

111. $f(x) = \left(\frac{1}{2}\right)^x$ at $x = \sqrt{2}$

112. $f(x) = \left(\frac{1}{2}\right)^x$ at $x = \sqrt{5}$

113. $f(x) = 4^x$ at $x = \pi$

114. $f(x) = 4^x$ at $x = -\pi$

Find the range of the function given by the equation.

115. $f(x) = 2^x$

116. $f(x) = \left(\frac{1}{2}\right)^x$

117. $f(x) = 3^{x-1}$

118. $f(x) = 2^x - 1$

Find the domain of the function given by the equation.

119. $f(x) = \log_2 x$

120. $f(x) = \log_2 x + 1$

121. $f(x) = 2 \log_3 x$

122. $f(x) = \log_2(x + 1)$

123. Evaluate $f(x) = 2^{\log_2 x}$ for $x = 1, 2, 4,$ and 8. Make a conjecture as to the value of $f(a)$, $a > 0$.

124. Evaluate $f(x) = \log_{10}10^x$ for $x = -2, -1, 0, 1,$ and 2. Make a conjecture as to the value of $f(a)$ for any real number a.

SECTION 11.2 The Properties of Logarithms

1 The Properties of Logarithms

Since a logarithm is an exponent, the Properties of Logarithms are similar to the Properties of Exponents.

The table at the right shows some powers of 2 and the equivalent logarithmic form.

$2^0 = 1$	$\log_2 1 = 0$
$2^1 = 2$	$\log_2 2 = 1$
$2^2 = 4$	$\log_2 4 = 2$
$2^3 = 8$	$\log_2 8 = 3$
$2^4 = 16$	$\log_2 16 = 4$
$2^5 = 32$	$\log_2 32 = 5$

The table can be used to show that $\log_2 4 + \log_2 8$ equals $\log_2 32$.

$$\log_2 4 + \log_2 8 = 2 + 3 = 5$$
$$\log_2 32 = 5$$
$$\log_2 4 + \log_2 8 = \log_2 32$$

Note that $\log_2 32 = \log_2(4 \times 8) = \log_2 4 + \log_2 8$.

The property of logarithms that states that the logarithm of the product of two numbers equals the sum of the logarithms of the two numbers is similar to the property of exponents that states that to multiply two exponential expressions with the same base, add the exponents.

The Logarithm Property of the Product of Two Numbers

For any positive real numbers, x, y, and b, $b \neq 1$,

$$\log_b xy = \log_b x + \log_b y.$$

Proof:
Let $\log_b x = m$ and $\log_b y = n$.

Write each equation in its equivalent exponential form. $\quad x = b^m \qquad y = b^n$

Use substitution and the Properties of Exponents.

$$xy = b^m b^n$$
$$xy = b^{m+n}$$

Write the equation in its equivalent logarithmic form. $\log_b xy = m + n$

Substitute $\log_b x$ for m and $\log_b y$ for n. $\log_b xy = \log_b x + \log_b y$

The Logarithm Property of Products is used to rewrite logarithmic expressions.

The $\log_b 6z$ is written in **expanded form** as $\log_b 6 + \log_b z$.

$$\log_b 6z = \log_b 6 + \log_b z$$

The $\log_b 12 + \log_b r$ is written as a single logarithm as $\log_b 12r$.

$$\log_b 12 + \log_b r = \log_b 12r$$

The Logarithm Property of Products can be extended to include the logarithm of the product of more than two factors. For example,

$$\log_b xyz = \log_b (xy)z = \log_b xy + \log_b z = \log_b x + \log_b y + \log_b z$$

To write $\log_b 5st$ in expanded form, use the Logarithm Property of Products. $\log_b 5st = \log_b 5 + \log_b s + \log_b t$

A second property of logarithms involves the logarithm of the quotient of two numbers. This property of logarithms is also based on the fact that a logarithm is an exponent and that to divide two exponential expressions with the same base, the exponents are subtracted.

The Logarithm Property of the Quotient of Two Numbers

For any positive real numbers, x, y, and b, $b \neq 1$, $\log_b \frac{x}{y} = \log_b x - \log_b y$.

Proof:
Let $\log_b x = m$ and $\log_b y = n$.

Write each equation in its equivalent exponential form. $x = b^m$ $\quad y = b^n$

Use substitution and the Properties of Exponents.

$$\frac{x}{y} = \frac{b^m}{b^n}$$

$$\frac{x}{y} = b^{m-n}$$

Write the equation in its equivalent logarithmic form. $\log_b \frac{x}{y} = m - n$

Substitute $\log_b x$ for m and $\log_b y$ for n. $\log_b \frac{x}{y} = \log_b x - \log_b y$

The Logarithm Property of Quotients is used to rewrite logarithmic expressions.

The $\log_b \frac{p}{8}$ is written in expanded form as $\log_b p - \log_b 8$.

$$\log_b \frac{p}{8} = \log_b p - \log_b 8$$

The $\log_b y - \log_b v$ is written as a single logarithm as $\log_b \frac{y}{v}$.

$$\log_b y - \log_b v = \log_b \frac{y}{v}$$

A third property of logarithms, which is especially useful in the computation of the power of a number, is based on the fact that a logarithm is an exponent, and the power of an exponential expression is found by multiplying the exponents.

The table of the powers of 2 shown on page 542 can be used to show that $\log_2 2^3$ equals $3 \log_2 2$.

$$\log_2 2^3 = \log_2 8 = 3$$
$$3 \log_2 2 = 3 \cdot 1 = 3$$
$$\log_2 2^3 = 3 \log_2 2$$

The Logarithm Property of the Power of a Number

For any positive real numbers x and b, $b \neq 1$, and for any real number r, $\log_b x^r = r \log_b x$.

Proof:
Let $\log_b x = m$.

Write the equation in its equivalent exponential form. $\quad x = b^m$

Raise each side to the r power. $\quad x^r = (b^m)^r$

$$x^r = b^{mr}$$

Write the equation in its equivalent logarithmic form. $\quad \log_b x^r = mr$

Substitute $\log_b x$ for m. $\quad \log_b x^r = r \log_b x$

The Logarithm Property of Powers is used to rewrite logarithmic expressions.

The $\log_b x^3$ is written in terms of $\log_b x$ as $3 \log_b x$.

$$\log_b x^3 = 3 \log_b x$$

$\frac{2}{3} \log_4 z$ is written with a coefficient of 1 as $\log_4 z^{\frac{2}{3}}$.

$$\frac{2}{3} \log_4 z = \log_4 z^{\frac{2}{3}}$$

The Properties of Logarithms can be used in combination to simplify expressions containing logarithms.

Example 1 Write the logarithm in expanded form.

A. $\log_b \frac{xy}{z}$ B. $\ln \frac{x^2}{y^3}$ C. $\log_8 \sqrt{x^3 y}$

Solution

A. $\log_b \frac{xy}{z} = \log_b(xy) - \log_b z$ ▶ Use the Logarithm Property of Quotients.

$= \log_b x + \log_b y - \log_b z$ ▶ Use the Logarithm Property of Products.

B. $\ln \frac{x^2}{y^3} = \ln x^2 - \ln y^3$ ▶ Use the Logarithm Property of Quotients.

$= 2 \ln x - 3 \ln y$ ▶ Use the Logarithm Property of Powers.

C. $\log_8 \sqrt{x^3 y} = \log_8 (x^3 y)^{\frac{1}{2}}$ ▶ Write the radical expression as an exponential expression.

$= \frac{1}{2} \log_8 x^3 y$ ▶ Use the Logarithm Property of Powers.

$= \frac{1}{2}(\log_8 x^3 + \log_8 y)$ ▶ Use the Logarithm Property of Products.

$= \frac{1}{2}(3 \log_8 x + \log_8 y)$ ▶ Use the Logarithm Property of Powers.

$= \frac{3}{2} \log_8 x + \frac{1}{2} \log_8 y$ ▶ Use the Distributive Property.

Problem 1 Write the logarithm in expanded form.

A. $\log_b \frac{x^2}{y}$ B. $\ln y^{\frac{1}{3}} z^3$ C. $\log_8 \sqrt[3]{xy^2}$

Solution See page A51.

Example 2 Express as a single logarithm with a coefficient of 1.

A. $3 \log_5 x + \log_5 y - 2 \log_5 z$ B. $\frac{1}{2}(\log_3 x - 3 \log_3 y + \log_3 z)$

C. $\frac{1}{3}(2 \ln x - 4 \ln y)$

Solution

A. $3 \log_5 x + \log_5 y - 2 \log_5 z =$

$\log_5 x^3 + \log_5 y - \log_5 z^2 =$ ▶ Use the Logarithm Property of Powers.

$\log_5 x^3 y - \log_5 z^2 =$ ▶ Use the Logarithm Property of Products.

$\log_5 \frac{x^3 y}{z^2}$ ▶ Use the Logarithm Property of Quotients.

B. $\frac{1}{2}(\log_3 x - 3\log_3 y + \log_3 z) =$

$\frac{1}{2}(\log_3 x - \log_3 y^3 + \log_3 z) =$ ▶ Use the Logarithm Property of Powers.

$\frac{1}{2}\left(\log_3 \frac{x}{y^3} + \log_3 z\right) =$ ▶ Use the Logarithm Property of Quotients.

$\frac{1}{2}\left(\log_3 \frac{xz}{y^3}\right) =$ ▶ Use the Logarithm Property of Products.

$\log_3\left(\frac{xz}{y^3}\right)^{\frac{1}{2}} = \log_3 \sqrt{\frac{xz}{y^3}}$ ▶ Use the Logarithm Property of Powers. Write the exponential expression as a radical expression.

C. $\frac{1}{3}(2\ln x - 4\ln y) =$

$\frac{1}{3}(\ln x^2 - \ln y^4) =$ ▶ Use the Logarithm Property of Powers.

$\frac{1}{3}\left(\ln \frac{x^2}{y^4}\right) =$ ▶ Use the Logarithm Property of Quotients.

$\ln\left(\frac{x^2}{y^4}\right)^{\frac{1}{3}} = \ln \sqrt[3]{\frac{x^2}{y^4}}$ ▶ Use the Logarithm Property of Powers. Write the exponential expression as a radical expression.

Problem 2 Express as a single logarithm with a coefficient of 1.

A. $2\log_b x - 3\log_b y - \log_b z$ B. $\frac{1}{3}(\log_4 x - 2\log_4 y + \log_4 z)$

C. $\frac{1}{2}(2\ln x - 5\ln y)$

Solution See page A51.

Two other properties of logarithmic functions can be established directly from the equivalence of the expressions $\log_b x = y$ and $x = b^y$.

For any positive real number b, $b \neq 1$, $\mathbf{\log_b 1 = 0}$.

For any positive real number b, $b \neq 1$, and any real number n, $\mathbf{\log_b b^n = n}$.

Note that for natural logs, $\ln e = 1$.

Example 3 Find $8\log_4 4$.

Solution $8\log_4 4 = \log_4 4^8 = 8$ ▶ Since $\log_b b^n = n$, $\log_4 4^8 = 8$.

Problem 3 Find $\log_9 1$.

Solution See page A51.

EXERCISES 11.2

1 Write the logarithm in expanded form.

1. $\log_8(xz)$

2. $\log_7(4y)$

3. $\log_3 x^5$

4. $\log_2 y^7$

5. $\ln\left(\frac{r}{s}\right)$

6. $\ln\left(\frac{z}{4}\right)$

7. $\log_3(x^2y^6)$

8. $\log_4(t^4u^2)$

9. $\log_7\left(\frac{u^3}{v^4}\right)$

10. $\log\left(\frac{s^5}{t^2}\right)$

11. $\log_2(rs)^2$

12. $\log_3(x^2y)^3$

13. $\log_9 x^2yz$

14. $\log_6 xy^2z^3$

15. $\ln\left(\frac{xy^2}{z^4}\right)$

16. $\ln\left(\frac{r^2s}{t^3}\right)$

17. $\log_8\left(\frac{x^2}{yz^2}\right)$

18. $\log_9\left(\frac{x}{y^2z^3}\right)$

19. $\log_7\sqrt{xy}$

20. $\log_8\sqrt[3]{xz}$

21. $\log_2\sqrt{\frac{x}{y}}$

22. $\log_3\sqrt[3]{\frac{r}{s}}$

23. $\ln\sqrt{x^3y}$

24. $\ln\sqrt{x^5y^3}$

25. $\log_7\sqrt{\frac{x^3}{y}}$

26. $\log_b\sqrt[3]{\frac{r^2}{t}}$

27. $\log_b x\sqrt{\frac{y}{z}}$

28. $\log_4 y\sqrt[3]{\frac{r}{s}}$

29. $\log_3 \frac{t}{\sqrt{x}}$

30. $\log_4\left(\frac{\sqrt{uv}}{x}\right)$

Express as a single logarithm with a coefficient of 1.

31. $\log_3 x^3 - \log_3 y$

32. $\log_7 t + \log_7 v^2$

33. $\log_8 x^4 + \log_8 y^2$

34. $\log_2 r^2 + \log_2 s^3$

35. $3 \ln x$

36. $4 \ln y$

37. $3 \log_5 x + 4 \log_5 y$

38. $2 \log_6 x + 5 \log_6 y$

39. $-2 \log_4 x$

40. $-3 \log_2 y$

41. $2 \log_3 x - \log_3 y + 2 \log_3 z$

42. $4 \log_5 r - 3 \log_5 s + \log_5 t$

43. $\log_b x - (2 \log_b y + \log_b z)$

44. $2 \log_2 x - (3 \log_2 y + \log_2 z)$

45. $2(\ln x + \ln y)$

46. $3(\ln r + \ln t)$

47. $\frac{1}{2}(\log_6 x - \log_6 y)$

48. $\frac{1}{3}(\log_8 x - \log_8 y)$

49. $2(\log_4 s - 2 \log_4 t + \log_4 r)$

50. $3(\log_9 x + 2 \log_9 y - 2 \log_9 z)$

51. $\log_5 x - 2(\log_5 y + \log_5 z)$

52. $\log_4 t - 3(\log_4 u + \log_4 v)$

53. $3 \ln t - 2(\ln r - \ln v)$

54. $2 \ln x - 3(\ln y - \ln z)$

55. $\frac{1}{2}(3 \log_4 x - 2 \log_4 y + \log_4 z)$

56. $\frac{1}{3}(4 \log_5 t - 3 \log_5 u - 3 \log_5 v)$

57. $\frac{1}{2}\log_b x - \frac{2}{3}\log_b y + \frac{1}{2}\log_b z$

58. $\frac{2}{3}\log_b x + \frac{1}{3}\log_b y - \frac{1}{2}\log_b z$

59. $\log_b x - \frac{2}{3}\log_b y + 4\log_b z$

60. $2\log_b x - 3\log_b y - 4\log_b z$

Simplify.

61. $9\log_3 3$

62. $5\log_6 6$

63. $\log_7 1$

64. $\log_4 1$

65. $8\log_5 5$

66. $2\log_7 7$

SUPPLEMENTAL EXERCISES 11.2

Use the Properties of Logarithms to solve for x.

67. $\log_8 x = 3\log_8 2$

68. $\log_5 x = 2\log_5 3$

69. $\log_4 x = \log_4 2 + \log_4 3$

70. $\log_3 x = \log_3 4 + \log_3 7$

71. $\log_7 x = \log_7 8 - \log_7 4$

72. $\log x = \log 64 - \log 8$

73. $\log_6 x = 3\log_6 2 - \log_6 4$

74. $\log_9 x = 5\log_9 2 - \log_9 8$

75. $\log x = \frac{1}{3}\log 27$

76. $\log_2 x = \frac{3}{2}\log_2 4$

77. Using the Properties of Logarithms, show that $\log_a a^x = x$, $a > 0$.

78. Using the Properties of Logarithms, show that $a^{\log_a x} = x$, $a > 0$, $x > 0$.

79. Using the Properties of Logarithms, show that $\log x^2 = 2\log x$. Graph $y = \log x^2$ and $y = 2\log x$. Are their graphs the same? Why or why not?

SECTION 11.3
Common Logarithms

1 Find common logarithms

Recall that logarithms to the base 10 are called **common logarithms** and that the base, 10, is omitted when writing the common logarithm of a number. Throughout this chapter, $\log x$ is used for $\log_{10} x$.

Using the definition of logarithms, the examples below show that the logarithm of a power of 10 is the exponent on 10.

$$\log 100 = \log 10^2 = 2$$
$$\log 10 = \log 10^1 = 1$$
$$\log 1 = \log 10^0 = 0$$
$$\log 0.1 = \log 10^{-1} = -1$$
$$\log 0.01 = \log 10^{-2} = -2$$

To find the base 10 logarithm of a number other than a power of 10, it is necessary to use a calculator or a table of logarithms. Since most logarithms are irrational numbers, a calculator or table of logarithms gives approximate values of logarithms. Despite the fact that the logarithms are approximate values, it is customary to use the equals sign (=) rather than the approximately equal sign (≈) when writing logarithms.

The base 10 logarithms of numbers from 1 to 9.99 can be found in the Table of Common Logarithms on pages A4–A5. The logarithms in this table have been rounded to the nearest ten-thousandth. A portion of this table is shown below.

x	0	1	2	3	4	5	6
2.0	.3010	.3032	.3054	.3075	.3096	.3118	.3139
2.1	.3222	.3243	.3263	.3284	.3304	.3324	.3345
2.2	.3424	.3444	.3464	.3483	.3502	.3522	.3541
2.3	.3617	.3636	.3655	.3674	.3692	.3711	.3729
2.4	.3802	.3820	.3838	.3856	.3874	.3892	.3909
2.5	.3979	.3997	.4014	.4031	.4048	.4065	.4082
2.6	.4150	.4166	.4183	.4200	.4216	.4232	.4249

To find log 2.35, locate 2.3 under x in the left-hand column of the table. Move right to the column headed by 5.

$$\log 2.35 = 0.3711$$

Remember that the common logarithm of a number, N, is the power to which 10 must be raised to equal N.

From the example above, $\log 2.35 = 0.3711$. Therefore, $2.35 = 10^{0.3711}$.

To find the common logarithm of a number not in the Table of Common Logarithms, first write the number in scientific notation. Then use the table and the Properties of Logarithms to find the logarithm of the number.

To find log 247, write 247 in scientific notation.	$\log 247 = \log (2.47 \times 10^2)$
Use the Logarithm Property of Products.	$= \log 2.47 + \log 10^2$
Locate log 2.47 in the table.	$= 0.3927 + \log 10^2$
By the definition of logarithms, the logarithm of a power of 10 is the exponent on 10.	$= 0.3927 + 2$
Add.	$= 2.3927$

To find log 2470, write 2470 in scientific notation.	$\log 2470 = \log (2.47 \times 10^3)$
Use the Logarithm Property of Products.	$= \log 2.47 + \log 10^3$
Use the table and the definition of logarithms.	$= 0.3927 + 3$
Add.	$= 3.3927$

Note that for log 247 and log 2470, the decimal part of the logarithm, 0.3927, is the same. The integer part is the exponent on 10 when the number is written in scientific notation.

The decimal part of the logarithm is called the **mantissa.** The integer part is called the **characteristic.**

To find log 0.00247, write 0.00247 in scientific notation.	$\log 0.00247 = \log (2.47 \times 10^{-3})$
Use the Logarithm Property of Products.	$= \log 2.47 + \log 10^{-3}$
Use the table and the definition of logarithms.	$= 0.3927 + (-3)$

In this last example, adding the characteristic to the mantissa would result in a negative logarithm. This form of a logarithm is inconvenient to use with logarithm tables, since these tables contain only positive logarithms. Therefore, when the characteristic of a logarithm is negative, it is customary to leave the logarithm as written or to rewrite the logarithm so that the mantissa remains unchanged.

For example: $\log 0.00247 = 0.3927 + (-3)$

$\log 0.00247 = 0.3927 + (7 - 10) = 7.3927 - 10$

$\log 0.00247 = 0.3927 + (12 - 15) = 12.3927 - 15$

Of these forms, $\log 0.00247 = 7.3927 - 10$ is most common.

Example 1 Find the logarithm.

A. $\log 8360$ B. $\log 0.0217$

Solution A. $\log 8360 = \log (8.36 \times 10^3)$ ▶ Write the number in scientific notation.

$= \log 8.36 + \log 10^3$ ▶ Use the Logarithm Property of Products.

$= 0.9222 + 3$ ▶ Use the table and the definition of logarithms.

$= 3.9222$

B. $\log 0.0217 = \log (2.17 \times 10^{-2})$ ▶ Write the number in scientific notation.

$= \log 2.17 + \log 10^{-2}$ ▶ Use the Logarithm Property of Products.

$= 0.3365 + (-2)$ ▶ Use the table and the definition of logarithms.

$= 0.3365 + (8 - 10)$ ▶ Rewrite -2 as $8 - 10$.

$= 8.3365 - 10$

Problem 1 Find the logarithm.

A. $\log 93{,}000$ B. $\log 0.0006$

Solution See page A52.

Since 3.257 is not included in the Table of Common Logarithms on pages A4–A5, it is not possible to find log 3.257 directly from the table. In the Appendix, page A2, the tables of common logarithms are used to obtain an approximation for log 3.257 by using linear interpolation. However, the [log] key on a calculator gives the logarithm base 10 of any number. Throughout the remainder of this chapter, a calculator will be used to find the logarithm of a number.

To find log 3.257 using a calculator, enter 3.257 and press [log].

The number in the display should be 0.5128178.

Rounded to the nearest ten-thousandth, $\log 3.257 = 0.5128$.

Example 2 Use a calculator to find log 649.8.

Solution $\log 649.8 = 2.8128$ ▶ Enter 649.8 and press [log]. Round to the nearest ten-thousandth.

Problem 2 Use a calculator to find log 0.0235.

Solution See page A52.

2 Find common antilogarithms

In the last objective, given a number, N, the logarithm of N was found. In this objective, the reverse process is examined. Given the *logarithm* of N, the number N will be found.

For example, to find N given $\log N = 0.3636$, locate 0.3636 in the body of the Table of Common Logarithms.

Find the number that corresponds to this mantissa.

The number is 2.31.

$\log 2.31 = 0.3636$

x	0	1	2	3
2.0	.3010	.3032	.3054	.3075
2.1	.3222	.3243	.3263	.3284
2.2	.3424	.3444	.3464	.3483
2.3	.3617	.3636	.3655	.3674
2.4	.3802	.3820	.3838	.3856
2.5	.3979	.3997	.4014	.4031
2.6	.4150	.4166	.4183	.4200

In the equation $\log 2.31 = 0.3636$, the number 2.31 is called the **antilogarithm** of 0.3636. This is written antilog $0.3636 = 2.31$, where the abbreviation antilog is used for antilogarithm.

The common logarithm of a number, N, is the power to which 10 must be raised to equal N.

$\log 2 = 0.3010$ because $2 = 10^{0.3010}$.

The common antilogarithm of a number, N, is the Nth power of 10.

antilog $0.3010 = 2$ because $10^{0.3010} = 2$.

To solve $\log N = 1.1206$, write 1.1206 as the sum of the mantissa and the characteristic.

$\log N = 1.1206$
$\log N = 0.1206 + 1$

Locate the mantissa in the Table of Common Logarithms. Find the number that corresponds to this mantissa. Since the characteristic is 1, the exponent on 10 is 1 when the number is written in scientific notation.

$N = 1.32 \times 10^1$

Simplify. $N = 13.2$

Check: $\log 13.2 =$
$\log (1.32 \times 10^1) =$
$\log 1.32 + \log 10^1 =$
$0.1206 + 1 = 1.1206$

Example 3 Find the antilogarithm.

A. antilog 1.9745 B. antilog (8.7332 − 10)

Solution A. antilog 1.9745 =
antilog (0.9745 + 1) = ▶ Write the number as the sum of the mantissa and the characteristic.
$9.43 \times 10^1 =$ ▶ Use the table to find the number that corresponds to the mantissa. The characteristic is the exponent on 10.
94.3 ▶ Simplify.

B. antilog (8.7332 − 10) =
antilog [0.7332 + (−2)] = ▶ Write the number as the sum of the mantissa and the characteristic.
$5.41 \times 10^{-2} =$ ▶ Use the table to find the number that corresponds to the mantissa. The characteristic is the exponent on 10.
0.0541 ▶ Simplify.

Problem 3 Find the antilogarithm.

A. antilog 2.3365 B. antilog (9.7846 − 10)

Solution See page A52.

Given $\log N = x$, then antilog $x = N$. Since $\log N = x$ is equivalent to $N = 10^x$ and to antilog $x = N$, 10^x can be substituted for N in the equation antilog $x = N$. The resulting equation is

$$\textbf{antilog } x = 10^x.$$

This equation is especially useful when using a calculator to find antilogarithms. Throughout the remainder of this chapter, a calculator will be used to find the antilogarithm of a number. To find antilog x, enter x. Then press the [10^x] key.

Find antilog 1.23514.

Enter: 1.23514 [10^x]

Round to the nearest ten-thousandth. antilog 1.23514 = 17.1846

Find antilog (−2.351).

Enter: 2.351 [+/−] [10ˣ]

Round to the nearest ten-thousandth. antilog (−2.351) = 0.0045

Note that in this last example, a negative logarithm can be used.

For those calculators that do not have a [10ˣ] key, the antilogarithm can be found by using the [INV] or [2nd] keys. The [INV] key will be used here, but the keystrokes for the [2nd] key would be the same.

Find antilog 2.9132.

Enter: 2.9132 [INV] [log]

Round to the nearest ten-thousandth. antilog 2.9132 = 818.8418

EXERCISES 11.3

1 Find the logarithm. Round to the nearest ten-thousandth.

1. 5.87	**2.** 6.42	**3.** 34.5
4. 10.9	**5.** 389	**6.** 592
7. 1030	**8.** 9060	**9.** 47,300
10. 68,200	**11.** 0.114	**12.** 0.209
13. 0.002	**14.** 0.034	**15.** 0.00175
16. 0.0806	**17.** 0.984	**18.** 0.79
19. 27,300	**20.** 10,800	**21.** 0.00571
22. 0.00367	**23.** 0.909	**24.** 20.3
25. 3.845	**26.** 4.965	**27.** 84.52
28. 67.38	**29.** 499.3	**30.** 699.6
31. 0.07071	**32.** 0.01453	**33.** 75,930
34. 83,090	**35.** 0.8407	**36.** 0.9813

2 Find the antilogarithm. Round to the nearest ten-thousandth.

37. 0.7396	**38.** 0.6474	**39.** 1.5353	**40.** 1.4829
41. 3.9015	**42.** 3.9557	**43.** 2.8007	**44.** 2.0128

45. −0.342	**46.** −0.2967	**47.** −1.6108
48. −1.1656	**49.** −2.1296	**50.** −2.0209

51. 2.6767
52. −0.8508
53. −0.308
54. 0.9006
55. 0.8466
56. 2.7360
57. 2.5375
58. −0.2592
59. −0.268
60. 2.6003
61. 2.7004
62. −1.6912
63. −1.4628
64. 1.6525
65. 1.6809

SUPPLEMENTAL EXERCISES 11.3

Complete. Use the fact that log 3.14 = 0.4969.

66. $10^{0.4969} =$ ______
67. antilog 0.4969 = ______
68. antilog (log 3.14) = ______
69. log (antilog 0.4969) = ______
70. $10^{\log 3.14} =$ ______
71. 10 antilog 0.4969 = ______

Complete each statement using the equation $\log_a b = c$.

72. $a^c =$ ______
73. $\text{antilog}_a(\log_a b) =$ ______
74. $a^{\log_a b} =$ ______
75. $\text{antilog}_a\, c =$ ______
76. $\log_a(\text{antilog}_a\, c) =$ ______
77. $\log_a a^b =$ ______

SECTION 11.4 Exponential and Logarithmic Equations

1 Solve exponential equations

An **exponential equation** is one in which the variable occurs in the exponent. The examples at the right are exponential equations.

$6^{2x+1} = 6^{3x-2}$
$4^x = 3$
$2^{x+1} = 7$

An exponential equation in which each side of the equation can be expressed in terms of the same base can be solved by using the Equality of Exponents Property.

Example 1 Solve and check: $9^{x+1} = 27^{x-1}$

Solution

$$9^{x+1} = 27^{x-1}$$
$$(3^2)^{x+1} = (3^3)^{x-1}$$ ▶ Rewrite each side of the equation using the same base.
$$3^{2x+2} = 3^{3x-3}$$
$$2x + 2 = 3x - 3$$ ▶ Use the Equality of Exponents Property to equate the exponents.
$$2 = x - 3$$ ▶ Solve the resulting equation.
$$5 = x$$

The solution is 5.

Check: $9^{x+1} = 27^{x-1}$

9^{x+1}	27^{x-1}
9^{5+1}	27^{5-1}
9^6	27^4
$(3^2)^6$	$(3^3)^4$
3^{12} =	3^{12}

Problem 1 Solve and check: $10^{3x+5} = 10^{x-3}$

Solution See page A52.

When each side of an exponential equation cannot easily be expressed in terms of the same base, logarithms are used to solve the exponential equation.

One-to-One Property of Logarithms

If $x = y$, then $\log_b x = \log_b y$.
If $\log_b x = \log_b y$, then $x = y$.

Example 2 Solve for x. Round to the nearest ten-thousandth.

A. $4^x = 7$ B. $3^{2x} = 4$

Solution A.

$$4^x = 7$$
$$\log 4^x = \log 7$$ ▶ Take the common logarithm of each side of the equation.
$$x \log 4 = \log 7$$ ▶ Rewrite using the Properties of Logarithms.
$$x = \frac{\log 7}{\log 4}$$ ▶ Solve for x.
$$= 1.4037$$ ▶ Calculator sequence: 7 [Log] [÷] [(] 4 [log] [)] [=]

The solution is 1.4037.

B.

$$3^{2x} = 4$$

$$\log 3^{2x} = \log 4$$

▶ Take the common logarithm of each side of the equation.

$$2x \log 3 = \log 4$$

▶ Rewrite using the Properties of Logarithms.

$$2x = \frac{\log 4}{\log 3}$$

▶ Solve for x.

$$x = \frac{\log 4}{2 \log 3}$$

$$x = 0.6309$$

The solution is 0.6309.

Problem 2 Solve for x. Round to the nearest ten-thousandth.

A. $4^{3x} = 25$ B. $(1.06)^x = 1.5$

Solution See page A52.

2 Solve logarithmic equations

A logarithmic equation can be solved by using the Properties of Logarithms.

To solve $\log_9 x + \log_9(x - 8) = 1$, use the Logarithm Property of Products to rewrite the left side of the equation.

$$\log_9 x + \log_9(x - 8) = 1$$

$$\log_9 x(x - 8) = 1$$

Write the equation in exponential form.

$$9^1 = x(x - 8)$$

Simplify and solve for x.

$$9 = x^2 - 8x$$

$$0 = x^2 - 8x - 9$$

$$0 = (x - 9)(x + 1)$$

$$x - 9 = 0 \qquad x + 1 = 0$$

$$x = 9 \qquad x = -1$$

When x is replaced by 9 in the original equation, 9 checks as a solution. Replacing x by -1, the original equation contains the expression $\log_9(-1)$. Because the logarithm of a negative number is not a real number, -1 does not check as a solution. Therefore, the solution of the equation is 9.

Some logarithmic equations can be solved by using the One-to-One Property of Logarithms. The use of this property is illustrated in Example 3B.

Example 3 Solve for x.

A. $\log_3(2x - 1) = 2$ B. $\log_2 x - \log_2(x - 1) = \log_2 2$

Solution A. $\log_3(2x - 1) = 2$

$3^2 = 2x - 1$ ▶ Rewrite in exponential form.

$9 = 2x - 1$ ▶ Solve for x.

$10 = 2x$

$5 = x$

The solution is 5.

B. $\log_2 x - \log_2(x - 1) = \log_2 2$

$\log_2\left(\frac{x}{x-1}\right) = \log_2 2$ ▶ Use the Logarithm Property of Quotients.

$\frac{x}{x-1} = 2$ ▶ Use the One-to-One Property of Logarithms.

$(x - 1)\left(\frac{x}{x-1}\right) = (x - 1)2$ ▶ Solve for x.

$x = 2x - 2$

$-x = -2$

$x = 2$

The solution is 2.

Problem 3 Solve for x.

A. $\log_4(x^2 - 3x) = 1$ B. $\log_3 x + \log_3(x + 3) = \log_3 4$

Solution See page A52.

3 Find the logarithm of a number other than base 10

It is possible to find the logarithm of a number to any positive base other than 10 by using a scientific calculator.

For example, to find $\log_2 7$, write an equation. $\log_2 7 = x$

Rewrite in exponential form. $2^x = 7$

Solve the exponential equation using common logarithms.

$$\log 2^x = \log 7$$
$$x \log 2 = \log 7$$

Use a calculator to divide log 7 by log 2. Round to the nearest ten-thousandth.

$$x = \frac{\log 7}{\log 2}$$
$$\log_2 7 = 2.8074$$

Using a procedure similar to the one shown above, a formula can be written that makes it possible to use any given table of logarithms to find the logarithm of a number with a different base.

The Change of Base Formula

$$\log_a N = \frac{\log_b N}{\log_b a}$$

Example 4 Find $\log_3 12$.

Solution $\log_3 12 = \dfrac{\log_{10} 12}{\log_{10} 3}$ ▶ Use the Change of Base Formula. $N = 12$, $a = 3$, $b = 10$.

$= 2.2619$

Problem 4 Find $\log_9 23$.

Solution See page A52.

Example 5 Find ln 5.

Solution Recall that e is the base for natural logarithms. The change of base formula, $\ln 5 = \dfrac{\log_{10} 5}{\log_{10} e}$, could be used to find the natural logarithm. However, all scientific calculators have a natural logarithm key, usually labeled [ln]. To find ln 5, enter 5. Then press the [ln] key. Round to the nearest ten-thousandth. $\ln 5 = 1.6094$

Problem 5 Find ln 3. Round to the nearest ten-thousandth.

Solution See page A52.

EXERCISES 11.4

1 Solve for x. Round to the nearest ten-thousandth.

1. $5^{4x-1} = 5^{x+2}$	**2.** $7^{4x-3} = 7^{2x+1}$	**3.** $8^{x-4} = 8^{5x+8}$
4. $10^{4x-5} = 10^{x+4}$	**5.** $5^x = 6$	**6.** $7^x = 10$
7. $12^x = 6$	**8.** $10^x = 5$	**9.** $\left(\frac{1}{2}\right)^x = 3$
10. $\left(\frac{1}{3}\right)^x = 2$	**11.** $(1.5)^x = 2$	**12.** $(2.7)^x = 3$
13. $10^x = 21$	**14.** $10^x = 37$	**15.** $2^{-x} = 7$
16. $3^{-x} = 14$	**17.** $2^{x-1} = 6$	**18.** $4^{x+1} = 9$
19. $3^{2x-1} = 4$	**20.** $4^{-x+2} = 12$	**21.** $9^x = 3^{x+1}$
22. $2^{x-1} = 4^x$	**23.** $8^{x+2} = 16^x$	**24.** $9^{3x} = 81^{x-4}$

25. $5^{x^2} = 21$
26. $3^{x^2} = 40$
27. $2^{4x-2} = 20$
28. $4^{3x+8} = 12$
29. $3^{-x+2} = 18$
30. $5^{-x+1} = 15$
31. $4^{2x} = 100$
32. $3^{3x} = 1000$
33. $2.5^{-x} = 4$
34. $3.25^{x+1} = 4.2$
35. $0.25^{x} = 0.125$
36. $0.1^{5x} = 10^{-2}$

2 Solve for x.

37. $\log_3(x+1) = 2$
38. $\log_5(x-1) = 1$
39. $\log_2(2x-3) = 3$
40. $\log_4(3x+1) = 2$
41. $\log_2(x^2+2x) = 3$
42. $\log_3(x^2+6x) = 3$
43. $\log_5\left(\frac{2x}{x-1}\right) = 1$
44. $\log_6\frac{3x}{x+1} = 1$
45. $\log_7 x = \log_7(1-x)$
46. $\frac{3}{4}\log x = 3$
47. $\frac{2}{3}\log x = 6$
48. $\log(x-2) - \log x = 3$
49. $\log_2(x-3) + \log_2(x+4) = 3$
50. $\log x - 2 = \log(x-4)$
51. $\log_3 x + \log_3(x-1) = \log_3 6$
52. $\log_4 x + \log_4(x-2) = \log_4 15$
53. $\log_2 8x - \log_2(x^2-1) = \log_2 3$
54. $\log_5 3x - \log_5(x^2-1) = \log_5 2$
55. $\log_9 x + \log_9(2x-3) = \log_9 2$
56. $\log_6 x + \log_6(3x-5) = \log_6 2$
57. $\log_8 6x = \log_8 2 + \log_8(x-4)$
58. $\log_7 5x = \log_7 3 + \log_7(2x+1)$
59. $\log_9 7x = \log_9 2 + \log_9(x^2-2)$
60. $\log_3 x = \log_3 2 + \log_3(x^2-3)$
61. $\log(x^2+3) - \log(x+1) = \log 5$
62. $\log(x+3) + \log(2x-4) = \log 3$

3 Find the logarithm. Round to the nearest ten-thousandth.

63. $\log_3 7$
64. $\log_2 9$
65. $\log_5 12$
66. $\log_5 10$
67. $\ln 4$
68. $\ln 6$
69. $\log_3 15$
70. $\log_{12} 9$
71. $\log_8 6$
72. $\ln 23$
73. $\ln 0.56$
74. $\log_6 28$
75. $\log_3 8.6$
76. $\log_5 9.2$
77. $\log_9 4$
78. $\log_8 5$
79. $\log_3(0.5)$
80. $\log_5(0.6)$
81. $\log_7(1.7)$
82. $\log_6(2.3)$
83. $\log_7(1.4)$

84. $\log_4 121$
85. $\log_7 432$
86. $\log_2 0.27$
87. $\log_5 0.38$
88. $\ln 0.069$
89. $\ln 6800$
90. $\log_6 0.0053$
91. $\log_8 0.0079$
92. $\log_6 72$
93. $\log_7 91$
94. $\log_3 121$
95. $\log_4 832$
96. $\log_2 0.3$
97. $\log_3 0.4$
98. $\log_2 17.3$
99. $\log_5 18.6$
100. $\log_3 1.79$
101. $\log_6 7.84$
102. $\log_9 43$
103. $\log_8 61$
104. $\log_5 44.2$

SUPPLEMENTAL EXERCISES 11.4

Solve for x. Round to the nearest ten-thousandth.

105. $8^{\frac{x}{2}} = 6$
106. $4^{\frac{x}{3}} = 2$
107. $5^{\frac{3x}{2}} = 7$
108. $9^{\frac{2x}{3}} = 8$
109. $1.2^{\frac{x}{2}-1} = 1.4$
110. $5.6^{\frac{x}{3}+1} = 7.8$

Solve the system of equations.

111. $2^x = 8^y$
$x + y = 4$

112. $3^{2y} = 27^x$
$y - x = 2$

113. $\log (x + y) = 3$
$x = y + 4$

114. $\log (x + y) = 3$
$x - y = 20$

115. $8^{3x} = 4^{2y}$
$x - y = 5$

116. $9^{3x} = 81^{3y}$
$x + y = 3$

The following "proof" shows that $0.04 < 0.008$. Find the error.

117.
$$\begin{aligned} 2 &< 3 \\ 2 \log (0.2) &< 3 \log (0.2) \\ \log (0.2)^2 &< \log (0.2)^3 \\ (0.2)^2 &< (0.2)^3 \\ 0.04 &< 0.008 \end{aligned}$$

SECTION 11.5
Applications of Exponential and Logarithmic Functions

1 Application problems

Solving problems that involve exponential or logarithmic equations can be quite tedious without a calculator. Each of the problems in this section was solved by using a calculator.

A biologist places one single-celled bacterium in a culture, and each hour that particular species of bacteria divides into two bacteria. After one hour, there will be two bacteria. After two hours, each of the two bacteria will divide, and there will be four bacteria. After three hours, each of the four bacteria will divide, and there will be eight bacteria.

The table at the right shows the number of bacteria in the culture after various intervals of time, t, in hours. Values in this table could also be found by using the exponential equation $N = 2^t$.

Time, t	Number of bacteria, N
0	1
1	2
2	4
3	8
4	16

The equation $N = 2^t$ is an example of an **exponential growth equation.** In general, any equation that can be written in the form $\boldsymbol{A = A_0 b^{kt}}$, where A is the size at time t, A_0 is the initial size, $b > 1$, and k is a positive real number, is an exponential growth equation. These equations play an important role not only in population growth studies but also in physics, chemistry, psychology, and economics.

Interest is the amount of money paid or received when borrowing or investing money. **Compound interest** is interest that is computed not only on the original principal but also on the interest already earned. The compound interest formula is an exponential equation.

Compound Interest Formula

The compound interest formula is $\boldsymbol{P = A(1 + i)^n}$, where A is the original value of an investment, i is the interest rate per compounding period, n is the total number of compounding periods, and P is the value of the investment after n periods.

An investment broker deposits \$1000 into an account that earns 12% annual interest compounded quarterly. What is the value of the investment after two years? Round to the nearest dollar.

Find i, the interest rate per quarter. The quarterly rate is the annual rate divided by 4, the number of quarters in one year.

$$i = \frac{12\%}{4} = \frac{0.12}{4} = 0.03$$

Find n, the number of compounding periods. The investment is compounded quarterly, 4 times a year, for 2 years.

$$n = 4 \cdot 2 = 8$$

Use the compound interest formula.

$$P = A(1 + i)^n$$

Replace A, i, and n by their values. $P = 1000(1 + 0.03)^8$

Solve for P. $P \approx 1267$

The value of the investment after two years is $1267.

Exponential decay is another important example of an exponential equation. One of the most common illustrations is the decay of a radioactive substance.

An isotope of cobalt has a half-life of approximately 5 years. This means that one half of any given amount of a cobalt isotope will disintegrate in 5 years.

The table at the right indicates the amount of the initial 10 mg of a cobalt isotope that remains after various intervals of time, t, in years. Values in this table could also be found by using the exponential equation $A = 10\left(\frac{1}{2}\right)^{\frac{t}{5}}$.

Time, t	Amount, A
0	10
5	5
10	2.5
15	1.25
20	0.625

The equation $A = 10\left(\frac{1}{2}\right)^{\frac{t}{5}}$ is an example of an **exponential decay equation.** Comparing this equation to the equation on exponential growth, note that for exponential growth, the base of the exponential equation is greater than 1, while for exponential decay, the base is between 0 and 1.

A method by which an archeologist can measure the age of a bone is called **carbon dating.** Carbon dating is based on a radioactive isotope of carbon called carbon 14, which has a half-life of approximately 5570 years. The exponential decay equation is given by $A = A_0\left(\frac{1}{2}\right)^{\frac{t}{5570}}$, where A_0 is the original amount of carbon 14 present in the bone, t is the age of the bone, and A is the amount present after t years.

A bone that originally contained 100 mg of carbon 14 now has 70 mg of carbon 14. What is the appproximate age of the bone? Round to the nearest year.

Use the exponential decay equation. $A = A_0\left(\frac{1}{2}\right)^{\frac{t}{5570}}$

Replace A_0 and A by their given values, and solve for t.

$$70 = 100\left(\frac{1}{2}\right)^{\frac{t}{5570}}$$

$$70 = 100(0.5)^{\frac{t}{5570}}$$

Divide each side of the equation by 100. $0.7 = (0.5)^{\frac{t}{5570}}$

Take the common logarithm of each side of the equation. Then simplify.

$$\log 0.7 = \log (0.5)^{\frac{t}{5570}}$$

$$\log 0.7 = \frac{t}{5570} \log 0.5$$

$$\frac{5570 \log 0.7}{\log 0.5} = t$$

$$2866 = t$$

The age of the bone is approximately 2866 years.

A chemist measures the acidity or alkalinity of a solution by the concentration of hydrogen ions, H^+, in the solution by the formula $\mathbf{pH = -\log (H^+)}$. A neutral solution such as distilled water has a pH of 7, acids have a pH of less than 7, and alkaline solutions (also called basic solutions) have a pH of greater than 7.

Find the pH of vinegar for which $H^+ = 1.26 \times 10^{-3}$. Round to the nearest tenth.

Use the pH equation.
$H^+ = 1.26 \times 10^{-3}$.

$$\begin{aligned} pH &= -\log (H^+) \\ &= -\log (1.26 \times 10^{-3}) \\ &= -(\log 1.26 + \log 10^{-3}) \\ &= -[0.1004 + (-3)] = 2.8996 \end{aligned}$$

The pH of vinegar is 2.9.

The **Richter scale** measures the magnitude, M, of an earthquake in terms of the intensity, I, of its shock waves. This can be expressed as the logarithmic equation $M = \log \frac{I}{I_0}$, where I_0 is a constant.

How many times stronger is an earthquake that has magnitude 4 on the Richter scale than one that has magnitude 2 on the scale?

Let I_1 represent the intensity of the earthquake that has magnitude 4, and let I_2 represent the intensity of the earthquake that has magnitude 2.

$$4 = \log \frac{I_1}{I_0}$$

$$2 = \log \frac{I_2}{I_0}$$

The ratio of I_1 to I_2, $\frac{I_1}{I_2}$, measures how much stronger I_1 is than I_2.

Use the Richter equation to write a system of equations, one equation for magnitude 4 and one for magnitude 2. Then rewrite the system using the Properties of Logarithms.

$$4 = \log I_1 - \log I_0$$
$$2 = \log I_2 - \log I_0$$

Use the addition method to eliminate $\log I_0$.

$$2 = \log I_1 - \log I_2$$

Rewrite the equations using the Properties of Logarithms.

$$2 = \log \frac{I_1}{I_2}$$

Solve for the ratio using the relationship between logarithms and exponents.

$$\frac{I_1}{I_2} = 10^2 = 100$$

An earthquake that has magnitude 4 on the Richter scale is 100 times stronger than an earthquake that has magnitude 2.

The percent of light that will pass through a substance is given by the equation **$\log P = -kd$**, where P is the percent of light passing through the substance, k is a constant that depends on the substance, and d is the thickness of the substance in centimeters.

Find the percent of light that will pass through opaque glass for which $k = 0.4$, and d is 0.5 cm.

Replace k and d in the equation by their given values, and solve for P.

$$\log P = -kd$$
$$\log P = -(0.4)(0.5)$$
$$\log P = -0.2$$

Use the relationship between the logarithmic and exponential function.

$$P = 10^{-0.2}$$
$$P = 0.6310$$

Approximately 63.1% of the light will pass through the glass.

Example 1 An investment of \$3000 is placed into an account that earns 12% annual interest compounded monthly. In approximately how many years will the investment be worth twice the original amount? Round to the nearest whole number.

Strategy To find the time, solve the compound interest formula for n. Use $P = 6000$, $A = 3000$, and $i = \frac{12\%}{12} = \frac{0.12}{12} = 0.01$.

Solution

$$P = A(1 + i)^n$$
$$6000 = 3000(1 + 0.01)^n$$
$$6000 = 3000(1.01)^n$$
$$2 = (1.01)^n$$
$$\log 2 = \log (1.01)^n$$
$$\log 2 = n \log 1.01$$
$$\frac{\log 2}{\log 1.01} = n$$
$$70 \approx n$$

► n is the number of months.

70 months ÷ 12 months ≈ 6

In approximately 6 years, the investment will be worth twice the original amount.

Problem 1 Find the pH of sodium carbonate for which $H^+ = 2.51 \times 10^{-12}$. Round to the nearest tenth.

Solution See page A53.

Example 2 The number of words per minute a student can type will increase with practice and can be approximated by the equation $N = 100[1 - (0.9)^t]$, where N is the number of words typed per minute after t days of instruction. Find the number of words a student will type per minute after 8 days of instruction.

Strategy To find the number of words per minute, replace t by its given value in the equation, and solve for N.

Solution $N = 100[1 - (0.9)^t] = 100[1 - (0.9)^8] = 56.95$

After 8 days, a student will type approximately 57 words per minute.

Problem 2 An earthquake that measures 5 on the Richter scale can cause serious damage to buildings. The San Francisco earthquake of 1906 would have measured about 7.8 on this scale. How many times stronger was the San Francisco earthquake than one that can cause serious damage? Round to the nearest whole number.

Solution See page A53.

EXERCISES 11.5

1 Solve. Round to the nearest whole number or percent.

Use the compound interest formula $P = A(1 + i)^n$, where A is the original value of an investment, i is the interest rate per compounding period, n is the total number of compounding periods, and P is the value of the investment after n periods.

1. An investment club deposits \$5000 into an account that earns 9% annual interest compounded monthly. What is the value of the investment after two years?

2. An investment advisor deposits \$8000 into an account that earns 8% annual interest compounded daily. What is the value of the investment after one year? (1 year = 365 days)

3. An investor deposits \$12,000 into an account that earns 10% annual interest compounded semi-annually. In approximately how many years will the investment double?

4. An insurance broker deposits \$4000 into an account that earns 7% annual interest compounded monthly. In approximately how many years will the investment be worth \$8000?

5. The shop foreman for a company estimates that in 4 years the company will need to purchase a new bottling machine at a cost of \$25,000. How much money must be deposited in an account that earns 10% annual interest compounded monthly so that the value of the account in four years will be \$25,000?

6. The comptroller of a company has determined that it will be necessary to purchase a new computer in three years. The estimated cost of the computer is \$10,000. How much money must be deposited in an account that earns 9% annual interest compounded quarterly so that the value of the account in three years will be \$10,000?

Use the exponential decay equation $A = A_0\left(\frac{1}{2}\right)^{\frac{t}{k}}$, where A is the amount of a radioactive material present after a time t, k is the half-life, and A_0 is the original amount of radioactive material.

7. An isotope of carbon has a half-life of approximately 1600 years. How long will it take an original sample of 15 mg of this isotope to decay to 10 mg?

8. An isotope has a half-life of 80 days. How many days are required for a 10 mg sample of this isotope to decay to 1 mg?

9. A laboratory assistant measures the amount of a radioactive material as 15 mg. Five hours later a second measurement shows that there are 12 mg of the material remaining. Find the half-life of this radioactive material.

10. A scientist measured the amount of a radioactive substance as 10 mg. Twenty-four hours later, another measurement showed that there were 8 mg remaining. What is the half-life of this substance?

The percent of correct welds a student welder can make will increase with practice and can be approximated by the equation $P = 100[1 - (0.75)^t]$, where P is the percent of correct welds and t is the number of weeks of practice.

11. How many weeks of practice are necessary before a student will make 80% of the welds correctly?

12. Find the percent of correct welds a student will make after four weeks of practice.

Use the pH equation $\text{pH} = -\log(\text{H}^+)$, where H^+ is the hydrogen ion concentration of a solution.

13. Find the pH of a sodium hydroxide solution for which the hydrogen ion concentration is 7.5×10^{-9}.

14. Find the pH of a hydrogen chloride solution for which the hydrogen ion concentration is 2.4×10^{-3}.

The percent of light that will pass through a material is given by the equation $\log P = -kd$, where P is the percent of light passing through the material, k is a constant which depends upon the material, and d is the thickness of the material in centimeters.

15. The constant k for a piece of opaque glass that is 0.5 cm thick is 0.2. Find the percent of light that will pass through the glass.

16. The constant k for a piece of tinted glass is 0.5. How thick a piece of this glass is needed so that 32% of the light incident to the glass will pass through it?

Use the Richter equation $M = \log \frac{I}{I_0}$, where M is the magnitude of an earthquake, I is the intensity of its shock waves, and I_0 is a constant.

17. An earthquake in China in 1976 measured 8.2 on the Richter scale and resulted in a death toll of 750,000. The greatest physical damage from an earthquake resulted from the 1988 earthquake in Armenia which measured 6.9 on the Richter scale. How much stronger was China's earthquake?

18. How many times stronger is an earthquake of magnitude 7.2 on the Richter scale than one of magnitude 3.2 on the scale?

The number of decibels, D, of a sound can be given by the equation $D = 10(\log I + 16)$, where I is the power of the sound measured in watts.

19. Find the number of decibels of normal conversation. The power of the sound of normal conversation is approximately 3.2×10^{-10} watts.

20. The loudest sound of any animal is made by the blue whale and can be heard over 500 miles away. The power of the sound made by the blue whale is 630 watts. Find the number of decibels from the sound of the blue whale.

SUPPLEMENTAL EXERCISES 11.5

Use the exponential growth equation $A = A_0 b^{kt}$, where A is the size at time t, A_0 is the initial size, and $b = 2$.

21. At 9 A.M., a culture of bacteria had a population of 1.5×10^6. At noon, the population was 3×10^6. If the population is growing exponentially, at what time will the population be 9×10^6? Round to the nearest hour.

22. The population of a colony of bacteria was 2×10^4 at 10 A.M. At noon, the population had grown to 3×10^4. If the population is growing exponentially, at what time will the population be 8×10^4? Round to the nearest hour.

Use the compound interest formula $P = A(1 + i)^n$.

23. If the average annual rate of inflation is 8%, in how many years will prices double? Round to the nearest whole number.

24. If the average annual rate of inflation is 5%, in how many years will prices double? Round to the nearest whole number.

25. An investment of $1000 earns $177.23 in interest in two years. If the interest is compounded annually, find the annual interest rate. Round to the nearest tenth of a percent.

26. An investment of $1000 earns $242.30 in interest in two years. If the interest is compounded annually, find the annual interest rate. Round to the nearest tenth of a percent.

27. Frequently exponential equations of the form $y = Ab^{kt}$ are rewritten in the form $y = Ae^{mt}$, where the base e is used rather than base b. Rewrite $y = A2^{0.14t}$ in the form $y = Ae^{mt}$.

28. Rewrite $y = A2^{kt}$ in the form $y = Ae^{mt}$. See Exercise 27.

29. The value of an investment in an account that earns an annual interest rate of 10% compounded daily grows according to the equation $A = A_0e^{0.10t}$. Find the time for the investment to double in value. Round to the nearest year.

Calculators and Computers

The [10^x] and [log] Keys on a Calculator

Here is an illustration of the use of the [10^x] and [log] keys.

An investor wishes to double an investment of $4000 in 5 years. What interest rate, compounded quarterly, is necessary to reach this goal?

Use the compound interest equation $P = A(1 + i)^n$, where $P = 8000$, $A = 4000$, and $n = 5 \cdot 4 = 20$.

$$\begin{aligned} P &= A(1+i)^n \\ 8000 &= 4000(1+i)^{20} \\ 2 &= (1+i)^{20} \\ \log 2 &= 20 \log (1+i) \\ \frac{\log 2}{20} &= \log (1+i) \end{aligned}$$

Use your calculator to find $\frac{\log 2}{20}$.

$0.0150515 = \log(1 + i)$	Enter: 2 [log] [÷] 20 [=]
$10^{0.0150515} = 1 + i$	Recall that $\log_{10}a = b$ means $10^b = a$. Use your calculator to find $10^{0.0150515}$.
$1.0352649 = 1 + i$	Enter: 0.0150515 [10ˣ].
$0.0352649 = i$	

The quarterly interest rate is 3.53%.
The annual interest rate is $4 \times 3.53\% = 14.12\%$.

Something Extra

Finding the Proper Dosage

The elimination of a drug from the body is at a rate that is usually proportional to the amount of the drug in the body. This relationship as a function of time can be given by the formula

$$f(t) = Ae^{kt}$$

where $k = -\frac{\ln 2}{H}$, A is the original dosage of the drug, and H is the time for $\frac{1}{2}$ of the drug to be eliminated.

The three graphs show one administration of a drug, repeated dosages in which the amount of drug in the body remains relatively constant, and dosages in which the amount of drug will increase with time.

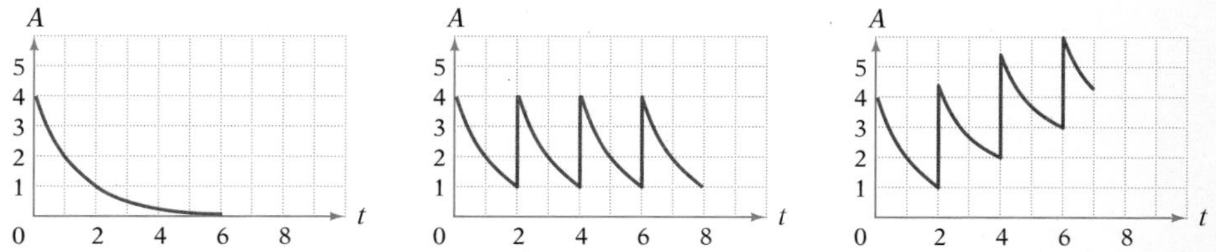

The amount of a drug in the body after n time periods is given by

$$A + Ae^{kt} + Ae^{2kt} + \ldots + Ae^{nkt},$$

where n is the number of dosages and t is the time between doses.

As will be shown in Chapter 12, the amount of a drug in the body at time t can be represented by

$$S = \frac{Ae^{kt}(1 - e^{nkt})}{1 - e^{kt}}.$$

Over a period of time, with continuing dosages at periods of time t, the amount S may increase, decrease, or reach an equilibrium state, as shown in the diagram below.

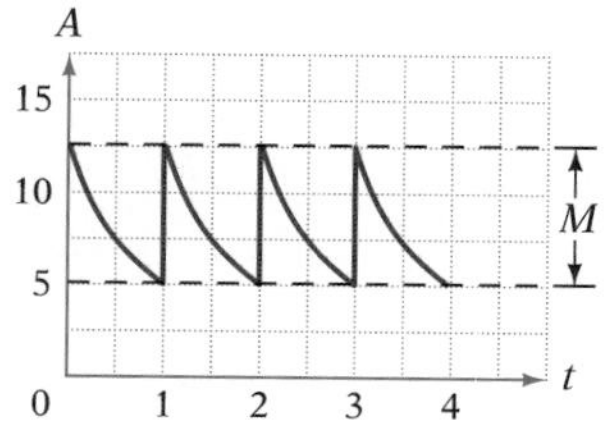

To reach this equilibrium state, the original dosage A is reduced to some maintenance dosage M, given by the expression

$$M = A(1 - e^{nkt}).$$

Solve.

1. Cancer cells have been exposed to x-rays. The number of surviving cells depends on the strength of the x-rays applied. Find the strength of the x-rays (in roentgens) for 40% of the cancer cells to be destroyed. Use the equation $A(r) = A_0e^{-0.2r}$.

2. Sodium pentobarbital is to be given to a patient for surgical anesthesia. The operation is assumed to last one-half hour and the half-life of the sodium pentobarbital is 2 h. The amount of anesthesia in the patient's system should not go below 500 mg. If only one dosage is to be given, find the initial dose of the sodium pentobarbital.

3. One dose of a drug increases the blood level of the drug by 0.5 mg/ml. The half-life of the drug is 8 h and the dose is given every 4 h. Find the concentration of the drug just before the fourth dose.

4. A dosage of 50 mg of medication every 2 h for 8 h will achieve a desired level of medication. (a) Find the level of medication after 8 h if the half-life of the medication is 5 h. (b) Find the maintenance dose after 8 h.

Chapter Summary

Key Words

A function of the form $f(x) = b^x$ is an *exponential function,* where b is a positive real number not equal to 1. The number b is the *base* of the exponential function.

For $b > 0$, $b \neq 1$, $y = \log_b x$ is equivalent to $x = b^y$. $\log_b x$ is the logarithm of x to the base b.

Common logarithms are logarithms to the base 10. (Usually the base 10 is omitted when writing the common logarithm of a number.)

Natural logarithms are logarithms to the base e. The number e is an irrational number approximately equal to 2.718281828. The natural logarithm is abbreviated $\ln x$.

The *mantissa* is the decimal part of a logarithm. The *characteristic* is the integer part of a logarithm.

The *common antilogarithm* of a number, N, is the Nth power of 10, written antilog $N = 10^N$.

An *exponential equation* is one in which the variable occurs in the exponent.

Essential Rules

The Logarithm Property of the Product of Two Numbers
For any positive real numbers x, y, and b, $b \neq 1$, $\log_b xy = \log_b x + \log_b y$.

The Logarithm Property of the Quotient of Two Numbers
For any positive real numbers x, y, and b, $b \neq 1$, $\log_b \frac{x}{y} = \log_b x - \log_b y$.

The Logarithm Property of the Power of a Number
For any positive real numbers x and b, $b \neq 1$, and any real number r, $\log_b x^r = r \log_b x$.

Additional Properties of Logarithms
For any positive real number b, $b \neq 1$, and any real number n, $\log_b b^n = n$.
For any positive real number b, $b \neq 1$, $\log_b 1 = 0$.

One-to-One Property of Logarithms
If $x = y$, then $\log_b x = \log_b y$.
If $\log_b x = \log_b y$, then $x = y$.

Equality of Exponents Property
If $b^x = b^y$, then $x = y$.

The Change of Base Formula
$$\log_a N = \frac{\log_b N}{\log_b a}$$

Chapter Review

1. Evaluate: $\log_4 16$

2. Write $\frac{1}{2}(\log_3 x - \log_3 y)$ as a single logarithm with a coefficient of 1.

3. Find antilog 2.8523.

4. Solve for x: $8^x = 2^{x-6}$

5. Evaluate $f(x) = \left(\frac{2}{3}\right)^x$ at $x = 0$.

6. Solve for x: $\log_3 x = -2$

7. Find log 0.00367.

8. Solve for x: $\log x + \log (x - 4) = \log 12$

9. Write $\log_6 \sqrt{xy^3}$ in expanded form.

10. Solve for x: $4^{5x-2} = 4^{3x+2}$

11. Solve for x: $3^{7x+1} = 3^{4x-5}$

12. Evaluate $f(x) = 3^{x+1}$ at $x = -2$.

13. Evaluate: $\log_2 16$

14. Find log 0.0491.

15. Find $\log_2 5$.

16. Find antilog -1.3936.

17. Solve for x: $4^x = 8^{x-1}$

18. Evaluate $f(x) = \left(\frac{1}{4}\right)^x$ at $x = -1$.

19. Write $\log_5 \sqrt{\frac{x}{y}}$ in expanded form.

20. Find log 86.52.

21. Solve for x: $\log_5 x = 3$

22. Solve for x: $\log x + \log (2x + 3) = \log 2$

23. Write $3 \log_b x - 5 \log_b y$ as a single logarithm with a coefficient of 1.

24. Evaluate $f(x) = 2^{-x-1}$ at $x = -3$.

25. Find $\log_3 19$.

26. Find antilog 1.7048.

27. Graph: $f(x) = 2^x - 3$

28. Graph: $f(x) = \left(\frac{1}{2}\right)^x + 1$

29. Graph: $f(x) = \log_2(2x)$

30. Graph: $f(x) = \log_2 x - 1$

31. Use the exponential decay equation $A = A_0\left(\frac{1}{2}\right)^{\frac{t}{k}}$, where A is the amount of a radioactive material present after time t, k is the half-life, and A_0 is the original amount of radioactive material, to find the half-life of a material that decays from 10 mg to 9 mg in 5 h. Round to the nearest whole number.

32. The percent of light that will pass through an opaque material is given by the equation $\log P = -0.5d$, where P is the percent of light that passes through the material, and d is the thickness of the material in centimeters. How thick must this opaque material be so that only 50% of the light that is incident to the material will pass through it? Round to the nearest thousandth.

Chapter Test

1. Evaluate $f(x) = \left(\frac{3}{4}\right)^x$ at $x = 0$.

2. Evaluate $f(x) = 4^{x-1}$ at $x = -2$.

3. Evaluate: $\log_4 64$

4. Solve for x: $\log_4 x = -2$

5. Write $\log_6 \sqrt[3]{x^2y^5}$ in expanded form.

6. Write $\frac{1}{2}(\log_5 x - \log_5 y)$ as a single logarithm with a coefficient of 1.

7. Find log 0.0421.

8. Find antilog -0.8794.

9. Find antilog 3.1277.

10. Find $\log_5 24$.

11. Solve for x: $\log_2 x + 3 = \log_2(x^2 - 20)$

12. Solve for x: $5^{6x-2} = 5^{3x+7}$

13. Solve for x: $4^x = 2^{3x+4}$

14. Solve for x: $\log(2x + 1) + \log x = \log 6$

15. Graph $f(x) = 2^x - 1$.

16. Graph $f(x) = 2^x + 2$.

17. Graph $f(x) = \log_2(3x)$.

18. Graph $f(x) = \log_3(x + 1)$.

19. Find the value of \$10,000 invested for 6 years at 7.5% compounded monthly. Use the compound interest formula $P = A(1 + i)^n$, where A is the original value of the investment, i is the interest rate per compounding period, and n is the number of compounding periods. Round to the nearest dollar.

20. Use the exponential decay equation $A = A_0\left(\frac{1}{2}\right)^{\frac{t}{k}}$, where A is the amount of a radioactive material present after time t, k is the half-life, and A_0 is the original amount of radioactive material, to find the half-life of a material that decays from 40 mg to 30 mg in 10 h. Round to the nearest whole number.

Cumulative Review

1. Solve: $4 - 2[x - 3(2 - 3x) - 4x] = 2x$

2. Solve $S = 2WH + 2WL + 2LH$ for L.

3. Solve: $|2x - 5| \leq 3$

4. Factor: $4x^{2n} + 7x^n + 3$

5. Solve: $x^2 + 4x - 5 \leq 0$

6. Simplify: $\dfrac{1 - \frac{5}{x} + \frac{6}{x^2}}{1 + \frac{1}{x} - \frac{6}{x^2}}$

7. Simplify: $\dfrac{\sqrt{xy}}{\sqrt{x} - \sqrt{y}}$

8. Simplify: $y\sqrt{18x^5y^4} - x\sqrt{98x^3y^6}$

9. Simplify: $\dfrac{i}{2 - i}$

10. Find the equation of the line containing the point $(2, -2)$ and parallel to the line $2x - y = 5$.

11. Write a quadratic equation that has integer coefficients and has as solutions $\frac{1}{3}$ and -3.

12. Solve: $x^2 - 4x - 6 = 0$

13. Find the range of $f(x) = x^2 - 3x - 4$ if the domain is $\{-1, 0, 1, 2, 3\}$.

14. Given $f(x) = x^2 + 2x + 1$ and $g(x) = 2x - 3$, find $f(g(0))$.

15. Solve by the addition method:

$$\begin{aligned} 3x - y + z &= 3 \\ x + y + 4z &= 7 \\ 3x - 2y + 3z &= 8 \end{aligned}$$

16. Solve: $y = x^2 + x - 3$
$y = 2x - 1$

17. Evaluate $f(x) = 3^{-x+1}$ at $x = -4$.

18. Solve for x: $\log_4 x = 3$

19. Solve for x: $2^{3x+2} = 4^{x+5}$

20. Solve for x: $\log x + \log (3x + 2) = \log 5$

21. Graph the set $\{x|x < 0\} \cap \{x|x > -4\}$.

22. Graph the solution set of $\dfrac{x + 2}{x - 1} \geq 0$.

23. Graph $y = -x^2 - 2x + 3$.

24. Graph $\dfrac{x^2}{4} - \dfrac{y^2}{16} = 1$.

25. Graph $f(x) = \left(\dfrac{1}{2}\right)^x - 1$.

26. Graph $f(x) = \log_2 x + 1$.

27. An alloy containing 25% tin is mixed with an alloy containing 50% tin. How much of each were used to make 2000 lb of an alloy containing 40% tin?

28. An account executive earns $500 per month plus 8% commission on the amount of sales. The executive's goal is to earn a minimum of $3000 per month. What amount of sales will enable the executive to earn $3000 or more per month?

29. A new printer can print checks three times faster than an older printer. The old printer can print the checks in 30 min. How long would it take to print the checks when both printers are operating?

30. For a constant temperature, the pressure (P) of a gas varies inversely as the volume (V). If the pressure is 50 lb/in.^{2} when the volume is 250 ft^3 find the pressure when the volume is 25 ft^3.

31. A contractor buys 45 yd of nylon carpet and 30 yd of wool carpet for $1170. A second purchase, at the same prices, includes 25 yd of nylon carpet and 80 yd of wool carpet for $1410. Find the cost per yard of the wool carpet.

32. An investor deposits $10,000 into an account that earns 9% interest compounded monthly. What is the value of the investment after 5 years? Use the compound interest formula $P = A(1 + i)^n$, where A is the original value of an investment, i is the interest rate per compounding period, n is the total number of compounding periods, and P is the value of the investment after n periods. Round to the nearest dollar.

12

Sequences and Series

Objectives

- Write the terms of a sequence
- Find the sum of a series
- Find the nth term of an arithmetic sequence
- Find the sum of an arithmetic series
- Application problems
- Find the nth term of a geometric sequence
- Finite geometric series
- Infinite geometric series
- Application problems
- Expand $(a + b)^n$

POINT OF INTEREST

Tower of Hanoi

The Tower of Hanoi is a puzzle that has the following form.

Three pegs are placed in a board. A number of disks, graded in size, are stacked on one of the pegs with the largest disk on the bottom and the succeeding smaller disks placed on top.

The disks are moved according to the following rules:

1) Only one disk at a time may be moved.
2) A larger disk cannot be placed over a smaller disk.

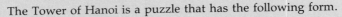

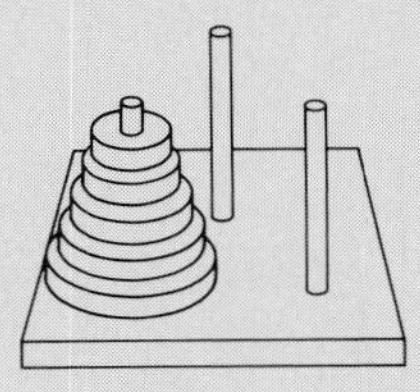

The object of the puzzle is to transfer all the disks from one peg to one of the other two pegs. If initially there is only one disk, then only one move would be required. With two disks initially, three moves are required, and with three disks, seven moves are required.

You can try this puzzle using playing cards. Select a number of playing cards, in sequence, from a deck. The number of cards corresponds to the number of disks and the number on the card corresponds to the size of the disk. For example, an "ace" would correspond to the smallest disk, a "two" would correspond to the next largest disk and so on for five cards. Now place three coins on a table to be used as the pegs. Pile the cards on one of the coins (in order) and try to move the pile to a second coin. Now a numerically larger card cannot be placed on a numerically smaller card. Try this for 5 cards. The minimum number of moves is 31.

Below is a chart that shows the minimum number of moves required for an initial number of disks. The difference between the number of moves for each succeeding disk is also given.

1		2		3		4		5		6		7		8
1		3		7		15		31		63		127		255
	2		4		8		16		32		64		128	

For this last list of numbers, each succeeding number can be found by multiplying the preceding number by a constant (in this case 2). Such a list of numbers is called a *geometric sequence.*

The formula for the minimum number of moves is given by $M = 2^n - 1$, where M is the number of moves and n is the number of disks. This equation is an exponential equation.

Here's a hint for solving the Tower of Hanoi puzzle with 5 disks. First solve the puzzle for three disks. Now solve the puzzle for 4 disks by first moving the top three disks to one peg. (You just did this when you solved the puzzle for three disks.) Now move the fourth disk to a peg and then again use your solution for 3 disks to move the disks back to the peg with the fourth disk. Now use this solution for 4 disks to solve the 5-disk problem.

SECTION **12.1**

Introduction to Sequences and Series

1 Write the terms of a sequence

An investor deposits \$100 in an account that earns 10% interest compounded annually. The amount of interest earned each year can be determined by using the compound interest formula used in Chapter 11. The amount of interest earned in each of the first four years of the investment is shown below.

Year	1	2	3	4
Interest Earned	\$10	\$11	\$12.10	\$13.31

The list of numbers 10, 11, 12.10, 13.31 is called a sequence. A **sequence** is an ordered list of numbers. The list 10, 11, 12.10, 13.31 is ordered because the position of a number in this list indicates the year in which that amount of interest was earned. Each of the numbers of a sequence is called a **term** of the sequence.

Examples of other sequences are shown at the right. These sequences are separated into two groups. A **finite sequence** contains a finite number of terms. An **infinite sequence** contains an infinite number of terms.

Finite Sequences

1, 1, 2, 3, 5, 8
1, 2, 3, 4, 5, 6, 7, 8
1, −1, 1, −1

Infinite Sequences

1, 3, 5, 7, . . .
$1, \frac{1}{2}, \frac{1}{4}, \frac{1}{8}, \ldots$
1, 1, 2, 3, 5, 8, . . .

For the sequence at the right, the first term is 2, the second term is 4, the third term is 6, and the fourth term is 8.

2, 4, 6, 8, . . .

A general sequence is shown at the right. The first term is a_1, the second term is a_2, the third term is a_3, and the nth term, also called the **general term** of the sequence, is a_n. Note that each term of the sequence is paired with a natural number.

$a_1, a_2, a_3, \ldots a_n, \ldots$

Frequently, a sequence has a definite pattern that can be expressed by a formula.

Each term of the sequence shown at the right is paired with a natural number by the formula $a_n = 3n$. The first term, a_1, is 3. The second term, a_2, is 6. The third term, a_3, is 9. The nth term, a_n, is $3n$.

$a_n = 3n$

$$\begin{array}{lllll} a_1, & a_2, & a_3, \ldots & a_n, \ldots \\ 3(1), & 3(2), & 3(3), \ldots & 3(n), \ldots \\ 3, & 6, & 9, \ldots & 3n, \ldots \end{array}$$

Example 1 Write the first three terms of the sequence whose nth term is given by the formula $a_n = 2n - 1$.

Solution
$a_n = 2n - 1$
$a_1 = 2(1) - 1 = 1$ ▶ Replace n by 1.
$a_2 = 2(2) - 1 = 3$ ▶ Replace n by 2.
$a_3 = 2(3) - 1 = 5$ ▶ Replace n by 3.

The first term is 1, the second term is 3, and the third term is 5.

Problem 1 Write the first four terms of the sequence whose nth term is given by the formula $a_n = n(n + 1)$.

Solution See page A53.

Example 2 Find the eighth and tenth terms of the sequence whose nth term is given by the formula $a_n = \frac{n}{n+1}$.

Solution
$a_n = \frac{n}{n+1}$

$a_8 = \frac{8}{8+1} = \frac{8}{9}$ ▶ Replace n by 8.

$a_{10} = \frac{10}{10+1} = \frac{10}{11}$ ▶ Replace n by 10.

The eighth term is $\frac{8}{9}$, and the tenth term is $\frac{10}{11}$.

Problem 2 Find the sixth and ninth terms of the sequence whose nth term is given by the formula $a_n = \frac{1}{n(n+2)}$.

Solution See page A54.

2 Find the sum of a series

12

On page 581, the sequence 10, 11, 12.10, 13.31 was shown to represent the amount of interest earned in each of 4 years of an investment.

10, 11, 12.10, 13.31

The sum of the terms of this sequence represents the total interest earned by the investment over the four-year period.

$10 + 11 + 12.10 + 13.31 = 46.41$

The total interest earned over the four-year period is \$46.41.

The indicated sum of the terms of a sequence is called a **series.** Given the sequence 10, 11, 12.10, 13.31, the series $10 + 11 + 12.10 + 13.31$ can be written.

S_n is used to indicate the sum of the first n terms of a sequence.

For the preceding example, the sums of the series S_1, S_2, S_3, and S_4 represent the total interest earned for 1, 2, 3, and 4 years, respectively.

$$S_1 = 10 = 10$$
$$S_2 = 10 + 11 = 21$$
$$S_3 = 10 + 11 + 12.10 = 33.10$$
$$S_4 = 10 + 11 + 12.10 + 13.31 = 46.41$$

For the general sequence $a_1, a_2, a_3, \ldots a_n$, the series S_1, S_2, S_3, and S_n are shown at the right.

$$S_1 = a_1$$
$$S_2 = a_1 + a_2$$
$$S_3 = a_1 + a_2 + a_3$$
$$S_n = a_1 + a_2 + a_3 + \cdots + a_n$$

It is convenient to represent a series in a compact form called **summation notation,** or **sigma notation.** The Greek letter sigma, Σ, is used to indicate a sum.

The first four terms of the sequence whose nth term is given by the formula $a_n = 2n$ are 2, 4, 6, 8. The corresponding series is shown at the right written in summation notation and is read "the summation from 1 to 4 of $2n$." The letter n is called the **index** of the summation.

$$\sum_{n=1}^{4} 2n$$

To write the terms of the series, replace n by the consecutive integers from 1 to 4.

$$\sum_{n=1}^{4} 2n = 2(1) + 2(2) + 2(3) + 2(4)$$

The series is $2 + 4 + 6 + 8$.

$$= 2 + 4 + 6 + 8$$

The sum of the series is 20.

$$= 20$$

Example 3 Find the sum of the series.

A. $\sum_{i=1}^{3} (2i - 1)$ B. $\sum_{n=3}^{6} \frac{1}{2}n$

Solution A. $\sum_{i=1}^{3} (2i - 1) =$

$[2(1) - 1] + [2(2) - 1] + [2(3) - 1] =$ ▶ Replace i by 1, 2, and 3.

$1 + 3 + 5 =$ ▶ Write the series.

9 ▶ Find the sum of the series.

B. $\sum_{n=3}^{6} \frac{1}{2}n =$

$\frac{1}{2}(3) + \frac{1}{2}(4) + \frac{1}{2}(5) + \frac{1}{2}(6) =$ ▶ Replace n by 3, 4, 5, and 6.

$\frac{3}{2} + 2 + \frac{5}{2} + 3 =$ ▶ Write the series.

9 ▶ Find the sum of the series.

Problem 3 Find the sum of the series.

A. $\sum_{n=1}^{4}(7-n)$ B. $\sum_{i=3}^{6}(i^2-2)$

Solution See page A54.

Example 4 Write $\sum_{i=1}^{5} x^i$ in expanded form.

Solution $\sum_{i=1}^{5} x^i =$ ▶ This is a variable series.

$x + x^2 + x^3 + x^4 + x^5$ ▶ Replace i by 1, 2, 3, 4, and 5.

Problem 4 Write $\sum_{n=1}^{5} nx$ in expanded form.

Solution See page A54.

EXERCISES 12.1

1 Write the first four terms of the sequence whose nth term is given by the formula.

1. $a_n = n + 1$
2. $a_n = n - 1$
3. $a_n = 2n + 1$
4. $a_n = 3n - 1$
5. $a_n = 2 - 2n$
6. $a_n = 1 - 2n$
7. $a_n = 2^n$
8. $a_n = 3^n$
9. $a_n = n^2 + 1$
10. $a_n = n^2 - 1$
11. $a_n = \dfrac{n}{n^2 + 1}$
12. $a_n = \dfrac{n^2 - 1}{n}$
13. $a_n = n - \dfrac{1}{n}$
14. $a_n = n^2 - \dfrac{1}{n}$
15. $a_n = (-1)^{n+1}n$

16. $a_n = \frac{(-1)^{n+1}}{n+1}$

17. $a_n = \frac{(-1)^{n+1}}{n^2+1}$

18. $a_n = (-1)^n(n^2 + 2n + 1)$

19. $a_n = (-1)^n 2^n$

20. $a_n = \frac{1}{3}n^3 + 1$

21. $a_n = 2\left(\frac{1}{3}\right)^{n+1}$

Find the indicated term of the sequence whose nth term is given by the formula.

22. $a_n = 3n + 4; a_{12}$

23. $a_n = 2n - 5; a_{10}$

24. $a_n = n(n - 1); a_{11}$

25. $a_n = \frac{n}{n+1}; a_{12}$

26. $a_n = (-1)^{n-1}n^2; a_{15}$

27. $a_n = (-1)^{n-1}(n - 1); a_{25}$

28. $a_n = \left(\frac{1}{2}\right)^n; a_8$

29. $a_n = \left(\frac{2}{3}\right)^n; a_5$

30. $a_n = (n + 2)(n + 3); a_{17}$

31. $a_n = (n + 4)(n + 1); a_7$

32. $a_n = \frac{(-1)^{2n-1}}{n^2}; a_6$

33. $a_n = \frac{(-1)^{2n}}{n+4}; a_{16}$

34. $a_n = \frac{3}{2}n^2 - 2; a_8$

35. $a_n = \frac{1}{3}n + n^2; a_6$

2 Find the sum of the series.

36. $\sum_{n=1}^{5} (2n + 3)$

37. $\sum_{i=1}^{7} (i + 2)$

38. $\sum_{i=1}^{4} 2i$

39. $\sum_{n=1}^{7} n$

40. $\sum_{i=1}^{6} i^2$

41. $\sum_{i=1}^{5} (i^2 + 1)$

42. $\sum_{n=1}^{6} (-1)^n$

43. $\sum_{n=1}^{4} \frac{1}{2n}$

44. $\sum_{i=3}^{6} i^3$

45. $\sum_{n=2}^{4} 2^n$

46. $\sum_{n=3}^{7} \frac{n}{n-1}$

47. $\sum_{i=3}^{6} \frac{i+1}{i}$

48. $\sum_{i=1}^{4} \frac{1}{2^i}$

49. $\sum_{i=1}^{5} \frac{1}{2i}$

50. $\sum_{n=1}^{4} (-1)^{n-1}n^2$

51. $\sum_{i=1}^{4} (-1)^{i-1}(i + 1)$

52. $\sum_{n=3}^{5} \frac{(-1)^{n-1}}{n-2}$

53. $\sum_{n=4}^{7} \frac{(-1)^{n-1}}{n-3}$

Write the series in expanded form.

54. $\sum_{n=1}^{5} 2x^n$

55. $\sum_{n=1}^{4} \frac{2n}{x}$

56. $\sum_{i=1}^{5} \frac{x^i}{i}$

57. $\sum_{i=1}^{4} \frac{x^i}{i+1}$

58. $\sum_{i=3}^{5} \frac{x^i}{2i}$

59. $\sum_{i=2}^{4} \frac{x^i}{2i-1}$

60. $\sum_{n=1}^{5} x^{2n}$

61. $\sum_{n=1}^{4} x^{2n-1}$

62. $\sum_{i=1}^{4} \frac{x^i}{i^2}$

63. $\sum_{n=1}^{4} (2n)x^n$

64. $\sum_{n=1}^{4} nx^{n-1}$

65. $\sum_{i=1}^{5} x^{-i}$

SUPPLEMENTAL EXERCISES 12.1

Write a formula for the nth term of the sequence.

66. The sequence of the natural numbers

67. The sequence of the odd natural numbers

68. The sequence of the negative even integers

69. The sequence of the negative odd integers

70. The sequence of the positive multiples of 7

71. The sequence of the positive integers that are divisible by 4

Find the sum of the series. Write your answer as a single logarithm.

72. $\sum_{n=1}^{5} \log n$

73. $\sum_{i=1}^{4} \log 2i$

Solve.

74. The first 22 numbers in the sequence 4, 44, 444, 4444, . . . are added together. What digit is in the thousands' place of the sum?

75. The first 31 numbers in the sequence 6, 66, 666, 6666, . . . are added together. What digit is in the hundreds' place of the sum?

A recursive sequence is one for which each term of the sequence is defined by using preceding terms. Find the first three terms of each recursively defined sequences.

76. $a_1 = 1,\ a_n = na_{n-1},\ n \geq 2$

77. $a_1 = 1,\ a_2 = 1,\ a_n = a_{n-1} + a_{n-2},\ n \geq 3$

SECTION 12.2 Arithmetic Sequences and Series

1 To find the *n*th term of an arithmetic sequence

A company's expenses for training a new employee are quite high. To encourage employees to continue their employment with the company, a company that has a six-month training program offers a starting salary of \$900 a month and then a \$100-per-month pay increase each month during the training period.

The sequence below shows the employee's monthly salaries during the training period. Each term of the sequence is found by adding \$100 to the previous term.

Month	1	2	3	4	5	6
Salary	900	1000	1100	1200	1300	1400

The sequence 900, 1000, 1100, 1200, 1300, 1400 is called an arithmetic sequence.

Definition of an Arithmetic Sequence

An **arithmetic sequence,** or **arithmetic progression,** is one in which the difference between any two consecutive terms is constant. The difference between consecutive terms is called the **common difference** of the sequence.

Each of the sequences shown below is an arithmetic sequence. To find the common difference of an arithmetic sequence, subtract the first term from the second term.

$2, 7, 12, 17, 22, \ldots$ Common difference: 5

$3, 1, -1, -3, -5, \ldots$ Common difference: -2

$1, \frac{3}{2}, 2, \frac{5}{2}, 3, \frac{7}{2}, \ldots$ Common difference: $\frac{1}{2}$

Consider an arithmetic sequence in which the first term is a_1 and the common difference is d. By adding the common difference to each successive term of the arithmetic sequence, a formula for the nth term can be found.

The first term is a_1. — $a_1 = a_1$

To find the second term, add the common difference d to the first term. — $a_2 = a_1 + d$

To find the third term, add the common difference d to the second term.

$$a_3 = a_2 + d = (a_1 + d) + d$$
$$a_3 = a_1 + 2d$$

To find the fourth term, add the common difference d to the third term.

$$a_4 = a_3 + d = (a_1 + 2d) + d$$
$$a_4 = a_1 + 3d$$

Note the relationship between the term number and the number that multiplies d. The multiplier of d is one less than the term number.

$$a_n = a_1 + (n - 1)d$$

The Formula for the *n*th Term of an Arithmetic Sequence

> The nth term of an arithmetic sequence with a common difference of d is given by $a_n = a_1 + (n - 1)d$.

Example 1 Find the 27th term of the arithmetic sequence $-4, -1, 2, 5, 8, \ldots$.

Solution

$d = a_2 - a_1 = -1 - (-4) = 3$ ▶ Find the common difference.

$$a_n = a_1 + (n - 1)d$$
$$a_{27} = -4 + (27 - 1)3$$
$$= -4 + (26)3 = -4 + 78$$
$$= 74$$

▶ Use the Formula for the nth Term of an Arithmetic Sequence to find the 27th term. $n = 27$, $a_1 = -4$, $d = 3$

Problem 1 Find the 15th term of the arithmetic sequence $9, 3, -3, -9, \ldots$.

Solution See page A54.

Example 2 Find the formula for the nth term of the arithmetic sequence $-5, -2, 1, 4, \ldots$.

Solution

$d = a_2 - a_1 = -2 - (-5) = 3$ ▶ Find the common difference.

$$a_n = a_1 + (n - 1)d$$
$$a_n = -5 + (n - 1)3$$
$$a_n = -5 + 3n - 3$$
$$a_n = 3n - 8$$

▶ Use the Formula for the nth Term of an Arithmetic Sequence. $a_1 = -5$, $d = 3$

Problem 2 Find the formula for the nth term of the arithmetic sequence $-3, 1, 5, 9, \ldots$.

Solution See page A54.

Example 3 Find the number of terms in the finite arithmetic sequence 7, 10, 13, . . . , 55.

Solution

$d = a_2 - a_1 = 10 - 7 = 3$ ▶ Find the common difference.

$a_n = a_1 + (n - 1)d$ ▶ Use the Formula for the nth Term of an Arithmetic Sequence. $a_n = 55$, $a_1 = 7$, $d = 3$

$55 = 7 + (n - 1)3$ ▶ Solve for n.

$55 = 7 + 3n - 3$

$55 = 3n + 4$

$51 = 3n$

$17 = n$

There are 17 terms in the sequence.

Problem 3 Find the number of terms in the finite arithmetic sequence 7, 9, 11, . . . , 59.

Solution See page A54.

2 Find the sum of an arithmetic series

The indicated sum of the terms of an arithmetic sequence is called an **arithmetic series.** The sum of an arithmetic series can be found by using a formula.

The Formula for the Sum of n Terms of an Arithmetic Series

Let a_1 be the first term of a finite arithmetic sequence, n the number of terms, and a_n the last term of the sequence. Then the sum of the series S_n is given by $S_n = \frac{n}{2}(a_1 + a_n)$.

Each term of the arithmetic sequence shown at the right was found by adding 3 to the previous term.

2, 5, 8, . . . , 17, 20

Each term of the reverse arithmetic sequence can be found by subtracting 3 from the previous term.

20, 17, 14, . . . , 5, 2

This idea is used in the following proof of the Formula for the Sum of n Terms of an Arithmetic Series:

Let S_n represent the sum of the sequence

$$S_n = a_1 + (a_1 + d) + (a_1 + 2d) + \ldots + a_n$$

Write the terms of the sum of the sequence in reverse order. The sum will be the same.

$$S_n = a_n + (a_n - d) + (a_n - 2d) + \ldots + a_1$$

Add the two equations.

$$2S_n = (a_1 + a_n) + (a_1 + a_n) + (a_1 + a_n) + \dots + (a_1 + a_n)$$

Simplify the right side of the equation by using the fact that there are n terms in the sequence.

$$2S_n = n(a_1 + a_n)$$

Solve for S_n.

$$S_n = \frac{n}{2}(a_1 + a_n)$$

Example 4 Find the sum of the first 10 terms of the arithmetic sequence 2, 4, 6, 8, ...

Solution

$d = a_2 - a_1 = 4 - 2 = 2$ ▶ Find the common difference.

$a_n = a_1 + (n-1)d$
$a_{10} = 2 + (10-1)2 = 2 + (9)2$
$= 2 + 18 = 20$

▶ Use the Formula for the nth Term of an Arithmetic Sequence to find the 10th term.

$S_n = \frac{n}{2}(a_1 + a_n)$

▶ Use the Formula for the Sum of n Terms of an Arithmetic Series. $n = 10$, $a_1 = 2$, $a_n = 20$

$S_{10} = \frac{10}{2}(2 + 20) = 5(22) = 110$

Problem 4 Find the sum of the first 25 terms of the arithmetic sequence $-4, -2, 0, 2, 4, \dots$.

Solution See page A54.

Example 5 Find the sum of the arithmetic series $\sum_{n=1}^{25} (3n + 1)$.

Solution

$a_n = 3n + 1$
$a_1 = 3(1) + 1 = 4$ ▶ Find the first term.
$a_{25} = 3(25) + 1 = 76$ ▶ Find the 25th term.

$S_n = \frac{n}{2}(a_1 + a_n)$

▶ Use the Formula for the Sum of n Terms of an Arithmetic Series. $n = 25$, $a_1 = 4$, $a_n = 76$

$S_{25} = \frac{25}{2}(4 + 76) = \frac{25}{2}(80)$
$= 1000$

Problem 5 Find the sum of the arithmetic series $\sum_{n=1}^{18} (3n - 2)$.

Solution See page A55.

12

3 Application problems

Example 6 The distance a ball rolls down a ramp each second is given by an arithmetic sequence. The distance in feet traveled by the ball during the nth second is given by $2n - 1$. Find the distance the ball will travel during the 10th second.

Strategy To find the distance:
- Find the first and second terms of the sequence.
- Find the common difference of the arithmetic sequence.
- Use the Formula for the nth Term of an Arithmetic Sequence to find the 10th term.

Solution
$a_n = 2n - 1$
$a_1 = 2(1) - 1 = 1$
$a_2 = 2(2) - 1 = 3$

$d = a_2 - a_1 = 3 - 1 = 2$
$a_n = a_1 + (n - 1)d$
$a_{10} = 1 + (10 - 1)2 = 1 + (9)2 = 19$

The ball will travel 19 ft during the 10th second.

Problem 6 A contest offers 20 prizes. The first prize is \$10,000, and each successive prize is \$300 less than the preceding prize. What is the value of the 20th-place prize? What is the total amount of prize money that is being awarded?

Solution See page A55.

EXERCISES 12.2

1 Find the indicated term of the arithmetic sequence.

1. $1, 11, 21, \ldots; a_{15}$
2. $3, 8, 13, \ldots; a_{20}$
3. $-6, -2, 2, \ldots; a_{15}$
4. $-7, -2, 3, \ldots; a_{14}$
5. $3, 7, 11, \ldots; a_{18}$
6. $-13, -6, 1, \ldots; a_{31}$
7. $-\frac{3}{4}, 0, \frac{3}{4}, \ldots; a_{11}$
8. $\frac{3}{8}, 1, \frac{13}{8}, \ldots; a_{17}$
9. $2, \frac{5}{2}, 3, \ldots; a_{31}$
10. $1, \frac{5}{4}, \frac{3}{2}, \ldots; a_{17}$
11. $6, 5.75, 5.50, \ldots; a_{10}$
12. $4, 3.7, 3.4, \ldots; a_{12}$

Find the formula for the nth term of the arithmetic sequence.

13. $1, 2, 3, \ldots$
14. $1, 4, 7, \ldots$
15. $6, 2, -2, \ldots$

16. $3, 0, -3, \ldots$

17. $2, \frac{7}{2}, 5, \ldots$

18. $7, 4.5, 2, \ldots$

19. $-8, -13, -18, \ldots$

20. $17, 30, 43, \ldots$

21. $26, 16, 6, \ldots$

Find the number of terms in the finite arithmetic sequence.

22. $-2, 1, 4, \ldots, 73$

23. $7, 11, 15, \ldots, 171$

24. $-\frac{1}{2}, \frac{3}{2}, \frac{7}{2}, \ldots, \frac{71}{2}$

25. $\frac{1}{3}, \frac{5}{3}, 3, \ldots, \frac{61}{3}$

26. $1, 5, 9, \ldots, 81$

27. $3, 8, 13, \ldots, 98$

28. $2, 0, -2, \ldots, -56$

29. $1, -3, -7, \ldots, -75$

30. $\frac{5}{2}, 3, \frac{7}{2}, \ldots, 13$

31. $\frac{7}{3}, \frac{13}{3}, \frac{19}{3}, \ldots, \frac{79}{3}$

32. $1, 0.75, 0.50, \ldots, -4$

33. $3.5, 2, 0.5, \ldots, -25$

2 Find the sum of the indicated number of terms of the arithmetic sequence.

34. $1, 3, 5, \ldots; n = 50$

35. $2, 4, 6, \ldots; n = 25$

36. $20, 18, 16, \ldots; n = 40$

37. $25, 20, 15, \ldots; n = 22$

38. $\frac{1}{2}, 1, \frac{3}{2}, \ldots; n = 27$

39. $2, \frac{11}{4}, \frac{7}{2}, \ldots; n = 10$

Find the sum of the arithmetic series.

40. $\sum_{i=1}^{15} (3i - 1)$

41. $\sum_{i=1}^{15} (3i + 4)$

42. $\sum_{n=1}^{17} \left(\frac{1}{2}n + 1\right)$

43. $\sum_{n=1}^{10} (1 - 4n)$

44. $\sum_{i=1}^{15} (4 - 2i)$

45. $\sum_{n=1}^{10} (5 - n)$

3 Solve.

46. The distance that an object dropped from a cliff will fall is 16 ft the first second, 48 ft the next second, 80 ft the third second, and so on in an arithmetic sequence. What is the total distance the object will fall in 6 s?

47. An exercise program calls for walking 12 min each day for a week. Each week thereafter, the amount of time spent walking increases by 6 min per day. In how many weeks will a person be walking 60 min each day?

48. A display of cans in a grocery store consists of 20 cans in the bottom row, 18 cans in the next row, and so on in an arithmetic sequence. The top row has 4 cans. Find the total number of cans in the display.

49. A theater in the round has 52 seats in the first row, 58 seats in the second row, 64 seats in the third row, and so on in an arithmetic sequence. Find the total number of seats in the theater if there are 20 rows of seats.

50. The loge seating section in a concert hall consists of 26 rows of chairs. There are 65 seats in the first row, 71 seats in the second row, 77 seats in the third row, and so on in an arithmetic sequence. How many seats are in the loge seating section?

51. The salary schedule for an engineering assistant is \$800 for the first month and a \$45-per-month salary increase for the next nine months. Find the monthly salary during the ninth month. Find the total salary for the nine-month period.

SUPPLEMENTAL EXERCISES 12.2

Solve.

52. Find the sum of the first 50 positive integers.

53. Find the sum of the first 100 natural numbers.

54. How many terms of the arithmetic sequence $-6, -2, 2, \ldots$ must be added together for the sum of the series to be 90?

55. How many terms of the arithmetic sequence $-3, 2, 7, \ldots$ must be added together for the sum of the series to be 116?

56. Given $a_1 = -9$, $a_n = 21$, and $S_n = 36$, find d and n.

57. Given $a_1 = -5$, $a_n = 19$, and $S_n = 49$, find d and n.

58. The third term of an arithmetic sequence is 4, and the eighth term is 30. Find the first term.

59. The fourth term of an arithmetic sequence is 9, and the ninth term is 29. Find the first term.

60. Show that $f(n) = mn + b$, n a natural number, is an arithmetic sequence.

61. The sum of the interior angles of a triangle is 180°. The sum is 360° for a quadrilateral and 540° for a pentagon. Assuming this pattern continues, find the sum of the angles of a dodecagon (12-sided figure). Find a formula for the sum of the angles of an n-sided polygon.

SECTION 12.3
Geometric Sequences and Series

1 Find the *n*th term of a geometric sequence

An ore sample contains 20 mg of a radioactive material with a half-life of one week. The amount of the radioactive material that the sample contains at the beginning of each week can be determined by using the exponential decay equation used in Chapter 11.

The sequence below represents the amount in the sample at the beginning of each week. Each term of the sequence is found by multiplying the preceding term by $\frac{1}{2}$.

Week	1	2	3	4	5
Amount	20	10	5	2.5	1.25

The sequence 20, 10, 5, 2.5, 1.25 is called a geometric sequence.

Definition of a Geometric Sequence

A **geometric sequence,** or **geometric progression,** is one in which each successive term of the sequence is the same nonzero constant multiple of the preceding term. The common multiple is called the **common ratio** of the sequence.

Each of the sequences shown at the right is a geometric sequence. To find the common ratio of a geometric sequence, divide the second term of the sequence by the first term.

$3, 6, 12, 24, 48, \ldots$ Common ratio: 2

$4, -12, 36, -108, 324, \ldots$ Common ratio: -3

$6, 4, \frac{8}{3}, \frac{16}{9}, \frac{32}{27}, \ldots$ Common ratio: $\frac{2}{3}$

Consider a geometric sequence in which the first term is a_1 and the common ratio is r. By multiplying each successive term of the geometric sequence by the common ratio, a formula for the *n*th term can be found.

The first term is a_1.	$a_1 = a_1$
To find the second term, multiply the first term by the common ratio r.	$a_2 = a_1 r$
To find the third term, multiply the second term by the common ratio r.	$a_3 = (a_1 r)r$ $a_3 = a_1 r^2$
To find the fourth term, multiply the third term by the common ratio r.	$a_4 = (a_1 r^2)r$ $a_4 = a_1 r^3$
Note the relationship between the term number and the number that is the exponent on r. The exponent on r is one less than the term number.	$a_n = a_1 r^{n-1}$

The Formula for the *n*th Term of a Geometric Sequence

The nth term of a geometric sequence with first term a_1 and common ratio r is given by $a_n = a_1 r^{n-1}$.

Example 1 Find the 6th term of the geometric sequence 3, 6, 12, ...

Solution

$r = \frac{a_2}{a_1} = \frac{6}{3} = 2$ ► Find the common ratio.

$a_n = a_1 r^{n-1}$
$a_6 = 3(2)^{6-1} = 3(2)^5 = 3(32)$
$= 96$

► Use the Formula for the nth Term of a Geometric Sequence. $n = 6$, $a_1 = 3$, $r = 2$

Problem 1 Find the 5th term of the geometric sequence, 5, 2, $\frac{4}{5}$, ...

Solution See page A55.

Example 2 Find a_3 for the geometric sequence 8, a_2, a_3, 27, ...

Solution

$a_n = a_1 r^{n-1}$
$a_4 = a_1 r^{4-1}$ ► Find the common ratio. $a_4 = 27$, $a_1 = 8$, $n = 4$
$27 = 8r^{4-1}$
$\frac{27}{8} = r^3$
$\frac{3}{2} = r$

$a_3 = 8\left(\frac{3}{2}\right)^{3-1} = 8\left(\frac{3}{2}\right)^2 = 8\left(\frac{9}{4}\right)$
$= 18$

► Use the Formula for the nth Term of a Geometric Sequence.

Problem 2 Find a_3 for the geometric sequence $3, a_2, a_3, -192, \ldots$.

Solution See page A56.

12

2 Finite geometric series

The indicated sum of the terms of a geometric sequence is called a **geometric series.** The sum of a geometric series can be found by a formula.

The Formula for the Sum of n Terms of a Finite Geometric Series

Let a_1 be the first term of a finite geometric sequence, n the number of terms, r the common ratio, and $r \neq 1$. The sum of the series S_n is given by $S_n = \dfrac{a_1(1 - r^n)}{1 - r}$.

Proof of the Formula for the Sum of n Terms of a Finite Geometric Series:

Let S_n represent the sum of n terms of the sequence.

$$S_n = a_1 + a_1r + a_1r^2 + \cdots + a_1r^{n-2} + a_1r^{n-1}$$

Multiply each side of the equation by r.

$$rS_n = a_1r + a_1r^2 + a_1r^3 + \cdots + a_1r^{n-1} + a_1r^n$$

Subtract the two equations.

$$S_n - rS_n = a_1 - a_1r^n$$

Assuming $r \neq 1$, solve for S_n.

$$(1 - r)S_n = a_1(1 - r^n)$$

$$S_n = \frac{a_1(1 - r^n)}{1 - r}$$

Example 3 Find the sum of the geometric sequence 2, 8, 32, 128, 512.

Solution

$$r = \frac{a_2}{a_1} = \frac{8}{2} = 4$$

▶ Find the common ratio.

$$S_n = \frac{a_1(1 - r^n)}{1 - r}$$

▶ Use the Formula for the Sum of n Terms of a Finite Geometric Series. $n = 5$, $a_1 = 2$, $r = 4$

$$S_5 = \frac{2(1 - 4^5)}{1 - 4} = \frac{2(1 - 1024)}{-3}$$

$$= \frac{2(-1023)}{-3} = \frac{-2046}{-3} = 682$$

Problem 3 Find the sum of the geometric sequence $1, -\frac{1}{3}, \frac{1}{9}, -\frac{1}{27}$.

Solution See page A56.

Example 4 Find the sum of the geometric series $\sum_{n=1}^{10} (-20)(-2)^{n-1}$.

Solution

$$a_n = (-20)(-2)^{n-1}$$

$$a_1 = (-20)(-2)^{1-1} = (-20)(-2)^0 = (-20)(1) = -20$$ ▶ Find the first term.

$$a_2 = (-20)(-2)^{2-1} = (-20)(-2)^1 = (-20)(-2) = 40$$ ▶ Find the second term.

$$r = \frac{a_2}{a_1} = \frac{40}{-20} = -2$$ ▶ Find the common ratio.

$$S_n = \frac{a_1(1 - r^n)}{1 - r}$$

▶ Use the Formula for the Sum of n Terms of a Finite Geometric Series. $n = 10$, $a_1 = -20$, $r = -2$

$$S_{10} = \frac{-20[1 - (-2)^{10}]}{1 - (-2)} = \frac{-20(1 - 1024)}{3}$$

$$= \frac{-20(-1023)}{3} = \frac{20{,}460}{3} = 6820$$

Problem 4 Find the sum of the geometric series $\sum_{n=1}^{5} \left(\frac{1}{2}\right)^n$.

Solution See page A56.

3 Infinite geometric series

When the absolute value of the common ratio of a geometric sequence is less than 1, $|r| < 1$, then as n becomes larger, r^n becomes closer to zero.

Examples of geometric sequences for which $|r| < 1$ are shown at the right. As the number of terms increases, the absolute value of the last term listed is closer to zero.

$$1, \frac{1}{3}, \frac{1}{9}, \frac{1}{27}, \frac{1}{81}, \frac{1}{243}, \ldots$$

$$1, -\frac{1}{2}, \frac{1}{4}, -\frac{1}{8}, \frac{1}{16}, -\frac{1}{32}, \ldots$$

The indicated sum of the terms of an infinite geometric sequence is called an **infinite geometric series.**

An example of an infinite geometric series is shown at the right. The first term is 1. The common ratio is $\frac{1}{3}$.

$$1 + \frac{1}{3} + \frac{1}{9} + \frac{1}{27} + \frac{1}{81} + \frac{1}{243} + \cdots$$

The sum of the first 5, 7, 12, and 15 terms, along with the values of r^n, are shown at the right. Note that as n increases, the sum of the terms is closer to 1.5, and the value of r^n is closer to zero.

n	S_n	r^n
5	1.4938272	0.0041152
7	1.4993141	0.0004572
12	1.4999972	0.0000019
15	1.4999999	0.0000001

Using the Formula for the Sum of n Terms of a Geometric Series and the fact that r^n approaches zero when $|r| < 1$ and n increases, a formula for an infinite geometric series can be found.

The sum of the first n terms of a geometric series is shown at the right. If $|r| < 1$, then r^n can be made very close to zero by using larger and larger values of n. Therefore, the sum of the first n terms is approximately $\frac{a_1}{1-r}$.

Approximately zero

$$S_n = \frac{a_1(1 - r^n)}{1 - r}$$

$$S_n \approx \frac{a_1(1 - 0)}{1 - r} \approx \frac{a_1}{1 - r}$$

The Formula for the Sum of an Infinite Geometric Series

> The sum of an infinite geometric series in which $|r| < 1$ and a_1 is the first term is given by $S = \frac{a_1}{1-r}$.

When $|r| \geq 1$, the infinite geometric series does not have a finite sum. For example, the sum of the infinite geometric series $1 + 2 + 4 + 8 + \cdots$ increases without limit.

Example 5 Find the sum of the infinite geometric sequence $1, -\frac{1}{2}, \frac{1}{4}, -\frac{1}{8}, \ldots$.

Solution $S = \frac{a_1}{1-r} = \frac{1}{1-\left(-\frac{1}{2}\right)} = \frac{1}{\frac{3}{2}} = \frac{2}{3}$

▶ The common ratio is $-\frac{1}{2}$. $\left|-\frac{1}{2}\right| < 1$.

Use the Formula for the Sum of an Infinite Geometric Series.

Problem 5 Find the sum of the infinite geometric sequence $3, -2, \frac{4}{3}, -\frac{8}{9}, \ldots$.

Solution See page A57.

The sum of an infinite geometric series can be used to find a fraction equivalent to a nonterminating repeating decimal.

The repeating decimal shown at the right has been rewritten as an infinite geometric series, with the first term $\frac{3}{10}$ and common ratio $\frac{1}{10}$.

$$0.33\overline{3} = 0.3 + 0.03 + 0.003 + \cdots$$
$$= \frac{3}{10} + \frac{3}{100} + \frac{3}{1000} + \cdots$$

Use the Formula for the Sum of an Infinite Geometric Series.

$$S = \frac{a_1}{1-r} = \frac{\frac{3}{10}}{1-\frac{1}{10}} = \frac{\frac{3}{10}}{\frac{9}{10}} = \frac{3}{9} = \frac{1}{3}$$

$\frac{1}{3}$ is equivalent to the nonterminating, repeating decimal $0.\overline{3}$.

Example 6 Find an equivalent fraction for $0.122\overline{2}$.

Solution $0.122\overline{2}$

$0.1 + 0.02 + 0.002 + 0.0002 + \cdots =$

$\frac{1}{10} + \frac{2}{100} + \frac{2}{1000} + \frac{2}{10{,}000} + \cdots$

▶ Write the decimal as an infinite geometric series. The geometric series does not begin with the first term.

The series begins with $\frac{2}{100}$.

The common ratio is $\frac{1}{10}$.

$$S = \frac{a_1}{1-r} = \frac{\frac{2}{100}}{1-\frac{1}{10}} = \frac{\frac{2}{100}}{\frac{9}{10}} = \frac{2}{90}$$

▶ Use the Formula for the Sum of an Infinite Geometric Series.

$$0.122\overline{2} = \frac{1}{10} + \frac{2}{90} = \frac{11}{90}$$

▶ Add $\frac{1}{10}$ to the sum of the series.

An equivalent fraction is $\frac{11}{90}$.

Problem 6 Find an equivalent fraction for $0.36\overline{36}$.

Solution See page A57.

4 Application problems

Example 7 On the first swing, the length of the arc through which a pendulum swings is 16 in. The length of each successive swing is $\frac{7}{8}$ of the preceding swing. Find the length of the arc on the fifth swing. Round to the nearest tenth.

Strategy To find the length of the arc on the fifth swing, use the Formula for the nth Term of a Geometric Sequence. $n = 5$, $a_1 = 16$, $r = \frac{7}{8}$

Solution $a_n = a_1 r^{n-1}$

$$a_5 = 16\left(\frac{7}{8}\right)^{5-1} = 16\left(\frac{7}{8}\right)^4 = 16\left(\frac{2401}{4096}\right) = \frac{38{,}416}{4096} \approx 9.4$$

The length of the arc on the fifth swing is 9.4 in.

Problem 7 You start a chain letter and send it to three friends. Each of the three friends sends the letter to three other friends, and the sequence is repeated. Assuming no one breaks the chain, how many letters will have been mailed from the first through the sixth mailings?

Solution See page A57.

EXERCISES 12.3

1 Find the indicated term of the geometric sequence.

1. $2, 8, 32, \ldots; a_9$
2. $4, 3, \frac{9}{4}, \ldots; a_8$
3. $6, -4, \frac{8}{3}, \ldots; a_7$
4. $-5, 15, -45, \ldots; a_7$
5. $1, \sqrt{2}, 2, \ldots; a_9$
6. $3, 3\sqrt{3}, 9, \ldots; a_8$

Find a_2 and a_3 for the geometric sequence.

7. $9, a_2, a_3, \frac{8}{3}, \ldots$
8. $8, a_2, a_3, \frac{27}{8}, \ldots$
9. $3, a_2, a_3, -\frac{8}{9}, \ldots$
10. $6, a_2, a_3, -48, \ldots$
11. $-3, a_2, a_3, 192, \ldots$
12. $5, a_2, a_3, 625, \ldots$

2 Find the sum of the indicated number of terms of the geometric sequence.

13. $2, 6, 18, \ldots; n = 7$
14. $-4, 12, -36, \ldots; n = 7$
15. $12, 9, \frac{27}{4}, \ldots; n = 5$
16. $3, 3\sqrt{2}, 6, \ldots; n = 12$

Find the sum of the geometric series.

17. $\sum_{i=1}^{5} (2)^i$
18. $\sum_{n=1}^{6} \left(\frac{3}{2}\right)^n$
19. $\sum_{i=1}^{5} \left(\frac{1}{3}\right)^i$
20. $\sum_{n=1}^{6} \left(\frac{2}{3}\right)^n$
21. $\sum_{i=1}^{5} (4)^i$
22. $\sum_{n=1}^{8} (3)^n$
23. $\sum_{i=1}^{4} (7)^i$
24. $\sum_{n=1}^{5} (5)^n$
25. $\sum_{i=1}^{5} \left(\frac{3}{4}\right)^i$
26. $\sum_{n=1}^{3} \left(\frac{7}{4}\right)^n$
27. $\sum_{i=1}^{4} \left(\frac{5}{3}\right)^i$
28. $\sum_{n=1}^{6} \left(\frac{1}{2}\right)^n$

3 Find the sum of the infinite geometric series.

29. $3 + 2 + \frac{4}{3} + \cdots$

30. $2 - \frac{1}{4} + \frac{1}{32} + \cdots$

31. $6 - 4 + \frac{8}{3} + \cdots$

32. $\frac{1}{10} + \frac{1}{100} + \frac{1}{1000} + \cdots$

33. $\frac{7}{10} + \frac{7}{100} + \frac{7}{1000} + \cdots$

34. $\frac{5}{100} + \frac{5}{10{,}000} + \frac{5}{1{,}000{,}000} + \cdots$

Find an equivalent fraction for the repeating decimal.

35. $0.88\overline{8}$ **36.** $0.55\overline{5}$ **37.** $0.22\overline{2}$ **38.** $0.99\overline{9}$

39. $0.4\overline{545}$ **40.** $0.1\overline{818}$ **41.** $0.16\overline{66}$ **42.** $0.83\overline{33}$

4 Solve.

43. A laboratory ore sample contains 500 mg of a radioactive material with a half-life of 1 day. Find the amount of radioactive material in the sample at the beginning of the seventh day.

44. On the first swing, the length of the arc through which a pendulum swings is 18 in. The length of each successive swing is $\frac{3}{4}$ of the preceding swing. What is the total distance the pendulum has traveled during the five swings? Round to the nearest tenth.

45. To test the bounce of a tennis ball, the ball is dropped from a height of 8 ft. The ball bounces to 80% of its previous height with each bounce. How high does the ball bounce on the fifth bounce? Round to the nearest tenth.

46. The temperature of a hot water spa is 75°F. Each hour the temperature is 10% higher than during the previous hour. Find the temperature of the spa after 3 h. Round to the nearest tenth.

47. A real estate broker estimates that a piece of land will increase in value at a rate of 12% each year. If the original value of the land is $15,000, what will be its value in 15 years?

48. Suppose an employee receives a wage of 1¢ for the first day of work, 2¢ the second day, 4¢ the third day, and so on in a geometric sequence. Find the total amount of money earned for working 30 days.

49. Assume the average value of a home increases 5% per year. How much would a house costing $100,000 be worth in 30 years?

50. A culture of bacteria doubles every 2 h. If there were 500 bacteria at the beginning, how many bacteria will there be after 24 h?

SUPPLEMENTAL EXERCISES 12.3

State whether the sequence is arithmetic (A), geometric (G), or neither (N), and write the next term in the sequence.

51. $4, -2, 1, \ldots$

52. $-8, 0, 8, \ldots$

53. $5, 6.5, 8, \ldots$

54. $-7, 14, -28, \ldots$

55. $1, 4, 9, 16, \ldots$

56. $\sqrt{1}, \sqrt{2}, \sqrt{3}, \sqrt{4}$

57. $x^8, x^6, x^4, \ldots$

58. $5a^2, 3a^2, a^2, \ldots$

59. $\log x, 2 \log x, 3 \log x, \ldots$

60. $\log x, 3 \log x, 9 \log x, \ldots$

Solve.

61. The third term of a geometric sequence is 3, and the sixth term is $\frac{1}{9}$. Find the first term.

62. The fourth term of a geometric sequence is -8, and the seventh term is -64. Find the first term.

63. Given $a_n = 5$, $r = \frac{1}{2}$, and $S_n = 155$ for a geometric sequence, find a_1 and n.

64. Given $a_n = 162$, $r = -3$, and $S_n = 122$ for a geometric sequence, find a_1 and n.

65. For the geometric sequence given by $a_n = 2^n$, show that the sequence $b_n = \log a_n$ is an arithmetic sequence.

66. For the geometric sequence given by $a_n = e^n$, show that the sequence $b_n = \ln a_n$ is an arithmetic sequence.

67. For the arithmetic sequence given by $a_n = 3n - 2$, show that the sequence $b_n = 2a^n$ is a geometric sequence.

68. For $f(n) = ab^n$, n a natural number, show that $f(n)$ is a geometric sequence.

SECTION 12.4
Binomial Expansions

1 Expand $(a + b)^n$

By carefully observing the series for each expansion of the binomial $(a + b)^n$ shown below, it is possible to identify some interesting patterns.

$$(a + b)^1 = a + b$$

$$(a + b)^2 = a^2 + 2ab + b^2$$

$$(a + b)^3 = a^3 + 3a^2b + 3ab^2 + b^3$$

$$(a + b)^4 = a^4 + 4a^3b + 6a^2b^2 + 4ab^3 + b^4$$

$$(a + b)^5 = a^5 + 5a^4b + 10a^3b^2 + 10a^2b^3 + 5ab^4 + b^5$$

Patterns for the Variable Part of the Expansion of $(a + b)^n$

1. The first term is a^n. The exponent on a decreases by 1 for each successive term.
2. The exponent on b increases by 1 for each successive term. The last term is b^n.
3. The degree of each term is n.

The variable parts of the terms of the expansion of $(a + b)^6$ are shown below.

$$a^6,\ a^5b,\ a^4b^2,\ a^3b^3,\ a^2b^4,\ ab^5,\ b^6$$

The first term is a^6. For each successive term, the exponent on a decreases by 1, and the exponent on b increases by 1. The last term is b^6.

The variable parts of the general expansion of $(a + b)^n$ are

$$\boldsymbol{a^n,\ a^{n-1}b,\ a^{n-2}b^2,\ \ldots\ ab^{n-1},\ b^n}$$

A pattern for the coefficients of the terms of the expanded binomial can be found by writing the coefficients in a triangular array, which is known as **Pascal's Triangle.**

Each row begins and ends with the number 1. Any other number in a row is the sum of the two closest numbers above it. For example, $4 + 6 = 10$.

For $(a + b)^1$:	1	1				
For $(a + b)^2$:	1	2	1			
For $(a + b)^3$:	1	3	3	1		
For $(a + b)^4$:	1	4	6	4	1	
For $(a + b)^5$:	1	5	10	10	5	1

To write the sixth row of Pascal's Triangle, first write the numbers of the fifth row. The first and last numbers of the sixth row are 1. Each of the other numbers of the sixth row can be obtained by finding the sum of the two closest numbers above it in the fifth row.

1 5 10 10 5 1
1 6 15 20 15 6 1

These numbers will be the coefficients of the terms of the expansion of $(a + b)^6$.

Using the numbers of the sixth row of Pascal's Triangle for the coefficients and the pattern for the variable part of each term, the expanded form of $(a + b)^6$ can be written

$$(a + b)^6 = a^6 + 6a^5b + 15a^4b^2 + 20a^3b^3 + 15a^2b^4 + 6ab^5 + b^6$$

Although Pascal's Triangle can be used to find the coefficients for the expanded form of the power of any binomial, this method is inconvenient when the power of the binomial is large. An alternative method for determining these coefficients is based on the concept of a **factorial.**

Definition of Factorial

$n!$ (read "n factorial") is the product of the first n consecutive natural numbers.

$$n! = n(n-1)(n-2)\cdots 3\cdot 2\cdot 1$$

0! is defined to be 1: $0! = 1$

$$5! = 5\cdot 4\cdot 3\cdot 2\cdot 1 = 120$$
$$7! = 7\cdot 6\cdot 5\cdot 4\cdot 3\cdot 2\cdot 1 = 5040$$

To evaluate 6!, write the factorial as a product. Then simplify.

$$6! = 6\cdot 5\cdot 4\cdot 3\cdot 2\cdot 1$$
$$= 720$$

Example 1 Evaluate: $\dfrac{7!}{4!3!}$

Solution $\dfrac{7!}{4!3!} = \dfrac{7\cdot 6\cdot 5\cdot 4\cdot 3\cdot 2\cdot 1}{(4\cdot 3\cdot 2\cdot 1)(3\cdot 2\cdot 1)}$ ▶ Write each factorial as a product.

$= 35$ ▶ Simplify.

Problem 1 Evaluate: $\dfrac{12!}{7!5!}$

Solution See page A57.

The coefficients in a binomial expansion can be given in terms of factorials. Note that in the expansion of $(a + b)^5$ shown below, the coefficient of a^2b^3 can be given by $\frac{5!}{2!3!}$. The numerator is the factorial of the power of the binomial. The denominator is the product of the factorials of the exponents on a and b.

$$(a + b)^5 = a^5 + 5a^4b + 10a^3b^2 + 10a^2b^3 + 5ab^4 + b^5$$

$$\frac{5!}{2!3!} = \frac{5 \cdot 4 \cdot 3 \cdot 2 \cdot 1}{(2 \cdot 1)(3 \cdot 2 \cdot 1)} = 10$$

In general, the coefficients of $(a + b)^n$ are given as the quotients of factorials. The **coefficient of $a^{n-r}b^r$ is** $\frac{n!}{(n-r)!r!}$. The symbol $\binom{n}{r}$ is used to express this quotient of factorials.

$$\binom{n}{r} = \frac{n!}{(n-r)!r!}$$

Example 2 Evaluate: $\binom{8}{5}$

Solution $\binom{8}{5} = \frac{8!}{(8-5)!5!} = \frac{8!}{3!5!}$ ▶ Write the quotient of the factorials.

$= \frac{8 \cdot 7 \cdot 6 \cdot 5 \cdot 4 \cdot 3 \cdot 2 \cdot 1}{(3 \cdot 2 \cdot 1)(5 \cdot 4 \cdot 3 \cdot 2 \cdot 1)} = 56$ ▶ Simplify.

Problem 2 Evaluate: $\binom{7}{0}$

Solution See page A57.

Using factorials and the pattern for the variable part of each term, a formula for any natural number power of a binomial can be written.

The Binomial Expansion Formula

$$(a + b)^n = \binom{n}{0}a^n + \binom{n}{1}a^{n-1}b + \binom{n}{2}a^{n-2}b^2 + \cdots + \binom{n}{r}a^{n-r}b^r + \cdots + \binom{n}{n}b^n$$

The Binomial Expansion Formula is used below to expand $(a + b)^7$.

$$(a + b)^7$$
$$= \binom{7}{0}a^7 + \binom{7}{1}a^6b + \binom{7}{2}a^5b^2 + \binom{7}{3}a^4b^3 + \binom{7}{4}a^3b^4 + \binom{7}{5}a^2b^5 + \binom{7}{6}ab^6 + \binom{7}{7}b^7$$
$$= a^7 + 7a^6b + 21a^5b^2 + 35a^4b^3 + 35a^3b^4 + 21a^2b^5 + 7ab^6 + b^7$$

Example 3 Write $(4x + 3y)^3$ in expanded form.

Solution
$$(4x + 3y)^3 = \binom{3}{0}(4x)^3 + \binom{3}{1}(4x)^2(3y) + \binom{3}{2}(4x)(3y)^2 + \binom{3}{3}(3y)^3$$
$$= 1(64x^3) + 3(16x^2)(3y) + 3(4x)(9y^2) + 1(27y^3)$$
$$= 64x^3 + 144x^2y + 108xy^2 + 27y^3$$

Problem 3 Write $(3m - n)^4$ in expanded form.

Solution See page A58.

Example 4 Find the first 3 terms in the expansion of $(x + 3)^{15}$.

Solution
$$(x + 3)^{15} = \binom{15}{0}x^{15} + \binom{15}{1}x^{14}(3) + \binom{15}{2}x^{13}(3)^2 + \cdots$$
$$= 1x^{15} + 15x^{14}(3) + 105x^{13}(9) + \cdots$$
$$= x^{15} + 45x^{14} + 945x^{13} + \cdots$$

Problem 4 Find the first 3 terms in the expansion of $(y - 2)^{10}$.

Solution See page A58.

The Binomial Theorem can also be used to write any term of a binomial expansion.

Note that in the expansion of $(a + b)^5$, the exponent on b is one less than the term number.

$$(a + b)^5 = a^5 + 5a^4b + 10a^3b^2 + 10a^2b^3 + 5ab^4 + b^5$$

The Formula for the rth Term in a Binomial Expansion

The rth term of $(a + b)^n$ is $\binom{n}{r-1}a^{n-r+1}b^{r-1}$.

Example 5 Find the 4th term in the expansion of $(x + 3)^7$.

Solution $\binom{n}{r-1}a^{n-r+1}b^{r-1}$

▶ Use the Formula for the rth Term in a Binomial Expansion. $r = 4$, $n = 7$

$$\binom{7}{4-1}x^{7-4+1}(3)^{4-1} = \binom{7}{3}x^4(3)^3$$
$$= 35x^4(27)$$
$$= 945x^4$$

Problem 5 Find the 3rd term in the expansion of $(t - 2s)^7$.

Solution See page A58.

EXERCISES 12.4

1 Evaluate.

1. $3!$ **2.** $4!$ **3.** $8!$

4. $9!$ **5.** $0!$ **6.** $1!$

7. $\frac{5!}{2!3!}$ **8.** $\frac{8!}{5!3!}$ **9.** $\frac{6!}{6!0!}$

10. $\frac{10!}{10!0!}$ **11.** $\frac{9!}{6!3!}$ **12.** $\frac{10!}{2!8!}$

Evaluate.

13. $\binom{7}{2}$ **14.** $\binom{8}{6}$ **15.** $\binom{10}{2}$ **16.** $\binom{9}{6}$

17. $\binom{9}{0}$ **18.** $\binom{10}{10}$ **19.** $\binom{6}{3}$ **20.** $\binom{7}{6}$

21. $\binom{11}{1}$ **22.** $\binom{13}{1}$ **23.** $\binom{4}{2}$ **24.** $\binom{8}{4}$

Write in expanded form.

25. $(x + y)^4$ **26.** $(r - s)^3$

27. $(x - y)^5$ **28.** $(y - 3)^4$

29. $(2m + 1)^4$ **30.** $(2x + 3y)^3$

31. $(2r - 3)^5$ **32.** $(x + 3y)^4$

Find the first three terms in the expansion.

33. $(a + b)^{10}$

34. $(a + b)^9$

35. $(a - b)^{11}$

36. $(a - b)^{12}$

37. $(2x + y)^8$

38. $(x + 3y)^9$

39. $(4x - 3y)^8$

40. $(2x - 5)^7$

41. $\left(x + \frac{1}{x}\right)^7$

42. $\left(x - \frac{1}{x}\right)^8$

43. $(x^2 + 3)^5$

44. $(x^2 - 2)^6$

Find the indicated term in the expansion.

45. $(2x - 1)^7$; 4th term

46. $(x + 4)^5$; 3rd term

47. $(x^2 - y^2)^6$; 2nd term

48. $(x^2 + y^2)^7$; 6th term

49. $(y - 1)^9$; 5th term

50. $(x - 2)^8$; 8th term

51. $\left(n + \frac{1}{n}\right)^5$; 2nd term

52. $\left(x + \frac{1}{2}\right)^6$; 3rd term

53. $\left(\frac{x}{2} + 2\right)^5$; 1st term

54. $\left(y - \frac{2}{3}\right)^6$; 3rd term

SUPPLEMENTAL EXERCISES 12.4

Solve.

55. Write the 7th row of Pascal's Triangle.

56. Write the 8th row of Pascal's Triangle.

57. Evaluate $\frac{n!}{(n-2)!}$ for $n = 50$.

58. Evaluate $\frac{n!}{(n-3)!}$ for $n = 20$.

59. Simplify $\frac{n!}{(n-1)!}$.

60. Simplify $\frac{(n+1)!}{(n-1)!}$.

61. Write the term that contains an x^3 in the expansion of $(x + a)^7$.

62. Write the term that contains a y^4 in the expansion of $(y - a)^8$.

Expand the binomial.

63. $(x^{\frac{1}{2}} + 2)^4$

64. $(x^{\frac{1}{3}} + 3)^3$

65. $(x^{-1} + y^{-1})^3$

66. $(x^{-1} - y^{-1})^3$

67. $(1 + i)^6$

68. $(1 - i)^6$

69. For $0 \leq k \leq n$, show that $\binom{n}{r} = \binom{n}{n-r}$.

70. For $n \geq 1$, evaluate $\dfrac{2 \cdot 4 \cdot 6 \cdot 8 \cdots (2n)}{2^n n!}$.

A mathematical theorem is used to expand $(a + b + c)^n$. According to this theorem, the coefficient of $a^r b^k c^{n-r-k}$ in the expansion of $(a + b + c)^n$ is $\dfrac{n!}{r!k!(n-r-k)!}$.

71. Use the mathematical theorem to find the coefficient of a^2b^3c in the expansion of $(a + b + c)^6$.

72. Use the mathematical theorem to find the coefficient of $a^4b^2c^3$ in the expansion of $(a + b + c)^9$.

Calculators and Computers

Continued Fractions

An example of a **continued fraction** is $1 + \cfrac{1}{2 + \cfrac{9}{2 + \cfrac{25}{2 + \cfrac{49}{2 + \cdots}}}}$

The pattern continues with each numerator the square of an odd integer Approximations to these fractions can be found by using a calculator.

1. To evaluate the continued fraction above, begin by dividing 49 by 2. — 49 [÷] 2
2. Now add 2. Store the result in memory. — [+] 2 [=] [M+]

3. Divide 25 by the result in memory. Then add 2 and store in memory. The MC clears the memory before the new result is stored. 25 [÷] [MR] [+] 2 [MC] [=] [M+]

4. Repeat step 3, but divide 9 by the result in memory.

5. Repeat step 3, but divide 1 by the result in memory and add 1 instead of 2.

The result is 1.197719. This value approximates $\frac{4}{\pi}$. In fact, as the fraction is continued, the approximation becomes closer and closer to $\frac{4}{\pi}$.

Here are two other continued fractions. Evaluate them by using your calculator.

$$1 + \cfrac{1}{2 + \cfrac{1}{2 + \cfrac{1}{2 + \cdots}}} \qquad 1 + \cfrac{1}{1 + \cfrac{1}{1 + \cfrac{1}{1 + \cdots}}}$$

The first continued fraction is approximately $\sqrt{2}$. As the fraction is continued, the approximation becomes closer to $\sqrt{2}$. The second continued fraction is approximately $\frac{1 + \sqrt{5}}{2}$. This number is called the golden ratio and has played an important role in art and architecture.

Something Extra

The Fibonacci Sequence

Assume that a pair of rabbits produces another pair of rabbits after two months. If the new rabbits produce another pair after two months and the parents produce other pairs every month, we get the sequence as shown below.

Number of Months	Rabbits	Number of Pairs
1	XX	1
2	XX	1
3	XX XX	2
4	XX XX XX	3
5	XX XX XX XX XX	5
6	XX XX XX XX XX XX XX XX	8
7	XX XX XX XX XX XX XX XX XX XX XX XX XX	13

The sequence produced by this, 1, 1, 2, 3, 5, 8, 13, . . . , was first studied by Fibonacci and is known as the **Finonacci sequence.** Applications of the Fibonacci sequence are found in many diverse fields, such as botany, physical science, business, and psychology.

Note that after the first two terms in the Fibonacci sequence, a term is the sum of the two preceding terms, or

$$a_n = a_{n-2} + a_{n-1}.$$

Solve.

1. Find the first 18 terms of the Fibonacci sequence.

2. **a.** Find the squares of the first 8 terms of the Fibonacci sequence.

b. Add each successive pair of the squares found in Exercise 2a. What do you observe?

3. Find the ratios $\frac{a_{10}}{a_9}$, $\frac{a_{13}}{a_{12}}$, and $\frac{a_{18}}{a_{17}}$ in decimal form of the Fibonacci sequence. What do you observe?

4. The golden rectangle favored by the early Greeks is shown at the right. Calculate AB and find the ratio of the length of the golden rectangle to the width. Compare the result with the ratios in Exercise 3.

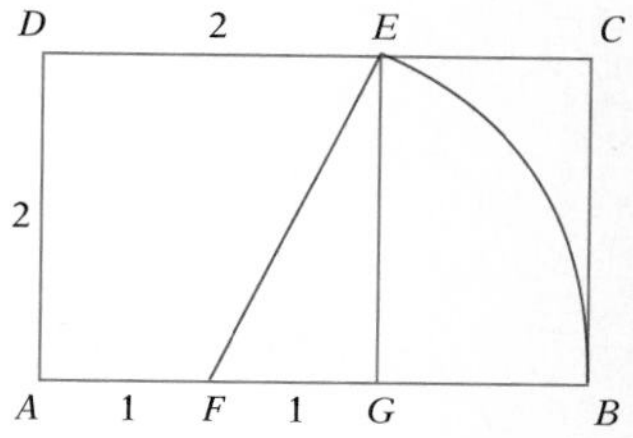

Chapter Summary

Key Words

A *sequence* is an ordered list of numbers.

Each of the numbers of a sequence is called a *term* of the sequence.

A *finite sequence* contains a finite number of terms.

An *infinite sequence* contains an infinite number of terms.

The indicated sum of a sequence is a *series*.

An *arithmetic sequence*, or arithmetic progression, is one in which the difference between any two consecutive terms is constant. The difference between consecutive terms is called the *common difference* of the sequence.

A *geometric sequence*, or geometric progression, is one in which each successive term of the sequence is the same nonzero constant multiple of the preceding term. The common multiple is called the *common ratio* of the sequence.

n factorial, written $n!$, is the product of the first n positive integers.

Essential Rules

The Formula for the *n*th Term of an Arithmetic Sequence The nth term of an arithmetic sequence with a common difference of d is given by $a_n = a_1 + (n - 1)d$.

The Formula for the Sum of *n* Terms of an Arithmetic Series Let a_1 be the first term of a finite arithmetic sequence, n the number of terms, and a_n the last term of the sequence. Then the sum of the series S_n is given by $S_n = \frac{n}{2}(a_1 + a_n)$.

The Formula for the *n*th Term of a Geometric Sequence The nth term of a geometric sequence with first term a_1 and common ratio r is given by $a_n = a_1r^{n-1}$.

The Formula for the Sum of *n* Terms of a Finite Geometric Series Let a_1 be the first term of a finite geometric sequence, n the number of terms, r the common ratio, and $r \neq 1$. Then the sum of the series S_n is given by $S_n = \frac{a_1(1 - r^n)}{1 - r}$.

The Formula for the Sum of an Infinite Geometric Series The sum of an infinite geometric series in which $|r| < 1$ and a_1 is the first term is given by $S = \frac{a_1}{1 - r}$.

The Binomial Expansion Formula

$$(a + b)^n = \binom{n}{0}a^n + \binom{n}{1}a^{n-1}b + \binom{n}{2}a^{n-2}b^2 + \cdots + \binom{n}{r}a^{n-r}b^r + \cdots + \binom{n}{n}b^n$$

The Formula for the *r*th Term in a Binomial Expansion

The rth term of $(a + b)^n$ is $\binom{n}{r - 1}a^{n-r+1}b^{r-1}$.

Chapter Review

1. Write $\sum_{i=1}^{4} 3x^i$ in expanded form.

2. Find the number of terms in the finite arithmetic sequence $-5, -8, -11, \ldots, -50$.

3. Find the 7th term of the geometric sequence $4, 4\sqrt{2}, 8, \ldots$.

4. Find the sum of the infinite geometric sequence $4, 3, \frac{9}{4}, \ldots$.

5. Evaluate: $\binom{9}{3}$

6. Write the 14th term of the sequence whose nth term is given by the formula $a_n = \frac{8}{n+2}$.

7. Find the 10th term of the arithmetic sequence $-10, -4, 2, \ldots$.

8. Find the sum of the first 18 terms of the arithmetic sequence $-25, -19, -13, \ldots$.

9. Find the sum of the first five terms of the geometric sequence $-6, 12, -24, \ldots$.

10. Evaluate: $\frac{8!}{4!4!}$

11. Find the 7th term in the expansion of $(3x + y)^9$.

12. Find the sum of the series $\sum_{n=1}^{4} (3n + 1)$.

13. Write the 6th term of the sequence whose nth term is given by the formula $a_n = \frac{n+1}{n}$.

14. Find the formula for the nth term of the arithmetic sequence $12, 9, 6, \ldots$.

15. Find the 5th term of the geometric sequence $6, 2, \frac{2}{3}, \ldots$.

16. Find an equivalent fraction for $0.23\overline{3}$.

17. Find the 35th term of the arithmetic sequence $-13, -16, -19, \ldots$.

18. Find the sum of the first six terms of the geometric sequence $1, \frac{3}{2}, \frac{9}{4}, \ldots$.

19. Find the sum of the first 21 terms of the arithmetic sequence $5, 12, 19, \ldots$.

20. Find the 4th term in the expansion of $(x - 2y)^7$.

21. Find the number of terms in the finite arithmetic sequence $1, 7, 13, \ldots, 121$.

22. Find the 8th term of the geometric sequence $\frac{3}{8}, \frac{3}{4}, \frac{3}{2}, \ldots$.

23. Find the sum of the series $\sum_{i=1}^{5} 2i$.

24. Find the sum of the first five terms of the geometric series 1, 4, 16, ...

25. Evaluate: 5!

26. Find the 3rd term in the expansion of $(x - 4)^6$.

27. Find the 30th term of the arithmetic sequence $-2, 3, 8, \ldots$.

28. Find the sum of the first 25 terms of the arithmetic sequence 25, 21, 17, ...

29. Write the 5th term of the sequence whose nth term is given by the formula $a_n = \frac{(-1)^{2n-1}n}{n^2 + 2}$.

30. Write $\sum_{i=1}^{4} 2x^{i-1}$ in expanded form.

31. Find an equivalent fraction for $0.23\overline{23}$.

32. Find the sum of the infinite geometric sequence $4 - 1 + \frac{1}{4} - \cdots$.

33. Find the sum of the geometric series $\sum_{n=1}^{5} 2(3)^n$.

34. Find the eighth term in the expansion of $(x - 2y)^{11}$.

35. Find the sum of the geometric series $\sum_{n=1}^{8} \left(\frac{1}{2}\right)^n$. Round to the nearest thousandth.

36. Find the sum of the infinite geometric sequence $2 + \frac{4}{3} + \frac{8}{9} + \cdots$.

37. Find an equivalent fraction for $0.63\overline{3}$.

38. Write $(x - 3y^2)^5$ in expanded form.

39. Find the number of terms in the finite arithmetic sequence $8, 2, -4, \ldots, -118$.

40. Evaluate: $\frac{12!}{5!8!}$

41. Write $\sum_{i=1}^{5} \frac{(2x)^i}{i}$ in expanded form.

42. Find the sum of the series $\sum_{n=1}^{4} \frac{(-1)^{n-1}n}{n+1}$.

43. The salary schedule for an apprentice electrician is \$1200 for the first month and a \$40-per-month salary increase for the next nine months. Find the total salary for the nine-month period.

44. The temperature of a hot water spa is 102°F. Each hour the temperature is 5% lower than during the previous hour. Find the temperature of the spa after 8 h. Round to the nearest tenth.

Chapter Test

1. Write the 14th term of the sequence whose nth term is given by the formula $a_n = \dfrac{6}{n+4}$.

2. Write the 9th and 10th terms of the sequence whose nth term is given by the formula $a_n = \dfrac{n-1}{n}$.

3. Find the sum of the series $\sum_{n=1}^{4} (2n + 3)$.

4. Write $\sum_{i=1}^{4} 2x^{2i}$ in expanded form.

5. Find the 28th term of the arithmetic sequence $-12, -16, -20, \ldots$.

6. Find the formula for the nth term of the arithmetic sequence $-3, -1, 1, \ldots$.

7. Find the number of terms in the finite arithmetic sequence $7, 3, -1, \ldots, -77$.

8. Find the sum of the first 15 terms of the arithmetic sequence $-42, -33, -24, \ldots$.

9. Find the sum of the first 24 terms of the arithmetic sequence $-4, 2, 8, \ldots$.

10. Evaluate: $\dfrac{10!}{5!5!}$

11. Find the 10th term of the geometric sequence $4, -4\sqrt{2}, 8, \ldots$.

12. Find the 5th term of the geometric sequence $5, 3, \dfrac{9}{5}, \ldots$.

13. Find the sum of the first five terms of the geometric sequence $1, \dfrac{3}{4}, \dfrac{9}{16}, \ldots$.

14. Find the sum of the first five terms of the geometric sequence $-5, 10, -20, \ldots$.

15. Find the sum of the infinite geometric sequence $2, 1, \frac{1}{2}, \ldots$.

16. Find an equivalent fraction for $0.233\overline{3}$.

17. Evaluate: $\binom{11}{4}$

18. Find the 5th term in the expansion of $(3x - y)^8$.

19. An inventory of supplies for a fabric manufacturer indicated that 7500 yd of material were in stock on January 1. On February 1, and on the first of the month for each successive month, the manufacturer sent 550 yd of material to retail outlets. How much material was in stock after the shipment on October 1?

20. An ore sample contains 320 mg of a radioactive substance with a half-life of one day. Find the amount of radioactive material in the sample at the beginning of the fifth day.

Cumulative Review

1. Simplify: $\frac{4x^2}{x^2 + x - 2} - \frac{3x - 2}{x + 2}$

2. Factor: $2x^6 + 16$

3. Simplify: $\sqrt{2y}(\sqrt{8xy} - \sqrt{y})$

4. Simplify: $\left(\frac{x^{-\frac{3}{4}} \cdot x^{\frac{3}{2}}}{x^{-\frac{5}{2}}}\right)^{-8}$

5. Solve: $5 - \sqrt{x} = \sqrt{x + 5}$

6. Solve: $2x^2 - x + 7 = 0$

7. Solve by the addition method:
$3x - 3y = 2$
$6x - 4y = 5$

8. Find the equation of the circle that passes through point $P(4, 2)$ and whose center is $C(-1, -1)$.

9. Solve: $2x - 1 > 3$ or $1 - 3x > 7$

10. Evaluate the determinant:
$\begin{vmatrix} -3 & 1 \\ 4 & 2 \end{vmatrix}$

11. Write $\log_5 \sqrt{\frac{x}{y}}$ in expanded form.

12. Solve for x: $4^x = 8^{x-1}$

13. Write the 5th and 6th terms of the sequence whose nth term is given by the formula $a_n = n(n-1)$.

14. Find the sum of the series $\sum_{n=1}^{7} (-1)^{n-1}(n+2)$.

15. Solve by the addition method:

$$x + 2y + z = 3$$
$$2x - y + 2z = 6$$
$$3x + y - z = 5$$

16. Solve for x: $\log_6 x = 3$

17. Simplify: $(4x^3 - 3x + 5) \div (2x + 1)$

18. For $g(x) = -3x + 4$, find $g(1 + h)$.

19. Find the range of $f(a) = \frac{a^3 - 1}{2a + 1}$ if the domain is $\{0, 1, 2\}$.

20. Find the equation of the circle that passes through point $P(2, 5)$ and whose center is $C(-1, 4)$.

21. Graph: $3x - 2y = -4$

22. Graph the solution set of $2x - 3y < 9$.

23. A new computer can complete a payroll in 16 min less time than it takes an older computer to complete the same payroll. Working together, both computers can complete the payroll in 15 min. How long would it take each computer working alone to complete the payroll?

24. A boat traveling with the current went 15 mi in 2 h. Against the current it took 3 h to travel the same distance. Find the rate of the boat in calm water and the rate of the current.

25. An 80-mg sample of a radioactive material decays to 55 mg in 30 days. Use the exponential decay equation $A = A_0\left(\frac{1}{2}\right)^{\frac{t}{k}}$ where A is the amount of radioactive material present after time t, k is the half-life, and A_0 is the original amount of radioactive material, to find the half-life of the 80-mg sample. Round to the nearest whole number.

26. A "theater in the round" has 62 seats in the first row, 74 seats in the second row, 86 seats in the third row, and so on in an arithmetic sequence. Find the total number of seats in the theater if there are 12 rows of seats.

27. To test the "bounce" of a ball, the ball is dropped from a height of 8 ft. The ball bounces to 80% of its previous height with each bounce. How high does the ball bounce on the fifth bounce? Round to the nearest tenth.

Final Exam

1. Simplify: $12 - 8[3 - (-2)]^2 \div 5 - 3$

2. Evaluate $\frac{a^2 - b^2}{a - b}$ when $a = 3$ and $b = -4$.

3. Simplify: $5 - 2[3x - 7(2 - x) - 5x]$

4. Solve: $\frac{3}{4}x - 2 = 4$

5. Solve: $\frac{2 - 4x}{3} - \frac{x - 6}{12} = \frac{5x - 2}{6}$

6. Solve: $8 - |5 - 3x| = 1$

7. Solve: $|2x + 5| < 3$

8. Solve: $2 - 3x < 6$ and $2x + 1 > 4$

9. Find the equation of the line containing the point $(-2, 1)$ and perpendicular to the line $3x - 2y = 6$.

10. Simplify: $2a[5 - a(2 - 3a) - 2a] + 3a^2$

11. Simplify: $\frac{3}{2 + i}$

12. Write a quadratic equation that has integer coefficients and has solutions $-\frac{1}{2}$ and 2.

13. Factor: $8 - x^3y^3$

14. Factor: $x - y - x^3 + x^2y$

15. Simplify: $(2x^3 - 7x^2 + 4) \div (2x - 3)$

16. Simplify: $\frac{x^2 - 3x}{2x^2 - 3x - 5} \div \frac{4x - 12}{4x^2 - 4}$

17. Simplify: $\frac{x - 2}{x + 2} - \frac{x + 3}{x - 3}$

18. Simplify: $\dfrac{\frac{3}{x} + \frac{1}{x + 4}}{\frac{1}{x} + \frac{3}{x + 4}}$

19. Solve: $\frac{5}{x - 2} - \frac{5}{x^2 - 4} = \frac{1}{x + 2}$

20. Solve $a_n = a_1 + (n - 1)d$ for d.

21. Simplify: $\left(\dfrac{4x^2y^{-1}}{3x^{-1}y}\right)^{-2}\left(\dfrac{2x^{-1}y^2}{9x^{-2}y^2}\right)^3$

22. Simplify: $\left(\dfrac{3x^{\frac{2}{3}}y^{\frac{1}{2}}}{6x^2y^{\frac{4}{3}}}\right)^6$

23. Simplify: $x\sqrt{18x^2y^3} - y\sqrt{50x^4y}$

24. Simplify: $\dfrac{\sqrt{16x^5y^4}}{\sqrt{32xy^7}}$

25. Solve by using the quadratic formula:
$2x^2 - 3x - 1 = 0$

26. Solve: $x^{\frac{2}{3}} - x^{\frac{1}{3}} - 6 = 0$

27. Find the equation of the line containing the points $(3, -2)$ and $(1, 4)$.

28. Solve: $\dfrac{2}{x} - \dfrac{2}{2x+3} = 1$

29. Solve by the addition method:
$3x - 2y = 1$
$5x - 3y = 3$

30. Evaluate the determinant:
$\begin{vmatrix} 3 & 4 \\ -1 & 2 \end{vmatrix}$

31. Solve for x: $\log_3 x - \log_3(x - 3) = \log_3 2$

32. Write $\sum_{i=1}^{5} 2y^i$ in expanded form.

33. Find an equivalent fraction for $0.5\overline{1}$.

34. Find the third term in the expansion of $(x - 2y)^9$.

35. Solve: $x^2 - y^2 = 4$
$x + y = 1$

36. Find the inverse function of $f(x) = \frac{2}{3}x - 4$.

37. Write $2(\log_2 a - \log_2 b)$ as a single logarithm with a coefficient of 1.

38. Graph $2x - 3y = 9$ by using the x- and y-intercepts.

39. Graph the solution set of $3x + 2y > 6$.

40. Graph: $f(x) = -x^2 + 4$

41. Graph: $\frac{x^2}{16} + \frac{y^2}{4} = 1$

42. Graph: $f(x) = \log_2(x + 1)$

43. An average score of 70–79 in a history class receives a C grade. A student has grades of 64, 58, 82, and 77 on four history tests. Find the range of scores on the fifth test that will give the student a C grade for the course.

44. A jogger and a cyclist set out at 8 A.M. from the same point headed in the same direction. The average speed of the cyclist is two and a half times the speed of the jogger. In 2 h, the cyclist is 24 mi ahead of the jogger. How far did the cyclist ride?

45. You have a total of $12,000 invested in two simple interest accounts. On one account, a money market fund, the annual simple interest rate is 8.5%. On the other account, a tax-free bond fund, the annual simple interest rate is 6.4%. The total annual interest earned by the two accounts is $936. How much do you have invested in each account?

46. The length of a rectangle is one foot less than three times the width. The area of the rectangle is 140 ft^2. Find the length and width of the rectangle.

47. Three hundred shares of a utility stock earn a yearly dividend of $486. How many additional shares of the utility stock would give a total dividend income of $810?

48. An account executive traveled 45 mi by car and then an additional 1050 mi by plane. The rate of the plane was seven times the rate of the car. The total time of the trip was $3\frac{1}{4}$ h. Find the rate of the plane.

49. An object is dropped from the top of a building. Find the distance the object has fallen when the speed reaches 75 ft/s. Use the equation $v = \sqrt{64d}$, where v is the speed of the object and d is the distance. Round to the nearest whole number.

50. A small plane made a trip of 660 mi in 5 h. The plane traveled the first 360 mi at a constant rate before increasing its speed by 30 mph. Another 300 mi was traveled at the increased speed. Find the rate of the plane for the first 360 mi.

51. The intensity (L) of a light source is inversely proportional to the square of the distance (d) from the source. If the intensity is 8 lumens at a distance of 20 ft, what is the intensity when the distance is 4 ft?

52. A motorboat traveling with the current can go 30 mi in 2 h. Against the current, it takes 3 h to go the same distance. Find the rate of the motorboat in calm water and the rate of the current.

53. An investor deposits $4000 into an account that earns 9% annual interest compounded monthly. Use the compound interest formula $P = A(1 + i)^n$, where A is the original value of the investment, i is the interest rate per compounding period, n is the total number of compounding periods, and P is the value of the investment after n periods, to find the value of the investment after two years. Round to the nearest cent.

54. Assume the average value of a home increases 6% per year. How much would a house costing $80,000 be worth in 20 years? Round to the nearest dollar.

APPENDIX

Interpolation: Logarithms and Antilogarithms

Since 3.257 is not included in the Table of Common Logarithms on pages A4–A5, it is not possible to find log 3.257 directly from the table. However, using the first number less than 3.257 that can be found in the table (3.250) and the first number greater than 3.257 (3.260), it is possible to obtain an approximation for log 3.257 by using linear interpolation. Linear interpolation is based on the fact that many functions can be approximated, over small intervals, by a straight line.

The graph of the equation $y = \log x$ is shown at the right. The units on the x- and y-axes have been distorted to illustrate linear interpolation.

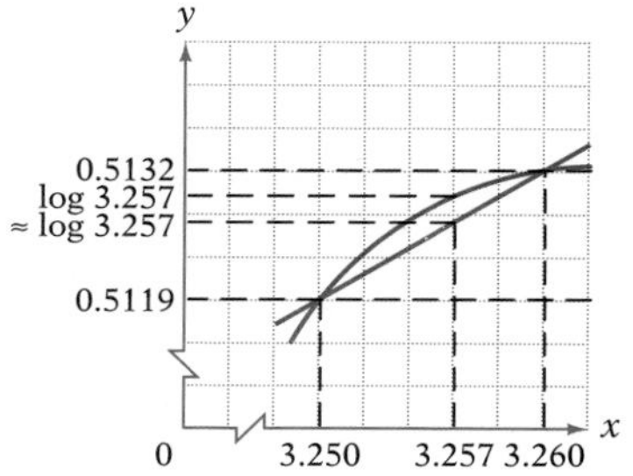

When $x = 3.250$, $y = 0.5119$. When $x = 3.260$, $y = 0.5132$. When $x = 3.257$, the value of y is greater than 0.5119 and less than 0.5132. The actual value of log 3.257 is on the curve. An approximation for this value is on the straight line.

A proportion can be written using the fact that between any two points on a straight line, the slope is the same.

$$\frac{y_2 - y_1}{x_2 - x_1} = \frac{y - y_1}{x - x_1}$$

Let $(x_1, y_1) = (3.250, 0.5119)$ and $(x_2, y_2) = (3.260, 0.5132)$. (x, y) is the point where $x = 3.257$.

$$\frac{0.5132 - 0.5119}{3.260 - 3.250} = \frac{y - 0.5119}{3.257 - 3.250}$$

$$\frac{0.0013}{0.01} = \frac{y - 0.5119}{0.007}$$

Solve the proportion for y.

$$0.00091 = y - 0.5119$$
$$0.51281 = y$$

$\log 3.257 \approx 0.5128$, rounded to the nearest ten-thousandth.

A more convenient method of using linear interpolation is illustrated in the following example:

Find log 32.57.

$$\log 32.57 = \log (3.257 \times 10^1)$$
$$= \log 3.257 + \log 10^1$$

Arrange the numbers 3.250, 3.257, and 3.260 and their corresponding mantissas in a table. Indicate the differences between the numbers and the mantissas, as shown at the right, using d to represent the unknown difference between the mantissa of 3.250 and the mantissa of 3.257.

		Number	Mantissa		
0.01	0.007	3.250	0.5119	d	0.0013
		3.257	$0.5119 + d$		
		3.260	0.5132		

Write a proportion.

$$\frac{0.007}{0.01} = \frac{d}{0.0013}$$

Solve for d, rounding to the nearest ten-thousandth.

$$0.0009 = d$$

Add the value of d to the mantissa 0.5119.

$$\begin{aligned} \log 3.257 + \log 10^1 &= (0.5119 + d) + 1 \\ &= (0.5119 + 0.0009) + 1 \\ &= 0.5128 + 1 \\ &= 1.5128 \end{aligned}$$

Linear interpolation can also be used to approximate an antilogarithm.

To find antilog (7.3431 − 10), use the table to find the first mantissa less than 0.3431 and the first mantissa greater than 0.3431. Arrange the mantissas and their corresponding antilogs in a table. Indicate the differences between the mantissas and the antilogs using d to represent the unknown difference.

		Mantissa	Antilog		
0.002	0.0007	0.3424	2.200	d	0.01
		0.3431	$2.200 + d$		
		0.3444	2.210		

Write a proportion.

$$\frac{0.0007}{0.002} = \frac{d}{0.01}$$

Solve for d, rounding to the nearest thousandth.

$$0.004 = d$$

Add the value of d to the antilog 2.20.

$$\begin{aligned} \text{antilog}\,(7.3431 - 10) &= (2.20 + d) \times 10^{-3} \\ &= (2.20 + 0.004) \times 10^{-3} \\ &= 2.204 \times 10^{-3} \\ &= 0.002204 \end{aligned}$$

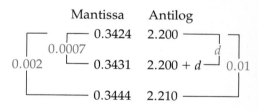

Table of Common Logarithms

Decimal approximations have been rounded to the nearest ten-thousandth.

x	0	1	2	3	4	5	6	7	8	9
1.0	.0000	.0043	.0086	.0128	.0170	.0212	.0253	.0294	.0334	.0374
1.1	.0414	.0453	.0492	.0531	.0569	.0607	.0645	.0682	.0719	.0755
1.2	.0792	.0828	.0864	.0899	.0934	.0969	.1004	.1038	.1072	.1106
1.3	.1139	.1173	.1206	.1239	.1271	.1303	.1335	.1367	.1399	.1430
1.4	.1461	.1492	.1523	.1553	.1584	.1614	.1644	.1673	.1703	.1732
1.5	.1761	.1790	.1818	.1847	.1875	.1903	.1931	.1959	.1987	.2014
1.6	.2041	.2068	.2095	.2122	.2148	.2175	.2201	.2227	.2253	.2279
1.7	.2304	.2330	.2355	.2380	.2405	.2430	.2455	.2480	.2504	.2529
1.8	.2553	.2577	.2601	.2625	.2648	.2672	.2695	.2718	.2742	.2765
1.9	.2788	.2810	.2833	.2856	.2878	.2900	.2923	.2945	.2967	.2989
2.0	.3010	.3032	.3054	.3075	.3096	.3118	.3139	.3160	.3181	.3201
2.1	.3222	.3243	.3263	.3284	.3304	.3324	.3345	.3365	.3385	.3404
2.2	.3424	.3444	.3464	.3483	.3502	.3522	.3541	.3560	.3579	.3598
2.3	.3617	.3636	.3655	.3674	.3692	.3711	.3729	.3747	.3766	.3784
2.4	.3802	.3820	.3838	.3856	.3874	.3892	.3909	.3927	.3945	.3962
2.5	.3979	.3997	.4014	.4031	.4048	.4065	.4082	.4099	.4116	.4133
2.6	.4150	.4166	.4183	.4200	.4216	.4232	.4249	.4265	.4281	.4298
2.7	.4314	.4330	.4346	.4362	.4378	.4393	.4409	.4425	.4440	.4456
2.8	.4472	.4487	.4502	.4518	.4533	.4548	.4564	.4579	.4594	.4609
2.9	.4624	.4639	.4654	.4669	.4683	.4698	.4713	.4728	.4742	.4757
3.0	.4771	.4786	.4800	.4814	.4829	.4843	.4857	.4871	.4886	.4900
3.1	.4914	.4928	.4942	.4955	.4969	.4983	.4997	.5011	.5024	.5038
3.2	.5051	.5065	.5079	.5092	.5105	.5119	.5132	.5145	.5159	.5172
3.3	.5185	.5198	.5211	.5224	.5237	.5250	.5263	.5276	.5289	.5302
3.4	.5315	.5328	.5340	.5353	.5366	.5378	.5391	.5403	.5416	.5428
3.5	.5441	.5453	.5465	.5478	.5490	.5502	.5514	.5527	.5539	.5551
3.6	.5563	.5575	.5587	.5599	.5611	.5623	.5635	.5647	.5658	.5670
3.7	.5682	.5694	.5705	.5717	.5729	.5740	.5752	.5763	.5775	.5786
3.8	.5798	.5809	.5821	.5832	.5843	.5855	.5866	.5877	.5888	.5899
3.9	.5911	.5922	.5933	.5944	.5955	.5966	.5977	.5988	.5999	.6010
4.0	.6021	.6031	.6042	.6053	.6064	.6075	.6085	.6096	.6107	.6117
4.1	.6128	.6138	.6149	.6160	.6170	.6180	.6191	.6201	.6212	.6222
4.2	.6232	.6243	.6253	.6263	.6274	.6284	.6294	.6304	.6314	.6325
4.3	.6335	.6345	.6355	.6365	.6375	.6385	.6395	.6405	.6415	.6425
4.4	.6435	.6444	.6454	.6464	.6474	.6484	.6493	.6503	.6513	.6522
4.5	.6532	.6542	.6551	.6561	.6571	.6580	.6590	.6599	.6609	.6618
4.6	.6628	.6637	.6646	.6656	.6665	.6675	.6684	.6693	.6702	.6712
4.7	.6721	.6730	.6739	.6749	.6758	.6767	.6776	.6785	.6794	.6803
4.8	.6812	.6821	.6830	.6839	.6848	.6857	.6866	.6875	.6884	.6893
4.9	.6902	.6911	.6920	.6928	.6937	.6946	.6955	.6964	.6972	.6981
5.0	.6990	.6998	.7007	.7016	.7024	.7033	.7042	.7050	.7059	.7067
5.1	.7076	.7084	.7093	.7101	.7110	.7118	.7126	.7135	.7143	.7152
5.2	.7160	.7168	.7177	.7185	.7193	.7202	.7210	.7218	.7226	.7235
5.3	.7243	.7251	.7259	.7267	.7275	.7284	.7292	.7300	.7308	.7316
5.4	.7324	.7332	.7340	.7348	.7356	.7364	.7372	.7380	.7388	.7396

Table of Common Logarithms

x	0	1	2	3	4	5	6	7	8	9
5.5	.7404	.7412	.7419	.7427	.7435	.7443	.7451	.7459	.7466	.7474
5.6	.7482	.7490	.7497	.7505	.7513	.7520	.7528	.7536	.7543	.7551
5.7	.7559	.7566	.7574	.7582	.7489	.7597	.7604	.7612	.7619	.7627
5.8	.7634	.7642	.7649	.7657	.7664	.7672	.7679	.7686	.7694	.7701
5.9	.7709	.7716	.7723	.7731	.7738	.7745	.7752	.7760	.7767	.7774
6.0	.7782	.7789	.7796	.7803	.7810	.7818	.7825	.7832	.7839	.7846
6.1	.7853	.7860	.7868	.7875	.7882	.7889	.7896	.7903	.7910	.7917
6.2	.7924	.7931	.7938	.7945	.7952	.7959	.7966	.7973	.7980	.7987
6.3	.7993	.8000	.8007	.8014	.8021	.8028	.8035	.8041	.8048	.8055
6.4	.8062	.8069	.8075	.8082	.8089	.8096	.8102	.8109	.8116	.8122
6.5	.8129	.8136	.8142	.8149	.8156	.8162	.8169	.8176	.8182	.8189
6.6	.8195	.8202	.8209	.8215	.8222	.8228	.8235	.8241	.8248	.8254
6.7	.8261	.8267	.8274	.8280	.8287	.8293	.8299	.8306	.8312	.8319
6.8	.8325	.8331	.8338	.8344	.8351	.8357	.8363	.8370	.8376	.8382
6.9	.8388	.8395	.8401	.8407	.8414	.8420	.8426	.8432	.8439	.8445
7.0	.8451	.8457	.8463	.8470	.8476	.8482	.8488	.8494	.8500	.8506
7.1	.8513	.8519	.8525	.8531	.8537	.8543	.8549	.8555	.8561	.8567
7.2	.8573	.8579	.8585	.8591	.8597	.8603	.8609	.8615	.8621	.8627
7.3	.8633	.8639	.8645	.8651	.8657	.8663	.8669	.8675	.8681	.8686
7.4	.8692	.8698	.8704	.8710	.8716	.8722	.8727	.8733	.8739	.8745
7.5	.8751	.8756	.8762	.8768	.8774	.8779	.8785	.8791	.8797	.8802
7.6	.8808	.8814	.8820	.8825	.8831	.8837	.8842	.8848	.8854	.8859
7.7	.8865	.8871	.8876	.8882	.8887	.8893	.8899	.8904	.8910	.8915
7.8	.8921	.8927	.8932	.8938	.8943	.8949	.8954	.8960	.8965	.8971
7.9	.8976	.8982	.8987	.8993	.8998	.9004	.9009	.9015	.9020	.9025
8.0	.9031	.9036	.9042	.9047	.9053	.9058	.9063	.9069	.9074	.9079
8.1	.9085	.9090	.9096	.9101	.9106	.9112	.9117	.9122	.9128	.9133
8.2	.9138	.9143	.9149	.9154	.9159	.9165	.9170	.9175	.9180	.9186
8.3	.9191	.9196	.9201	.9206	.9212	.9217	.9222	.9227	.9232	.9238
8.4	.9243	.9248	.9253	.9258	.9263	.9269	.9274	.9279	.9284	.9289
8.5	.9294	.9299	.9304	.9309	.9315	.9320	.9325	.9330	.9335	.9340
8.6	.9345	.9350	.9355	.9360	.9365	.9370	.9375	.9380	.9385	.9390
8.7	.9395	.9400	.9405	.9410	.9415	.9420	.9425	.9430	.9435	.9440
8.8	.9445	.9450	.9455	.9460	.9465	.9469	.9474	.9479	.9484	.9489
8.9	.9494	.9499	.9504	.9509	.9513	.9518	.9523	.9528	.9533	.9538
9.0	.9542	.9547	.9552	.9557	.9562	.9566	.9571	.9576	.9581	.9586
9.1	.9590	.9595	.9600	.9605	.9609	.9614	.9619	.9624	.9628	.9633
9.2	.9638	.9643	.9647	.9652	.9657	.9661	.9666	.9671	.9675	.9680
9.3	.9685	.9689	.9694	.9699	.9703	.9708	.9713	.9717	.9722	.9727
9.4	.9731	.9736	.9741	.9745	.9750	.9754	.9759	.9763	.9768	.9773
9.5	.9777	.9782	.9786	.9791	.9795	.9800	.9805	.9809	.9814	.9818
9.6	.9823	.9827	.9832	.9836	.9841	.9845	.9850	.9854	.9859	.9863
9.7	.9868	.9872	.9877	.9881	.9886	.9890	.9894	.9899	.9903	.9908
9.8	.9912	.9917	.9921	.9926	.9930	.9934	.9939	.9943	.9948	.9952
9.9	.9956	.9961	.9965	.9969	.9974	.9978	.9983	.9987	.9991	.9996

Table of Square and Cube Roots

Decimal approximations have been rounded to the nearest thousandth.

N	$\sqrt{N}$	$\sqrt[3]{N}$	N	$\sqrt{N}$	$\sqrt[3]{N}$	N	$\sqrt{N}$	$\sqrt[3]{N}$
1	1	1	36	6	3.302	71	8.426	4.141
2	1.414	1.260	37	6.083	3.332	72	8.485	4.160
3	1.732	1.442	38	6.164	3.362	73	8.544	4.179
4	2	1.587	39	6.245	3.391	74	8.602	4.198
5	2.236	1.710	40	6.325	3.420	75	8.660	4.217
6	2.449	1.817	41	6.403	3.448	76	8.718	4.236
7	2.646	1.913	42	6.481	3.476	77	8.775	4.254
8	2.828	2	43	6.557	3.503	78	8.832	4.273
9	3	2.080	44	6.633	3.530	79	8.888	4.291
10	3.162	2.154	45	6.708	3.557	80	8.944	4.309
11	3.317	2.224	46	6.782	3.583	81	9	4.327
12	3.464	2.289	47	6.856	3.609	82	9.055	4.344
13	3.606	2.351	48	6.928	3.634	83	9.110	4.362
14	3.742	2.410	49	7	3.659	84	9.165	4.380
15	3.873	2.466	50	7.071	3.684	85	9.220	4.397
16	4	2.520	51	7.141	3.708	86	9.274	4.414
17	4.123	2.571	52	7.211	3.733	87	9.327	4.431
18	4.243	2.621	53	7.280	3.756	88	9.381	4.448
19	4.359	2.668	54	7.348	3.780	89	9.434	4.465
20	4.472	2.714	55	7.416	3.803	90	9.487	4.481
21	4.583	2.759	56	7.483	3.826	91	9.539	4.498
22	4.690	2.802	57	7.550	3.849	92	9.592	4.514
23	4.796	2.844	58	7.616	3.871	93	9.644	4.531
24	4.899	2.884	59	7.681	3.893	94	9.695	4.547
25	5	2.924	60	7.746	3.915	95	9.747	4.563
26	5.099	2.962	61	7.810	3.936	96	9.798	4.579
27	5.196	3	62	7.874	3.958	97	9.849	4.595
28	5.292	3.037	63	7.937	3.979	98	9.899	4.610
29	5.385	3.072	64	8	4	99	9.950	4.626
30	5.477	3.107	65	8.062	4.021	100	10	4.642
31	5.568	3.141	66	8.124	4.041	101	10.050	4.657
32	5.657	3.175	67	8.185	4.062	102	10.100	4.672
33	5.745	3.208	68	8.246	4.082	103	10.149	4.688
34	5.831	3.240	69	8.307	4.102	104	10.198	4.703
35	5.916	3.271	70	8.367	4.121	105	10.247	4.718

SOLUTIONS to Chapter 1 Problems

SECTION 1.1 *pages 3–11*

Problem 1 A. -18 B. 5.2

Problem 2 A. 11 B. $-\frac{4}{5}$

Problem 3 A. $\frac{5}{8} \div \left(-\frac{15}{40}\right) = \frac{5}{8} \cdot \left(-\frac{40}{15}\right) = -\frac{5 \cdot 40}{8 \cdot 15} = -\frac{\overset{1}{\cancel{5}} \cdot \overset{1}{\cancel{2}} \cdot \overset{1}{\cancel{2}} \cdot \overset{1}{\cancel{2}} \cdot 5}{\underset{1}{\cancel{2}} \cdot \underset{1}{\cancel{2}} \cdot \underset{1}{\cancel{2}} \cdot 3 \cdot \underset{1}{\cancel{5}}} = -\frac{5}{3}$

B.
$$\begin{array}{r} 12.094 \\ -\ 8.729 \\ \hline 3.365 \end{array}$$

$-8.729 + 12.094 = 3.365$

Problem 4
$$\begin{array}{r} 4.027 \\ \times\ 0.49 \\ \hline 36243 \\ 16108 \\ \hline 1.97323 \approx 1.97 \end{array}$$

$-4.027(0.49) \approx -1.97$

Problem 5 $(-2)^4 = (-2)(-2)(-2)(-2) = 16$

$-2^4 = -(2 \cdot 2 \cdot 2 \cdot 2) = -16$

Problem 6 $-\left(\frac{2}{5}\right)^3 \cdot 5^2 = -\left(\frac{2}{5}\right)\left(\frac{2}{5}\right)\left(\frac{2}{5}\right) \cdot (5)(5) = -\frac{8}{5}$

Problem 7
$(3.81 - 1.41)^2 \div 0.036 - 1.89$
$(2.40)^2 \div 0.036 - 1.89$
$5.76 \div 0.036 - 1.89$
$160 - 1.89$
158.11

Problem 8
$\frac{1}{3} + \frac{5}{8} \div \frac{15}{16} - \frac{7}{12}$

$\frac{1}{3} + \frac{5}{8} \cdot \frac{16}{15} - \frac{7}{12}$

$\frac{1}{3} + \frac{2}{3} - \frac{7}{12}$

$1 - \frac{7}{12}$

$\frac{5}{12}$

Problem 9
$\frac{11}{12} - \frac{\frac{5}{4}}{2 - \frac{7}{2}} \cdot \frac{3}{4}$

$\frac{11}{12} - \frac{\frac{5}{4}}{-\frac{3}{2}} \cdot \frac{3}{4}$

$\frac{11}{12} - \left[\frac{5}{4} \cdot \left(-\frac{2}{3}\right)\right] \cdot \frac{3}{4}$

$\frac{11}{12} - \left(-\frac{5}{6}\right) \cdot \frac{3}{4}$

$\frac{11}{12} - \left(-\frac{5}{8}\right)$

$\frac{37}{24}$

SECTION 1.2 *pages 14–20*

Problem 1
$(b - c)^2 \div ab$
$[2 - (-4)]^2 \div (-3)(2)$
$[6]^2 \div (-3)(2)$
$36 \div (-3)(2)$
$-12(2)$
-24

Problem 2 $(x)\left(\frac{1}{4}\right) = \left(\frac{1}{4}\right)(x)$

Problem 3 The Associative Property of Addition

Problem 4
$(2x + xy - y) - (5x - 7xy + y)$
$2x + xy - y - 5x + 7xy - y$
$-3x + 8xy - 2y$

Problem 5
$2x - 3[y - 3(x - 2y + 4)]$
$2x - 3[y - 3x + 6y - 12]$
$2x - 3[7y - 3x - 12]$
$2x - 21y + 9x + 36$
$11x - 21y + 36$

Problem 6
the unknown number: n
the difference between 8 and twice the unknown number: $8 - 2n$

$n - (8 - 2n)$
$n - 8 + 2n$
$3n - 8$

Problem 7
the unknown number: n

three eighths of the number: $\frac{3}{8}n$

five twelfths of the number: $\frac{5}{12}n$

$\frac{3}{8}n + \frac{5}{12}n$

$\frac{9}{24}n + \frac{10}{24}n$

$\frac{19}{24}n$

SECTION 1.3 *pages 25–29*

Problem 1 $A = \{2, 4, 6, 8, 10\}$

Problem 2 $A = \{1, 3, 5, 7, 9\}$

Problem 3 0.35 is less than 2. Yes, 0.35 is an element of the set.

Problem 4 $A \cup C = \{-5, -2, -1, 0, 1, 2, 5\}$

Problem 5 There are no integers that are both odd and even. $E \cap F = \varnothing$

Problem 6 The set is $\{x|x > -3\}$.

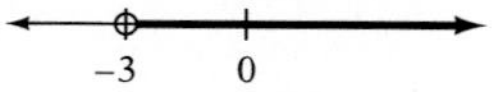

Problem 7 The set is $\{x|x \geq 1 \text{ or } x \leq -3\}$.

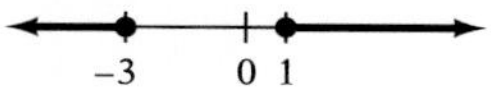

SOLUTIONS to Chapter 2 Problems

SECTION 2.1 *pages 41–47*

Problem 1

$$\begin{aligned} 6x - 5 - 3x &= 14 - 5x \\ 3x - 5 &= 14 - 5x \\ 3x - 5 + 5x &= 14 - 5x + 5x \\ 8x - 5 &= 14 \\ 8x - 5 + 5 &= 14 + 5 \\ 8x &= 19 \\ \frac{8x}{8} &= \frac{19}{8} \\ x &= \frac{19}{8} \end{aligned}$$

The solution is $\frac{19}{8}$.

Problem 2

$$\begin{aligned} 6(5 - x) - 12 &= 2x - 3(4 + x) \\ 30 - 6x - 12 &= 2x - 12 - 3x \\ 18 - 6x &= -x - 12 \\ 18 - 5x &= -12 \\ -5x &= -30 \\ x &= 6 \end{aligned}$$

The solution is 6.

Problem 3

The LCM of 3, 5, and 30 is 30.

$$\begin{aligned} \frac{2x-7}{3} - \frac{5x+4}{5} &= \frac{-x-4}{30} \\ 30\left(\frac{2x-7}{3} - \frac{5x+4}{5}\right) &= 30\left(\frac{-x-4}{30}\right) \\ \frac{30(2x-7)}{3} - \frac{30(5x+4)}{5} &= \frac{30(-x-4)}{30} \\ 10(2x - 7) - 6(5x + 4) &= -x - 4 \\ 20x - 70 - 30x - 24 &= -x - 4 \\ -10x - 94 &= -x - 4 \\ -9x - 94 &= -4 \\ -9x &= 90 \\ x &= -10 \end{aligned}$$

The solution is -10.

Problem 4

Strategy

- Next year's salary: s
- Next year's salary is the sum of this year's salary and the raise.

Solution

$$\begin{aligned} s &= 14{,}500 + 0.08(14{,}500) \\ &= 14{,}500 + 1160 \\ &= 15{,}660 \end{aligned}$$

Next year's salary is \$15,660.

SECTION 2.2 *pages 51–55*

Problem 1

Strategy

- Number of 3¢ stamps: x
  Number of 10¢ stamps: $2x + 2$
  Number of 15¢ stamps: $3x$

Stamps	*Number*	*Value*	*Total Value*
3¢	x	3	$3x$
10¢	$2x + 2$	10	$10(2x + 2)$
15¢	$3x$	15	$45x$

- The sum of the total values of each type of stamp equals the total value of all the stamps (156 cents).

Solution

$$3x + 10(2x + 2) + 45x = 156$$
$$3x + 20x + 20 + 45x = 156$$
$$68x + 20 = 156$$
$$68x = 136$$
$$x = 2$$

$3x = 3(2) = 6$

There are six 15¢ stamps in the collection.

Problem 2

Strategy

- The first number: n
  The second number: $2n$
  The third number: $4n - 3$
- The sum of the numbers is 81.

Solution

$$n + 2n + (4n - 3) = 81$$
$$7n - 3 = 81$$
$$7n = 84$$
$$n = 12$$

$2n = 2(12) = 24$
$4n - 3 = 4(12) - 3 = 48 - 3 = 45$

The numbers are 12, 24, and 45.

Problem 3

Strategy

- First odd integer: n
  Second odd integer: $n + 2$
  Third odd integer: $n + 4$
- Three times the sum of the first two integers is ten more than the product of the third integer and four.

Solution

$$3[n + (n + 2)] = (n + 4)4 + 10$$
$$3[2n + 2] = 4n + 16 + 10$$
$$6n + 6 = 4n + 26$$
$$2n + 6 = 26$$
$$2n = 20$$
$$n = 10$$

Since 10 is not an odd integer, there is no solution.

SECTION 2.3 *pages 57–61*

Problem 1

Strategy ▬ Pounds of \$3.00 hamburger: x
Pounds of \$1.80 hamburger: $75 - x$

	Amount	*Cost*	*Value*
\$3.00 hamburger	x	3.00	$3.00x$
\$1.80 hamburger	$75 - x$	1.80	$1.80(75 - x)$
Mixture	75	2.20	75(2.20)

▬ The sum of the values before mixing equals the value after mixing.

Solution

$$\begin{aligned} 3.00x + 1.80(75 - x) &= 75(2.20) \\ 3x + 135 - 1.80x &= 165 \\ 1.2x + 135 &= 165 \\ 1.2x &= 30 \\ x &= 25 \end{aligned}$$

$75 - x = 75 - 25 = 50$

The mixture must contain 25 lb of the \$3.00 hamburger and 50 lb of the \$1.80 hamburger.

Problem 2

Strategy ▬ Rate of the second plane: r
Rate of the first plane: $r + 30$

	Rate	*Time*	*Distance*
First plane	$r + 30$	4	$4(r + 30)$
Second plane	r	4	$4r$

▬ The total distance traveled by the two planes is 1160 mi.

Solution

$$\begin{aligned} 4(r + 30) + 4r &= 1160 \\ 4r + 120 + 4r &= 1160 \\ 8r + 120 &= 1160 \\ 8r &= 1040 \\ r &= 130 \end{aligned}$$

$r + 30 = 130 + 30 = 160$

The first plane is traveling 160 mph.
The second plane is traveling 130 mph.

SECTION 2.4 *pages 64–68*

Problem 1

Strategy ▬ Amount invested at 11.5%: x

	Principal	*Rate*	*Interest*
Amount at 13.2%	3500	0.132	0.132(3500)
Amount at 11.5%	x	0.115	$0.115x$

▬ The sum of the interest earned by the two investments equals the total annual interest earned ($1037).

Solution

$$0.132(3500) + 0.115x = 1037$$
$$462 + 0.115x = 1037$$
$$0.115x = 575$$
$$x = 5000$$

The amount invested at 11.5% is $5000.

Problem 2

Strategy ▬ Pounds of 22% fat hamburger: x
Pounds of 12% fat hamburger: $80 - x$

	Amount	*Percent*	*Quantity*
22%	x	0.22	$0.22x$
12%	$80 - x$	0.12	$0.12(80 - x)$
18%	80	0.18	0.18(80)

▬ The sum of the quantities before mixing is equal to the quantity after mixing.

Solution

$$0.22x + 0.12(80 - x) = 0.18(80)$$
$$0.22x + 9.6 - 0.12x = 14.4$$
$$0.10x + 9.6 = 14.4$$
$$0.10x = 4.8$$
$$x = 48$$

$$80 - x = 80 - 48 = 32$$

The butcher needs 48 lb of the hamburger that is 22% fat and 32 lb of the hamburger that is 12% fat.

SECTION 2.5 *pages 71–77*

Problem 1

$$\begin{aligned} 2x - 1 &< 6x + 7 \\ -4x - 1 &< 7 \\ -4x &< 8 \\ \frac{-4x}{-4} &> \frac{8}{-4} \\ x &> -2 \end{aligned}$$

$\{x|x > -2\}$

Problem 2

$$\begin{aligned} 5x - 2 &\le 4 - 3(x - 2) \\ 5x - 2 &\le 4 - 3x + 6 \\ 5x - 2 &\le 10 - 3x \\ 8x - 2 &\le 10 \\ 8x &\le 12 \\ \frac{8x}{8} &\le \frac{12}{8} \\ x &\le \frac{3}{2} \end{aligned}$$

$\left\{x|x \le \frac{3}{2}\right\}$

Problem 3

$$\begin{aligned} -2 \le 5x + 3 &\le 13 \\ -2 - 3 \le 5x + 3 - 3 &\le 13 - 3 \\ -5 \le 5x &\le 10 \\ \frac{-5}{5} \le \frac{5x}{5} &\le \frac{10}{5} \\ -1 \le x &\le 2 \end{aligned}$$

$\{x| -1 \le x \le 2\}$

Problem 4

$5 - 4x > 1$ and $6 - 5x < 11$

$-4x > -4$ $\quad -5x < 5$

$x < 1$ $\quad x > -1$

$\{x|x < 1\}$ $\quad \{x|x > -1\}$

$\{x|x < 1\} \cap \{x|x > -1\} = \{x|-1 < x < 1\}$

Problem 5

$2 - 3x > 11$ or $5 + 2x > 7$

$-3x > 9$ $\quad 2x > 2$

$x < -3$ $\quad x > 1$

$\{x|x < -3\}$ $\quad \{x|x > 1\}$

$\{x|x < -3\} \cup \{x|x > 1\} = \{x|x < -3 \text{ or } x > 1\}$

Problem 6

Strategy To find the maximum height, substitute the given values in the inequality $\frac{1}{2}bh < 50$ and solve for x.

Solution

$$\frac{1}{2}bh < 50$$
$$\frac{1}{2}(12)(x + 2) < 50$$
$$6(x + 2) < 50$$
$$6x + 12 < 50$$
$$6x < 38$$
$$x < \frac{19}{3}$$

The largest integer less than $\frac{19}{3}$ is 6.

$x + 2 = 6 + 2 = 8$

The maximum height of the triangle is 8 in.

Problem 7

Strategy To find the range of scores, write and solve an inequality using N to represent the score on the fifth exam.

Solution

$$80 \le \frac{72 + 94 + 83 + 70 + N}{5} \le 89$$
$$80 \le \frac{319 + N}{5} \le 89$$
$$5(80) \le 5\left(\frac{319 + N}{5}\right) \le 5(89)$$
$$400 \le 319 + N \le 445$$
$$400 - 319 \le 319 + N - 319 \le 445 - 319$$
$$81 \le N \le 126$$

Since 100 is a maximum score, the range of scores that will give the student a B for the course is $81 \le N \le 100$.

SECTION 2.6 *pages 83–88*

Problem 1

A. $|x| = 25$

$x = 25$ $\quad$ $x = -25$

The solutions are 25 and -25.

B. $|2x - 3| = 5$

$$2x - 3 = 5 \qquad 2x - 3 = -5$$
$$2x = 8 \qquad 2x = -2$$
$$x = 4 \qquad x = -1$$

The solutions are 4 and -1.

C. $|x - 3| = -2$

The absolute value of a number must be nonnegative.
There is no solution to this equation.

D. $5 - |3x + 5| = 3$
$-|3x + 5| = -2$
$|3x + 5| = 2$

$3x + 5 = 2$ $\quad$ $3x + 5 = -2$
$3x = -3$ $\quad$ $3x = -7$
$x = -1$ $\quad$ $x = -\frac{7}{3}$

The solutions are -1 and $-\frac{7}{3}$.

Problem 2 $|3x + 2| < 8$

$-8 < 3x + 2 < 8$
$-8 - 2 < 3x + 2 - 2 < 8 - 2$
$-10 < 3x < 6$
$\frac{-10}{3} < \frac{3x}{3} < \frac{6}{3}$
$-\frac{10}{3} < x < 2$
$\left\{x \middle| -\frac{10}{3} < x < 2\right\}$

Problem 3 $|3x - 7| < 0$

The absolute value of a number must be nonnegative. The solution set is the empty set.
$\varnothing$

Problem 4 $|5x + 3| > 8$

$5x + 3 < -8$ or $5x + 3 > 8$
$5x < -11$ $\quad$ $5x > 5$
$x < -\frac{11}{5}$ $\quad$ $x > 1$
$\left\{x \middle| x < -\frac{11}{5}\right\}$ $\quad$ $\{x | x > 1\}$
$\left\{x \middle| x < -\frac{11}{5}\right\} \cup \{x | x > 1\} = \left\{x \middle| x < -\frac{11}{5} \text{ or } x > 1\right\}$

Problem 5

Strategy Let b represent the diameter of the bushing, T the tolerance, and d the lower and upper limits of the diameter. Solve the absolute value inequality $|d - b| \le T$ for d.

Solution $|d - b| \le T$
$|d - 2.55| \le 0.003$

$-0.003 \le d - 2.55 \le 0.003$
$-0.003 + 2.55 \le d - 2.55 + 2.55 \le 0.003 + 2.55$
$2.547 \le d \le 2.553$

The lower and upper limits of the diameter of the bushing are 2.547 in. and 2.553 in.

SOLUTIONS to Chapter 3 Problems

SECTION 3.1 *pages 103–113*

Problem 1

$$\begin{array}{r} 5x^2 + 3x - 1 \\ 2x^2 + 4x - 6 \\ \underline{x^2 - 7x + 8} \\ 8x^2 \qquad + 1 \end{array}$$

Problem 2

$(4x^2 + 3x - 5) + (6x^3 - 2 + x^2)$
$6x^3 + (4x^2 + x^2) + 3x + (-5 - 2)$
$6x^3 + 5x^2 + 3x - 7$

Problem 3

$(-5x^2 + 2x - 3) - (6x^2 + 3x - 7)$

$$\begin{array}{r} -5x^2 + 2x - 3 \\ \underline{-6x^2 - 3x + 7} \\ -11x^2 - \quad x + 4 \end{array}$$

Problem 4

$(5x^{2n} - 3x^n - 7) - (-2x^{2n} - 5x^n + 8)$
$(5x^{2n} - 3x^n - 7) + (2x^{2n} + 5x^n - 8)$
$7x^{2n} + 2x^n - 15$

Problem 5 $(7xy^3)(-5x^2y^2)(-xy^2) = 35x^4y^7$

Problem 6 **A.** $(y^3)^6 = y^{18}$ **B.** $(x^n)^3 = x^{3n}$

Problem 7 $(-2ab^3)^4 = (-2)^4a^{1 \cdot 4}b^{3 \cdot 4} = 16a^4b^{12}$

Problem 8

$$\begin{aligned} 6a(2a)^2 + 3a(2a^2) &= 6a(2^2a^2) + 6a^3 \\ &= 6a(4a^2) + 6a^3 \\ &= 24a^3 + 6a^3 = 30a^3 \end{aligned}$$

Problem 9

A. $(2x^{-5}y)(5x^4y^{-3}) = 10x^{-5+4}y^{1-3} = 10x^{-1}y^{-2} = \dfrac{10}{xy^2}$

B. $\dfrac{a^{-1}b^4}{a^{-2}b^{-2}} = a^{-1-(-2)}b^{4-(-2)} = ab^6$

C. $\left(\dfrac{2^{-1}x^2y^{-3}}{4x^{-2}y^{-5}}\right)^{-2} = \left(\dfrac{x^4y^2}{8}\right)^{-2} = \dfrac{x^{-8}y^{-4}}{8^{-2}} = \dfrac{8^2}{x^8y^4} = \dfrac{64}{x^8y^4}$

D. $[(a^{-1}b)^{-2}]^3 = [a^2b^{-2}]^3 = a^6b^{-6} = \dfrac{a^6}{b^6}$

Problem 10 $942{,}000{,}000 = 9.42 \times 10^8$

Problem 11 $2.7 \times 10^{-5} = 0.000027$

Problem 12

$$\frac{5{,}600{,}000 \times 0.000000081}{900 \times 0.000000028} = \frac{5.6 \times 10^6 \times 8.1 \times 10^{-8}}{9 \times 10^2 \times 2.8 \times 10^{-8}} = \frac{(5.6)(8.1) \times 10^{6+(-8)-2-(-8)}}{(9)(2.8)} = 1.8 \times 10^4$$

Problem 13

Strategy To find the number of arithmetic operations:

- Find the reciprocal of 1×10^{-7}, which is the number of operations performed in one second.
- Write the number of seconds in one minute (60) in scientific notation.
- Multiply the number of arithmetic operations per second by the number of seconds in one minute.

Solution $\dfrac{1}{1 \times 10^{-7}} = 10^7$

$60 = 6 \times 10$

$6 \times 10 \times 10^7 = 6 \times 10^8$

The computer can perform 6×10^8 operations in one minute.

SECTION 3.2 *pages 118–123*

Problem 1 A. $-4y(y^2 - 3y + 2) = -4y(y^2) - (-4y)(3y) + (-4y)(2) = -4y^3 + 12y^2 - 8y$

B.
$$x^2 - 2x[x - x(4x - 5) + x^2]$$
$$x^2 - 2x[x - 4x^2 + 5x + x^2]$$
$$x^2 - 2x[6x - 3x^2]$$
$$x^2 - 12x^2 + 6x^3$$
$$6x^3 - 11x^2$$

C.
$$y^{n+3}(y^{n-2} - 3y^2 + 2)$$
$$y^{n+3}(y^{n-2}) - (y^{n+3})(3y^2) + (y^{n+3})(2)$$
$$y^{n+3+(n-2)} - 3y^{n+3+2} + 2y^{n+3}$$
$$y^{2n+1} - 3y^{n+5} + 2y^{n+3}$$

Problem 2

$$\begin{array}{r} -2b^2 + 5b - 4 \\ -3b + 2 \\ \hline -4b^2 + 10b - 8 \\ 6b^3 - 15b^2 + 12b \phantom{{}-8} \\ \hline 6b^3 - 19b^2 + 22b - 8 \end{array}$$

Problem 3 $(5a - 3b)(2a + 7b) = 10a^2 + 35ab - 6ab - 21b^2 = 10a^2 + 29ab - 21b^2$

Problem 4 A. $(3x - 7)(3x + 7) = 9x^2 - 49$ B. $(3x - 4y)^2 = 9x^2 - 24xy + 16y^2$

C. $(2x^n + 3)(2x^n - 3) = 4x^{2n} - 9$ D. $(2x^n - 8)^2 = 4x^{2n} - 32x^n + 64$

Problem 5

Strategy To find the area, replace the variables b and h in the equation $A = \frac{1}{2}bh$ with the given values and solve for A.

Solution
$$A = \frac{1}{2}bh$$
$$A = \frac{1}{2}(2x + 6)(x - 4)$$
$$A = (x + 3)(x - 4)$$
$$A = x^2 - 4x + 3x - 12$$
$$A = x^2 - x - 12$$

The area is $(x^2 - x - 12)$ ft^2.

Problem 6

Strategy To find the volume, subtract the volume of the small rectangular solid from the volume of the large rectangular solid.

Large rectangular solid: Length = $L_1 = 12x$
Width = $w_1 = 7x + 2$
Height = $h_1 = 5x - 4$
Small rectangular solid: Length = $L_2 = 12x$
Width = $w_2 = x$
Height = $h_2 = 2x$

Solution V = Volume of large rectangular solid − volume of small rectangular solid
$$V = (L_1 \cdot w_1 \cdot h_1) - (L_2 \cdot w_2 \cdot h_2)$$
$$V = (12x)(7x + 2)(5x - 4) - (12x)(x)(2x)$$
$$V = (84x^2 + 24x)(5x - 4) - (12x^2)(2x)$$
$$V = (420x^3 - 336x^2 + 120x^2 - 96x) - (24x^3)$$
$$V = 396x^3 - 216x^2 - 96x$$

The volume is $(396x^3 - 216x^2 - 96x)$ ft^3.

SECTION 3.3 *pages 127–130*

Problem 1 **A.**

$$\begin{array}{r}
5x \;-\; 1 \\
3x+4\overline{)15x^2+17x-20} \\
\underline{15x^2+20x} \\
-3x-20 \\
\underline{-3x-\;\;4} \\
-16
\end{array}$$

$$\frac{15x^2+17x-20}{3x+4}=5x-1-\frac{16}{3x+4}$$

B.

$$\begin{array}{r}
x^2+3x\;-1 \\
3x-1\overline{)3x^3+8x^2-6x+2} \\
\underline{3x^3-\;x^2}\qquad\quad \\
9x^2-6x\quad \\
\underline{9x^2-3x}\quad \\
-3x+2 \\
\underline{-3x+1} \\
1
\end{array}$$

$$\frac{3x^3+8x^2-6x+2}{3x-1}=x^2+3x-1+\frac{1}{3x-1}$$

Problem 2 **A.**

$$\begin{array}{r|rrr}
-2 & 6 & 8 & -5 \\
 & & -12 & 8 \\
\hline
 & 6 & -4 & 3
\end{array}$$

$$(6x^2+8x-5)\div(x+2)=6x-4+\frac{3}{x+2}$$

B.

$$\begin{array}{r|rrrrr}
3 & 2 & -3 & -8 & 0 & -2 \\
 & & 6 & 9 & 3 & 9 \\
\hline
 & 2 & 3 & 1 & 3 & 7
\end{array}$$

$$(2x^4-3x^3-8x^2-2)\div(x-3)=2x^3+3x^2+x+3+\frac{7}{x-3}$$

SECTION 3.4 *pages 133–143*

Problem 1 **A.** The GCF of $3x^3y-6x^2y^2-3xy^3$ is $3xy$.

$3x^3y-6x^2y^2-3xy^3=3xy(x^2-2xy-y^2)$

B. The GCF of $6t^{2n}$ and $9t^n$ is $3t^n$.

$6t^{2n}-9t^n=3t^n(2t^n-3)$

Problem 2 $6a(2b-5)+7(5-2b)=6a(2b-5)-7(2b-5)=(2b-5)(6a-7)$

Problem 3 $3rs-2r-3s+2=(3rs-2r)-(3s-2)=r(3s-2)-(3s-2)=(3s-2)(r-1)$

Problem 4 **A.** $x^2 + 13x + 42 = (x + 6)(x + 7)$

B. $x^2 - x - 20 = (x + 4)(x - 5)$

C. $x^2 + 5xy + 6y^2 = (x + 2y)(x + 3y)$

Problem 5 $x^2 + 5x - 1$

There are no factors of -1 whose sum is 5.

The trinomial is nonfactorable over the integers.

Problem 6 **A.** $4x^2 + 15x - 4$

Factors of 4	Factors of -4
1, 4	1, -4
2, 2	-1, 4
	2, -2

Trial Factors	Middle Term
$(4x + 1)(x - 4)$	$-16x + x = -15x$
$(4x - 1)(x + 4)$	$16x - x = 15x$

$4x^2 + 15x - 4 = (4x - 1)(x + 4)$

B. $10x^2 + 39x + 14$

Factors of 10	Factors of 14
1, 10	1, 14
2, 5	2, 7

Trial Factors	Middle Term
$(x + 2)(10x + 7)$	$7x + 20x = 27x$
$(2x + 1)(5x + 14)$	$28x + 5x = 33x$
$(10x + 1)(x + 14)$	$140x + x = 141x$
$(5x + 2)(2x + 7)$	$35x + 4x = 39x$

$10x^2 + 39x + 14 = (5x + 2)(2x + 7)$

Problem 7 **A.** $6x^2 + 7x - 20$

$a \cdot c = -120$

Factors of -120	Sum
120, -1	119
60, -2	58
40, -3	37
30, -4	26
24, -5	19
20, -6	14
15, -8	7

$$\begin{aligned} 6x^2 + 7x - 20 &= 6x^2 + 15x - 8x - 20 \\ &= 3x(2x + 5) - 4(2x + 5) \\ &= (2x + 5)(3x - 4) \end{aligned}$$

B. $2 - x - 6x^2$

$a \cdot c = -12$

Factors of -12	Sum
-12, 1	-11
-6, 2	-4
-4, 3	-1

$$\begin{aligned} 2 - x - 6x^2 &= 2 - 4x + 3x - 6x^2 \\ &= 2(1 - 2x) + 3x(1 - 2x) \\ &= (1 - 2x)(2 + 3x) \end{aligned}$$

Problem 8 **A.** $3a^3b^3 + 3a^2b^2 - 60ab = 3ab(a^2b^2 + ab - 20) = 3ab(ab + 5)(ab - 4)$

B. $40a - 10a^2 - 15a^3 = 5a(8 - 2a - 3a^2) = 5a(2 + a)(4 - 3a)$

SECTION 3.5 *pages 146–151*

Problem 1 $x^2 - 36y^4 = x^2 - (6y^2)^2$
$= (x + 6y^2)(x - 6y^2)$

Problem 2 $9x^2 + 12x + 4 = (3x + 2)^2$

Problem 3
A. $8x^3 + y^3z^3 = (2x)^3 + (yz)^3 = (2x + yz)(4x^2 - 2xyz + y^2z^2)$
B. $(x - y)^3 + (x + y)^3 =$
$[(x - y) + (x + y)][(x - y)^2 - (x - y)(x + y) + (x + y)^2] =$
$2x[x^2 - 2xy + y^2 - (x^2 - y^2) + x^2 + 2xy + y^2] =$
$2x[x^2 - 2xy + y^2 - x^2 + y^2 + x^2 + 2xy + y^2] = 2x(x^2 + 3y^2)$

Problem 4
A. $6x^2y^2 - 19xy + 10 = (3xy - 2)(2xy - 5)$
B. $3x^4 + 4x^2 - 4 = (x^2 + 2)(3x^2 - 2)$

Problem 5
A. $4x - 4y - x^3 + x^2y = (4x - 4y) - (x^3 - x^2y) = 4(x - y) - x^2(x - y) =$
$(x - y)(4 - x^2) = (x - y)(2 + x)(2 - x)$
B. $x^{4n} - x^{2n}y^{2n} = x^{2n+2n} - x^{2n}y^{2n} = x^{2n}(x^{2n} - y^{2n}) = x^{2n}[(x^n)^2 - (y^n)^2] = x^{2n}(x^n + y^n)(x^n - y^n)$

SECTION 3.6 *pages 155–157*

Problem 1

A.
$4x^2 + 11x = 3$
$4x^2 + 11x - 3 = 0$
$(4x - 1)(x + 3) = 0$

$4x - 1 = 0$ $x + 3 = 0$
$4x = 1$ $x = -3$
$x = \frac{1}{4}$

The solutions are $\frac{1}{4}$ and -3.

B.
$(x - 2)(x + 5) = 8$
$x^2 + 3x - 10 = 8$
$x^2 + 3x - 18 = 0$
$(x + 6)(x - 3) = 0$

$x + 6 = 0$ $x - 3 = 0$
$x = -6$ $x = 3$

The solutions are -6 and 3.

C.
$x^3 + 4x^2 - 9x - 36 = 0$
$x^2(x + 4) - 9(x + 4) = 0$
$(x + 4)(x^2 - 9) = 0$
$(x + 4)(x + 3)(x - 3) = 0$

$x + 4 = 0$ $x + 3 = 0$ $x - 3 = 0$
$x = -4$ $x = -3$ $x = 3$

The solutions are -4, -3, and 3.

Problem 2

Strategy
- Width of the rectangle: W
  Length of the rectangle: $W + 5$
- Use the equation $A = LW$.

Solution

$$A = LW$$
$$66 = (W + 5)(W)$$
$$66 = W^2 + 5W$$
$$0 = W^2 + 5W - 66$$
$$0 = (W + 11)(W - 6)$$

$$W + 11 = 0 \qquad\qquad W - 6 = 0$$
$$W = -11 \qquad\qquad W = 6 \qquad \text{The width cannot be a negative number.}$$

$$\text{Length} = W + 5 = 6 + 5 = 11$$

The length is 11 in., and the width is 6 in.

SOLUTIONS to Chapter 4 Problems

SECTION 4.1 *pages 173–174*

Problem 1 A. $\dfrac{6x^4 - 24x^3}{12x^3 - 48x^2} = \dfrac{6x^3(x - 4)}{12x^2(x - 4)} = \dfrac{6x^3\overset{1}{\cancel{(x - 4)}}}{12x^2\underset{1}{\cancel{(x - 4)}}} = \dfrac{x}{2}$

B. $\dfrac{20x - 15x^2}{15x^3 - 5x^2 - 20x} = \dfrac{5x(4 - 3x)}{5x(3x^2 - x - 4)} = \dfrac{5x(4 - 3x)}{5x(3x - 4)(x + 1)} = \dfrac{5x\overset{-1}{\cancel{(4 - 3x)}}}{5x\underset{1}{\cancel{(3x - 4)}}(x + 1)} = -\dfrac{1}{x + 1}$

C. $\dfrac{x^{2n} + x^n - 12}{x^{2n} - 3x^n} = \dfrac{(x^n + 4)(x^n - 3)}{x^n(x^n - 3)} = \dfrac{(x^n + 4)\overset{1}{\cancel{(x^n - 3)}}}{x^n\underset{1}{\cancel{(x^n - 3)}}} = \dfrac{x^n + 4}{x^n}$

SECTION 4.2 *pages 176–181*

Problem 1 A. $\dfrac{12 + 5x - 3x^2}{x^2 + 2x - 15} \cdot \dfrac{2x^2 + x - 45}{3x^2 + 4x} = \dfrac{(4 + 3x)(3 - x)}{(x + 5)(x - 3)} \cdot \dfrac{(2x - 9)(x + 5)}{x(3x + 4)} =$

$\dfrac{(4 + 3x)(3 - x)(2x - 9)(x + 5)}{(x + 5)(x - 3) \cdot x(3x + 4)} = \dfrac{\overset{1}{\cancel{(4 + 3x)}}\overset{-1}{\cancel{(3 - x)}}(2x - 9)\overset{1}{\cancel{(x + 5)}}}{\underset{1}{\cancel{(x + 5)}}\underset{1}{\cancel{(x - 3)}} \cdot x\underset{1}{\cancel{(3x + 4)}}} = -\dfrac{2x - 9}{x}$

B. $\dfrac{2x^2 - 13x + 20}{x^2 - 16} \cdot \dfrac{2x^2 + 9x + 4}{6x^2 - 7x - 5} = \dfrac{(2x - 5)(x - 4)}{(x - 4)(x + 4)} \cdot \dfrac{(2x + 1)(x + 4)}{(3x - 5)(2x + 1)} =$

$\dfrac{(2x - 5)(x - 4)(2x + 1)(x + 4)}{(x - 4)(x + 4)(3x - 5)(2x + 1)} = \dfrac{(2x - 5)\overset{1}{\cancel{(x - 4)}}\overset{1}{\cancel{(2x + 1)}}\overset{1}{\cancel{(x + 4)}}}{\underset{1}{\cancel{(x - 4)}}\underset{1}{\cancel{(x + 4)}}(3x - 5)\underset{1}{\cancel{(2x + 1)}}} = \dfrac{2x - 5}{3x - 5}$

Problem 2 A. $\dfrac{6x^2 - 3xy}{10ab^4} \div \dfrac{16x^2y^2 - 8xy^3}{15a^2b^2} = \dfrac{6x^2 - 3xy}{10ab^4} \cdot \dfrac{15a^2b^2}{16x^2y^2 - 8xy^3} = \dfrac{3x(2x - y)}{10ab^4} \cdot \dfrac{15a^2b^2}{8xy^2(2x - y)} =$

$\dfrac{45a^2b^2x(2x - y)}{80ab^4xy^2(2x - y)} = \dfrac{45a^2b^2x\overset{1}{\cancel{(2x - y)}}}{80ab^4xy^2\underset{1}{\cancel{(2x - y)}}} = \dfrac{9a}{16b^2y^2}$

B. $$\frac{6x^2-7x+2}{3x^2+x-2} \div \frac{4x^2-8x+3}{5x^2+x-4} = \frac{6x^2-7x+2}{3x^2+x-2} \cdot \frac{5x^2+x-4}{4x^2-8x+3} =$$

$$\frac{(2x-1)(3x-2)}{(x+1)(3x-2)} \cdot \frac{(x+1)(5x-4)}{(2x-1)(2x-3)} = \frac{(2x-1)(3x-2)(x+1)(5x-4)}{(x+1)(3x-2)(2x-1)(2x-3)} =$$

$$\frac{\overset{1}{\cancel{(2x-1)}}\overset{1}{\cancel{(3x-2)}}\overset{1}{\cancel{(x+1)}}(5x-4)}{\underset{1}{\cancel{(x+1)}}\underset{1}{\cancel{(3x-2)}}\underset{1}{\cancel{(2x-1)}}(2x-3)} = \frac{5x-4}{2x-3}$$

Problem 3 The LCM is $a(a-5)(a+5)$.

$$\frac{a-3}{a^2-5a} + \frac{a-9}{a^2-25} = \frac{a-3}{a(a-5)} \cdot \frac{a+5}{a+5} + \frac{a-9}{(a-5)(a+5)} \cdot \frac{a}{a} =$$

$$\frac{a^2+2a-15}{a(a-5)(a+5)} + \frac{a^2-9a}{a(a-5)(a+5)} = \frac{(a^2+2a-15)+(a^2-9a)}{a(a-5)(a+5)} =$$

$$\frac{a^2+2a-15+a^2-9a}{a(a-5)(a+5)} = \frac{2a^2-7a-15}{a(a-5)(a+5)} = \frac{(2a+3)(a-5)}{a(a-5)(a+5)} =$$

$$\frac{(2a+3)\overset{1}{\cancel{(a-5)}}}{a\underset{1}{\cancel{(a-5)}}(a+5)} = \frac{2a+3}{a(a+5)}$$

Problem 4 The LCM is $(x-2)(2x-3)$.

$$\frac{x-1}{x-2} - \frac{7-6x}{2x^2-7x+6} + \frac{4}{2x-3} = \frac{x-1}{x-2} \cdot \frac{2x-3}{2x-3} - \frac{7-6x}{(x-2)(2x-3)} + \frac{4}{2x-3} \cdot \frac{x-2}{x-2} =$$

$$\frac{2x^2-5x+3}{(x-2)(2x-3)} - \frac{7-6x}{(x-2)(2x-3)} + \frac{4x-8}{(x-2)(2x-3)} =$$

$$\frac{(2x^2-5x+3)-(7-6x)+(4x-8)}{(x-2)(2x-3)} = \frac{2x^2-5x+3-7+6x+4x-8}{(x-2)(2x-3)} =$$

$$\frac{2x^2+5x-12}{(x-2)(2x-3)} = \frac{(x+4)(2x-3)}{(x-2)(2x-3)} = \frac{(x+4)\overset{1}{\cancel{(2x-3)}}}{(x-2)\underset{1}{\cancel{(2x-3)}}} = \frac{x+4}{x-2}$$

SECTION 4.3 *pages 186–188*

Problem 1 **A.** The LCM of x and x^2 is x^2.

$$\frac{3+\frac{16}{x}+\frac{16}{x^2}}{6+\frac{5}{x}-\frac{4}{x^2}} = \frac{3+\frac{16}{x}+\frac{16}{x^2}}{6+\frac{5}{x}-\frac{4}{x^2}} \cdot \frac{x^2}{x^2} = \frac{3\cdot x^2+\frac{16}{x}\cdot x^2+\frac{16}{x^2}\cdot x^2}{6\cdot x^2+\frac{5}{x}\cdot x^2-\frac{4}{x^2}\cdot x^2} = \frac{3x^2+16x+16}{6x^2+5x-4} =$$

$$\frac{(3x+4)(x+4)}{(2x-1)(3x+4)} = \frac{\overset{1}{\cancel{(3x+4)}}(x+4)}{(2x-1)\underset{1}{\cancel{(3x+4)}}} = \frac{x+4}{2x-1}$$

B. The LCM is $x-3$.

$$\frac{2x+5+\frac{14}{x-3}}{4x+16+\frac{49}{x-3}} = \frac{2x+5+\frac{14}{x-3}}{4x+16+\frac{49}{x-3}} \cdot \frac{x-3}{x-3} = \frac{(2x+5)(x-3)+\frac{14}{x-3}(x-3)}{(4x+16)(x-3)+\frac{49}{x-3}(x-3)} =$$

$$\frac{2x^2-x-15+14}{4x^2+4x-48+49} = \frac{2x^2-x-1}{4x^2+4x+1} = \frac{(2x+1)(x-1)}{(2x+1)(2x+1)} = \frac{\overset{1}{\cancel{(2x+1)}}(x-1)}{\underset{1}{\cancel{(2x+1)}}(2x+1)} = \frac{x-1}{2x+1}$$

Problem 2 $3 + \dfrac{3}{3 + \frac{3}{y}} = 3 + \dfrac{3}{3 + \frac{3}{y}} \cdot \dfrac{y}{y} = 3 + \dfrac{3y}{3y + 3} = 3 + \dfrac{3y}{3(y + 1)} = 3 + \dfrac{y}{y + 1} =$

$\dfrac{3(y + 1)}{y + 1} + \dfrac{y}{y + 1} = \dfrac{3y + 3 + y}{y + 1} = \dfrac{4y + 3}{y + 1}$

SECTION 4.4 *pages 191–198*

Problem 1 **A.**

$$\frac{5}{2x - 3} = \frac{-2}{x + 1}$$
$$(x + 1)(2x - 3)\frac{5}{2x - 3} = (x + 1)(2x - 3)\frac{-2}{x + 1}$$
$$5(x + 1) = -2(2x - 3)$$
$$5x + 5 = -4x + 6$$
$$9x + 5 = 6$$
$$9x = 1$$
$$x = \frac{1}{9}$$

The solution is $\frac{1}{9}$.

B.

$$\frac{4x + 1}{2x - 1} = 2 + \frac{3}{x - 3}$$
$$(2x - 1)(x - 3)\frac{4x + 1}{2x - 1} = (2x - 1)(x - 3)\left(2 + \frac{3}{x - 3}\right)$$
$$(x - 3)(4x + 1) = (2x - 1)(x - 3)2 + (2x - 1)3$$
$$4x^2 - 11x - 3 = 4x^2 - 14x + 6 + 6x - 3$$
$$-11x - 3 = -8x + 3$$
$$-3x = 6$$
$$x = -2$$

The solution is -2.

Problem 2

Strategy To find the cost, write and solve a proportion using x to represent the cost.

Solution

$$\frac{2}{3.10} = \frac{15}{x}$$
$$x(3.10)\,\frac{2}{3.10} = x(3.10)\,\frac{15}{x}$$
$$2x = 15(3.10)$$
$$2x = 46.50$$
$$x = 23.25$$

The cost of 15 lb of cashews is \$23.25.

Problem 3

Strategy ▬ Time required for the small pipe to fill the tank: x

	Rate	*Time*	*Part*
Large pipe	$\frac{1}{9}$	6	$\frac{6}{9}$
Small pipe	$\frac{1}{x}$	6	$\frac{6}{x}$

▬ The sum of the part of the task completed by the large pipe and the part of the task completed by the small pipe is 1.

Solution

$$\frac{6}{9}+\frac{6}{x}=1$$
$$\frac{2}{3}+\frac{6}{x}=1$$
$$3x\left(\frac{2}{3}+\frac{6}{x}\right)=3x\cdot 1$$
$$2x+18=3x$$
$$18=x$$

The small pipe working alone will fill the tank in 18 h.

Problem 4

Strategy ▬ Rate of the wind: r

	Distance	*Rate*	*Time*
With wind	700	$150+r$	$\frac{700}{150+r}$
Against wind	500	$150-r$	$\frac{500}{150-r}$

▬ The time flying with the wind equals the time flying against the wind.

Solution

$$\frac{700}{150+r}=\frac{500}{150-r}$$
$$(150+r)(150-r)\left(\frac{700}{150+r}\right)=(150+r)(150-r)\left(\frac{500}{150-r}\right)$$
$$(150-r)700=(150+r)500$$
$$105{,}000-700r=75{,}000+500r$$
$$30{,}000=1200r$$
$$25=r$$

The rate of the wind is 25 mph.

SECTION 4.5 *pages 205–206*

Problem 1 **A.**

$$\frac{1}{R_1} + \frac{1}{R_2} = \frac{1}{R}$$
$$RR_1R_2\left(\frac{1}{R_1} + \frac{1}{R_2}\right) = RR_1R_2\left(\frac{1}{R}\right)$$
$$RR_1R_2\left(\frac{1}{R_1}\right) + RR_1R_2\left(\frac{1}{R_2}\right) = R_1R_2$$
$$RR_2 + RR_1 = R_1R_2$$
$$R(R_2 + R_1) = R_1R_2$$
$$R = \frac{R_1R_2}{R_2 + R_1}$$

B.

$$\frac{r}{r+1} = t$$
$$(r+1)\left(\frac{r}{r+1}\right) = t(r+1)$$
$$r = tr + t$$
$$r - tr = t$$
$$r(1-t) = t$$
$$r = \frac{t}{1-t}$$

SOLUTIONS to Chapter 5 Problems

SECTION 5.1 *pages 219–224*

Problem 1 **A.** $64^{\frac{2}{3}} = (2^6)^{\frac{2}{3}} = 2^4 = 16$ **B.** $16^{-\frac{3}{4}} = (2^4)^{-\frac{3}{4}} = 2^{-3} = \frac{1}{2^3} = \frac{1}{8}$

C. $(-81)^{\frac{3}{4}}$

The base of the exponential expression is negative, while the denominator of the exponent is a positive even number.

Therefore, $(-81)^{\frac{3}{4}}$ is not a real number.

Problem 2 **A.** $\dfrac{x^{\frac{1}{2}}y^{-\frac{5}{4}}}{x^{-\frac{4}{3}}y^{\frac{1}{3}}} = \dfrac{x^{\frac{1}{2}+\frac{4}{3}}}{y^{\frac{1}{3}+\frac{5}{4}}} = \dfrac{x^{\frac{11}{6}}}{y^{\frac{19}{12}}}$ **B.** $(x^{\frac{3}{4}}y^{\frac{1}{2}}z^{-\frac{2}{3}})^{-\frac{4}{3}} = x^{-1}y^{-\frac{2}{3}}z^{\frac{8}{9}} = \dfrac{z^{\frac{8}{9}}}{xy^{\frac{2}{3}}}$

C. $\left(\dfrac{16a^{-2}b^{\frac{4}{3}}}{9a^4b^{-\frac{2}{3}}}\right)^{-\frac{1}{2}} = \left(\dfrac{2^4a^{-6}b^2}{3^2}\right)^{-\frac{1}{2}} = \dfrac{2^{-2}a^3b^{-1}}{3^{-1}} = \dfrac{3a^3}{2^2b} = \dfrac{3a^3}{4b}$

Problem 3 **A.** $(2x^3)^{\frac{3}{4}} = \sqrt[4]{(2x^3)^3} = \sqrt[4]{8x^9}$

B. $-5a^{\frac{5}{6}} = -5(a^5)^{\frac{1}{6}} = -5\sqrt[6]{a^5}$

Problem 4 **A.** $\sqrt[3]{3ab} = (3ab)^{\frac{1}{3}}$

B. $\sqrt[4]{x^4 + y^4} = (x^4 + y^4)^{\frac{1}{4}}$

Problem 5 **A.** $-\sqrt[4]{x^{12}} = -(x^{12})^{\frac{1}{4}} = -x^3$

B. $\sqrt{121x^{10}y^4} = \sqrt{11^2x^{10}y^4} = 11x^5y^2$

C. $\sqrt[3]{-125a^6b^9} = \sqrt[3]{(-5)^3a^6b^9} = -5a^2b^3$

SECTION 5.2 *pages 228–234*

Problem 1 $\sqrt[5]{x^7} = \sqrt[5]{x^5 \cdot x^2} = \sqrt[5]{x^5}\sqrt[5]{x^2} = x\sqrt[5]{x^2}$

Problem 2 **A.** $3xy\sqrt[3]{81x^5y} - \sqrt[3]{192x^8y^4} = 3xy\sqrt[3]{3^4x^5y} - \sqrt[3]{2^6 \cdot 3x^8y^4} =$
$3xy\sqrt[3]{3^3x^3}\sqrt[3]{3x^2y} - \sqrt[3]{2^6x^6y^3}\sqrt[3]{3x^2y} = 3xy \cdot 3x\sqrt[3]{3x^2y} - 2^2x^2y\sqrt[3]{3x^2y} =$
$9x^2y\sqrt[3]{3x^2y} - 4x^2y\sqrt[3]{3x^2y} = 5x^2y\sqrt[3]{3x^2y}$

B. $4a\sqrt[3]{54a^7b^9} + a^2b\sqrt[3]{128a^4b^6} = 4a\sqrt[3]{3^3 \cdot 2a^7b^9} + a^2b\sqrt[3]{2^6 \cdot 2a^4b^6} =$
$4a\sqrt[3]{3^3a^6b^9}\sqrt[3]{2a} + a^2b\sqrt[3]{2^6a^3b^6}\sqrt[3]{2a} = 4a \cdot 3a^2b^3\sqrt[3]{2a} + a^2b \cdot 2^2ab^2\sqrt[3]{2a} =$
$12a^3b^3\sqrt[3]{2a} + 4a^3b^3\sqrt[3]{2a} = 16a^3b^3\sqrt[3]{2a}$

Problem 3 $\sqrt{5b}(\sqrt{3b} - \sqrt{10}) = \sqrt{15b^2} - \sqrt{50b} = \sqrt{3 \cdot 5b^2} - \sqrt{2 \cdot 5^2b} =$
$\sqrt{b^2}\sqrt{3 \cdot 5} - \sqrt{5^2}\sqrt{2b} = b\sqrt{15} - 5\sqrt{2b}$

Problem 4 **A.** $(2\sqrt[3]{2x} - 3)(\sqrt[3]{2x} - 5) = 2\sqrt[3]{4x^2} - 10\sqrt[3]{2x} - 3\sqrt[3]{2x} + 15 = 2\sqrt[3]{4x^2} - 13\sqrt[3]{2x} + 15$

B. $(2\sqrt{x} - 3)(2\sqrt{x} + 3) = 2^2\sqrt{x^2} - 9 = 4x - 9$

Problem 5 **A.** $\frac{y}{\sqrt{3y}} = \frac{y}{\sqrt{3y}} \cdot \frac{\sqrt{3y}}{\sqrt{3y}} = \frac{y\sqrt{3y}}{\sqrt{3^2y^2}} = \frac{y\sqrt{3y}}{3y} = \frac{\sqrt{3y}}{3}$

B. $\frac{3}{\sqrt[3]{3x^2}} = \frac{3}{\sqrt[3]{3x^2}} \cdot \frac{\sqrt[3]{3^2x}}{\sqrt[3]{3^2x}} = \frac{3\sqrt[3]{9x}}{\sqrt[3]{3^3x^3}} = \frac{3\sqrt[3]{9x}}{3x} = \frac{\sqrt[3]{9x}}{x}$

Problem 6 **A.** $\frac{4 + \sqrt{2}}{3 - \sqrt{3}} = \frac{4 + \sqrt{2}}{3 - \sqrt{3}} \cdot \frac{3 + \sqrt{3}}{3 + \sqrt{3}} = \frac{12 + 4\sqrt{3} + 3\sqrt{2} + \sqrt{6}}{9 - (\sqrt{3})^2}$
$= \frac{12 + 4\sqrt{3} + 3\sqrt{2} + \sqrt{6}}{6}$

B. $\frac{\sqrt{2} + \sqrt{x}}{\sqrt{2} - \sqrt{x}} = \frac{\sqrt{2} + \sqrt{x}}{\sqrt{2} - \sqrt{x}} \cdot \frac{\sqrt{2} + \sqrt{x}}{\sqrt{2} + \sqrt{x}} = \frac{\sqrt{2^2} + \sqrt{2x} + \sqrt{2x} + \sqrt{x^2}}{(\sqrt{2})^2 - (\sqrt{x})^2} = \frac{2 + 2\sqrt{2x} + x}{2 - x}$

SECTION 5.3 *pages 237–243*

Problem 1 $\sqrt{-45} = i\sqrt{45} = i\sqrt{3^2 \cdot 5} = 3i\sqrt{5}$

Problem 2 $\sqrt{98} - \sqrt{-60} = \sqrt{98} - i\sqrt{60} = \sqrt{2 \cdot 7^2} - i\sqrt{2^2 \cdot 3 \cdot 5} = 7\sqrt{2} - 2i\sqrt{15}$

Problem 3 $(-4 + 2i) - (6 - 8i) = -10 + 10i$

Problem 4 $(16 - \sqrt{-45}) - (3 + \sqrt{-20}) = (16 - i\sqrt{45}) - (3 + i\sqrt{20}) =$
$(16 - i\sqrt{3^2 \cdot 5}) - (3 + i\sqrt{2^2 \cdot 5}) = (16 - 3i\sqrt{5}) - (3 + 2i\sqrt{5}) = 13 - 5i\sqrt{5}$

Problem 5 $\sqrt{-3}(\sqrt{27} - \sqrt{-6}) = i\sqrt{3}(\sqrt{27} - i\sqrt{6}) = i\sqrt{81} - i^2\sqrt{18} =$
$i\sqrt{3^4} - (-1)\sqrt{2 \cdot 3^2} = 9i + 3\sqrt{2} = 3\sqrt{2} + 9i$

Problem 6 **A.** $(4 - 3i)(2 - i) = 8 - 4i - 6i + 3i^2 = 8 - 10i + 3i^2 = 8 - 10i + 3(-1) = 5 - 10i$

B. $(3 - i)\left(\frac{3}{10} + \frac{1}{10}i\right) = \frac{9}{10} + \frac{3}{10}i - \frac{3}{10}i - \frac{1}{10}i^2 = \frac{9}{10} - \frac{1}{10}i^2 = \frac{9}{10} - \frac{1}{10}(-1) = \frac{9}{10} + \frac{1}{10} = 1$

C. $(3 + 6i)(3 - 6i) = 3^2 + 6^2 = 9 + 36 = 45$

Problem 7 $\frac{2-3i}{4i} = \frac{2-3i}{4i}\cdot\frac{i}{i} = \frac{2i-3i^2}{4i^2} = \frac{2i-3(-1)}{4(-1)} = \frac{3+2i}{-4} = -\frac{3}{4} - \frac{1}{2}i$

Problem 8 $\frac{2+5i}{3-2i} = \frac{(2+5i)}{(3-2i)}\cdot\frac{(3+2i)}{(3+2i)} = \frac{6+4i+15i+10i^2}{3^2+2^2} = \frac{6+19i+10(-1)}{13} = \frac{-4+19i}{13} = -\frac{4}{13} + \frac{19}{13}i$

SECTION 5.4 *pages 246–250*

Problem 1 A.

$$\begin{aligned} \sqrt{4x+5} - 12 &= -5 \\ \sqrt{4x+5} &= 7 \\ (\sqrt{4x+5})^2 &= 7^2 \\ 4x+5 &= 49 \\ 4x &= 44 \\ x &= 11 \end{aligned}$$

Check:

$$\begin{array}{r|l} \multicolumn{2}{c}{\sqrt{4x+5} - 12 = -5} \\ \hline \sqrt{4\cdot 11+5} - 12 & -5 \\ \sqrt{44+5} - 12 & \\ \sqrt{49} - 12 & \\ 7 - 12 & \\ \multicolumn{2}{r}{-5 = -5} \end{array}$$

The solution is 11.

B.

$$\begin{aligned} \sqrt[4]{x-8} &= 3 \\ (\sqrt[4]{x-8})^4 &= 3^4 \\ x-8 &= 81 \\ x &= 89 \end{aligned}$$

Check:

$$\begin{array}{r|l} \multicolumn{2}{c}{\sqrt[4]{x-8} = 3} \\ \hline \sqrt[4]{89-8} & 3 \\ \sqrt[4]{81} & \\ \multicolumn{2}{r}{3 = 3} \end{array}$$

The solution is 89.

Problem 2 A.

$$\begin{aligned} x + 3\sqrt{x+2} &= 8 \\ 3\sqrt{x+2} &= 8 - x \\ (3\sqrt{x+2})^2 &= (8-x)^2 \\ 9(x+2) &= 64 - 16x + x^2 \\ 9x + 18 &= 64 - 16x + x^2 \\ 0 &= x^2 - 25x + 46 \\ 0 &= (x-2)(x-23) \end{aligned}$$

$x - 2 = 0 \qquad x - 23 = 0$

$x = 2 \qquad x = 23$

Check:

$$\begin{array}{r|l} \multicolumn{2}{c}{x + 3\sqrt{x+2} = 8} \\ \hline 2 + 3\sqrt{2+2} & 8 \\ 2 + 3\sqrt{4} & \\ 2 + 3\cdot 2 & \\ 2 + 6 & \\ \multicolumn{2}{r}{8 = 8} \end{array} \qquad \begin{array}{r|l} \multicolumn{2}{c}{x + 3\sqrt{x+2} = 8} \\ \hline 23 + 3\sqrt{23+2} & 8 \\ 23 + 3\sqrt{25} & \\ 23 + 3\cdot 5 & \\ 23 + 15 & \\ \multicolumn{2}{r}{38 \neq 8} \end{array}$$

23 does not check as a solution.
The solution is 2.

B.

$$\begin{aligned} \sqrt{x+5} &= 5 - \sqrt{x} \\ (\sqrt{x+5})^2 &= (5 - \sqrt{x})^2 \\ x + 5 &= 25 - 10\sqrt{x} + x \\ -20 &= -10\sqrt{x} \\ 2 &= \sqrt{x} \\ 2^2 &= (\sqrt{x})^2 \\ 4 &= x \end{aligned}$$

Check: $\sqrt{x+5} = 5 - \sqrt{x}$

$$\begin{array}{r|l} \sqrt{4+5} & 5 - \sqrt{4} \\ \sqrt{9} & 5 - 2 \\ \multicolumn{2}{c}{3 = 3} \end{array}$$

The solution is 4.

Problem 3

Strategy To find the diagonal, use the Pythagorean Theorem. One leg is the length of the rectangle. The second leg is the width of the rectangle. The hypotenuse is the diagonal of the rectangle.

Solution

$$c^2 = a^2 + b^2$$
$$c^2 = (6)^2 + (3)^2$$
$$c^2 = 36 + 9$$
$$c^2 = 45$$
$$(c^2)^{\frac{1}{2}} = (45)^{\frac{1}{2}}$$
$$c = \sqrt{45}$$
$$c \approx 6.7$$

The diagonal is 6.7 cm.

Problem 4

Strategy To find the height above water, replace d in the equation with the given value, and solve for h.

Solution

$$d = 1.4\sqrt{h}$$
$$5.5 = 1.4\sqrt{h}$$
$$\frac{5.5}{1.4} = \sqrt{h}$$
$$\left(\frac{5.5}{1.4}\right)^2 = (\sqrt{h})^2$$
$$\frac{30.25}{1.96} = h$$
$$15.434 \approx h$$

The periscope must be 15.434 ft above the water.

SOLUTIONS to Chapter 6 Problems

SECTION 6.1 *pages 265–270*

Problem 1 A.

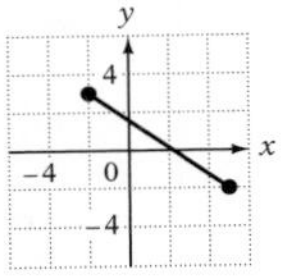

B.

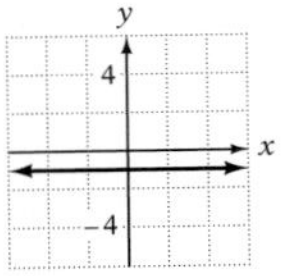

Problem 2

$$y = -2x + 5$$
$$y = -2\left(\frac{1}{3}\right) + 5$$
$$y = -\frac{2}{3} + 5$$
$$y = \frac{13}{3}$$

The ordered pair solution is $\left(\frac{1}{3}, \frac{13}{3}\right)$.

Problem 3

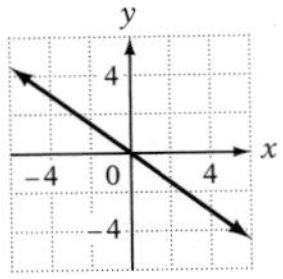

Problem 4

$$-3x + 2y = 4$$
$$2y = 3x + 4$$
$$y = \frac{3}{2}x + 2$$

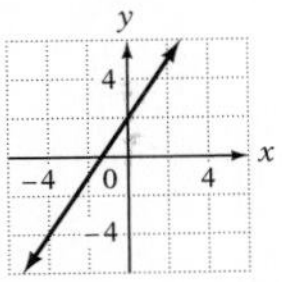

Problem 5

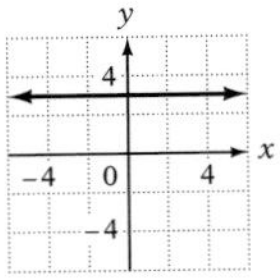

SECTION 6.2 *pages 274–280*

Problem 1 Let $P_1 = (4, -3)$ and $P_2 = (2, 7)$.

$$m = \frac{y_2 - y_1}{x_2 - x_1} = \frac{7 - (-3)}{2 - 4} = \frac{10}{-2} = -5$$

The slope is -5.

Problem 2 x-intercept:

$$3x - y = 2$$
$$3x - 0 = 2$$
$$3x = 2$$
$$x = \frac{2}{3}$$

$\left(\frac{2}{3}, 0\right)$

y-intercept:

$$3x - y = 2$$
$$3(0) - y = 2$$
$$-y = 2$$
$$y = -2$$

$(0, -2)$

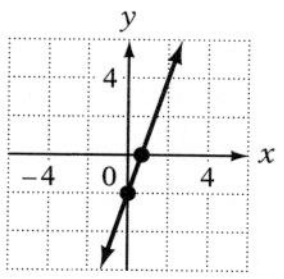

Problem 3 x-intercept:

$$y = \frac{1}{4}x + 1$$
$$0 = \frac{1}{4}x + 1$$
$$-\frac{1}{4}x = 1$$
$$x = -4$$

$(-4, 0)$

y-intercept:

$(0, b)$

$b = 1$

$(0, 1)$

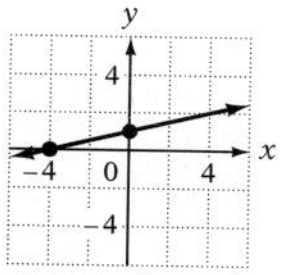

Problem 4

$$2x + 3y = 6$$
$$3y = -2x + 6$$
$$y = -\frac{2}{3}x + 2$$

$$m = -\frac{2}{3} = \frac{-2}{3}$$

y-intercept: $(0, 2)$

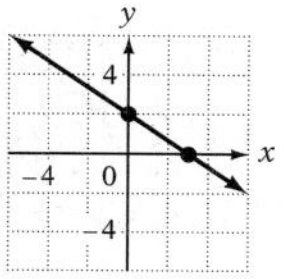

Problem 5 $(x_1, y_1) = (-3, -2)$

$$m = 3 = \frac{3}{1}$$

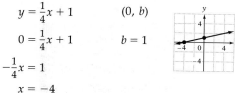

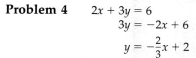

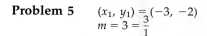

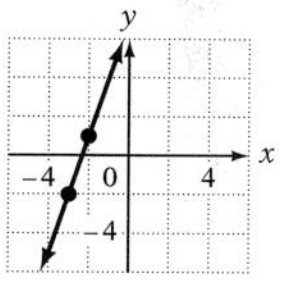

SECTION 6.3 *pages 284–290*

Problem 1 $m = -\frac{5}{4}$ $b = 3$

$$y = mx + b$$
$$y = -\frac{5}{4}x + 3$$

The equation of the line is $y = -\frac{5}{4}x + 3$.

Problem 2 $m = -3$ $(x_1, y_1) = (4, -3)$

$$\begin{aligned} y - y_1 &= m(x - x_1) \\ y - (-3) &= -3(x - 4) \\ y + 3 &= -3x + 12 \\ y &= -3x + 9 \end{aligned}$$

The equation of the line is $y = -3x + 9$.

Problem 3 **A.** Let $(x_1, y_1) = (4, -2)$ and $(x_2, y_2) = (-1, -7)$.

$$m = \frac{y_2 - y_1}{x_2 - x_1} = \frac{-7 - (-2)}{-1 - 4} = \frac{-5}{-5} = 1$$
$$\begin{aligned} y - y_1 &= m(x - x_1) \\ y - (-2) &= 1(x - 4) \\ y + 2 &= x - 4 \\ y &= x - 6 \end{aligned}$$

The equation of the line is $y = x - 6$.

B. Let $(x_1, y_1) = (2, 3)$ and $(x_2, y_2) = (-5, 3)$.

$$m = \frac{y_2 - y_1}{x_2 - x_1} = \frac{3 - 3}{-5 - 2} = \frac{0}{-7} = 0$$

The line has zero slope.
The line is a horizontal line.
All points on the line have an ordinate of 3.
The equation of the line is $y = 3$.

Problem 4

$$\begin{aligned} 5x + 2y &= 2 \\ 2y &= -5x + 2 \\ y &= -\frac{5}{2}x + 1 \end{aligned} \qquad \begin{aligned} 5x + 2y &= -6 \\ 2y &= -5x - 6 \\ y &= -\frac{5}{2}x - 3 \end{aligned}$$

$$m_1 = m_2 = -\frac{5}{2}$$

The slopes of the lines are equal. The lines are parallel.

Problem 5

$$x - 4y = 3$$
$$-4y = -x + 3$$
$$y = \frac{1}{4}x - \frac{3}{4}$$
$$m_1 = \frac{1}{4}$$
$$m_1 \cdot m_2 = -1$$
$$\frac{1}{4} \cdot m_2 = -1$$
$$m_2 = -4$$
$$y - y_1 = m(x - x_1)$$
$$y - 2 = -4[x - (-2)]$$
$$y - 2 = -4(x + 2)$$
$$y - 2 = -4x - 8$$
$$y = -4x - 6$$

The equation of the line is $y = -4x - 6$.

SECTION 6.4 *pages 295–297*

Problem 1

Strategy To write the equation:

- Use two points on the graph to find the slope of the line.
- Locate the y-intercept of the line on the graph.
- Use the slope-intercept form of an equation to write the equation of the line.

To find the Fahrenheit temperature, substitute 40° for x in the equation, and solve for y.

Solution $(x_1, y_1) = (0, 32)$ $(x_2, y_2) = (100, 212)$

$$m = \frac{y_2 - y_1}{x_2 - x_1} = \frac{212 - 32}{100 - 0} = \frac{180}{100} = \frac{9}{5}$$

The y-intercept is (0, 32).

$$y = mx + b$$
$$y = \frac{9}{5}x + 32$$

The equation of the line is $y = \frac{9}{5}x + 32$.

$$y = \frac{9}{5}x + 32$$
$$y = \frac{9}{5}(40) + 32 = 72 + 32 = 104$$

The Fahrenheit temperature is 104°.

SECTION 6.5 *pages 300–301*

Problem 1 **A.** $x + 3y > 6$

$3y > -x + 6$

$y > -\frac{1}{3}x + 2$

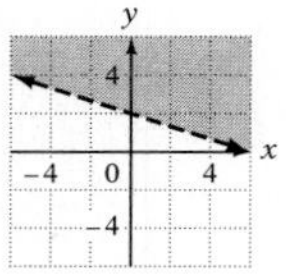

B. $y < 2$

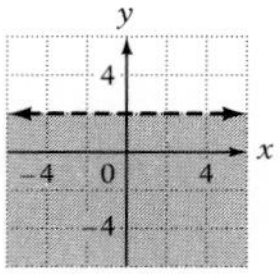

SOLUTIONS to Chapter 7 Problems

SECTION 7.1 *pages 317–320*

Problem 1

$x^2 - 3ax - 4a^2 = 0$

$(x + a)(x - 4a) = 0$

$x + a = 0 \qquad x - 4a = 0$

$x = -a \qquad x = 4a$

The solutions are $-a$ and $4a$.

Problem 2

$(x - r_1)(x - r_2) = 0$

$\left[x - \left(-\frac{2}{3}\right)\right]\left(x - \frac{1}{6}\right) = 0$

$\left(x + \frac{2}{3}\right)\left(x - \frac{1}{6}\right) = 0$

$x^2 + \frac{3}{6}x - \frac{2}{18} = 0$

$18\left(x^2 + \frac{3}{6}x - \frac{2}{18}\right) = 0$

$18x^2 + 9x - 2 = 0$

Problem 3

$2(x + 1)^2 + 24 = 0$

$2(x + 1)^2 = -24$

$(x + 1)^2 = -12$

$\sqrt{(x + 1)^2} = \sqrt{-12}$

$x + 1 = \pm\sqrt{-12} = \pm 2i\sqrt{3}$

$x + 1 = 2i\sqrt{3} \qquad x + 1 = -2i\sqrt{3}$

$x = -1 + 2i\sqrt{3} \qquad x = -1 - 2i\sqrt{3}$

The solutions are $-1 + 2i\sqrt{3}$ and $-1 - 2i\sqrt{3}$.

SECTION 7.2 *pages 324–331*

Problem 1

A.
$$4x^2 - 4x - 1 = 0$$
$$4x^2 - 4x = 1$$
$$\frac{4x^2 - 4x}{4} = \frac{1}{4}$$
$$x^2 - x = \frac{1}{4}$$

Complete the square.
$$x^2 - x + \frac{1}{4} = \frac{1}{4} + \frac{1}{4}$$
$$\left(x - \frac{1}{2}\right)^2 = \frac{2}{4}$$
$$\sqrt{\left(x - \frac{1}{2}\right)^2} = \sqrt{\frac{2}{4}}$$
$$x - \frac{1}{2} = \pm\frac{\sqrt{2}}{2}$$

$$x - \frac{1}{2} = \frac{\sqrt{2}}{2} \qquad x - \frac{1}{2} = -\frac{\sqrt{2}}{2}$$
$$x = \frac{1}{2} + \frac{\sqrt{2}}{2} \qquad x = \frac{1}{2} - \frac{\sqrt{2}}{2}$$

The solutions are $\frac{1 + \sqrt{2}}{2}$ and $\frac{1 - \sqrt{2}}{2}$.

B.
$$2x^2 + x - 5 = 0$$
$$2x^2 + x = 5$$
$$\frac{2x^2 + x}{2} = \frac{5}{2}$$
$$x^2 + \frac{1}{2}x = \frac{5}{2}$$

Complete the square.
$$x^2 + \frac{1}{2}x + \frac{1}{16} = \frac{5}{2} + \frac{1}{16}$$
$$\left(x + \frac{1}{4}\right)^2 = \frac{41}{16}$$
$$\sqrt{\left(x + \frac{1}{4}\right)^2} = \sqrt{\frac{41}{16}}$$
$$x + \frac{1}{4} = \pm\frac{\sqrt{41}}{4}$$

$$x + \frac{1}{4} = \frac{\sqrt{41}}{4} \qquad x + \frac{1}{4} = -\frac{\sqrt{41}}{4}$$
$$x = -\frac{1}{4} + \frac{\sqrt{41}}{4} \qquad x = -\frac{1}{4} - \frac{\sqrt{41}}{4}$$

The solutions are $\frac{-1 + \sqrt{41}}{4}$ and $\frac{-1 - \sqrt{41}}{4}$.

Problem 2

A. $x^2 + 6x - 9 = 0$

$a = 1,\ b = 6,\ c = -9$
$$x = \frac{-b \pm \sqrt{b^2 - 4ac}}{2a}$$
$$= \frac{-6 \pm \sqrt{6^2 - 4(1)(-9)}}{2 \cdot 1}$$
$$= \frac{-6 \pm \sqrt{36 + 36}}{2}$$
$$= \frac{-6 \pm \sqrt{72}}{2} = \frac{-6 \pm 6\sqrt{2}}{2}$$
$$= -3 \pm 3\sqrt{2}$$

The solutions are $-3 + 3\sqrt{2}$ and $-3 - 3\sqrt{2}$.

B.
$$4x^2 = 4x - 1$$
$$4x^2 - 4x + 1 = 0$$

$a = 4,\ b = -4,\ c = 1$
$$x = \frac{-b \pm \sqrt{b^2 - 4ac}}{2a}$$
$$= \frac{-(-4) \pm \sqrt{(-4)^2 - 4(4)(1)}}{2 \cdot 4}$$
$$= \frac{4 \pm \sqrt{16 - 16}}{8} = \frac{4 \pm \sqrt{0}}{8}$$
$$= \frac{4}{8} = \frac{1}{2}$$

The solution is $\frac{1}{2}$.

Problem 3 $3x^2 - x - 1 = 0$

$a = 3,\ b = -1,\ c = -1$

$b^2 - 4ac$
$(-1)^2 - 4(3)(-1) = 1 + 12 = 13$
$13 > 0$

Since the discriminant is greater than zero, the equation has two real number solutions.

SECTION 7.3 *pages 335–341*

Problem 1

A.

$$x - 5x^{\frac{1}{2}} + 6 = 0$$
$$\left(x^{\frac{1}{2}}\right)^2 - 5\left(x^{\frac{1}{2}}\right) + 6 = 0$$
$$u^2 - 5u + 6 = 0$$
$$(u - 2)(u - 3) = 0$$

$u - 2 = 0$, $u = 2$ $\qquad$ $u - 3 = 0$, $u = 3$

Replace u with $x^{\frac{1}{2}}$.

$x^{\frac{1}{2}} = 2$, $\left(x^{\frac{1}{2}}\right)^2 = 2^2$, $x = 4$ $\qquad$ $x^{\frac{1}{2}} = 3$, $\left(x^{\frac{1}{2}}\right)^2 = 3^2$, $x = 9$

4 and 9 check as solutions.
The solutions are 4 and 9.

B.

$$4x^4 + 35x^2 - 9 = 0$$
$$4(x^2)^2 + 35(x^2) - 9 = 0$$
$$4u^2 + 35u - 9 = 0$$
$$(4u - 1)(u + 9) = 0$$

$4u - 1 = 0$, $4u = 1$, $u = \frac{1}{4}$ $\qquad$ $u + 9 = 0$, $u = -9$

Replace u with x^2.

$x^2 = \frac{1}{4}$, $\sqrt{x^2} = \sqrt{\frac{1}{4}}$, $x = \pm\frac{1}{2}$ $\qquad$ $x^2 = -9$, $\sqrt{x^2} = \sqrt{-9}$, $x = \pm 3i$

The solutions are $\frac{1}{2}$, $-\frac{1}{2}$, $3i$, and $-3i$.

Problem 2

$$\sqrt{2x + 1} + x = 7$$
$$\sqrt{2x + 1} = 7 - x$$
$$(\sqrt{2x + 1})^2 = (7 - x)^2$$
$$2x + 1 = 49 - 14x + x^2$$
$$0 = x^2 - 16x + 48$$
$$0 = (x - 4)(x - 12)$$

$x - 4 = 0$, $x = 4$ $\qquad$ $x - 12 = 0$, $x = 12$

4 checks as a solution.
12 does not check as a solution.

The solution is 4.

Problem 3

$$\sqrt{2x - 1} + \sqrt{x} = 2$$

Solve for one of the radical expressions.

$$\sqrt{2x - 1} = 2 - \sqrt{x}$$
$$(\sqrt{2x - 1})^2 = (2 - \sqrt{x})^2$$
$$2x - 1 = 4 - 4\sqrt{x} + x$$
$$x - 5 = -4\sqrt{x}$$

Square each side of the equation.

$$(x - 5)^2 = (-4\sqrt{x})^2$$
$$x^2 - 10x + 25 = 16x$$
$$x^2 - 26x + 25 = 0$$
$$(x - 1)(x - 25) = 0$$

$x - 1 = 0$, $x = 1$ $\qquad$ $x - 25 = 0$, $x = 25$

1 checks as a solution.
25 does not check as a solution.

The solution is 1.

Problem 4 **A.**

$$3y + \frac{25}{3y-2} = -8$$
$$(3y-2)\left(3y + \frac{25}{3y-2}\right) = (3y-2)(-8)$$
$$(3y-2)(3y) + (3y-2)\left(\frac{25}{3y-2}\right) = (3y-2)(-8)$$
$$9y^2 - 6y + 25 = -24y + 16$$
$$9y^2 + 18y + 9 = 0$$
$$9(y^2 + 2y + 1) = 0$$
$$9(y+1)(y+1) = 0$$

$y + 1 = 0$ $\quad$ $y + 1 = 0$

$y = -1$ $\quad$ $y = -1$

The solution is -1.

B.

$$\frac{5}{x+2} = 2x - 5$$
$$(x+2)\left(\frac{5}{x+2}\right) = (x+2)(2x-5)$$
$$5 = 2x^2 - x - 10$$
$$0 = 2x^2 - x - 15$$
$$0 = (2x+5)(x-3)$$

$2x + 5 = 0$ $\quad$ $x - 3 = 0$

$2x = -5$ $\quad$ $x = 3$

$x = -\frac{5}{2}$

The solutions are $-\frac{5}{2}$ and 3.

SECTION 7.4 *pages 344–350*

Problem 1 **A.**

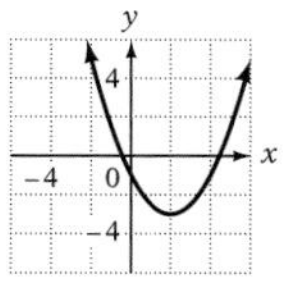

B.

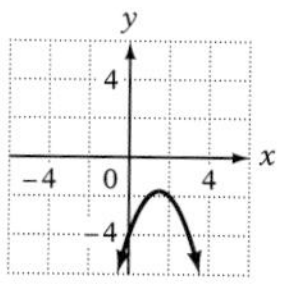

Problem 2 x-coordinate: $-\frac{b}{2a} = -\frac{0}{2(1)} = 0$

The axis of symmetry is the line $x = 0$.

$y = x^2 - 2$

$= 0^2 - 2$

$= -2$

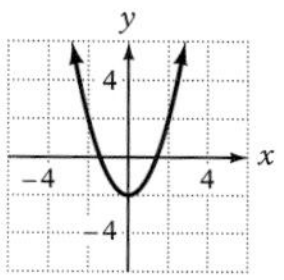

The vertex is $(0, -2)$.

Problem 3

A. $y = 2x^2 - 5x + 2$
$0 = 2x^2 - 5x + 2$
$0 = (2x - 1)(x - 2)$

$2x - 1 = 0 \qquad x - 2 = 0$
$2x = 1 \qquad x = 2$
$x = \frac{1}{2}$

The x-intercepts are $\left(\frac{1}{2}, 0\right)$ and $(2, 0)$.

B. $y = x^2 + 4x + 4$
$0 = x^2 + 4x + 4$
$0 = (x + 2)(x + 2)$

$x + 2 = 0 \qquad x + 2 = 0$
$x = -2 \qquad x = -2$

The x-intercept is $(-2, 0)$.

Problem 4

$y = x^2 - x - 6$

$a = 1,\ b = -1,\ c = -6$

$b^2 - 4ac$
$(-1)^2 - 4(1)(-6) = 1 + 24 = 25$

Since the discriminant is greater than zero, the parabola has two x-intercepts.

SECTION 7.5 *pages 355–357*

Problem 1

Strategy

- This is a geometry problem.
- Width of the rectangle: w
  Length of the rectangle: $w + 3$
- Use the equation $A = L \cdot w$.

Solution

$A = L \cdot w$
$54 = (w + 3)(w)$
$54 = w^2 + 3w$
$0 = w^2 + 3w - 54$
$0 = (w + 9)(w - 6)$

$w + 9 = 0 \qquad w - 6 = 0$
$w = -9 \qquad w = 6$

The solution -9 is not possible.

length $= w + 3 = 6 + 3 = 9$

The length is 9 m.

SECTION 7.6 *pages 360–362*

Problem 1

$$2x^2 - x - 10 \le 0$$
$$(2x - 5)(x + 2) \le 0$$

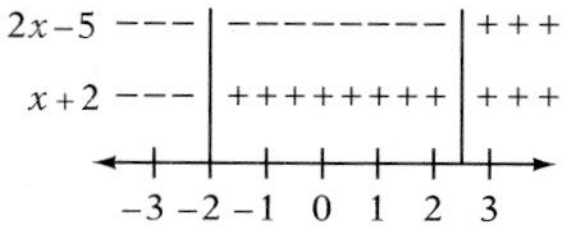

$$\left\{x \middle| -2 \le x \le \frac{5}{2}\right\}$$

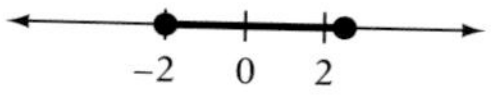

Problem 2

$$\frac{x}{x-2} \le 0$$

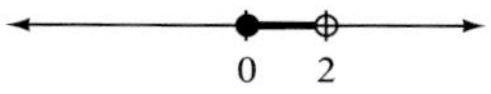

$$\{x \mid 0 \le x < 2\}$$

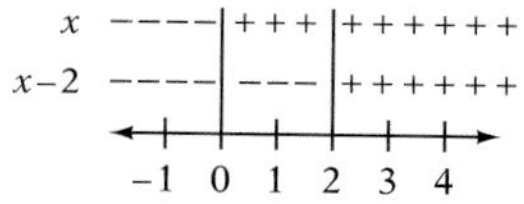

SOLUTIONS to Chapter 8 Problems

SECTION 8.1 *pages 375–380*

Problem 1

$$\begin{aligned} s(t) &= 2t^2 + 3t - 4 \\ s(-3) &= 2(-3)^2 + 3(-3) - 4 \\ &= 2(9) + 3(-3) - 4 \\ &= 18 + (-9) - 4 \\ &= 9 - 4 \\ &= 5 \end{aligned}$$

Problem 2

$$\begin{aligned} f(x) &= x^2 \\ f(a+1) &= (a+1)^2 \\ &= a^2 + 2a + 1 \end{aligned}$$

Problem 3

$$\begin{aligned} f(x) - g(x) &= 2x^2 - (4x - 1) \\ f(-1) - g(-1) &= 2(-1)^2 - [4(-1) - 1] \\ &= 2(1) - (-4 - 1) \\ &= 2 - (-5) \\ &= 7 \end{aligned}$$

Problem 4

The domain is {0, 1, 2, 3, 4}.
The range is {1, 3, 5, 7, 9}.

Problem 5

$$\begin{aligned} f(x) &= x^2 - 2x + 1 \\ f(-2) &= (-2)^2 - 2(-2) + 1 = 4 - (-4) + 1 = 9 \\ f(-1) &= (-1)^2 - 2(-1) + 1 = 1 - (-2) + 1 = 4 \\ f(0) &= (0)^2 - 2(0) + 1 = 0 - 0 + 1 = 1 \\ f(1) &= (1)^2 - 2(1) + 1 = 1 - 2 + 1 = 0 \\ f(2) &= (2)^2 - 2(2) + 1 = 4 - 4 + 1 = 1 \end{aligned}$$

The range is {0, 1, 4, 9}.

Problem 6

$$\begin{aligned} f(x) &= -4 \\ 2x - 1 &= -4 \\ 2x &= -3 \\ x &= -\frac{3}{2} \end{aligned}$$

$$\left(-\frac{3}{2}, -4\right)$$

Problem 7

$$\begin{aligned} 2x - 5 &< 0 \\ 2x &< 5 \\ x &< \frac{5}{2} \end{aligned}$$

$$\left\{x \middle| x < \frac{5}{2}\right\}$$

SECTION 8.2 *pages 384–394*

Problem 1

A.

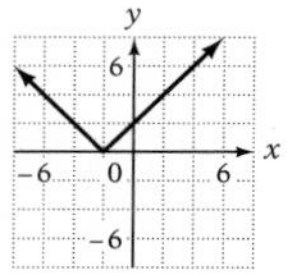

The domain is all real numbers.
The range is $\{y|y \geq 0\}$.

B.

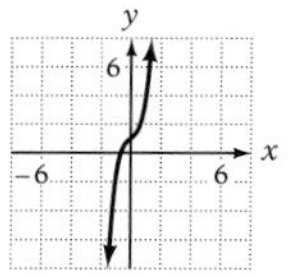

The domain is all real numbers.
The range is all real numbers.

Problem 2 Since any vertical line would intersect the graph at no more than one point, the graph is the graph of a function.

Problem 3

A. Since any vertical line will intersect the graph at no more than one point, and any horizontal line will intersect the graph at no more than one point, the graph is the graph of a 1–1 function.

B. Since any vertical line will intersect the graph at no more than one point, and any horizontal line will intersect the graph at no more than one point, the graph is the graph of a 1–1 function.

SECTION 8.3 *pages 399–404*

Problem 1

A.

$$\begin{aligned} h(x) &= x^2 + 1 \\ h(0) &= 0 + 1 = 1 \\ g(x) &= 3x - 2 \\ g(1) &= 3(1) - 2 = 1 \\ g(h(0)) &= 1 \end{aligned}$$

B.

$$\begin{aligned} h(g(x)) &= h(3x - 2) \\ &= (3x - 2)^2 + 1 \\ &= 9x^2 - 12x + 4 + 1 \\ &= 9x^2 - 12x + 5 \end{aligned}$$

Problem 2

$$\begin{aligned} f(x) &= 4x + 2 \\ y &= 4x + 2 \\ x &= 4y + 2 \\ 4y &= x - 2 \\ y &= \frac{1}{4}x - \frac{1}{2} \\ f^{-1}(x) &= \frac{1}{4}x - \frac{1}{2} \end{aligned}$$

Problem 3 $h(g(x)) = 4\left(\frac{1}{4}x - \frac{1}{2}\right) + 2 = x - 2 + 2 = x$

$$g(h(x)) = \frac{1}{4}(4x + 2) - \frac{1}{2} = x + \frac{1}{2} - \frac{1}{2} = x$$

The functions are inverses of each other.

SECTION 8.4 *pages 408–413*

Problem 1

Strategy To find the distance:

- Write the basic direct variation equation, replace the variables by the given values, and solve for k.
- Write the direct variation equation, replacing k by its value. Substitute 5 for t, and solve for s.

Solution

$$\begin{aligned} s &= kt^2 \\ 64 &= k(2)^2 \\ 64 &= k \cdot 4 \\ 16 &= k \end{aligned}$$

$$\begin{aligned} s &= 16t^2 \\ &= 16(5)^2 = 400 \end{aligned}$$

The object will fall 400 ft in 5 s.

Problem 2

Strategy To find the resistance:

- Write the basic inverse variation equation, replace the variables by the given values, and solve for k.
- Write the inverse variation equation, replacing k by its value. Substitute 0.02 for d, and solve for R.

Solution

$$\begin{aligned} R &= \frac{k}{d^2} \\ 0.5 &= \frac{k}{(0.01)^2} \\ 0.5 &= \frac{k}{0.0001} \\ 0.00005 &= k \end{aligned}$$

$$\begin{aligned} R &= \frac{0.00005}{d^2} \\ &= \frac{0.00005}{(0.02)^2} = 0.125 \end{aligned}$$

The resistance is 0.125 ohms.

Problem 3

Strategy To find the strength of the beam:

- Write the basic combined variation equation, replace the variables by the given values, and solve for k.
- Write the basic combined variation equation, replacing k by its value and substituting 4 for w and 8 for d. Solve for s.

Solution

$$s = \frac{kw}{d^2} \qquad s = \frac{kw}{d^2}$$

$$1200 = \frac{2k}{12^2} \qquad s = \frac{86{,}400(4)}{8^2}$$

$$1200 = \frac{2k}{144} \qquad s = \frac{345{,}600}{64}$$

$$172{,}800 = 2k \qquad s = 5400$$

$$86{,}400 = k$$

The strength of the beam is 5400 lb.

Problem 4

$$x = -\frac{b}{2a} = -\frac{-3}{2(2)} = \frac{3}{4}$$

$$f(x) = 2x^2 - 3x + 1$$

$$f\left(\frac{3}{4}\right) = 2\left(\frac{3}{4}\right)^2 - 3\left(\frac{3}{4}\right) + 1 = \frac{9}{8} - \frac{9}{4} + 1 = -\frac{1}{8}$$

Since a is positive, the function has a minimum value.

The minimum value of the function is $-\frac{1}{8}$.

Problem 5

Strategy

- To find the time it takes the ball to reach its maximum height, find the t-coordinate of the vertex.
- To find the maximum height, evaluate the function at the t-coordinate of the vertex.

Solution $t = -\frac{b}{2a} = -\frac{64}{2(-16)} = 2$

The ball reaches its maximum height in 2 s.

$$s(t) = -16t^2 + 64t$$

$$s(2) = -16(2)^2 + 64(2) = -64 + 128 = 64$$

The maximum height is 64 ft.

Problem 6

Strategy The perimeter is 44 ft.

$$44 = 2L + 2W$$

$$22 = L + W$$

$$22 - L = W$$

The area is $L \cdot W = L(22 - L) = 22L - L^2$.

- To find the length, find the L-coordinate of the vertex of the function $f(L) = -L^2 + 22L$.
- To find the width, replace L in $22 - L$ by the L-coordinate of the vertex and evaluate.

Solution $L = -\frac{b}{2a} = -\frac{22}{2(-1)} = 11$

The length is 11 ft.

$22 - L = 22 - 11 = 11$

The width is 11 ft.

SOLUTIONS to Chapter 9 Problems

SECTION 9.1 *pages 431–433*

Problem 1 $-\frac{b}{2a} = -\frac{1}{2(-1)} = \frac{1}{2}$

$$y = -x^2 + x + 3$$
$$= -\left(\frac{1}{2}\right)^2 + \frac{1}{2} + 3 = \frac{13}{4}$$

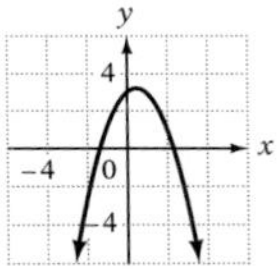

The vertex is $\left(\frac{1}{2}, \frac{13}{4}\right)$.

The axis of symmetry is the line $x = \frac{1}{2}$.

Problem 2 y-coordinate: $-\frac{b}{2a} = -\frac{-4}{2(-2)} = -1$

$$x = -2y^2 - 4y - 3$$
$$= -2(-1)^2 - 4(-1) - 3$$
$$= -1$$

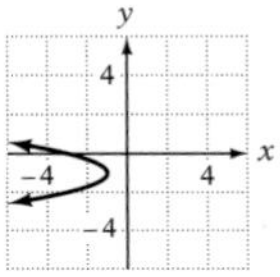

The vertex is $(-1, -1)$.

The axis of symmetry is the line $y = -1$.

SECTION 9.2 *pages 435–441*

Problem 1 $(x_1, y_1) = (-1, -5)$ $(x_2, y_2) = (3, -2)$

$$d = \sqrt{(x_2 - x_1)^2 + (y_2 - y_1)^2}$$
$$= \sqrt{[3 - (-1)]^2 + [(-2) - (-5)]^2}$$
$$= \sqrt{4^2 + 3^2} = \sqrt{16 + 9} = \sqrt{25} = 5$$

Problem 2 $x_m = \frac{x_1 + x_2}{2}$ $y_m = \frac{y_1 + y_2}{2}$

$$= \frac{-8 - 4}{2} \qquad = \frac{4 + 5}{2}$$
$$= -6 \qquad = \frac{9}{2}$$

The midpoint is $\left(-6, \frac{9}{2}\right)$.

Problem 3

$$(x-h)^2+(y-k)^2=r^2$$
$$(x-2)^2+[y-(-3)]^2=4^2$$
$$(x-2)^2+(y+3)^2=16$$

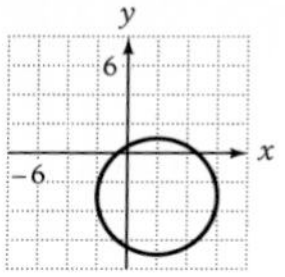

Problem 4

$$x_m=\frac{x_1+x_2}{2} \qquad y_m=\frac{y_1+y_2}{2}$$
$$x_m=\frac{-2+4}{2}=1 \qquad y_m=\frac{1-1}{2}=0;\ \text{Center } (1,\ 0)$$
$$r=\sqrt{(x_1-x_m)^2+(y_1-y_m)^2}$$
$$=\sqrt{(-2-1)^2+(1-0)^2}=\sqrt{9+1}=\sqrt{10};\ r=\sqrt{10}$$
$$(x-h)^2+(y-k)^2=r^2$$
$$(x-1)^2+(y-0)^2=10$$
$$(x-1)^2+y^2=10$$

Problem 5

$$x^2+y^2-4x+8y+15=0$$
$$(x^2-4x)+(y^2+8y)=-15$$
$$(x^2-4x+4)+(y^2+8y+16)=-15+4+16$$
$$(x-2)^2+(y+4)^2=5$$

Center: $(2,\ -4)$
Radius: $\sqrt{5}$

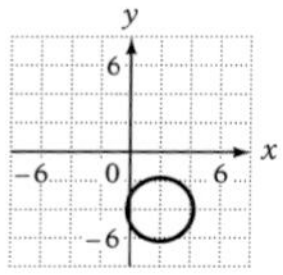

SECTION 9.3 *pages 444–449*

Problem 1 **A.** x-intercepts:
$(2,\ 0)$ and $(-2,\ 0)$

y-intercepts:
$(0,\ 5)$ and $(0,\ -5)$

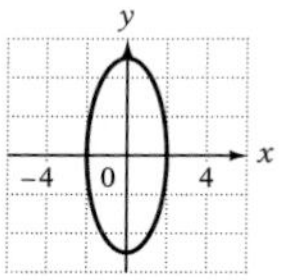

B. x-intercepts:
$(3\sqrt{2},\ 0)$ and $(-3\sqrt{2},\ 0)$

y-intercepts:
$(0,\ 3)$ and $(0,\ -3)$

$\left(3\sqrt{2}\approx 4\frac{1}{4}\right)$

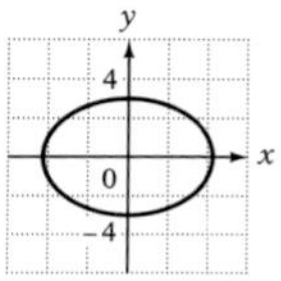

Problem 2 **A.** Axis of symmetry:
x-axis

Vertices:
(3, 0) and (−3, 0)

Asymptotes:
$y = \frac{5}{3}x$ and $y = -\frac{5}{3}x$

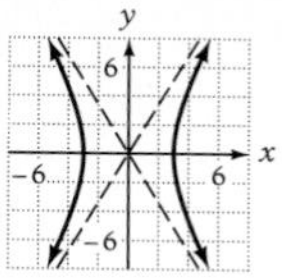

B. Axis of symmetry:
y-axis

Vertices:
(0, 3) and (0, −3)

Asymptotes:
$y = x$ and $y = -x$

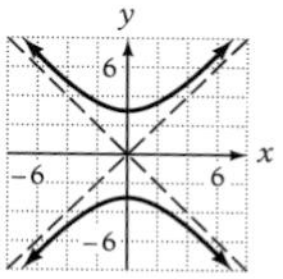

SECTION 9.4 *pages 452–453*

Problem 1 **A.** Graph the ellipse $\frac{x^2}{9} + \frac{y^2}{16} = 1$ as a solid line.
Shade the region of the plane that includes (0, 0).

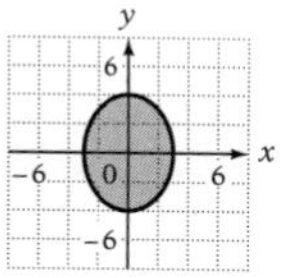

B. Graph the hyperbola $\frac{x^2}{9} - \frac{y^2}{4} = 1$ as a solid line.
Shade the region that includes (0, 0).

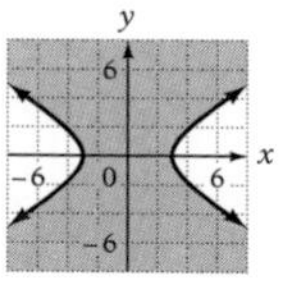

SOLUTIONS to Chapter 10 Problems

SECTION 10.1 *pages 465–469*

Problem 1 **A.**

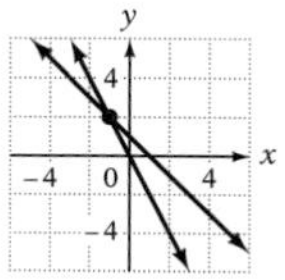

The solution is (−1, 2).

B.

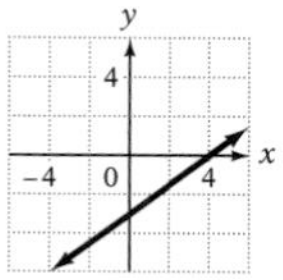

The system of equations is dependent. The solutions are the ordered pairs that are solutions of the equation $y = \frac{3}{4}x - 3$.

Problem 2 **A.** (1) $3x - y = 3$
(2) $6x + 3y = -4$

Solve equation (1) for y.

$$\begin{aligned} 3x - y &= 3 \\ -y &= -3x + 3 \\ y &= 3x - 3 \end{aligned}$$

Substitute into equation (2).

$$\begin{aligned} 6x + 3y &= -4 \\ 6x + 3(3x - 3) &= -4 \\ 6x + 9x - 9 &= -4 \\ 15x - 9 &= -4 \\ 15x &= 5 \\ x &= \frac{5}{15} = \frac{1}{3} \end{aligned}$$

Substitute into equation (1).

$$\begin{aligned} 3x - y &= 3 \\ 3\left(\frac{1}{3}\right) - y &= 3 \\ 1 - y &= 3 \\ -y &= 2 \\ y &= -2 \end{aligned}$$

The solution is $\left(\frac{1}{3}, -2\right)$.

B. (1) $6x - 3y = 6$
(2) $2x - y = 2$

Solve equation (2) for y.

$$\begin{aligned} 2x - y &= 2 \\ -y &= -2x + 2 \\ y &= 2x - 2 \end{aligned}$$

Substitute into equation (1).

$$\begin{aligned} 6x - 3y &= 6 \\ 6x - 3(2x - 2) &= 6 \\ 6x - 6x + 6 &= 6 \\ 6 &= 6 \end{aligned}$$

This is a true equation. The system of equations is dependent. The solutions are the ordered pairs that are solutions of the equation $2x - y = 2$.

SECTION 10.2 *pages 472–479*

Problem 1 **A.** (1) $2x + 5y = 6$
(2) $3x - 2y = 6x + 2$

Write equation (2) in the form $Ax + By = C$.

$$\begin{aligned} 3x - 2y &= 6x + 2 \\ -3x - 2y &= 2 \end{aligned}$$

Solve the system $2x + 5y = 6$
$-3x - 2y = 2$.

Eliminate y.

$$\begin{aligned} 2(2x + 5y) &= 2(6) \\ 5(-3x - 2y) &= 5(2) \end{aligned}$$

$$\begin{aligned} 4x + 10y &= 12 \\ -15x - 10y &= 10 \end{aligned}$$

Add the equations.

$$\begin{aligned} -11x &= 22 \\ x &= -2 \end{aligned}$$

Replace x in equation (1).

$$\begin{aligned} 2x + 5y &= 6 \\ 2(-2) + 5y &= 6 \\ -4 + 5y &= 6 \\ 5y &= 10 \\ y &= 2 \end{aligned}$$

The solution is $(-2, 2)$.

B. $2x + y = 5$
$4x + 2y = 6$

Eliminate y.
$-2(2x + y) = -2(5)$
$4x + 2y = 6$

$-4x - 2y = -10$
$4x + 2y = 6$

Add the equations.
$0x + 0y = -4$
$0 = -4$

This is not a true equation. The system is inconsistent and therefore has no solution.

Problem 2

(1) $x - y + z = 6$
(2) $2x + 3y - z = 1$
(3) $x + 2y + 2z = 5$

Eliminate z. Add equations (1) and (2).
$x - y + z = 6$
$2x + 3y - z = 1$
(4) $3x + 2y = 7$

Multiply equation (2) by 2 and add to equation (3).
$4x + 6y - 2z = 2$
$x + 2y + 2z = 5$
(5) $5x + 8y = 7$

Solve the system of two equations.
(4) $3x + 2y = 7$
(5) $5x + 8y = 7$

Multiply equation (4) by -4 and add to equation (5).
$-12x - 8y = -28$
$5x + 8y = 7$
$-7x = -21$
$x = 3$

Replace x by 3 in equation (4).
$3x + 2y = 7$
$3(3) + 2y = 7$
$9 + 2y = 7$
$2y = -2$
$y = -1$

Replace x by 3 and y by -1 in equation (1).
$x - y + z = 6$
$3 - (-1) + z = 6$
$4 + z = 6$
$z = 2$

The solution is $(3, -1, 2)$.

SECTION 10.3 *pages 483–496*

Problem 1 **A.** $\begin{vmatrix} -1 & -4 \\ 3 & -5 \end{vmatrix} = -1(-5) - 3(-4) = 5 + 12 = 17$

The value of the determinant is 17.

B. Expand by cofactors of the first column.

$$\begin{vmatrix} 1 & 4 & -2 \\ 3 & 1 & 1 \\ 0 & -2 & 2 \end{vmatrix} = 1\begin{vmatrix} 1 & 1 \\ -2 & 2 \end{vmatrix} - 3\begin{vmatrix} 4 & -2 \\ -2 & 2 \end{vmatrix} + 0$$

$$= 1(2 + 2) - 3(8 - 4)$$
$$= 4 - 12$$
$$= -8$$

The value of the determinant is -8.

Problem 2 $D = \begin{vmatrix} 6 & -6 \\ 2 & -10 \end{vmatrix} = -48$ $D_x = \begin{vmatrix} 5 & -6 \\ -1 & -10 \end{vmatrix} = -56$ $D_y = \begin{vmatrix} 6 & 5 \\ 2 & -1 \end{vmatrix} = -16$

$x = \frac{D_x}{D} = \frac{-56}{-48} = \frac{7}{6}$ $y = \frac{D_y}{D} = \frac{-16}{-48} = \frac{1}{3}$

The solution is $\left(\frac{7}{6}, \frac{1}{3}\right)$.

Problem 3 $D = \begin{vmatrix} 2 & -1 & 1 \\ 3 & 2 & -1 \\ 1 & 3 & 1 \end{vmatrix} = 21$ $D_x = \begin{vmatrix} -1 & -1 & 1 \\ 3 & 2 & -1 \\ -2 & 3 & 1 \end{vmatrix} = 9$

$D_y = \begin{vmatrix} 2 & -1 & 1 \\ 3 & 3 & -1 \\ 1 & -2 & 1 \end{vmatrix} = -3$ $D_z = \begin{vmatrix} 2 & -1 & -1 \\ 3 & 2 & 3 \\ 1 & 3 & -2 \end{vmatrix} = -42$

$x = \frac{D_x}{D} = \frac{9}{21} = \frac{3}{7}$ $y = \frac{D_y}{D} = \frac{-3}{21} = -\frac{1}{7}$ $z = \frac{D_z}{D} = \frac{-42}{21} = -2$

The solution is $\left(\frac{3}{7}, -\frac{1}{7}, -2\right)$.

Problem 4 $\begin{bmatrix} 3 & -5 & -12 \\ 4 & -3 & -5 \end{bmatrix}$

$\begin{bmatrix} 1 & -\frac{5}{3} & -4 \\ 4 & -3 & -5 \end{bmatrix}$ ▬ Multiply row 1 by $\frac{1}{3}$.

$$\begin{bmatrix} 1 & -\frac{5}{3} & -4 \\ 0 & \frac{11}{3} & 11 \end{bmatrix}$$

▬ Multiply row 1 by −4 and add to row 2. Replace row 2 by the sum.

$$\begin{bmatrix} 1 & -\frac{5}{3} & -4 \\ 0 & 1 & 3 \end{bmatrix}$$

▬ Multiply row 2 by $\frac{3}{11}$.

$$\begin{aligned} x - \tfrac{5}{3}y &= -4 \\ y &= 3 \end{aligned} \qquad \begin{aligned} x - \tfrac{5}{3}(3) &= -4 \\ x - 5 &= -4 \\ x &= 1 \end{aligned}$$

The solution is (1, 3).

Problem 5

$$\begin{bmatrix} 3 & -2 & -3 & 5 \\ 1 & 3 & -2 & -4 \\ 2 & 6 & 3 & 6 \end{bmatrix}$$

$$\begin{bmatrix} 1 & 3 & -2 & -4 \\ 3 & -2 & -3 & 5 \\ 2 & 6 & 3 & 6 \end{bmatrix}$$

▬ Interchange rows 1 and 2.

$$\begin{bmatrix} 1 & 3 & -2 & -4 \\ 0 & -11 & 3 & 17 \\ 0 & 0 & 7 & 14 \end{bmatrix}$$

▬ Multiply row 1 by −3 and add to row 2.
▬ Multiply row 1 by −2 and add to row 3.

$$\begin{bmatrix} 1 & 3 & -2 & -4 \\ 0 & 1 & -\frac{3}{11} & -\frac{17}{11} \\ 0 & 0 & 7 & 14 \end{bmatrix}$$

▬ Multiply row 2 by $-\frac{1}{11}$.

$$\begin{bmatrix} 1 & 3 & -2 & -4 \\ 0 & 1 & -\frac{3}{11} & -\frac{17}{11} \\ 0 & 0 & 1 & 2 \end{bmatrix}$$

▬ Multiply row 3 by $\frac{1}{7}$.

$$\begin{aligned} x + 3y - 2z &= -4 \\ y - \tfrac{3}{11}z &= -\tfrac{17}{11} \\ z &= 2 \end{aligned} \qquad \begin{aligned} y - \tfrac{3}{11}(2) &= -\tfrac{17}{11} \\ y - \tfrac{6}{11} &= -\tfrac{17}{11} \\ y &= -1 \end{aligned} \qquad \begin{aligned} x + 3(-1) - 2(2) &= -4 \\ x - 3 - 4 &= -4 \\ x - 7 &= -4 \\ x &= 3 \end{aligned}$$

The solution is (3, −1, 2).

SECTION 10.4 *pages 498–503*

Problem 1

Strategy ■ Rate of the rowing team in calm water: t
Rate of the current: c

	Rate	*Time*	*Distance*
With current	$t + c$	2	$2(t + c)$
Against current	$t - c$	2	$2(t - c)$

■ The distance traveled with the current is 18 mi.
The distance traveled against the current is 10 mi.

Solution $2(t + c) = 18$ $\quad \frac{1}{2} \cdot 2(t + c) = \frac{1}{2} \cdot 18$

$2(t - c) = 10$ $\quad \frac{1}{2} \cdot 2(t - c) = \frac{1}{2} \cdot 10$

$$t + c = 9$$
$$t - c = 5$$
$$2t = 14$$
$$t = 7$$

$$t + c = 9$$
$$7 + c = 9$$
$$c = 2$$

The rate of the rowing team in calm water is 7 mph.
The rate of the current is 2 mph.

Problem 2

Strategy ■ Cost of an orange tree: x
Cost of a grapefruit tree: y

First purchase:

	Amount	*Unit Cost*	*Value*
Orange trees	25	x	$25x$
Grapefruit trees	20	y	$20y$

Second purchase:

	Amount	*Unit Cost*	*Value*
Orange trees	20	x	$20x$
Grapefruit trees	30	y	$30y$

■ The total of the first purchase was \$290.
The total of the second purchase was \$330.

Solution

$$\begin{aligned} 25x + 20y &= 290 \\ 20x + 30y &= 330 \end{aligned} \qquad \begin{aligned} 4(25x + 20y) &= 4 \cdot 290 \\ -5(20x + 30y) &= -5 \cdot 330 \end{aligned}$$

$$\begin{aligned} 100x + 80y &= 1160 \\ -100x - 150y &= -1650 \\ -70y &= -490 \\ y &= 7 \end{aligned}$$

$$\begin{aligned} 25x + 20y &= 290 \\ 25x + 20(7) &= 290 \\ 25x + 140 &= 290 \\ 25x &= 150 \\ x &= 6 \end{aligned}$$

The cost of an orange tree is \$6.
The cost of a grapefruit tree is \$7.

SECTION 10.5 *pages 506–511*

Problem 1

A.

$$\begin{aligned} y &= 2x^2 + x - 3 \quad (1) \\ y &= 2x^2 - 2x + 9 \quad (2) \end{aligned}$$

Use the substitution method.

$$\begin{aligned} 2x^2 - 2x + 9 &= 2x^2 + x - 3 \\ -3x + 9 &= -3 \\ -3x &= -12 \\ x &= 4 \end{aligned}$$

Substitute into equation (1).

$$\begin{aligned} y &= 2x^2 + x - 3 \\ y &= 2(4)^2 + 4 - 3 \\ y &= 32 + 4 - 3 \\ y &= 33 \end{aligned}$$

The solution is (4, 33).

B.

$$\begin{aligned} x^2 - y^2 &= 10 \quad (1) \\ x^2 + y^2 &= 8 \quad (2) \end{aligned}$$

Use the addition method.

$$\begin{aligned} 2x^2 &= 18 \\ x^2 &= 9 \\ x &= \pm\sqrt{9} = \pm 3 \end{aligned}$$

Substitute into equation (2).

$$\begin{aligned} x^2 + y^2 &= 8 \\ 3^2 + y^2 &= 8 \\ 9 + y^2 &= 8 \\ y^2 &= -1 \\ y &= \pm\sqrt{-1} \end{aligned} \qquad \begin{aligned} x^2 + y^2 &= 8 \\ (-3)^2 + y^2 &= 8 \\ 9 + y^2 &= 8 \\ y^2 &= -1 \\ y &= \pm\sqrt{-1} \end{aligned}$$

y is not a real number. Therefore, the system of equations has no real solution. The graphs do not intersect.

Problem 2

A. $x^2 + y^2 < 16$
$y^2 > x$

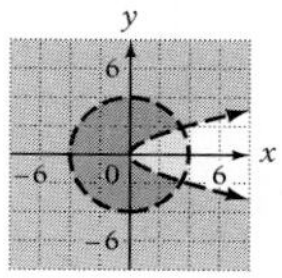

B. $y \geq x - 1$
$y < -2x$

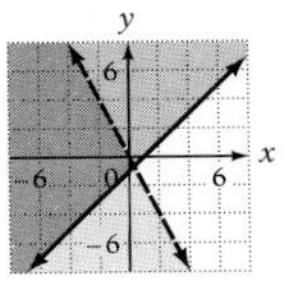

SOLUTIONS to Chapter 11 Problems

SECTION 11.1 *pages 529–537*

Problem 1

$$f(x) = \left(\frac{2}{3}\right)^x$$
$$f(3) = \left(\frac{2}{3}\right)^3 = \frac{8}{27}$$
$$f(-2) = \left(\frac{2}{3}\right)^{-2} = \left(\frac{3}{2}\right)^2 = \frac{9}{4}$$

Problem 2

$$f(x) = 2^{2x+1}$$
$$f(0) = 2^{2(0)+1} = 2^1 = 2$$
$$f(-2) = 2^{2(-2)+1} = 2^{-3} = \frac{1}{2^3} = \frac{1}{8}$$

Problem 3 $f(x) = \pi^x$

A. $f(3) = \pi^3 \approx 31.0063$

B. $f(-2) = \pi^{-2} \approx 0.1013$

C. $f(\pi) = \pi^\pi \approx 36.4622$

Problem 4 $f(x) = e^x$

A. $f(1.2) = e^{1.2} \approx 3.3201$

B. $f(-2.5) = e^{-2.5} \approx 0.0821$

C. $f(e) = e^e \approx 15.1543$

Problem 5

A.

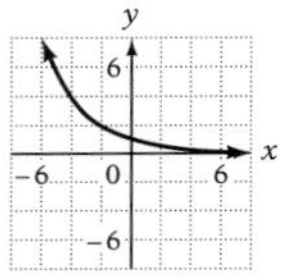

B.

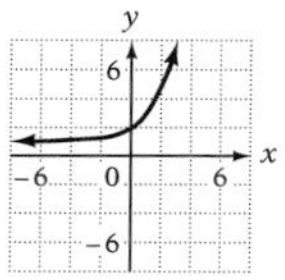

Problem 6 $7^3 = 343$ is equivalent to $\log_7 343 = 3$.

Problem 7 $\log_{\frac{1}{2}}\left(\frac{1}{8}\right) = 3$ is equivalent to $\left(\frac{1}{2}\right)^3 = \frac{1}{8}$.

Problem 8 $\log 0.1 = -1$ is equivalent to $10^{-1} = 0.1$.

Problem 9 $\ln 7.389 = 2$ is equivalent to $e^2 = 7.389$.

Problem 10

$$\log_4 64 = x$$
$$64 = 4^x$$
$$4^3 = 4^x$$
$$3 = x$$

$$\log_4 64 = 3$$

Problem 11

$$\log_2 x = -4$$
$$2^{-4} = x$$
$$\frac{1}{2^4} = x$$
$$\frac{1}{16} = x$$

The solution is $\frac{1}{16}$.

Problem 12 **A.** $f(x) = \log_2(x-1)$
$y = \log_2(x-1)$

$y = \log_2(x-1)$ is equivalent to $2^y = x - 1$.
$2^y + 1 = x$

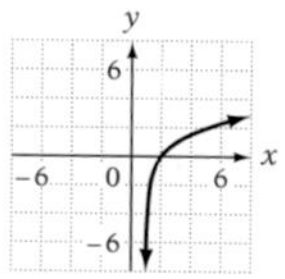

B. $f(x) = \log_3 2x$
$y = \log_3 2x$

$y = \log_3 2x$ is equivalent to $3^y = 2x$.
$\frac{3^y}{2} = x$

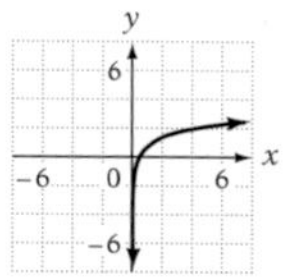

SECTION 11.2 *pages 542–546*

Problem 1 **A.** $\log_b \frac{x^2}{y} = \log_b x^2 - \log_b y = 2\log_b x - \log_b y$

B. $\ln y^{\frac{1}{3}}z^3 = \ln y^{\frac{1}{3}} + \ln z^3 =$
$\frac{1}{3}\ln y + 3\ln z$

C. $\log_8 \sqrt[3]{xy^2} = \log_8(xy^2)^{\frac{1}{3}} = \frac{1}{3}\log_8 xy^2 =$
$\frac{1}{3}(\log_8 x + \log_8 y^2) = \frac{1}{3}(\log_8 x + 2\log_8 y) =$
$\frac{1}{3}\log_8 x + \frac{2}{3}\log_8 y$

Problem 2 **A.** $2\log_b x - 3\log_b y - \log_b z = \log_b x^2 - \log_b y^3 - \log_b z =$
$\log_b \frac{x^2}{y^3} - \log_b z = \log_b \frac{x^2}{y^3 z}$

B. $\frac{1}{3}(\log_4 x - 2\log_4 y + \log_4 z) =$
$\frac{1}{3}(\log_4 x - \log_4 y^2 + \log_4 z) =$
$\frac{1}{3}\left(\log_4 \frac{x}{y^2} + \log_4 z\right) = \frac{1}{3}\left(\log_4 \frac{xz}{y^2}\right) =$
$\log_4\left(\frac{xz}{y^2}\right)^{\frac{1}{3}} = \log_4 \sqrt[3]{\frac{xz}{y^2}}$

C. $\frac{1}{2}(2\ln x - 5\ln y) =$
$\frac{1}{2}(\ln x^2 - \ln y^5) =$
$\frac{1}{2}\left(\ln \frac{x^2}{y^5}\right) =$
$\ln\left(\frac{x^2}{y^5}\right)^{\frac{1}{2}} =$
$\ln \sqrt{\frac{x^2}{y^5}}$

Problem 3 Since $\log_b 1 = 0$, $\log_9 1 = 0$.

SECTION 11.3 *pages 550–555*

Problem 1 **A.** $\log 93{,}000 = \log(9.3 \times 10^4) = \log 9.3 + \log 10^4 = 0.9685 + 4 = 4.9685$

B. $\log 0.0006 = \log(6 \times 10^{-4}) = \log 6 + \log 10^{-4} = 0.7782 + (-4) = 6.7782 - 10$

Problem 2 $\log 0.0235 = -1.6289$

Problem 3 **A.** $\text{antilog } 2.3365 = 2.17 \times 10^2 = 217$

B. $\text{antilog } (9.7846 - 10) = 6.09 \times 10^{-1} = 0.609$

SECTION 11.4 *pages 556–560*

Problem 1

$$\begin{aligned} 10^{3x+5} &= 10^{x-3} \\ 3x + 5 &= x - 3 \\ 2x + 5 &= -3 \\ 2x &= -8 \\ x &= -4 \end{aligned}$$

Check:

$$\begin{array}{c|c} \multicolumn{2}{c}{10^{3x+5} = 10^{x-3}} \\ \hline 10^{3(-4)+5} & 10^{-4-3} \\ 10^{-12+5} & 10^{-7} \\ \multicolumn{2}{c}{10^{-7} = 10^{-7}} \end{array}$$

The solution is −4.

Problem 2 **A.**

$$\begin{aligned} 4^{3x} &= 25 \\ \log 4^{3x} &= \log 25 \\ 3x \log 4 &= \log 25 \\ 3x &= \frac{\log 25}{\log 4} \\ x &= \frac{\log 25}{3 \log 4} \\ x &= 0.7740 \end{aligned}$$

The solution is 0.7740.

B.

$$\begin{aligned} (1.06)^x &= 1.5 \\ \log (1.06)^x &= \log 1.5 \\ x \log 1.06 &= \log 1.5 \\ x &= \frac{\log 1.5}{\log 1.06} \\ x &= 6.9585 \end{aligned}$$

The solution is 6.9585.

Problem 3 **A.** $\log_4(x^2 - 3x) = 1$

Rewrite in exponential form.

$$\begin{aligned} 4^1 &= x^2 - 3x \\ 4 &= x^2 - 3x \\ 0 &= x^2 - 3x - 4 \\ 0 &= (x + 1)(x - 4) \end{aligned}$$

$$\begin{aligned} x + 1 &= 0 & \qquad x - 4 &= 0 \\ x &= -1 & \qquad x &= 4 \end{aligned}$$

The solutions are −1 and 4.

B.

$$\begin{aligned} \log_3 x + \log_3(x + 3) &= \log_3 4 \\ \log_3[x(x + 3)] &= \log_3 4 \end{aligned}$$

Use the One-to-One Property of Logarithms.

$$\begin{aligned} x(x + 3) &= 4 \\ x^2 + 3x &= 4 \\ x^2 + 3x - 4 &= 0 \\ (x + 4)(x - 1) &= 0 \end{aligned}$$

$$\begin{aligned} x + 4 &= 0 & \qquad x - 1 &= 0 \\ x &= -4 & \qquad x &= 1 \end{aligned}$$

−4 does not check as a solution.
The solution is 1.

Problem 4 $\log_9 23 = \dfrac{\log 23}{\log 9} = 1.4270$

Problem 5 $\ln 3 = 1.0986$

SECTION 11.5 *pages 562–567*

Problem 1

Strategy To find the pH, replace H^+ by 2.51×10^{-12} in the equation $pH = -\log (H^+)$ and solve for pH.

Solution

$$\begin{aligned} pH &= -\log (H^+) = -\log (2.51 \times 10^{-12}) \\ &= -(\log 2.51 + \log 10^{-12}) \\ &= -[0.3997 + (-12)] = 11.6003 \end{aligned}$$

The pH of sodium carbonate is 11.6.

Problem 2

Strategy To find how many times stronger the San Francisco earthquake was, use the Richter equation to write a system of equations. Solve the system of equations for the ratio $\frac{I_1}{I_2}$.

Solution

$$7.8 = \log \frac{I_1}{I_0}$$

$$5 = \log \frac{I_2}{I_0}$$

$$\begin{aligned} 7.8 &= \log I_1 - \log I_0 \\ 5 &= \log I_2 - \log I_0 \end{aligned}$$

$$2.8 = \log I_1 - \log I_2$$

$$2.8 = \log \frac{I_1}{I_2}$$

$$\frac{I_1}{I_2} = 10^{2.8} = 630.957$$

The San Francisco earthquake was 631 times stronger than one that can cause serious damage.

SOLUTIONS to Chapter 12 Problems

SECTION 12.1 *pages 581–584*

Problem 1 $a_n = n(n + 1)$

$a_1 = 1(1 + 1) = 2$ The first term is 2.

$a_2 = 2(2 + 1) = 6$ The second term is 6.

$a_3 = 3(3 + 1) = 12$ The third term is 12.

$a_4 = 4(4 + 1) = 20$ The fourth term is 20.

Problem 2 $a_n = \frac{1}{n(n+2)}$

$a_6 = \frac{1}{6(6+2)} = \frac{1}{48}$ The sixth term is $\frac{1}{48}$.

$a_9 = \frac{1}{9(9+2)} = \frac{1}{99}$ The ninth term is $\frac{1}{99}$.

Problem 3 **A.** $\sum_{n=1}^{4}(7-n) = (7-1)+(7-2)+(7-3)+(7-4) = 6+5+4+3 = 18$

B. $\sum_{i=3}^{6}(i^2-2) = (3^2-2)+(4^2-2)+(5^2-2)+(6^2-2)$

$= 7+14+23+34 = 78$

Problem 4 $\sum_{n=1}^{5} nx = x+2x+3x+4x+5x$

SECTION 12.2 *pages 587–591*

Problem 1 $9, 3, -3, -9, \ldots$

$d = a_2 - a_1 = 3 - 9 = -6$

$a_n = a_1 + (n-1)d$
$a_{15} = 9 + (15-1)(-6) = 9 + (14)(-6) = 9 - 84$
$a_{15} = -75$

Problem 2 $-3, 1, 5, 9, \ldots$

$d = a_2 - a_1 = 1 - (-3) = 4$

$a_n = a_1 + (n-1)d$
$a_n = -3 + (n-1)4$
$a_n = -3 + 4n - 4$
$a_n = 4n - 7$

Problem 3 $7, 9, 11, \ldots, 59$

$d = a_2 - a_1 = 9 - 7 = 2$

$a_n = a_1 + (n-1)d$
$59 = 7 + (n-1)2$
$59 = 7 + 2n - 2$
$59 = 5 + 2n$
$54 = 2n$
$27 = n$

There are 27 terms in the sequence.

Problem 4 $-4, -2, 0, 2, 4, \ldots$

$d = a_2 - a_1 = -2 - (-4) = 2$

$a_n = a_1 + (n-1)d$
$a_{25} = -4 + (25-1)2 = -4 + (24)2 = -4 + 48$
$a_{25} = 44$

$S_n = \frac{n}{2}(a_1 + a_n)$

$S_{25} = \frac{25}{2}(-4 + 44) = \frac{25}{2}(40) = 25(20)$

$S_{25} = 500$

Problem 5 $\sum_{n=1}^{18} (3n - 2)$

$$a_n = 3n - 2$$
$$a_1 = 3(1) - 2 = 1$$
$$a_{18} = 3(18) - 2 = 52$$

$$S_n = \frac{n}{2}(a_1 + a_n)$$

$$S_{18} = \frac{18}{2}(1 + 52) = 9(53) = 477$$

Problem 6

Strategy To find the value of the 20th-place prize:

- Write the equation for the nth-place prize.
- Find the 20th term of the sequence.

To find the total amount of prize money being awarded, use the Formula for the Sum of n Terms of an Arithmetic Sequence.

Solution 10,000, 9700, . . .

$$d = a_2 - a_1 = 9700 - 10{,}000 = -300$$

$$\begin{aligned} a_n &= a_1 + (n - 1)d \\ &= 10{,}000 + (n - 1)(-300) \\ &= 10{,}000 - 300n + 300 \\ &= -300n + 10{,}300 \end{aligned}$$
$$a_{20} = -300(20) + 10{,}300 = -6000 + 10{,}300 = 4300$$

$$S_n = \frac{n}{2}(a_1 + a_n)$$

$$S_{20} = \frac{20}{2}(10{,}000 + 4300) = 10(14{,}300) = 143{,}000$$

The value of the 20th-place price is $4300.

The total amount of prize money being awarded is $143,000.

SECTION 12.3 *pages 594–600*

Problem 1 $5, 2, \frac{4}{5}, \ldots$

$$r = \frac{a_2}{a_1} = \frac{2}{5}$$
$$a_n = a_1 r^{n-1}$$
$$a_5 = 5\left(\frac{2}{5}\right)^{5-1} = 5\left(\frac{2}{5}\right)^4 = 5\left(\frac{16}{625}\right)$$
$$a_5 = \frac{16}{125}$$

Problem 2 $3, a_2, a_3, -192, \ldots$

$$a_n = a_1 r^{n-1}$$
$$a_4 = 3r^{4-1}$$
$$-192 = 3r^{4-1}$$
$$-192 = 3r^3$$
$$-64 = r^3$$
$$-4 = r$$

$$a_n = a_1 r^{n-1}$$
$$a_3 = 3(-4)^{3-1} = 3(-4)^2 = 3(16) = 48$$

Problem 3 $1, -\frac{1}{3}, \frac{1}{9}, -\frac{1}{27}$

$$r = \frac{a_2}{a_1} = \frac{-\frac{1}{3}}{1} = -\frac{1}{3}$$

$$S_n = \frac{a_1(1 - r^n)}{1 - r}$$

$$S_4 = \frac{1\left[1 - \left(-\frac{1}{3}\right)^4\right]}{1 - \left(-\frac{1}{3}\right)} = \frac{1 - \frac{1}{81}}{\frac{4}{3}} = \frac{\frac{80}{81}}{\frac{4}{3}}$$

$$= \frac{80}{81} \cdot \frac{3}{4} = \frac{20}{27}$$

Problem 4 $\sum_{n=1}^{5} \left(\frac{1}{2}\right)^n$

$$a_n = \left(\frac{1}{2}\right)^n$$

$$a_1 = \left(\frac{1}{2}\right)^1 = \frac{1}{2}$$

$$a_2 = \left(\frac{1}{2}\right)^2 = \frac{1}{4}$$

$$r = \frac{a_2}{a_1} = \frac{\frac{1}{4}}{\frac{1}{2}} = \frac{1}{4} \cdot \frac{2}{1} = \frac{1}{2}$$

$$S_n = \frac{a_1(1 - r^n)}{1 - r}$$

$$S_5 = \frac{\frac{1}{2}\left[1 - \left(\frac{1}{2}\right)^5\right]}{1 - \frac{1}{2}} = \frac{\frac{1}{2}\left(1 - \frac{1}{32}\right)}{\frac{1}{2}} = \frac{\frac{1}{2}\left(\frac{31}{32}\right)}{\frac{1}{2}} = \frac{\frac{31}{64}}{\frac{1}{2}} = \frac{31}{64} \cdot \frac{2}{1} = \frac{31}{32}$$

Problem 5 $3, -2, \frac{4}{3}, -\frac{8}{9}, \ldots$

$$r = \frac{a_2}{a_1} = -\frac{2}{3}$$

$$S = \frac{a_1}{1 - r} = \frac{3}{1 - \left(-\frac{2}{3}\right)} = \frac{3}{1 + \frac{2}{3}}$$

$$= \frac{3}{\frac{5}{3}} = \frac{9}{5}$$

Problem 6 $0.36\overline{36} = 0.36 + 0.0036 + 0.000036 + \ldots$

$$S = \frac{a_1}{1 - r} = \frac{\frac{36}{100}}{1 - \frac{1}{100}} = \frac{\frac{36}{100}}{\frac{99}{100}} = \frac{36}{99} = \frac{4}{11}$$

An equivalent fraction is $\frac{4}{11}$.

Problem 7

Strategy To find the total number of letters mailed, use the Formula for the Sum of n Terms of a Finite Geometric Series.

Solution $n = 6,\ a_1 = 3,\ r = 3$

$$S_n = \frac{a_1(1 - r^n)}{1 - r}$$

$$S_6 = \frac{3(1 - 3^6)}{1 - 3} = \frac{3(1 - 729)}{1 - 3}$$

$$= \frac{3(-728)}{-2} = \frac{-2184}{-2} = 1092$$

From the first through the sixth mailings, 1092 letters will have been mailed.

SECTION 12.4 *pages 603–607*

Problem 1 $\frac{12!}{7!5!} = \frac{12 \cdot 11 \cdot 10 \cdot 9 \cdot 8 \cdot 7 \cdot 6 \cdot 5 \cdot 4 \cdot 3 \cdot 2 \cdot 1}{(7 \cdot 6 \cdot 5 \cdot 4 \cdot 3 \cdot 2 \cdot 1)(5 \cdot 4 \cdot 3 \cdot 2 \cdot 1)} = 792$

Problem 2 $\binom{7}{0} = \frac{7!}{(7 - 0)!0!} = \frac{7!}{7!0!} = \frac{7 \cdot 6 \cdot 5 \cdot 4 \cdot 3 \cdot 2 \cdot 1}{(7 \cdot 6 \cdot 5 \cdot 4 \cdot 3 \cdot 2 \cdot 1)(1)} = 1$

Problem 3 $(3m - n)^4 =$

$\binom{4}{0}(3m)^4 + \binom{4}{1}(3m)^3(-n) + \binom{4}{2}(3m)^2(-n)^2 + \binom{4}{3}(3m)(-n)^3 + \binom{4}{4}(-n)^4 =$

$1(81m^4) + 4(27m^3)(-n) + 6(9m^2)(n^2) + 4(3m)(-n^3) + 1(n^4) =$

$81m^4 - 108m^3n + 54m^2n^2 - 12mn^3 + n^4$

Problem 4 $(y - 2)^{10} =$

$\binom{10}{0}y^{10} + \binom{10}{1}y^9(-2) + \binom{10}{2}y^8(-2)^2 + \cdots =$

$1(y^{10}) + 10y^9(-2) + 45y^8(4) + \cdots = y^{10} - 20y^9 + 180y^8 + \cdots$

Problem 5 $(t - 2s)^7$

$n = 7,\ a = t,\ b = -2s,\ r = 3$

$\binom{7}{3-1}(t)^{7-3+1}(-2s)^{3-1} = \binom{7}{2}(t)^5(-2s)^2 = 21t^5(4s^2) = 84t^5s^2$

ANSWERS to Chapter 1 Exercises

SECTION 1.1 *pages 12–14*

1. -83 **3.** 75 **5.** -9.3 **7.** 6.4 **9.** $\frac{11}{12}$ **11.** -126 **13.** 436 **15.** -16 **17.** 4.93
19. $-\frac{7}{8}$ **21.** $\frac{43}{48}$ **23.** $-\frac{67}{45}$ **25.** $-\frac{13}{36}$ **27.** $\frac{11}{24}$ **29.** $\frac{13}{24}$ **31.** $-\frac{3}{56}$ **33.** $-\frac{2}{3}$ **35.** $-\frac{11}{14}$
37. $-\frac{1}{24}$ **39.** -12.974 **41.** -6.008 **43.** 1.9215 **45.** -6.02 **47.** -6.7 **49.** 35
51. -7 **53.** -7 **55.** -64 **57.** 64 **59.** 432 **61.** -1125 **63.** 512 **65.** -160
67. 24 **69.** -36 **71.** -11 **73.** $\frac{1}{4}$ **75.** $\frac{1}{4}$ **77.** 12 **79.** 25 **81.** 44 **83.** $\frac{109}{150}$
85. $\frac{91}{36}$ **87.** 4.4 **89.** integer, rational number, real number **91.** rational number, real number
93. irrational number, real number **95.** irrational number, real number
97. irrational number, real number **99.** positive **101.** negative **103.** 1, -1
105. 3 integers; 1, 4, and 9 **107.** 1 **109.** no

SECTION 1.2 *pages 21–24*

1. 10 **3.** 4 **5.** 0 **7.** $-\frac{1}{7}$ **9.** $\frac{9}{2}$ **11.** $\frac{1}{2}$ **13.** 2 **15.** 3 **17.** -12 **19.** 2
21. 6 **23.** -2 **25.** 6.51 **27.** -19.71 **29.** 15 **31.** 4 **33.** 0 **35.** 5
37. $[-(x + y)]$ **39.** x **41.** ab **43.** The Inverse Property of Addition
45. The Commutative Property of Multiplication **47.** The Associative Property of Addition
49. The Distributive Property **51.** The Multiplication Property of Zero
53. The Commutative Property of Addition **55.** $13x$ **57.** $-4x$ **59.** $7a + 7b$ **61.** x
63. $-3x + 6$ **65.** $5x + 10$ **67.** $x + y$ **69.** $3x - 6y - 5$ **71.** $-11a + 21$ **73.** $-x + 6y$
75. $-30a + 140$ **77.** $30x - 10y$ **79.** $-10a + 2b$ **81.** $-12a + b$ **83.** $-2x - 144y - 96$
85. $5x + 3y - 32$ **87.** $x + 6$ **89.** $n - (5 - n)$; $2n - 5$ **91.** $\frac{3}{8}n - \frac{1}{6}n$; $\frac{5}{24}n$ **93.** $n + \frac{2}{3}n$; $\frac{5}{3}n$
95. $\frac{1}{2}(6n + 22)$; $3n + 11$ **97.** $16 - (5n - 4)$; $-5n + 20$ **99.** $\frac{2}{3}n + \frac{5}{8}n$; $\frac{31}{24}n$ **101.** $3\left(\frac{2n}{6}\right)$; n
103. $2(n + 11) - 4$; $2n + 18$ **105.** $20 - (4 + n)12$; $-12n - 28$ **107.** $n + (n - 12)3$; $4n - 36$
109. a. The Distributive Property **b.** The Commutative Property of Addition
c. The Associative Property of Addition **d.** The Distributive Property
111. a. The Distributive Property **b.** The Associative Property of Multiplication
c. The Multiplication Property of One **113.** $3c$, where c = the amount of cashews
115. $\frac{1}{3}x$, where x = the measure of the largest angle **117.** $2s - 15$, where s = the age of the silver coin

SECTION 1.3 *pages 29–31*

1. $\{-2, -1, 0, 1, 2, 3, 4\}$ **3.** $\{2, 4, 6, 8, 10, 12\}$ **5.** $\{3\}$ **7.** $\{3, 6, 9, 12, 15, 18\}$
9. $\{x|x > 4, x \text{ is an integer}\}$ **11.** $\{x|x \geq 1, x \in \text{real numbers}\}$ **13.** $\{x|-2 < x < 5, x \text{ is an integer}\}$
15. $\{x|0 < x < 1, x \in \text{real numbers}\}$ **17.** $A \cup B = \{1, 2, 4, 6, 9\}$ **19.** $A \cup B = \{2, 3, 5, 8, 9, 10\}$
21. $A \cup B = \{-4, -2, 0, 2, 4, 8\}$ **23.** $A \cup B = \{1, 2, 3, 4, 5\}$ **25.** $A \cap B = \{6\}$

27. $A \cap B = \{5, 10, 20\}$ **29.** $A \cap B = \varnothing$ **31.** $A \cap B = \{4, 6\}$

33. 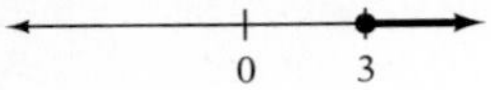

35. 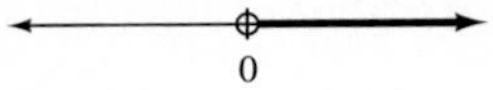

37. 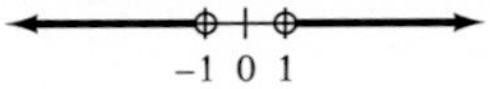

39. 

41. 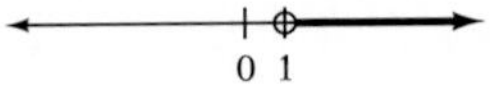

43. 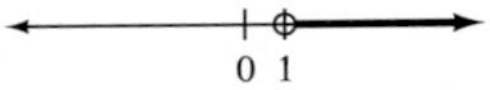

45. 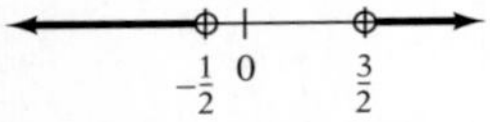

47. 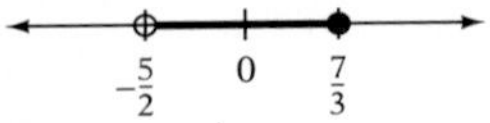

49. 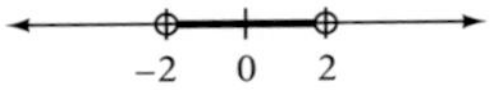

51. 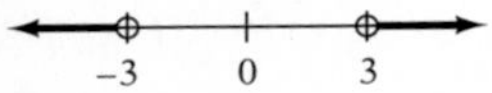

53. $A \cup B = \{x|x > 0, x \text{ is an integer}\}$

55. $A \cap B = \{x|x \geq 15, x \text{ is an odd integer}\}$ **57.** a **59.** b, c

VENN DIAGRAMS *page 34*

1. a. $\{6, 7, 8\}$ **b.** $\varnothing$ **c.** $\{1\}$

CHAPTER REVIEW *pages 36–37*

1. −23 (Objective 1.1.1) **2.** 15 (Objective 1.1.1) **3.** 7 (Objective 1.1.1)
4. −5 (Objective 1.1.1) **5.** −15 (Objective 1.1.2) **6.** −72 (Objective 1.1.2)
7. 12 (Objective 1.1.2) **8.** 4 (Objective 1.1.2) **9.** −4 (Objective 1.1.2)
10. 14 (Objective 1.1.2) **11.** 72 (Objective 1.1.3) **12.** −108 (Objective 1.1.3)
13. −288 (Objective 1.1.3) **14.** 2560 (Objective 1.1.3) **15.** $\frac{7}{120}$ (Objective 1.1.2)
16. $-\frac{7}{6}$ (Objective 1.1.2) **17.** $\frac{2}{15}$ (Objective 1.1.2) **18.** $-\frac{2}{21}$ (Objective 1.1.2)
19. $-\frac{5}{8}$ (Objective 1.1.2) **20.** $\frac{13}{9}$ (Objective 1.1.2) **21.** −2.84 (Objective 1.1.2)
22. −4.41 (Objective 1.1.2) **23.** 3.1 (Objective 1.1.2) **24.** 44.2 (Objective 1.1.2)
25. 0 (Objective 1.1.4) **26.** $\frac{312}{5}$ (Objective 1.1.4) **27.** $\frac{11}{24}$ (Objective 1.1.4)
28. $\frac{28}{5}$ (Objective 1.1.4) **29.** 20 (Objective 1.2.1) **30.** −19 (Objective 1.2.1)
31. −20 (Objective 1.2.1) **32.** $\frac{7}{6}$ (Objective 1.2.1) **33.** 3 (Objective 1.2.2)
34. y (Objective 1.2.2) **35.** ab (Objective 1.2.2) **36.** 4 (Objective 1.2.2)
37. The Inverse Property of Addition (Objective 1.2.2) **38.** The Distributive Property (Objective 1.2.2)
39. The Associative Property of Multiplication (Objective 1.2.2)
40. The Commutative Property of Addition (Objective 1.2.2) **41.** $-3a + 9$ (Objective 1.2.3)
42. $-6x + 14$ (Objective 1.2.3) **43.** $9x - 3y + 12$ (Objective 1.2.3) **44.** $-15x + 18$ (Objective 1.2.3)
45. $5 - 2(n + 2)$; $-2n + 1$ (Objective 1.2.4) **46.** $2(n + 5) - 3$; $2n + 7$ (Objective 1.2.4)
47. $\frac{5}{8}n - \frac{3}{4}n$; $-\frac{1}{8}n$ (Objective 1.2.4) **48.** $12 - \frac{n+3}{n}$; $\frac{11n-3}{n}$ (Objective 1.2.4)
49. $A \cup B = \{1, 2, 3, 4, 5, 6, 7, 8\}$ (Objective 1.3.1) **50.** $A \cup B = \{-2, -1, 0, 1, 2, 3\}$ (Objective 1.3.1)
51. $A \cap B = \varnothing$ (Objective 1.3.1) **52.** $A \cap B = \{2, 3\}$ (Objective 1.3.1)

53. 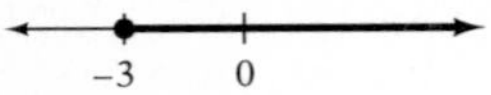

(Objective 1.3.2) **54.** 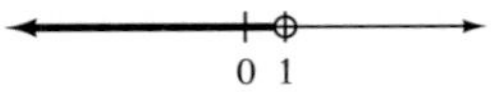

(Objective 1.3.2)

55. 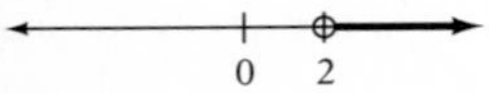

(Objective 1.3.2) **56.** 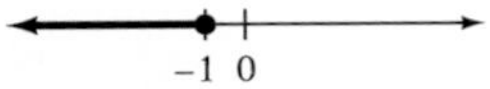

(Objective 1.3.2)

57. 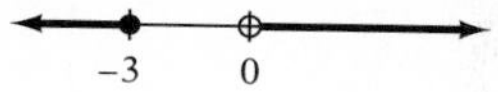

(Objective 1.3.2)

58. 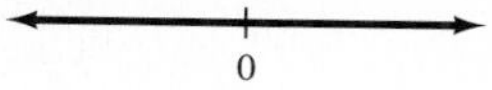

(Objective 1.3.2)

59. 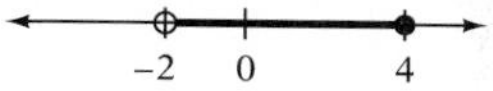

(Objective 1.3.2)

60. 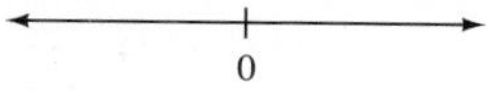

(Objective 1.3.2)

CHAPTER TEST *page 38*

1. 6 (Objective 1.1.4) **2.** 2 (Objective 1.1.2) **3.** $\frac{25}{36}$ (Objective 1.1.2) **4.** 2 (Objective 1.2.1)
5. −36 (Objective 1.1.3) **6.** $-\frac{4}{27}$ (Objective 1.1.2) **7.** −270 (Objective 1.1.4)
8. 10 (Objective 1.1.4) **9.** −5 (Objective 1.2.1) **10.** 1 (Objective 1.2.1)
11. The Distributive Property (Objective 1.2.2) **12.** 4 (Objective 1.2.2)
13. $13x - y$ (Objective 1.2.3) **14.** $14x + 48y$ (Objective 1.2.3) **15.** $-5x - 2y$ (Objective 1.2.3)
16. $A \cup B = \{-3, -1, 0, 1, 3, 5\}$ (Objective 1.3.1) **17.** $A \cup B = \{2, 3, 4, 5, 6, 8\}$ (Objective 1.3.1)
18. $A \cap B = \{-4, 0\}$ (Objective 1.3.1) **19.** 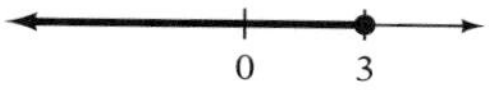

(Objective 1.3.2)

20. 

(Objective 1.3.2)

21. 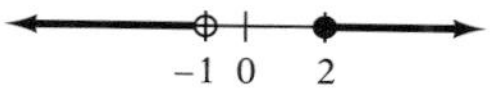

(Objective 1.3.2)

22. 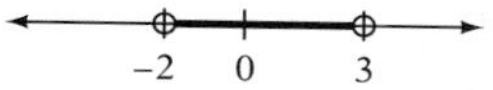

(Objective 1.3.2) **23.** $13 - (n - 3)9$; $-9n + 40$ (Objective 1.2.4)
24. $8 - (4n - 12)$; $-4n + 20$ (Objective 1.2.4) **25.** $\frac{7}{8}n - \frac{3}{5}n$; $\frac{11}{40}n$ (Objective 1.2.4)

ANSWERS to Chapter 2 Exercises

SECTION 2.1 *pages 47–50*

1. 9 **3.** −10 **5.** 4 **7.** $\frac{11}{21}$ **9.** $-\frac{1}{8}$ **11.** 20 **13.** −49 **15.** $-\frac{21}{20}$ **17.** −3.73
19. $\frac{3}{2}$ **21.** 8 **23.** no solution **25.** −3 **27.** 1 **29.** $\frac{7}{4}$ **31.** 24 **33.** $\frac{5}{3}$ **35.** $-\frac{3}{2}$
37. no solution **39.** $\frac{11}{2}$ **41.** −1.25 **43.** 0.433 **45.** 6 **47.** $\frac{1}{2}$ **49.** $\frac{2}{3}$ **51.** −6
53. $-\frac{4}{3}$ **55.** $\frac{9}{4}$ **57.** $\frac{6}{7}$ **59.** $\frac{35}{12}$ **61.** −1 **63.** −33 **65.** 6 **67.** 6 **69.** $\frac{25}{14}$ **71.** $\frac{3}{4}$
73. $-\frac{4}{29}$ **75.** 3 **77.** −10 **79.** $\frac{5}{11}$ **81.** 9 **83.** −2 **85.** 57 **87.** 1.35 **89.** 9.4
91. 24 **93.** \$394.40 **95.** 15 oz **97.** 17 min **99.** 10 days **101.** 10 min **103.** $-\frac{2}{5}$
105. −9 **107.** $-\frac{15}{2}$ **109.** no solution **111.** no solution **113.** −1 **115.** 52
117. all real numbers

SECTION 2.2 *pages 55–57*

1. 24 dimes **3.** 8 dimes **5.** 26 twenty-dollar bills **7.** 20¢ stamps: 8; 15¢ stamps: 16
9. 5 stamps **11.** 7 stamps **13.** 3 **15.** 3 and 7 **17.** 21 and 29 **19.** 21, 44, and 58
21. −20, −19, and −18 **23.** no solution **25.** 5, 7, 9 **27.** \$4.30 **29.** 14 stamps **31.** −32

SECTION 2.3 *pages 61–64*

1. \$2.95/lb **3.** 320 adult tickets **5.** 225 L **7.** 20 oz of pure gold; 30 oz of the alloy **9.** \$4.04/lb **11.** 37.5 gal **13.** 28 mi **15.** 1st plane: 420 mph; 2nd plane: 500 mph **17.** 43.2 mi **19.** 1st plane: 225 mph; 2nd plane: 305 mph **21.** 2.8 mi **23.** 1050 mi **25.** 12:10 P.M. **27.** 480 mi

SECTION 2.4 *pages 68–71*

1. \$6000 at 7.2%; \$6500 at 9.8% **3.** \$2000 **5.** \$24,500 at 7%; \$17,500 at 9.8% **7.** \$25,000 **9.** \$4000 **11.** 3 qt **13.** 22% **15.** 30 oz **17.** 25 L of 65% solution; 25 L of 15% solution **19.** $2\frac{2}{3}$ gal **21.** 8 qt **23.** 31.1% **25.** \$5000 at 9%; \$6000 at 8%; \$9000 at 9.5% **27.** \$4.84/lb **29.** 60 g

SECTION 2.5 *pages 77–82*

1. $\{x|x < 5\}$ **3.** $\{x|x \le 2\}$ **5.** $\{x|x < -4\}$ **7.** $\{x|x > 3\}$ **9.** $\{x|x > 4\}$ **11.** $\{x|x \le 2\}$ **13.** $\{x|x > -2\}$ **15.** $\{x|x \ge 2\}$ **17.** $\{x|x > -2\}$ **19.** $\{x|x \le 3\}$ **21.** $\{x|x < 2\}$ **23.** $\{x|x < -3\}$ **25.** $\{x|x \le 5\}$ **27.** $\{x|x \ge 1\}$ **29.** $\{x|x < -5\}$ **31.** $\{x|x < -24\}$ **33.** $\left\{x\middle|x < \frac{23}{16}\right\}$ **35.** $\left\{x\middle|x \ge \frac{8}{3}\right\}$ **37.** $\{x|x < 1\}$ **39.** $\{x|x < 3\}$ **41.** $\{x|x > -3\}$ **43.** $\{x|-1 < x < 2\}$ **45.** $\{x|x \ge 3 \text{ or } x \le 1\}$ **47.** $\{x|-2 < x < 4\}$ **49.** $\{x|x < -3 \text{ or } x > 0\}$ **51.** $\{x|x \ge 3\}$ **53.** $\varnothing$ **55.** $\varnothing$ **57.** $\{x|x < 1 \text{ or } x > 3\}$ **59.** $\{x|-3 < x < 4\}$ **61.** $\{x|3 < x < 5\}$ **63.** $\left\{x\middle|x > 3 \text{ or } x \le -\frac{5}{2}\right\}$ **65.** $\{x|x > 4\}$ **67.** $\varnothing$ **69.** the set of real numbers **71.** $\{x|-4 \le x \le 2\}$ **73.** $\{x|-4 < x < -2\}$ **75.** $\{x|x < -4 \text{ or } x > 3\}$ **77.** $\{x|-4 \le x < 1\}$ **79.** $\left\{x\middle|x > \frac{27}{2} \text{ or } x < \frac{5}{2}\right\}$ **81.** $\left\{x\middle|7 < x < \frac{15}{2}\right\}$ **83.** $\left\{x\middle|-10 \le x \le \frac{11}{2}\right\}$ **85.** 3 **87.** 2 ft **89.** 15, 16, 17, 18; 16, 17, 18, 19; or 17, 18, 19, 20 **91.** 5, 6, or 7 in. **93.** 160 min **95.** between 351 mi and 384.75 mi **97.** less than 167 checks **99.** less than 700 mi **101.** $86 \le N \le 100$ **103.** {1, 2, 3} **105.** {1, 2} **107.** {1, 2, 3, 4, 5, 6} **109.** {1, 2, 3, 4} **111.** {1, 2} **113.** {1, 2, 3, 4, 5, 6} **115.** between 25°C and 30°C **117.** 44 **119.** 10 min

SECTION 2.6 *pages 88–91*

1. 6 **3.** 5 **5.** −7 and 7 **7.** −3 and 3 **9.** no solution **11.** −5 and 1 **13.** 8 and 2 **15.** 2 **17.** no solution **19.** $\frac{1}{2}$ and $\frac{9}{2}$ **21.** 0 and $\frac{4}{5}$ **23.** $-\frac{1}{5}$ and 1 **25.** −1 **27.** no solution **29.** 4 and 14 **31.** −4 and −5 **33.** 4 and 12 **35.** $\frac{7}{4}$ **37.** no solution **39.** −3 and $\frac{3}{2}$ **41.** $\frac{4}{5}$ and 0 **43.** no solution **45.** no solution **47.** −2 and 1 **49.** $-\frac{3}{5}$ **51.** $-\frac{1}{3}$ and $\frac{5}{3}$ **53.** $\{x|x > 3 \text{ or } x < -3\}$ **55.** $\{x|x > 1 \text{ or } x < -3\}$ **57.** $\{x|4 \le x \le 6\}$ **59.** $\{x|x \le -1 \text{ or } x \ge 5\}$ **61.** $\{x|-3 < x < 2\}$ **63.** $\left\{x\middle|x > 2 \text{ or } x < -\frac{14}{5}\right\}$ **65.** $\varnothing$

67. the set of real numbers **69.** $\left\{x|x \le -\frac{1}{3} \text{ or } x \ge 3\right\}$ **71.** $\left\{x|-2 \le x \le \frac{9}{2}\right\}$ **73.** 2

75. $\left\{x|x < -2 \text{ or } x > \frac{22}{9}\right\}$ **77.** lower limit: 1.742 in.; upper limit: 1.758 in.

79. lower limit: 2.3 cc; upper limit: 2.7 cc **81.** lower limit: 195 volts; upper limit: 245 volts

83. lower limit: $3\frac{19}{64}$ in.; upper limit: $3\frac{21}{64}$ in. **85.** lower limit: 13,500 ohms; upper limit: 16,500 ohms

87. lower limit: 53.2 ohms; upper limit: 58.8 ohms **89.** $7, -\frac{11}{2}$ **91.** $\frac{17}{5}, -\frac{19}{5}$

93. $\{x|x \ge 15 \text{ or } x \le -12\}$ **95.** $\left\{x|-3 < x < \frac{19}{3}\right\}$ **97.** $\{y|y \ge -6\}$ **99.** $\{b|b \le 7\}$

ABSOLUTE VALUE EQUATIONS AND INEQUALITIES *page 93*

1. $\frac{3}{2}, 2$ **3.** $\frac{1}{3}, 1$ **5.** $\frac{13}{7}, \frac{7}{3}$ **7.** $\{x|x > 2\}$ **9.** $\{x|x > 0\}$

CHAPTER REVIEW *pages 95–97*

1. -9 (Objective 2.1.1) **2.** $-\frac{1}{12}$ (Objective 2.1.1) **3.** 7 (Objective 2.1.1)

4. $\frac{2}{3}$ (Objective 2.1.1) **5.** $\frac{8}{5}$ (Objective 2.1.1) **6.** 6 (Objective 2.1.1) **7.** $-\frac{40}{7}$ (Objective 2.1.1)

8. $-\frac{23}{27}$ (Objective 2.1.1) **9.** $\frac{26}{17}$ (Objective 2.1.2) **10.** $\frac{5}{2}$ (Objective 2.1.2)

11. $-\frac{17}{2}$ (Objective 2.1.2) **12.** $-\frac{9}{19}$ (Objective 2.1.2) **13.** $\left\{x|x > \frac{5}{3}\right\}$ (Objective 2.5.1)

14. $\{x|x > -4\}$ (Objective 2.5.1) **15.** $\left\{x|x \le -\frac{87}{14}\right\}$ (Objective 2.5.1)

16. $\left\{x|x \ge \frac{16}{13}\right\}$ (Objective 2.5.1) **17.** $\{x|-1 < x < 2\}$ (Objective 2.5.2)

18. $\{x|x > 2 \text{ or } x < -2\}$ (Objective 2.5.2) **19.** $\left\{x|-3 < x < \frac{4}{3}\right\}$ (Objective 2.5.2)

20. The set of real numbers. (Objective 2.5.2) **21.** $-\frac{5}{2}$ and $\frac{11}{2}$ (Objective 2.6.1)

22. no solution (Objective 2.6.1) **23.** $-\frac{8}{5}$ (Objective 2.6.1) **24.** no solution (Objective 2.6.1)

25. $\{x|1 \le x \le 4\}$ (Objective 2.6.2) **26.** $\left\{x|x \le \frac{1}{2} \text{ or } x \ge 2\right\}$ (Objective 2.6.2) **27.** $\varnothing$ (Objective 2.6.2)

28. $\{x|1 < x < 4\}$ (Objective 2.6.2) **29.** lower limit: 2.747 in.; upper limit: 2.753 in. (Objective 2.6.3)
30. lower limit: 1.75 cc; upper limit: 2.25 cc (Objective 2.6.3) **31.** 6 and 14 (Objective 2.2.2)
32. $-1, 0, 1$ (Objective 2.2.2) **33.** 7 quarters (Objective 2.2.1) **34.** 7 stamps (Objective 2.2.1)
35. $4.25/oz (Objective 2.3.1) **36.** 52 gal (Objective 2.3.1)
37. 1st plane: 440 mph; 2nd plane: 520 mph (Objective 2.3.2) **38.** 36 mi (Objective 2.3.2)
39. $3000 at 10.5%; $5000 invested at 6.4% (Objective 2.4.1) **40.** $6000 (Objective 2.4.1)
41. 30 oz (Objective 2.4.2) **42.** 375 lb of 30% tin; 125 lb of 70% tin (Objective 2.4.2)
43. $55,000 or more (Objective 2.5.3) **44.** $82 \le N \le 100$ (Objective 2.5.3)

CHAPTER TEST *pages 97–98*

1. -2 (Objective 2.1.1) **2.** $-\frac{1}{8}$ (Objective 2.1.1) **3.** $\frac{5}{6}$ (Objective 2.1.1)
4. 4 (Objective 2.1.1) **5.** $\frac{32}{3}$ (Objective 2.1.1) **6.** $-\frac{1}{5}$ (Objective 2.1.1) **7.** 1 (Objective 2.1.2)
8. -24 (Objective 2.1.2) **9.** $\frac{12}{7}$ (Objective 2.1.2) **10.** $\{x|x \le -3\}$ (Objective 2.5.1)
11. $\{x|x > -1\}$ (Objective 2.5.1) **12.** $\{x|x > -2\}$ (Objective 2.5.2) **13.** $\varnothing$ (Objective 2.5.2)
14. 3 and $-\frac{9}{5}$ (Objective 2.6.1) **15.** 7 and -2 (Objective 2.6.1)
16. $\left\{x\middle|-\frac{1}{3} \le x \le 1\right\}$ (Objective 2.6.2) **17.** $\{x|x > 2 \text{ or } x < -1\}$ (Objective 2.6.2)
18. no solution (Objective 2.6.1) **19.** less than 120 mi (Objective 2.5.3)
20. lower limit: 2.9 cc; upper limit: 3.1 cc (Objective 2.6.3) **21.** 6 stamps (Objective 2.2.1)
22. \$2.20 (Objective 2.3.1) **23.** 12 mi (Objective 2.3.2)
24. \$5000 at 7.8%; \$7000 at 9% (Objective 2.4.1) **25.** 100 oz (Objective 2.4.2)

CUMULATIVE REVIEW *pages 99–100*

1. -108 (Objective 1.1.3) **2.** 3 (Objective 1.1.4) **3.** -64 (Objective 1.1.4)
4. -8 (Objective 1.2.1) **5.** The Commutative Property of Addition (Objective 1.2.2)
6. $A \cap B = \{3, 9\}$ (Objective 1.3.1) **7.** $-17x + 2$ (Objective 1.2.3) **8.** $25y$ (Objective 1.2.3)
9. 2 (Objective 2.1.1) **10.** $\frac{1}{2}$ (Objective 2.1.1) **11.** 1 (Objective 2.1.1)
12. 24 (Objective 2.1.1) **13.** 2 (Objective 2.1.2) **14.** 2 (Objective 2.1.2)
15. $-\frac{13}{5}$ (Objective 2.1.2) **16.** $\{x|x \le -3\}$ (Objective 2.5.1) **17.** $\varnothing$ (Objective 2.5.2)
18. $\{x|x > -2\}$ (Objective 2.5.2) **19.** -1 and 4 (Objective 2.6.1) **20.** -4 and 7 (Objective 2.6.1)
21. $\left\{x\middle|\frac{1}{3} \le x \le 3\right\}$ (Objective 2.6.2) **22.** $\left\{x\middle|x > 2 \text{ or } x < -\frac{1}{2}\right\}$ (Objective 2.6.2)
23. 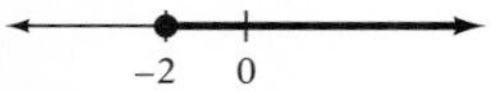

(Objective 1.3.2) **24.** 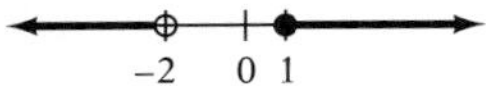

(Objective 1.3.2)
25. $(3n + 6) + 3n$; $6n + 6$ (Objective 1.2.4) **26.** 1 (Objective 2.2.2) **27.** 11 stamps (Objective 2.2.1)
28. 48 adult tickets (Objective 2.3.1) **29.** 340 mph (Objective 2.3.2) **30.** 3 L (Objective 2.4.2)
31. \$6500 (Objective 2.4.1)

ANSWERS to Chapter 3 Exercises

SECTION 3.1 *pages 113–118*

1. $6x^2 - 6x + 5$ **3.** $3x^2 - 3xy - 2y^2$ **5.** $-x^2 + 1$ **7.** $9x^n + 5$ **9.** $3y^3 + 2y^2 - 15y + 2$
11. $7a^2 - a + 2$ **13.** $3x^4 - 8x^2 + 2x$ **15.** $-5a^2 - 6a - 4$ **17.** $-b^{2n} + 2b^n - 7$
19. $-3x^2 + 10x^2 - 18x - 11$ **21.** $2x^2 + 4x - 7$ **23.** $-3x^3 + 7x^2 - 5x - 10$
25. $7x^4 - 9x^3 - 5x^2 + 9x - 2$ **27.** $3x^{2n} + x^n + 5$ **29.** $6a^3b^8$ **31.** x^4y^2 **33.** $-8a^3b^6$ **35.** $64a^6b^9$

37. x^9y^5 **39.** $256x^8$ **41.** $x^{40}y^{20}$ **43.** $64a^6b^6$ **45.** $81x^8y^{12}$ **47.** y^{3n} **49.** y^{6n+1} **51.** a^{2n^2}
53. y^{6n-3} **55.** b^{2n^2-n} **57.** $-6x^6y^3z^8$ **59.** $6a^4b^2c^4$ **61.** $72a^8b^{11}$ **63.** $-128a^5b^8c^9$ **65.** $\frac{1}{8}$
67. x^4 **69.** $\frac{2}{x^2y^4}$ **71.** 1 **73.** $\frac{9}{x^4}$ **75.** $\frac{1}{x^5}$ **77.** a^2 **79.** $\frac{x^4}{y^8}$ **81.** $\frac{4}{a^2}$ **83.** $\frac{1}{x^6}$
85. $\frac{x}{16y^2}$ **87.** $\frac{y^4}{243x^7}$ **89.** $\frac{12}{a^3b}$ **91.** $-\frac{3b^7}{2a^3}$ **93.** $2x$ **95.** x^2y^{10} **97.** $-x^{3n}$ **99.** $\frac{a^{n-3}}{b^{n+1}}$
101. $\frac{8}{a^{17}b^8}$ **103.** $\frac{x^{24}}{64y^{16}}$ **105.** $\frac{y^{12}}{x^6}$ **107.** $\frac{y^{12}}{x^6}$ **109.** $\frac{a^2+b}{a}$ **111.** $\frac{1+x^2y^2}{xy}$ **113.** 4.67×10^{-6}
115. 1.7×10^{-10} **117.** 2×10^{11} **119.** 0.000000123 **121.** 8,200,000,000,000,000 **123.** 0.039
125. 1.5×10^5 **127.** 2.08×10^7 **129.** 1.5×10^{-8} **131.** 1.78×10^{-11} **133.** 1.4×10^8
135. 1.14567901×10^7 **137.** 8×10^{-6} **139.** 7.2×10^9 operations **141.** 8.64×10^{12} m
143. 1.5×10^{-7} s **145.** 3.3898305×10^5 times **147.** 1.97311828×10^4 s **149.** $2.52224928 \times 10^{13}$ mi
151. 2.82743339×10^9 mi^2 **153.** $1.66030218 \times 10^{-24}$ g **155.** yes **157.** no **159.** no
161. -2 **163.** $8x^n$ **165.** $24a^2b^2$ **167.** $5m^4n^2$ **169.** ab^2 **171.** $\frac{1}{25} \neq \frac{13}{36}$ **173.** $\frac{1}{25} \neq \frac{1}{13}$

SECTION 3.2 *pages 123–127*

1. $2x^2 - 6x$ **3.** $6x^4 - 3x^3$ **5.** $6x^2y - 9xy^2$ **7.** $x^{n+1} + x^n$ **9.** $x^{2n} + x^ny^n$ **11.** $-4b^2 + 10b$
13. $-6a^4 + 4a^3 - 6a^2$ **15.** $9b^5 - 9b^3 + 24b$ **17.** $-3y^4 - 4y^3 + 2y^2$ **19.** $-20x^5 - 15x^4 + 15x^3 - 20x^2$
21. $-2x^4y + 6x^3y^2 - 4x^2y^3$ **23.** $x^{3n} + x^{2n} + x^{n+1}$ **25.** $a^{2n+1} - 3a^{n+2} + 2a^{n+1}$ **27.** $5y^2 - 11y$
29. $6y^2 - 31y$ **31.** $24n^3 + 16n^2 - n - 12$ **33.** $y^2 + 11y + 24$ **35.** $15x^2 - 61x + 56$
37. $14x^2 - 69xy + 27y^2$ **39.** $3a^2 + 16ab - 35b^2$ **41.** $15x^2 + 37xy + 18y^2$ **43.** $30x^2 - 79xy + 45y^2$
45. $2x^2y^2 - 3xy - 35$ **47.** $x^4 - 10x^2 + 24$ **49.** $x^4 + 2x^2y^2 - 8y^4$ **51.** $x^{2n} - 9x^n + 20$
53. $10b^{2n} + 18b^n - 4$ **55.** $3x^{2n} + 7x^nb^n + 2b^{2n}$ **57.** $x^3 + 8x^2 + 7x - 24$ **59.** $a^4 - a^3 - 6a^2 + 7a + 14$
61. $6a^3 - 13a^2b - 14ab^2 - 3b^3$ **63.** $6b^4 - 6b^3 + 3b^2 + 9b - 18$ **65.** $6a^5 - 15a^4 - 6a^3 + 19a^2 - 20a + 25$
67. $x^4 - 5x^3 + 14x^2 - 23x + 7$ **69.** $3b^3 - 14b^2 + 17b - 6$ **71.** $a^{7n} - 3a^{5n} - a^{4n} + a^{3n} + 3a^{2n} - 3a^n$
73. $x^{3n} - 4x^{2n}y^n + 2x^ny^{2n} + y^{3n}$ **75.** $b^2 - 49$ **77.** $b^2 - 121$ **79.** $25x^2 - 40xy + 16y^2$
81. $x^4 + 2x^2y^2 + y^4$ **83.** $4a^2 - 9b^2$ **85.** $x^4 - 1$ **87.** $4x^{2n} + 4x^ny^n + y^{2n}$ **89.** $25a^2 - 81b^2$
91. $4x^{2n} - 25$ **93.** $y^2 + 4y + 4$ **95.** $4x^2 - 4xy + y^2$ **97.** $x^2 - y^2z^2$ **99.** $36 - x^2$
101. $9a^2 - 24ab + 16b^2$ **103.** $9x^{2n} + 12x^n + 4$ **105.** $x^{2n} - 9$ **107.** $x^{2n} - 2x^n + 1$
109. $4x^{2n} + 20x^ny^n + 25y^{2n}$ **111.** $\left(x^2 + \frac{x}{2} - 3\right)$ ft^2 **113.** $(x^2 + 12x + 16)$ ft^2
115. $(2x^3 - 7x^2 - 15x)$ cm^3 **117.** $(4x^3 + 32x^2 + 48x)$ cm^3 **119.** $(3.14x^2 - 12.56x + 12.56)$ in.^{2}
121. $9x^2 - 30x + 25$ **123.** $2x^2 + 4xy$ **125.** $6x^5 - 5x^4 + 8x^3 + 13x^2$ **127.** $5x^2 - 5y^2$
129. $4x^4y^2 - 4x^4y + x^4$ **131.** 5 **133.** -1 **135.** a^2 **137.** $2x^2 + 11x - 21$ **139.** $x^2 - 2xy + y^2$
141. 256

SECTION 3.3 *pages 130–133*

1. $x + 8$ **3.** $x^2 + \frac{2}{x-3}$ **5.** $3x + 5 + \frac{3}{2x+1}$ **7.** $5x + 7 + \frac{2}{2x-1}$ **9.** $4x^2 + 6x + 9 + \frac{18}{2x-3}$
11. $3x^2 + 1 + \frac{1}{2x^2-5}$ **13.** $x^2 - 3x - 10$ **15.** $-x^2 + 2x - 1 + \frac{1}{x-3}$ **17.** $2x^3 - 3x^2 + x - 4$
19. $x - 4 + \frac{x+3}{x^2+1}$ **21.** $2x - 3 + \frac{1}{x-1}$ **23.** $3x + 1 + \frac{10x+7}{2x^2-3}$ **25.** $2x - 8$ **27.** $3x - 8$
29. $3x + 3 - \frac{1}{x-1}$ **31.** $x - 4 + \frac{7}{x+4}$ **33.** $x - 2 + \frac{16}{x+2}$ **35.** $4x - 12 + \frac{15}{x+1}$ **37.** $2x^2 - 3x + 9$

39. $x^2 - 3x + 2$ 41. $x^2 - 5x + 16 - \frac{41}{x+2}$ 43. $x^2 - x + 2 - \frac{4}{x+1}$ 45. $4x^2 + 8x + 15 + \frac{12}{x-2}$
47. $2x^2 - 3x + 7 - \frac{8}{x+4}$ 49. $2x^3 - 3x^2 + x - 4$ 51. $3x^3 + 2x^2 + 12x + 19 + \frac{33}{x-2}$
53. $3x^3 - x + 4 - \frac{2}{x+1}$ 55. $2x^3 + 6x^2 + 17x + 51 + \frac{155}{x-3}$ 57. $x^2 - 5x + 25$ 59. $x - y$
61. $2a - 2b - \frac{3b^2}{2a+b}$ 63. $a^2 + ab + b^2$ 65. $x^4 - x^3y + x^2y^2 - xy^3 + y^4$ 67. 3 69. −1
71. $x - 3$

SECTION 3.4 *pages 143–146*

1. $3a(2a - 5)$ 3. $x^2(4x - 3)$ 5. nonfactorable 7. $x(x^4 - x^2 - 1)$ 9. $4(4x^2 - 3x + 6)$
11. $5b^2(1 - 2b + 5b^2)$ 13. $x^n(x^n - 1)$ 15. $x^{2n}(x^n - 1)$ 17. $a^2(a^{2n} + 1)$ 19. $6x^2y(2y - 3x + 4)$
21. $4a^2b^2(-4b^2 - 1 + 6a)$ 23. $y^2(y^{2n} + y^n - 1)$ 25. $(a + 2)(x - 2)$ 27. $(x - 2)(a + b)$
29. $(x + 3)(x + 2)$ 31. $(x + 4)(y - 2)$ 33. $(a + b)(x - y)$ 35. $(y - 3)(x^2 - 2)$ 37. $(3 + y)(2 + x^2)$
39. $(2a + b)(x^2 - 2y)$ 41. $(y - 5)(x^n + 1)$ 43. $(x + 1)(x^2 + 2)$ 45. $(2x - 1)(x^2 + 2)$
47. $(x - 5)(x - 3)$ 49. $(a + 11)(a + 1)$ 51. $(b + 7)(b - 5)$ 53. $(y - 3)(y - 13)$ 55. $(b + 8)(b - 4)$
57. $(a - 7)(a - 8)$ 59. $(y + 12)(y + 1)$ 61. $(x + 5)(x - 1)$ 63. $(a + 6b)(a + 5b)$
65. $(x - 12y)(x - 2y)$ 67. $(y + 9x)(y - 7x)$ 69. $(7 + x)(3 - x)$ 71. $(5 + a)(10 - a)$
73. nonfactorable 75. $(2x + 5)(x - 8)$ 77. $(4y - 3)(y - 3)$ 79. $(2a + 1)(a + 6)$
81. nonfactorable 83. $(5x + 1)(x + 5)$ 85. $(11x - 1)(x - 11)$ 87. $(6x - 1)(2x + 3)$
89. $(4a + 1)(3a - 5)$ 91. $(3x + 2)(5x + 3)$ 93. $(12x - 5)(x - 1)$ 95. $(4y - 3)(2y - 3)$
97. nonfactorable 99. nonfactorable 101. $(2x - 3y)(3x + 7y)$ 103. $(4a + 7b)(a + 9b)$
105. $(5x - 4y)(2x - 3y)$ 107. $(8 - x)(3 + 2x)$ 109. nonfactorable 111. $(3 - 4a)(5 + 2a)$
113. $(4 - 7a)(3 + 4a)$ 115. $(3 - 10a)(5 + 2a)$ 117. $3a(3a - 2)(a + 4)$ 119. $y^2(5y - 4)(y - 5)$
121. $2x(2x - 3y)(x + 4y)$ 123. $5(5 + x)(4 - x)$ 125. $4x(10 + x)(8 - x)$ 127. $2x^2(5 + 3x)(2 - 5x)$
129. $a^2b^2(ab - 5)(ab + 2)$ 131. $5(18a^2b^2 + 9ab + 2)$ 133. nonfactorable 135. $2a(a^2 + 5)(a^2 + 2)$
137. $3y^2(x^2 - 5)(x^2 - 8)$ 139. $3b^2(3a + 4)(5a - 6)$ 141. $3xy(6x - 5y)(2x + 3y)$
143. $6b^2(2a - 3b)(4a + 3b)$ 145. $5(2x^n - 3)(x^n + 4)$ 147. $(8b + 3)(b - 4)$ 149. $p(2p - 1)(2p + 3)$
151. $xy(3x + y)(x - y)$ 153. $3b(b^2 + 5)(b^2 - 2)$ 155. $(4a - 5)(a + 1)$ 157. $(3y + 5)(4y + 11)$
159. $b(5a - 6)(a - 5)$ 161. 5, −5, 1, −1 163. 29, −29, 7, −7, 13, −13, 1, −1 165. 3, −3, 9, −9
167. 1, −1, 5, −5 169. 0, 3, −3 171. 5, −5, 1, −1

SECTION 3.5 *pages 151–154*

1. $(x + 4)(x - 4)$ 3. $(2x + 1)(2x - 1)$ 5. $(b - 1)^2$ 7. $(4x - 5)^2$ 9. $(xy + 10)(xy - 10)$
11. nonfactorable 13. $(x + 3y)^2$ 15. $(2x + y)(2x - y)$ 17. $(a^n + 1)(a^n - 1)$ 19. $(a + 2)^2$
21. $(x - 6)^2$ 23. $(4x + 11)(4x - 11)$ 25. $(1 + 3a)(1 - 3a)$ 27. nonfactorable 29. nonfactorable
31. $(5 + ab)(5 - ab)$ 33. $(5a - 4b)^2$ 35. $(x^n + 3)^2$ 37. $(x - 3)(x^2 + 3x + 9)$
39. $(2x - 1)(4x^2 + 2x + 1)$ 41. $(x - y)(x^2 + xy + y^2)$ 43. $(m + n)(m^2 - mn + n^2)$
45. $(4x + 1)(16x^2 - 4x + 1)$ 47. $(3x - 2y)(9x^2 + 6xy + 4y^2)$ 49. $(xy + 4)(x^2y^2 - 4xy + 16)$
51. nonfactorable 53. nonfactorable 55. $(a - 2b)(a^2 - ab + b^2)$ 57. $(x^{2n} + y^n)(x^{4n} - x^{2n}y^n + y^{2n})$
59. $(x^n + 2)(x^{2n} - 2x^n + 4)$ 61. $(xy - 3)(xy - 5)$ 63. $(xy - 5)(xy - 12)$ 65. $(x^2 - 6)(x^2 - 3)$
67. $(b^2 - 18)(b^2 + 5)$ 69. $(x^2y^2 - 6)(x^2y^2 - 2)$ 71. $(x^n + 2)(x^n + 1)$ 73. $(3xy - 5)(xy - 3)$
75. $(2ab - 3)(3ab - 7)$ 77. $(2x^2 - 15)(x^2 + 1)$ 79. $(2x^n - 1)(x^n - 3)$ 81. $(2a^n + 5)(3a^n + 2)$
83. $3(2x - 3)^2$ 85. $a(3a - 1)(9a^2 + 3a + 1)$ 87. $5(2x + 1)(2x - 1)$ 89. $y^3(y + 11)(y - 5)$
91. $(4x^2 + 9)(2x + 3)(2x - 3)$ 93. $2a(2 - a)(4 + 2a + a^2)$ 95. $(x + 2)(x + 1)(x - 1)$
97. $(x + 2)(2x^2 - 3)$ 99. $(x + 1)(x + 4)(x - 4)$ 101. $b^3(ab - 1)(a^2b^2 + ab + 1)$ 103. $x^2(x - 7)(x + 5)$

105. nonfactorable **107.** $2x^3(3x + 1)(x + 12)$ **109.** $(4a^2 + b^2)(2a + b)(2a - b)$ **111.** nonfactorable
113. $3b^2(b - 2)(b^2 + 2b + 4)$ **115.** $x^2y^2(x - 3y)(x - 2y)$ **117.** $2xy(4x + 7y)(2x - 3y)$
119. $(x - 2)(x + 1)(x - 1)$ **121.** $4(b - 1)(2x - 1)$ **123.** $(y + 1)(y - 1)(2x + 3)(2x - 3)$
125. $(x + 2)(x - 2)^2(x^2 + 2x + 4)$ **127.** $a^2(a^n - 3)^2$ **129.** $x^n(2x - 1)(x - 3)$ **131.** 12, −12
133. 20, −20 **135.** 8, −8 **137.** $(x - 1)(x^2 + x + 1)(a - b)$ **139.** $(p - 4)^2$ **141.** $-5(2a - 1)$
143. $(y^{2n} + 1)^2(y^n - 1)^2(y^n + 1)^2$ **145.** $-4(y + 2)^2(y - 2)$ **147.** 16 **149.** 1, 4, 9, 121, 484

SECTION 3.6 *pages 158–161*

1. −4 and −6 **3.** 0 and 7 **5.** 0 and $-\frac{5}{2}$ **7.** $-\frac{3}{2}$ and 7 **9.** 7 and −7 **11.** $\frac{4}{3}$ and $-\frac{4}{3}$
13. −5 and 1 **15.** $-\frac{3}{2}$ and 4 **17.** 0 and 9 **19.** 0 and 4 **21.** 7 and −4 **23.** $\frac{2}{3}$ and −5
25. $\frac{2}{5}$ and 3 **27.** $\frac{3}{2}$ and $-\frac{1}{4}$ **29.** 7 and −5 **31.** 9 and 3 **33.** $-\frac{4}{3}$ and 2 **35.** $\frac{4}{3}$ and −5
37. 5 and −3 **39.** −11 and −4 **41.** 4 and 2 **43.** 2 and 1 **45.** 2 and 3 **47.** −4, −1, and 1
49. −1, $-\frac{3}{2}$, and $\frac{3}{2}$ **51.** −12 or 11 **53.** 15 or −13 **55.** 11 and 13 **57.** 0, 1, or 7
59. length: 20 ft; width: 15 ft **61.** 14 cm **63.** 6 s **65.** length: 12 cm; width: 8 cm
67. $-7a$ and $3a$ **69.** $4a$ and $-4a$ **71.** $6a$ **73.** $\frac{2a}{3}$ and $-2a$ **75.** $8a$ and $-3a$ **77.** 236 or −44
79. length: 12 m; width: 10 m **81.** length: 22 in.; width: 10 in. **83.** 160 cm^3

REVERSE POLISH NOTATION *pages 162–163*

1. 12 [ENTER] 6 [÷] **3.** 9 [ENTER] 3 [+] 4 [÷] **5.** 6 [ENTER] 7 [x] 10 [x]
7. 9 **9.** 19 **11.** 2

CHAPTER REVIEW *pages 164–167*

1. $2x^2 - 5x - 2$ (Objective 3.1.1) **2.** $4x^2 - 8xy + 5y^2$ (Objective 3.1.1)
3. $2x^{2n} - 11x^n + 13$ (Objective 3.1.1) **4.** $70xy^2z^6$ (Objective 3.1.2) **5.** $-24a^7b^{10}$ (Objective 3.1.2)
6. $-\frac{48y^9z^{17}}{x}$ (Objective 3.1.2) **7.** $\frac{x^2 + y^2}{xy}$ (Objective 3.1.3) **8.** $-\frac{x^3}{4y^2z^3}$ (Objective 3.1.3)
9. $\frac{c^{10}}{2b^{17}}$ (Objective 3.1.3) **10.** 9.3×10^7 (Objective 3.1.4) **11.** 0.00254 (Objective 3.1.4)
12. 2×10^{-6} (Objective 3.1.4) **13.** $5x + 4 + \frac{6}{3x - 2}$ (Objective 3.3.1)
14. $2x - 3 - \frac{4}{6x + 1}$ (Objective 3.3.1) **15.** $4x + 1 + \frac{7}{5x - 7}$ (Objective 3.3.1)
16. $x + 4 + \frac{13}{x - 3}$ (Objective 3.3.1, 3.3.2) **17.** $4x^2 + 3x - 8 + \frac{50}{x + 6}$ (Objective 3.3.1, 3.3.2)
18. $x^3 + 4x^2 + 16x + 64 + \frac{252}{x - 4}$ (Objective 3.3.1, 3.3.2) **19.** $12x^5y^3 + 8x^3y^2 - 28x^2y^4$ (Objective 3.2.1)
20. $a^{3n+3} - 5a^{2n+4} + 2a^{2n+3}$ (Objective 3.2.1) **21.** $9x^2 + 8x$ (Objective 3.2.1)
22. $x^{3n+1} - 3x^{2n} - x^{n+2} + 3x$ (Objective 3.2.2) **23.** $x^4 + 3x^3 - 23x^2 - 29x + 6$ (Objective 3.2.2)
24. $6x^3 - 29x^2 + 14x + 24$ (Objective 3.2.2) **25.** $25a^2 - 4b^2$ (Objective 3.2.3)
26. $16x^2 - 24xy + 9y^2$ (Objective 3.2.3) **27.** $6a^2b(3a^3b - 2ab^2 + 5)$ (Objective 3.4.1)

28. $x^{3n}(x^{2n} - 3x^n + 12)$ (Objective 3.4.1) 29. $x^2(5x^{n+3} + x^{n+1} + 4)$ (Objective 3.4.1)
30. $(y - 3)(x - 4)$ (Objective 3.4.2) 31. $(a + 2b)(2x - 3y)$ (Objective 3.4.2)
32. $(x + 5)(x + 7)$ (Objective 3.4.3) 33. $(3 + x)(4 - x)$ (Objective 3.4.3)
34. $(x - 7)(x - 9)$ (Objective 3.4.3) 35. $(3x - 2)(2x - 9)$ (Objective 3.4.4)
36. $(8x - 1)(3x + 8)$ (Objective 3.4.4) 37. $(xy + 3)(xy - 3)$ (Objective 3.5.1)
38. $(2x + 3y)^2$ (Objective 3.5.1) 39. $(x^n - 6)^2$ (Objective 3.5.1)
40. $(6 + a^n)(6 - a^n)$ (Objective 3.5.1) 41. $(4a - 3b)(16a^2 + 12ab + 9b^2)$ (Objective 3.5.2)
42. $(5x^n + y^n)(25x^{2n} - 5x^ny^n + y^{2n})$ (Objective 3.5.2)
43. $(a + b + 1)(a^2 + 2ab + b^2 - a - b + 1)$ (Objective 3.5.2) 44. $(2 - y^n)(4 + 2y^n + y^{2n})$ (Objective 3.5.2)
45. $(3x^2 + 2)(5x^2 - 3)$ (Objective 3.5.3) 46. $(6x^4 - 5)(6x^4 - 1)$ (Objective 3.5.3)
47. $(3x^2y^2 + 2)(7x^2y^2 + 3)$ (Objective 3.5.3) 48. $3a^2(a^2 - 6)(a^2 + 1)$ (Objective 3.5.4)
49. $(x^n + 2)^2(x^n - 2)^2$ (Objective 3.5.4) 50. $3ab(a - b)(a^2 + ab + b^2)$ (Objective 3.5.4)
51. $-\frac{9}{2}$ and $\frac{7}{5}$ (Objective 3.6.1) 52. -2, 0, and 3 (Objective 3.6.1)
53. $\frac{5}{2}$ and 4 (Objective 3.6.1) 54. -4, 0, and 4 (Objective 3.6.1)
55. -1, -6, and 6 (Objective 3.6.1) 56. -1, 1, and 8 (Objective 3.6.1)
57. 1.137048×10^{23} mi (Objective 3.1.5) 58. 1.09×10^{21} hp (Objective 3.1.5)
59. $(10x^2 - 29x - 21)$ cm^2 (Objective 3.2.4) 60. $(27x^3 - 27x^2 + 9x - 1)$ ft^3 (Objective 3.2.4)
61. $(5x^2 + 8x - 8)$ in.^{2} (Objective 3.2.4) 62. -6 and -4, or 4 and 6 (Objective 3.6.2)
63. -8 or 7 (Objective 3.6.2) 64. 12 m (Objective 3.6.2)

CHAPTER TEST *pages 167–168*

1. $2x^3 - 4x^2 + 6x - 14$ (Objective 3.1.1) 2. $64a^7b^7$ (Objective 3.1.2) 3. $\frac{2b^7}{a^{10}}$ (Objective 3.1.3)
4. 5.01×10^{-7} (Objective 3.1.4) 5. 6.048×10^5 (Objective 3.1.5) 6. $\frac{x}{1 + xy}$ (Objective 3.1.3)
7. $35x^2 - 55x$ (Objective 3.2.1) 8. $6a^2 - 13ab - 28b^2$ (Objective 3.2.2)
9. $6t^5 - 8t^4 - 15t^3 + 22t^2 - 5$ (Objective 3.2.2) 10. $9z^2 - 30z + 25$ (Objective 3.2.3)
11. $2x^2 + 3x + 5$ (Objective 3.3.1) 12. $x^2 - 2x - 1 + \frac{2}{x - 3}$ (Objective 3.3.1, 3.3.2)
13. $2ab^2(3a^2 - 2a + 2b^2)$ (Objective 3.4.1) 14. $(3 - 2x)(4 - 3x)$ (Objective 3.4.4)
15. $(2a^2 - 5)(3a^2 + 1)$ (Objective 3.5.3) 16. $3x(2x - 3)(2x + 5)$ (Objective 3.5.4)
17. $(4x - 5)(4x + 5)$ (Objective 3.5.1) 18. $(4t + 3)^2$ (Objective 3.5.1)
19. $(3x - 2)(9x^2 + 6x + 4)$ (Objective 3.5.2) 20. $(3x - 2)(2x - a)$ (Objective 3.4.2)
21. $(3x^2 + 4)(x + 3)(x - 3)$ (Objective 3.5.4) 22. $-\frac{1}{3}$ and $\frac{1}{2}$ (Objective 3.6.1)
23. -1, $-\frac{1}{6}$, and 1 (Objective 3.6.1) 24. $(10x^2 - 3x - 1)$ ft^2 (Objective 3.2.4)
25. 12 h (Objective 3.1.5)

CUMULATIVE REVIEW *pages 168–170*

1. 6 (Objective 1.1.4) 2. $-\frac{5}{4}$ (Objective 1.2.1) 3. Inverse Property of Addition (Objective 1.2.2)
4. $-18x + 8$ (Objective 1.2.3) 5. $-\frac{1}{6}$ (Objective 2.1.1) 6. $-\frac{11}{4}$ (Objective 2.1.1)
7. $\frac{35}{3}$ (Objective 2.1.2) 8. -1 and $\frac{7}{3}$ (Objective 2.6.1) 9. $\frac{49}{26}$ (Objective 2.1.2)

10. -44 (Objective 2.1.2) **11.** $\left\{x \mid x > -\frac{5}{4}\right\}$ (Objective 2.5.1)
12. $\left\{x \mid x > \frac{5}{2} \text{ or } x < 1\right\}$ (Objective 2.5.2) **13.** $\left\{x \mid 0 < x < \frac{4}{3}\right\}$ (Objective 2.6.2)
14. $\frac{21}{8}$ (Objective 3.1.3) **15.** $-3x^2 + 60x + 8$ (Objective 3.2.1) **16.** $4x^3 - 7x + 3$ (Objective 3.2.2)
17. $x^{2n} + 2x^n + 1$ (Objective 3.2.3) **18.** $(4x - 3)(2x - 5)$ (Objective 3.4.4)
19. $-2x(2x - 3)(x - 2)$ (Objective 3.5.4) **20.** $(2x - 5y)^2$ (Objective 3.5.1)
21. $(x - y)(a + b)$ (Objective 3.4.2) **22.** $(x^2 + 4)(x + 2)(x - 2)$ (Objective 3.5.4)
23. $2(x - 2)(x^2 + 2x + 4)$ (Objective 3.5.4) **24.** $(a + 4)(a^2 - 4a + 16)$ (Objective 3.5.2)
25. 9 and 15 (Objective 2.2.2) **26.** 40 oz (Objective 2.3.1)
27. slower cyclist: 5 mph; faster cyclist: 7.5 mph (Objective 2.3.2) **28.** \$4500 (Objective 2.4.1)
29. 8, 10, and 12; or 10, 12, and 14 (Objective 2.2.2) **30.** 12 in. (Objective 3.6.2)

ANSWERS to Chapter 4 Exercises

SECTION 4.1 *pages 175–176*

1. $1 - 2x$ **3.** $3x - 1$ **5.** $2x$ **7.** $-\frac{2}{x}$ **9.** $\frac{x}{2}$ **11.** The expression is in simplest form.
13. $4x^2 - 2x + 3$ **15.** $a^2 + 2a - 3$ **17.** $\frac{x^n - 3}{4}$ **19.** $\frac{x^n}{x^n - y^n}$ **21.** $\frac{x - 3}{x - 5}$ **23.** $\frac{x + y}{x - y}$
25. $-\frac{x + 3}{3x - 4}$ **27.** $-\frac{x + 7}{x - 7}$ **29.** The expression is in simplest form. **31.** $\frac{a - b}{a^2 - ab + b^2}$
33. $\frac{4x^2 + 2xy + y^2}{2x + y}$ **35.** $\frac{x + y}{3x}$ **37.** $\frac{x - 2}{2x(x + 1)}$ **39.** $\frac{x + 2y}{3x + 4y}$ **41.** $-\frac{2x - 3}{2(2x + 3)}$
43. $\frac{x - 2}{a - b}$ **45.** $\frac{x^2 + 2}{(x + 1)(x - 1)}$ **47.** $\frac{xy + 7}{xy - 7}$ **49.** $\frac{a^n - 2}{a^n + 2}$ **51.** $\frac{a^n + 1}{a^n - 1}$ **53.** $-\frac{x - 3}{x + 3}$ **55.** $-3, 3$
57. $-6, 1$ **59.** $0, 3$ **61.** $-\frac{3}{2}, \frac{1}{3}$ **63.** $3, -4$ **65.** $-2, -1, 2$

SECTION 4.2 *pages 181–186*

1. $\frac{15b^4xy}{4}$ **3.** 1 **5.** $\frac{2x - 3}{x^2(x + 1)}$ **7.** $-\frac{x - 1}{x - 5}$ **9.** $\frac{2(x + 2)}{x - 1}$ **11.** $\frac{5x + 1}{5x - 1}$ **13.** $-\frac{x - 3}{2x + 3}$
15. $\frac{x^n - 6}{x^n - 1}$ **17.** $x + y$ **19.** $\frac{a^2y}{10x}$ **21.** $\frac{3}{2}$ **23.** $\frac{5(x - y)}{xy(x + y)}$ **25.** $\frac{(x - 3)^2}{(x + 3)(x - 4)}$ **27.** $\frac{2x + 5}{2x - 5}$
29. $-\frac{x + 2}{x + 5}$ **31.** $\frac{x^{2n}}{8}$ **33.** $\frac{(x - 3)(3x - 5)}{(2x + 5)(x + 4)}$ **35.** -1 **37.** 1 **39.** $-\frac{2(x^2 + x + 1)}{(x + 1)^2}$ **41.** $-\frac{13}{2xy}$
43. $\frac{1}{x - 1}$ **45.** $\frac{15 - 16xy - 9x}{10x^2y}$ **47.** $-\frac{5y + 21}{30xy}$ **49.** $-\frac{6x + 19}{36x}$ **51.** $\frac{10x + 6y + 7xy}{12x^2y^2}$
53. $-\frac{x^2 + x}{(x - 3)(x - 5)}$ **55.** -1 **57.** $\frac{5x - 8}{(x + 5)(x - 5)}$ **59.** $-\frac{3x^2 - 4x + 8}{x(x - 4)}$ **61.** $\frac{4x^2 - 14x + 15}{2x(2x - 3)}$
63. $\frac{2x^2 + x + 3}{(x + 1)^2(x - 1)}$ **65.** $\frac{x + 1}{x + 3}$ **67.** $\frac{x + 2}{x + 3}$ **69.** $-\frac{x^n - 2}{(x^n + 1)(x^n - 1)}$ **71.** $\frac{2}{x^n + 2}$ **73.** $\frac{12x + 13}{(2x + 3)(2x - 3)}$
75. $\frac{2x + 3}{2x - 3}$ **77.** $\frac{2x + 5}{(x + 3)(x - 2)(x + 1)}$ **79.** $\frac{3x^2 + 4x + 3}{(2x + 3)(x + 4)(x - 3)}$ **81.** $\frac{x - 2}{x - 3}$ **83.** $-\frac{2x^2 - 9x - 9}{(x + 6)(x - 3)}$

85. $\frac{3x^2 - 14x + 24}{(3x-2)(2x+5)}$ **87.** $\frac{x-12}{2x-5}$ **89.** $\frac{5x-1}{2x-3}$ **91.** $\frac{4}{(x-2)(x^2+2x+4)}$ **93.** $\frac{2}{x^2+1}$ **95.** $-x-1$
97. $\frac{108a}{b}$ **99.** $\frac{y-2}{x^4}$ **101.** $\frac{4y}{(y-1)^2}$ **103.** $\frac{4}{3(a-2)}$ **105.** $\frac{5(x+4)(x-4)}{2(x+2)(x-2)(x-3)}$
107. $-\frac{(a+3)(a-2)(5a-2)}{(3a+1)(a+2)(2a-3)}$ **109.** $\frac{6(x-1)}{x(2x+1)}$ **111.** $\frac{8}{15} \neq \frac{1}{8}$ **113.** $\frac{3}{y} + \frac{6}{x}$ **115.** $\frac{4}{b^2} + \frac{3}{ab}$

SECTION 4.3 *pages 188–190*

1. $\frac{5}{23}$ **3.** $\frac{2}{5}$ **5.** $\frac{1-y}{y}$ **7.** $\frac{5-a}{a}$ **9.** $\frac{b}{2(2-b)}$ **11.** $\frac{4}{5}$ **13.** $-\frac{1}{2}$ **15.** $\frac{2a^2-3a+3}{3a-2}$
17. $3x+2$ **19.** $\frac{x+2}{x-1}$ **21.** $-\frac{x+4}{2x+3}$ **23.** $\frac{2(x+2)}{x-4}$ **25.** $\frac{1}{3x+8}$ **27.** $\frac{x-2}{x+2}$ **29.** $\frac{x-3}{x+4}$
31. $\frac{3x+1}{x-5}$ **33.** $\frac{2(a+1)}{7a-4}$ **35.** $\frac{x+y}{x-y}$ **37.** $-\frac{2x}{x^2+1}$ **39.** $\frac{6(x-2)}{2x-3}$ **41.** $-\frac{a^2}{1-2a}$ **43.** $-\frac{3x+2}{x-2}$
45. $\frac{a^2-3a+1}{a-2}$ **47.** $\frac{2y}{x}$ **49.** $\frac{1+xy}{1-xy}$ **51.** $\frac{1}{a-1}$ **53.** $\frac{c-2}{c-3}$ **55.** $-\frac{2x+h}{x^2(x+h)^2}$ **57.** $-\frac{6z-1}{6z+1}$
59. $-\frac{4a-1}{6a}$

SECTION 4.4 *pages 198–205*

1. -5 **3.** -1 **5.** $\frac{5}{2}$ **7.** 0 **9.** -3 and 3 **11.** 8 **13.** 5 **15.** -3 **17.** no solution
19. 4 and 10 **21.** -1 **23.** -3 and 4 **25.** 2 **27.** 2 and 5 **29.** -3 and 1 **31.** $\frac{7}{2}$
33. no solution **35.** 4000 ducks **37.** \$165,000 **39.** 17 ft by 22 ft **41.** 4 diodes
43. 0.5 additional ounces **45.** 34 additional ounces **47.** 175 additional acres **49.** \$937.50
51. 160 s **53.** 2.5 mi **55.** 32.4 min **57.** 18 min **59.** 3 h **61.** 10 h **63.** 12 h
65. 80 min **67.** 20 h **69.** 10 min **71.** 30 min **73.** 30 h **75.** 64 mph
77. passenger train: 59 mph; freight train: 45 mph **79.** 16.25 mph **81.** 12 mph **83.** 4 mph
85. 70 mph **87.** jet plane: 720 mph; single-engined plane: 180 mph **89.** 4 mph **91.** 25 mph
93. 2 mph **95.** 60.80 mph **97.** 175 mi **99.** 120 dentists **101.** $\frac{4}{6}$ **103.** $5\frac{1}{3}$ min
105. 60 mph **107.** $\frac{5}{8}$ h

SECTION 4.5 *pages 207–208*

1. $W = \frac{P-2L}{2}$ **3.** $C = \frac{S}{1-r}$ **5.** $R = \frac{PV}{nT}$ **7.** $m_2 = \frac{Fr^2}{Gm_1}$ **9.** $R = \frac{E-Ir}{I}$ **11.** $b_2 = \frac{2A-hb_1}{h}$
13. $R_2 = \frac{RR_1}{R_1-R}$ **15.** $d = \frac{a_n-a_1}{n-1}$ **17.** $H = \frac{S-2WL}{2W+2L}$ **19.** $x = \frac{-by-c}{a}$ **21.** $x = \frac{d-b}{a-c}$
23. $x = \frac{ac}{b}$ **25.** $x = \frac{ab}{a+b}$ **27.** $5+2y$ **29.** $0, y$ **31.** $\frac{y}{2}, 5y$ **33.** $f = \frac{w_2f_2 + w_1f_1}{w_1+w_2}$

ERRORS IN ALGEBRAIC OPERATIONS *pages 209–210*

1. a^8, Rule for Simplifying Powers of Exponential Expressions **3.** $\frac{y+x}{xy}$, Rule of Negative Exponents
5. $2ab$ is in simplest form. **7.** $x^2 + x^3$ is in simplest form.
9. $\frac{1}{x}$, Rule for Multiplying Exponential Expressions **11.** $x^3 - x$ is in simplest form.

CHAPTER REVIEW *pages 211–213*

1. $3a^{2n} + 2a^n - 1$ (Objective 4.1.1) **2.** $-\frac{x+4}{x(x+2)}$ (Objective 4.1.1) **3.** $\frac{3x^2-1}{3x^2+1}$ (Objective 4.1.1)
4. $\frac{x^2+3x+9}{x+3}$ (Objective 4.1.1) **5.** $\frac{1}{a-1}$ (Objective 4.2.1) **6.** x (Objective 4.2.1)
7. $\frac{3x-1}{x-1}$ (Objective 4.2.1) **8.** $-\frac{(x+4)(x+3)(x+2)}{6(x-1)(x-4)}$ (Objective 4.2.1) **9.** $\frac{x^n+6}{x^n-3}$ (Objective 4.2.1)
10. $\frac{3x+2}{x}$ (Objective 4.2.1) **11.** $\frac{1}{x}$ (Objective 4.2.1) **12.** $-\frac{x-3}{x+3}$ (Objective 4.2.1)
13. $\frac{21a+40b}{24a^2b^4}$ (Objective 4.2.2) **14.** $\frac{4x-1}{x+2}$ (Objective 4.2.2) **15.** $\frac{3x+26}{(3x-2)(3x+2)}$ (Objective 4.2.2)
16. $\frac{9x^2+x+4}{(3x-1)(x-2)}$ (Objective 4.2.2) **17.** $-\frac{5x^2-17x-9}{(x-3)(x+2)}$ (Objective 4.2.2) **18.** $\frac{x-4}{x+5}$ (Objective 4.3.1)
19. $\frac{x-1}{4}$ (Objective 4.3.1) **20.** $\frac{x}{x-3}$ (Objective 4.3.1) **21.** $\frac{7x+4}{2x+1}$ (Objective 4.3.1)
22. $-\frac{9x^2+16}{24x}$ (Objective 4.3.1) **23.** 12 (Objective 4.4.2) **24.** 5 (Objective 4.4.2)
25. $\frac{15}{13}$ (Objective 4.4.1) **26.** −2 (Objective 4.4.1) **27.** 12 (Objective 4.4.1)
28. −3 and 5 (Objective 4.4.2) **29.** 6 (Objective 4.4.1) **30.** no solution (Objective 4.4.1)
31. 1 (Objective 4.4.1) **32.** 10 (Objective 4.4.1) **33.** $R = \frac{V}{I}$ (Objective 4.5.1)
34. $N = \frac{S}{1-Q}$ (Objective 4.5.1) **35.** $v = \frac{Ft}{m}$ (Objective 4.5.1) **36.** $r = \frac{S-a}{S}$ (Objective 4.5.1)
37. $6\frac{2}{3}$ tanks (Objective 4.4.2) **38.** 375 min (Objective 4.4.2) **39.** 48 mi (Objective 4.4.2)
40. 104 min (Objective 4.4.3) **41.** 40 min (Objective 4.4.3) **42.** 24 min (Objective 4.4.3)
43. 4 h (Objective 4.4.3) **44.** 2 mph (Objective 4.4.4) **45.** 45 mph (Objective 4.4.4)
46. 180 mph (Objective 4.4.4) **47.** 30 mph (Objective 4.4.4)

CHAPTER TEST *pages 213–214*

1. $\frac{v(v+2)}{2v-1}$ (Objective 4.1.1) **2.** $-\frac{2(a-2)}{3a+2}$ (Objective 4.1.1) **3.** $\frac{6(x-2)}{5}$ (Objective 4.2.1)
4. $\frac{x+2}{x-1}$ (Objective 4.2.1) **5.** $\frac{x+1}{3x-4}$ (Objective 4.2.1) **6.** $\frac{x^n-1}{x^n+2}$ (Objective 4.2.1)
7. $\frac{4y^2+6x^2-5xy}{2x^2y^2}$ (Objective 4.2.2) **8.** $\frac{2(5x+2)}{(x-2)(x+2)}$ (Objective 4.2.2) **9.** $\frac{x^2-9x+3}{(x+2)(x-3)}$ (Objective 4.2.2)
10. $-\frac{x^2+5x-2}{(x+1)(x+4)(x-1)}$ (Objective 4.2.2) **11.** $\frac{x-4}{x+3}$ (Objective 4.3.1) **12.** $\frac{x+4}{x+2}$ (Objective 4.3.1)
13. 2 (Objective 4.4.1) **14.** 1, 2 (Objective 4.4.1) **15.** no solution (Objective 4.4.1)
16. $x = \frac{c}{a-b}$ (Objective 4.5.1) **17.** 80 min (Objective 4.4.3) **18.** 14 rolls (Objective 4.4.2)
19. 10 min (Objective 4.4.3) **20.** 10 mph (Objective 4.4.4)

CUMULATIVE REVIEW *pages 215–216*

1. $\frac{36}{5}$ (Objective 1.1.4) **2.** -52 (Objective 1.2.1) **3.** $\frac{2}{3}$ (Objective 2.1.1)
4. $\frac{15}{2}$ (Objective 2.1.2) **5.** 7 and 1 (Objective 2.6.1) **6.** -2 and 3 (Objective 3.6.1)
7. $\frac{a^5}{b^6}$ (Objective 3.1.3) **8.** $\{x \mid x \leq 4\}$ (Objective 2.5.1) **9.** $-4a^4 + 6a^3 - 2a^2$ (Objective 3.2.1)
10. $(2x^n - 1)(x^n + 2)$ (Objective 3.4.4) **11.** $(xy - 3)(x^2y^2 + 3xy + 9)$ (Objective 3.5.2)
12. $x + 3y$ (Objective 4.1.1) **13.** $4x^2 + 10x + 10 + \frac{21}{x-2}$ (Objective 3.3.1 or 3.3.2)
14. $\frac{3xy}{x-y}$ (Objective 4.2.1) **15.** $2x^2 - 3x + 4 + \frac{2}{3x+2}$ (Objective 3.3.1)
16. $-\frac{x}{(3x+2)(x+1)}$ (Objective 4.2.2) **17.** $\frac{x-3}{x+3}$ (Objective 4.3.1) **18.** 9 (Objective 4.4.1)
19. 13 (Objective 4.4.1) **20.** $r = \frac{E - IR}{I}$ (Objective 4.5.1) **21.** $\frac{x}{x-1}$ (Objective 3.1.3)
22. $(y + 4)(2x - 3)$ (Objective 3.4.2) **23.** 50 lb (Objective 2.3.1)
24. 75,000 people (Objective 4.4.2) **25.** 14 min (Objective 4.4.3) **26.** 60 mph (Objective 4.4.4)

ANSWERS to Chapter 5 Exercises

SECTION 5.1 *pages 224–228*

1. 2 **3.** 27 **5.** $\frac{1}{9}$ **7.** 4 **9.** not a real number **11.** $\frac{343}{125}$ **13.** x **15.** $y^{\frac{1}{2}}$ **17.** $x^{\frac{1}{12}}$
19. $a^{\frac{7}{12}}$ **21.** $\frac{1}{a}$ **23.** $\frac{1}{y}$ **25.** $y^{\frac{3}{2}}$ **27.** $\frac{1}{x}$ **29.** $\frac{1}{x^4}$ **31.** a **33.** $x^{\frac{3}{10}}$ **35.** a^3
37. $\frac{1}{x^{\frac{1}{2}}}$ **39.** $y^{\frac{2}{3}}$ **41.** x^4y **43.** $x^6y^3z^9$ **45.** $\frac{x}{y^2}$ **47.** $\frac{x^{\frac{3}{2}}}{y^{\frac{1}{4}}}$ **49.** x^2y^8 **51.** $\frac{1}{x^{\frac{11}{12}}}$ **53.** $\frac{1}{y^{\frac{5}{2}}}$
55. $\frac{1}{b^{\frac{7}{8}}}$ **57.** a^5b^{13} **59.** $\frac{m^2}{4n^{\frac{3}{2}}}$ **61.** $\frac{y^{\frac{17}{2}}}{x^3}$ **63.** $\frac{16b^2}{a^{\frac{1}{3}}}$ **65.** $\frac{y^2}{x^3}$ **67.** $a - a^2$ **69.** $y + 1$
71. $a - \frac{1}{a}$ **73.** $\frac{1}{a^{10}}$ **75.** $a^{\frac{n}{6}}$ **77.** $\frac{1}{b^{\frac{2m}{3}}}$ **79.** x^{10n^2} **81.** $x^{3n}y^{2n}$ **83.** $\sqrt{5}$ **85.** $\sqrt[3]{b^4}$
87. $\sqrt[3]{9x^2}$ **89.** $-3\sqrt[5]{a^2}$ **91.** $\sqrt[4]{x^6y^9}$ **93.** $\sqrt{a^9b^{21}}$ **95.** $\sqrt[3]{3x-2}$ **97.** $14^{\frac{1}{2}}$ **99.** $x^{\frac{1}{3}}$
101. $x^{\frac{4}{3}}$ **103.** $b^{\frac{3}{5}}$ **105.** $(2x^2)^{\frac{1}{3}}$ **107.** $-(3x^5)^{\frac{1}{2}}$ **109.** $3xy^{\frac{2}{3}}$ **111.** $(a^2 + 2)^{\frac{1}{2}}$ **113.** y^7
115. $-a^3$ **117.** a^7b^3 **119.** $11y^6$ **121.** a^2b^4 **123.** $-a^3b^3$ **125.** $5b^5$ **127.** $-a^2b^3$
129. $5x^4y$ **131.** not a real number **133.** $2a^7b^2$ **135.** $-3ab^5$ **137.** y^3 **139.** $3a^5$
141. $-a^4b$ **143.** ab^5 **145.** $2a^2b^5$ **147.** $-2x^3y^4$ **149.** $\frac{4x}{y^7}$ **151.** $\frac{3b}{a^3}$ **153.** $2x + 3$
155. $x + 1$ **157.** c **159.** a **161.** a **163.** x **165.** a **167.** $2x^3$ **169.** ab^2
171. $2a^2b^4$ **173.** $\frac{3}{5}$ **175.** 3 **177.** $\frac{5}{4}$ **179.** $-\frac{1}{3}$ **181.** $16^{\frac{1}{4}}$, 2 **183.** $32^{-\frac{1}{5}}$, $\frac{1}{2}$

SECTION 5.2 *pages 234–237*

1. $x^2yz^2\sqrt{yz}$ **3.** $2ab^4\sqrt{2a}$ **5.** $3xyz^2\sqrt{5yz}$ **7.** $-5y\sqrt[3]{x^2y}$ **9.** $-6xy^3\sqrt[3]{x^2}$ **11.** $ab^2\sqrt[3]{a^2b^2}$
13. $-6\sqrt{x}$ **15.** $-2\sqrt{2}$ **17.** $\sqrt{2x}$ **19.** $3\sqrt{3a}-2\sqrt{2a}$ **21.** $10x\sqrt{2x}$ **23.** $2x\sqrt{3xy}$
25. $xy\sqrt{2y}$ **27.** $5a^2b^2\sqrt{ab}$ **29.** $9\sqrt[3]{2}$ **31.** $-7a\sqrt[3]{3a}$ **33.** $-5xy^2\sqrt[3]{x^2y}$ **35.** $16ab\sqrt[4]{ab}$
37. $6\sqrt{3}-6\sqrt{2}$ **39.** $6x^3y^2\sqrt{xy}$ **41.** $-9a^2\sqrt{3ab}$ **43.** $10xy\sqrt[3]{3y}$ **45.** 16 **47.** $2\sqrt[3]{4}$
49. $xy^3\sqrt{x}$ **51.** $8xy\sqrt{x}$ **53.** $2x^2y\sqrt[3]{2}$ **55.** $2ab\sqrt[4]{3a^2b}$ **57.** 6 **59.** $x-\sqrt{2x}$ **61.** $4x-8\sqrt{x}$
63. $x-6\sqrt{x}+9$ **65.** $84+16\sqrt{5}$ **67.** $672x^2y^2$ **69.** $4a^3b^3\sqrt[3]{a}$ **71.** $-8\sqrt{5}$ **73.** $x-y^2$
75. $12x-y$ **77.** $x-3\sqrt{x}-28$ **79.** $\sqrt[3]{x^2}+\sqrt[3]{x}-20$ **81.** $4\sqrt{x}$ **83.** $b^2\sqrt{3a}$
85. $\frac{\sqrt{5}}{5}$ **87.** $\frac{\sqrt{2x}}{2x}$ **89.** $\frac{\sqrt{5x}}{x}$ **91.** $\frac{\sqrt{5x}}{5}$ **93.** $\frac{3\sqrt[3]{4}}{2}$ **95.** $\frac{3\sqrt[3]{2x}}{2x}$ **97.** $\frac{\sqrt{2xy}}{2y}$ **99.** $\frac{2\sqrt{3ab}}{3b^2}$
101. $2\sqrt{5}-4$ **103.** $\frac{3\sqrt{y}+6}{y-4}$ **105.** $-5+2\sqrt{6}$ **107.** $8+4\sqrt{3}-2\sqrt{2}-\sqrt{6}$
109. $\frac{\sqrt{15}+\sqrt{10}-\sqrt{6}-5}{3}$ **111.** $\frac{3\sqrt[4]{2x}}{2x}$ **113.** $\frac{2\sqrt[5]{2a^3}}{a}$ **115.** $\frac{\sqrt[5]{16x^2}}{2}$ **117.** $\frac{1+2\sqrt{ab}+ab}{1-ab}$
119. $\frac{5x\sqrt{y}+5y\sqrt{x}}{x-y}$ **121.** $2\sqrt{2}$ **123.** $14\sqrt{2}-20$ **125.** $29\sqrt{2}-45$ **127.** $\frac{3\sqrt{y+1}-3}{y}$
129. $\frac{a+4\sqrt{a+4}+8}{a}$ **131.** $\sqrt{x+3}+\sqrt{x}$ **133.** $\sqrt[6]{x+y}$ **135.** $\sqrt[4]{2y(x+3)^2}$ **137.** $\sqrt[6]{a^3(a+3)^2}$
139. $\frac{1}{\sqrt{9+h}+3}$ **141.** $\frac{\sqrt[3]{4}-\sqrt[3]{2}+1}{3}$

SECTION 5.3 *pages 243–246*

1. $2i$ **3.** $7i\sqrt{2}$ **5.** $3i\sqrt{3}$ **7.** $4+2i$ **9.** $2\sqrt{3}-3i\sqrt{2}$ **11.** $4\sqrt{10}-7i\sqrt{3}$ **13.** $2ai$
15. $7x^6i$ **17.** $12ab^2i\sqrt{ab}$ **19.** $2\sqrt{a}+2ai\sqrt{3}$ **21.** $3b^2\sqrt{2b}-3bi\sqrt{3b}$ **23.** $5x^3y\sqrt{y}+5xyi\sqrt{2xy}$
25. $2a^2bi\sqrt{a}$ **27.** $2a\sqrt{3a}+3bi\sqrt{3b}$ **29.** $10-7i$ **31.** $11-7i$ **33.** $-6+i$ **35.** $-4-8i\sqrt{3}$
37. $-\sqrt{10}-11i\sqrt{2}$ **39.** $6-4i$ **41.** 0 **43.** $11+6i$ **45.** -24 **47.** -15 **49.** $-5\sqrt{2}$
51. $-15-12i$ **53.** $3\sqrt{2}+6i$ **55.** $-10i$ **57.** $29-2i$ **59.** 1 **61.** 1 **63.** 89 **65.** 50
67. $80-18i$ **69.** $-\frac{4}{5}i$ **71.** $-\frac{5}{3}+\frac{16}{3}i$ **73.** $\frac{30}{29}-\frac{12}{29}i$ **75.** $\frac{20}{17}+\frac{5}{17}i$ **77.** $\frac{11}{13}+\frac{29}{13}i$
79. $-\frac{1}{5}+\frac{\sqrt{6}}{10}i$ **81.** $-1+4i$ **83.** -1 **85.** i **87.** 1 **89.** 1 **91.** -1 **93.** -1
95. $(y+i)(y-i)$ **97.** $(x+5i)(x-5i)$ **99.** $(7x+4i)(7x-4i)$ **101. a.** yes **b.** yes
103. a. yes **b.** yes **105.** i **107.** $-\frac{\sqrt{2}}{2}+i\frac{\sqrt{2}}{2}$

SECTION 5.4 *pages 251–253*

1. 25 **3.** 27 **5.** 48 **7.** -2 **9.** no solution **11.** 9 **13.** -23 **15.** -16 **17.** $\frac{9}{4}$
19. 14 **21.** 7 **23.** 7 **25.** -122 **27.** 35 **29.** $-5, -2$ **31.** 45 **33.** 23
35. -4 **37.** 10 **39.** 1 **41.** 3 **43.** 21 **45.** $\frac{25}{72}$ **47.** 5 **49.** $-2, 4$ **51.** 6
53. 0, 1 **55.** $2, -2$ **57.** $-2, 1$ **59.** 2, 11 **61.** 3 **63.** 11.3 ft **65.** 10 ft **67.** 6.25 ft
69. 225 ft **71.** 96 ft **73.** 4.67 ft **75.** 27 **77.** 243 **79.** $s=\frac{v^2}{2a}$ **81.** $r=\frac{\sqrt{A\pi}}{\pi}$
83. $r=\frac{\sqrt[3]{6V\pi^2}}{2\pi}$ **85.** -1 **87.** first cyclist: 40 mi; second cyclist: 30 mi

DIOPHANTINE EQUATIONS *page 256*

1. 6, 13, 20, 27, 34, or 41 tickets **3.** (0, 11), (3, 7), or (6, 3)

CHAPTER REVIEW *pages 257–259*

1. $\frac{1}{3}$ (Objective 5.1.1) **2.** $\frac{1}{x^5}$ (Objective 5.1.1) **3.** $\frac{1}{a^{10}}$ (Objective 5.1.1)
4. $\frac{160y^{11}}{x}$ (Objective 5.1.1) **5.** $3\sqrt[4]{x^3}$ (Objective 5.1.2) **6.** $\frac{1}{\sqrt[3]{5a+2}}$ (Objective 5.1.2)
7. $x^{\frac{5}{2}}$ (Objective 5.1.2) **8.** $7x^{\frac{2}{3}}y$ (Objective 5.1.2) **9.** $3a^2b^3$ (Objective 5.1.3)
10. $-7x^3y^8$ (Objective 5.1.3) **11.** $-2a^2b^4$ (Objective 5.1.3) **12.** xy^3 (Objective 5.1.3)
13. $3ab^3\sqrt{2a}$ (Objective 5.2.1) **14.** $3x^2y^2\sqrt[3]{3y^2}$ (Objective 5.2.1) **15.** $-2ab^2\sqrt[5]{2a^3b^2}$ (Objective 5.2.1)
16. $xy^2z^2\sqrt{xy}$ (Objective 5.2.1) **17.** $5\sqrt{6}$ (Objective 5.2.2)
18. $4x^2\sqrt{3xy} - 4x^2\sqrt{5xy}$ (Objective 5.2.2) **19.** $2a^2b\sqrt{2b}$ (Objective 5.2.2)
20. $5x^3y^3\sqrt[3]{2x^2y}$ (Objective 5.2.2) **21.** $6x^2\sqrt{3y}$ (Objective 5.2.2) **22.** 40 (Objective 5.2.3)
23. $4xy^2\sqrt[3]{x^2}$ (Objective 5.2.3) **24.** $3x + 3\sqrt{3x}$ (Objective 5.2.3) **25.** $31 - 10\sqrt{6}$ (Objective 5.2.3)
26. $-13 + 6\sqrt{3}$ (Objective 5.2.3) **27.** $5x\sqrt{x}$ (Objective 5.2.4) **28.** $\frac{8\sqrt{3y}}{3y}$ (Objective 5.2.4)
29. $\frac{12\sqrt{x} + 12\sqrt{7}}{x-7}$ (Objective 5.2.4) **30.** $\frac{x\sqrt{x} - x\sqrt{2} + 2\sqrt{x} - 2\sqrt{2}}{x-2}$ (Objective 5.2.4)
31. $\frac{x + 2\sqrt{xy} + y}{x-y}$ (Objective 5.2.4) **32.** $6i$ (Objective 5.3.1) **33.** $5i\sqrt{2}$ (Objective 5.3.1)
34. $7 - 4i$ (Objective 5.3.1) **35.** $10\sqrt{2} + 2i\sqrt{3}$ (Objective 5.3.1) **36.** $4\sqrt{2} - 3i\sqrt{5}$ (Objective 5.3.1)
37. $9 - i$ (Objective 5.3.2) **38.** $-12 + 10i$ (Objective 5.3.2) **39.** $14 + 2i$ (Objective 5.3.2)
40. $-4\sqrt{2} + 8i\sqrt{2}$ (Objective 5.3.2) **41.** $10 - 9i$ (Objective 5.3.2) **42.** -16 (Objective 5.3.3)
43. $7 + 3i$ (Objective 5.3.3) **44.** $-6\sqrt{2}$ (Objective 5.3.3) **45.** $39 - 2i$ (Objective 5.3.3)
46. 13 (Objective 5.3.3) **47.** $6i$ (Objective 5.3.4) **48.** $\frac{2}{3} - \frac{5}{3}i$ (Objective 5.3.4)
49. $\frac{14}{5} + \frac{7}{5}i$ (Objective 5.3.4) **50.** $1 + i$ (Objective 5.3.4) **51.** $-2 + 7i$ (Objective 5.3.4)
52. -24 (Objective 5.4.1) **53.** $\frac{13}{3}$ (Objective 5.4.1) **54.** 20 (Objective 5.4.1)
55. -2 (Objective 5.4.1) **56.** 30 (Objective 5.4.1) **57.** 5 in. (Objective 5.4.2)
58. 120 watts (Objective 5.4.2) **59.** 242 ft (Objective 5.4.2) **60.** 6.63 ft (Objective 5.4.2)

CHAPTER TEST *pages 259–260*

1. $r^{\frac{1}{6}}$ (Objective 5.1.1) **2.** $\frac{64x^3}{y^6}$ (Objective 5.1.1) **3.** $\frac{b^3}{8a^6}$ (Objective 5.1.1)
4. $3\sqrt[5]{y^2}$ (Objective 5.1.2) **5.** $\frac{1}{2}x^{\frac{3}{4}}$ (Objective 5.1.2) **6.** $2xy^2$ (Objective 5.1.3)
7. $4x^2y^3\sqrt{2y}$ (Objective 5.2.1) **8.** $3abc^2\sqrt[3]{ac}$ (Objective 5.2.1) **9.** $8a\sqrt{2a}$ (Objective 5.2.2)
10. $-2x^2y\sqrt[3]{2x}$ (Objective 5.2.2) **11.** $-4x\sqrt{3}$ (Objective 5.2.3) **12.** $14 + 10\sqrt{3}$ (Objective 5.2.3)
13. $2a - \sqrt{ab} - 15b$ (Objective 5.2.3) **14.** $4x + 4\sqrt{xy} + y$ (Objective 5.2.3)
15. $\frac{4x^2}{y}$ (Objective 5.2.4) **16.** 2 (Objective 5.2.4) **17.** $\frac{x + \sqrt{xy}}{x-y}$ (Objective 5.2.4)
18. -4 (Objective 5.3.1) **19.** $-3 + 2i$ (Objective 5.3.2) **20.** $18 + 16i$ (Objective 5.3.3)

21. $-\frac{4}{5}+\frac{7}{5}i$ (Objective 5.3.4) **22.** $10+2i$ (Objective 5.3.2, 5.3.3) **23.** 4 (Objective 5.4.1)
24. -3 (Objective 5.4.1) **25.** 576 ft (Objective 5.4.2)

CUMULATIVE REVIEW *pages 260–262*

1. The Distributive Property (Objective 1.2.2) **2.** $-x-24$ (Objective 1.2.3)
3. $A\cap B=\varnothing$ (Objective 1.3.1) **4.** 16 (Objective 5.4.1) **5.** $\frac{3}{2}$ (Objective 2.1.1)
6. $\frac{2}{3}$ (Objective 2.1.2) **7.** $\{x|x\ge -1\}$ (Objective 2.5.1) **8.** $\{x|4<x<5\}$ (Objective 2.5.2)
9. $\frac{1}{3}$ and $\frac{7}{3}$ (Objective 2.6.1) **10.** $\left\{x\middle|x<2 \text{ or } x>\frac{8}{3}\right\}$ (Objective 2.6.2)
11. $(8a+b)(8a-b)$ (Objective 3.5.1) **12.** $x(x^2+3)(x+1)(x-1)$ (Objective 3.5.4)
13. $\frac{2}{3}$ and -5 (Objective 3.6.1) **14.** -2 and 4 (Objective 3.6.1) **15.** $\frac{4}{a-1}$ (Objective 4.1.1)
16. $y-2$ (Objective 4.3.1) **17.** $R=Pn+C$ (Objective 4.5.1) **18.** -12 (Objective 4.4.1)
19. $\frac{3x^3}{y}$ (Objective 3.1.3) **20.** $\frac{y^8}{x^2}$ (Objective 5.1.1) **21.** $-x\sqrt{5x}$ (Objective 5.2.2)
22. $11-5\sqrt{5}$ (Objective 5.2.3) **23.** $\frac{x\sqrt[3]{4y^2}}{2y}$ (Objective 5.2.4) **24.** $22-7i$ (Objective 5.3.3)
25. $-\frac{3}{5}+\frac{6}{5}i$ (Objective 5.3.4) **26.** 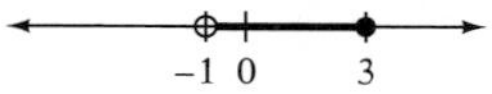

(Objective 1.3.2)
27. 19 stamps (Objective 2.2.1) **28.** \$4000 (Objective 2.4.1)
29. length: 12 ft; width: 6 ft (Objective 3.6.2) **30.** 250 mph (Objective 4.4.4)
31. 1.25 s (Objective 3.1.5) **32.** 25 ft (Objective 5.4.2)

ANSWERS to Chapter 6 Exercises

SECTION 6.1 *pages 270–273*

1. 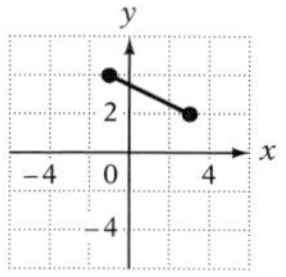

3. 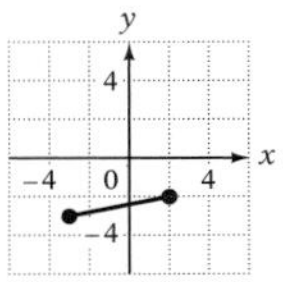

5. A is (0, 3), B is (1, 1), C is (3, −4), D is (−4, 4)

7. A is (4, 0), B is (−1, 1), C is (−4, −2), D is (1, 2)

9. 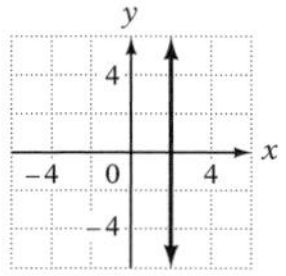

11. 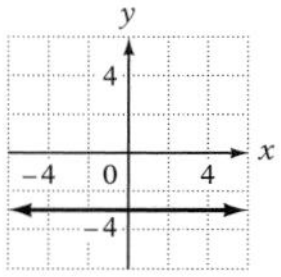

13. (−3, −6) **15.** (−6, 8) **17.** (−4, −3) **19.** (−2, 1)

21. $\left(-2, -\frac{1}{3}\right)$

23. 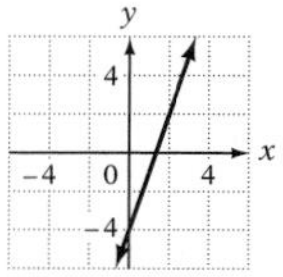

25. 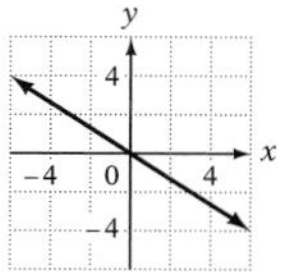

27. 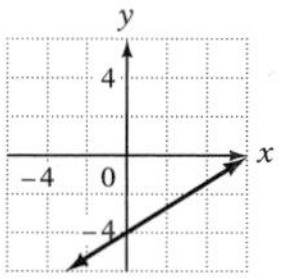

29. 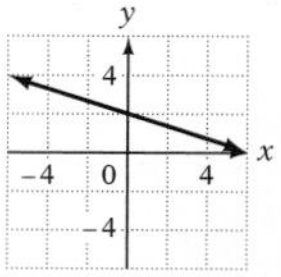

31. 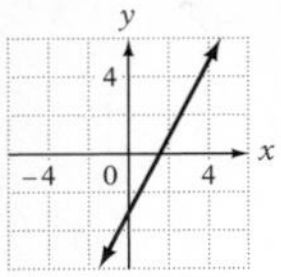 **33.** 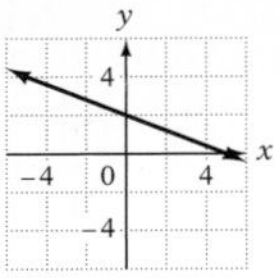 **35.** 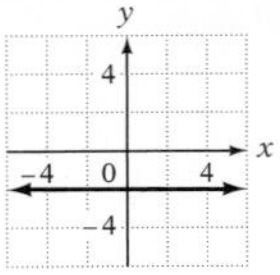 **37.** 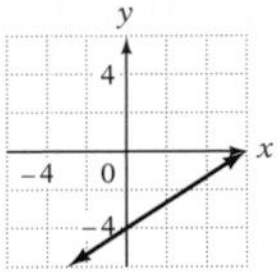 **39.** (0, 5) **41.** (−3, 5)

43. (−1, 8) **45.** 2 **47.** 4 **49.** 2 **51.** 10 square units **53.** $(8 + \sqrt{34})$ units **55.** 3

SECTION 6.2 *pages 280–284*

1. −1 **3.** $\frac{1}{3}$ **5.** $-\frac{2}{3}$ **7.** $-\frac{3}{4}$ **9.** undefined **11.** $\frac{7}{5}$ **13.** 0 **15.** $-\frac{1}{2}$ **17.** −3

19. undefined **21.** −2 **23.** 0 **25.** 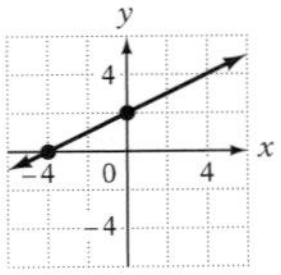 **27.** 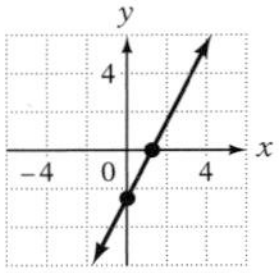 **29.** 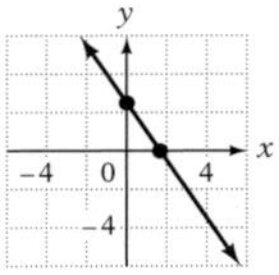

31. 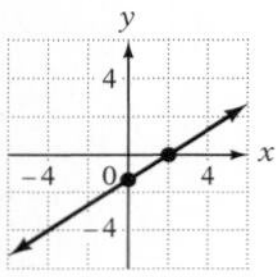 **33.** 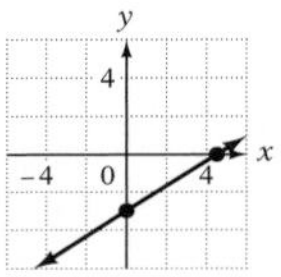 **35.** 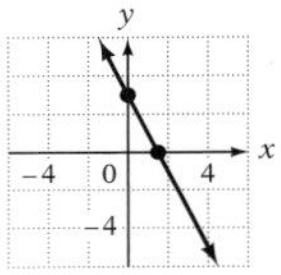 **37.** 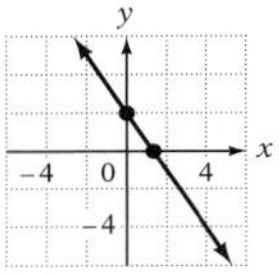 **39.** 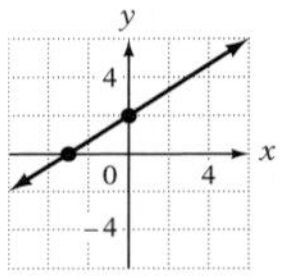

41. 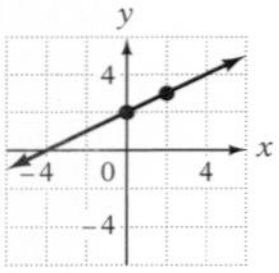 **43.** 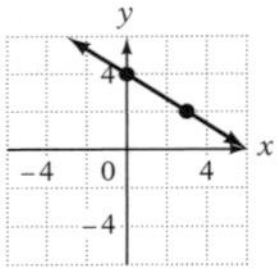 **45.** 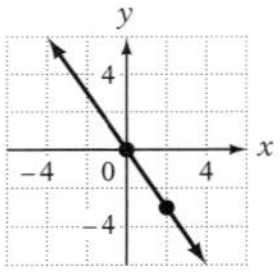 **47.** 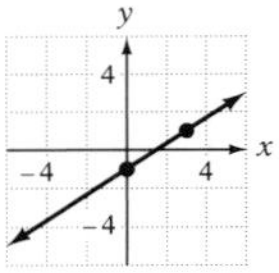 **49.** 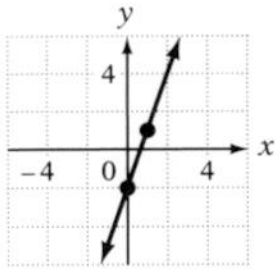

51. 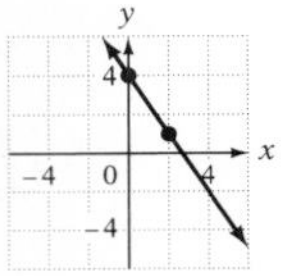 **53.** 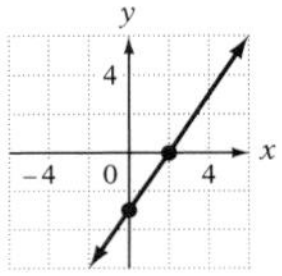 **55.** 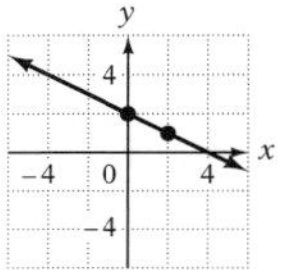 **57.** 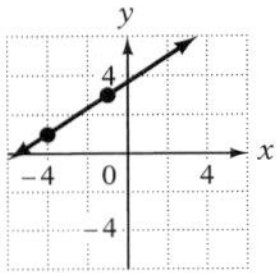 **59.** 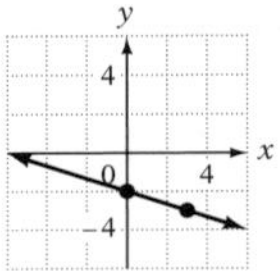

61. 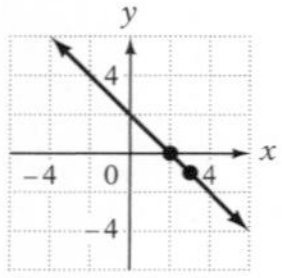

63. $y = \frac{1}{2}x - 2$ **65.** $y = -3x - 3$ **67. a.** $y = \frac{5}{2}x - 5$ **b.** (2, 0) and (0, −5)

69. a. $y = 8x - 8$ **b.** (1, 0) and (0, −8)

71. The graph of the line rotates counterclockwise.

73. The graph of the line moves up. **75.** m

SECTION 6.3 *pages 290–294*

1. $y = 2x + 5$ **3.** $y = \frac{1}{2}x + 2$ **5.** $y = \frac{5}{4}x + \frac{21}{4}$ **7.** $y = -\frac{5}{3}x + 5$ **9.** $y = -3x + 9$ **11.** $y = -3x + 4$

13. $y = \frac{2}{3}x - \frac{7}{3}$ **15.** $y = \frac{1}{2}x$ **17.** $y = 3x - 9$ **19.** $y = -\frac{2}{3}x + 7$ **21.** $y = 0$ **23.** $y = \frac{3}{2}x$

25. $x = 0$ **27.** $y = 3x - 3$ **29.** $y = -\frac{5}{2}x - \frac{25}{2}$ **31.** $y = \frac{4}{3}x + 5$ **33.** $x = -\frac{2}{5}x + \frac{3}{5}$ **35.** $x = 3$

37. $y = -\frac{5}{4}x - \frac{15}{2}$ **39.** $y = -3$ **41.** $y = x + 2$ **43.** $y = -2x - 3$ **45.** $y = \frac{2}{3}x + \frac{5}{3}$ **47.** $y = \frac{1}{3}x + \frac{10}{3}$

49. $y = \frac{3}{2}x - \frac{1}{2}$ **51.** $y = -\frac{3}{2}x + 3$ **53.** $y = -1$ **55.** $y = x - 1$ **57.** $y = -2x - 1$ **59.** $x = -2$

61. $y = -\frac{3}{4}x + \frac{17}{4}$ **63.** $y = 3x - 10$ **65.** $y = -x + 1$ **67.** $y = \frac{3}{7}x - \frac{15}{7}$ **69.** $y = \frac{1}{2}x - 1$

71. $y = -4$ **73.** $y = \frac{3}{4}x$ **75.** $y = -\frac{4}{3}x + \frac{5}{3}$ **77.** $x = -2$ **79.** $y = x - 1$ **81.** yes **83.** no

85. no **87.** yes **89.** yes **91.** yes **93.** no **95.** yes **97.** yes **99.** $y = \frac{2}{3}x - \frac{8}{3}$

101. $y = \frac{1}{3}x - \frac{1}{3}$ **103.** $y = -\frac{5}{3}x - \frac{14}{3}$ **105.** $y = \frac{2}{3}x + 3$ **107.** $y = -\frac{5}{3}x - \frac{13}{3}$ **109.** $y = \frac{5}{2}x + 8$

111. $y = -\frac{1}{5}x - \frac{11}{5}$ **113.** $y = -\frac{2}{3}x - 5$ **115.** -3 **117.** -2 **119.** -3 **121.** 5

123. $y = -\frac{6}{5}x + \frac{12}{5}$; $y = \frac{1}{4}x - \frac{1}{2}$; $y = -7x - 15$

125. Let $P_1 = (1, 4)$ and $P_2 = (-3, -2)$; $m = \frac{-2-4}{-3-1} = \frac{-6}{-4} = \frac{3}{2}$.

Let $P_3 = (-3, -2)$ and $P_4 = (3, -6)$; $m = \frac{-6-(-2)}{3-(-3)} = \frac{-4}{6} = -\frac{2}{3}$; $m_1 \cdot m_2 = \frac{3}{2}\left(-\frac{2}{3}\right) = -1$.

127. Let $P_1 = (1, 6)$ and $P_2 = (3, 2)$; $m_1 = \frac{2-6}{3-1} = \frac{-4}{2} = -2$.

Let $P_3 = (-3, -2)$ and $P_4 = (-1, -6)$; $m_2 = \frac{-6-(-2)}{-1-(-3)} = \frac{-4}{2} = -2$; $m_1 = m_2$.

Let $P_5 = (-3, -2)$ and $P_6 = (1, 6)$; $m_3 = \frac{6-(-2)}{1-(-3)} = \frac{8}{4} = 2$.

Let $P_7 = (-1, -6)$ and $P_8 = (3, 2)$; $m_4 = \frac{2-(-6)}{3-(-1)} = \frac{8}{4} = 2$, $m_3 = m_4$.

129. $y = -x + 4$ **131.** $y = \frac{1}{2}x + 1$ **133.** 0 **135.** 6

SECTION 6.4 *pages 297–299*

1. $y = 0.08x$; \$320 **3.** $y = 10x$; 600 gal **5.** $y = -\frac{25{,}000}{3}x + 250{,}000$; \$150,000

7. $y = 120x + 3000$; \$7200 **9.** $y = -\frac{1}{2}x + 20$; 24 units **11.** $y = -\frac{1}{4}x + 5$; 8 units

13.

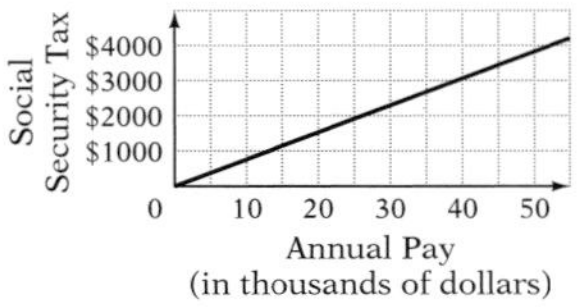

15. The slope represents the Social Security tax per \$1. The y-intercept represents the amount of tax paid when \$0 is earned.

SECTION 6.5 *pages 302–304*

1.

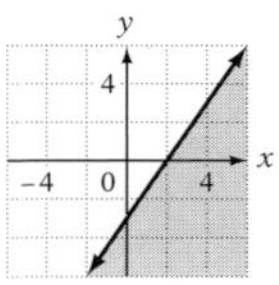

3.

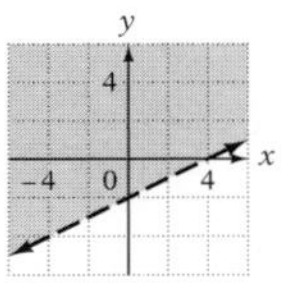

5.

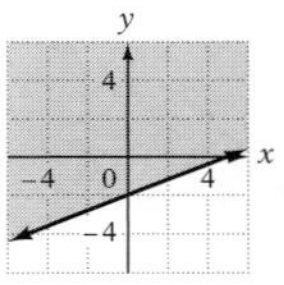

7.

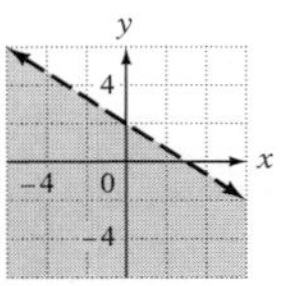

9.

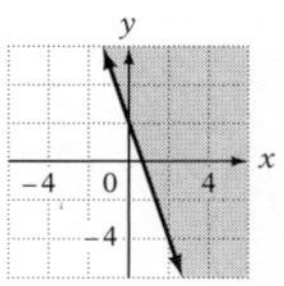

11. 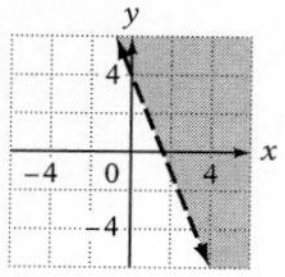

13. 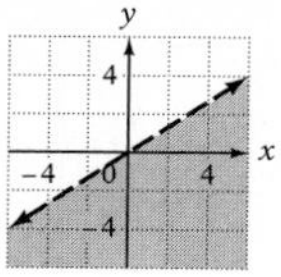

15. 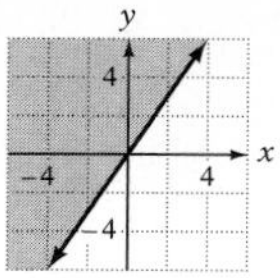

17. 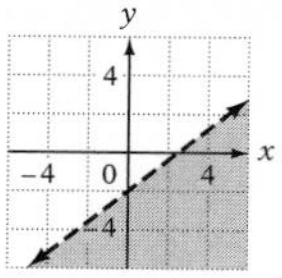

19. 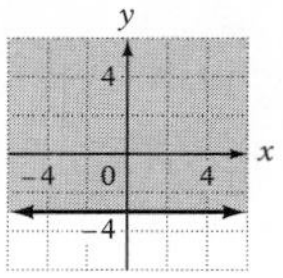

21. 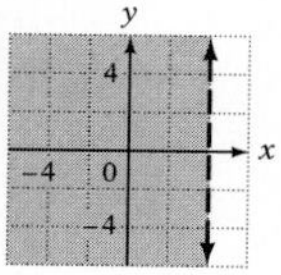

23. 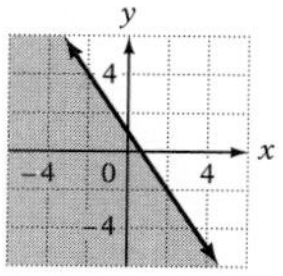

25. 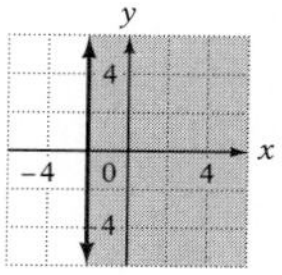

27. 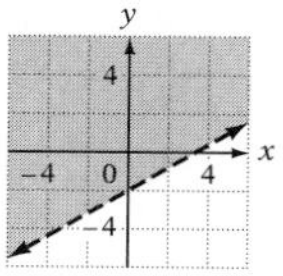

29. b 31. a and b

33. a 35. x = ounces of Symx A; y = ounces of Symx B; $7x + 4y \geq 18$; yes

CALCULATORS AND COMPUTERS *page 305*

1. $x \approx -4.971745$ 3. $(-1.004216, -6.391558)$ 5. $(-0.3700658, -1.912202)$

APPLICATION OF SLOPE *page 307*

1. 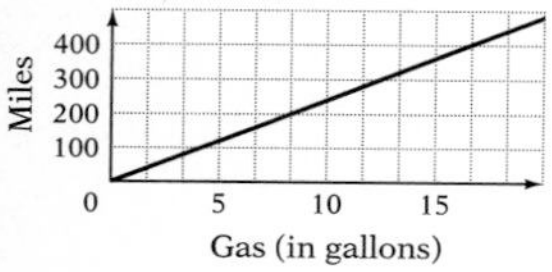

Use the points $P_1(0, 0)$ and $P_2(15, 360)$.

$$m = \frac{y_2 - y_1}{x_2 - x_1} = \frac{360 - 0}{15 - 0} = 24$$

CHAPTER REVIEW *pages 308–310*

1. $(-3, 0)$ (Objective 6.1.2) 2. $y = \frac{2}{5}x + 4$ (Objective 6.3.1)

3. $y = -\frac{2}{3}x + 2$ (Objective 6.3.2) 4. $y = 2x + 1$ (Objective 6.3.2) 5. $-\frac{1}{6}$ (Objective 6.2.1)

6. $y = -\frac{7}{5}x + \frac{1}{5}$ (Objective 6.3.1) 7. $x = 2$ (Objective 6.3.1) 8. $y = -\frac{3}{4}x + 2$ (Objective 6.3.1)

9. $(3, 0)$ (Objective 6.2.2) 10. 0 (Objective 6.2.1) 11. $y = -\frac{3}{2}x$ (Objective 6.3.2)

12. $\left(3, -\frac{1}{4}\right)$ (Objective 6.1.2) 13. undefined (Objective 6.2.1) 14. $y = -\frac{3}{2}x$ (Objective 6.3.2)

15. $y = 4$ (Objective 6.3.1) 16. undefined (Objective 6.2.1) 17. $y = -3x + 7$ (Objective 6.3.2)

18. $y = \frac{3}{2}x + 2$ (Objective 6.3.2) 19. $y = \frac{5}{2}x - 9$ (Objective 6.3.1)

20. 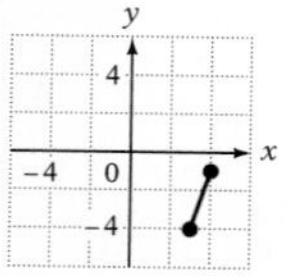

(Objective 6.1.1)

21. 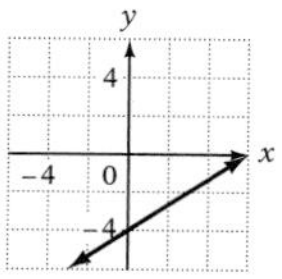

(Objective 6.1.3)

22. 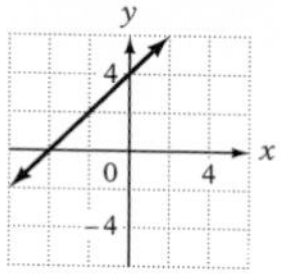

(Objective 6.2.3)

23. 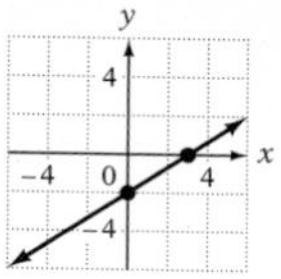

(Objective 6.2.2)

24. 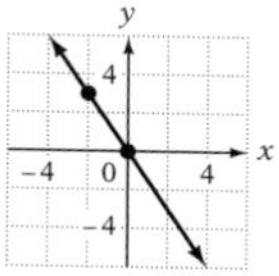

(Objective 6.2.3)

25. 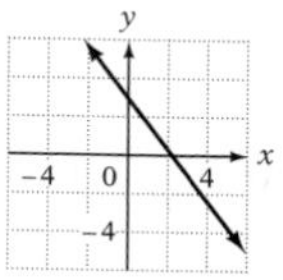

(Objective 6.1.3)

26. 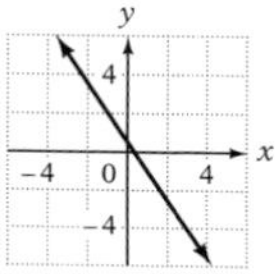

(Objective 6.1.3)

27. 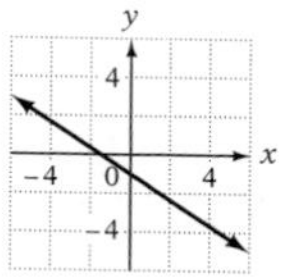

(Objective 6.1.3)

28. 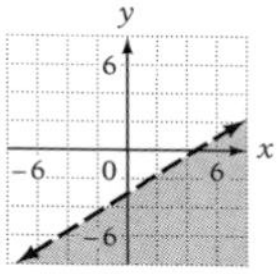

(Objective 6.5.1)

29. 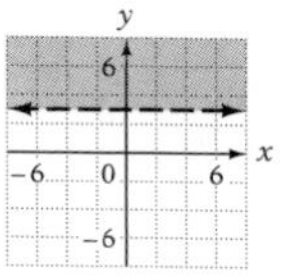

(Objective 6.5.1)

30. 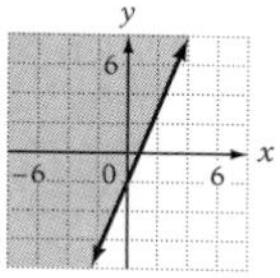

(Objective 6.5.1)

31. 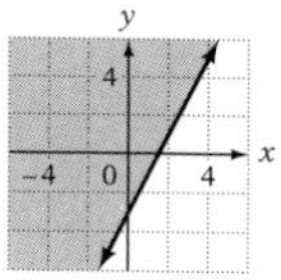

(Objective 6.5.1)

32. 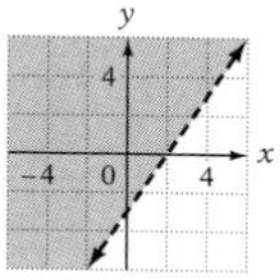

(Objective 6.5.1)

33. $y = 20x + 2000$; \$4500 (Objective 6.4.1)

34. $y = 400x$; 1000 mi (Objective 6.4.1)

CHAPTER TEST *pages 310–312*

1. $x = -2$ (Objective 6.3.1) **2.** $(-6, -6)$ (Objective 6.1.2) **3.** $\frac{2}{3}$ (Objective 6.2.1)

4. 0 (Objective 6.2.1) **5.** $y = -\frac{3}{2}x - 1$ (Objective 6.3.1) **6.** $y = -\frac{4}{3}x - 3$ (Objective 6.3.1)

7. $y = -\frac{5}{6}x + \frac{4}{3}$ (Objective 6.3.1) **8.** $y = -\frac{1}{4}x + \frac{7}{4}$ (Objective 6.3.1) **9.** $y = 3$ (Objective 6.3.2)

10. $y = -\frac{3}{2}x + \frac{7}{2}$ (Objective 6.3.2) **11.** $y = -2x - 4$ (Objective 6.3.2) **12.** $y = -3$ (Objective 6.3.1)

13. $y = \frac{3}{2}x - \frac{19}{2}$ (Objective 6.3.2) **14.**

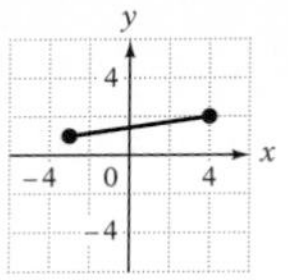

(Objective 6.1.1) **15.**

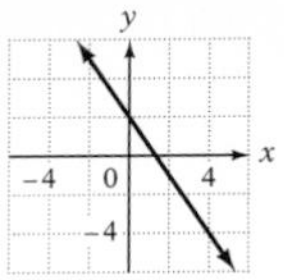

(Objective 6.1.3)

16.

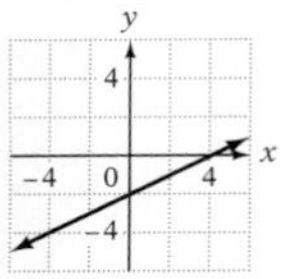

(Objective 6.1.3) **17.**

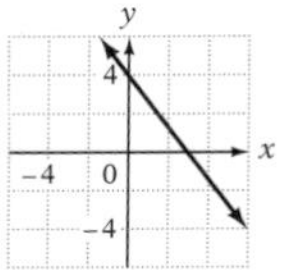

(Objective 6.2.2) **18.**

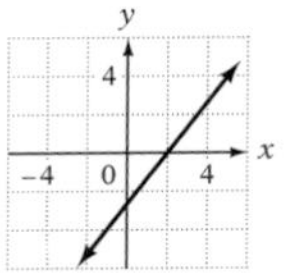

(Objective 6.2.3)

19. $y = -\frac{10{,}000}{3}x + 50{,}000$ (Objective 6.4.1) **20.**

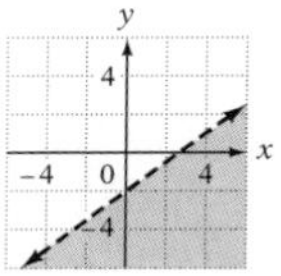

(Objective 6.5.1)

CUMULATIVE REVIEW *pages 312–314*

1. 432 (Objective 1.1.3) **2.** 0 (Objective 1.1.4) **3.** $-\frac{13}{2}$ (Objective 1.2.1)

4. $\frac{9}{2}$ (Objective 2.1.1) **5.** $\frac{8}{9}$ (Objective 1.2.3) **6.** $\{x|1 < x < 3\}$ (Objective 2.5.2)

7. $\frac{5}{2}$ and $-\frac{3}{2}$ (Objective 2.6.1) **8.** $\left\{x\middle|-\frac{2}{5} < x < 2\right\}$ (Objective 2.6.2) **9.** $\frac{3b^8}{8a^2}$ (Objective 3.1.3)

10. $a^4 - a^3 + 7a^2 - 2a$ (Objective 3.2.1) **11.** $8xy(x + 2xy + 3y)$ (Objective 3.4.1)

12. $(2x - 3b)(3x + 2a)$ (Objective 3.4.2) **13.** 5 and −1 (Objective 3.6.1) **14.** $\frac{9}{4}$ (Objective 5.4.1)

15. $\frac{x-8}{x-5}$ (Objective 4.2.2) **16.** $4ab\sqrt[3]{2ab}$ (Objective 5.2.2) **17.** $14 - 27i$ (Objective 5.3.3)

18. 4 (Objective 5.4.1) **19.** (2, −1) (Objective 6.1.2) **20.** 2 (Objective 6.2.1)

21. $y = -2x - 8$ (Objective 6.3.1) **22.** $y = -2x$ (Objective 6.3.1) **23.** $y = \frac{2}{3}x + 6$ (Objective 6.3.2)

24. $y = -4x - 2$ (Objective 6.3.2) **25.** $y = -\frac{3}{2}x + 7$ (Objective 6.3.2)

26. $y = -\frac{2}{3}x + \frac{8}{3}$ (Objective 6.3.2) **27.**

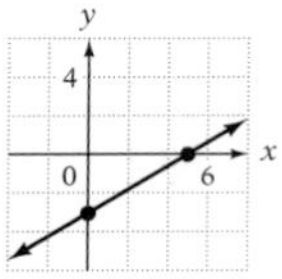

(Objective 6.2.2)

28.

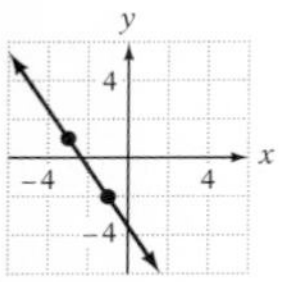

(Objective 6.2.3) **29.**

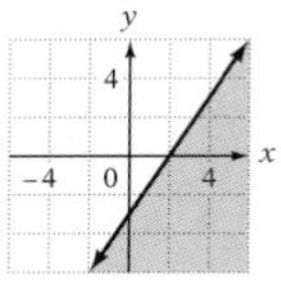

(Objective 6.5.1)

30. 7 dimes (Objective 2.2.1) **31.** 32 lb of \$8 coffee; 48 lb of \$3 coffee (Objective 2.3.1)
32. 1st plane: 200 mph; 2nd plane: 400 mph (Objective 2.3.2)
33. $y = -2500x + 15{,}000$; \$6250 (Objective 6.4.1)

ANSWERS to Chapter 7 Exercises

SECTION 7.1 *pages 320–324*

1. 0 and 4 **3.** −5 and 5 **5.** −2 and 3 **7.** 3 **9.** 0 and 2 **11.** −2 and 5
13. 2 and 5 **15.** 6 and $-\frac{3}{2}$ **17.** $\frac{1}{4}$ and 2 **19.** −4 and $\frac{1}{3}$ **21.** $-\frac{2}{3}$ and $\frac{9}{2}$ **23.** −4 and $\frac{1}{4}$
25. −2 and 9 **27.** −2 and $-\frac{3}{4}$ **29.** 2 and −5 **31.** $-\frac{3}{2}$ and −4 **33.** $2b$ and $7b$
35. $-c$ and $7c$ **37.** $-\frac{b}{2}$ and $-b$ **39.** $\frac{2a}{3}$ and $4a$ **41.** $-\frac{a}{3}$ and $3a$ **43.** $-\frac{3y}{2}$ and $-\frac{y}{2}$
45. $-\frac{a}{2}$ and $-\frac{4a}{3}$ **47.** $x^2 - 7x + 10 = 0$ **49.** $x^2 + 6x + 8 = 0$ **51.** $x^2 - 5x - 6 = 0$ **53.** $x^2 - 9 = 0$
55. $x^2 - 8x + 16 = 0$ **57.** $x^2 - 5x = 0$ **59.** $x^2 - 3x = 0$ **61.** $2x^2 - 7x + 3 = 0$
63. $4x^2 - 5x - 6 = 0$ **65.** $3x^2 + 11x + 10 = 0$ **67.** $9x^2 - 4 = 0$ **69.** $6x^2 - 5x + 1 = 0$
71. $10x^2 - 7x - 6 = 0$ **73.** $8x^2 + 6x + 1 = 0$ **75.** $50x^2 - 25x - 3 = 0$ **77.** 7 and −7
79. $2i$ and $-2i$ **81.** 2 and −2 **83.** $\frac{9}{2}$ and $-\frac{9}{2}$ **85.** $7i$ and $-7i$ **87.** $4\sqrt{3}$ and $-4\sqrt{3}$
89. $5\sqrt{3}$ and $-5\sqrt{3}$ **91.** $3i\sqrt{2}$ and $-3i\sqrt{2}$ **93.** 7 and −5 **95.** 0 and −6 **97.** −7 and 3
99. $-5 + 5i$ and $-5 - 5i$ **101.** $4 + 2i$ and $4 - 2i$ **103.** $9 + 3i$ and $9 - 3i$ **105.** 0 and 1
107. $-\frac{1}{5}$ and 1 **109.** $-\frac{3}{2}$ and 0 **111.** $\frac{7}{3}$ and 1 **113.** $-5 + \sqrt{6}$ and $-5 - \sqrt{6}$
115. $2 + 2\sqrt{6}$ and $2 - 2\sqrt{6}$ **117.** $-1 + 2i\sqrt{3}$ and $-1 - 2i\sqrt{3}$ **119.** $3 + 3i\sqrt{5}$ and $3 - 3i\sqrt{5}$
121. $\frac{-2 + 9\sqrt{2}}{3}$ and $\frac{-2 - 9\sqrt{2}}{3}$ **123.** $\frac{3 + 12\sqrt{3}}{4}$ and $\frac{3 - 12\sqrt{3}}{4}$ **125.** $-\frac{1}{2} + 2i\sqrt{10}$ and $-\frac{1}{2} - 2i\sqrt{10}$
127. $\frac{2}{3} + \frac{5}{3}i$ and $\frac{2}{3} - \frac{5}{3}i$ **129.** $x^2 - 2 = 0$ **131.** $x^2 + 1 = 0$ **133.** $x^2 - 8 = 0$ **135.** $x^2 - 12 = 0$
137. $x^2 + 12 = 0$ **139.** $-\frac{4b}{a}$ and $\frac{4b}{a}$ **141.** $-\frac{5z}{y}$ and $\frac{5z}{y}$ **143.** $b + 1$ and $b - 1$ **145.** $-\frac{1}{2}$
147.
$$ax^2 + bx = 0$$
$$x(ax + b) = 0$$
$$x = 0 \qquad ax + b = 0$$
$$ax = -b$$
$$x = -\frac{b}{a}$$
The solutions are 0 and $-\frac{b}{a}$.

SECTION 7.2 *pages 331–335*

1. 5 and −1 **3.** −9 and 1 **5.** 3 **7.** $-2 + \sqrt{11}$ and $-2 - \sqrt{11}$ **9.** $3 + \sqrt{2}$ and $3 - \sqrt{2}$
11. $1 + i$ and $1 - i$ **13.** 8 and −3 **15.** 4 and −9 **17.** $\frac{3 + \sqrt{5}}{2}$ and $\frac{3 - \sqrt{5}}{2}$

19. $\frac{1+\sqrt{5}}{2}$ and $\frac{1-\sqrt{5}}{2}$ **21.** $3+\sqrt{13}$ and $3-\sqrt{13}$ **23.** 3 and 5 **25.** $2+3i$ and $2-3i$

27. $-3+2i$ and $-3-2i$ **29.** $1+3\sqrt{2}$ and $1-3\sqrt{2}$ **31.** $\frac{1+\sqrt{17}}{2}$ and $\frac{1-\sqrt{17}}{2}$

33. $1+2i\sqrt{3}$ and $1-2i\sqrt{3}$ **35.** $-\frac{1}{2}$ and -1 **37.** $\frac{1}{2}$ and $\frac{3}{2}$ **39.** $-\frac{1}{2}$ and $\frac{4}{3}$

41. $\frac{1}{2}+i$ and $\frac{1}{2}-i$ **43.** $\frac{1}{3}+\frac{1}{3}i$ and $\frac{1}{3}-\frac{1}{3}i$ **45.** $\frac{2+\sqrt{14}}{2}$ and $\frac{2-\sqrt{14}}{2}$ **47.** $-\frac{3}{2}$ and 1

49. $1+\sqrt{5}$ and $1-\sqrt{5}$ **51.** $\frac{1}{2}$ and 5 **53.** $2+\sqrt{5}$ and $2-\sqrt{5}$ **55.** -3.236 and 1.236

57. 1.707 and 0.293 **59.** 0.309 and -0.809 **61.** 5 and -2 **63.** 4 and -9

65. $4+2\sqrt{22}$ and $4-2\sqrt{22}$ **67.** 3 and -8 **69.** $\frac{1}{2}$ and -3 **71.** $-\frac{1}{4}$ and $\frac{3}{2}$

73. $1+2\sqrt{2}$ and $1-2\sqrt{2}$ **75.** 10 and -2 **77.** $6+2\sqrt{3}$ and $6-2\sqrt{3}$ **79.** $\frac{1+2\sqrt{2}}{2}$ and $\frac{1-2\sqrt{2}}{2}$

81. $\frac{1}{2}$ and 1 **83.** $\frac{-5+\sqrt{7}}{3}$ and $\frac{-5-\sqrt{7}}{3}$ **85.** $\frac{2}{3}$ and $\frac{5}{2}$ **87.** $2+i$ and $2-i$

89. $-3+2i$ and $-3-2i$ **91.** $3+i$ and $3-i$ **93.** $-\frac{3}{2}$ and $-\frac{1}{2}$ **95.** $-\frac{1}{2}+\frac{5}{2}i$ and $-\frac{1}{2}-\frac{5}{2}i$

97. $\frac{-3+\sqrt{6}}{3}$ and $\frac{-3-\sqrt{6}}{3}$ **99.** $1+\frac{\sqrt{6}}{3}i$ and $1-\frac{\sqrt{6}}{3}i$ **101.** $-\frac{3}{2}$ and -1 **103.** $-\frac{3}{2}$ and 4

105. two complex **107.** one real **109.** two real **111.** two real **113.** 0.873 and -6.873

115. 3.236 and -1.236 **117.** 2.468 and -0.135 **119.** $\frac{\sqrt{2}}{2}$ and $-2\sqrt{2}$ **121.** $-3\sqrt{2}$ and $\frac{\sqrt{2}}{2}$

123. $\frac{\sqrt{3}}{2}+\frac{1}{2}i$ and $\frac{\sqrt{3}}{2}-\frac{1}{2}i$ **125.** $2a$ and $-a$ **127.** a and $-4a$ **129.** $2a$ and $4a$ **131.** $\frac{a}{2}$ and $-2a$

133. $1+\sqrt{y+1}$ and $1-\sqrt{y+1}$ **135.** $\{p \mid p<9, p \in \text{real numbers}\}$ **137.** $\{p \mid p>1, p \in \text{real numbers}\}$

139. $b^2+4>0$ for any real number b

SECTION 7.3 *pages 341–343*

1. 3, -3, 2, -2 **3.** 2, -2, $\sqrt{2}$, $-\sqrt{2}$ **5.** 1 and 4 **7.** 16 **9.** $2i$, $-2i$, 1, -1

11. $4i$, $-4i$, 2, -2 **13.** 16 **15.** 512 and 1 **17.** 2, 1, $-1+i\sqrt{3}$, $-1-i\sqrt{3}$, $-\frac{1}{2}+\frac{\sqrt{3}}{2}i$, $-\frac{1}{2}-\frac{\sqrt{3}}{2}i$

19. 1, -1, 2, -2, i, $-i$, $2i$, $-2i$ **21.** -64 and 8 **23.** $\frac{1}{4}$ and 1 **25.** 3 **27.** 9 **29.** 2 and -1

31. 0 and 2 **33.** $-\frac{1}{2}$ and 2 **35.** -2 **37.** 1 **39.** 1 **41.** -3 **43.** 10 and -1

45. $-\frac{1}{2}+\frac{\sqrt{7}}{2}i$ and $-\frac{1}{2}-\frac{\sqrt{7}}{2}i$ **47.** 1 and -3 **49.** 0 and -1 **51.** $\frac{1}{2}$ and $-\frac{1}{3}$ **53.** 6 and $-\frac{2}{3}$

55. $\frac{4}{3}$ and 3 **57.** $-\frac{1}{4}$ and 3 **59.** $\frac{-5+\sqrt{33}}{2}$ and $\frac{-5-\sqrt{33}}{2}$ **61.** 4 and -6

63. $-1+6\sqrt{2}$ and $-1-6\sqrt{2}$ **65.** 5 and 4 **67.** 2, -2, 1, -1 **69.** 3, -3, i, $-i$

71. $\frac{-2+\sqrt{2}}{2}$ and $\frac{-2-\sqrt{2}}{2}$ **73.** $i\sqrt{6}$, $-i\sqrt{6}$, 2, -2 **75.** $\sqrt{2}$ and i **77.** 9, 36 **79.** 4

SECTION 7.4 *pages 350–355*

1. 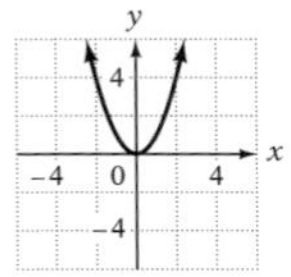

Vertex: (0, 0)
Axis of symmetry: $x = 0$

3. 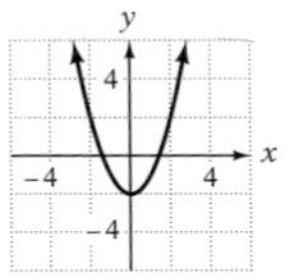

Vertex: (0, −2)
Axis of symmetry: $x = 0$

5. 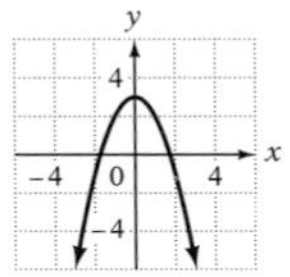

Vertex: (0, 3)
Axis of symmetry: $x = 0$

7. 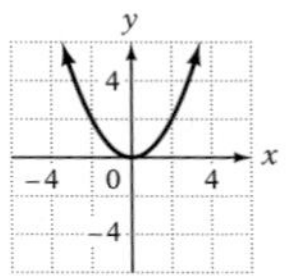

Vertex: (0, 0)
Axis of symmetry: $x = 0$

9. 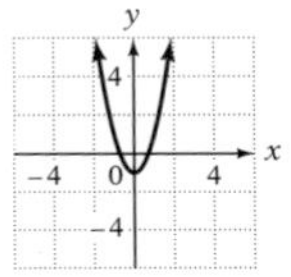

Vertex: (0, −1)
Axis of symmetry: $x = 0$

11. 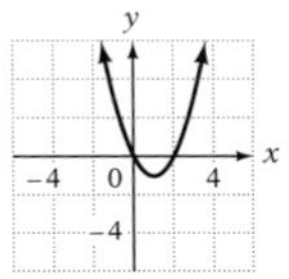

Vertex: (1, −1)
Axis of symmetry: $x = 1$

13. 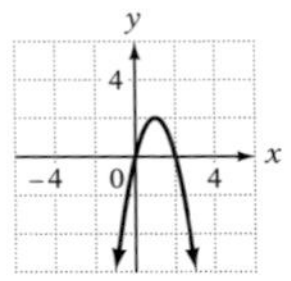

Vertex: (1, 2)
Axis of symmetry: $x = 1$

15. 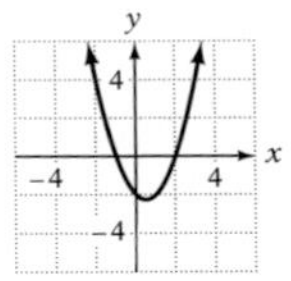

Vertex: $\left(\frac{1}{2}, -\frac{9}{4}\right)$
Axis of symmetry: $x = \frac{1}{2}$

17. 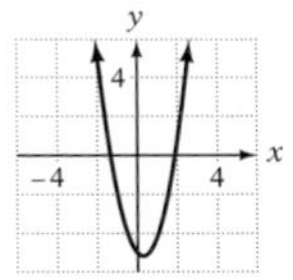

Vertex: $\left(\frac{1}{4}, -\frac{41}{8}\right)$
Axis of symmetry: $x = \frac{1}{4}$

19. 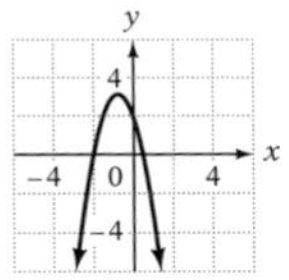

Vertex: $\left(-\frac{3}{4}, \frac{25}{8}\right)$
Axis of symmetry: $x = -\frac{3}{4}$

21. 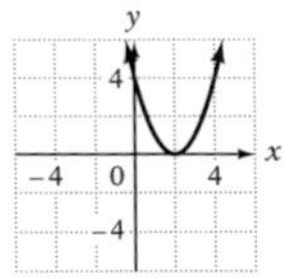

Vertex: (2, 0)
Axis of symmetry: $x = 2$

23. 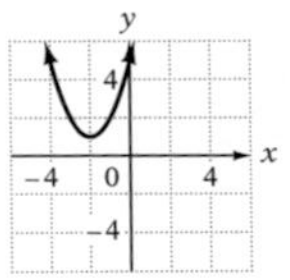

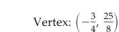

Vertex: (−2, 1)
Axis of symmetry: $x = -2$

25. (2, 0) and (−2, 0) **27.** (0, 0) and (2, 0) **29.** (2, 0) and (−1, 0) **31.** (3, 0) and $\left(-\frac{1}{2}, 0\right)$
33. $\left(-\frac{2}{3}, 0\right)$ and (7, 0) **35.** (5, 0) and $\left(\frac{4}{3}, 0\right)$ **37.** $\left(\frac{2}{3}, 0\right)$ **39.** $\left(\frac{\sqrt{2}}{3}, 0\right)$ and $\left(-\frac{\sqrt{2}}{3}, 0\right)$
41. $\left(-\frac{1}{2}, 0\right)$ and (3, 0) **43.** $(-2 + \sqrt{7}, 0)$ and $(-2 - \sqrt{7}, 0)$ **45.** no x-intercepts
47. $(-1 + \sqrt{2}, 0)$ and $(-1 - \sqrt{2}, 0)$ **49.** $(-2 + \sqrt{2}, 0)$ and $(-2 - \sqrt{2}, 0)$
51. $(-4 + \sqrt{2}, 0)$ and $(-4 - \sqrt{2}, 0)$ **53.** $(1 + \sqrt{5}, 0)$ and $(1 - \sqrt{5}, 0)$ **55.** no x-intercepts
57. two **59.** one **61.** no x-intercepts **63.** two **65.** no x-intercepts **67.** one **69.** two
71. two **73.** no x-intercepts **75.** no x-intercepts **77.** no x-intercepts **79.** two **81.** two
83. −2 **85.** −10 **87.** −4 **89.** 1 **91.** −4 **93.** −3 **95.** The graph becomes wider.
97. The graph is lower on the rectangular coordinate system. **99.** $y = (x - 1)^2 - 3$; vertex: $(1, -3)$
101. $y = (x + 2)^2 - 5$; vertex: $(-2, -5)$ **103.** $y = \left(x - \frac{1}{2}\right)^2 - \frac{13}{4}$; vertex: $\left(\frac{1}{2}, -\frac{13}{4}\right)$ **105.** $y = \frac{1}{9}x^2 + \frac{4}{9}x + \frac{22}{9}$
107. $y = \frac{1}{9}x^2 - 3$

SECTION 7.5 *pages 357–360*

1. height: 3 cm; base: 14 cm **3.** length: 13 ft; width: 5 ft **5.** 3 and 5 or −5 and −3 **7.** 3 or −8 **9.** 12.5 s **11.** 5 s **13.** larger air conditioner: 8 min; smaller air conditioner: 24 min **15.** new sorter: 14 min; old sorter: 35 min **17.** 10 mph **19.** 20 mph **21.** 6 mph **23.** $\frac{1}{4}$ or 8 **25.** $\frac{2}{4}$ **27.** 6 and 7, or −7 and −6 **29.** 2 cm by 8 cm by 16 cm **31.** 9 cm

SECTION 7.6 *pages 363–365*

1. $\{x|x < -2 \text{ or } x > 4\}$ 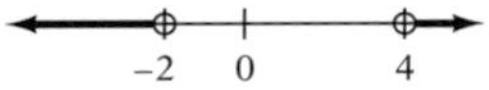

3. $\{x|x \le 1 \text{ or } x \ge 2\}$ 

5. $\{x|-3 < x < 4\}$ 

7. $\{x|x < -2 \text{ or } 1 < x < 3\}$ 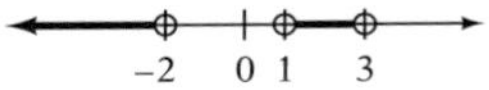

9. $\{x|-4 \le x \le 1 \text{ or } x \ge 2\}$ 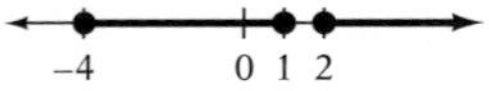

11. $\{x|x < -4 \text{ or } x > 4\}$ **13.** $\{x|x < 2 \text{ or } x > 2\}$ **15.** $\{x|-3 \le x \le 12\}$ **17.** $\left\{x|x \le \frac{1}{2} \text{ or } x \ge 2\right\}$ **19.** $\left\{x|\frac{1}{2} < x < \frac{3}{2}\right\}$ **21.** $\{x|x \le -3 \text{ or } 2 \le x \le 6\}$ **23.** $\left\{x|-\frac{3}{2} < x < \frac{1}{2} \text{ or } x > 4\right\}$ **25.** $\left\{x|x < -1 \text{ or } \frac{7}{2} < x < 5\right\}$ **27.** $\{x|x \le -3 \text{ or } -1 \le x \le 1\}$ **29.** $\{x|-2 \le x \le 1 \text{ or } x \ge 2\}$
31. $\{x|x < -2 \text{ or } x > 4\}$ 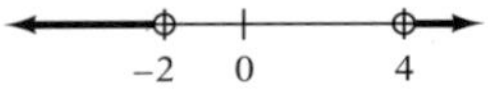

33. $\{x|-1 < x \le 3\}$ 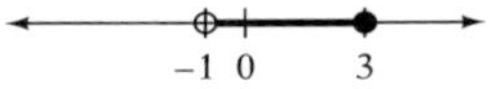

35. $\{x|x \le -2 \text{ or } 1 \le x < 3\}$ 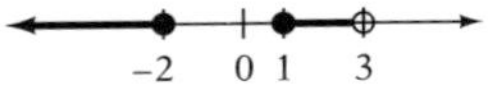

37. $\{x|x < -1 \text{ or } x > 2\}$ **39.** $\{x|-1 < x \le 0\}$ **41.** $\{x|-2 < x \le 0 \text{ or } x > 1\}$ **43.** $\left\{x|x < 0 \text{ or } x > \frac{1}{2}\right\}$
45. 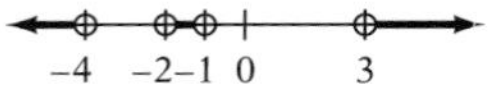

47. 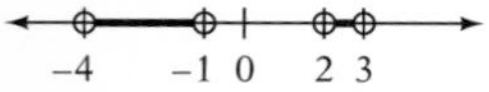

49. 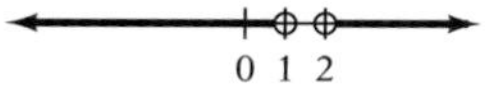

51. 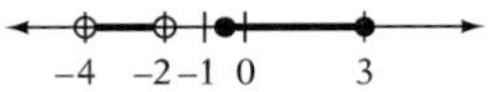

53. 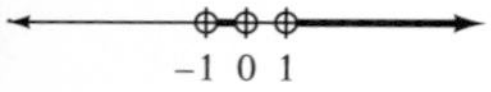

55. 

TRAJECTORIES *pages 366–367*

1. $h = 111.64$ ft
t (max. height) = 1.4 s
t (to ground) = 4.05 s

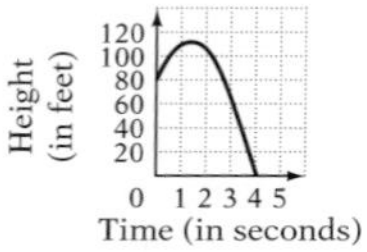

CHAPTER REVIEW *pages 367–370*

1. 0 and $\frac{3}{2}$ (Objective 7.1.1) **2.** $-2c$ and $\frac{c}{2}$ (Objective 7.1.1) **3.** $-4\sqrt{3}$ and $4\sqrt{3}$ (Objective 7.1.3) **4.** $-\frac{1}{2} - 2i$ and $-\frac{1}{2} + 2i$ (Objective 7.1.3) **5.** $-\frac{2}{3}$ and $\frac{3}{2}$ (Objective 7.1.1)

6. $2 + 2\sqrt{2}$ and $2 - 2\sqrt{2}$ (Objective 7.1.3) **7.** $\frac{3 - \sqrt{15}}{3}$ and $\frac{3 + \sqrt{15}}{3}$ (Objective 7.2.1)
8. $-2 - 2i\sqrt{2}$ and $-2 + 2i\sqrt{2}$ (Objective 7.2.1) **9.** $3x^2 + 8x - 3 = 0$ (Objective 7.1.2)
10. $2x^2 + 7x - 4 = 0$ (Objective 7.1.2) **11.** -5 and $\frac{1}{2}$ (Objective 7.1.1)
12. $-1 - 3\sqrt{2}$ and $-1 + 3\sqrt{2}$ (Objective 7.1.3) **13.** $-3 - i$ and $-3 + i$ (Objective 7.2.1)
14. 2 and 3 (Objective 7.3.3) **15.** $-\sqrt{2}, \sqrt{2}, -2, 2$ (Objective 7.3.1)
16. $2 + \sqrt{10}$ and $2 - \sqrt{10}$ (Objective 7.2.1) **17.** $\frac{25}{18}$ (Objective 7.3.2)
18. -8 and $\frac{1}{8}$ (Objective 7.3.1) **19.** 2 (Objective 7.3.2)
20. $-\sqrt{3}, \sqrt{3}, -1, 1$ (Objective 7.3.1) **21.** -4 and $\frac{2}{3}$ (Objective 7.1.1)
22. $3 - \sqrt{11}$ and $3 + \sqrt{11}$ (Objective 7.2.2) **23.** 5 and $-\frac{4}{9}$ (Objective 7.3.3)
24. $\frac{1 + \sqrt{3}}{2}$ and $\frac{1 - \sqrt{3}}{2}$ (Objective 7.2.2) **25.** 27 and -64 (Objective 7.3.1) **26.** $\frac{5}{4}$ (Objective 7.3.2)
27. 4 (Objective 7.3.2) **28.** 5 (Objective 7.3.2) **29.** 3 and -1 (Objective 7.3.3)
30. -1 (Objective 7.3.3) **31.** $\frac{-3 + \sqrt{249}}{10}$ and $\frac{-3 - \sqrt{249}}{10}$ (Objective 7.3.3)
32. $\frac{-11 - \sqrt{129}}{2}$ and $\frac{-11 + \sqrt{129}}{2}$ (Objective 7.3.3) **33.** $x = 3$ (Objective 7.4.1)
34. $\left(\frac{3}{2}, \frac{1}{4}\right)$ (Objective 7.4.1) **35.** no x-intercepts (Objective 7.4.2)
36. two x-intercepts (Objective 7.4.2) **37.** $\left(\frac{-3 - \sqrt{5}}{2}, 0\right)$ and $\left(\frac{-3 + \sqrt{5}}{2}, 0\right)$ (Objective 7.4.2)
38. $(-2, 0)$ and $\left(\frac{1}{2}, 0\right)$ (Objective 7.4.2) **39.** two x-intercepts (Objective 7.4.2)
40. no x-intercepts (Objective 7.4.2) **41.** $(0, 0)$ and $(-3, 0)$ (Objective 7.4.2)
42. $\left(\frac{-7 + \sqrt{145}}{4}, 0\right)$ and $\left(\frac{-7 - \sqrt{145}}{4}, 0\right)$ (Objective 7.4.2) **43.** $\left\{x \mid -3 < x < \frac{5}{2}\right\}$ (Objective 7.6.1)
44. $\left\{x \mid x \le -4 \text{ or } -\frac{3}{2} \le x \le 2\right\}$ (Objective 7.6.1)
45. $\left\{x \mid x < \frac{3}{2} \text{ or } x \ge 2\right\}$ 

(Objective 7.6.2)
46. $\left\{x \mid x \le -3 \text{ or } \frac{1}{2} \le x < 4\right\}$ 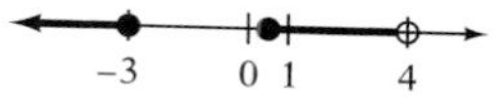

(Objective 7.6.2) **47.** 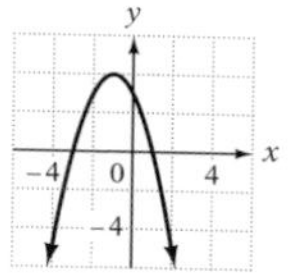

(Objective 7.4.1)
48. 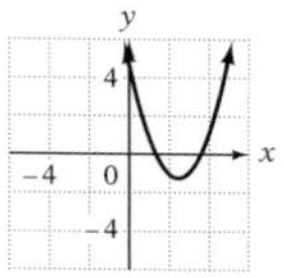

(Objective 7.4.1) **49.** length: 12 cm; width: 5 cm (Objective 7.5.1)
50. 2, 4, and 6 or -6, -4, and -2 (Objective 7.5.1) **51.** 12 min (Objective 7.5.1)
52. 1st car: 40 mph; 2nd car: 50 mph (Objective 7.5.1)

CHAPTER TEST *pages 370–371*

1. $\frac{3}{2}$ and -2 (Objective 7.1.1) **2.** $\frac{3}{4}$ and $-\frac{4}{3}$ (Objective 7.1.1) **3.** $x^2 - 9 = 0$ (Objective 7.1.2)
4. $3x^2 - 8x - 3 = 0$ (Objective 7.1.2) **5.** $-3 + 3\sqrt{2}$ and $-3 - 3\sqrt{2}$ (Objective 7.1.3)
6. $-2 + \sqrt{5}$ and $-2 - \sqrt{5}$ (Objective 7.2.1) **7.** $-1 + i\sqrt{7}$ and $-1 - i\sqrt{7}$ (Objective 7.2.2)
8. $\frac{1}{6} - \frac{\sqrt{95}}{6}i$ and $\frac{1}{6} + \frac{\sqrt{95}}{6}i$ (Objective 7.2.2) **9.** $-3 - \sqrt{10}$ and $-3 + \sqrt{10}$ (Objective 7.3.3)
10. $\frac{1}{4}$ (Objective 7.3.1) **11.** $-3, 3, -\sqrt{2}, \sqrt{2}$ (Objective 7.3.1) **12.** 4 (Objective 7.3.2)
13. no solution (Objective 7.3.2) **14.** 2 and -9 (Objective 7.3.3)
15. $(-4, 0)$ and $\left(\frac{3}{2}, 0\right)$ (Objective 7.4.2) **16.** $x = -\frac{3}{2}$ (Objective 7.4.1)
17. 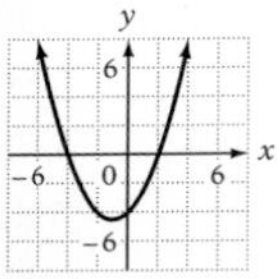

(Objective 7.4.1) **18.** $\left\{x \mid -4 < x \le \frac{3}{2}\right\}$ 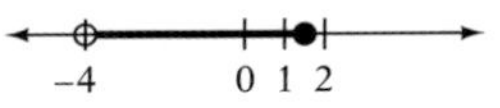

(Objective 7.6.2)
19. base: 15 ft; height: 4 ft (Objective 7.5.1) **20.** 4 mph (Objective 7.5.1)

CUMULATIVE REVIEW *pages 371–372*

1. 14 (Objective 1.2.1) **2.** -28 (Objective 2.1.2) **3.** $-\frac{3}{2}$ (Objective 6.2.1)
4. $y = x + 1$ (Objective 6.3.2) **5.** $-3xy(x^2 - 2xy + 3y^2)$ (Objective 3.4.1)
6. $(2x - 5)(3x + 4)$ (Objective 3.4.4) **7.** $(x + y)(a^n - 2)$ (Objective 3.4.2)
8. $x^2 - 3x - 4 - \frac{6}{3x - 4}$ (Objective 3.3.1) **9.** $\frac{x}{2}$ (Objective 4.2.1) **10.** $-\frac{1}{2}$ (Objective 4.4.1)
11. $b = \frac{2S - an}{n}$ (Objective 4.5.1) **12.** $-8 - 14i$ (Objective 5.3.3) **13.** $1 - a$ (Objective 5.1.1)
14. $\frac{x\sqrt[3]{4y^2}}{2y}$ (Objective 5.2.4) **15.** $\frac{2}{3}$ and -3 (Objective 7.1.1)
16. $-3 + i$ and $-3 - i$ (Objective 7.2.2) **17.** $-2, 2, -\sqrt{2}, \sqrt{2}$ (Objective 7.3.1)
18. 0 and 1 (Objective 7.3.2) **19.** $-\frac{3}{2}$ and -1 (Objective 7.3.3)
20. $\left(\frac{5}{2}, 0\right)$ and $(0, -3)$ (Objective 6.2.2) **21.** 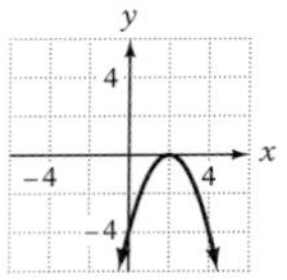

(Objective 7.4.1)
22. $\{x \mid -3 < x \le 1 \text{ or } x \ge 5\}$ 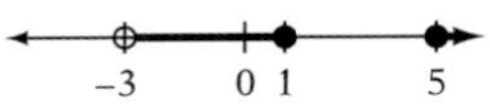

(Objective 7.6.2)
23. lower limit: $9\frac{23}{64}$ in.; upper limit: $9\frac{25}{64}$ in. (Objective 2.6.3) **24.** $(x^2 + 6x - 16)$ ft^2 (Objective 3.2.4)
25. 5.25 qt (Objective 2.4.2) **26.** 50 mph (Objective 7.5.1)

ANSWERS to Chapter 8 Exercises

SECTION 8.1 *pages 381–384*

1. 12 **3.** 3 **5.** $3a^2$ **7.** 1 **9.** 7 **11.** $t^2 - t + 1$ **13.** -11 **15.** $4h + 5$ **17.** $4h$
19. 3 **21.** $h^2 + 2h$ **23.** $h^2 + 6h$ **25.** 6 **27.** 0 **29.** $2h^2 + 4h$ **31.** -2 **33.** $2h + 12$
35. -2 **37.** domain: {1, 2, 3, 4, 5}; range: {1, 4, 7, 10, 13} **39.** domain: {0, 2, 4, 6}; range: {1, 2, 3, 4}
41. domain: {1, 3, 5, 7, 9}; range: {0} **43.** domain: {−2, −1, 0, 1, 2}; range: {0, 1, 2}
45. {−3, 1, 5, 9, 13} **47.** {1, 2, 3, 4, 5} **49.** $\left\{\frac{2}{5}, \frac{1}{2}, \frac{2}{3}, 1, 2\right\}$ **51.** $\left\{-1, -\frac{1}{2}, 1\right\}$ **53.** -8; $(-8, -3)$
55. -1; $(-1, -1)$ **57.** 1; (1, 1) **59.** ± 2; (2, 7) or (−2, 7) **61.** 34; (34, 7) **63.** none
65. 0, $\frac{1}{3}$ **67.** -3, 2 **69.** $\left\{x \mid x < \frac{1}{3}\right\}$ **71.** none **73.** 18 **75.** $f(2, 5) + g(2, 5) = 17$
77. \$160 **79.** $f(14) = 8$ **81.** $\{x \mid -1 < x < 1\}$ **83.** -3

SECTION 8.2 *pages 394–399*

1.
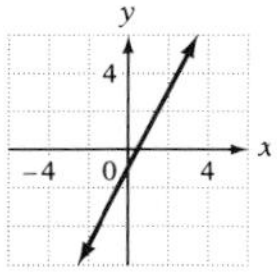

D: all reals
R: all reals

3.
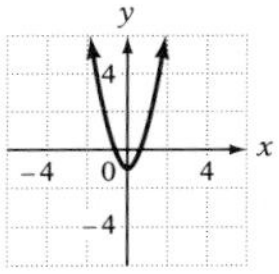

D: all reals
R: $y \geq -1$

5.
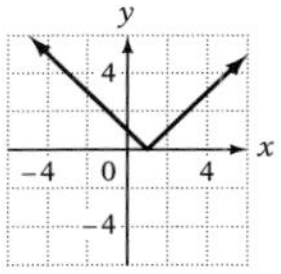

D: all reals
R: $y \geq 0$

7.
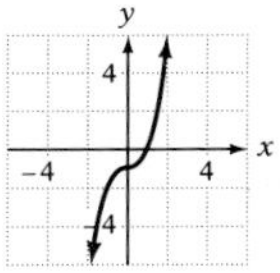

D: all reals
R: all reals

9.
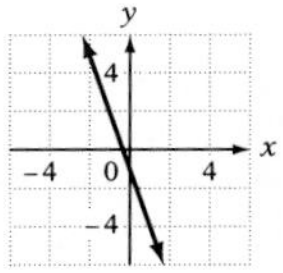

D: all reals
R: all reals

11.
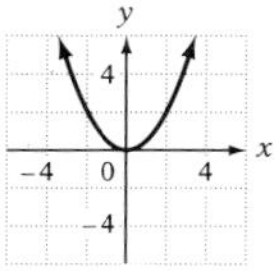

D: all reals
R: $y \geq 0$

13.
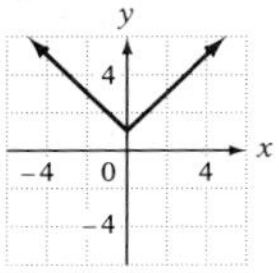

D: all reals
R: $y \geq 1$

15.
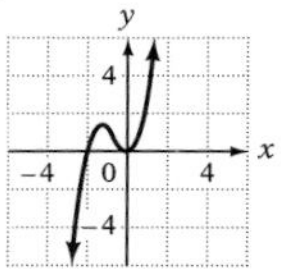

D: all reals
R: all reals

17.
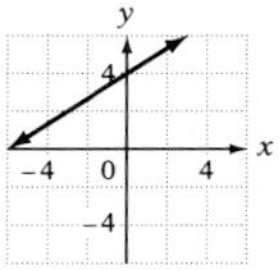

D: all reals
R: all reals

19.
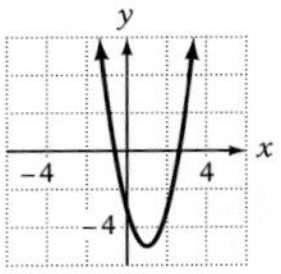

D: all reals
R: $y \geq -5$

21.
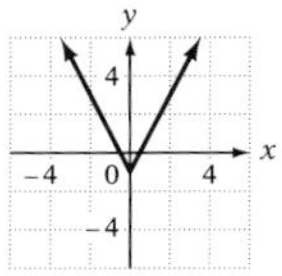

D: all reals
R: $y \geq -1$

23.
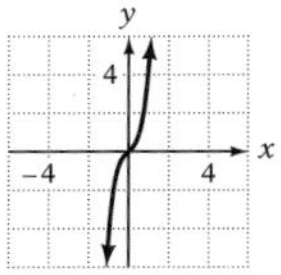

D: all reals
R: all reals

25.
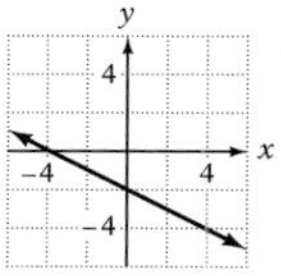

D: all reals
R: all reals

27.
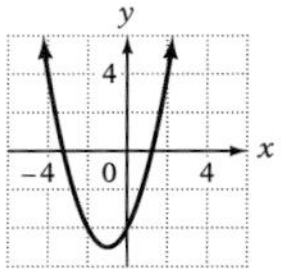

D: all reals
R: $y \geq -5$

29.
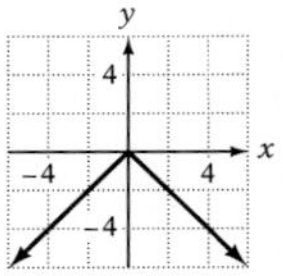

D: all reals
R: $y \leq 0$

31.
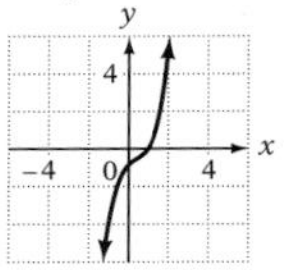

D: all reals
R: all reals

33. yes **35.** yes **37.** no **39.** yes **41.** yes **43.** no **45.** yes
47. yes **49.** no **51.** yes **53.** no **55.** c **57.** $f(-1) = -5$ **59.** $f(1) = 0$ **61.** $f(-1) = 5$
63. $f(0) = -4$ **65. a.** 0 **b.** (0, 0)

SECTION 8.3 *pages 405–408*

1. $f(g(0)) = -5$ **3.** $f(g(2)) = 11$ **5.** $f(g(x)) = 8x - 5$ **7.** $h(f(0)) = 8$ **9.** $h(f(2)) = 10$ **11.** $h(f(x)) = x + 8$ **13.** $g(h(0)) = 7$ **15.** $g(h(4)) = 7$ **17.** $g(h(x)) = x^2 - 4x + 7$ **19.** $f(h(0)) = 7$ **21.** $f(h(-1)) = 1$ **23.** $f(h(x)) = 9x^2 + 15x + 7$ **25.** $\{(0, 1), (3, 2), (8, 3), (15, 4)\}$ **27.** no inverse **29.** $f^{-1}(x) = \frac{1}{4}x + 2$ **31.** no inverse **33.** $f^{-1}(x) = x + 5$ **35.** $f^{-1}(x) = 3x - 6$ **37.** $f^{-1}(x) = -\frac{1}{3}x - 3$ **39.** $f^{-1}(x) = \frac{3}{2}x - 6$ **41.** $f^{-1}(x) = -3x + 3$ **43.** $f^{-1}(x) = \frac{1}{2}x + \frac{5}{2}$ **45.** no inverse **47.** $f^{-1}(x) = \frac{1}{4}x + \frac{1}{2}$ **49.** $f^{-1}(x) = -\frac{1}{8}x + \frac{1}{2}$ **51.** $f^{-1}(x) = \frac{1}{8}x - \frac{3}{4}$ **53.** yes **55.** no **57.** yes **59.** yes **61.** range **63.** 3 **65.** $f^{-1}(5) = 0$ **67.** $f^{-1}(9) = 2$ **69.** $f^{-1}(5) = 2$ **71.** $f^{-1}(2) = 2$ **73.** $f^{-1}(6) = -2$ **75.** $f^{-1}(2) = \frac{7}{3}$ **77.** $f(g(0)) = -3$ **79.** $f(g(4)) = 1$ **81.** $f(g(-4)) = 1$ **83.** $g(f(4)) = -2$

85.

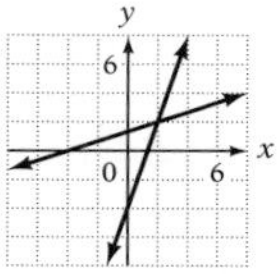

87.

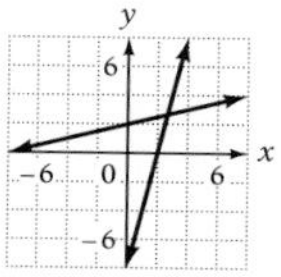

89.

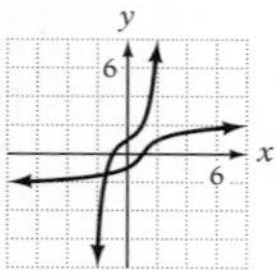

91.

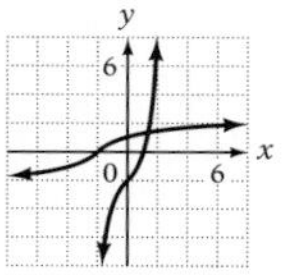

93. 19 **95.** $-2x^2 - 1$

SECTION 8.4 *pages 413–417*

1. \$37,500 **3.** 13.5 lb/in.^{2} **5.** 24.04 mi **7.** 80 ft **9.** 8 ft **11.** $66\frac{2}{3}$ lb/in.^{2} **13.** 22.5 lb/in.^{2} **15.** 112.5 lb **17.** 56.25 ohms **19.** 180 lb **21.** minimum: 2 **23.** maximum: -1 **25.** minimum: $-\frac{11}{4}$ **27.** maximum: $-\frac{2}{3}$ **29.** 114 ft **31.** 5 days **33.** \$250 **35.** 10 and 10 **37.** 12 and -12 **39. a.**

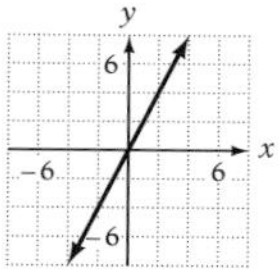

b. a linear function

41. a.

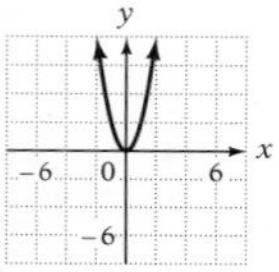

b. a quadratic function **43. a.**

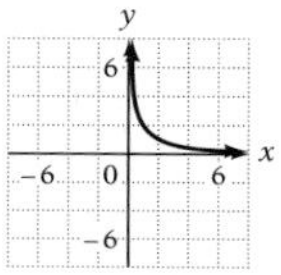

b. yes **45.** x is halved

47. directly, inversely **49.** directly

CALCULATORS AND COMPUTERS *pages 418–419*

1.

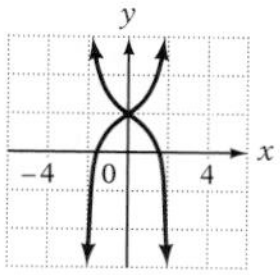

3.

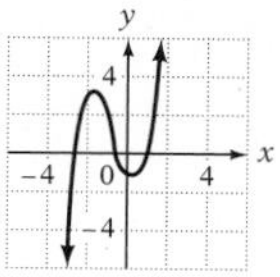

3 x-intercepts: $(-0.34, 0.04)$, $(1.03, -0.04)$, $(-2.63, -0.08)$

5.

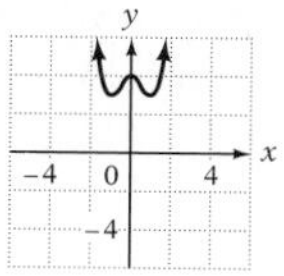

none, no

7.

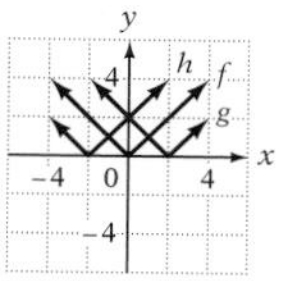

It moves the graph horizontally along the x axis.

9.

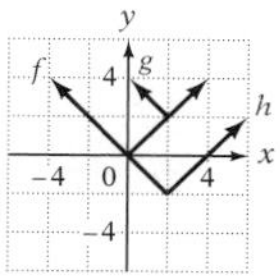

RSA PUBLIC-KEY CRYPTOGRAPHY *page 421*

1. 6 **3.** 2 **5.** 6 **7.** 2 **9.** 3

CHAPTER REVIEW *pages 422–425*

1. -16 (Objective 8.1.1) **2.** $h^2 + 9h$ (Objective 8.1.1) **3.** $3h + 12$ (Objective 8.1.1)
4. domain: $\{3, 5, 7\}$; range: $\{8\}$ (Objective 8.1.2) **5.** $\{0, 6, 12, 18\}$ (Objective 8.1.2)
6. -2 or 2 (Objective 8.1.2) **7.** -1 and 3 (Objective 8.1.2) **8.** $\{x|x < 3\}$ (Objective 8.1.2)
9. no (Objective 8.2.2) **10.** yes (Objective 8.2.3) **11.** $12x^2 + 12x - 1$ (Objective 8.3.1)
12. 5 (Objective 8.3.1) **13.** 10 (Objective 8.3.1) **14.** $6x^2 + 3x - 16$ (Objective 8.3.1)
15. $f^{-1}(x) = 2x - 16$ (Objective 8.3.2) **16.** $f^{-1}(x) = -\frac{1}{6}x + \frac{2}{3}$ (Objective 8.3.2)
17. $-\frac{7}{2}$ (Objective 8.4.2) **18.** yes (Objective 8.3.2) **19.** yes (Objective 8.2.3)
20. no (Objective 8.2.3) **21.** no (Objective 8.2.3) **22.** no (Objective 8.2.2)

23. (Objective 8.2.1)

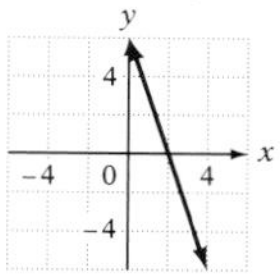

D: real numbers
R: real numbers

24. (Objective 8.2.1)

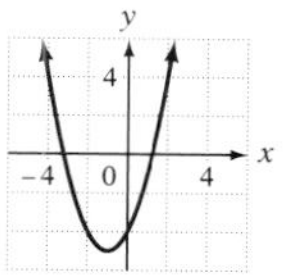

D: real numbers
R: $y \geq -5$

25. (Objective 8.2.1)

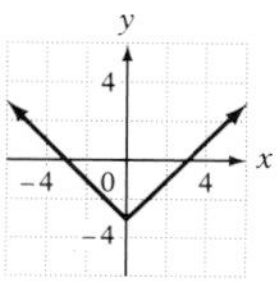

D: real numbers
R: $y \geq -3$

26. yes (Objective 8.2.2) **27.** 40 lb (Objective 8.4.1)
28. 300 lumens (Objective 8.4.1) **29.** 0.4 ohm (Objective 8.4.1)
30. 90 items (Objective 8.4.2)

CHAPTER TEST *pages 425–426*

1. $f(4) = 15$ (Objective 8.1.1) **2.** $g(-2) = -12$ (Objective 8.1.1) **3.** yes (Objective 8.2.2)
4. 0 and 8 (Objective 8.1.2) **5.** $\{-11, -7, -3, 1, 5\}$ (Objective 8.1.2)
6. $\{3, 5, 7, 9\}$ (Objective 8.1.2) **7.** $g(1 + h) = 1 - 2h$ (Objective 8.1.1)
8. $\{x|x < 7\}$ (Objective 8.1.2) **9.** 4 (Objective 8.4.2) **10.** $g(0) + h(0) = 5$ (Objective 8.1.1)
11. $f(g(0)) = -1$ (Objective 8.3.1) **12.** $f^{-1}(x) = 2x + 2$ (Objective 8.3.2)
13. $g(h(x)) = -3x + 14$ (Objective 8.3.1) **14.** no (Objective 8.3.2) **15.** no (Objective 8.2.2)
16. yes (Objective 8.2.3) **17.**

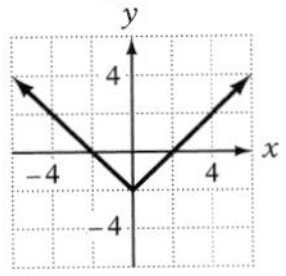

(Objective 8.2.1)

D: real numbers
R: $y \geq -2$

18.

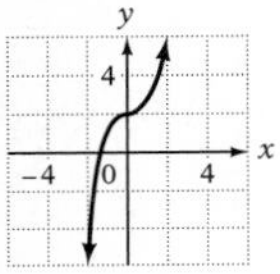

(Objective 8.2.1)

D: real numbers
R: real numbers

19.

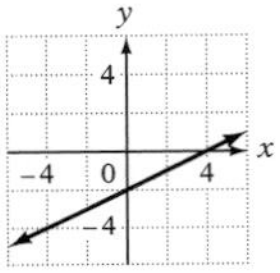

(Objective 8.2.1)

20.

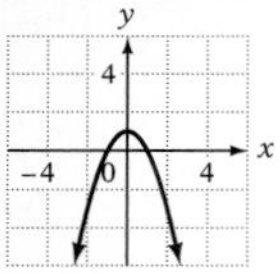

(Objective 8.2.1)

21.

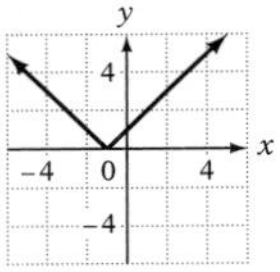

(Objective 8.2.1)

D: real numbers
R: $y \geq 0$

22.

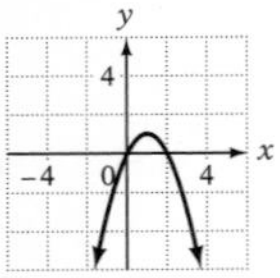

(Objective 8.2.1)

D: real numbers
R: $y \leq 1$

23. 61.2 ft (Objective 8.4.1)
24. 2 amps (Objective 8.4.1) **25.** 12 cm by 12 cm (Objective 8.4.2)

CUMULATIVE REVIEW *pages 426–428*

1. $-\frac{23}{4}$ (Objective 1.2.1) **2.** $x^2 - x + \frac{1}{x-2}$ (Objective 3.3.2) **3.** 3 (Objective 2.1.2)
4. $\{x|x < -2 \text{ or } x > 3\}$ (Objective 2.5.2) **5.** all real numbers (Objective 2.6.2)
6. $-\frac{a^{10}}{12b^4}$ (Objective 3.1.3) **7.** $2x^3 - 4x^2 - 17x + 4$ (Objective 3.2.2)
8. $(a + 2)(a - 2)(a^2 + 2)$ (Objective 3.5.4) **9.** $xy(x - 2y)(x + 3y)$ (Objective 3.5.4)

10. −3 and 8 (Objective 3.6.1) **11.** $\{x|x < -3 \text{ or } x > 5\}$ (Objective 7.6.1)

12. $\frac{5}{2x-1}$ (Objective 4.2.2) **13.** −2 (Objective 4.4.1) **14.** $-3 - 2i$ (Objective 5.3.4)

15. $y = -2x - 2$ (Objective 6.3.1) **16.** $y = -\frac{3}{2}x - \frac{7}{2}$ (Objective 6.3.2)

17. $\frac{1}{2} + \frac{\sqrt{3}}{6}i$ and $\frac{1}{2} - \frac{\sqrt{3}}{6}i$ (Objective 7.2.2) **18.** 3 (Objective 7.3.2) **19.** $f(-2) = 5$ (Objective 8.1.1)

20. {1, 2, 4, 5} (Objective 8.1.2) **21.** yes (Objective 8.2.2) **22.** 2 (Objective 5.4.1)

23. $f^{-1}(x) = -\frac{1}{3}x + 3$ (Objective 8.3.2) **24.**

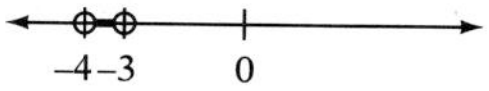

(Objective 1.3.2)

25.

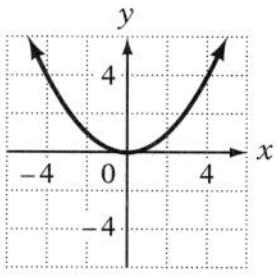

(Objective 8.2.1)

26.

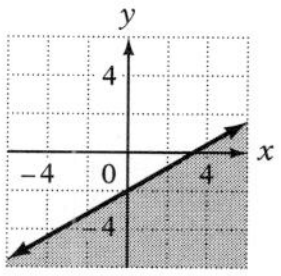

(Objective 6.5.1)

27. \$3.96 (Objective 2.3.1) **28.** 25 lb (Objective 2.4.2) **29.** 4.5 oz (Objective 4.4.2)
30. 4 min (Objective 4.4.3) **31.** 24 in. (Objective 8.4.1) **32.** 80 vibrations/min (Objective 8.4.1)

ANSWERS to Chapter 9 Exercises

SECTION 9.1 *pages 433–435*

1.

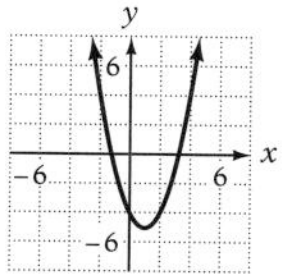

Vertex: (1, −5)
Axis of symmetry:
$x = 1$

3.

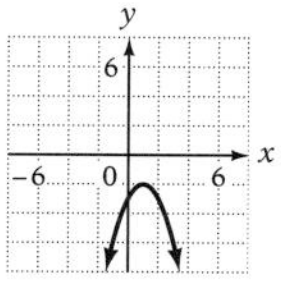

Vertex: (1, −2)
Axis of symmetry:
$x = 1$

5.

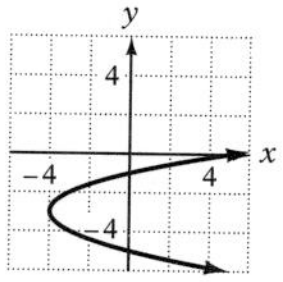

Vertex: (−4, −3)
Axis of symmetry:
$y = -3$

7.

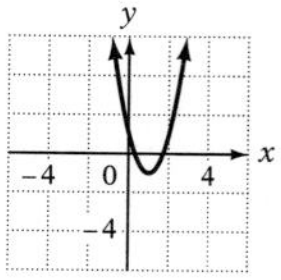

Vertex: (1, −1)
Axis of symmetry:
$x = 1$

9.

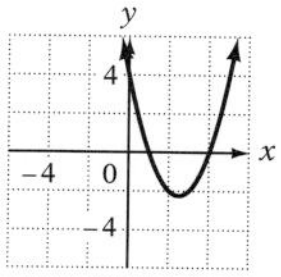

Vertex: $\left(\frac{5}{2}, -\frac{9}{4}\right)$
Axis of symmetry:
$x = \frac{5}{2}$

11.

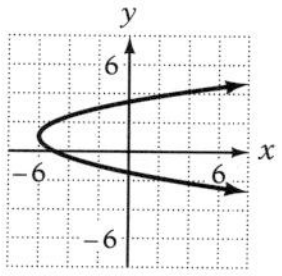

Vertex: (−6, 1)
Axis of symmetry:
$y = 1$

13.

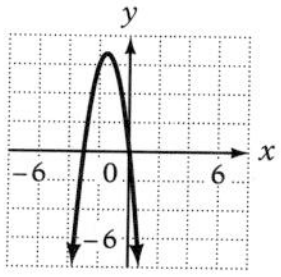

Vertex: $\left(-\frac{3}{2}, \frac{27}{4}\right)$
Axis of symmetry:
$x = -\frac{3}{2}$

15.

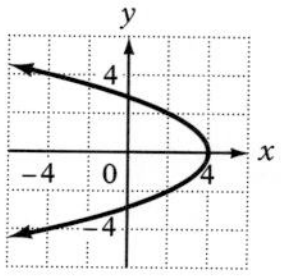

Vertex: (4, 0)
Axis of symmetry:
$y = 0$

17. 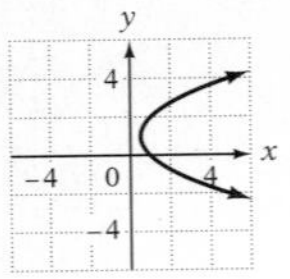

Vertex: $\left(\frac{1}{2}, 1\right)$
Axis of symmetry:
$y = 1$

19. 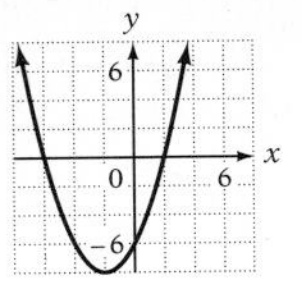

Vertex: $(-2, -8)$
Axis of symmetry:
$x = -2$

21. D: the real numbers; R: $y \geq -6$
23. D: the real numbers; R: $y \leq -2$
25. D: $x \geq -14$; R: the real numbers
27. D: $x \leq 7$; R: the real numbers
29. $\left(1, -\frac{7}{8}\right)$
31. $(-2, -1)$

SECTION 9.2 *pages 441–444*

1. $\sqrt{10}$ 3. $\sqrt{34}$ 5. $\sqrt{53}$ 7. 5 9. $\left(\frac{3}{2}, 2\right)$ 11. $\left(\frac{3}{2}, 1\right)$ 13. $(0, -5)$ 15. $\left(-\frac{7}{2}, \frac{1}{2}\right)$

17. 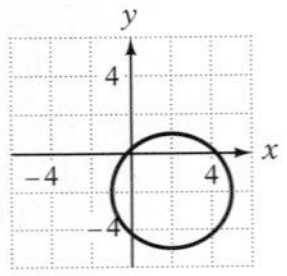

19. 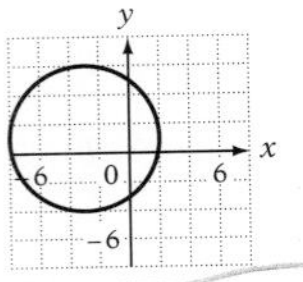

21. 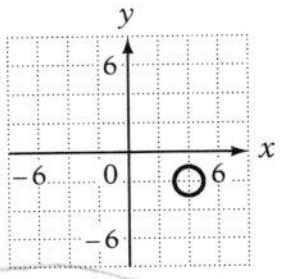

23. 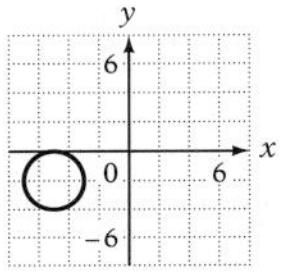

25. $(x - 2)^2 + (y + 1)^2 = 4$

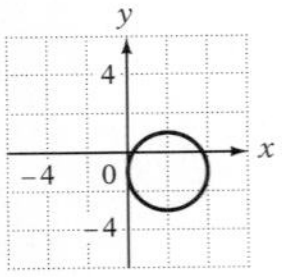

27. $(x + 1)^2 + (y - 1)^2 = 5$

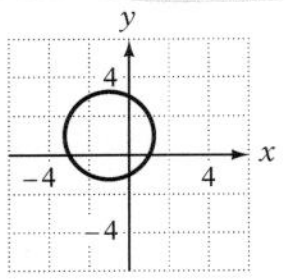

29. $(x + 3)^2 + (y - 6)^2 = 8$

31. $(x + 2)^2 + (y - 1)^2 = 5$

33. $(x - 1)^2 + (y + 2)^2 = 25$

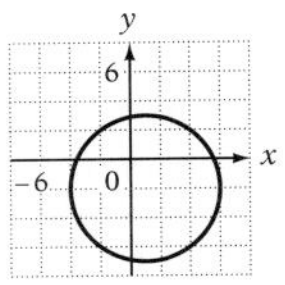

35. $(x + 3)^2 + (y + 4)^2 = 16$

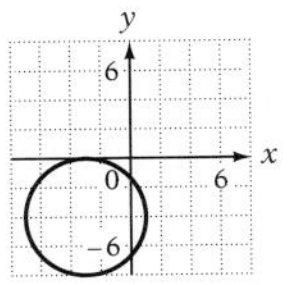

37. $\left(x - \frac{1}{2}\right)^2 + (y + 2)^2 = 1$

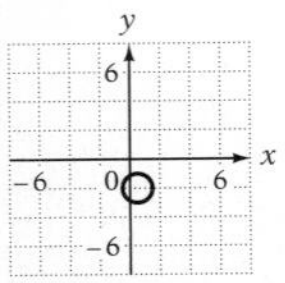

39. $(x - 3)^2 + (y + 2)^2 = 9$

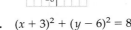

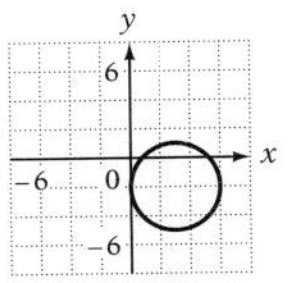

41. yes 43. no 45. $(x - 3)^2 + y^2 = 9$

47. $(x - 2)^2 + (y - 4)^2 = 10$ 49. $(x + 1)^2 + (y - 1)^2 = 1$

SECTION 9.3 *pages 449–451*

1. 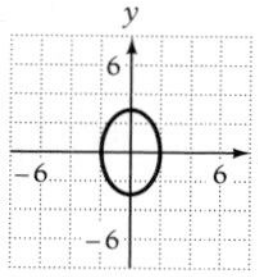

3. 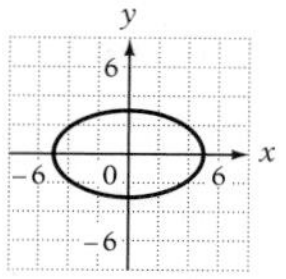

5. 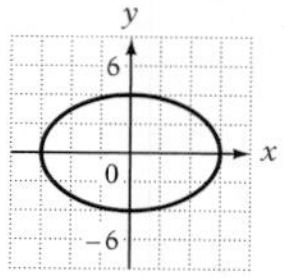

7. 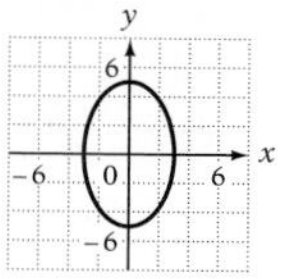

9. 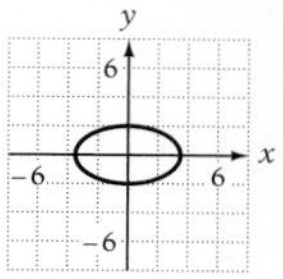

11. 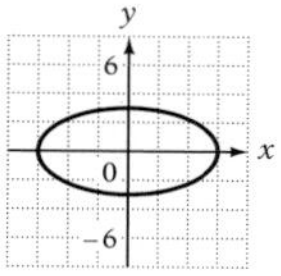

13. 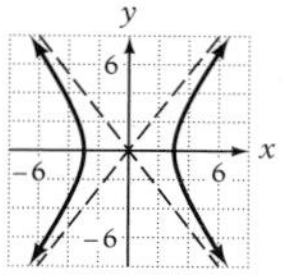

15. 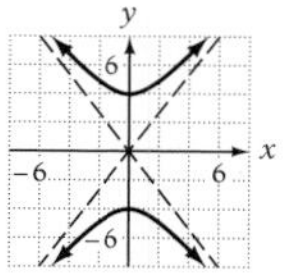

17. 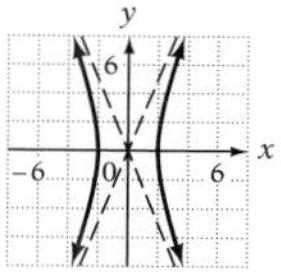

19. 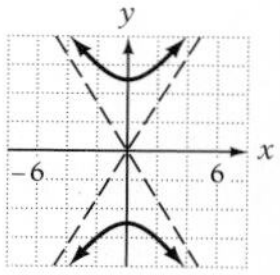

21. 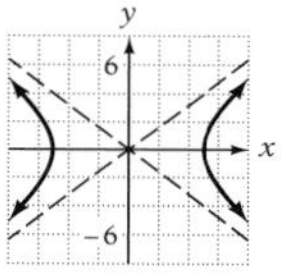

23. 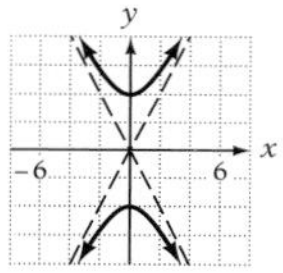

25. 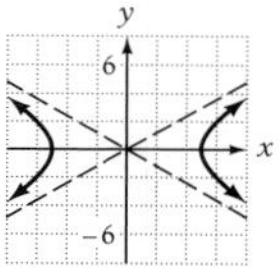

27. 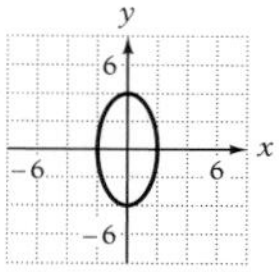

29. 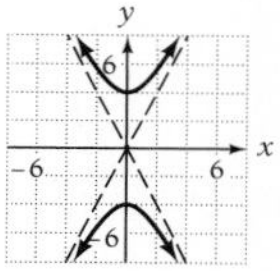

31. 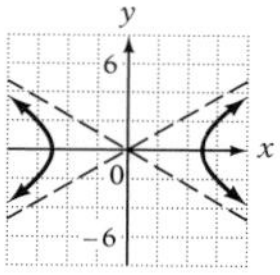

33. 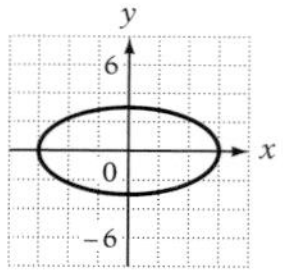

35. $F_1(3, 0)$, $F_2(-3, 0)$

37. $F_1(5, 0)$, $F_2(-5, 0)$

SECTION 9.4 *pages 453–454*

1. 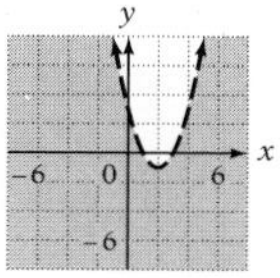

3. 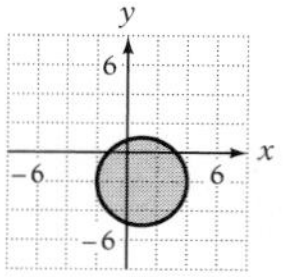

5. 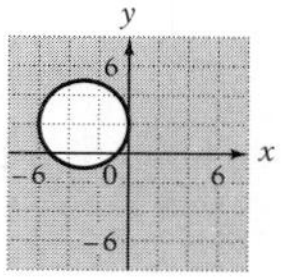

7. 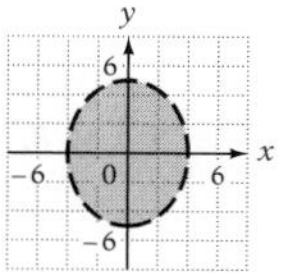

9. 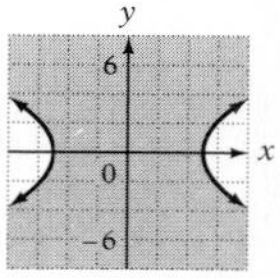

11. 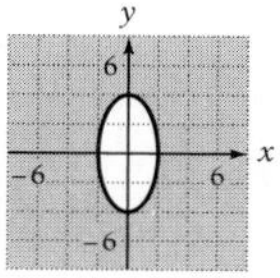

13. 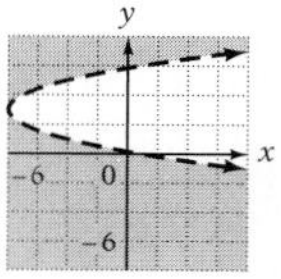

15. 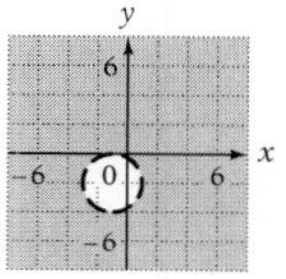

17. 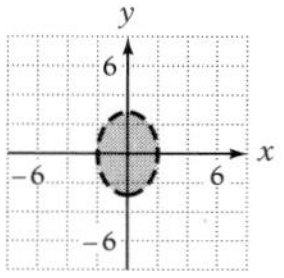

19. all points in the plane 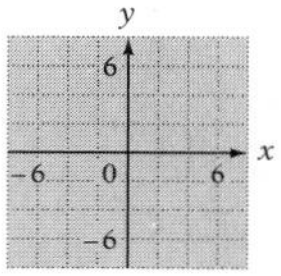

THE DIFFERENCE QUOTIENT *page 456*

1. −3; They are the same. 3. $2x + h$ 5. $4x - 2 + 2h$

CHAPTER REVIEW *pages 457–459*

1. $y = 1$ (Objective 9.1.1) **2.** $\left(\frac{-3-\sqrt{17}}{4}, 0\right)$ and $\left(\frac{-3+\sqrt{17}}{4}, 0\right)$ (Objective 9.1.1)

3. $\left(\frac{11}{4}, \frac{3}{2}\right)$ (Objective 9.1.1) **4.** $(x+3)^2 + (y-7)^2 = 4$ (Objective 9.2.3) **5.** $x = \frac{3}{4}$ (Objective 9.1.1)

6. $(x-2)^2 + (y+1)^2 = 34$ (Objective 9.2.3) **7.** $\sqrt{58}$ (Objective 9.2.1) **8.** $\left(4, -\frac{9}{2}\right)$ (Objective 9.2.2)

9. $(-2, 3)$ (Objective 9.2.2) **10.** $(x-3)^2 + (y-2)^2 = 26$ (Objective 9.2.3)
11. $(4, 1)$ (Objective 9.2.2) **12.** $x^2 + (y-3)^2 = 17$ (Objective 9.2.3)
13. $(x+2)^2 + (y-4)^2 = 9$ (Objective 9.2.3) **14.** $(x+2)^2 + (y-1)^2 = 32$ (Objective 9.2.3)
15. $(4, -2)$ (Objective 9.2.2) **16.**

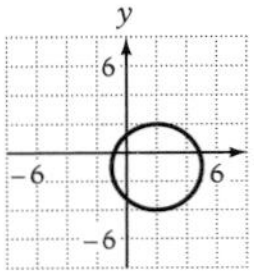

(Objective 9.2.3)

17. $(x-2)^2 + (y+1)^2 = 4$ (Objective 9.2.4)

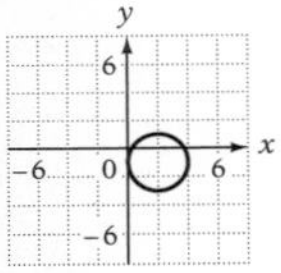

18.

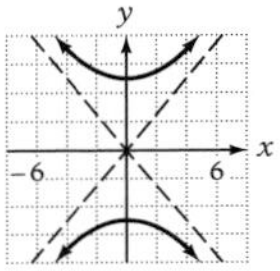

(Objective 9.3.2)

19.

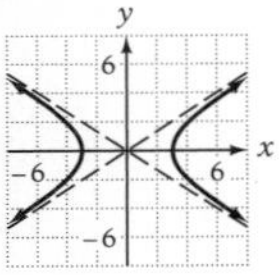

(Objective 9.3.2)

20.

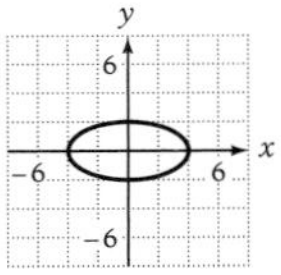

(Objective 9.3.1)

21.

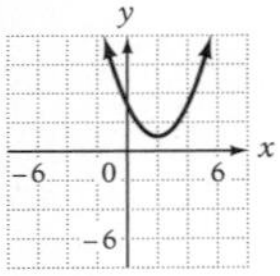

(Objective 9.1.1)

22.

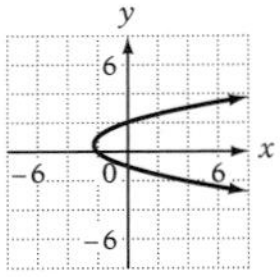

(Objective 9.1.1)

23.

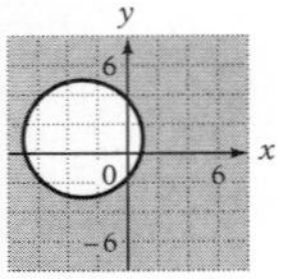

(Objective 9.4.1)

24.

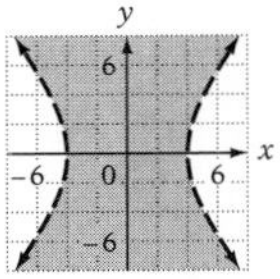

(Objective 9.4.1)

25.

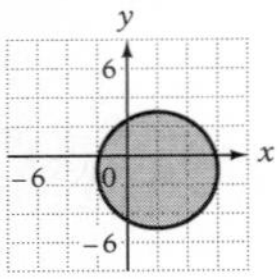

(Objective 9.4.1)

26.

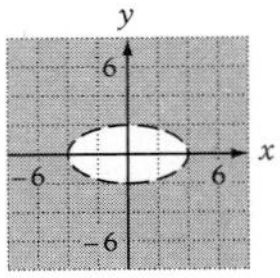

(Objective 9.4.1)

27. 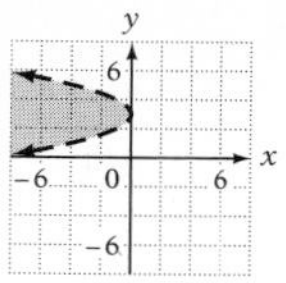 (Objective 9.4.1)

28. 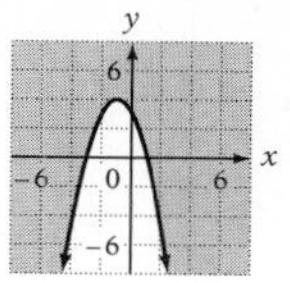 (Objective 9.4.1)

29. 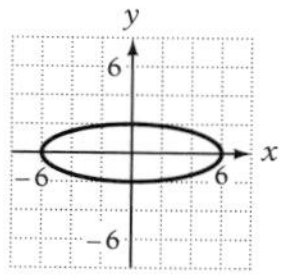 (Objective 9.3.1)

30. 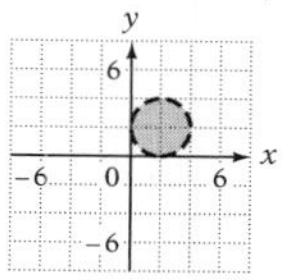 (Objective 9.4.1)

CHAPTER TEST *pages 459–461*

1. $\left(\frac{3}{2}, 0\right)$ and $\left(-\frac{1}{2}, 0\right)$ (Objective 9.1.1) **2.** $(x-4)^2 + (y-1)^2 = 18$ (Objective 9.2.3)

3. $(-4, -2)$ (Objective 9.2.2) **4.** $\left(\frac{7}{8}, \frac{3}{4}\right)$ (Objective 9.1.1) **5.** $x = \frac{1}{3}$ (Objective 9.1.1)

6. $(-3, 2)$ (Objective 9.1.1) **7.** $y = -\frac{3}{4}$ (Objective 9.1.1)

8. $(x-3)^2 + (y-1)^2 = 9$ (Objective 9.2.4) **9.** $\left(\frac{1}{2}, -\frac{5}{2}\right)$ (Objective 9.2.2) **10.** $5\sqrt{2}$ (Objective 9.2.1)

11. $(x-2)^2 + (y+2)^2 = 10$ (Objective 9.2.3) **12.** $(x-2)^2 + (y+1)^2 = 9$ (Objective 9.2.4)

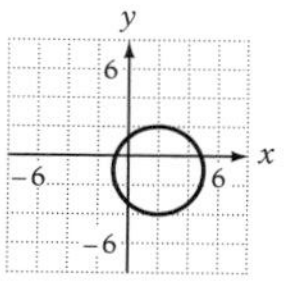

13. 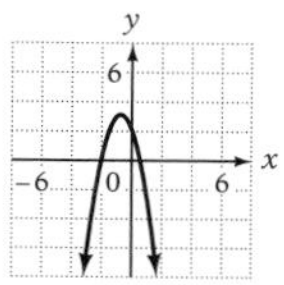 (Objective 9.1.1)

14. 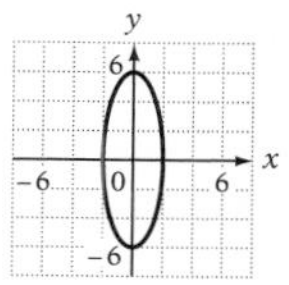 (Objective 9.3.1)

15. 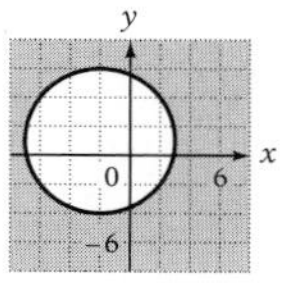 (Objective 9.4.1)

16. 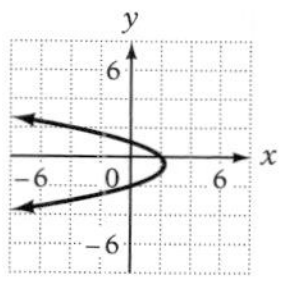 (Objective 9.1.1)

17. 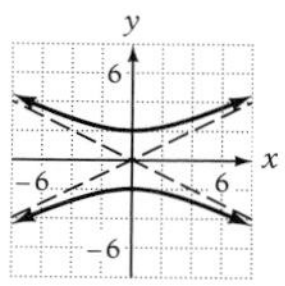 (Objective 9.3.2)

18. 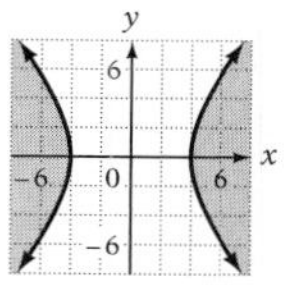 (Objective 9.4.1)

19. 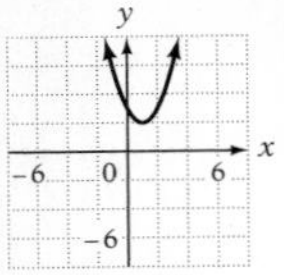

(Objective 9.1.1)

20. 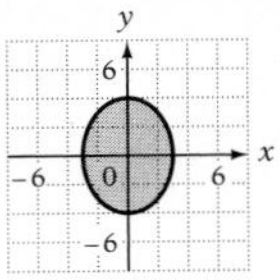

(Objective 9.4.1)

CUMULATIVE REVIEW *pages 461–462*

1. $\frac{38}{53}$ (Objective 2.1.2) **2.** $\frac{5}{2}$ (Objective 4.4.1) **3.** $\left\{x \middle| -\frac{4}{3} < x < 0\right\}$ (Objective 2.6.2)

4. $y = -\frac{3}{2}x$ (Objective 6.3.1) **5.** $y = x - 6$ (Objective 6.3.2)

6. $x^{4n} + 2x^{3n} - 3x^{2n+1}$ (Objective 3.2.1) **7.** $(x - y - 1)(x^2 - 2x + 1 + xy - y + y^2)$ (Objective 3.5.2)

8. $\{x \mid -4 < x \leq 3\}$ (Objective 7.6.2) **9.** $\frac{x}{x+y}$ (Objective 4.1.1) **10.** $\frac{x-5}{3x-2}$ (Objective 4.2.2)

11. $\frac{4a}{3b^8}$ (Objective 3.1.3) **12.** $2x^{\frac{3}{4}}$ (Objective 5.1.2) **13.** $3\sqrt{2} - 5i$ (Objective 5.3.1)

14. $\frac{-1+\sqrt{7}}{2}$ and $\frac{-1-\sqrt{7}}{2}$ (Objective 7.2.2) **15.** 27 and −1 (Objective 7.3.1)

16. 6 (Objective 7.3.2) **17.** −20 (Objective 8.1.1) **18.** $f^{-1}(x) = \frac{1}{4}x - 2$ (Objective 8.3.2)

19. 0 (Objective 8.4.2) **20.** 400 (Objective 8.4.2) **21.** 5 (Objective 9.2.1)

22. $(x + 1)^2 + (y - 2)^2 = 17$ (Objective 9.2.3) **23.** 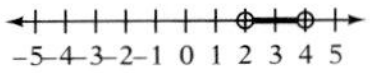

(Objective 1.3.2)

24. 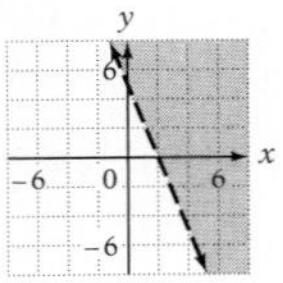

(Objective 6.5.1)

25. 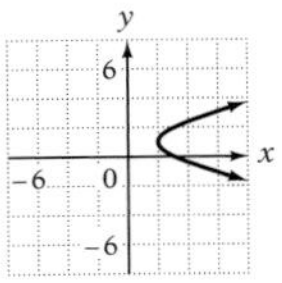

(Objective 9.1.1)

26. 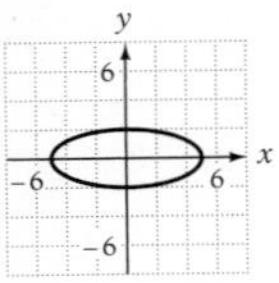

(Objective 9.3.1)

27. 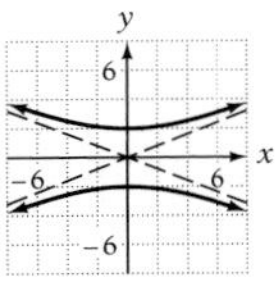

(Objective 9.3.2)

28. 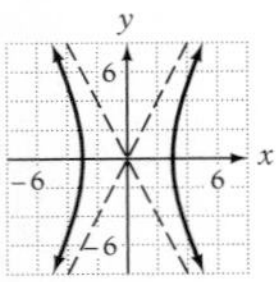

(Objective 9.3.2)

29. 82 tickets (Objective 2.3.1) **30.** 60 mph (Objective 4.4.4)

31. 4.5 mph (Objective 7.5.1) **32.** 18 revolutions (Objective 8.4.1)

ANSWERS to Chapter 10 Exercises

SECTION 10.1 *pages 469-472*

1.
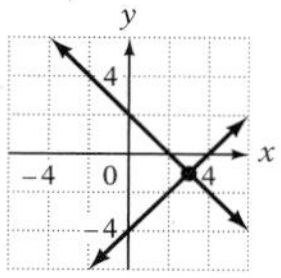

The solution is (3, −1).

3.
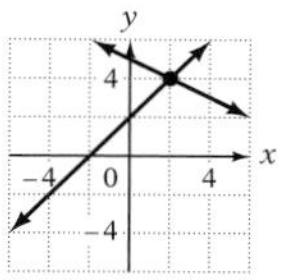

The solution is (2, 4).

5.
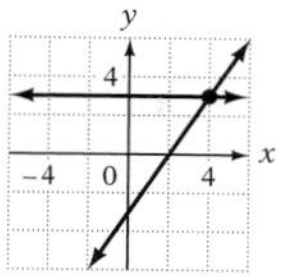

The solution is (4, 3).

7.
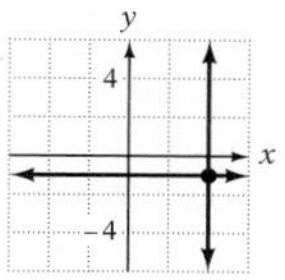

The solution is (4, −1).

9.
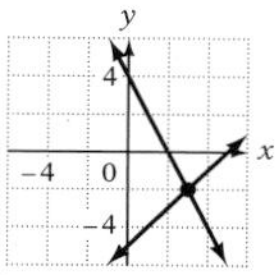

The solution is (3, −2).

11.
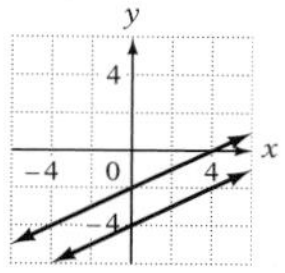

inconsistent

13.
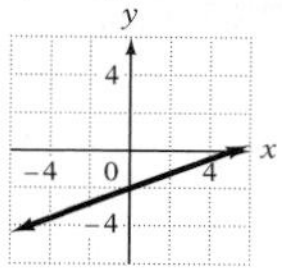

dependent

15.
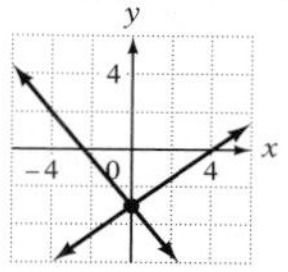

The solution is (0, −3).

17. (2, 1) **19.** (1, 1) **21.** (16, 5) **23.** (2, 1) **25.** (−1, −7) **27.** (−2, −3) **29.** (3, −4) **31.** (3, 3) **33.** (0, −1) **35.** $\left(\frac{1}{2}, 3\right)$ **37.** (1, 2) **39.** (−1, 2) **41.** (2, −4) **43.** (−2, 5) **45.** (0, 0) **47.** (−4, −6) **49.** (1, 5) **51.** $\left(\frac{2}{3}, 3\right)$ **53.** (5, 2) **55.** (1, 4) **57.** (−1, 6) **59.** 2 **61.** $\frac{1}{2}$ **63.** 26 and 18 **65.** 56 and 72 **67.** 8 and 11 **69.** (2, 1) **71.** $\left(\frac{13}{11}, \frac{13}{5}\right)$ **73.** (−3, 1)

SECTION 10.2 *pages 480–483*

1. (6, 1) **3.** (1, 1) **5.** (2, 1) **7.** (−2, 1) **9.** dependent **11.** $\left(-\frac{1}{2}, 2\right)$ **13.** inconsistent **15.** (−1, −2) **17.** (−5, 4) **19.** (2, 5) **21.** $\left(\frac{1}{2}, \frac{3}{4}\right)$ **23.** (0, 0) **25.** (−1, 3) **27.** $\left(\frac{2}{3}, -\frac{2}{3}\right)$ **29.** (1, −1) **31.** dependent **33.** (5, 3) **35.** $\left(\frac{1}{3}, -1\right)$ **37.** $\left(\frac{5}{3}, \frac{1}{3}\right)$ **39.** inconsistent **41.** (−1, 2, 1) **43.** (6, 2, 4) **45.** (4, 1, 5) **47.** (3, 1, 0) **49.** (−1, −2, 2) **51.** inconsistent **53.** (2, 1, −3) **55.** (2, −1, 3) **57.** (6, −2, 2) **59.** (0, −2, 0) **61.** (2, 3, 1) **63.** (1, 1, 3) **65.** (2, 1) **67.** (1, −3) **69.** (2, −1) **71.** (2, 0, −1) **73.** (2, −1, 0) **75.** $A = 4$, $B = 5$ **77.** $A = 2$, $B = 3$, $C = -3$ **79.** $a = 2$, $b = -1$, $c = 1$ **81.** (nickels, dimes, quarters) → $(3z - 5, -4z + 35, z)$ where $z = 2, 3, 4, 5, 6, 7, 8$

SECTION 10.3 *pages 496–498*

1. 11 **3.** 18 **5.** 0 **7.** 15 **9.** −30 **11.** 0 **13.** (3, −4) **15.** (4, −1) **17.** $\left(\frac{11}{14}, \frac{17}{21}\right)$ **19.** $\left(\frac{1}{2}, 1\right)$ **21.** inconsistent **23.** (−1, 0) **25.** (1, −1, 2) **27.** (2, −2, 3) **29.** inconsistent

31. $\left(\frac{68}{25}, \frac{56}{25}, -\frac{8}{25}\right)$ 33. (1, 3) 35. $(-1, -3)$ 37. (3, 2) 39. inconsistent 41. $(-2, 2)$

43. $(0, 0, -3)$ 45. $(1, -1, -1)$ 47. inconsistent 49. $\left(\frac{1}{3}, \frac{1}{2}, 0\right)$ 51. $\left(\frac{1}{5}, \frac{2}{5}, -\frac{3}{5}\right)$

53. $\left(\frac{1}{4}, 0, -\frac{2}{3}\right)$ 55. 3 57. -14 59. -3 61. 0

SECTION 10.4 *pages 503–506*

1. plane: 150 mph; wind: 10 mph 3. cabin cruiser: 14 mph; current: 2 mph
5. plane: 165 mph; wind: 15 mph 7. boat: 19 km/h; current: 3 km/h
9. plane: 105 mph; wind: 15 mph 11. boat: 16.5 mph; current: 1.5 mph
13. cabin cruiser: 18 mph; current: 2.5 mph 15. pine: \$.18/ft; redwood: \$.30/ft 17. \$.08
19. 18 quarters 21. 60 color TV's 23. 1st powder: 200 mg; 2nd powder: 450 mg 25. 9° and 81°
27. gold coin: 25 years; silver coin: 50 years 29. 25 nickels, 10 dimes, and 5 quarters 31. 84

SECTION 10.5 *pages 511–514*

1. $(-2, 5)$ and $(5, 19)$ 3. $(-1, -2)$ and $(2, 1)$ 5. $(2, -2)$ 7. $(2, 2)$ and $\left(-\frac{2}{9}, -\frac{22}{9}\right)$

9. (3, 2) and (2, 3) 11. $\left(\frac{\sqrt{3}}{2}, 3\right)$ and $\left(-\frac{\sqrt{3}}{2}, 3\right)$ 13. no solution

15. $(1, 2)$, $(1, -2)$, $(-1, 2)$, and $(-1, -2)$ 17. $(3, 2)$, $(3, -2)$, $(-3, 2)$, and $(-3, -2)$
19. $(2, 7)$ and $(-2, 11)$ 21. $(3, \sqrt{2})$, $(3, -\sqrt{2})$, $(-3, \sqrt{2})$, and $(-3, -\sqrt{2})$ 23. no solution
25. $(\sqrt{2}, 3)$, $(\sqrt{2}, -3)$, $(-\sqrt{2}, 3)$, and $(-\sqrt{2}, -3)$ 27. $(2, -1)$ and $(8, 11)$ 29. $(1, 0)$

31.
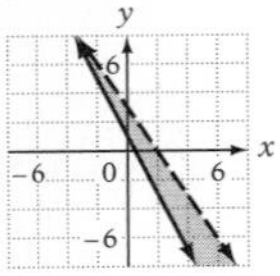

33.
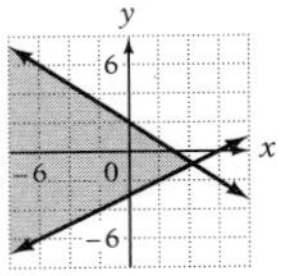

35.
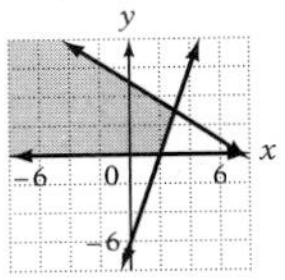

37.
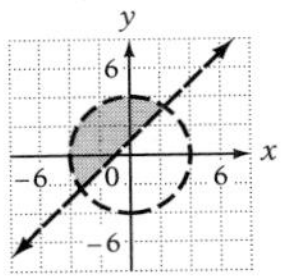

39.
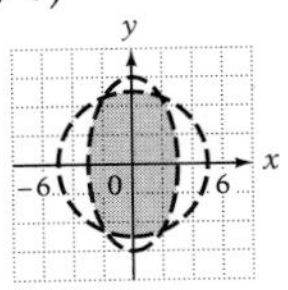

41.
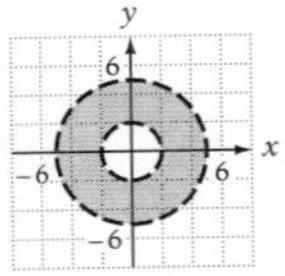

43.
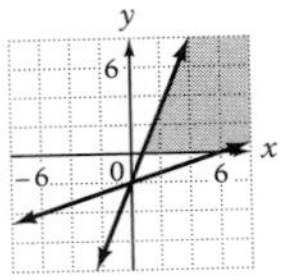

45.
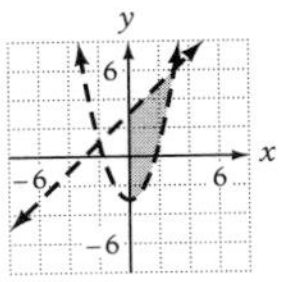

47.
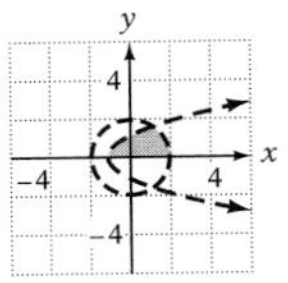

49.
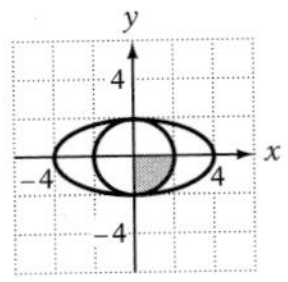

51. no 53.
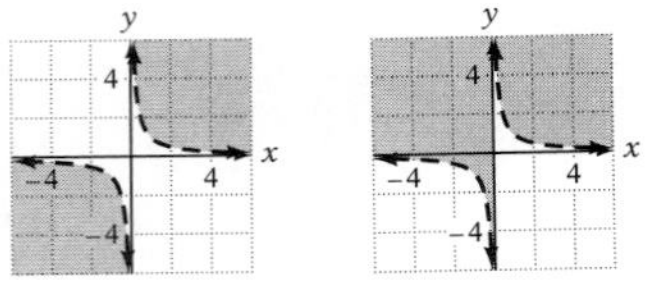

Changing the inequality sign depends on whether x is positive or negative.

CALCULATORS AND COMPUTERS *page 516*

1. (2.7977839, 0.6122449) 3. $(0.45761773, -0.6424792)$
5. $(-2.236069, -2.000064)$, $(2.236069, 2.000064)$ 7.
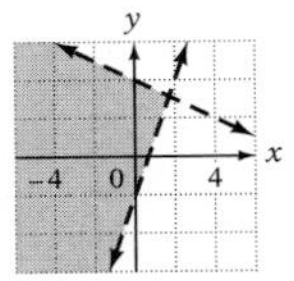

9.
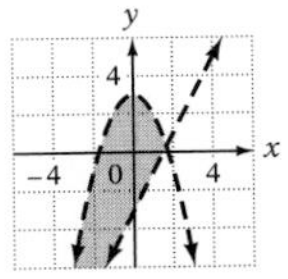

LINEAR PROGRAMMING *page 518*

1. Solve the system of equations $\begin{aligned} x &= 2y. \\ x + y &= 12 \end{aligned}$
3. Profit = \$2400 for the ordered pair (8, 4).

CHAPTER REVIEW *pages 520–523*

1. inconsistent (Objective 10.1.2) **2.** dependent (Objective 10.1.2) **3.** (−4, 7) (Objective 10.2.1)
4. dependent (Objective 10.2.1) **5.** $\left(\frac{1}{4}, -3\right)$ (Objective 10.2.1) **6.** (−1, 2, 3) (Objective 10.2.2)
7. (5, −2, 3) (Objective 10.2.2) **8.** (3, −1, −2) (Objective 10.2.2) **9.** 28 (Objective 10.3.1)
10. 0 (Objective 10.3.1) **11.** (3, −1) (Objective 10.3.2) **12.** $\left(\frac{110}{23}, \frac{25}{23}\right)$ (Objective 10.3.2)
13. (−1, −3, 4) (Objective 10.3.2) **14.** (2, 3, −5) (Objective 10.3.2)
15. (−2, −12) (Objective 10.5.1) **16.** (4, 2), (−4, 2), (4, −2), (−4, −2) (Objective 10.5.1)
17. (5, 1), (−5, 1), (5, −1), (−5, −1) (Objective 10.5.1)
18. $(\sqrt{5}, 3), (-\sqrt{5}, 3), (\sqrt{5}, -3), (-\sqrt{5}, -3)$ (Objective 10.5.1) **19.** 26 (Objective 10.3.1)
20. (3, −2) (Objective 10.3.3) **21.** (−1, 2, 1) (Objective 10.3.2)
22. (1, 1), (1, −1), (−1, 1), (−1, −1) (Objective 10.5.1) **23.** $\left(\frac{7}{3}, -\frac{10}{3}\right)$ (Objective 10.2.1)
24. (0, −2) (Objective 10.2.1) **25.** $\left(\frac{81}{16}, -\frac{9}{8}\right)$, (4, 1) (Objective 10.5.1)
26. (−1, −5) (Objective 10.1.2) **27.** (1, −1, 4) (Objective 10.2.2) **28.** $\left(\frac{8}{5}, \frac{7}{5}\right)$ (Objective 10.3.2)
29. $\left(\frac{1}{2}, -1, \frac{1}{3}\right)$ (Objective 10.3.3) **30.** 12 (Objective 10.3.1) **31.** (2, −3) (Objective 10.3.2)
32. (2, −3, 1) (Objective 10.3.3)

33.
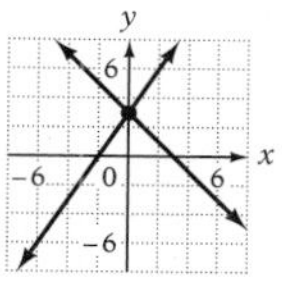

(Objective 10.1.1)
The solution is (0, 3).

34.
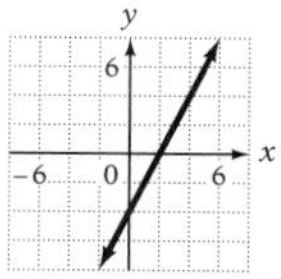

(Objective 10.1.1)
dependent

35.
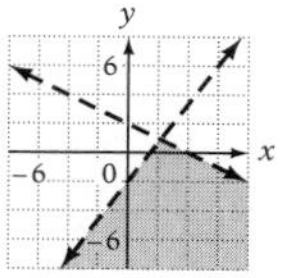

(Objective 10.5.2)

36.
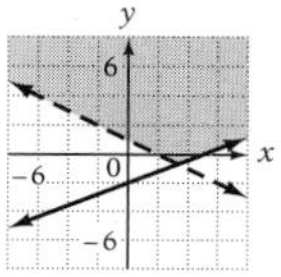

(Objective 10.5.2)

37.
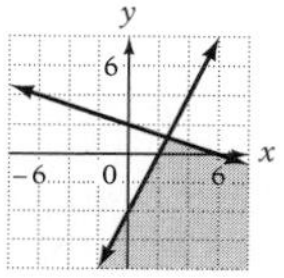

(Objective 10.5.2)

38.
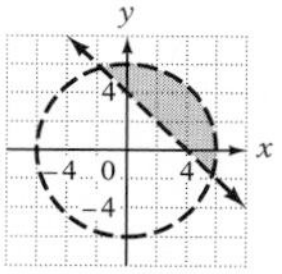

(Objective 10.5.2)

39.
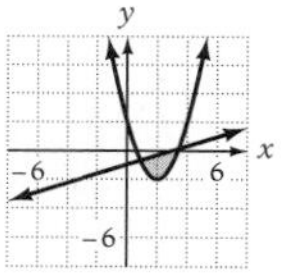

(Objective 10.5.2)

40. 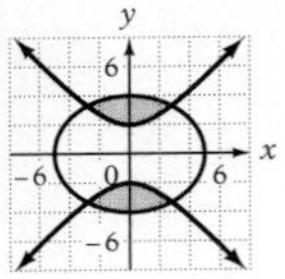

(Objective 10.5.2) **41.** cabin cruiser: 16 mph; current: 4 mph (Objective 10.4.1)

42. plane: 175 mph; wind: 25 mph (Objective 10.4.1) **43.** 100 children (Objective 10.4.2)
44. milk chocolate: \$5.50/lb; semi-sweet chocolate: \$4.50/lb (Objective 10.4.2)
45. \$.20 (Objective 10.4.2) **46.** 14 one-dollar bills (Objective 10.4.2)

CHAPTER TEST *pages 523–524*

1. $\left(\frac{3}{4}, \frac{7}{8}\right)$ (Objective 10.1.2) **2.** $(-3, -4)$ (Objective 10.1.2) **3.** $(2, -1)$ (Objective 10.1.2)
4. $(-2, 1)$ (Objective 10.3.3) **5.** inconsistent (Objective 10.2.1) **6.** $(1, 1)$ (Objective 10.2.1)
7. inconsistent (Objective 10.2.2) **8.** $(2, -1, -2)$ (Objective 10.3.3) **9.** 10 (Objective 10.3.1)
10. -32 (Objective 10.3.1) **11.** $\left(\frac{59}{19}, -\frac{62}{19}\right)$ (Objective 10.3.2) **12.** $(0, -2, 3)$ (Objective 10.3.2)
13. $(4, -1), \left(-\frac{7}{2}, -\frac{19}{4}\right)$ (Objective 10.5.1) **14.** $(2, 2), (2, -2), (-2, 2), (-2, -2)$ (Objective 10.5.1)

15. 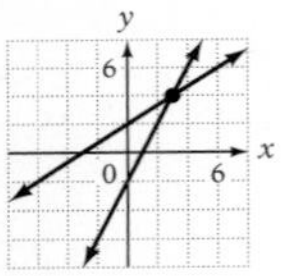

(Objective 10.1.1)
The solution is (3, 4).

16. 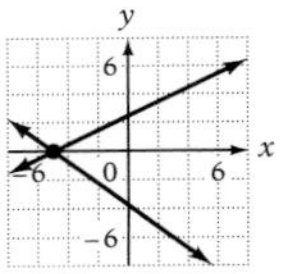

(Objective 10.1.1)
The solution is (−5, 0).

17. 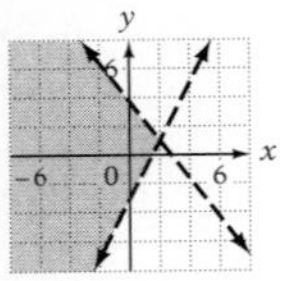

(Objective 10.5.2)

18. 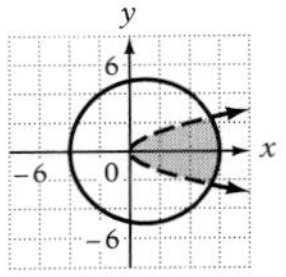

(Objective 10.5.2)

19. plane: 150 mph; wind: 25 mph (Objective 10.4.1) **20.** cotton: \$4.50; wool: \$7 (Objective 10.4.2)

CUMULATIVE REVIEW *pages 524–526*

1. $-\frac{11}{28}$ (Objective 2.1.2) **2.** $\{x \mid -4 \le x \le 1\}$ (Objective 2.5.2) **3.** $y = 5x - 11$ (Objective 6.3.1)
4. $(2x - 5)(3x - 2)$ (Objective 3.4.4) **5.** $\frac{2x^2 - x + 6}{(x - 2)(x - 3)(x + 1)}$ (Objective 4.2.2)
6. -3 (Objective 4.4.1) **7.** $\frac{1 + b}{a}$ (Objective 5.1.1) **8.** $-2ab^2\sqrt[3]{ab^2}$ (Objective 5.2.3)
9. -3 (Objective 5.4.1) **10.** $\frac{1}{2}$ and -5 (Objective 7.1.1)
11. $\frac{1 + \sqrt{7}}{3}$ and $\frac{1 - \sqrt{7}}{3}$ (Objective 7.2.1/7.2.2) **12.** 8 and 27 (Objective 7.3.1)
13. no x-intercepts (Objective 7.4.2) **14.** 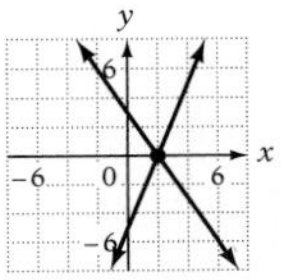

(Objective 10.1.1)
The solution is (2, 0).

15. $f^{-1}(x) = \frac{3}{2}x + \frac{3}{2}$ (Objective 8.3.2) **16.** $2\sqrt{10}$ (Objective 9.2.1) **17.** $(-5, -11)$ (Objective 10.1.2)
18. $(1, 0, -1)$ (Objective 10.2.2) **19.** 3 (Objective 10.3.1) **20.** $\left(\frac{7}{6}, \frac{1}{2}\right)$ (Objective 10.3.2)
21. $(-1, 0, 2)$ (Objective 10.3.3) **22.** $(1, 2)$ and $(1, -2)$ (Objective 10.5.1)
23. $(x - 2)^2 + (y + 1)^2 = 9$ (Objective 9.2.4)

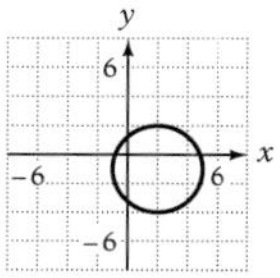

24.

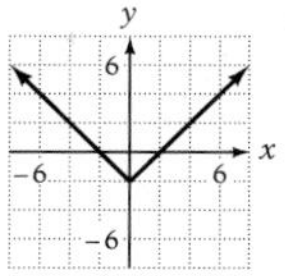

(Objective 8.2.1)

25. 16 nickels (Objective 2.2.1) **26.** 36 in. (Objective 8.4.1) **27.** 6 m (Objective 7.5.1)
28. 60 ml (Objective 2.4.2) **29.** 77 or better (Objective 2.5.3)
30. rate of the rowboat: 6 mph; rate of the current: 1.5 mph (Objective 10.4.1)

ANSWERS to Chapter 11 Exercises

SECTION 11.1 *pages 538–542*

1. 9 **3.** $\frac{1}{9}$ **5.** 1 **7.** 16 **9.** $\frac{1}{4}$ **11.** 1 **13.** 1 **15.** 16 **17.** $\frac{1}{8}$ **19.** $\frac{1}{3}$ **21.** 9
23. 3 **25.** 2 **27.** 16 **29.** 2 **31.** 2 **33.** 1.4142 **35.** 2.7183 **37.** 8.1548

39.

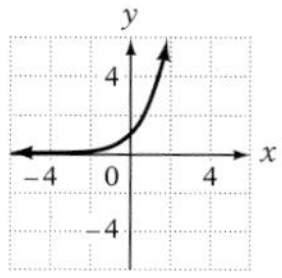

41.

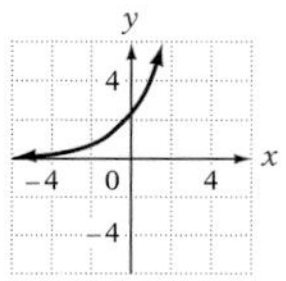

43.

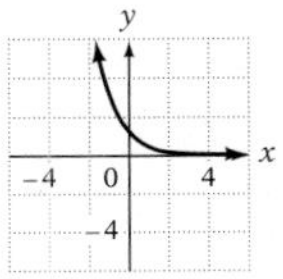

45.

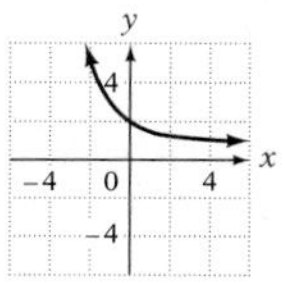

47.

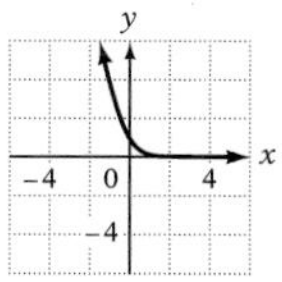

49. $\log_2 32 = 5$ **51.** $\log_5 25 = 2$ **53.** $\log_4 \frac{1}{16} = -2$ **55.** $\log_{\frac{1}{2}} \frac{1}{4} = 2$

57. $\log_3 1 = 0$ **59.** $\log_a w = x$ **61.** $3^2 = 9$ **63.** $4^1 = 4$ **65.** $10^0 = 1$ **67.** $10^{-2} = 0.01$
69. $\left(\frac{1}{3}\right)^2 = \frac{1}{9}$ **71.** $b^v = u$ **73.** $10^3 = 1000$ **75.** $e^0 = 1$ **77.** 2 **79.** 5 **81.** 2 **83.** 3
85. 9 **87.** 64 **89.** $\frac{1}{7}$ **91.** 216

93.

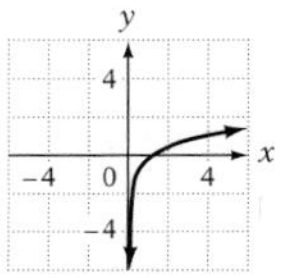

95.

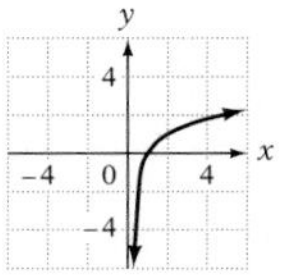

97.

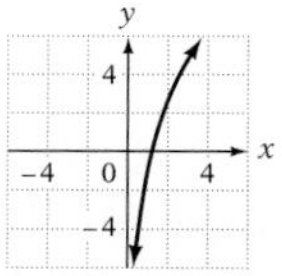

99.

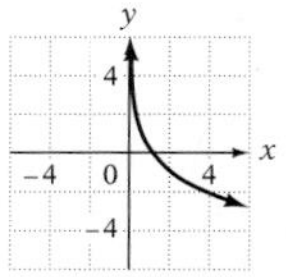

101.

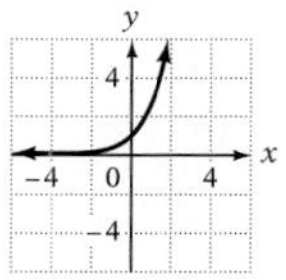

103.

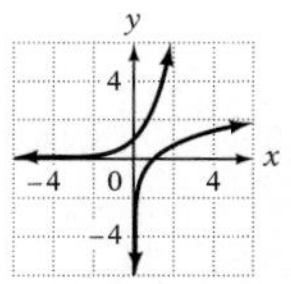

105. 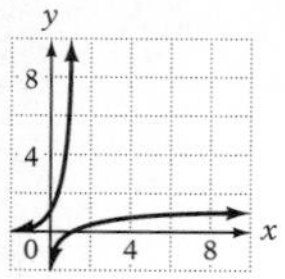

107. 2.67 **109.** 4.73 **111.** 0.38 **113.** 77.88 **115.** $y > 0$

117. $y > 0$ **119.** $x > 0$ **121.** $x > 0$ **123.** $f(1) = 1, f(2) = 2, f(4) = 4, f(8) = 8, f(a) = a$

SECTION 11.2 *pages 547–549*

1. $\log_8 x + \log_8 z$ **3.** $5 \log_3 x$ **5.** $\ln r - \ln s$ **7.** $2 \log_3 x + 6 \log_3 y$ **9.** $3 \log_7 u - 4 \log_7 v$
11. $2 \log_2 r + 2 \log_2 s$ **13.** $2 \log_9 x + \log_9 y + \log_9 z$ **15.** $\ln x + 2 \ln y - 4 \ln z$
17. $2 \log_8 x - \log_8 y - 2 \log_8 z$ **19.** $\frac{1}{2} \log_7 x + \frac{1}{2} \log_7 y$ **21.** $\frac{1}{2} \log_2 x - \frac{1}{2} \log_2 y$ **23.** $\frac{3}{2} \ln x + \frac{1}{2} \ln y$
25. $\frac{3}{2} \log_7 x - \frac{1}{2} \log_7 y$ **27.** $\log_b x + \frac{1}{2} \log_b y - \frac{1}{2} \log_b z$ **29.** $\log_3 t - \frac{1}{2} \log_3 x$ **31.** $\log_3 \frac{x^3}{y}$
33. $\log_8 x^4 y^2$ **35.** $\ln x^3$ **37.** $\log_5 x^3 y^4$ **39.** $\log_4 \frac{1}{x^2}$ **41.** $\log_3 \frac{x^2 z^2}{y}$ **43.** $\log_b \frac{x}{y^2 z}$ **45.** $\ln x^2 y^2$
47. $\log_6 \sqrt{\frac{x}{y}}$ **49.** $\log_4 \frac{s^2 r^2}{t^4}$ **51.** $\log_5 \frac{x}{y^2 z^2}$ **53.** $\ln \frac{t^3 v^2}{r^2}$ **55.** $\log_4 \sqrt{\frac{x^3 z}{y^2}}$ **57.** $\log_b \frac{\sqrt{xz}}{\sqrt[3]{y^2}}$ **59.** $\log_b \frac{xz^4}{\sqrt[3]{y^2}}$
61. 9 **63.** 0 **65.** 8 **67.** 8 **69.** 6 **71.** 2 **73.** 2 **75.** 3

77. $\log_a a^x = x$ Use the Logarithm Property of the
$x \log_a a = x$ Power of a Number. $\text{Log}_a a = 1$, since
$x = x$ $\log_b b^n = n$.

79. 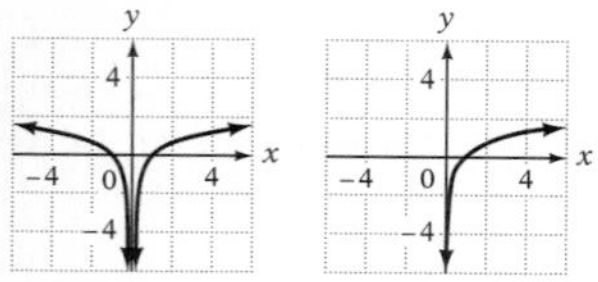

$y = \log x^2$ $y = 2 \log x$

The graphs are not the same. The logarithm of a negative number is not defined. Therefore $y = 2 \log x$ is not defined when x represents a negative number. However, $y = \log x^2$ is defined for all values of x, $x \neq 0$, since x^2 is positive.

SECTION 11.3 *pages 555–556*

1. 0.7686 **3.** 1.5378 **5.** 2.5899 **7.** 3.0128 **9.** 4.6749 **11.** −0.9431 **13.** −2.6990
15. −2.7570 **17.** −0.0070 **19.** 4.4362 **21.** −2.2434 **23.** −0.0414 **25.** 0.5849
27. 1.9270 **29.** 2.6984 **31.** −1.1505 **33.** 4.8804 **35.** −0.0754 **37.** 5.4903 **39.** 34.3005
41. 7970.7649 **43.** 631.9751 **45.** 0.4550 **47.** 0.0245 **49.** 0.0074 **51.** 475.0070
53. 0.4920 **55.** 7.0243 **57.** 344.7466 **59.** 0.5395 **61.** 501.6491 **63.** 0.0345 **65.** 47.9623
67. 3.14 **69.** 0.4969 **71.** 31.4 **73.** *b* **75.** *b* **77.** *b*

SECTION 11.4 *pages 560–562*

1. 1 **3.** −3 **5.** 1.1133 **7.** 0.7211 **9.** −1.5850 **11.** 1.7095 **13.** 1.3222 **15.** −2.8074
17. 3.5850 **19.** 1.1309 **21.** 1 **23.** 6 **25.** 1.3754 and −1.3754 **27.** 1.5805 **29.** −0.6309
31. 1.6610 **33.** −1.5129 **35.** 1.5 **37.** 8 **39.** $\frac{11}{2}$ **41.** 2 and −4 **43.** $\frac{5}{3}$ **45.** $\frac{1}{2}$
47. 1,000,000,000 **49.** 4 **51.** 3 **53.** 3 **55.** 2 **57.** no solution **59.** 4 **61.** $\frac{5 + \sqrt{33}}{2}$

63. 1.7712 **65.** 1.5440 **67.** 1.3863 **69.** 2.4650 **71.** 0.8617 **73.** −0.5798 **75.** 1.9586
77. 0.6309 **79.** −0.6309 **81.** 0.2727 **83.** 0.1729 **85.** 3.1186 **87.** −0.6012 **89.** 8.8247
91. −2.3280 **93.** 2.3181 **95.** 4.8502 **97.** −0.8340 **99.** 1.8163 **101.** 1.1493
103. 1.9769 **105.** 1.7233 **107.** 0.8060 **109.** 5.6910 **111.** (3, 1) **113.** (502, 498)
115. (−4, −9) **117.** The error is in step 2, $\log 0.2 < 0$. When both sides of an inequality are multiplied by a negative number, the inequality symbol must be reversed.

SECTION 11.5 *pages 567–570*

1. \$5982 **3.** 7 years **5.** \$16,786 **7.** 936 years **9.** 16 h **11.** 6 weeks **13.** 8
15. 79% **17.** 20 times **19.** 65 decibels **21.** 5 P.M. **23.** 9 years **25.** 8.5%
27. $y = Ae^{0.09704t}$ **29.** 7 years

FINDING THE PROPER DOSAGE *page 572*

1. 4.58 roentgens **3.** 0.7803 mg/ml

CHAPTER REVIEW *pages 574–575*

1. 2 (Objective 11.1.3) **2.** $\log_3 \sqrt{\frac{x}{y}}$ (Objective 11.2.1) **3.** 711.7050 (Objective 11.3.2)

4. −3 (Objective 11.4.1) **5.** 1 (Objective 11.1.1) **6.** $\frac{1}{9}$ (Objective 11.4.2)

7. −2.4353 (Objective 11.3.1) **8.** 6 (Objective 11.4.2) **9.** $\frac{1}{2}\log_6 x + \frac{3}{2}\log_6 y$ (Objective 11.2.1)

10. 2 (Objective 11.4.1) **11.** −2 (Objective 11.4.1) **12.** $\frac{1}{3}$ (Objective 11.1.1)

13. 4 (Objective 11.1.3) **14.** −1.3089 (Objective 11.3.1) **15.** 2.3219 (Objective 11.4.3)
16. 0.0404 (Objective 11.3.2) **17.** 3 (Objective 11.4.1) **18.** 4 (Objective 11.1.1)

19. $\frac{1}{2}\log_5 x - \frac{1}{2}\log_5 y$ (Objective 11.2.1) **20.** 1.9371 (Objective 11.3.1) **21.** 125 (Objective 11.1.3)

22. $\frac{1}{2}$ (Objective 11.4.2) **23.** $\log_b \frac{x^2}{y^5}$ (Objective 11.2.1) **24.** 4 (Objective 11.1.1)

25. 2.6801 (Objective 11.4.3) **26.** 50.6757 (Objective 11.3.2)

27.
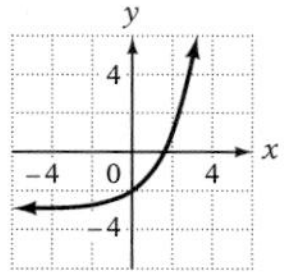

(Objective 11.1.2)

28.
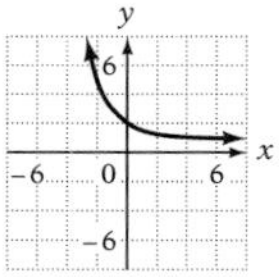

(Objective 11.1.2)

29.
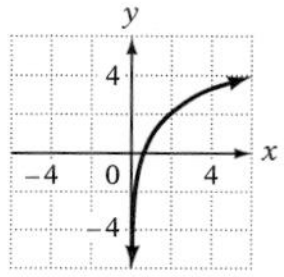

(Objective 11.1.4)

30.
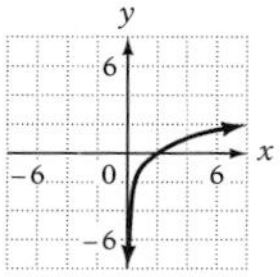

(Objective 11.1.4)

31. 33 h (Objective 11.5.1) **32.** 0.602 cm (Objective 11.5.1)

CHAPTER TEST *pages 575–576*

1. $f(0) = 1$ (Objective 11.1.1) **2.** $f(-2) = \frac{1}{64}$ (Objective 11.1.1) **3.** 3 (Objective 11.1.3)

4. $\frac{1}{16}$ (Objective 11.4.2) **5.** $\frac{1}{3}(2\log_6 x + 5\log_6 y)$ (Objective 11.2.1) **6.** $\log_5\sqrt{\frac{x}{y}}$ (Objective 11.2.1)

7. -1.3757 (Objective 11.3.1) **8.** 0.1320 (Objective 11.3.2) **9.** 1341.8377 (Objective 11.3.2)

10. 1.9746 (Objective 11.4.3) **11.** 10 (Objective 11.4.2) **12.** 3 (Objective 11.4.1)

13. -4 (Objective 11.4.1) **14.** $\frac{3}{2}$ (Objective 11.4.2) **15.**

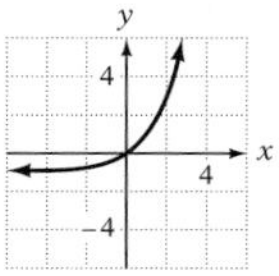

(Objective 11.1.2)

16.

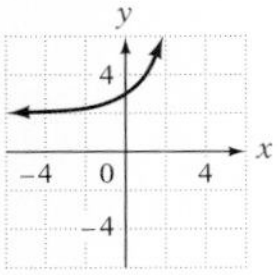

Objective 11.1.2)

17.

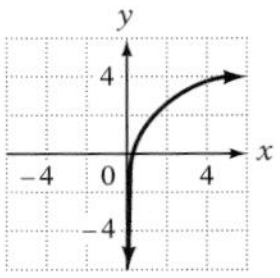

(Objective 11.1.4)

18.

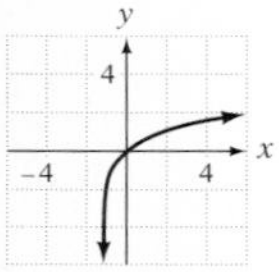

(Objective 11.1.4)

19. \$15,661 (Objective 11.5.1)

20. 24 h (Objective 11.5.1)

CUMULATIVE REVIEW *pages 576–578*

1. $\frac{8}{7}$ (Objective 2.1.2) **2.** $L = \frac{S - 2WH}{2W + 2H}$ (Objective 4.5.1) **3.** $\{x \mid 1 \le x \le 4\}$ (Objective 2.6.2)

4. $(4x^n + 3)(x^n + 1)$ (Objective 3.5.3) **5.** $\{x \mid -5 \le x \le 1\}$ (Objective 7.6.1) **6.** $\frac{x-3}{x+3}$ (Objective 4.3.1)

7. $\frac{x\sqrt{y} + y\sqrt{x}}{x - y}$ (Objective 5.2.4) **8.** $-4x^2y^3\sqrt{2x}$ (Objective 5.2.2) **9.** $-\frac{1}{5} + \frac{2}{5}i$ (Objective 5.3.4)

10. $y = 2x - 6$ (Objective 6.3.2) **11.** $3x^2 + 8x - 3 = 0$ (Objective 7.1.2)

12. $2 + \sqrt{10}$ and $2 - \sqrt{10}$ (Objective 7.2.1, 7.2.2) **13.** $\{-6, -4, 0\}$ (Objective 8.1.2)

14. 4 (Objective 8.3.1) **15.** $(0, -1, 2)$ (Objective 10.2.2)

16. $(2, 3)$ and $(-1, -3)$ (Objective 10.5.1) **17.** 243 (Objective 11.1.1) **18.** 64 (Objective 11.1.3)

19. 8 (Objective 11.4.1) **20.** 1 (Objective 11.4.2) **21.**

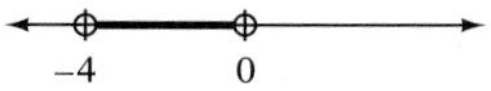

(Objective 1.3.2)

22.

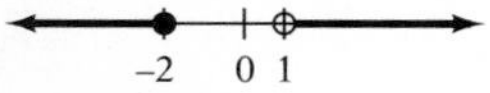

(Objective 7.6.2)

23.

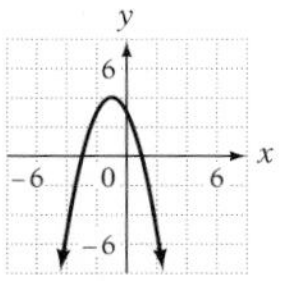

(Objective 9.1.1)

24. 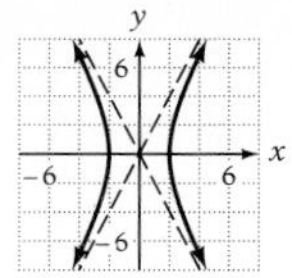

(Objective 9.3.2)

25. 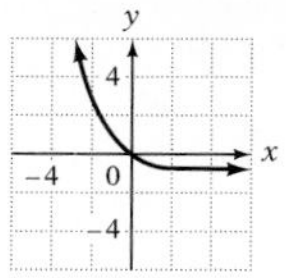

(Objective 11.1.2)

26. 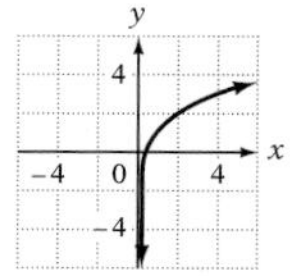

(Objective 11.1.4)

27. 25% alloy: 800 lb; 50% alloy: 1200 lb (Objective 2.4.2)

28. \$31,250 or more (Objective 2.5.3) **29.** 7.5 min (Objective 4.4.3)
30. 500 lb/in.^{2} (Objective 8.4.1) **31.** \$12 (Objective 10.4.2) **32.** \$15,657 (Objective 11.5.1)

ANSWERS to Chapter 12 Exercises

SECTION 12.1 *pages 584–586*

1. 2, 3, 4, 5 **3.** 3, 5, 7, 9 **5.** 0, −2, −4, −6 **7.** 2, 4, 8, 16 **9.** 2, 5, 10, 17
11. $\frac{1}{2}, \frac{2}{5}, \frac{3}{10}, \frac{4}{17}$ **13.** $0, \frac{3}{2}, \frac{8}{3}, \frac{15}{4}$ **15.** 1, −2, 3, −4 **17.** $\frac{1}{2}, -\frac{1}{5}, \frac{1}{10}, -\frac{1}{17}$
19. −2, 4, −8, 16 **21.** $\frac{2}{9}, \frac{2}{27}, \frac{2}{81}, \frac{2}{243}$ **23.** 15 **25.** $\frac{12}{13}$ **27.** 24 **29.** $\frac{32}{243}$ **31.** 88
33. $\frac{1}{20}$ **35.** 38 **37.** 42 **39.** 28 **41.** 60 **43.** $\frac{25}{24}$ **45.** 28 **47.** $\frac{99}{20}$ **49.** $\frac{137}{120}$
51. −2 **53.** $-\frac{7}{12}$ **55.** $\frac{2}{x} + \frac{4}{x} + \frac{6}{x} + \frac{8}{x}$ **57.** $\frac{x}{2} + \frac{x^2}{3} + \frac{x^3}{4} + \frac{x^4}{5}$ **59.** $\frac{x^2}{3} + \frac{x^3}{5} + \frac{x^4}{7}$
61. $x + x^3 + x^5 + x^7$ **63.** $2x + 4x^2 + 6x^3 + 8x^4$ **65.** $\frac{1}{x} + \frac{1}{x^2} + \frac{1}{x^3} + \frac{1}{x^4} + \frac{1}{x^5}$ **67.** $a_n = 2n - 1$
69. $a_n = -2n + 1$ **71.** $a_n = 4n$ **73.** log 384 **75.** 3 **77.** 2, 3, 5

SECTION 12.2 *pages 591–593*

1. 141 **3.** 50 **5.** 71 **7.** $\frac{27}{4}$ **9.** 17 **11.** 3.75 **13.** $a_n = n$ **15.** $a_n = -4n + 10$
17. $a_n = \frac{3n + 1}{2}$ **19.** $a_n = -5n - 3$ **21.** $a_n = -10n + 36$ **23.** 42 **25.** 16 **27.** 20 **29.** 20
31. 13 **33.** 20 **35.** 650 **37.** −605 **39.** $\frac{215}{4}$ **41.** 420 **43.** −210 **45.** −5
47. 9 weeks **49.** 2180 seats **51.** \$1160; \$8820 **53.** 5050 **55.** 8 **57.** $d = 4$; $n = 7$
59. −3 **61.** 1800°; $180(n - 2)$

SECTION 12.3 *pages 600–602*

1. 131,072 **3.** $\frac{128}{243}$ **5.** 16 **7.** 6, 4 **9.** $-2, \frac{4}{3}$ **11.** 12, −48 **13.** 2186 **15.** $\frac{2343}{64}$
17. 62 **19.** $\frac{121}{243}$ **21.** 1364 **23.** 2800 **25.** $\frac{2343}{1024}$ **27.** $\frac{1360}{81}$ **29.** 9 **31.** $\frac{18}{5}$ **33.** $\frac{7}{9}$

35. $\frac{8}{9}$ **37.** $\frac{2}{9}$ **39.** $\frac{5}{11}$ **41.** $\frac{1}{6}$ **43.** 7.8125 mg **45.** 2.6 ft **47.** $82,103.49
49. $432,194.24 **51.** (*G*), $-\frac{1}{2}$ **53.** (*A*), 9.5 **55.** (*N*), 25 **57.** (*G*), x^2 **59.** (*A*), $4 \log x$
61. 27 **63.** $a_1 = 80$; $n = 5$ **65.** The common difference is log 2. **67.** The common ratio is 8.

SECTION 12.4 *pages 607–609*

1. 6 **3.** 40,320 **5.** 1 **7.** 10 **9.** 1 **11.** 84 **13.** 21 **15.** 45 **17.** 1 **19.** 20
21. 11 **23.** 6 **25.** $x^4 + 4x^3y + 6x^2y^2 + 4xy^3 + y^4$ **27.** $x^5 - 5x^4y + 10x^3y^2 - 10x^2y^3 + 5xy^4 - y^5$
29. $16m^4 + 32m^3 + 24m^2 + 8m + 1$ **31.** $32r^5 - 240r^4 + 720r^3 - 1080r^2 + 810r - 243$
33. $a^{10} + 10a^9b + 45a^8b^2$ **35.** $a^{11} - 11a^{10}b + 55a^9b^2$ **37.** $256x^8 + 1024x^7y + 1792x^6y^2$
39. $65{,}536x^8 - 393{,}216x^7y + 1{,}032{,}192x^6y^2$ **41.** $x^7 + 7x^5 + 21x^3$ **43.** $x^{10} + 15x^8 + 90x^6$ **45.** $-560x^4$
47. $-6x^{10}y^2$ **49.** $126y^5$ **51.** $5n^3$ **53.** $\frac{x^5}{32}$ **55.** 1 7 21 35 35 21 7 1 **57.** 2450
59. n **61.** $35x^3a^4$ **63.** $x^2 + 8x^{\frac{3}{2}} + 24x + 32x^{\frac{1}{2}} + 16$ **65.** $\frac{1}{x^3} + \frac{3}{x^2y} + \frac{3}{xy^2} + \frac{1}{y^3}$ **67.** $-8i$
69. $\binom{n}{r} = \binom{n}{(n-r)!r!}$ **71.** 60

$$\binom{n}{n-r} = \binom{n}{[n-(n-r)]!(n-r)!}$$
$$= \binom{n}{r!(n-r)!}$$
$$\binom{n}{r} = \binom{n}{n-r}$$

THE FIBONACCI SEQUENCE *page 611*

1. 1, 1, 2, 3, 5, 8, 13, 21, 34, 55, 89, 144, 233, 377, 610, 987, 1597, 2584
3. 1.6176471, 1.6180556, 1.6180338; each is approximately 1.6

CHAPTER REVIEW *pages 612–615*

1. $3x + 3x^2 + 3x^3 + 3x^4$ (Objective 12.1.2) **2.** 16 (Objective 12.2.1) **3.** 32 (Objective 12.3.1)
4. 16 (Objective 12.3.3) **5.** 84 (Objective 12.4.1) **6.** $\frac{1}{2}$ (Objective 12.1.1)
7. 44 (Objective 12.2.1) **8.** 468 (Objective 12.2.2) **9.** −66 (Objective 12.3.2)
10. 70 (Objective 12.4.1) **11.** $2268x^3y^6$ (Objective 12.4.1) **12.** 34 (Objective 12.2.2)
13. $\frac{7}{6}$ (Objective 12.1.1) **14.** $a_n = -3n + 15$ (Objective 12.2.1) **15.** $\frac{2}{27}$ (Objective 12.3.1)
16. $\frac{7}{30}$ (Objective 12.3.3) **17.** −115 (Objective 12.2.1) **18.** $\frac{665}{32}$ (Objective 12.3.2)
19. 1575 (Objective 12.2.2) **20.** $-280x^4y^3$ (Objective 12.4.1) **21.** 21 (Objective 12.2.1)
22. 48 (Objective 12.3.1) **23.** 30 (Objective 12.2.2) **24.** 341 (Objective 12.3.2)
25. 120 (Objective 12.4.1) **26.** $240x^4$ (Objective 12.4.1) **27.** 143 (Objective 12.2.1)
28. −575 (Objective 12.2.2) **29.** $-\frac{5}{27}$ (Objective 12.1.1) **30.** $2 + 2x + 2x^2 + 2x^3$ (Objective 12.1.2)
31. $\frac{23}{99}$ (Objective 12.3.3) **32.** $\frac{16}{5}$ (Objective 12.3.3) **33.** 726 (Objective 12.3.2)
34. $-42{,}240x^4y^7$ (Objective 12.4.1) **35.** 0.996 (Objective 12.3.2) **36.** 6 (Objective 12.3.3)

37. $\frac{19}{30}$ (Objective 12.3.3) **38.** $x^5 - 15x^4y^2 + 90x^3y^4 - 270x^2y^6 + 405xy^8 - 243y^{10}$ (Objective 12.4.1)
39. 22 (Objective 12.2.1) **40.** 99 (Objective 12.4.1)
41. $2x + 2x^2 + \frac{8x^3}{3} + 4x^4 + \frac{32x^5}{5}$ (Objective 12.1.2) **42.** $-\frac{13}{60}$ (Objective 12.1.2)
43. $12,240 (Objective 12.2.3) **44.** 67.7°F (Objective 12.3.4)

CHAPTER TEST *pages 615–616*

1. $\frac{1}{3}$ (Objective 12.1.1) **2.** $\frac{8}{9}, \frac{9}{10}$ (Objective 12.1.1) **3.** 32 (Objective 12.1.2)
4. $2x^2 + 2x^4 + 2x^6 + 2x^8$ (Objective 12.1.2) **5.** -120 (Objective 12.2.1)
6. $a_n = 2n - 5$ (Objective 12.2.1) **7.** 22 (Objective 12.2.1) **8.** 315 (Objective 12.2.2)
9. 1560 (Objective 12.2.2) **10.** 252 (Objective 12.4.1) **11.** $-64\sqrt{2}$ (Objective 12.3.1)
12. $\frac{81}{125}$ (Objective 12.3.1) **13.** $\frac{781}{256}$ (Objective 12.3.2) **14.** -55 (Objective 12.3.2)
15. 4 (Objective 12.3.3) **16.** $\frac{7}{30}$ (Objective 12.3.3) **17.** 330 (Objective 12.4.1)
18. $5670x^4y^4$ (Objective 12.4.1) **19.** 2550 yd (Objective 12.2.3) **20.** 20 mg (Objective 12.3.4)

CUMULATIVE REVIEW *pages 616–617*

1. $\frac{x^2 + 5x - 2}{(x + 2)(x - 1)}$ (Objective 4.2.2) **2.** $2(x^2 + 2)(x^4 - 2x^2 + 4)$ (Objective 3.5.4)
3. $4y\sqrt{x} - y\sqrt{2}$ (Objective 5.2.3) **4.** $\frac{1}{x^{26}}$ (Objective 5.1.1) **5.** 4 (Objective 7.3.2)
6. $\frac{1}{4} - \frac{\sqrt{55}}{4}i$ and $\frac{1}{4} + \frac{\sqrt{55}}{4}i$ (Objective 7.2.2) **7.** $\left(\frac{7}{6}, \frac{1}{2}\right)$ (Objective 10.2.1)
8. $(x + 1)^2 + (y + 1)^2 = 34$ (Objective 9.2.3) **9.** $\{x \mid x < -2 \text{ or } x > 2\}$ (Objective 2.5.2)
10. -10 (Objective 10.3.1) **11.** $\frac{1}{2}\log_5 x - \frac{1}{2}\log_5 y$ (Objective 11.2.1) **12.** 3 (Objective 11.4.1)
13. $a_5 = 20;\ a_6 = 30$ (Objective 12.1.1) **14.** 6 (Objective 12.1.2) **15.** (2, 0, 1) (Objective 10.2.2)
16. 216 (Objective 11.1.3) **17.** $2x^2 - x - 1 + \frac{6}{2x + 1}$ (Objective 3.3.1) **18.** $-3h + 1$ (Objective 8.1.1)
19. $\left\{-1, 0, \frac{7}{5}\right\}$ (Objective 8.1.2) **20.** $(x + 1)^2 + (y - 4)^2 = 10$ (Objective 9.2.3)
21. 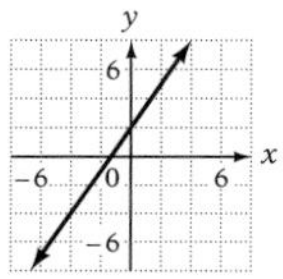

(Objective 6.1.3) **22.** 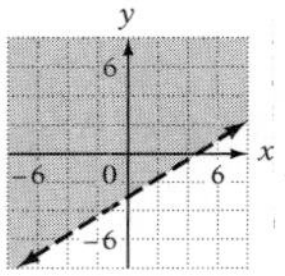

(Objective 6.5.1)

23. new computer: 24 min; old computer: 40 min (Objective 4.4.3)
24. boat: 6.25 mph; current: 1.25 mph (Objective 7.5.1) **25.** 55 days (Objective 11.5.1)
26. 1536 seats (Objective 12.2.3) **27.** 2.6 ft (Objective 12.3.4)

FINAL EXAM *pages 618–621*

1. -31 (Objective 1.1.4) **2.** -1 (Objective 1.2.1) **3.** $33 - 10x$ (Objective 1.2.3)

4. $x = 8$ (Objective 2.1.1) **5.** $\frac{2}{3}$ (Objective 2.1.2) **6.** $4, -\frac{2}{3}$ (Objective 2.6.1)

7. $\{x \mid -4 < x < -1\}$ (Objective 2.6.2) **8.** $\left\{x \mid x > \frac{3}{2}\right\}$ (Objective 2.5.2)

9. $y = -\frac{2}{3}x - \frac{1}{3}$ (Objective 6.3.2) **10.** $6a^3 - 5a^2 + 10a$ (Objective 3.2.1) **11.** $\frac{6}{5} - \frac{3}{5}i$ (Objective 5.3.4)

12. $2x^2 - 3x - 2 = 0$ (Objective 7.1.2) **13.** $(2 - xy)(4 + 2xy + x^2y^2)$ (Objective 3.5.2)

14. $(x - y)(1 + x)(1 - x)$ (Objective 3.5.4) **15.** $x^2 - 2x - 3 - \frac{5}{2x - 3}$ (Objective 3.3.1)

16. $\frac{x(x - 1)}{2x - 5}$ (Objective 4.2.1) **17.** $-\frac{10x}{(x + 2)(x - 3)}$ (Objective 4.2.2) **18.** $\frac{x + 3}{x + 1}$ (Objective 4.3.1)

19. $-\frac{7}{4}$ (Objective 4.4.1) **20.** $d = \frac{a_n - a_1}{n - 1}$ (Objective 4.5.1) **21.** $\frac{y^4}{162x^3}$ (Objective 3.1.3)

22. $\frac{1}{64x^8y^5}$ (Objective 5.1.1) **23.** $-2x^2y\sqrt{2y}$ (Objective 5.2.2) **24.** $\frac{x^2\sqrt{2y}}{2y^2}$ (Objective 5.2.4)

25. $\frac{3 + \sqrt{17}}{4}$ and $\frac{3 - \sqrt{17}}{4}$ (Objective 7.2.2) **26.** $27, -8$ (Objective 7.3.1)

27. $y = -3x + 7$ (Objective 6.3.1) **28.** $\frac{3}{2}, -2$ (Objective 7.3.3) **29.** $(3, 4)$ (Objective 10.2.1)

30. 10 (Objective 10.3.1) **31.** 6 (Objective 11.4.2)

32. $2y + 2y^2 + 2y^3 + 2y^4 + 2y^5$ (Objective 12.1.2) **33.** $\frac{23}{45}$ (Objective 12.3.3)

34. $144x^7y^2$ (Objective 12.4.1) **35.** $\left(\frac{5}{2}, -\frac{3}{2}\right)$ (Objective 10.5.1)

36. $f^{-1}(x) = \frac{3}{2}x + 6$ (Objective 8.3.2) **37.** $\log_2 \frac{a^2}{b^2}$ (Objective 11.2.1)

38. x-intercept: $\left(\frac{9}{2}, 0\right)$
y-intercept: $(0, -3)$

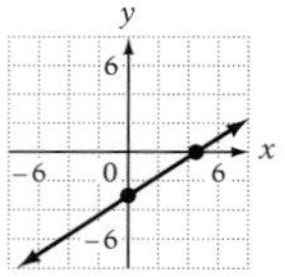

(Objective 6.2.2)

39.

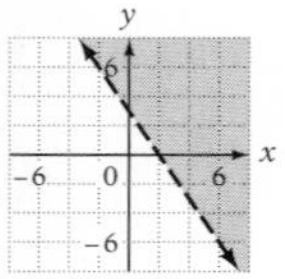

(Objective 6.5.1)

40.

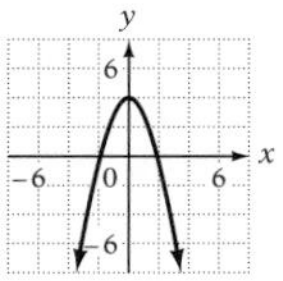

(Objective 8.2.1)

41.

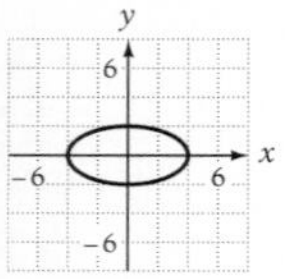

(Objective 9.3.1)

42.

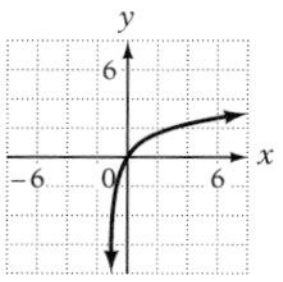

(Objective 11.1.4)

43. 69 or better (Objective 2.5.3) **44.** 40 mi (Objective 2.3.2)
45. \$8000 at 8.5%; \$4000 at 6.4% (Objective 2.4.1) **46.** length: 20 ft; width: 7 ft (Objective 7.5.1)
47. 200 additional shares (Objective 4.4.2) **48.** 420 mph (Objective 7.5.1)
49. 88 ft (Objective 5.4.2) **50.** 120 mph (Objective 7.5.1) **51.** 200 lumens (Objective 8.4.1)
52. boat: 12.5 mph; current: 2.5 mph (Objective 10.4.1) **53.** \$4785.65 (Objective 11.5.1)
54. \$256,571 (Objective 12.3.4)

INDEX

Table of Properties

Properties of Real Numbers

The Associative Property of Addition
If a, b, and c are real numbers, then
$(a + b) + c = a + (b + c)$.

The Commutative Property of Addition
If a and b are real numbers, then
$a + b = b + a$.

The Addition Property of Zero
If a is a real number, then
$a + 0 = 0 + a = a$.

The Inverse Property of Addition
If a is a real number, then
$a + (-a) = (-a) + a = 0$.

The Multiplication Property of Zero
If a is a real number, then
$a \cdot 0 = 0 \cdot a = 0$.

The Associative Property of Multiplication
If a, b, and c are real numbers, then
$(a \cdot b) \cdot c = a \cdot (b \cdot c)$.

The Commutative Property of Multiplication
If a and b are real numbers, then
$a \cdot b = b \cdot a$.

The Multiplication Property of One
If a is a real number, then
$a \cdot 1 = 1 \cdot a = a$.

The Inverse Property of Multiplication
If a is a nonzero real number, then
$a \cdot \frac{1}{a} = \frac{1}{a} \cdot a = 1$.

Distributive Property

If a, b, and c are real numbers, then $a(b + c) = ab + ac$ or $(b + c)a = ba + ca$.

Properties of Equations

Addition Property of Equations
If $a = b$, then $a + c = b + c$.

Multiplication Property of Equations
If $a = b$ and $c \neq 0$, then $a \cdot c = b \cdot c$.

Properties of Exponents

If m and n are integers, then
$x^m \cdot x^n = x^{m+n}$.
If m and n are integers, then
$(x^m)^n = x^{mn}$.
If $x \neq 0$, then
$x^0 = 1$.
If m, n, and p are integers, then
$\left(\frac{x^m}{y^n}\right)^p = \frac{x^{mp}}{y^{np}}$.

If m and n are integers and $x \neq 0$, then
$\frac{x^m}{x^n} = x^{m-n}$.
If m, n, and p are integers, then
$(x^m \cdot y^n)^p = x^{mp}y^{np}$.
If n is a positive integer and $x \neq 0$, then
$x^{-n} = \frac{1}{x^n}$ and $\frac{1}{x^{-n}} = x^n$.

One-to-One Property of Exponents
If $b^x = b^y$, then $x = y$.